40th Riso International Symposium on Materials Science 2019

Metal Microstructures in 2D, 3D and 4D

IOP Conference Series: Materials Science and Engineering
Volume 580

Kongens Lyngby, Denmark
2 - 6 September 2019

ISBN: 978-1-7138-0898-5
ISSN: 1757-8981

Printed from e-media with permission by:

Curran Associates, Inc.
57 Morehouse Lane
Red Hook, NY 12571

Some format issues inherent in the e-media version may also appear in this print version.

Institute of Physics
Dirac House, Temple Back
Bristol BS1 6BE UK

Phone: 44 1 17 929 7481
Fax: 44 1 17 920 0979

techtracking@iop.org

Additional copies of this publication are available from:

Curran Associates, Inc.
57 Morehouse Lane
Red Hook, NY 12571 USA
Phone: 845-758-0400
Fax: 845-758-2633
Email: curran@proceedings.com
Web: www.proceedings.com

TABLE OF CONTENTS

KEYNOTE PAPERS

INVITED PAPERS

CONTRIBUTED PAPERS

IOP Conf. Series: Materials Science and Engineering **580** (2019) 011001 doi:10.1088/1757-899X/580/1/011001

Preface

This year the Risø International Symposium on Materials Science is celebrating its 40th year jubilee. During this 40-year period, significant progress has taken place and new discoveries and understanding in the field of material science and engineering have materialized in advanced metals and alloys with wide applications in society. In parallel, there has been an unprecedented development of 2D, 3D and 4D characterization techniques operating on multiple length scales down to the atomic dimension. The development has led to revision and foundation of fundamental principles and models for example in the areas of deformation, recovery and recrystallization of metals and alloys. This in-turn has led to new and demanding challenges that are addressed in the Proceedings. The challenges include:

- Transitional stages in the evolution of deformation microstructure and strengthening mechanisms.
- Effects of local microstructural variations on the evolution of microstructures during annealing
- 2D, 3D and 4D characterization techniques based on for example x-rays and electron microscope techniques in the laboratory and at international facilities.
- Microstructural variations at different length scales which is of critical importance for the properties and performance of metals and alloys for engineering applications.
- Advanced and new processing techniques of bulk materials and by application of powder metallurgy in additive manufacturing and spark plasma sintering processes.
- Experimental findings in 3D and 4D used as input to as well as validation of analytical and numerical models.

The metals and alloys investigated are of fcc, bcc and hcp structure, in bulk form or present as gradients or multilayer structures.

The Proceedings contain 15 keynote, 11 invited and 34 contributed presentations covering the above-mentioned key themes. It is our hope that Proceedings can inspire the future development of this field.

IOP Conf. Series: Materials Science and Engineering **580** (2019) 011002 doi:10.1088/1757-899X/580/1/011002

Peer review statement

All papers published in this volume of *IOP Conference Series: Materials Science and Engineering* have been peer reviewed through processes administered by the proceedings Editors. Reviews were conducted by expert referees to the professional and scientific standards expected of a proceedings journal published by IOP Publishing.

Determination of the dislocation-based mechanism(s) responsible for the anomalous ductility of a class of B2 intermetallic alloys

S R Agnew[1] and T Ungar[2,3]

[1] Department of Materials Science and Engineering, University of Virginia, Charlottesville, Virginia, 22904, U.S.A.
[2] Eötvös University Budapest, Budapest, Hungary
[3] School of Materials, The University of Manchester, Manchester, M1 3BB, United Kingdom

Corresponding e-mail: agnew@virginia.edu

Abstract. A class of stoichiometric, ordered MR alloys (where M denotes a metal and R either a rare earth or group IV refractory metal) with B2 (CsCl) crystal structure, were discovered to possess anomalous ductility by researchers at Ames Laboratory in the early 2000's. Despite active research on the topic for over 15 years, a holistic explanation for the anomalous ductility of these metal alloys has not yet been accepted by the scientific community. The reason for this appears to be a failure to account for all relevant length scales. Researchers either focus on the *atomistic* length scale relevant to dislocation core structures, the *microscopic* length scale of dislocations and dislocation ensembles, the *mesoscopic* length scale of grains and their interactions, or the *macroscopic* scale of mechanical testing samples with millimetre dimensions or larger. Focused studies at each of these length scales have provided essential, but insufficient information to provide a complete answer to the question. The insight provided by in-situ diffraction studies, along with crystal plasticity modelling as an interpretive tool, and a novel grain-by-grain line profile analysis technique for assessing dislocation densities in polycrystals is highlighted. In addition to the primary <100> dislocations, which are most easily activated in all of these anomalous MR alloys, the results show that dislocations with large Burgers vectors (e.g., <110> and <111>) are present in large numbers *beyond a transition* in the strain hardening response. The motion of these large Burgers vector dislocations is experimentally observed to occur at stress levels consistent with atomistic modelling predictions of the Peierls resistance and simultaneously provides a satisfying explanation for the anomalous ductility and other experimental results.

1. Introduction

The practical application of many intermetallic compounds, despite numerous attractive properties (e.g., NiAl has a very high melting point, low density, and good oxidation behaviour) is limited by poor low temperature ductility. In particular, many B2 (CsCl prototypical structure) compounds are known to exhibit only the <001>{110} "cube" slip mode, which provides only 3 independent slip systems and fails to satisfy the von Mises criterion for polycrystalline ductility (Figure 1). Nearly a century ago, von Mises [1] noted that the ability to accommodate arbitrary strain boundary conditions enforced on a grain within a polycrystalline aggregate would require 5 independent easy slip systems. This criterion has since been viewed as a necessary, though not sufficient, requirement for polycrystalline ductility. The discovery of a number of ductile B2 (CsCl) structured intermetallic compounds [2] raised questions about a possible violation of the von Mises criterion. In particular, it raised the possibility that other deformation mechanisms could intervene and thus relax (or alter) this criterion. The discovered alloys are stoichiometric, ordered MR compounds (where M denotes a metal and (R) either a rare earth or group IV refractory metal). The discoverers pointed out that the class of alloys is large (over 100 compounds) and that many of them had already been shown to be ductile (e.g., AgY, CuY, CuDy). Although the present researchers investigated a number of these rare earth element containing compounds, the research also highlighted the previously noted anomalous ductility of related alloys, CoTi and CoZr [e.g., 3]. In what follows, primary focus will be placed on four anomalous systems: CoTi, CoZr, CuY, and AgY. CoTi and CuY appear to be borderline cases, whereas CoZr and AgY exhibit more ductility (Figure 2).

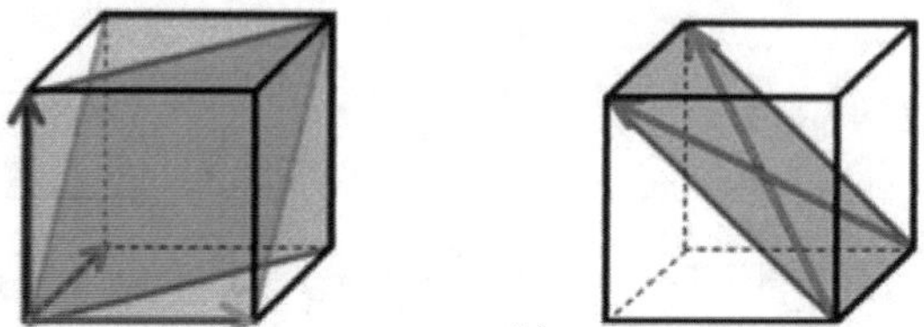

Figure 1. Schematic illustrations of two possible slip modes in the crystals with the B2 (CsCl) structure. a. The <100>{011} "cube" slip mode, which is common to many B2 alloys (note that the third possible {011} slip plane is not shown for clarity). b. The <111>{1$\bar{1}$0} slip (only two slip systems with a common slip plane are shown for clarity) (adapted from Wollmershauser [4]).

Previous researchers who examined this ductility anomaly tended to focus on either the *atomistic* length scale relevant to dislocation core structures, the *microscopic* length scale of dislocations and dislocation ensembles, the *mesoscopic* length scale of grains and their interactions, or the *macroscopic* scale of mechanical testing samples with millimeter dimensions or larger. It will be shown that these previous, focused studies have provided essential, but insufficient information to provide a complete answer to the question. In the present article, the findings of researchers who examined the problem at each of these length scales is reviewed, and in the process, a holistic explanation that is consistent with the results of these prior studies is presented.

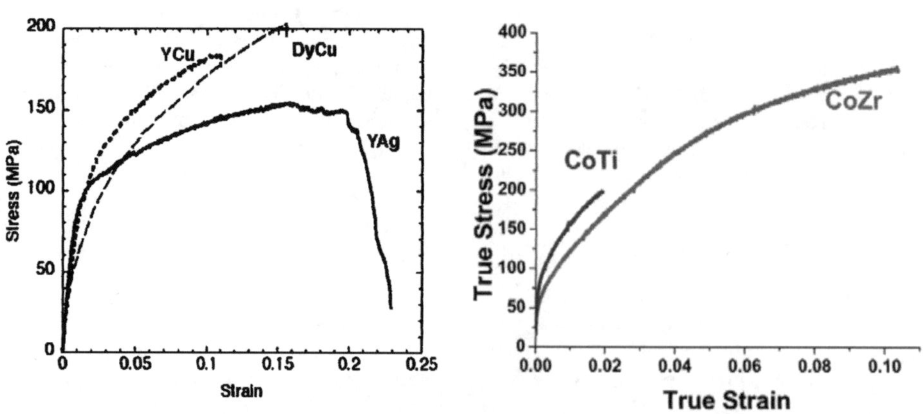

Figure 2. Tensile stress strain curves illustrating the anomalous ductility of AgY, CuY, and CuDy (left, after Zhang et al. [5], with permission, Elsevier) and CoTi and CoZr (right, reproduced from Wollmershauser [4]).

IOP Conf. Series: Materials Science and Engineering **580** (2019) 012001 doi:10.1088/1757-899X/580/1/012001

2. Macroscale phenomena

Although the MR alloys exhibit a level of ductility which is unusual for B2 compounds, it must be clarified that they do not exhibit the extended level of ductility (large reduction in area) which we expect from pure fcc metals (Al, Cu, Au, etc.) or their solid solution alloys (brass, austenitic stainless steels, etc.). Rather, it is a limited sort of ductility which is more akin to that exhibited by many noncubic metals like Mg, Sn, or Ti and some of their alloys. In fact, the fracture surfaces appear as brittle, transgranular cleavage [e.g., 6]. Macroscale mechanical testing has also revealed that the initial yield stresses of the anomalous MR are rather low (Table 1). Each of the alloys undergoes very rapid initial strain hardening, which will later be shown to be primarily due to the composite effect typical of the "elastoplastic transition" rather than dislocation density increase (i.e., Taylor strain hardening). Then, there typically is a transition in strain hardening behaviour which is more apparent during compression tests where deformation far beyond the transition is possible. It is noted that the earlier this transition from rapid hardening to slower hardening occurs (for a given alloy type), the higher the ductility.

Table 1. Mechanical behaviour of select MR alloys

Alloy	0.2% offset yield (MPa)	Transition stress (MPa)	Max. tensile ductility (%)
AgY [2]	95	100	20
CuY [2]	95-200	145	11
CuDy [7]	55	125	16
CoTi [3]	100	300	2
CoZr [3]	70	250	18

2.1. Slip trace analyses

Two surface slip trace analysis performed on single crystals is an accepted method to determine both the slip plane and slip (Burgers vector) direction. Since the advent of electron backscattered diffraction (EBSD), which allows ready determination of the crystallographic orientation of a grain, single-surface trace analysis combined with computational assessment of the most likely slip systems based upon Schmid factor (or tensor) calculations has become widespread [e.g., 8]. However, it is important to note that, unlike two surface slip trace analysis which can provide a mathematically unique answer, single surface slip trace analysis only provides a plausible answer selected from a list of candidate slip planes and directions. This plausible answer is based upon the assumption that the stress state is uniform throughout the polycrystal (which will be strongly challenged by the in-situ diffraction results which follow), and that the active slip system is the one with the highest resolved shear stress on it (i.e., that other components of stress do not matter).

For the anomalous B2 alloys, all slip trace analyses suggest that only <100>{011} or <100>{010} "cube" slip systems were active in these materials. For example, single-crystal slip traces analyses performed on CoTi reveal <100>{011} slip mode activity, regardless of crystal orientation [9]. Similarly, slip trace studies of the most ductile MR compound, AgY, and similar CuY and CuDy revealed only cube type slip systems with no evidence of slip in the <111> or <110> directions [2,10].

2.2. Thermal activation analysis

Schaarschuch et al. [11] recently performed an in-depth thermal activation analysis of the deformation of AgY, the most ductile of the anomalous compounds. They learned that polycrystalline AgY is not only ductile at RT, but maintains some limited ductility even at 4 K. The higher than previously observed yield strength in this study is reasonably attributed by the authors to the fine grain (or subgrain) size induced by their thermomechanical processing treatment. They highlight the fact that the active slip mode exhibits a low thermally activated lattice resistance, which is atypical for materials with a bcc-type structure. Furthermore, adherence to the Cottrell-Stokes relation suggested that forest dislocation cutting is the rate controlling deformation mechanism all the way to 4K, typical of pure fcc metals. They ultimately suggest that this similarity to fcc behaviour and the known low elastic anisotropy of the single crystal may be responsible for the anomalous ductility. However, these are also signatures exhibited by some non-cubic metals, such as hcp Mg [12], which are not noted for good ductility without the activation of secondary slip or deformation twinning mechanisms. In short, these observations do not provide a satisfactory observation for the dramatic ductility anomaly. Furthermore, the samples examined did not exhibit the transition in strain hardening response that has been observed in other experimental studies of coarse-grained samples. The interested reader is referred to the paper of Schaarschuch et al. [11] because it contains one of the most up-to-date summaries of the large number of investigations of the ductility anomaly which have been performed.

3. Alternative mechanisms

This list of mechanisms does not include diffusional mechanisms which are typically relegated to high homologous temperatures (dislocation climb or grain boundary sliding). The anomalous ductility under consideration occurs at temperatures well below ~$0.3T_m$ for the systems which we will consider most deeply. For example, AgY has the highest reported room temperature (RT) ductility and a melting point of 1160°C[13], placing RT at $0.2T_m$. Before enumerating these mechanisms, it is noted that Mulay et al. [14,15] examined a number of the candidate ductile MR alloys (AgY, AgEr, CoTi, CoZr, CuY, CuDy, MgCe and MgY) as well as NiAl for comparison and found that neither phase purity, strong crystallographic texture, nor disorder were prerequisites of the ductility anomaly.

3.1. Deformation twinning

Deformation twinning can also be revealed by slip trace analysis, in which the twin variant information (twinning plane and direction) can be revealed in the same was as slip system information. In cases where deformation twinning is active, it has been broadly acknowledged as a relaxation of von Mises criterion. Because twinning is a polar mechanism which can accommodate shear only in the twinning (but not anti-twinning) direction, it has been suggested that twinning modes may represent ½ of an independent slip mode. For example, Kocks and Westlake [16] suggested that pure Ti exhibits 4 ½ independent slip modes due to the importance of deformation twinning in that material. However, as noted by Schaarschuch et al. [11], "the ordinary bcc twinning mode becomes a pseudo[twin] mode [in B2 structured crystals] that produces incorrect ordering in the sheared lattice. A true twin mode would then require atoms to change the site occupation, in addition to the normal shear. Such interchange shuffles are virtually impossible at low… temperatures. Indeed, in B2 structured intermetallics, twinning has not been observed so far."

3.2. Kink banding

Kink banding is another phenomenon that was observed during early studies of single crystal deformation of metals [17,18]. In fact, it has often observed as a mechanism of surface relaxation which helped to accommodate deformation twinning. There have recently been extensive studies of kink banding in non-cubic crystal structured materials with high aspect ratio unit cells. These are typically layered crystal structures like the long-period ordered structure (LPSO) observed in some Mg alloys [19] and the MAX (metal-aluminum-carbide/nitride) phases [20]. However, it was long ago pointed out that kinking does not produce deformations other than those which the dislocations which compose their boundaries could produce on their own [21]. All crystallographic kinking of which the authors are aware involve the motion of dislocations of very easy slip systems (e.g., basal dislocations in LPSO Mg alloys or MAX phases). In short, kink bands do not produce new types of deformation, which are independent of the intrinsic dislocation slip systems of the material. At the risk of redundancy, kink bands represent localizations of deformation accommodated by intrinsic dislocation slip mechanisms. Therefore, neither deformation twinning nor kink banding are considered as important options to alleviate the von Mises criterion in what follows. The emphasis will be placed solely on exploring the possibility that other slip modes are active in these materials.

3.3. Martensitic phase transformation

Different authors have shown that CuY undergoes martensitic phase transformation to the B27 phase, especially at cryogenic temperatures [6,22], but the most recent study has shown that it leads to degradation, not improvement in the ductility [23]. In the case of CoZr, a transformation to B33 phase was observed at crack tips, which may have extended the toughness of that material [24]. However, other samples with significant tensile ductility showed no evidence for this phase transformation, nor did the in-situ diffraction studies outlined below reveal any phase transformation during RT deformation. In short, martensitic phase transformation does not seem to be a viable mechanism to explain the anomalous ductility of this class of MR alloys, though it may play a role in specific cases.

4. Atomistic-scale investigations of dislocation core structures

It is noted that slip of dislocations with a <111> Burgers vector direction, like body-centred cubic (bcc) metals, provides the 5 independent systems required by the von Mises criterion and that numerous bcc metals exhibit extensive ductility. In B2 compounds, the shortest lattice translation vector (and therefore expected Burgers vector of dislocations, if one ignores the role of elastic anisotropy and dislocation core structure) is <100>. In disordered bcc alloys the shortest lattice translation vector parallel to this direction is ½ <111>, whereas in ordered, B2 intermetallic compounds, the shortest lattice translation direction parallel to this direction is <111>, i.e., twice as long and (assuming isotropic elasticity) approximately four times as costly, energetically. In fact, B2 intermetallic alloys known to adopt <111> type slip (Figure 1b) as their primary slip mode, such as β-brass CuZn [25] and FeAl [26], can exhibit some tensile ductility. The preference for one type of slip mode or the other (<100> vs. <111>) has recently been the subject of intense investigation by atomistic modellers. Many of these studies had the goal of explaining the anomalous ductility of MR alloys, given the fact that all experimental evidence suggested that the primary slip mode in these alloys was <100>{011}.

Historically, the question of <100> vs. <111> slip has been addressed by experimental or simulation-based assessments of the antiphase domain boundary (APB) energy on the {110} planes. If the APB energy is low, it was presumed that the <111> "super dislocation" can dissociate into two "super partial dislocations" with an APB between them. If it is high, the <111> dislocation cannot dissociate and, therefore, will generally not be observed. For example, the APB energy in NiAl has been estimated at 500-900 mJ/m^2 [27, 28] and that of β brass is estimated at 60 mJ/m^2 [25]. Researchers have used this to explain why NiAl does not exhibit <111> slip at low homologous temperatures and is brittle, whereas β brass behaves similarly to a disordered bcc metal, despite a strong propensity for ordering.

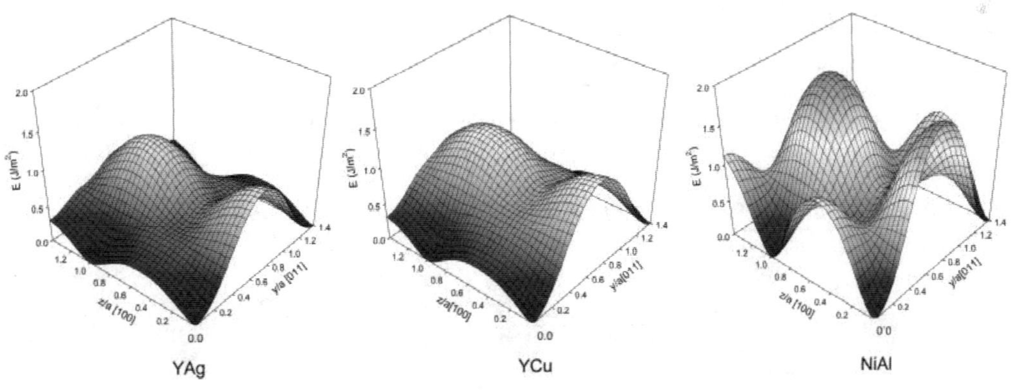

Figure 3. $(01\bar{1})$ plane γ-surface calculations illustrating the similar overall shape for ductile AgY, CuY, and NiAl (after Chen et al. [29], with permission, Elsevier).

Recently, researchers have emphasized that both thermodynamic (elastic self-energy and generalized stacking fault energy) and kinetic (Peierls lattice resistance) aspects need to be considered [30]. Using first-principles density functional theory (DFT), they have computed the γ-surface (or generalized stacking fault energy) on the {110} planes (Figure 3), and have shown that <111> dislocations tend to dissociate into, not just two, but three (or even four) partial dislocations with faults between them. For example, the following dissociation reaction for each half of the <111> dislocation has been proposed: $\frac{1}{2}[111] = \frac{1}{2}[xx1] + \frac{1}{2}(1-x)[110]$, where the value of x and the fault width, d, is determined by an energy minimization scheme [31]. In any event, the fault (anti-phase domain boundary) energies

IOP Conf. Series: Materials Science and Engineering **580** (2019) 012001 doi:10.1088/1757-899X/580/1/012001

predicted for the MR alloys are large and the dissociation core widths are still rather compact. Similarly, recent atomistic studies agree that CoTi has a high APB energy > 550 mJ/m^2 [32,33]. Note that there is little experimental evidence for the proposed dissociation reactions because such high fault energies will lead to very little spreading, making imaging by electron microscopy extremely challenging.

In short, <100> dislocations are thermodynamically preferred over <111>, for the anomalous MR alloys. Indeed, this is the attribute which renders their behavior anomalous! Chen et al. [29] showed that even the <100> perfect dislocations in the ductile yttrium-containing compounds have a low fault energy, permitting them to dissociate into ½<100> partials. This may help to explain the experimental results of Schaarschuch et al. [11], who found the thermally activated barrier associated with lattice friction to be quite low. Curiously, the Peierls barrier predicted by Chen et al. for <100>{011} slip has a strong character dependence, suggesting <100> dislocation motion will not be easy in all directions.

Another important aspect which has recently been emphasized by those who have modeled the detailed core structure of dislocations in B2 compounds is the effect of elastic anisotropy [32,33]. Although some authors have emphasized the near isotropic elastic constants of AgY, it is here emphasized that the Zener anisotropy parameter is greater than one for all of the B2 alloys presently in view, including the brittle NiAl, ductile CuZn and FeAl, and those with anomalous ductility AgY, CuY, CoTi and CoZr. The implication of this fact is that the elastic strain energy difference between <111> and <100> dislocations is not so great as that in isotropic crystals (where the difference would be a factor 3). Typically, the ratio in the self-energy of <111> and <100> screw dislocations predicted to be is about 1.5. Hence, there is a driving force for junction formation, whereby multiple <100> type dislocations meet and "zip up" to form these large Burgers vector dislocations. This has two important implications for what follows. First, the mere observation of dislocations with <111> (or <110>) Burgers vectors in a deformed microstructure is not proof that these dislocations are active participants in strain accommodation. A number of researchers have observed such hard-mode dislocations via TEM (and EBSD). In light of the present facts regarding strain energy, this is not viewed as conclusive evidence that these hard mode dislocations have been involved in significant strain accommodation.

Finally, first principles DFT-based predictions of the Peierls barriers for the motion of <111>$\{1\bar{1}0\}$ screw dislocations (the least mobile) are estimated to be 110 and 230 MPa, for CuY and AgY, respectively. It has been noted numerous times that estimates of the Peierls barrier made using DFT and MD simulations employing empirical potentials can vary significantly [29, 34-36]. However, the fact that these barriers are an order of magnitude smaller than that predicted for NiAl (~2 GPa) is viewed as significant. There is no doubt that <100>{011} slip is the primary slip mode in the anomalous MR crystals. What remains to be described is the experimental evidence of low temperature activation of <111>$\{1\bar{1}0\}$ slip in these MR crystals, which is not observed in the archetypal brittle B2 alloy, NiAl.

5. Mesoscopic investigations of MR alloy deformation behaviour

5.1. In situ neutron diffraction studies

Five systems were examined by this group: NiAl, CuZn, and CeAg [37] as well as CoTi and CoZr [3]. Williams performed a similar in-situ neutron diffraction study of CuY [6, 38], which behaved similarly to CeAg, CoTi and CoZr (Figure 4). The compression samples employed were right cylinders with diameters ~ 10 mm and aspect ratios of 1.5 (details in the original papers). NiAl was selected as archetypal of a cube-slip dominated B2 alloy and CuZn was selected as it undergoes bcc-type <111> slip, and the others are examples of the anomalous MR compounds. In-situ measurement of the internal strains were obtained via single peak fits of the time-of-flight neutron diffraction data, as follows:

$$\varepsilon(hkl) = \frac{d(hkl) - d_0(hkl)}{d_0(hkl)} \qquad (1)$$

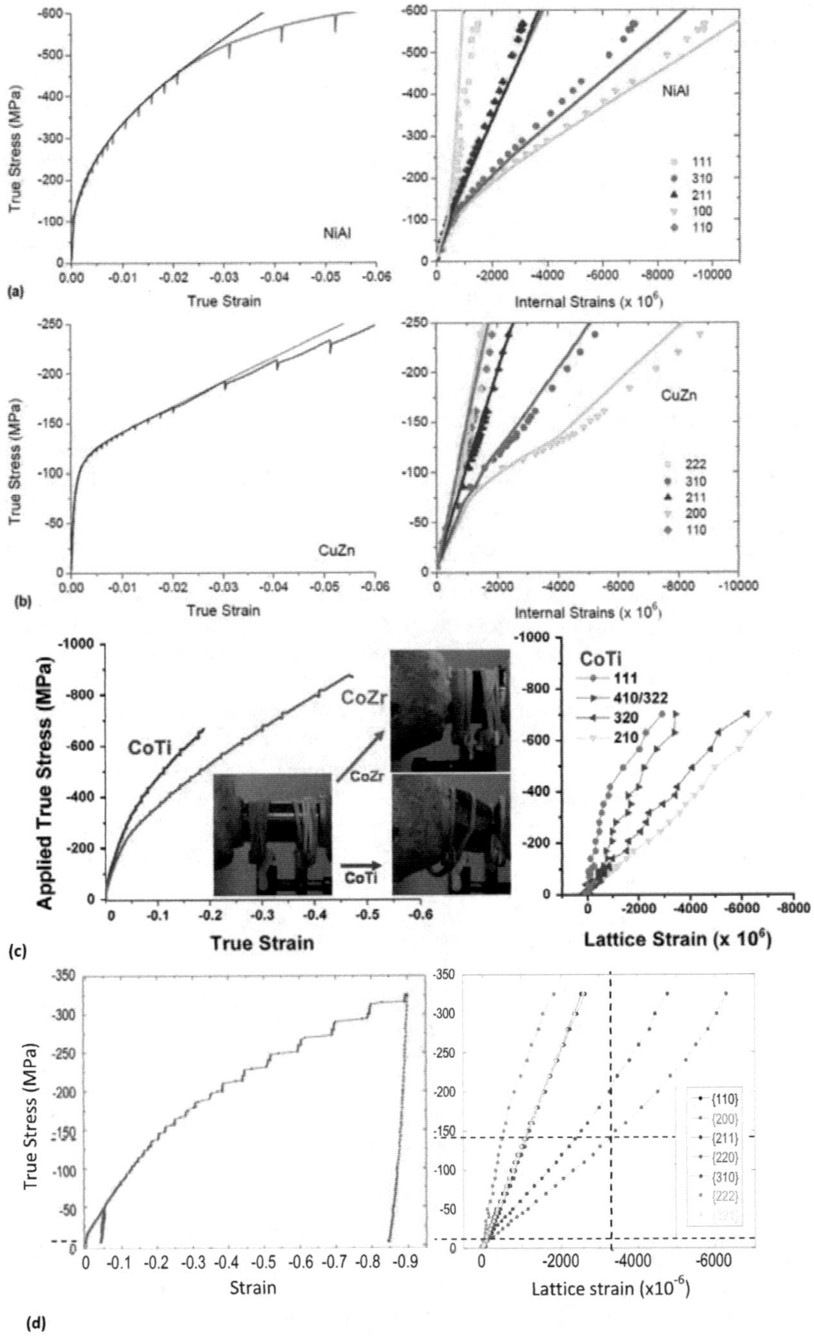

Figure 4. In-situ compression curves and internal strain data obtained from neutron diffraction: a) NiAl, b) CuZn, c) CoTi, (after Wollmershauser et al. with permission, Springer [3] & Elsevier [37]) and d) CuY (adapted from Williams [6]). The insets show images of the in-situ samples with extensometers attached. The serrations in the flow curves are due to interruptions in the straining, while the diffraction data was being collected (10's of minutes). Note the three distinct patterns in the flow curves and internal strain evolutions between the three sample types in (a), (b), & (c)/(d).

The studies revealed three distinct patterns in the flow curve and internal strain evolutions between the three sample types. NiAl exhibits steeply parabolic hardening behavior. The behavior at strains beyond about 3% should be ignored because it was observed that the sample buckled at higher strains. While elastic, all of the grain orientations develop similar levels of internal strain. Upon yielding, the internal strains within variously oriented grains sharply changed their slope but then remained essentially linear all the way up to fracture. As will be discussed in more detail in the modelling section below, the sharp change in the slopes of the internal strain evolutions at yielding is due to the fact that soft-oriented grains (such as those labelled (111), (211), and (110)) begin accommodating strain plastically, due to the easy-glide of dislocations with <100> Burgers vectors. Thus, the elastic strain accumulation in these grains slows down. On the other hand, the hard-orientated grains labelled (100) and (310) cannot easily undergo plasticity and continue loading essentially elastically, thus compensating for the lower load carrying capacity of the soft-oriented grains.

In contrast with this behavior, the ductile CuZn undergoes a much sharper elastoplastic transition followed by essentially linear strain hardening. While elastic, the various grain orientations develop distinct levels of internal strain, a reflection of the strong elastic anisotropy of CuZn [e.g., 33]. Upon yielding, the grains whose orientations are soft with respect to <111> slip, e.g., those labelled (222), (211), and (110) continue to slowly develop internal strain, whereas the hard-oriented crystals, again labelled (100) and (310) begin developing internal strain much more rapidly. Then, we can observe a second transition beyond which all of the grain orientations develop internal strain similarly. The remaining distinction in the slopes is again simply due to the strong elastic anisotropy of CuZn.

The final pattern, exhibited by CoTi, CoZr and CuY (Figure 5c and d) as well as AgCe (not shown), is a response which is intermediate to that of NiAl and CuZn. Initially, these materials behave similarly to NiAl. Following yield at a rather low stress level around 50 MPa, there is a very steep initial strain hardening in the flow curve and protracted build-up of internal strains within hard-oriented crystals, e.g., those labelled (210). Then, at an applied stress level of ~350 MPa or ~150 MPa for CoTi and CuY, respectively, the slope of the strain hardening slows and the internal strain developments (slopes) of the different grain orientations become more similar. Because these alloys are less elastically anisotropic [3,32] we do not observe the continued distinction in the internal strain evolutions that was observed in CuZn. It is hypothesized that the transition in the strain hardening rate is a signature of the activation of a secondary deformation mode which prevents the internal stress levels from continuing to build, and which would otherwise lead to intergranular fracture.

5.2. Elastoplastic Self-Consistent (EPSC) Modeling

In the context of the present Risø Symposium, it may be interesting for readers to note that the first application of an EPSC model for the interpretation of in-situ neutron diffraction straining data was performed by Risø National Laboratory researchers Clausen and Lorentzen [39]. The EPSC code employed in the current work is well-described by its developers, Turner and Tome [40]. In this work, the incorporated slip modes and the critical resolved shear stresses (CRSS) responsible for activating them are employed as fitting parameters used to describe not only the flow curves, but also the internal strain evolutions within the variously oriented grains in the polycrystal. Although all of the alloys examined by neutron diffraction were modelled, only CoTi data is presented as representative.

The aggregate was approximated as an ensemble of ~2900 discrete "grains" with specific orientations and volume fractions that well-approximate the initially weak crystallographic texture of the actual polycrystal that is being modelled. Analogous to the neutron diffraction experiment, where diffraction conditions serve to select the grains which contribute to a given diffraction peak, the model selects the N_{hkl} grains which would contribute to a given virtual diffraction peak (*hkl*). Knowing the stress level within the ith "grain," the elastic lattice strain can be computed by multiplying that stress times the known single crystal elastic compliance tensor, **S**.

$$\boldsymbol{\varepsilon}_i = \mathbf{S} \cdot \boldsymbol{\sigma}_i \tag{2}$$

Thus, the strain value equivalent to the measured value described by Eq. (1) can be modelled as the volume weighted average of the normal strain parallel to the (*hkl*) plane normal $\mathbf{n}_{hkl}$ in these grains

IOP Conf. Series: Materials Science and Engineering **580** (2019) 012001 doi:10.1088/1757-899X/580/1/012001

$$\varepsilon(hkl) = \frac{\sum_{i=1}^{N_{hkl}} f_i(\boldsymbol{\varepsilon}_i \cdot \mathbf{n}_{hkl}) \cdot \mathbf{n}_{hkl}}{\sum_{i=1}^{N_{hkl}} f_i} \quad (3)$$

where f_i is the volume fraction of the ith grain. By comparing the experimental data with the results of simulations with various slip systems (Figure 5), a number of conclusions may be drawn.

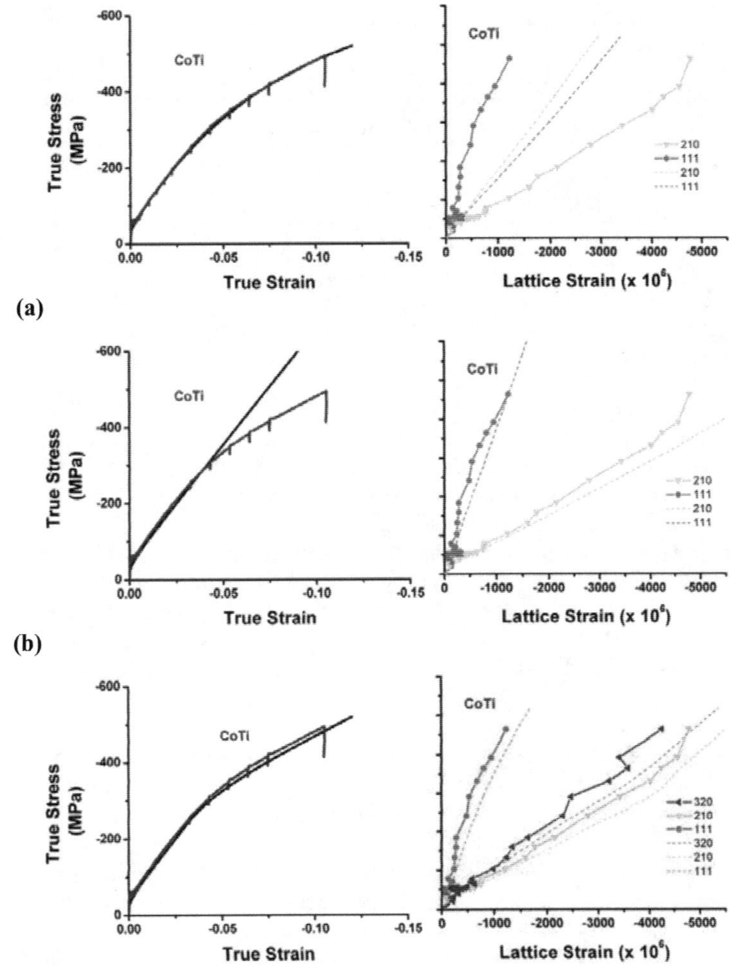

Figure 5: (left) Comparison between experimentally measured and EPSC modelled deformation invoking solely (a) <111> $\{1\bar{1}0\}$ or (b) <100>{011} slip, or (c) <100>{011} primary slip and <111> $\{1\bar{1}0\}$ secondary slip, respectively. (right) Comparison of the corresponding experimental (solid symbols and lines) and predicted (dashed lines) internal strain evolution. Only a small number of experimental and predicted internal strain evolution curves are presented in these plots, in the interest of clarity (after Wollmershauser et al. [3], with permission from Elsevier).

For example, Figure 5(a) shows that, while the flow curve can be simulated using solely <111> slip, the corresponding internal strain evolution predicted by the model is completely wrong (showing very little distinction between the internal strain evolutions of grains with hard and soft orientations). Conversely, Figure 5(b) shows that most of the internal strain evolution can be predicted by the model, if only <100> slip is assumed, whereas only the initial portion of the flow curve can be predicted (regardless of the critical resolve shear stress and hardening parameters that are used to describe the behavior of that slip system.) Finally, Figure 5(c) shows the results of model simulations which simultaneously capture both the stress strain curve and the internal strain evolution. This was accomplished by selecting a low CRSS for <100> slip (which becomes active during the early stages of plasticity), and a much higher CRSS for <110> or <111> slip, which only becomes active after about 5% strain, where a transition in strain hardening is observed (see Table 2). It is important to note that either <110> or <111> or both secondary slip modes can equally well describe the observe stress-strain and internal strain evolutions.

Table 2. Predicted critical resolved shear stresses (CRSS) and initial strain hardening rates (θ) for CoTi and CoZr (all in units of MPa)

	CoTi		CoZr	
Slip system	**CRSS**	θ	**CRSS**	θ
<001>{110}	10	310	8	150
<110> $\{1\bar{1}0\}$	305	210	250	185
<111> $\{1\bar{1}0\}$	235	600	185	520

5.3. The role of intergranular backstresses

A subsequent in-situ neutron diffraction and EPSC modeling study of CoZr performed by Mulay et al [41] provides additional evidence in support of the above understanding. It was shown that strong Bauschinger effects were exhibited by the material such that it would undergo reverse yielding, even prior to completely unloading. This is due to the fact that the primary <100> slip mode cannot accommodate all strain components demanded by the polycrystalline aggregate. Hence, the hard-oriented grains build up large internal stresses, which enforce a "backstress" on their softer surroundings. The amount of backstress-induced hysteresis exhibited by the material strongly increases up until the transition in strain hardening behavior. Beyond this transition, the activation of hard slip modes within the hard-oriented grains relaxes the backstresses. The EPSC model equipped with parameters similar to those in Table 2 predicts not only the monotonic deformation behavior but also the loading-unloading response and associated internal strain evolutions.

There are two effects of backstresses which must be emphasized. First, as emphasized by Wollmershauser et al. [37], they are a major contributor to the pre-transition strain hardening rate in all of the alloys with <100> primary slip modes (including NiAl and the anomalous MR alloys). This sort of composite (or *mesoscale*) effect is distinct from single-phase alloys in which the von Mises criterion is satisfied, where the increase in dislocation density (Taylor hardening) is the dominant source of strain hardening. Secondly, such *mesoscale* intergranular sources of backstress contribute to the athermal portion of the flow stress, in addition to Hall-Petch-type grain size strengthening. It is proposed that these intergranular backstresses are responsible for the large athermal contribution to the strength of YAg, which remained unexplained in the work of Schaarschuch et al. [11].

6. Microscopic investigations of dislocations in MR alloys

6.1. TEM investigations of dislocation types

It has already been mentioned that numerous researchers have examined the rare-earth containing alloys using TEM, and $\mathbf{g} \cdot \mathbf{b}$ diffraction contrast analysis has been used to determine what types of dislocations are present. In many cases, a mixture of primary <100> and secondary <111> (or <110>) dislocations were observed. In the case of AgY, one study reported that the majority of dislocations observed were of the <111> type despite the fact that only <100> slip traces were observed for this alloy [2, 7, 42, 43] In general, the samples were reported as "lightly deformed." As such, the appearance of hard mode dislocation is viewed as potentially incidental (i.e., the result of junction formation), rather than firm evidence of secondary slip mode activity. During the study of CuDy [44], small amounts of <111>{011} slip was activated at high stress levels, which supports the findings of the present study.

Similarly, there have been extensive studies of CoTi and CoZr, especially by Japanese researchers. Though kink banding and associated serrations in the stress-strain curve are seen in single crystal samples of CoTi deformed parallel to the <100> directions, the dislocations present in these samples are still <100> type [45]. Polycrystal studies only observe such serrations and kinking at elevated temperatures [46, 47]. Similar to CoTi, polycrystalline TEM studies of CoZr find the only active slip system to be <100>{011} [24, 48]. In short, despite intense TEM-based research, a mechanistic understanding for the greater than expected ductility of CoTi and CoZr was not provided.

Mulay et al. [49] reexamined the dislocation structures in CoTi in detail. Despite the fact that previous TEM studies had been performed, there were two reasons to look again. First, those previous studies were performed on single crystals in which hard deformation modes may not have activated in part due to a lack of sources. Secondly, previous studies had been performed at low strain levels, where the dislocation densities are lower, enabling clearer images of individual dislocations and simpler identification of the Burgers vectors and other salient information about the dislocations. Because the authors had already shown that hard slip modes are most likely activated only after the transition in strain hardening (at around 5% compressive strain) samples were examined after 7 and 9% compressive strain. Most of the dislocations in the samples deformed to 7 and 9% compressive strain were unambiguously determined to have <100> Burgers vectors, per the $\mathbf{g} \cdot \mathbf{b} = 0$ invisibility criterion. Figure 6 includes a series of two-beam bright field images in which dislocations B and D are clearly visible when $\mathbf{g} = (020)$ and $\mathbf{g} = (\bar{2}00)$, thus ruling out the possibility that these dislocations are also of <100> type. By analysing the images obtained with other diffraction conditions, it is possible to identify the Burgers vector of dislocation D as $\mathbf{b} = [\bar{1}10]$. Dislocation F can also be identified as this type, $\mathbf{b} = [\bar{1}01]$, and appears to be formed by the junction of [100] and [001] dislocations, E and G. There were other cases where dislocations were determined to be either <110> or <111> type.

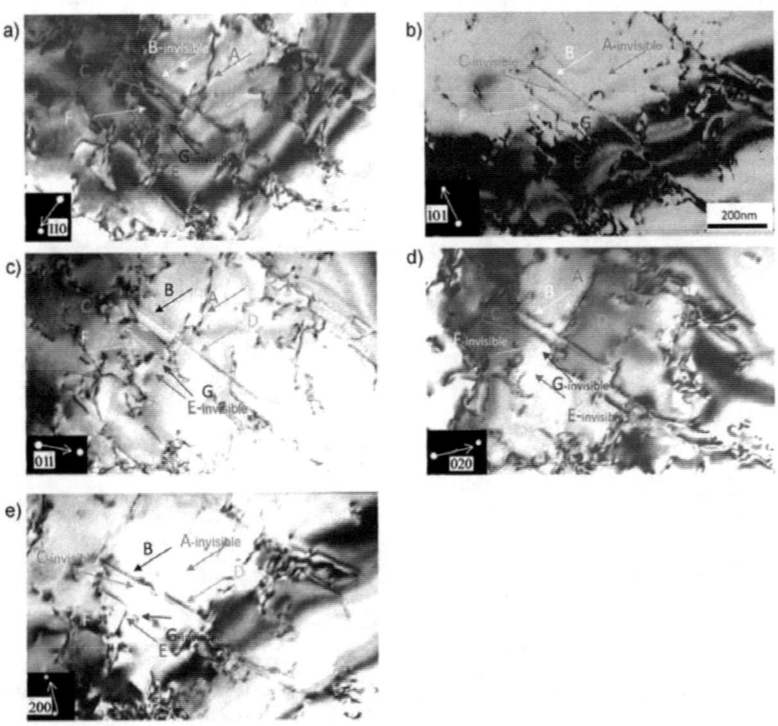

Figure 6. Series of two-beam bright field images of dislocations in CoTi deformed to 9 % strain. The electron beam direction is near [111] for images a, b, c and near [001] for images d and e (after Mulay et al. [49], reproduced with permission).

To summarize the results of the present TEM analysis, in addition to the overwhelming majority of <100> dislocations present in the deformed MR samples, some clear evidence of dislocations with <110> and <111> Burgers vectors has been obtained. Unfortunately, the mere presence of such non-cube type dislocations in TEM images does not provide a satisfying explanation for the anomalous ductility of these alloys. In particular, if it were not for the additional information provided by the in-situ neutron scattering experiments, these observations could indeed be dismissed as incidental. However, together with the in-situ neutron diffraction and EPSC modelling, a case for <110> or <111> hard slip modes which can accommodate strain incompatibilities between grains begins to take form. Nevertheless, more direct, statistically significant evidence of <110> or <111> hard slip modes is still desirable, and the last two techniques discussed in this paper address this point.

6.2. Intragranular misorientation axis (IGMA) measurements

Intragranular misorientation axis (IGMA) measurement using electron back-scattered diffraction (EBSD) can be used to determine the predominant type of geometrically necessary dislocations (GNDs), which are present in numerous grains within a material. This is based upon the fact that unpaired dislocations cause the lattice to rotate about an axis perpendicular to the Burger's vector and slip plane normal, often denoted the Taylor axis [50,51]:

$$\mathbf{IGMA} = \mathbf{b} \times \mathbf{n} \tag{4}$$

The **IGMA** for different slip modes under consideration for B2 alloys are shown in Table 3.

Mulay et al. [49] determined the misorientation axes for neighboring pixels, which have a misorientation of 0.5-2°. Misorientations less than 0.5° may be the result of uncertainty and are subject to large errors. Larger misorientations could include low angle grain boundaries. Table 3 contains the IGMA for 22 grains examined in a sample deformed 2% in compression (prior to the transition in the strain hardening response). Of the 82% which showed a clear IGMA, all of them could be associated with dislocations having a <100> (or <110>) Burgers vector. Table 3 also contains the IGMA for 30 grains in a sample deformed 16% in compression (well beyond the transition). Of the 83% which showed a clear IGMA, 66% of them could still be associated with dislocations having a <100> (or <110>) Burgers vector. However, 17% provided evidence for the activation of the <111>$\{1\bar{1}0\}$ slip mode. These results are in good agreement with the results of the aforementioned polycrystal plasticity modeling study, which showed that the activation of the <111>$\{1\bar{1}0\}$ slip mode provides an explanation for the transition in strain hardening and associated change in internal strain evolution.

Table 3. Intragranular misorientation (Taylor) Axes for common slip systems in B2 (CsCl) crystals and percentage of grains showing a particular IGMA at two different strain levels

Slip system	IGMA	% grains – prior to transition (2% strain)	% grains – beyond the transition (16% strain)
<001>{110}	$<1\bar{1}0>$ - 6 variants	32	43
<001>{100} or <110> $\{1\bar{1}0\}$	<010>- 3 variants	50	23
<111>$\{1\bar{1}0\}$	$<11\bar{2}>$-12 variants	0	17
N/A	No clear IGMA	18	17

6.3. Synchrotron x-ray diffraction-based assessment of dislocation densities

This final section of the paper describes a new high energy x-ray diffraction (HEXD) technique which enables one to obtain single crystal diffraction patterns from large numbers individual grains within a polycrystal [52]. Using these data, one can estimate the dislocation densities of different slip modes, slip systems, and dislocation characters within each of the grains which are interrogated. In short, information which was formerly relegated to TEM studies, such as those described in section 6.1 is now available via x-ray diffraction. More significantly, a statistical advantage is obtained, since it is in principle possible to evaluate hundreds of grains within a given sample. In the present study, 125 grains were analysed within a single sample of deformed CoTi (and about 30 were analysed in CoZr). Clearly, such an evaluation would be virtually impossible by TEM.

The same arc cast and hot-extruded CoTi and CoZr alloys described above were electro-discharge machined into small tensile specimens with a gauge diameter of 1.1 mm and a gauge length of 15.75 mm which was parallel to the extrusion axis. More details about alloy and tensile sample preparation can be found in the paper by Wollmershauser et al. [3]. The tensile samples were strained to failure, as shown in Figure 1. The failed tensile samples were then electropolished to a tapered needle shape such that a varying number of grains appear within the cross-section of the sample. During the high energy x-ray diffraction experiment, which was performed in transmission, the sample was be moved up and down along the tensile axis to obtain the desired number of grains within the illuminated region. The limiting goals are to get as many grains as possible, to improve the statistics of the microstructure assessment, but not so many as to render it impossible to discriminate single crystal diffraction peaks from individual grains.

The complete details of the experimental measurements and subsequent data analysis are provided in the original paper and only a summary is provided here. The sample was rotated on its tensile axis (ω) and two-dimensional detector data (see Figure 7) was collected every 0.5° over a range of ±90° using both intermediate (500 mm, yielding a large solid angle) and far (2500 mm, high resolution) sample-to-detector distances. Using the intermediate-detector, the FABLE and GRAINSPOTTER software packages [53, 54] utilized the fact that diffraction spots go in and out of contrast with sample rotation (ω) to determine which peaks belong to which grains. The x, y (or 2θ, η) and ω positions of the peaks on the intermediate detector were associated with the peaks on the far detector by the CORRELATION software, which was produced as a part of the research project. Once this task of associating (and indexing) individual (*hkl*) peaks with grains on the far-detector was complete, the individual diffraction peaks were analysed in the following fashion.

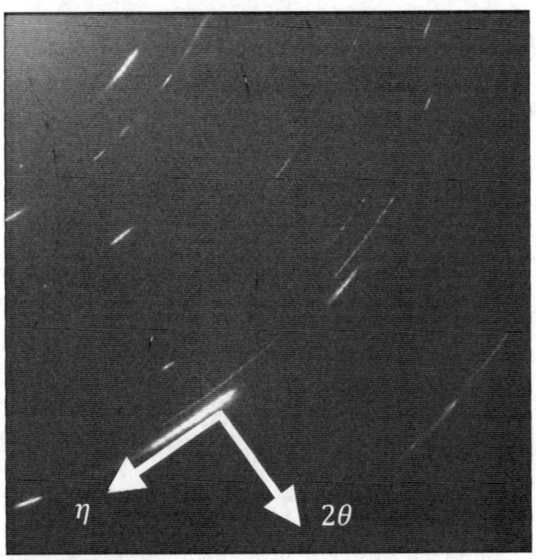

Figure 7. High resolution 2D x-ray diffraction data illustrating spreading of a diffraction peak along the radial (2θ) and azimuthal (η). (adapted from Ungar et al. [52], with permission, Elsevier)

For each diffraction spot within the 3D space of ω, η, and 2θ, the intensity spread along ω and η (rocking curve spread) was integrated by the INTEGRATE software [55, 56] to produce I(2θ) intensity distributions. Notably, the selection of the ω and η ranges over which this integration takes place is critical. If it is too small, the I(2θ) profiles will not be correctly estimated and if the range is too large data from other peaks (from other grains) will be included. Note that the ω and η (rocking curve) spread can be 100 times greater than the spread along the 2θ (radial) direction. The instrumental broadening in the I(2θ) profiles were estimated by analysing the shape of the narrowest of the ~3000 peaks analysed in the course of the research. The instrument profile was described by a pseudo-Voigt comprised of 80%

Gaussian and 20% Lorentzian contributions and a FWHM of ΔK = 0.0083 1/nm. This peak breadth corresponds to about 2 pixels on the far-detector.

Each peak profile is associated with its signed (*hkl*) indices and the corresponding K values (where K = 2sinθ/λ and λ is the wavelength of the incident monochromatic x-rays). Again, about 125 and 30 grains were identified for CoTi and CoZr, respectively, with about 12 to 40 different (*hkl*) reflections for each grain. Fewer grains were successfully indexed for the CoZr material because it was strained to a higher level, and thus, the diffraction peaks were highly spread in the ω and η (rocking curve) directions, rendering them difficult to discriminate from each other. Once the integration procedure is completed, all of the profiles corresponding to a single grain were assembled in a single data for the convolutional multiple whole profile (CMWP) analysis described in the original paper. All diffraction profiles comprising the diffraction pattern are simultaneously fit by the following convolution function:

$$I^M(2\theta) = \sum_{hkl} I^d_{hkl} \cdot I^D_{hkl} \cdot I^{SF}_{hkl} \cdot I^{INST}_{hkl} + I_{BG} \quad (5)$$

where, $I^M(2\theta)$ is the measured line profile, and I^S_{hkl}, I^D_{hkl}, and I^{SF}_{hkl} are the physically based, theoretical profile functions related to coherently scattering domain size (d), distortions caused by dislocations (D) and planar faults (stacking faults and twinning, SF) as well as the aforementioned instrumental profile I^{INST}_{hkl}, and the background (BG) intensity. In this case, a Monte Carlo-type algorithm was employed to minimize the residual difference between the measured and fitted multiple whole peak profiles [57].

The size and strain contributions are parsed similarly to the Fourier method originally introduced by Warren [58]. Admittedly, the procedure for extracting the dislocation densities of individual slip systems within individual grains is complex. However, three checks of self-consistency were employed. The first relates to the goodness of fit, even at very small fractions (10^{-2} to 10^{-3}) of the peak intensity (essentially the tails of the peaks), where it is known that the presence of dislocations can still have a profound effect. Secondly, although the data appear very irregular on the traditional Williamson-Hall plot of the integral peak breadth β vs. K (the magnitude of the diffraction vector), it is linearized on the *modified* Williamson-Hall plot[59] of β vs. $K\sqrt{C_{calc}(\boldsymbol{K})}$. This indicates that the contrast factors properly account for the different effects a given dislocation population will have on the integral breadth of different peaks. Finally, the average dislocation densities were also determined in the two alloys by laboratory, powder x-ray diffraction experiments using a dedicated high-resolution diffractometer as in [16,20,27]. For CoZr and CoTi the values were obtained to be: $2.3(\pm0.5)\times10^{14}$ m^{-2} and $1.7(\pm0.5)\times10^{14}$ m^{-2}, respectively, in good correlation with the average of the values obtained for the individual grains.

In CoTi deformed to about 2% tensile strain, dislocations with <100> Burgers were observed in in 99 (80%) of the 123 analysed grains. In fact, there were a number of grains in which the soft mode dislocations were the overwhelming majority (>70 or 80%). However, hard mode dislocations with <110> or <111> Burgers vectors were required to describe the data, even in this moderately deformed sample. In many grains, they were a significant fraction of hard <110> and <111> dislocations. Nevertheless, these are considered potentially incidental, since there is a driving force for dislocations to form due to reactions between the primary <100> dislocations. In CoZr, which was deformed well beyond the transition in strain hardening, where in-situ neutron diffraction and EPSC modelling results clearly suggest a significant activity of <110> or <111> slip modes, only a single grain of the more than 30 which were analysed exhibited >70% <100> dislocations. In all of the other grains, there was a significant presence of hard mode dislocations with <110> or <111> Burgers vectors.

With such a large fraction of the dislocations belonging to hard slip modes, in many of the analysed grains, it seems unlikely that these dislocations arise only from interactions between soft mode dislocations. Rather, such a high fraction of these dislocations leads to the obvious conclusion that these were active slip modes during the deformation process. There is also clear, positive correlation between the presence soft mode <100> dislocations and the Schmid factor of the grain in question. Curiously, there was no such trend could describe the relative fraction of dislocations with <110> or <111> Burgers vectors. Perhaps these slip modes are subject to non-Schmid effects, or perhaps their source operation relies heavily on junction formations (and hence preceding <100> slip).

As was recently articulated by Zihaly, Ungar, and Tichy [60], there is distinct information about the dislocation content within the 2θ (radial) spread as compared with the ω and η (rocking curve) spread.

The former is influenced by the variations in local lattice parameter induced by both redundant (statistically stored, SSD) and geometrically necessary (GND) dislocations, whereas the rocking curve is only sensitive to GND content, which contributes to lattice curvature. In many ways, the information obtained from rocking curve spread is analogous to that which may be obtained from EBSD (as described in section 6.2) or X-ray micro-Laue measurements of local misorientation, e.g. [61]. Unfortunately, and unlike those spatially resolved measurements, rocking curve spread of the present HEDM technique provides incomplete information regarding the GND content. Novel, forward-modelling techniques may help aid in the interpretation of such diffraction data [62, 63].

7. Discussion

As explained in the previous sections, the activation of hard slip modes, i.e., <111>$\{1\bar{1}0\}$ or <110>$\{1\bar{1}0\}$, is ultimately the mechanism held responsible for the anomalous ductility by the present researchers, as well as some others like Cao et al. [7, 43, 44]. It has been shown that this explanation is consistent with the results of most other researchers, who focused their efforts at specific length-scales such as the atomistic length scale of dislocation cores.

Even in cases where there are apparent contradictions, there are reasonable explanations. It has already been mentioned that prior TEM studies were often performed at low strain (stress) levels which are now known to have been insufficient to activate these hard modes. Therefore, it is not surprising the many researchers failed to observe evidence of significant hard deformation mode activity. Most troublesome are the apparently contradictory single crystal studies of AgY [e.g, 64], in which cracking intervened before any plasticity initiated in a sample pulled in tension along <100> with the intention of inducing <111>$\{1\bar{1}0\}$ slip. However, it is posited that these dislocations with large <111> Burgers vectors can become active in polycrystals, even when single crystal deformation fails to induce the mechanism. For example, given the high self-energy and Peierls resistance of these dislocations, it is likely that <111> slip is limited by both mobility and source operation. It may be necessary for some <100> slip to precede the activation of <111> slip, as a means toward source formation. Indeed, atomistic studies of core structure have shown that there is a driving force for <110> and <111> junction formation [33] and such junctions have been observed by TEM [49]. Ironically, another explanation for why there are few observations of <111> slip during single crystal studies is kink banding, which is a plastic instability involving the primary <100> slip mode, which can lead to premature failure when compression parallel to a hard direction. Again, the conditions required for <111> slip are not met.

Finally, it is admitted that <111> slip is not required to explain all observations of "ductility" in B2 compounds. Schaarschuch et al. [11] observed limited (1-3.5%) ductility in YAg over a wide range of temperatures between 4K and RT. The strain hardening rate exhibited in their tests was quite steep with no transition to lower hardening, which is observed in the published studies of more extensive (6 – 20%) ductility in MR compounds. It is suggested that the fine microstructure of their material led to an increase in the strength of both the soft and hard dislocation modes. In fact, studies of Mg alloys have suggested that slip modes involving dislocations with large Burgers vectors may exhibit a larger strain hardening effect from grain size refinement [65, 66]. The highlighted intergranular backstress (section 6.3) provides a self-consistent explanation for the high strain hardening rate (despite the fact that dislocation densities do not rapidly in materials with such a fine microstructure) and the large athermal contribution to the flow stress.

8. Summary

- There exists a broad class of fully ordered intermetallic alloys with the B2 structure, denoted MR alloys, which exhibit anomalous ductility confirmed by numerous research groups around the world. The behavior is considered anomalous, because such B2 alloys with a <100>{011} primary slip mode are expected to be completely brittle (like NiAl).
- These MR alloys do not exhibit extended ductility (large reduction in area) typical of pure fcc metals. Rather, it is a limited sort of ductility which is more akin to that of noncubic metals. The fracture surfaces typically appear brittle, either transgranular cleavage or intergranular.

- At least one representative (AgY) of this class of alloys exhibits some ductility down to liquid He temperatures (4 K), emphasizing that these alloys do not exhibit a sharp ductile-to-brittle-transition (DBTT) of bcc metals at low temperatures.
- Atomistic simulations of the core structure have universally concluded that all of the "anomalous" MR alloys bear more in common with NiAl than ductile CuZn (β-brass) or FeAl, which exhibit <111>$\{1\bar{1}0\}$ as the primary slip mode. In short, <100>{011} slip is shown to be the primary slip mode, both theoretically – the <100> dislocations have the lowest self-energy and experimentally – slip trace analyses universally highlight the activity of <100> dislocations.
- Experimental studies reveal the <100>{011} slip mode of the anomalous B2 alloys to be quite soft, in comparison with other alloy B2 intermetallic alloys.
- TEM analyses of dislocations in deformed MR alloys have also revealed dislocations with <110> or <111> Burgers vectors. The elastic anisotropy of the anomalous B2 compounds is such that the <111> screw dislocations are typically only 3/2 times as energetic as <100>. Hence, there is a driving force for junction formation. This means that the mere occurrence of dislocations with <111> (or <110>) Burgers vectors in a deformed microstructure is not proof that these dislocations are active participants in strain accommodation. However, it also provides a mechanism for <111> dislocation source creation.
- First-principles calculations predict the Peierls barrier for <111>$\{1\bar{1}0\}$ slip of screw dislocations in CuY and AgY to be (~100-200 MPa), much lower that the barrier predicted for NiAl (~2 GPa).
- In-situ neutron compression testing of anomalous MR alloys revealed a sharp transition in the strain hardening behavior and an associated break in the internal strain evolution. Prior to this transition, large intergranular strain development can be associated with the activity of <100>{011} alone.
- Beyond the transition in strain hardening behavior, the subsequent internal strain evolution and the reduction in strain hardening rate observed in in-situ neutron diffraction experiments is well explained by the activation of <111>$\{1\bar{1}0\}$ (or <110>$\{1\bar{1}0\}$) slip. Additionally, the CRSS level inferred from polycrystal plasticity modeling of the hard modes is in keeping with the estimated Peierls barriers of the anomalous MR alloys.
- A new synchrotron-based, high energy X-ray diffraction technique for measuring dislocation densities within individual grains by convolution multiple whole profile (CMWP) analysis was performed on a CoTi sample deformed to a strain level below the aforementioned transition in strain hardening. It revealed <100> dislocations in all grains, which constituted the vast majority of the dislocations in numerous grains. However, <110> and <111> dislocations were also observed in significant fractions in most grains.
- Grain-by-grain CMWP analysis was also performed on a CoZr sample deformed to a strain level well beyond the aforementioned transition in strain hardening. It revealed a combination of <100>, <110>, and <111> dislocations, no longer having a strong preference for <100>. This is viewed as strong evidence that motion of the hard dislocations has become an important mechanism of strain accommodation, thus explaining the anomalous ductility in this class of alloys.

9. Acknowledgments

The authors would like to thank their present institutions, University of Virginia and University of Manchester, for supporting this review. They would also like to thank Jishnu Bhattacharyya for valuable discussions and support in the article preparation.

10. References

[1] von Mises R 1928 *Z. Ang. Math. Mech.* **8** 161-85
[2] Gschneidner Jr K, Russell A, Pecharsky A, Morris J, Zhang Z, Lograsso T, Hsu D, Lo CC, Ye Y, Slager A and Kesse D 2003 *Nature Mater.* **2** 587
[3] Wollmershauser JA, Neil CJ and Agnew SR 2010 *Metall. Mater. Trans. A* **41** 1217-29

[4] Wollmershauser JA 2011 *Toward a more complete mechanistic understanding of the mechanical behavior of anisotropic single-phase polycrystalline materials* (Charlottesville, Virginia: University of Virginia)
[5] Zhang Z, Russell AM, Biner SB, Gschneidner Jr KA and Lo CC 2005 *Intermetallics* **13** 559-64
[6] Williams SH 2009 *In-situ neutron diffraction analysis of deformation behavior of ductile rare-earth intermetallic YCu* (Ames, Iowa: Iowa State University)
[7] Cao GH, Shechtman D, Wu DM, Becker AT, Chumbley LS, Lograsso TA, Russell AM and Gschneidner Jr KA 2007 *Acta Mater.* **55** 3765-70
[8] Li H, Mason DE, Bieler TR, Boehlert CJ and Crimp MA 2013 *Acta Mater.* **61** 7555-67
[9] Takasugi T, Tsurisaki K, Izumi O and Ono S 1990 *Philos. Mag. A* **61** 785 - 800
[10] Russell AM, Zhang Z, Gschneidner Jr KA, Lograsso TA, Pecharsky AO, Slager AJ and Kesse DC 2005 *Intermetallics* **13** 565-71
[11] Schaarschuch R, Oertel CG, Cao G, Freudenberger J, Brokmeier HG and Skrotzki W 2018 *Acta Mater.* **151** 149-58
[12] Conrad H, Hays L, Schoeck G and Wiedersich H 1961 *Acta Metall.* **9** 367-78
[13] Gschneidner K and Calderwood FW. 1983 *Binary Alloy Phase Diagrams,* 2nd Ed. vol 1 ed Massalski TB (Materials Park, Ohio: ASM International) p. 114
[14] Mulay RP, Wollmershauser JA, Heisel MA, Bei H, Russell AM and Agnew SR 2010 *Acta Mater.* **58** 2788-96
[15] Mulay RP 2011 *A diffraction based study of the deformation mechanisms in anomalously ductile B2 intermetallics* (Charlottesville, VA: University of Virginia)
[16] Kocks UF and Westlake DG 1967 *AIME Met. Soc. Trans.* **239** 1107-9
[17] Orowan E 1942 *Nature* **149** 643
[18] Hess JB and Barrett CS 1949 *JOM* **1** 599-606.
[19] Yamasaki M, Hagihara K, Inoue SI, Hadorn JP and Kawamura Y 2013 *Acta Mater.* **61** 2065-76
[20] Barsoum MW, Farber, L and El-Raghy T 1999 *Metall. Mater. Trans. A* **30** 1727-38
[21] Fraser HL, Smallman RE and Loretto MH 1973 *Phil. Mag.* **28** 651-65
[22] Ritter C, Ibarra MR and Ibberson RM 1994 *J. Phys.: Condens. Matter* **4** L339
[23] Oertel CG, Schaarschuch R, Cao GH, Tian HN, Freudenberger J, Brokmeier HG and Skrotzki W 2011 *Scripta Mater.* **65** 779-82
[24] François A and Veyssiere P 1994 *Intermetallics* **2** 9-22
[25] Potter DI 1970 *Mater. Sci. Eng.* **5** 201-9
[26] Fu CL and Yoo MH 1991 *Acta Metall. Mater.* **40** 703-11
[27] Fu CL and Yoo MH 1991 *High-Temperature Ordered Intermetallic Alloys IV* ed Johnson LA, et al. (Pittsburgh, PA: Materials Research Society) p. 667-72
[28] Veyssiere P and Noebe R 1992 *Phil. Mag. A* **65** 1-13
[29] Chen Q, Ji M, Wang CZ, Ho KM and Biner SB 2010 *Intermetallics* **18** 312-8
[30] Paidar V and Čák M 2016 *Intermetallics* **73** 21-5
[31] Yan JX, Zhang ZJ, Li KQ, Xia ZY, Yang JB and Zhang ZF 2019 *Intermetallics* **110** 106470
[32] Paidar V, Lin YS, Čák M, Vitek V 2010 *Intermetallics* **18** 1285-87
[33] Lin YS, Čák M, Paidar V and Vitek V 2012 *Acta Mater.* **60** 881-8
[34] Schoeck G 2001 Acta Mater. **49** 1179-87
[35] Schoeck G 2001 *Comp. Mater. Sci.* **21** 124-34
[36] Lu G, Kioussis N, Bulatov VV and Kaxiras E 2000 *Philos. Mag. Lett.* **80** 675-82
[37] Wollmershauser JA, Kabra S and Agnew SR 2009 *Acta Mater.* **57** 213-23
[38] Williams SH, Brown DW, Clausen B, Russell A and Gschneidner Jr KA 2014 *Acta Mater.* **70** 307-15
[39] Clausen B and Lorentzen T 1997 *Metall. Mater. Trans. A* **28** 2537-41
[40] Turner PA and Tomé CN 1994 *Acta Metall. Mater.* **42** 4143-53
[41] Mulay RP, Clausen B and Agnew SR 2011 *Metall. Mater. Trans. A* **42** 60–70
[42] Xie S, Russell AM, Becker AT and Gschneidner Jr KA 2008 *Scripta Mater.* **58** 1066-9

IOP Conf. Series: Materials Science and Engineering **580** (2019) 012001 doi:10.1088/1757-899X/580/1/012001

[43] Cao GH and Yu Z 2009 J. *Alloys Comp.* **488** 44-6.
[44] Cao GH, Yu Z and Russell AM 2011 *Mater. Sci. Eng. A*, ***528*** 7173-77
[45] Takasugi T, Yoshida M and Kawabata T 1992 *Philos. Mag. A* **65** 29 - 40
[46] Takasugi T and Izumi O 1988 *J. Mater. Sci.* **23** 1265-73
[47] Wittmann M and Baker I 2002 *Mater. Sci. Eng. A* **329-331** 206-12
[48] Yoshida M and Takasugi T 1993 *Philos. Mag. A* **68** 401-17
[49] Mulay RP and Agnew SR 2012 *Acta Mater.* **60** 1784-94
[50] Rapperport EJ and Hartley CS 1960 *Trans. AIME* **218** 869-77
[51] Chun YB, Battaini M, Davies CH and Hwang SK 2010 *Metall. Mater. Trans. A* **41A** 3473-87
[52] Ungár T, Ribárik G, Zilahi G, Mulay R, Lienert U, Balogh L and Agnew S 2014 *Acta Mater.* **71** 264-82
[53] Wright J http://sourceforge.net/projects/fable/files/ImageD11/
[54] Schmidt S http://sourceforge.net/apps/trac/fable/wiki/grainspotter
[55] Ungár T, Ribárik G, Balogh L, Salem AA, Semiatin SL and Vaughan GB 2010 *Scripta Mater.* **63** 69-72
[56] Nisr C, Ribárik G, Ungár T, Vaughan G, Cordier P and Merkel S 2012 *J Geophys. Res.* **117** B03201
[57] Ribárik G, Jóni B and Ungár T 2019 *J. Mater. Sci. Technol.*, **35** 1508–14
[58] Warren BE 1959 *Prog. Metal Phys.* **8** 147–202
[59] Ungár T and Borbely A 1996 *J. Appl. Phys.* **69** 3173-75
[60] Zilahi G, Ungár T and Tichy G 2015 *J. Appl. Cryst.* **48** 418-30
[61] Barabash RI, Ice GE, Liu W and Barabash OM 2009 *Micron* **40** 28-36
[62] Balogh L, Capolungo L and Tomé CN 2012 *Acta Mater.* **60**1467-77.
[63] Bertin N and Cai W 2018 *Comp. Mater. Sci.* **146** 268-77
[64] Russell AM, Zhang Z, Lograsso TA, Lo CC, Pecharsky AO, Morris JR, Ye Y, Gschneidner Jr KA and Slager AJ 2004 *Acta Mater.* **52** 4033-40
[65] Jain A, Duygulu O, Brown DW, Tomé CN and Agnew SR 2008 *Mater. Sci. Eng. A* **486** 545-55.
[66] Raeisinia B and Agnew SR 2010 *Scripta Mater.* **63** 731-6.

Length scales and alloys of iron

H K D H Bhadeshia

University of Cambridge, Materials Science and Metallurgy, CB3 0FS, UK

E-mail: `hkdb@cam.ac.uk`

Abstract. Many of the structures common within steels are not geometrically smooth, which means that their stereological parameters, such as the amount of interfacial area per unit volume, are resolution dependent. The tendency of brittle phases to crack is size dependent, as is the strength. The ability of steel to resist cleavage can be influenced by the size and distribution of coherent domains within the structure. A different diffusion theory applies when dealing with steep concentration gradients. All of these aspects are of importance in the design, characterisation and theory of steels, illustrated here with a few examples, where the neglect of length scales is improper.

1. Introduction

We do not as yet attach sufficient significance to length scales in the context of engineering materials. The problem is compounded by the plethora of characterisation methods and theoretical treatments, that cover many orders of magnitude of resolution. The purpose of the work presented here is to describe a few phenomena related to steels, where neglect of the coupling between length scale and theory or experiment, can become problematic.

2. Rough objects

It is intriguing that platelets of bainite (α_b) are arrested in their growth before they encounter a hard obstacle, such as an austenite grain boundary [1]. There is a shape deformation [2, 3], too large to accommodate elastically in the temperature regime where bainite grows. This causes the adjacent austenite to relax by plastic deformation [4, 5], creating a large density of tangled dislocations in the vicinity of the transformation interface [6]. This blocks the motion of the glissile α_b/γ interface [7], causing *mechanical stabilisation* [8, 9]. The platelets therefore grow to a limited size, so the transformation must propagate by the formation of new platelets, each of a limited size. The best location for a new platelet to form is either at the austenite grain boundary or at the tip of a prior platelet. Both of these sites activate to generate clusters of platelets that are known as sheaves.

The structure of a sheaf is illustrated in figure 1 at a variety of resolutions. Platelets of bainitic ferrite are separated by films of retained austenite (γ). These platelets are in fact contiguous and hence in the same crystallographic orientation in space. This means that the shape of the austenite/ferrite interface is complex, making any attempt to determine the amount of interface per unit volume (S_V) dependent on the spatial resolution of the measurement technique used.

In the terminology of fractal dimensions, the interface is said to be *rough*. This contrasts with smooth objects whose stereological properties do not depend on resolution. A perfect circle would have a precise perimeter, a perfect sphere an exact surface area, both quantities

IOP Conf. Series: Materials Science and Engineering **580** (2019) 012002 doi:10.1088/1757-899X/580/1/012002

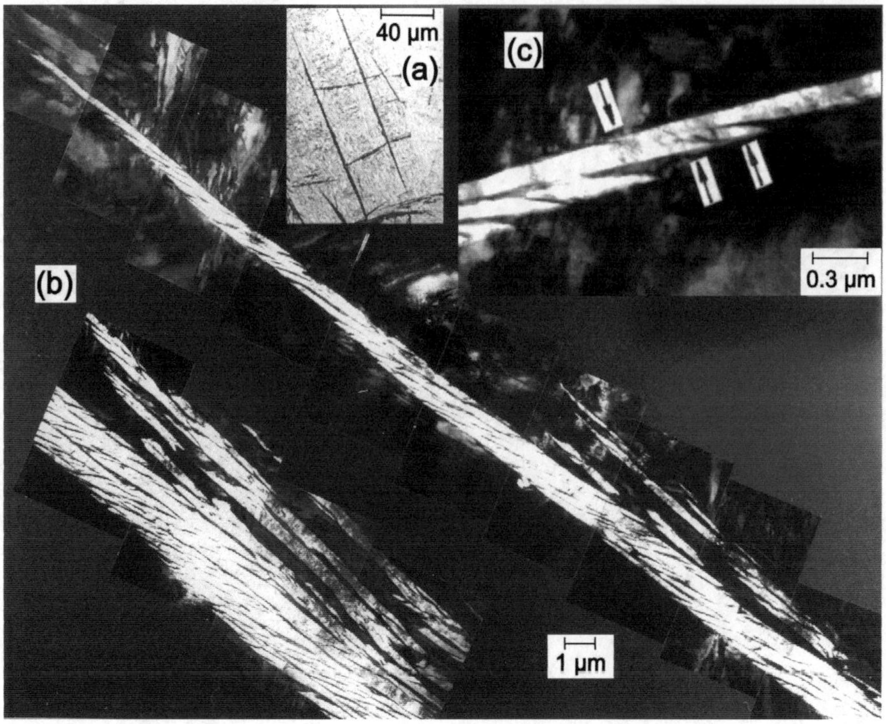

Figure 1. Structure of a bainite sheaf at a variety of resolutions [10]. (a) Optical micrograph, the dark plate-like features are bainite sheaves. (b) Transmission electron micrograph montage showing detailed structure of sheaf, consisting of very many platelets, each about 10 μm in length. (c) Another generation of much shorter sub-critical platelets.

defined by explicit equations independent of resolution. There is a theory for roughness [11], in which a fractal dimension D determines how, for example, the perimeter of a rough object such as the outline map of Scotland, scales with resolution. The dimension can be determined from the gradient of a plot of $\ln N_i$ against $\ln \varepsilon_i^{-1}$, where $i = 1, 2, 3, \ldots$ is the number of different resolutions at which measurements are made, $N_i \times \varepsilon_i$ is the total length measured at a particular resolution ε_i.

In the case of bainite, the α_b/γ interfacial area per unit volume is an important parameter in determining a variety of properties, such as toughness, the ability to trap hydrogen [12] etc. [13].

Toughness is related to the size of the plastic zone at a crack tip [14], so the appropriate resolution for a typical steel would be of the order of 10 μm, whereas the scale is much smaller when it comes to hydrogen absorption at an interface. It would not be unreasonable in the latter case to measure S_V with a resolution that is close to atomic, since there are multiple interstices per iron atom where hydrogen can reside [15]. The interfacial area per unit volume is given by:

$$\ln S_V = S_o \varepsilon^{(D_T - D)}, \tag{1}$$

where S_o is the surface to volume ratio of a smooth object with a topological dimension D_T, and D is the fractal dimension. By measuring S_V at a variety of resolutions, the fractal dimension for the bainite-austenite interfacial area has been determined as follows:

$$\ln S_V = 0.59 \ln \varepsilon^{-1} + 5.4 \tag{2}$$

Figure 2. Measuring the perimeter of Scotland with callipers, set to a resolution ε. The nooks and crannies are not captured at the resolution illustrated, but the finer features would enlarge the perimeter if the distance ε between its ends is reduced.

where 0.59 is the slope of the line in figure 3, i.e., $D - D_T$, with $D_T = 3$ for the three-dimensional object that a bainite sheaf is. Using this, it is possible to calculate the amount of surface per unit volume as a function of resolution, and the information used in order to address particular properties. It is emphasised however, that the bainite sheaf is not in fact a true fractal object because the self-similarity of platelets does not extend to indefinitely finer scales. So the change in S_V with resolution is smaller than if the structure represented a true fractal.

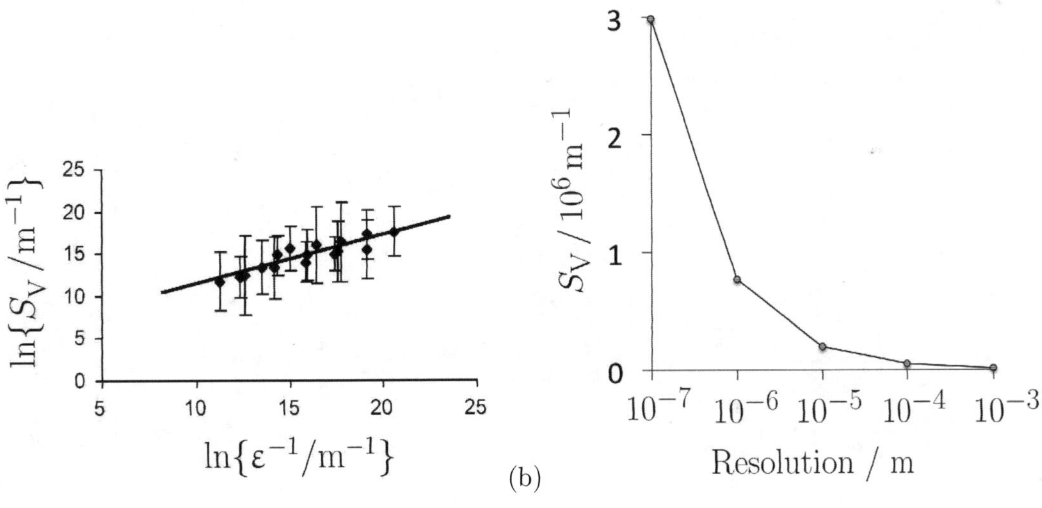

Figure 3. (a) Plot to determine the fractal dimension of α_b/γ interfaces; (b) variation in S_V as a function of resolution [16].

3. Cracking and complexity

It has been possible to create the largest ever density of strong-interfaces, some 100 million m^2/m^3, in steel by heat treatment alone [17–19]. The structure consists of platelets of bainitic ferrite, just 20-40 nm in thickness, embedded in carbon-enriched austenite. Unlike many methods for the production of nanostructured metals, it becomes possible to create components that are large in all three dimensions, with uniform properties throughout. The fundamental mechanisms behind this nanostructured steel are now fairly well-understood, along with the factors determining its strength, ductility and fracture toughness [20].

Nevertheless, there are difficulties which prevent the wider exploitation of the nanostructure. The large carbon concentration of the steel (about 1 wt%) prevents it from being welded because of the brittle, untempered martensite induced in the heat-affected zone of the joint [21, 22]. This dramatically reduces the utility of the steel to objects such as shafts, bearings and armour. The second difficulty is the poor impact toughness of just 5 J of absorbed energy at 20°C, in spite of the more than acceptable fracture toughness (K_{IC}) at a strength of more than 2 GPa.

Weldability in the context of strong steels can in principle be achieved by reducing the carbon concentration to less than ≈ 0.3 wt% [23] but this would compromise the strength if the structure remains bainitic [24, 25]. A martensitic structure can achieve the required strength at that concentration of carbon, but the toughness would not then be impressive. And toughness is essential for weldability.

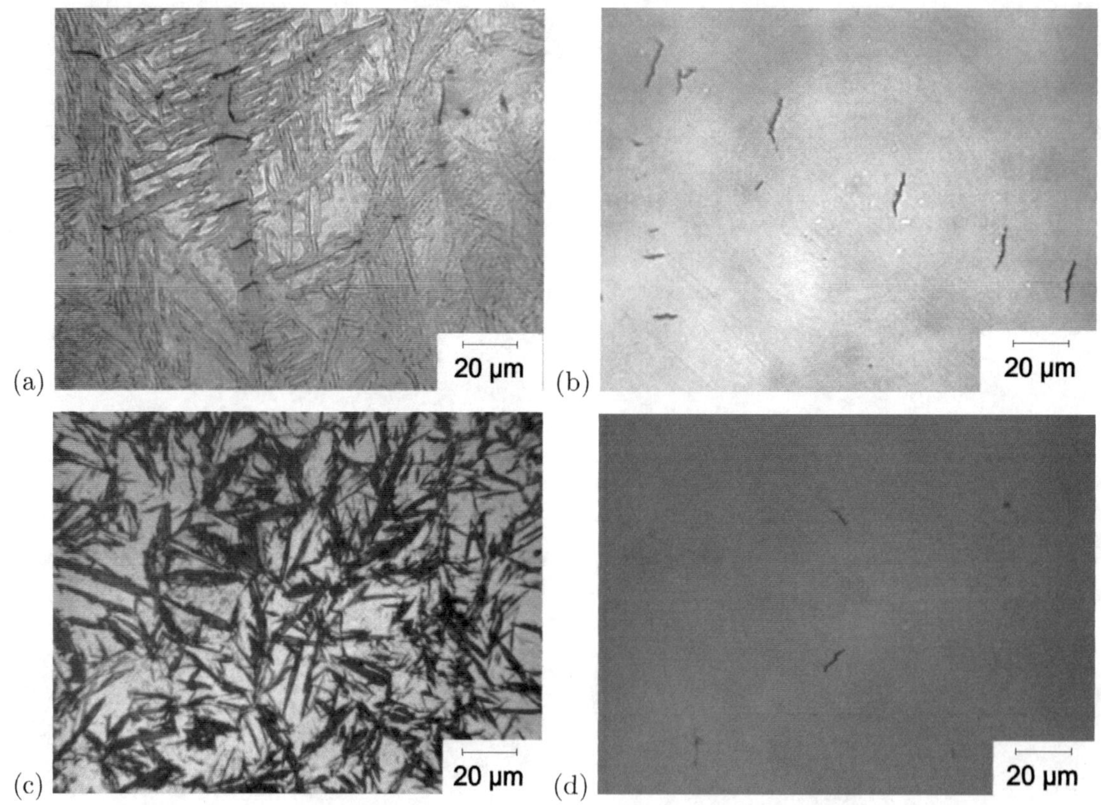

Figure 4. Microscopic cracking of martensite plates. (a,b) Etched and unetched samples respectively, showing martensite plates in austenite with a grain size (mean lineal intercept) of 410±53 μm. (c,d) Etched and unetched samples respectively, showing martensite plates in austenite with a grain size (mean lineal intercept) of 65±12 μm [26].

The general impression that high-carbon martensite is brittle may not be correct. Untempered high-carbon martensite is induced during the deformation of TRIP-assisted steels, and yet, they are formable. The induced-martensite does not crack even at large plastic strains. In fact, it is the large blocks of martensite that have been demonstrated to be brittle [27, 28]. A study of the microscopic cracking of martensite in a 1 wt% carbon alloy-steel has shown that the tendency for cracking is sensitive to the absolute size of martensite plates [26]. This is because load transfer onto the hard martensite becomes difficult as the scale of the martensite decreases. figure 4 illustrates the experiments in which the size of the martensite plates was controlled by altering the austenite grain size. There are two key observations, first, that the extent of cracking diminishes as the austenite grain size (and hence martensite plate size) is reduced. Secondly, the spacing of cracks in a fragmented plate is approximately constant and corresponds to a stress-transfer length of about 10 µm. Therefore, reducing the austenite grain size to a number less than 10 µm should dramatically increase the fracture toughness.

The approach is presented in figure 5, with the aim of generating fine martensite by rendering the austenite into a pancaked state using a particularly low finish-rolling temperature, combined with vanadium microalloying to ensure that the austenite, and its deformation defects, do not recover or recrystallise before martensitic transformation.

Figure 5. (a) Methodology for achieving tough and strong martensite. (b) Pancaked prior austenite grain structure, also containing deformation bands, after Chintha et al. [29].

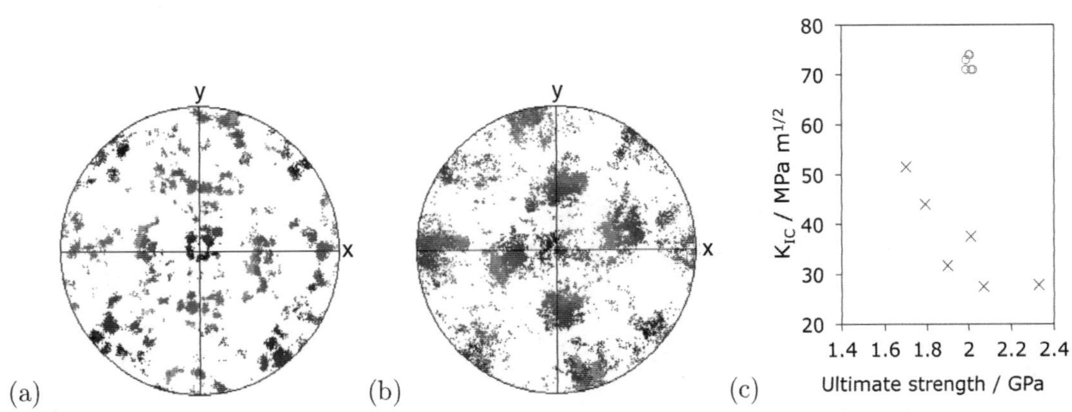

Figure 6. Stereographic projections relative to sample axes, showing martensite $\{100\}_{\alpha'}$ poles. (a) Martensite poles from a single, undeformed, prior austenite grain. (b) Martensite poles from a severely deformed, pancaked, prior austenite grain, showing an enormous spread of orientations [29]. (c) Mechanical properties of new steel (red circles - data courtesy of Dr S. Kundu, Tata Steel India) contrasted against a bulk nanostructured bainitic steel [30].

It is established that a refinement of the "crystallographic grain size", i.e., the size over which crystals are coherent, improves toughness by providing frequent intersections where a propagating cleavage crack must change orientation or adopt a complex topology as it traverses the coherent space [31–36]. The caveat is that the material must not be so brittle that boundaries are of little consequence [33]. The boundaries themselves do not contribute much to the cleavage resistance [37]. figure 6 shows $\{100\}_{\alpha'}$ martensite pole figures, each taken from a single prior-austenite grain, in both the deformed-austenitic and undeformed-austenitic states. The deformation of the austenite results in a huge spread in the orientations of the martensite plates that form subsequently, in contrast to the martensite that grows from undeformed austenite. This crystallographic *chaos* in the former case is because there is a spread of orientations within a defect-rich austenite grain. This, combined with the fine scale of the martensite plates, causes a large increase in the toughness of the steel; comparative data are presented in figure 6c.

To paraphrase, untempered martensite in Fe-C need not be brittle if its length scale is reduced and if the plates are crystallographically dispersed to induce complex cleavage.

4. Limit of detection

It often is taken for granted the austenite to martensite transformation occurs at a unique temperature, designated M_S. The fact that M_S depends on the sensitivity of the measuring technique is neglected in reaching this conclusion. Acoustic emission measurements record a greater M_S than the monitoring of electrical resistivity [38]. The martensite-start temperature depends also on the austenite grain size (figure 7) but curiously, this is a manifestation of the sensitivity described above. The size of the largest martensite plate that can form is related to the austenite grain size ($\overline{L}_\gamma$). The thickness to length ratio of a martensite plate is constant as the strain energy due to the shape deformation [39] balances with the chemical driving force. Therefore, the volume per plate of martensite will increase with $\overline{L}_\gamma$. If the same number of plates form in a large and small austenite grain, the latter would be more difficult to detect because the absolute volume of martensite would be smaller. Based on this concept, the classical Koistinen and Marburger equation [40]:

$$1 - V_V^{\alpha'} = \exp\{-\beta[T - M_S]\} \tag{3}$$

where β is a constant, becomes [41]:

$$M_S^\circ - M_S = \frac{1}{b}\ln\left[\frac{1}{\overline{L}_\gamma^3}\left\{\exp\left(-\frac{\ln(1 - V_V^{\alpha'})}{m}\right) - 1\right\} + 1\right] \tag{4}$$

where $b = 0.2689$, $\overline{L}_\gamma$ is in mm, $m = 0.05$ is the aspect ratio of martensite plates, and M_S is defined when a fraction $V_V^{\alpha'} = 0.01$ of martensite is obtained. The term M_S^o is defined as a fundamental martensite-start temperature for an austenite grain size that is so large that the formation of just one martensite plate is detectable using routine methods. It is given by the point where $\Delta G^{\gamma\rightarrow\alpha} = 700\,\mathrm{J\,mol^{-1}}$, i.e. the stored energy of martensite due to the shape deformation and twin interfaces. M_S^o is therefore purely a thermodynamic quantity with no consideration given to kinetic effects. figure 7 shows that the effect becomes insignificant beyond about 40µm so in conventional experiments, the issue is likely lost in the ±20 °C noise present in typical experimental data [42].

5. Steep diffusion gradients

It could be argued that the interpretation of growth rate data on the formation of ferrite in steels, is in a state of crisis. There is a failure to account for the limitations of experimental techniques and difficulties with the theory of diffusion-controlled growth in multicomponent steels. One might be forgiven in deducing from the published literature that whenever there is a gap between theory and experiment, it is explained by appealing to free energy dissipations that have little in the way of supporting evidence [43].

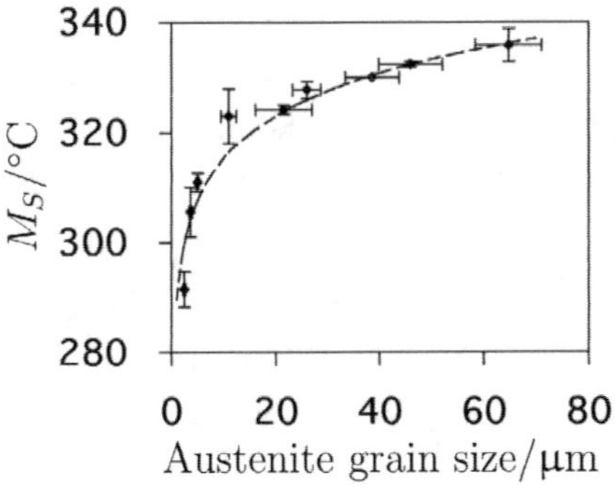

Figure 7. Measured variation in the martensite–start temperature as a function of austenite grain size, determined from dilatometric data (after Yang et al. [41]). This entire variation can be explained by the resolution of the experimental technique coupled with the spatial constraints that exist during the formation of elastically accommodated martensite plates in a given size of austenite grain.

A particular difficulty is that kinetic models for the growth of ferrite in multicomponent steels at large supersaturations leads to extremely steep gradients of concentration, recognised to be unrealistic in the original theoretical work by Coates [44]. There are publications in which the diffusion spike within the austenite, required to maintain local equilibrium at the α/γ interface while permitting the fluxes of substitutional and interstitials to keep pace, is just 0.03-0.003 nm in width. This clearly is physically impossible.

There are additional terms in diffusion theory when gradients are steep [reviewed by Hilliard in 45], which ensure, for example, that the composition waves that develop during spinodal decomposition have finite wavelengths. Whereas a homogeneous chemical solution of composition $\overline{c}$ can be ascribed a particular free energy per atom of $g\{\overline{c}\}$, that of a heterogeneous solution containing concentration gradients, g_{ih} is expressed as a multivariate Taylor expansion:

$$\begin{aligned} g_{\mathrm{ih}}\{a,b,\ldots\} &= g\{\overline{c}\} + a\frac{\partial g}{\partial a} + b\frac{\partial g}{\partial b} + \ldots \\ &\quad + \frac{1}{2}\left[a^2\frac{\partial^2 g}{\partial a^2} + b^2\frac{\partial^2 g}{\partial b^2} + 2ab\frac{\partial^2 g}{\partial a\,\partial b} + \ldots\right] + \ldots \end{aligned} \tag{5}$$

in which the variables, $a, b, \ldots$ in our context are the spatial composition derivatives ($\mathrm{d}c/\mathrm{d}z$, $\mathrm{d}^2c/\mathrm{d}z^2$, etc). After considerable mathematical manipulation [45] this becomes

$$g_{\mathrm{ih}} = \int \left[g\{\overline{c}\} + v_{\mathrm{a}}^3\kappa(\nabla c)^2\right] dV \tag{6}$$

where v_{a} is the volume per atom and κ is known as the gradient energy coefficient. Gradients of concentration lead to an increase in the free energy, so that the formation of steep gradients will in general be opposed. The application of this to ferrite growth is described in detail elsewhere [43], but figure 8 shows that the cost of the steep gradient becomes overwhelming at the sort of concentration spikes typical in the analysis of ferrite growth involving the negligible partitioning of substitutional solutes. This mode of growth is physically unrealistic, the alternative being that local equilibrium breaks down. In summary, any process that involves concentration profiles over very short distances must account for the gradient energy term. This includes, for example, models of solute drag that often are called upon to wile away discrepancies between experiments and theory.

6. Strength

The strength of crystals increases as they are made smaller because the chances of avoiding defects become greater as the volume of the sample decreases. Imperfections in the form of

IOP Conf. Series: Materials Science and Engineering **580** (2019) 012002 doi:10.1088/1757-899X/580/1/012002

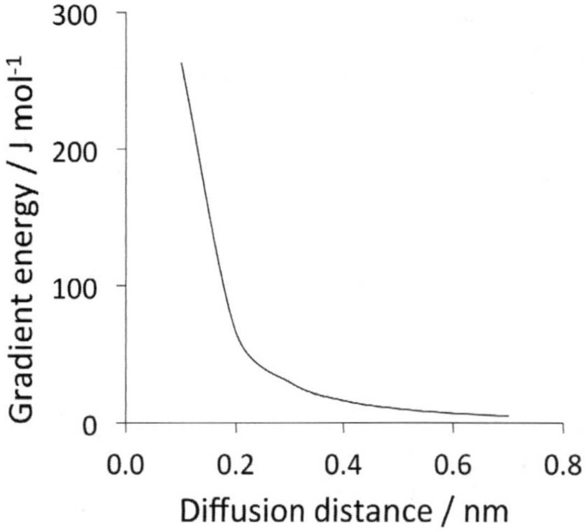

Figure 8. Estimate of the penalty on free energy due to the gradient of concentration in the austenite ahead of the α/γ interface. The diffusion distance in many calculations of growth with local equilibrium are much less than 0.01 nm, in which case, the penalty would be intolerable. This penalty essentially rules out any mechanism of diffusion-controlled growth that necessitates the existence of steep gradients, a case in hand being the so-called negligible-partitioning local equilibrium mode.

dislocations are able to facilitate shear at much lower stresses than would be the case if whole planes of atoms had to collectively slide across each other.

In an ideal, defect-crystal, the tensile strength $\sigma_t \simeq 0.1E$ where E is the Young's modulus. The corresponding ideal shear-strength is $\sigma_s \simeq b\mu/2\pi a$, where μ is the shear modulus, b is a repeat period along the displacement direction and a is the spacing of the slip planes. For ferritic iron, $\mu = 80.65$ GPa and $E \simeq 208.2$ GPa. It follows that the ideal values of tensile and shear strength should be about 21 and 11 GPa, respectively. Strengths approaching the theoretical values were achieved by Brenner as long ago as 1956 (figure 9a) during the testing of single-crystals of iron with diameters less than 2 µm.

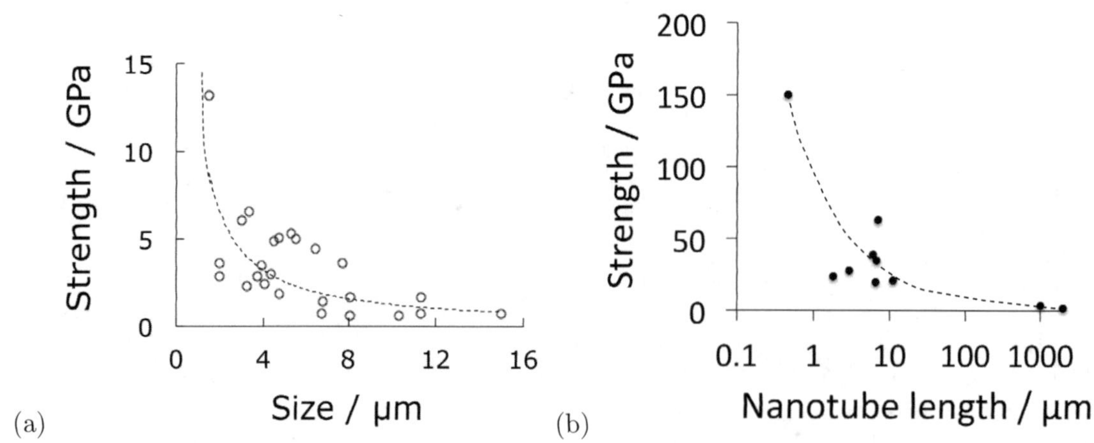

Figure 9. (a) The tensile strength of whiskers of iron [46, 47]. (b) Variation in the strength of carbon nanotubes as a function of length. Data collated from [48–50].

The experiments showed that the strength decreased sharply as the dimensions of the test-sample were increased (figure 9a), because the chances of finding defects increase as the sample gets bigger. It is therefore not wise to rely on perfection as a method of designing strong materials.

IOP Conf. Series: Materials Science and Engineering **580** (2019) 012002 doi:10.1088/1757-899X/580/1/012002

Carbon nanotubes can be imagined to be constructed from sheets of graphene consisting of sp^2 carbon arranged in a two-dimensional hexagonal lattice. The breaking strength of such a tube has been estimated to be an extraordinary 130 GPa, leading to many exaggerated comparisons against steel. However, this is the strength of an invisibly small nanotube. Larger tubes will contain defects that lead to a gross deterioration of strength, rather like the behaviour of single-crystals of iron. Some of these defects will be there at equilibrium and hence are unavoidable. For example, it is known that metals contain an *equilibrium* concentration of vacancies. The enthalpy change associated with the formation of a vacancy opposes its existence, whereas the change in configurational entropy due to the formation of a vacancy favours its formation.

The total change in free energy on forming n vacancies in a crystal is given by [51]:

$$\Delta G = n\Delta g - kT[(N+n)\ln\{N+n\} - N\ln\{N\} - n\ln\{n\}], \tag{7}$$

where k is the Boltzmann constant, T is the absolute temperature, N is the number of atoms, $\Delta g = \Delta h - T\Delta s$, Δh is the enthalpy of formation of one vacancy and Δs is the entropy of formation of a vacancy excluding any contribution from configurational entropy, which is the second term in the equation. The equilibrium mole fraction of vacancies (x) is obtained by writing $\partial\Delta G/\partial n = 0$ giving:

$$x = n/N \simeq \exp\{-\Delta g/kT\}. \tag{8}$$

On this basis, taking the energy of a vacancy in a nanotube as 7 eV, and setting T to be the manufacturing temperature of the tubes (2000-4000 K), it is possible to show that a carbon nanotube strand appropriate for a space elevator, weighing 5000 kg, would contain approximately 10^{10}-10^{20} defects. In fact, we have the less than surprising result the strength collapses as the length of the tube increases figure 9b. Systems which rely on perfection in order to achieve strength necessarily fail on scaling to engineering dimensions. Indeed, there is no carbon tube which can match the strength of iron beyond a scale of 2 mm. These considerations apply also to graphene sheets.

Graphene has been claimed to be between 100-300 times stronger than steel. Lee and co-workers determined the strength of a monolayer of graphene about 1 µm in diameter by a nanoindentation method [52]. They measured the intrinsic breaking strength of the perfect layer as $42\,\mathrm{N\,m^{-1}}$ and converted this into a strength of 130 GPa, the same value as reported for carbon nanotubes [53], which may not be surprising given that the data are in both cases for carbon-carbon bonds in perfect samples. Suppose that 130 GPa represents the true strength of two-dimensional graphene. Brenner [46] has measured the tensile strength of a 1.6 µm whisker to be 13.4 GPa so the intrinsic tensile strength of iron, along its weakest crystallographic direction is likely to be 14.2 GPa [54]. If follows that that pristine graphene is at best about 9 times stronger than steel. When scaled to sizes greater than the micrometre dimensions, it is likely to suffer the same fate as carbon nanotubes, i.e. lose its integrity. More detail in [55].

7. Summary

The study of scaling effects presents a fertile ground for revealing research, but it has only been possible to describe a few of the effects here. A few further examples in the context of steels, not described here but available in the open literature, are as follows:

- small concentrations of stainless steel nanoparticles in fluids disproportionately influencing their thermal conductivity [56];
- comparison of carbon concentration analysis using the atom probe, transmission electron microscopy and X-ray diffraction [57];
- the thermodynamics of the evolution of a solid solution beginning with large particles that are forced to mix at ever decreasing sizes by mechanical alloying [58].

IOP Conf. Series: Materials Science and Engineering **580** (2019) 012002 doi:10.1088/1757-899X/580/1/012002

A particular difficulty that has yet to addressed quantitatively, is how promising studies based on the characterisation of small laboratory samples (60 g-100 kg) can be scaled to much larger commercially viable quantities for component level testing. This is an Achilles heal of many alloy design programmes.

Acknowledgments: Many congratulations to Risø DTU National Laboratories on this 40th anniversary of the international symposia that have contributed regularly to our understanding of materials.

References

[1] Hehemann R F 1970 The bainite transformation *Phase Transformations* ed Aaronson H I and Zackay V F (Materials Park, Ohio, USA: American Society of Materials) pp 397–432

[2] Ko T and Cottrell S A 1952 *Journal of the Iron and Steel Institute* **172** 307–313

[3] Srinivasan G R and Wayman C M 1968 *Acta Metallurgica* **16** 621–636

[4] Swallow E and Bhadeshia H K D H 1996 *Materials Science and Technology* **12** 121–125

[5] Peet M and Bhadeshia H K D H 2011 *Metallurgical & Materials Transactions A* **42** 3344–3348

[6] Bhadeshia H K D H and Edmonds D V 1979 *Metallurgical Transactions A* **10A** 895–907

[7] Chatterjee S, Wang H S, Yang J R and Bhadeshia H K D H 2006 *Materials Science and Technology* **22** 641–644

[8] Machlin E S and Cohen M 1951 *Trans. Metall. Soc. AIME* **191** 746–754

[9] Fiedler H C, Averbach B L and Cohen M 1955 *Transactions of the American Society for Metals* **47** 267–290

[10] Bhadeshia H K D H 1978 *Theory and Significance of Retained Austenite in Steels* Ph.D. thesis University of Cambridge Cambridge, U. K.

[11] Mandelbrot B B 1983 *The fractal geometry of nature* (New York, USA: W. H. Freeman)

[12] Turnbull A and Hutchings R B 1994 *Materials Science & Engineering A* **177** 161–171

[13] Bhadeshia H K D H 2016 *ISIJ International* **56** 24–36

[14] Knott J F 1981 *Fundamentals of Fracture Mechanics* (London: Butterworths)

[15] Song E J, Suh D W and Bhadeshia H K D H 2013 *Computational Materials Science* **79** 36–44

[16] Kang Y and Bhadeshia H K D H 2006 *Materials Science and Technology* **22** 645–652

[17] Brown P M and Baxter D P 2004 Hyper-strength bainitic steels *Materials Science and Technology 2004* (Warrendale, Pennsylvania, USA: TMS) pp 433–438

[18] Secretary of State for Defence, Bhadeshia H K D H, Mateo C and Brown P 2010 Bainite steel and methods of manufacture thereof Tech. Rep. GB2462197 Intellectual Property Office London, U.K.

[19] Bhadeshia H K D H 2010 *Proceedings of the Royal Society of London A* **466** 3–18

[20] Bhadeshia H K D H 2013 *Science and Technology of Advanced Materials* **14** 014202

[21] Hong S G, Cho M H and Lee J S 2010 Effect of post weld rapid heat treatment on microstructural evolution in high-carbon bainitic steel welds *63rd Annual Assembly and International Conference of the International Institute of Welding* vol AWST-10/99 (France: International Institute of Welding) pp 47–51

[22] Fang K, Yang J, Zhao D, Song K, Yan Z and Fang H 2012 *Advanced Materials Research* **482-484** 2405–2408

[23] Lancaster J F 1970 *The metallurgy of welding brazing and soldering* (London, U.K.: George Allen and Unwin Ltd, eds C. R. Tottle and J. C. Wright)

[24] Pickering F B 1978 *Physical Metallurgy and the Design of Steels* (Essex, U.K.: Applied Science Publishers, page 104)

[25] Khare S, Lee K Y and Bhadeshia H K D H 2010 *Metallurgical & Materials Transactions A* **41A** 922–928

[26] Chatterjee S and Bhadeshia H K D H 2006 *Materials Science and Technology* **22** 645–649

[27] Bhadeshia H K D H and Edmonds D V 1983 *Metal Science* **17** 411–419

[28] Bhadeshia H K D H and Edmonds D V 1983 *Metal Science* **17** 420–425

[29] Chintha A R, Peet M, Kundu S, Bhadeshia H K D H, Valtonen K and Kuokkala V T 2019 *Wear* **428-429** 430–437

[30] Bhadeshia H K D H 2015 *Bainite in steels: theory and practice* 3rd ed (Leeds, U.K.: Maney Publishing)

[31] Gourgues A F, Flower H M and Lindley T C 2000 *Materials Science and Technology* **16** 26–40

[32] Lambert-Perlade A 2001 *Rupture par clivage de microstructures d'aciers bainitiques obtenues en conditions de soudage* Ph.D. thesis Ecole de Mines de Paris Paris, France

[33] Qiao Y and Argon A S 2003 *Mechanics of Materials* **35** 313–331

[34] Nohava J, Hausild P, Karlik M and Bompard P 2003 *Materials Characterization* **49** 211–217

[35] Lambert-Perlade A, Gourgues A F and Pineau A 2004 *Acta Materialia* **52** 2337–2348

[36] Lambert-Perlade A, Gourgues A F, Besson J, Sturel T and Pineau A 2004 *Metallurgical & Materials Transactions A* **35** 1039–1053

[37] Qiao R and Argon A S 2003 *Mechanics of Materials* **35** 129–154

[38] Olson G B, Tsuzaki K and Cohen M 1987 *Mat. Res. Soc. Symposium Proc.* **57** 129–148

[39] Christian J W 1958 *Acta Metallurgica* **6** 377–379

[40] Koistinen D P and Marburger R E 1959 *Acta Metallurgica* **7** 59–60

[41] Yang H S and Bhadeshia H K D H 2009 *Scripta Materialia* **60** 493–495

[42] Yang H S and Bhadeshia H K D H 2007 *Materials Science and Technology* **23** 556–560

[43] Bhadeshia H K D H 2016 *Current Opinion in Solid State and Materials Science* **20** 396–400

[44] Coates D E 1973 *Metallurgical Transactions* **4** 395–396

[45] Hilliard J E 1970 Spinodal decomposition *Phase Transformations* ed Zackay V F and Aaronson H I (Metals Park, Ohio, USA: ASM International) pp 497–560

[46] Brenner S S 1956 *Journal of Applied Physics* **27** 1484–1491

[47] Brenner S S 1956 *Acta Metallurgica* **4** 62–74

[48] Yu M F, Lourie O, Dyer M J, Moloni K, Kelly T F and Rouff R S 2000 *Science* **287** 637–640

[49] Xie S, Li W, Pan Z, Chang B and Sun L 2000 *Journal of Physics and Chemistry of Solids* **61** 1153–1158

[50] Pan Z W, Xie S S, Lu L, Chang B H, Sun L F, Zhou W Y, Wang G and Zhang D L 1999 *Applied Physics Letters* **74** 3152–3154

[51] Christian J W 2003 *Theory of Transformations in Metals and Alloys, Part I* 3rd ed (Oxford, U. K.: Pergamon Press)

[52] Lee C, Wei X, Kysar J W and Hone J 2008 *Science* **321** 385–388

[53] Yakobson B I and Smalley R E 1997 *American Scientist* **85** 324–337

[54] Clatterbuck D M, Chrzan D C and J W Morris Jr 2002 *Philosophical Magazine Letters* **82** 141–147

[55] Bhadeshia H K D H and Honeycombe R W K 2017 *Steels: Microstructure and Properties* 4th ed (Elsevier)

[56] Song Y Y, Bhadeshia H K D H and Suh D W 2015 *Powder Technology* **272** 34–44

[57] Bhadeshia H K D H 2015 *Materials Science and Technology* **31** 758–763

[58] Badmos A Y and Bhadeshia H K D H 2000 *Metallurgical & Materials Transactions A* **28** 2189–2193

IOP Conf. Series: Materials Science and Engineering **580** (2019) 012003 doi:10.1088/1757-899X/580/1/012003

Modelling creep induced by internal stresses in freestanding submicron Cu film

Laurent Delannay, Guerric Lemoine, Michaël Coulombier, Thomas Pardoen

iMMC, UCLouvain, av. G. Lemaitre 4, 1348 Louvain-la-Neuve, Belgium

E-mail: laurent.delannay@uclouvain.be

Abstract. In a lab-on-chip experiment, lithography and selective chemical etching are used to pattern microscopic tensile test samples within a thin metallic layer hosting large internal stresses. After partial release of the layer from the substrate on which it was deposited, the free-standing beam-like structures are stretched by the actuator to which they are connected. The lab-on-chip also comprises cantilever beams which shorten freely upon release from the substrate. Experimental observations of both the instantaneous and the delayed deformations in a 170 nm thick copper film were simulated using a theoretical model. The model properly reproduced the experiments only when accounting for both plasticity and significant kinematic hardening occurring already during the deposition of the polycristalline film. Once released from the substrate, cantilever beams contracted well beyond the elastic range because the amplitudes of back-stresses were sufficient to cause reverse plastic yielding. Large tensile stresses inside the actuated beams led to delayed uniform elongations (creep) exceeding 16%. Such values are much larger than the uniform strain of 5-6% that was observed in beams that underwent necking as soon as the film was released from the substrate, i.e. directly after etching of the sacrificial layer.

1. Introduction

Reducing the grain size of metallic alloys contributes to increasing not only their strength, but also the heterogeneity of internal stresses and the strain rate sensitivity [1, 2, 3, 4]. Under tensile loads, rate-dependent viscoplastic processes, which are promoted among others in grain boundary regions [5, 6], tend to postpone necking and they may hence impede that the gain in strength results in significant loss of ductility [7]. Under cyclic loading, nanocrystalline alloys undergo elastic-plastic transients which demonstrate large kinematic hardening: backstresses due to dislocation pile ups cause early yielding upon load reversal [8, 9].

Mechanical testing of submicron-thick freestanding films [10, 11, 12, 13, 14, 15] allows a rather convenient characterization of the latter phenomena. The experimental results reported here were obtained using a testing method specifically designed in order to probe the creep/relaxation response of e-beam evaporated metallic films. Such a lab-on-chip technique, which is illustrated in figure 1 and described in Section 2, has already been applied to aluminum [16], palladium [17] and copper [18, 19]. Over the years, the technique has been improved for instance by annealing of the deposited films prior to release in order to tune the amplitude of internal stresses [20]. However, it is an early experiment, dating from 2014, that is revisited here because it demonstrated phenomena which ceased to operate with the new lab-on-chip designs.

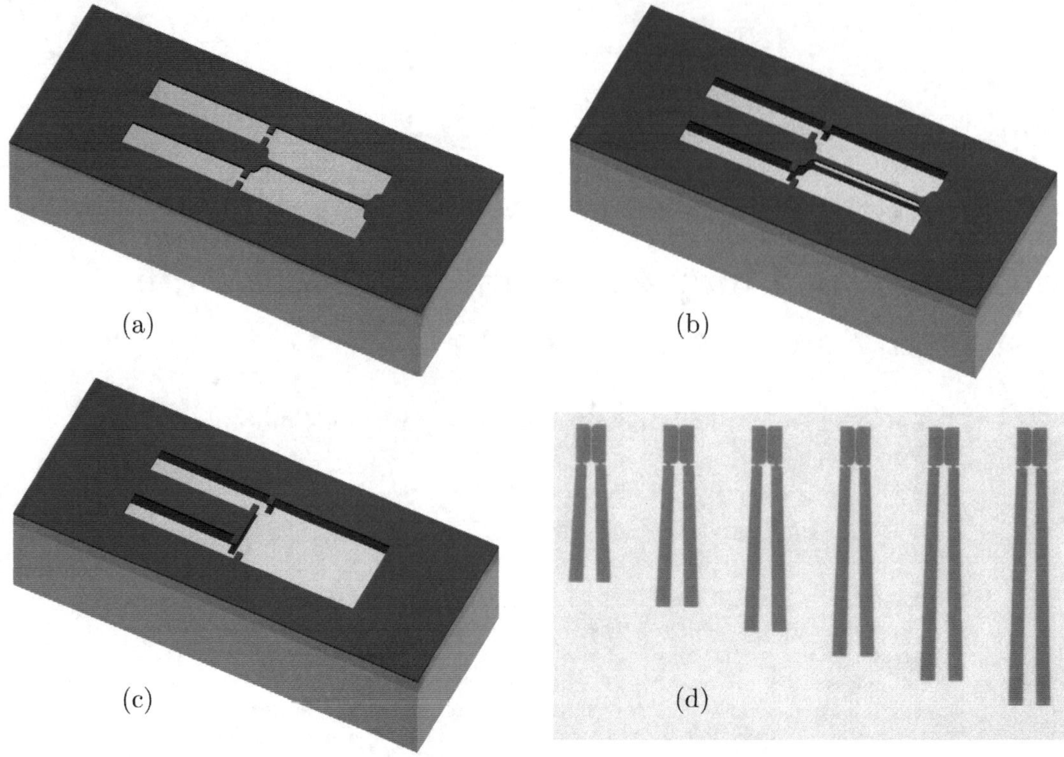

Figure 1. Principle of the lab-on-chip experiment: (a) patterned tensile test structure still tied to the substrate; (b) partial release of the Cu film by chemical etching of the sacrificial layer; (c) unconstrained shortening of a cantilever beam after release; (d) Adjacent tensile test structures with (tapered) actuators of increasing lengths ensuring that the amplitude of post-release elongation of the narrow beams also increases.

In recent on-chip experiments, the ultimate tensile strength of the test material is much lower than the strength of the actuators which are made of silicon nitride, brittle at more than 5 Gpa. As a result, the actuator behaves elastically and interpretation of the measurements is rather straightforward [20]. In the experiment considered here, actuators were patterned directly in the in Cu film that was tested. Both types of structural elements hence had the same (a priori unknown) non-linear elastic-viscoplastic response. The present contribution shows how to determine the latter material response by inverse modeling of such experiments, in which the post-release creep is dominated by significant backstress levels. The parameters of the viscoplastic model are identified iteratively, by repeated computational simulations aiming to reproduce the experimentally observed deformations of a large number of tensile test structures within the lab-on-chip. The theoretical model is described in Section 3 and a comparison of numerical predictions to experimental data is discussed in Section 4.

2. Description and main results of the lab-on-chip experiment

The fabrication of the lab-on-chip proceeded as follows, using a single mask. The first deposited layer was 1 μm thick and covered the whole surface of the silicon substrate. Such sacrificial layer consisted of a plasma enhanced chemical vapor deposited (PECVD) silicon dioxide layer. The precursor gases were composed of 100 sccm of SiH_4, 700 sccm of N_2O and 300 sccm of N_2. After deposition, the wafer was densified at 800°C for 20 min and then covered with a negative photoresist (AZ5214) allowing patterning by UV light. Copper was then deposited on the patterned substrate at room temperature, by e-beam evaporation at a pressure of 0.1 mTorr and a deposition rate of 0.1 nm/s. The photoresist was then removed in acetone (lift-off) revealing copper tests structures still bounded to the sacrificial layer. The thickness of the Cu film was measured to be 170 ± 10 nm by profilometry.

The test structures were released by wet etching of the sacrificial layer using hydrofluoric acid (HF 73%), which ensures a high etching selectivity between PECVD SiO_2 and the Cu layer. This led to an etching rate of 0.181 μm/s. As the release was performed in liquid, supercritical drying was performed to avoid sticking of the freestanding structures to the substrate underneath.

Once released, the cantilever beams shortened whereas the narrow beam of every auto-actuated tensile test structure was stretched. Such changes of lengths of the various beams were probed in the SEM by measuring the relative positions of two cursors which faced one another before etching of the sacrificial layer (figure 1a) but were then shifted due to the change of lengths of the beams (figure 1a and figure 2). Cursors patterned in regions of the Cu film still anchored to the substrate were considered to hold fixed positions. The displacements of the shifted cursors were measured using high magnification micrographs from a field emission gun scanning electron microscope (FEG-SEM). Displacements were measured with an accuracy of about 50 nm, limited by the photolithography process rather than by the image resolution achieved by FEG-SEM. The first observations of the deformed lab-on-chip were performed 7 hours after release. In order to probe not only the instantaneous, but also the delayed, response of tensile tests structures undergoing creep, electron microscopy observations of the lab-on-chip were performed 25 times, scheduled from one day up to four months after the release.

In the tensile test structures, the constant width of the beam was 1.5 μm, whereas the actuator had a tapered shape with minimum and maximum widths of 7.5 and 11.5 μm, respectively. The length of the narrow beam was systematically equal to 25 μm, whereas the length of the actuator varied from one structure to the next, ranging from 100 to 1200 μm. The initial lengths of freely contracting cantilever beams ranged from 100 to 400 μm.

Before the release, such patterned Cu films host uniform equibiaxial tensile stresses of unknown amplitude which, in the absence of an annealing treatment, vary a lot depending on the deposition conditions and the delay before release. After the release, equilibrium of axial force in the tensile test structures imposes that the stress in the narrow beam be 6.4 times larger than the average stress in the actuator (due to the different cross-section areas). This means that the actuator contracts and unloads elastically (figure 1b), whereas the beam undergoes plastic elongation. Its strain hardening causes tensile stresses much larger than the amplitude of the initial internal stress. During elastic unloading of the actuator, the change of length (measured by the cursor displacement) increases proportionally to the actuator length. Hence, mechanical equilibrium is reached for a cursor displacement, i.e. for an elongation of the test beam, that increases with the actuator length. On the other hand, as the unloading of the cantilever beams is unconstrained (figure 1c), their contraction is directly proportional to their length.

In the experiment on the 170 nm thick Cu film considered here, the instantaneous unconstrained contraction of cantilever beams was 0.4%. Necking and failure of the test beam were observed whenever the actuator was longer than 850 μm. In the measurements performed right after release, the largest unbroken beam was stretched by 5.1%. We may thus assume that the (yet unknown) ultimate tensile stress was reached for a uniform elongation lower than 6%.

Figure 3 shows the creep responses of four tensile test structures in which the actuator length was 100, 200, 400 and 800 μm, respectively. During the four months of measurements, the displacement Δu of the cursor was found to increase linearly with time. Hence, the creep rate was constant. After four months, among the unbroken beams, the largest elongation among all actuated test beams was 16%. Surprisingly, free beams underwent creep strains too. After 1750 hours (73 days), free beams had contracted by -2.2% (whereas the largest contraction of an actuator with a test beam attached to it was -1.3%).

IOP Conf. Series: Materials Science and Engineering **580** (2019) 012003 doi:10.1088/1757-899X/580/1/012003

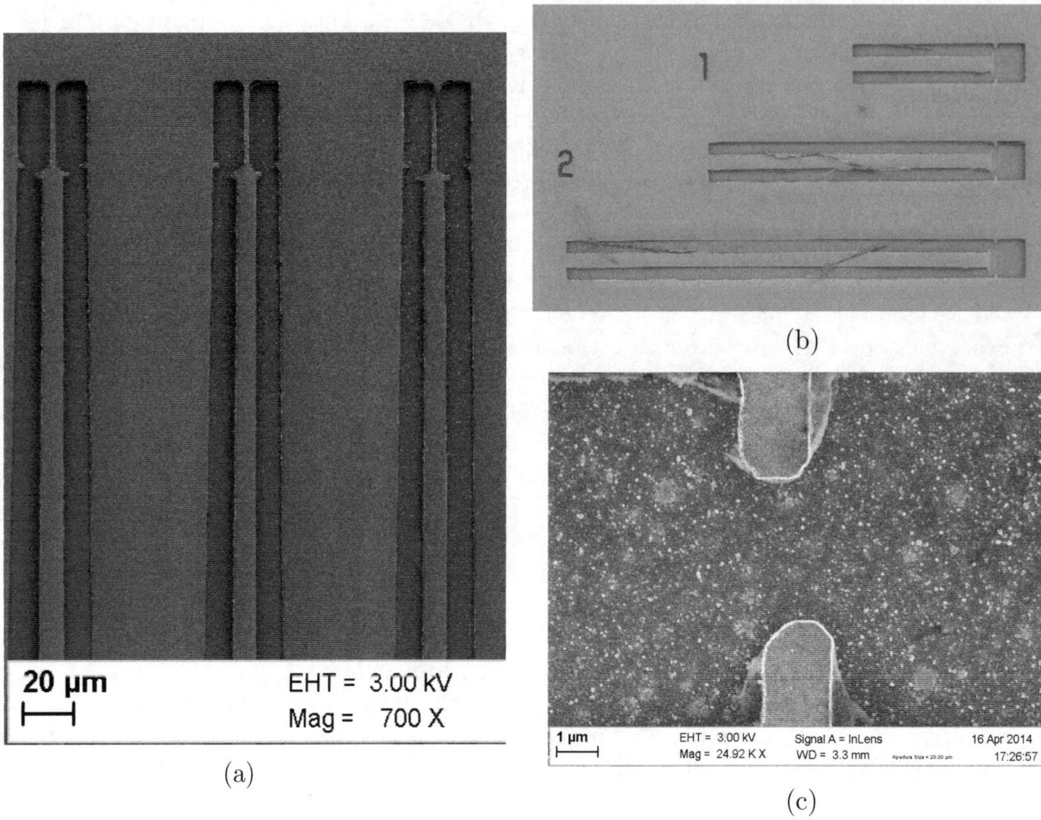

Figure 2. Micrographs showing two regions of the lab-on-chip three months after release. In the tensile test structures shown in (a), the actuator initial length increases from left to right and the test beam on the right underwent necking and failure. The initial lengths of the cantilever beams shown in (b) were 100, 200 and 300 μm, respectively. (c) Close-up of two cursors used to measure a beam contraction.

3. Description and main results of the model

The theoretical analysis is greatly simplified if we assume that all beams are subjected to a uniaxial stress once released from the substrate and if the deformation of each actuator is considered uniform (in spite of its tapered shape). Indeed, under uniaxial tension a scalar description of the stress-strain relationship suffices.

In reality, a uniaxial stress state is valid only when the beams are fully released from the substrate. Until full release, there is a transition from the equibiaxial stress state (resulting from the film deposition) to the uniaxial situation. Elastic unloading of the in-plane transverse stress causes a decrease of the axial stress in all tensile test structures. The latter axial stress is multiplied by $1 - \nu$ where $\nu = 0.3$ is the Poisson ratio of copper. As etching of the sacrificial layer underneath the beams starts from the edges, the central axis of the beams is released last and it is likely that the in-plane transverse stress is largely relaxed when the beam starts to deform along the longitudinal direction.

It is further assumed that the internal stresses were sufficiently large to induce plasticity (and kinematic hardening) already during the film deposition. Otherwise the amplitude of contraction of the free beams would not exceed that expected from elastic unloading. However, such pre-deformation is not modeled here. The initial state, just before the release, is characterized by a yield stress σ_{y0} and a positive backstress ζ_0 which are determined by inverse modelling.

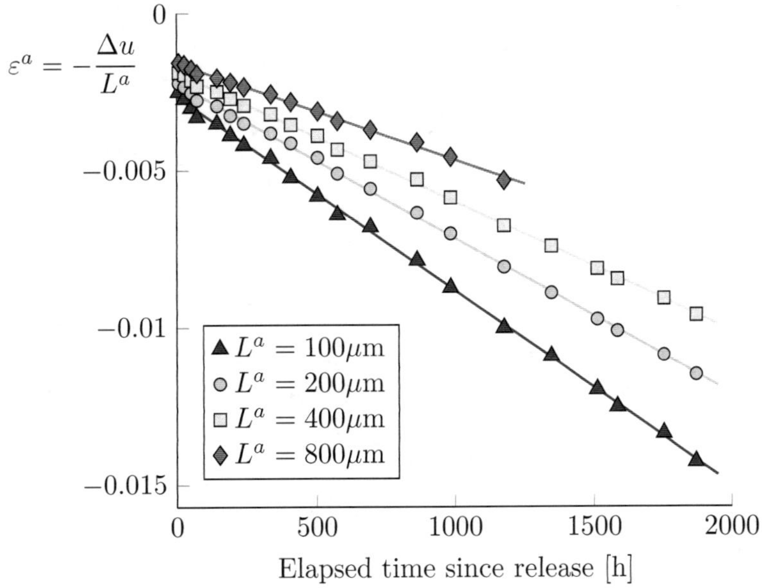

Figure 3. Illustration of the delayed deformation of tensile test structures by the progressive shortening of actuators having lengths L^a equal to 100, 200, 400 and 800 μm. For comparison, free beams underwent an instantaneous contraction of -0.004 and a delayed contraction of -0.022 after 1750 h.

3.1. Simulation of the instantaneous deformation

The instantaneous response, observed immediately after etching, is considered elastic-plastic in both the actuator and the stretched test beam (denoted, respectively, with superscripts a and b). As the total length $L^a + L^b$ of the tensile test structure remains constant, the cursor displacement Δu defines the axial strains of both components: $\Delta\varepsilon^a = -\Delta u/L^a$ and $\Delta\varepsilon^b = ln(1 + \Delta u/L^b)$.

In the model, Δu is computed by enforcing that $\sigma^b = 6.4\,\sigma^a$, which ensures force equilibrium (Section 2). The value of Δu is computed iteratively, leading to a different result in every tensile test structure since the actuator lengths L^a are different. The response is linear elastic if

$$\begin{aligned} &\sigma^a = (1-\nu)\big(\zeta_0 + \sigma_{y0}\big) + E\Delta\varepsilon^a > \zeta_0 - \sigma_{y0} && \text{in the actuator} \\ &\sigma^b = (1-\nu)\big(\zeta_0 + \sigma_{y0}\big) + E\Delta\varepsilon^b < \zeta_0 + \sigma_{y0} && \text{in the test beam} \end{aligned} \tag{1}$$

where E is the Young modulus. Otherwise, plastic yielding occurs and the increments of equivalent plastic strain Δp^a and Δp^b must fulfill the following conditions:

$$\begin{aligned} &\sigma^a = (1-\nu)\big(\zeta_0 + \sigma_{y0}\big) + E\big(\Delta\varepsilon^a + \Delta p^a\big) = \zeta^a(\Delta p^a) - \sigma_y(\Delta p^a) && \text{in the actuator} \\ &\sigma^b = (1-\nu)\big(\zeta_0 + \sigma_{y0}\big) + E\big(\Delta\varepsilon^b - \Delta p^b\big) = \zeta^b(\Delta p^b) + \sigma_y(\Delta p^b) && \text{in the test beam} \end{aligned} \tag{2}$$

Experimental measurements are best reproduced by relying on the following representation of strain hardening:

$$\begin{aligned} &\sigma_y(\Delta p \le p_l) = \sigma_{y0} + H\Delta p \\ &\sigma_y(\Delta p > p_l) = \sigma_{sat} - \big(\sigma_{sat} - \sigma_{y0} - Hp_l\big)\, exp\left(\frac{p_l - \Delta p}{p_s}\right) \end{aligned} \tag{3}$$

This implies that the hardening rate H is constant until $\Delta p = p_l$, after which it decreases until the yield stress reaches its saturation value σ_{sat}. Setting $p_s = (\sigma_{sat} - \sigma_{y0})/H - p_l$ ensures that the hardening rate decreases continuously.

The backstress evolves differently in the two components of the tensile test structure. Indeed, ζ^a undergoes a fast decay because yielding switches from elongation (before the release) to contraction. On the other hand, stretching of the beam causes ζ^b to increase beyond the initial value ζ_0:

$$\begin{aligned} \zeta^a(\Delta p^a) &= \zeta_0 - H_\zeta \Delta p^a \\ \zeta^b(\Delta p^b) &= \zeta_{sat} - \big(\zeta_{sat} - \zeta_0\big)\ exp\big(- \Delta p^b/p_\zeta\big) \end{aligned} \quad (4)$$

The selected values of the various hardening parameters are listed in table 1. They ensure that the free contraction of cantilever beams is -0.4%, as observed experimentally, and also that the onset of necking of the stretched beam occurs at $\Delta p^b = 6\%$ according to the Considère criterion.

Table 1. Fitted values of the parameters of the elastic-plastic model

σ_{y0}	H	p_l	p_s	σ_{sat}	ζ_0	H_ζ	p_ζ	ζ_{sat}
86 MPa	7 GPa	0.028	0.0126	370 MPa	207 MPa	54 GPa	0.01	222 MPa

Figure 4 shows that the model closely reproduces the instantaneous contraction of the actuators (which was probed 7 h after release). The predicted stress strain curve of stretched beams is shown in figure 5a. For comparison, figure 5b shows the stress-strain curve that would be predicted if, instead of relying on the elastic-plastic model, one would consider that the unloading of all actuators is purely elastic, which would imply that the initial backstress ζ_0 is much lower. The latter prediction of the Cu film response is unrealistic because the hardening of stretched beams is excessive, postponing necking and failure later than $\Delta p^b \simeq 6\%$.

3.2. Simulation of the delayed deformation

According to figure 3, the post-release displacement of the cursor occurs at constant velocity. This can be reproduced by the theoretical model if the behavior of the Cu film is considered viscoplastic. Inside the tensile test structures, the stress ratio σ^b/σ^a increases beyond the initial value of 6.4 since the cross-sections of stretched beams decrease and the opposite applies within axially contracting actuators. As the response is highly viscous, large variations of the axial stress are impeded and the delayed deformation must be driven by constant creep rates inside the actuators. For simplicity, the present model considers that both the yield stress and the backstress hold constant values inside every actuator. The viscoplastic creep rate, corresponding to the slope of the linear creep strains reported in figure 3, is computed using the following power law relationship:

$$\dot{\varepsilon}^a_{vp} = \frac{-\dot{u}}{L^a} = -\dot{\gamma}_0 \big|\sigma^a - \zeta^a(\Delta p^a)\big|^{1/m} = -\dot{\gamma}_0 \big(\sigma_y(\Delta p^a)\big)^{1/m} \quad (5)$$

Predicted creep rates correspond best to experimental observations when $\dot{\gamma}_0 = 2,74 \times 10^{-32}/\mathrm{h}$ and $m = 0.075$. The creep strains predicted within 24 actuators with different lengths are compared to experiments in figure 4.

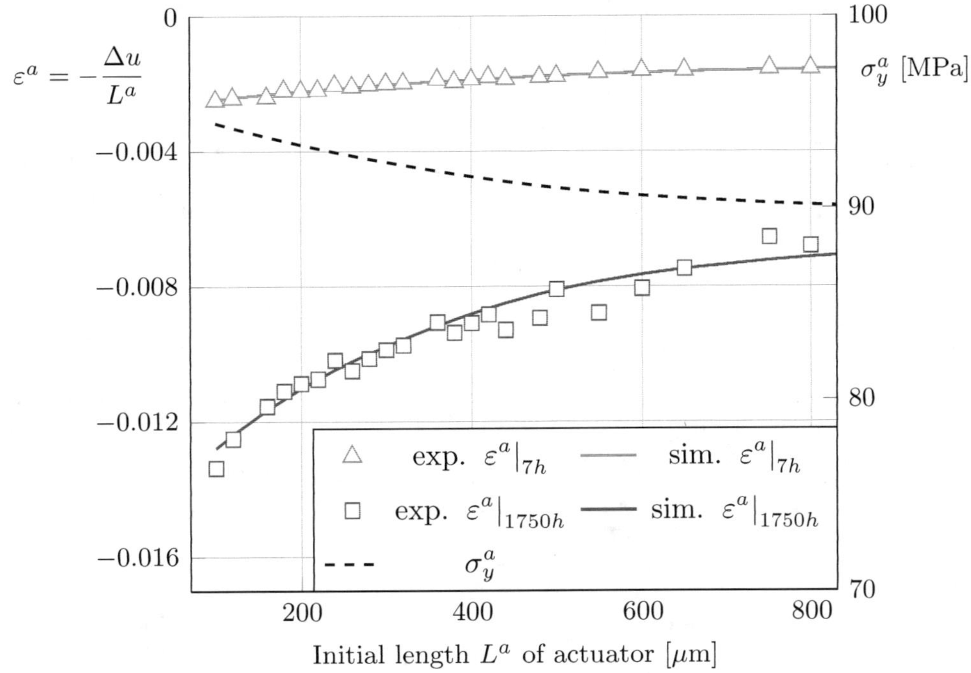

Figure 4. Assessment the model prediction of the shortening of actuators resulting from the instantaneous deformation (probed 7 h after release) and the delayed deformation (probed 1750 h after release). The delayed viscoplastic response depends on the yield stress after release which is shown too.

The viscoplastic deformation of free beams is likely to be described by the same law since the only difference with regard to the actuators is that free beams underwent larger instantaneous contraction and hence larger reverse plastic yielding, leading to $\sigma_y^{fb} = 102$ MPa after release. The power law predicts that, after 1750 h, free beams had contracted by -3.4% whereas the experimentally measured value is only -2.2%. Such a discrepancy might be due to the fact that free beams tend to deflect and may undergo friction if they are in contact with the substrate.

4. Discussion

According to the inverse modeling of the lab-on-chip experiment, the amplitude of internal stresses prior to release is 293 MPa. This is in the lower range of the levels of internal stresses (300-450 MPa) which were deduced from the wafer curvature (Stoney method) after annealing of e-beam evaporated copper [20]. However, experience has shown that internal stresses may vary significantly after deposition unless the microstructure is stabilized by annealing.

Considering that the grain size is of the order of the film thickness (170 nm), the predicted value of the ultimate tensile stress (570 MPa) also is realistic. It compares well to previous investigations on free-standing Cu films with comparable thicknesses [8, 20].

The amplitude of contraction of both the free beams and the actuators is much larger than the amplitude expected in case of linear elastic unloading. The fact that the contraction continues at a constant rate long after the release can be explained only in the case of a viscoplastic process driven by the heterogeneity of internal stresses, which is commonly referred to as kinematic hardening. The backstress developed before the release which means that plastic yielding

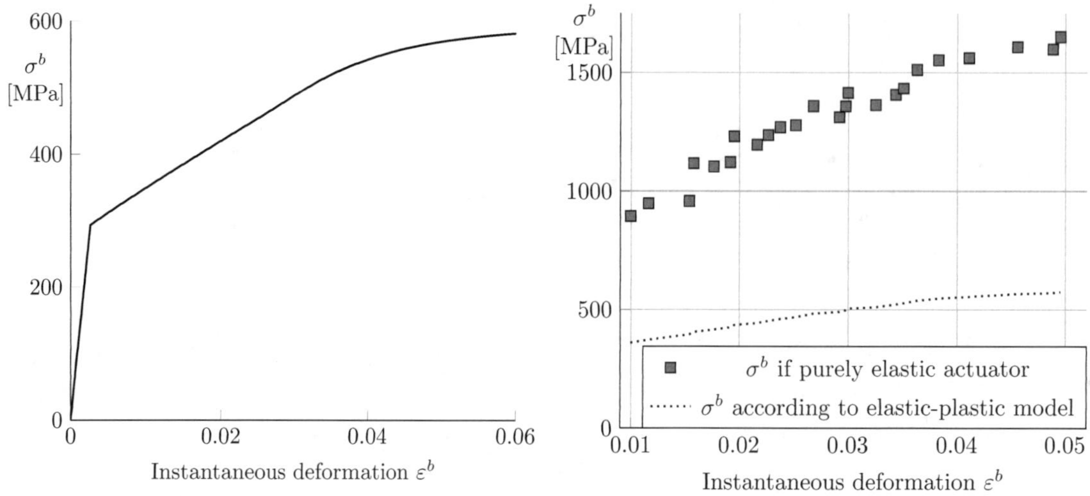

(a) (b)

Figure 5. (a) Uniaxial stress-strain curve of the Cu film as predicted through the inverse analysis. (b) Comparison of the axial stress in stretched beams when assuming either a purely elastic unloading of all actuators, or an elastic-plastic unloading as predicted by the model.

occurred already during the film deposition. Similar observations of reverse creep induced by a strong kinematic hardening were reported in other nanocrystalline films [8, 9].

In the present analysis, it was assumed that the strengthening of the beams stretched after release was due to an increase of the yield stress σ_y^b, whereas the backstress ζ_y^b rapidly reached a saturation value slightly larger than the initial backstress ζ_0. In reality, the experiment does not allow determining which proportion of the axial stress should be considered a backstress. If the heterogeneity of internal stresses increases during the instantaneous stretching of the beams, the backstress also increases. Probing kinematic hardening in thin films is possible using other experimental set ups in which the material undergoes cyclic loading [8, 10].

Tensile test samples subjected to an instantaneous elongation of more than 5% underwent necking upon release, but a much larger elongation was achieved during creep. It is well known that viscous processes delay plastic instability [21, 22] and hence increase ductility. When relying on a viscoplastic power law, necking under uniaxial tension occurs after a uniform strain $\varepsilon_u = m\,ln(f^{1/m} - 1)$ where f represents the reduction of cross-sectional area due to an initial imperfection. Using $f = 0.99$ and $m = 0.075$ (as identified by inverse modeling), a uniform strain $\varepsilon_u = 15.5\%$ is obtained. This corresponds to the maximum strain measured in the stretched beams, whereas many of the stretched beams underwent necking at lower amplitudes of the creep strain, presumably due to larger initial imperfections.

In a previous study involving lab-on-chip experiments on free-standing Pd films with different thicknesses [23], the influence of grain size on the heterogeneity of internal stresses was investigated using a microscopic model accounting for the crystalline anisotropy of individual grains and for thermally-activated dislocation glide which reproduced the observed creep rates. Adapting the model to the Cu film investigated here is under progress.

5. Conclusion

When a patterned submicron thick Cu film is released from the substrate on which it was deposited, both the instantaneous (elastic-plastic) and the delayed (viscoplastic) responses are strongly influenced by the amplitude and the heterogeneity of the internal stresses resulting from film deposition. In a lab-on-chip experiment, some of the patterned beam-like elements contract whereas others are stretched by an actuator. In the experiment investigated here, the amplitude of the experimentally observed contractions was much larger than that expected from a purely elastic unloading of the initial biaxial tensile stress. Moreover most of the shortening of the beams occurred at constant rate for up to four months after film release, which is characteristic of a viscoplastic process driven by internal stresses. The same viscoplastic process was active in stretched beams, which demonstrated unexpectedly high levels of ductility.

Most of the experimental observations were properly reproduced by the theoretical model and the identification of parameter values by an inverse modeling analysis led to realistic predictions of the strength and strain rate sensitivity of the Cu film.

Acknowledgements

LD is mandated by the FRS-FNRS. Financial support was obtained from the Communauté Française de Belgique (Action de Recherche Concertée $n°15/20-066$).

References

[1] May J, Höppel HW, Göken M 2005 *Scripta Mater.* **53** 189–94
[2] Meyers Ma, Mishra A, Benson DJ 2006 *Prog. Mater. Sci.* **51** 427–556
[3] Miyamoto H, Ota K, Mimaki T 2006 *Scripta Mater.* **54** 1721–25
[4] Izadi E, Rajagopalan J 2016 *Scripta Mater.* **114** 65–9
[5] Wei Y, Bower AF, Gao H 2008 *Acta Mater.* **56** 1741–52
[6] Kim HS, Estrin Y 2008 *Mater. Sci. Eng.* A **483-484** 127–30
[7] Sharon Ja, Padilla Ha, Boyce BL 2013 *J. Mater. Res.* **28** 1539–52
[8] Xiang Y, Vlassak JJ 2006 *Acta Mater.* **54** 5449–60
[9] Rajagopalan J, Rentenberger C, Peter Karnthaler H, Dehm G, Saif MTa 2010 *Acta Mater.* **58** 4772–82
[10] Haque MA, Saif MTA 2002 *Sens. Actuators* A **97-98** 239–45
[11] Nicola L, Xiang Y, Vlassak JJ, Van der Giessen E, Needleman A 2006 *J. Mech. Phys. Sol.* **54** 2089–110
[12] de Boer MP, Corwin AD, Kotula PG, Baker MS, Michael JR, Subhash G, et al. 2008 *Acta Mater.* **56** 3313–26
[13] Karanjgaokar NJ, Oh CS, Lambros J, Chasiotis I 2012 *Acta Mater.* **60** 5352–61
[14] Karanjgaokar N, Stump F, Geubelle P, Chasiotis I 2013 *Scripta Mater.* **68** 551–54
[15] Wang L, Xin T, Kong D, Shu X, Chen Y, Zhou H, et al. 2017 *Scripta Mater.* **134** 95–99.
[16] Coulombier M, Boé A, Brugger C, Raskin JPP, Pardoen T 2010 *Scripta Mater.* **62** 742–5
[17] Colla M, Idrissi H, Malet L, Godet S, Raskin J, Schryvers D, et al. 2014 *Nat. Commun.* **6** 1–8
[18] Vayrette R, Raskin JP, Pardoen T 2015 *Eng. Fract. Mech.* **150** 222–38
[19] Lapouge P, Onimus F, Vayrette R, Raskin JP, Pardoen T, Bréchet Y 2016 *J. Nucl. Mater.* **476** 20–9
[20] Lapouge P, Onimus F, Coulombier M, Raskin JP, Pardoen T, Bréchet Y 2017 *Acta Mater.* **131** 77–87
[21] Hutchinson JW, Neale KW 1977 *Acta Metal. Mater.* **25** 839–46
[22] Hosford WF, Caddell RM 2011 *Metal forming: mechanics and metallurgy* Cambridge University Press.
[23] Lemoine G, Delannay L, Idrissi H, Colla MS, Pardoen T 2016 *Acta Mater.* **111** 10–21

IOP Conf. Series: Materials Science and Engineering **580** (2019) 012004 doi:10.1088/1757-899X/580/1/012004

Monitoring microstructural evolution in-situ during cyclic loading with high-resolution reciprocal space mapping

Annika Martina Diederichs[1], Ulrich Lienert[2] and Wolfgang Pantleon[1]

[1] Department of Mechanical Engineering, Technical University of Denmark, Produktionstorvet 425, 4000 Kongens Lyngby, Denmark
[2] DESY Photon Science, Deutsches Elektronen Synchrotron, Notkestraße 85, 22607 Hamburg, Germany

E-mail pawo@dtu.dk

Abstract. High-resolution reciprocal space mapping using high-energy hard x-rays has been developed to investigate the microstructure of grains located in the bulk of a metallic sample in a non-destructive way *in-situ* during different loading conditions. The technique allows to identify and follow individual grains and subgrains during ongoing deformation such as loading in tension and compression, during repeated cyclic deformation or even individual load cycles while simultaneously monitoring macroscopic stress and strain. Insight in the structural reorganization within single grains is gained by *in-situ* monitoring of the characteristic intensity distribution of Bragg reflections from individual grains during cyclic deformation of commercially pure polycrystalline aluminium. By reciprocal space mapping with high angular resolution and combined analysis of the radial and azimuthal information, individual subgrains are tracked during single load cycles with different strain amplitudes. Additionally, changes in mean peak position, peak width and asymmetry of integrated radial profiles from individual grains are analyzed as well as their orientation spread. In this manner, the microstructural evolution in grains embedded in the bulk of polycrystalline specimens is traced and linked to the changing mechanical loads during cyclic deformation.

1. Introduction

Material failure due to fatigue as a consequence of repeatedly changing load conditions is the most dominant failure reason of structural components. Details about the microstructural changes occurring in metallic materials during periodically varying loads remained largely unidentified due to limitations of the available characterization techniques as well as challenges in performing *in-situ* experiments.

High-resolution reciprocal space mapping (HRRSM) using hard x-rays has been developed to investigate the microstructure of grains located in the bulk of a metallic sample in a non-destructive manner *in-situ* during mechanical loading [1-6]. The synchrotron technique enables investigating the characteristic intensity distribution of diffraction peaks in a crystalline material in large detail and to identify and follow individual grains and subgrains during ongoing deformation [1-6]. The technique is based on the acquisition of three-dimensional reciprocal space maps of Bragg reflections and reveals the orientation spread as well as the occurring elastic strains within an individual grain. As the time for acquisition of a single map is of the order of 20 min, changes in the microstructure of individual grains can be followed during mechanical loading – even during continued tensile deformation [7]. The acquired reciprocal space maps are analyzed and presented complementarily as radial profiles or azimuthal maps. This allows tracking the microstructural evolution through the development of subgrains and the internal stresses associated with the substructure formation in individual grains during deformation. Simultaneous monitoring of macroscopic stress and strain enables to establish a link between the resolved evolution of the substructure and the mechanical loading.

In the past, HRRSM was applied successfully to unidirectional loading [1-6], stress relaxation [8], strain path changes [9-14], unloading and strain reversal [15] as well as cyclic deformation [16,17]. Previous investigations on cyclic deformation focused on application of HRRSM either on repeated tension-tension cycling for many cycles or during an individual cycle [16,17]. Here, HRRSM is employed on cyclic deformation in (almost symmetric) tension and compression with focus on the effect of a change in strain amplitude.

2. Experimental details

High-resolution reciprocal space mapping was carried out at beam line 1-ID-E at the Advanced Photon Source (APS) at Argonne National Laboratory with a monochromatic beam of 52 keV while loading an undeformed sample in position control. Figure 1 shows a sketch of the experimental set-up.

2.1. Material

Uniaxial test specimens were manufactured from a commercially pure aluminium sheet (AA1050) homogeneously cold-rolled by 90% thickness reduction to a final thickness of 1 mm. Dog bone-shaped specimens with a gauge section of 15 mm in length and 5 mm in width were designed to fit to a custom-made load frame. After spark-cutting, tensile specimens were annealed at 600 °C for 2 h to ensure complete and homogeneous recrystallization. After annealing, the microstructure was investigated metallographically with light optical and scanning electron microscopy. The micro-structure was found to be homogeneous throughout the entire cross section of the gauge showing a pronounced cube texture; grain sizes were estimated to be between 30 µm and 100 µm. Tensile testing of recrystallized specimens with a nominal strain rate of 10^{-3} s^{-1} revealed a yield strength of 16 MPa.

2.2. Experimental set-up at APS

For the synchrotron investigation, the sample was equipped with two pre-wired strain gauges (Omega KFG-3 350 Ω), one on each side at the center of the gauge section and aligned with the loading axis to monitor the axial strain *in-situ*. The sample was mounted in a custom-made screw-driven load frame equipped with a 1 kN load cell as presented in the insert of figure 1. Using flat grips, this load frame enables mechanical loading in tension and compression while monitoring the local microstructure within the test specimen *in-situ* using synchrotron radiation. The load frame was placed with the load axis horizontally on a *xy*-translation stage allowing to translate the sample and with this the chosen grain of interest – which gets displaced with respect to the load frame during mechanical loading – to the center of the beam after each load step. The *xy*-translation stage is mounted on top of a rotation stage enabling rotation of the entire load frame around the vertical *z*-axis to obtain the desired reciprocal space maps by rocking in small intervals of the angle ω around this axis. An additional *z*-translation stage provides adjustment for possible changes in the height of the selected grain of interest due to mechanical loading (cf. figure 1).

2.3. Experimental procedure for high-resolution reciprocal space mapping

Suitable grains are identified with the help of a large area detector (detector 1), an amorphous silicon flat panel from General Electrics, placed 86 cm behind the sample on a horizontal translation to cover the first 6 diffraction rings of aluminium. Individual grains are selected by finding isolated 400 diffraction peaks (i.e. peaks without any overlap with peaks from other grains) from a set of diffraction patterns (acquired at different ω angles but at same sample position to determine grain size and neighbourhood). Four bulk grains are chosen which have their 400 diffraction vector close to the loading axis. (Grains with their crystallographic [100] direction along the loading axis develop cell structures with subgrains separated by dislocation walls in tension [18] and cyclic deformation [19].)

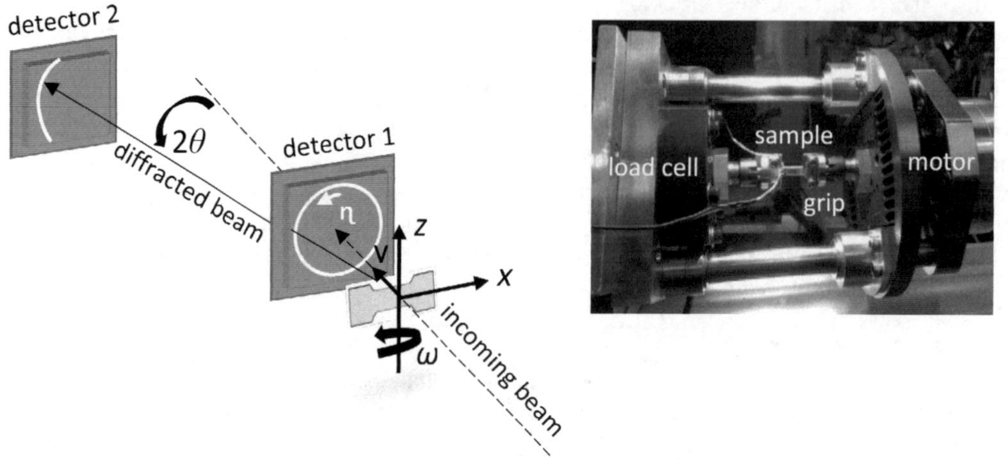

Figure 1. Principal set-up for High-Resolution Reciprocal Space Mapping (HRRSM) at the beam line 1-ID-E at the Advanced Photon Source at Argonne National Laboratory. The insert shows the custom-made screw-driven load frame with a dog bone-shaped specimen of aluminium.

After this, detector 1 is moved out of the beam and the diffraction peaks are investigated with significantly higher angular resolution by a Mar165 CCD (detector 2). This second detector is placed 4.65 m behind the sample at the location of 400 diffraction peaks with their diffraction vectors close to the loading axis (i.e. in the horizontal diffraction plane at a diffraction angle $2\theta_{400}$ of 13.53° for aluminium). An entire sequence of two-dimensional detector images for each selected diffraction peak is acquired with the far detector 2 while rocking the sample around the vertical z-axis perpendicular to the scattering plane in small intervals $\Delta\omega$ of 0.015° of the rocking angle ω. In this way three-dimensional distributions of the diffracted intensity (consisting of two dimensions, η and 2θ, on the far detector and additionally the rocking angle ω as third dimension) are obtained by stacking the images recorded for several adjacent ω intervals.

Two different projections of the acquired three-dimensional reciprocal space maps are conveniently used for analysis: azimuthal projections and radial profiles. Azimuthal projections characterize the intensity distribution of the diffraction peak in dependence on the two azimuthal angles (η,ω). They represent the pole density in dependence on the orientation as given by the angles (η,ω) and are essentially pole figures of quite high resolution. Radial profiles, on the other hand, are azimuthally integrated intensity distributions and represent the integrated intensity in dependence on either the diffraction angle 2θ (related to wave length λ and spacing d_{hkl} between the diffracting (hkl) lattice planes by Bragg's law $2\,d_{\mathrm{hkl}}\sin\theta = \lambda$) or the length of the diffraction vector $q = 2\pi/d_{\mathrm{hkl}} = 4\pi\sin\theta/\lambda$. The latter is the preferred means of representation as (relative) differences in the diffraction vector Δq relate directly to the elastic strain

$$\varepsilon_{\mathrm{el}} = \frac{d-d_0}{d_0} = \frac{q_0}{q} - 1 = -\frac{q-q_0}{q} = -\frac{\Delta q}{q}\,.$$

From such three-dimensional reciprocal space maps (and their two complementary projections), microstructural features such as dislocation walls and subgrains within individual grains can be identified from intensity variations in the map [1-6]: These intensity variations are caused by the slightly different and unique orientations of individual subgrains (e.g. [1-3,6,20]) in the deformation structure as the crystalline lattice becomes locally distorted by the emerging dislocation structures. Sharp peaks of high intensity in the high-resolution reciprocal space map correspond to individual subgrains, whereas a smooth cloud of lower intensity originates from the dislocations walls. From the characteristic intensity distribution of ordered dislocation structures, both the variation of elastic strains within single grains as well as the elastic strains of the embedded individual subgrains can be resolved.

3. Experimental procedure and results

3.1. Sample deformation

An undeformed and annealed dog bone-shaped tensile sample equipped with two strain gauges as described in section 2.2 was mounted in the load frame. After pre-deformation to 1% tensile strain by a macroscopic stress of 42 MPa, the specimen was cyclically deformed into saturation by 10000 tension-compression cycles with a strain amplitude $\varepsilon_{a1} = 0.4\cdot10^{-3}$. After this initial cycling, four grains were selected for further investigations and the sample was cyclically deformed further – for in total 30010 cycles. The last ten cycles where monitored in detail to choose the displacement for the desired stress-strain conditions along the stress-strain hysteresis curve for which data acquisition was intended. Based on this, a first tension-compression load cycle with the small strain amplitude $\varepsilon_{a1} = 0.4\cdot10^{-3}$ was performed for which reciprocal space mappings were undertaken at specific stress-strain conditions in such a manner that both the elastic region, as well as the yield point (or at least close to it), the plastic region and the state of maximum compression and tension were covered.

Thereafter, the strain amplitude was increased over 10 cycles to the intended strain amplitude of $0.82\cdot10^{-3}$. The sample was cycled in displacement control for another 10010 cycles, during which the strain amplitude decreased to a final value of $\varepsilon_{a2} = 0.68\cdot10^{-3}$. Again the last ten load cycles with the new strain amplitude of $\varepsilon_{a2} = 0.68\cdot10^{-3}$ were recorded in detail to select the displacements for the desired stress-strain conditions, before a second load cycle at the increased strain amplitude ε_{a2} was performed and reciprocal space maps were acquired. (Finally, the sample was cycled for another 8500 cycles with the large strain amplitude ε_{a2} before a further mapping of the grains was undertaken. The sample was then unloaded to zero stress and a final high-resolution reciprocal space map acquired.)

During acquisition of each high-resolution reciprocal space map, applied load and resulting strains were recorded by the load cell and the strain gauge, respectively. The average macroscopic stress and strain of all acquisitions are displayed in figure 2. Note, that during pausing of the motors for HRRSM after each load step, the position of the cross heads is fixed, while stresses and strains do not remain constant and change slightly. From the two hysteresis curves in figure 2, a clear increase in the width of the obtained hysteresis loop is obvious for load cycle 2 in comparison to load cycle 1. It also becomes evident that the average macroscopic strain decreased during cycling deformation.

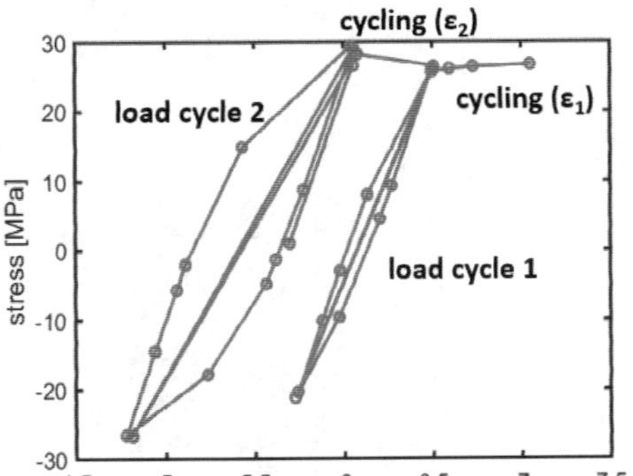

Figure 2. Macroscopic stress and strain measured during all acquisitions of high-resolution reciprocal space maps (highlighted with circles) showing the typical hysteresis curve during both load cycles. Before the two load cycles were investigated in detail, the sample was once unloaded and re-loaded without any intermediate step (represented by the lines in the middle of the hysteresis loops) to verify the strain amplitude.

3.2. Azimuthal maps

Grains were selected after pre-deformation to 1% tensile strain and 10000 tension-compression cycles. Four grains were chosen for further investigations with up to 29 HRRSMs per grain. Figure 3 shows azimuthal projections of the four grains measured always at maximum tension: (i) after the initial 10000 cycles, when the grains were initially selected, (ii) after cycling for 40010 cycles with ε_{a1} and (iii) after an increase of the strain amplitude to ε_{a2} followed by another 10020 cycles. Already when identifying the grains, the intensity distribution of their diffraction peaks showed a wide spread due to the 1% tensile pre-deformation. A clear increase of their extent in reciprocal space after the increase in strain amplitude is visible for grains 1 and 3, but less pronounced for grains 2 and 4.

IOP Conf. Series: Materials Science and Engineering **580** (2019) 012004 doi:10.1088/1757-899X/580/1/012004

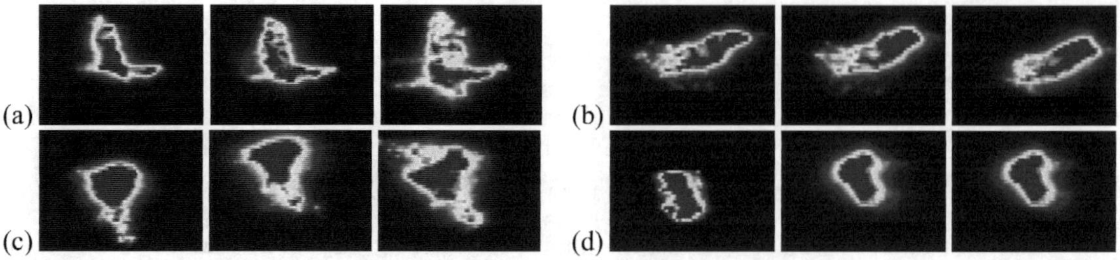

Figure 3. Azimuthal maps (η horizontally, ω vertically) of the 400 diffraction peak after grain selection (left), before (middle, L5a see below) and after (right, L1b) increase of strain amplitude from ε_{a1} to ε_{a2} for four different grains: (a) grain 1, (b) grain 2, (c) grain 3, and (d) grain 4.

The two grains 1 and 2 were mapped in detail during the two load cycles with different strain amplitudes with 9 (load cycle 1 designated with a, figure 4a) and 11 (load cycle 2 designated with b, figure 4b) acquisitions along the hysteresis curve. The acquisitions were selected to be at the maximum tensile load before (L1a, L1b) and after the load cycle (L5a, L6b), in the elastic region during unloading (U2a; U2b, U3b) and loading (L2a; L2b, L3b), around the yield point (U3a, L3a; U4b, L4b), at the maximum compressive load (U5a, U6b) and at several interesting conditions in the plastic region along the cycle (U4a, L4a; U5b, L5b). All measurements during unloading from maximum tension and loading to maximum compression are designated with U and all measurements during unloading from compression and loading to maximum tension are designated with L.

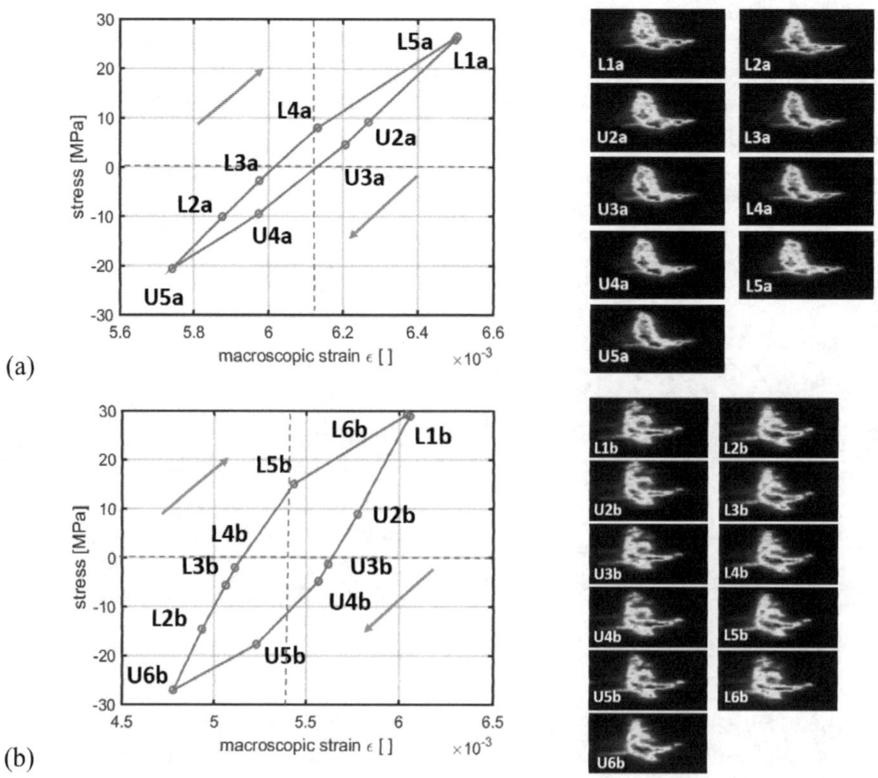

Figure 4. Hysteresis loops (left) and azimuthal maps of grain 1 (right) during the two load cycles with different strain amplitudes ε_{a1} (a) and ε_{a2} (b). Red lines indicate zero load and average macroscopic strain of the total strain variation during the cycle. The designation of the acquisitions starts with L1a (L1b) at maximum tensile load, following the arrows during unloading into compression to U5a (U6b) and loading in tension with a final map at maximum tension L5a (L6b).

Both load cycles interrupted for HRRSM were nearly symmetrical in their stress amplitude around zero load (indicated by the crossing red lines). In both cases, the absolute maximal stresses in compression were only slightly smaller than the absolute maximal stresses in tension. The stress ratio $R = \sigma_{max}/\sigma_{min}$ for the hysteresis loop changed from the value $R = -0.84$ at small strain amplitude $\varepsilon_{a1} = 0.4 \cdot 10^{-3}$ to an even higher value of $R = -0.97$ for the larger strain amplitude $\varepsilon_{a2} = 0.68 \cdot 10^{-3}$.

The azimuthal maps for grain 1 shown in figure 4a and b for all acquisitions during the two load cycles demonstrate exemplarily that no major changes are detectable during a hysteresis loop, but that significant changes are caused by the increase in strain amplitude and the performed 10020 cycles.

3.3. Radial profiles

Radial profiles are analysed for both load cycles. In figure 5, the radial profiles for the first load cycle are presented for grain 1. For a better overview, the profiles are split up into two sequences of unloading into compression (figure 5a) and loading into tension (figure 5b). The radial profiles shift to higher values of the diffraction vector $q = 2\pi/d_{hkl} = 4\pi \sin\theta / \lambda$ during compressive loading and to lower values during tensile loading in accordance with the applied load.

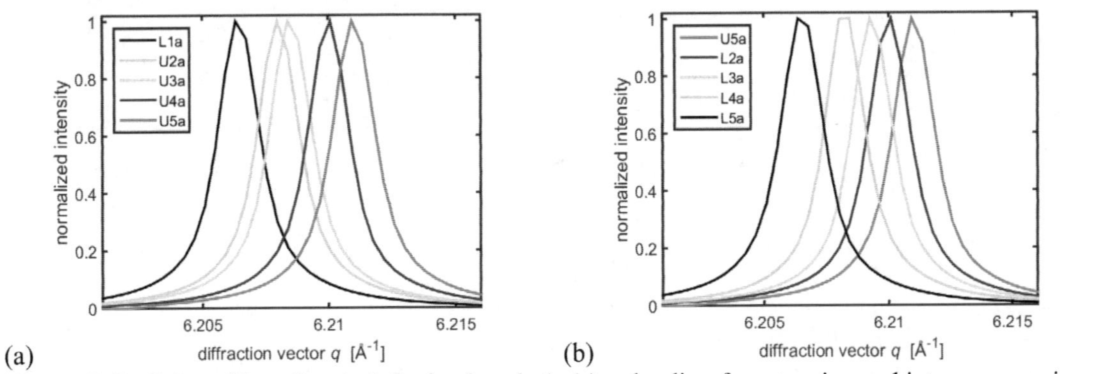

Figure 5. Radial profiles of grain 1 for load cycle 1: (a) unloading from tension and into compression (L1a, U1a-U5a), (b) unloading from compression and loading into tension (U5a, L2a-L5a).

Figure 6 shows the mean profile position of each of the four grains during the first load cycle compared to the strain-free diffraction vector $q_0 = 2\pi/d_{400} = 6.2064$ Å^{-1} (based on the lattice constant $a = 4.0495$ Å of pure aluminium and marked by a horizontal line). Grains 1 and 3 have peak positions generally above the strain-free diffraction vector q_0, whereas the average diffraction vectors of grains 2 and 4 remain mainly below q_0 – indicating quite different neighbourhoods between the four grains.

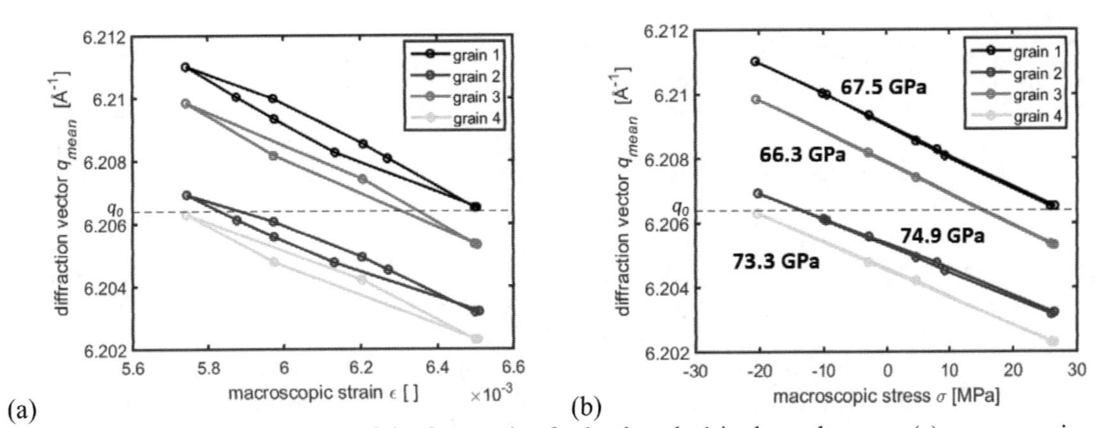

Figure 6. Mean profile position of the four grains for load cycle 1 in dependence on (a) macroscopic strain and (b) macroscopic stress. Apparent Young's moduli are noted next to the data of each grain.

The radial profiles are further analysed with regard to their integral profile width β and asymmetry κ as shown in figure 7 where the changes of both parameters during the two load cycles is illustrated for grain 1. During the first load cycle, the peak width (see figure 7a) is initially decreasing until U4a and then increasing towards the maximum compression (U5a) and further during tensile loading towards L5a. The second load cycle reveals more details due to the two additional measurements. Here it becomes obvious that the profile width is in general decreasing towards its lowest value at U5b and again strongly increasing towards the maximum compression U6b. After U6b it decreases first towards L5b, before increasing again towards L6b. The behaviour – in particular for the second load cycle – resembles a so-called butterfly pattern observed earlier [21,22] and discussed in detail in [23].

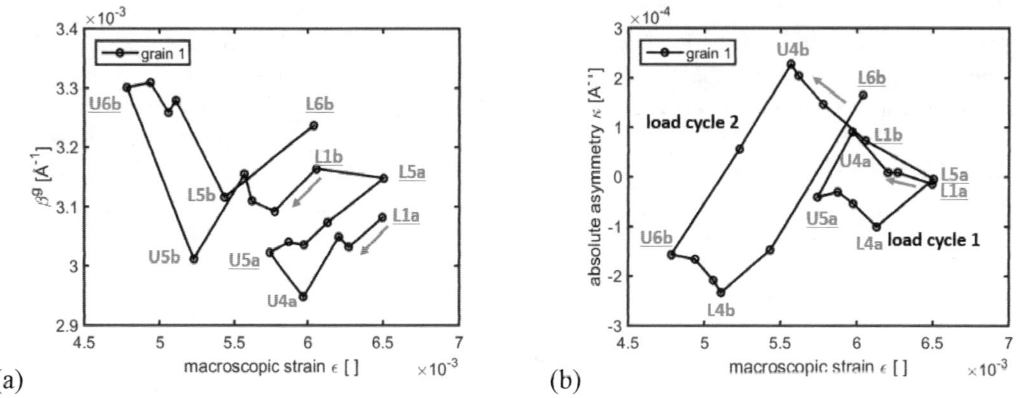

(a) (b)

Figure 7. Changes in (a) integral width β and (b) absolute asymmetry κ of grain 1 during the two load cycles (in dependence on macroscopic strain). The arrows indicate the sequence of the mappings.

The absolute asymmetry κ (i.e. the difference between the widths at half of the maximum intensity on both sides of the position of maximum intensity) follows a distinct pattern as well (cf. figure 7b). During the first load cycle, the asymmetry is first increasing from L1a until U4a and then decreasing towards U5a. The asymmetry is in general decreasing further until L4a and then increasing again towards maximum tensile load L5a. For the second load cycle the asymmetry is first increasing until U4b and then decreasing towards U6b. It behaves similar (but with opposite sign) during tensile loading with first decreasing until L4b and then increasing until the tensile point L6b is reached. The asymmetry is changing sign from positive to negative during loading into compression as expected from the composite model (see section 4.2), with a similar absolute value ± 1.6 Å^{-1} at maximum tension and compression during load cycle 2. The other three grains behave quite similar (cf. [23]).

3.4. Subgrains

In high-resolution reciprocal space maps, the existence of high-intensity peaks on a cloud of enhanced intensity reflects the developing ordered dislocation structure with perfect, slightly disoriented subgrains separated by dense dislocation walls. Each almost perfect subgrain gives rise to one individual peak of high intensity, whereas the dislocation walls contribute collectively a cloud of enhanced intensity to the diffraction peak [1-6]. For analysing subgrains, the two different contributions from subgrains and walls are separated by a mathematical approach [7,10]. Figure 8 shows the separated azimuthal map originating solely from all subgrains in grain 1 at the specific loading condition L1b just before the second load cycle. Circles mark the ten most intense peaks corresponding to the ten largest subgrains. For grain 1, up to 92 subgrains have been identified in this manner in the seperated subgrain contributions of the different acquired reciprocal space maps. No clear relation between the number of identified subgrains and the stress-strain condition was found, though a tendency for fewer identified peaks around the maximum compressive load was observed for load cycle 2. With such a large number of identified subgrains, a statistical analysis can be performed (as illustrated for unidirectional tension [2], strain path changes [13,14], unloading and strain reversal [15] or cyclic deformation in tension-tension [17] – revealing a Gaussian distribution of the elastic strains of subgrains in all cases). Alternatively, individual subgrains can be identified by their azimuthal position and followed during the loading sequence. The latter is illustrated here for the second load cycle by the largest subgrain in the azimuthal map in figure 8 marked with a red circle which could be followed through all acquisitions of grain 1.

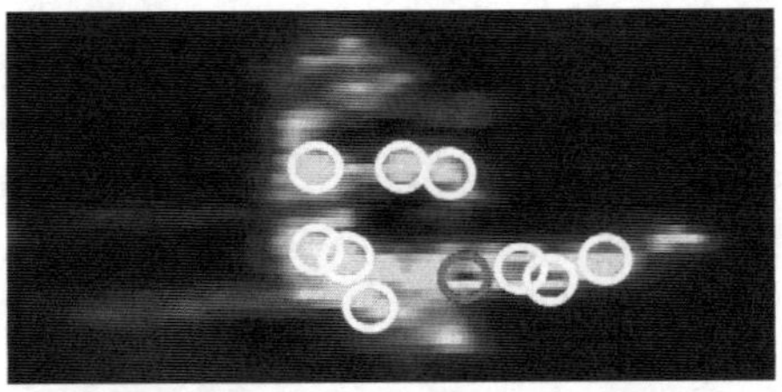

Figure 8. Azimuthal map of grain 1 at maximum tensile stress after 10020 cycles with larger strain amplitude ε_{a2} (load step L1b). Only the seperated part formed by the high-intensity peaks of all subgrains is illustrated. The ten most intense peaks corresponding to the ten largest subgrains are marked with circles.

The radial profiles of this subgrain are shown together with the radial profiles of grain 1 in figure 9a at the maximum tensile and maximum compressive load (L1b in black and U6b in red, respectively), where the mean profile positions are marked with straight lines additionally. It is apparent, that the grain profile at maximum compression U6b looks in general mirrored to the grain profile measured under maximum tensile load L1b. The subgrain profile shifts similar to the one of grain 1 to higher q-values during loading into compression and to lower q-values during loading into tension. Small differences can be identified between the mean profile positions of grain and subgrain. The mean position of the grain is at slightly higher q-values than the one for the subgrain at maximum tension and at slightly lower q-values at the maximum compression, meaning that the subgrain experiences slightly less tensile strain at maximum tensile load and slightly less compressive strain at maximum compressive load than the grain, i.e. the subgrain experiences back strains in both cases.

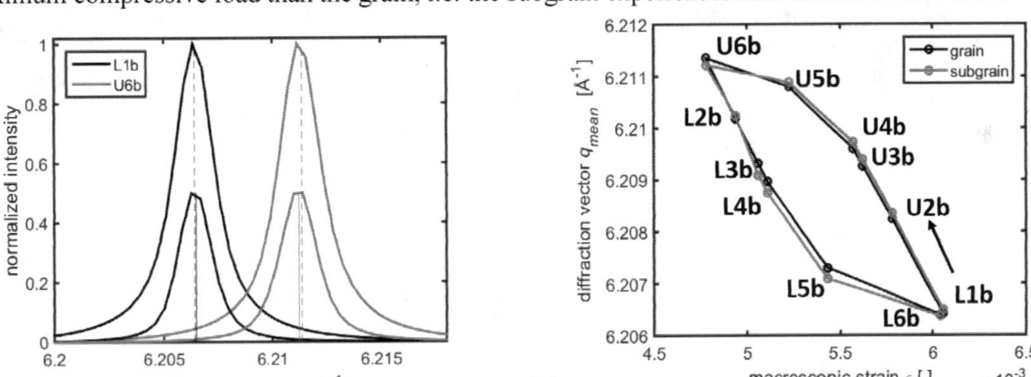

(a) (b)

Figure 9. (a) Radial profiles of grain 1 at maximum tension (L1b) and maximum compression (U6b) during load cycle 2. The additional profiles of the particular subgrain (which has been identified in all load steps and marked in the azimuthal map in figure 8 by a red circle) are scaled to a maximum of 0.5. The mean profile positions (of the fitted profiles) are marked with vertical dashed lines for the grain profiles and full lines for the subgrain profiles. (b) Average profile positions q_{mean} for grain 1 (black) and the particular subgrain (red) in dependence on macroscopic strain.

The mean positions of grain and subgrain throughout load cycle 2 are displayed in detail in figure 9b, where the shift of the subgrain profile clearly reproduces the hysteresis loop as observed for the grain. Obviously, the subgrain shows slightly higher q_{mean} than the entire grain at maximum tension, unloading from there and loading into compression (meaning the subgrain experiences slightly less tensile stress) and slightly lower q_{mean} than the grain at maximum compression, unloading from there and loading into tension (meaning the subgrain experiences slightly less compressive stresses than the grain). This is in accordance with previous findings for tension-tension cycling [17] that there is a size effect governing the stresses of individual subgrains with the largest subgrains always experiencing the largest back stresses.

IOP Conf. Series: Materials Science and Engineering **580** (2019) 012004 doi:10.1088/1757-899X/580/1/012004

3.5. *Increase in strain amplitude*

The strain amplitude was increased in between the detailed acquisition of the two load cycles from $\varepsilon_{a1} = 0.4\cdot10^{-3}$ to $\varepsilon_{a2} = 0.68\cdot10^{-3}$. Before the second load cycle was measured, 10020 cycles with the increased strain amplitude ε_{a2} were performed. Figure 10 visualizes the changes in the Bragg reflection along the azimuthal ω direction and with this the appearance of the grain in orientation space before and after the increase of the strain amplitude. The rocking curves of all four grains change slightly between before (blue) and after (red) the increase in ε_a. The intensity distributions become broader in width, most clearly seen for grains 1 and 3 which has earlier been indicated by the azimuthal maps in figure 3. The rocking curve of grain 4 seems to become slightly broader as well. The observed movements in ω are due to a rotation of the grains when the sample and the grains adjust to the higher strain amplitude and the different level of macroscopic strain of the second load cycle.

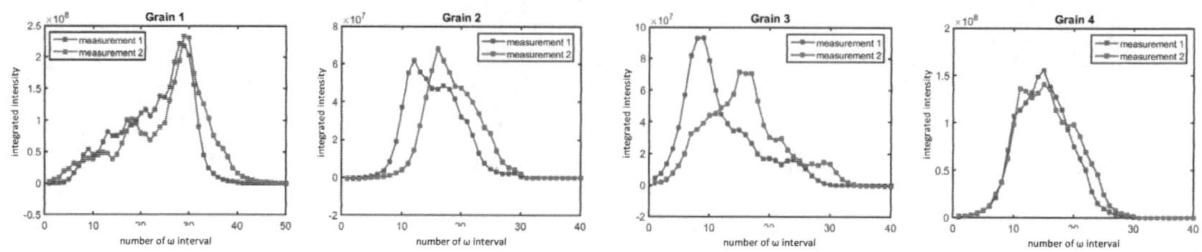

Figure 10. Rocking curves for four grains before (measurement 1, L5a) and after (measurement 2, L1b) increase of strain amplitude from ε_{a1} to ε_{a2}. Each ω interval corresponds to $\Delta\omega = 0.015°$.

Figure 11 presents an overview how radial profile position, width and asymmetry were affected by the change in strain amplitude. The radial profiles were analysed at maximum tensile load at the end of the first load cycle (L5a, measurement 1) and at maximum tensile load at the beginning of the second load cycle (L1b, measurement 2), i.e. after additional 10020 cycles with $\varepsilon_{a2.}$ The profile positions decrease only slightly due to the lower macroscopic stress at L5a than at L1b (cf. figure 2). The radial profile width is increasing in general for all grains, though less pronounced for grain 1. The absolute asymmetry is increasing for all grains in a similar manner. After increasing the strain amplitude, three grains (grain 1, 3 and 4) show a positive asymmetry at maximum tension and a negative asymmetry at maximum compression as expected by the composite model (cf. section 4.2); in contrast to all other grains, grain 2 has negative asymmetry throughout. Nevertheless, profile width and asymmetry increase both with an increase in strain amplitude as expected due to increasing internal stresses.

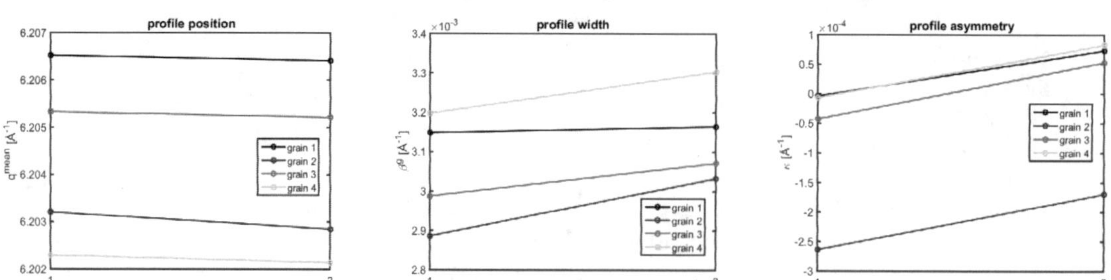

Figure 11. Comparison of (a) average profile position, (b) integral width, and (c) absolute asymmetry at maximum tensile load before the increase of strain amplitude after 40010 cycles with ε_{a1} (L5a, measurement 1) and after additional 10020 cycles with ε_{a2}. (L1b, measurement 2).

4. Discussion

4.1. *Peak positions*

The different average peak positions of the four grains seen in figure 6 reveal substantial internal stresses within the specimen. Grains 1 and 3 show in general compressive strain with respect to the strain-free lattice spacing, whereas grains 2 and 4 show tensile strains. These internal elastic strains must be caused by corresponding internal stresses arising during the tensile pre-deformation as a result of load transfer between the grains under investigation and their immediate surrounding of neighbouring grains. As grains 1 and 3 experience compressive stresses, i.e. back stresses compared to the initial tensile loading, they are expected to be located in rather strong (or stiff) neighbourhoods, so they will yield first (or elastically deform more) when under the same load. In order to deform compatibly with the stronger (or stiffer) neighbourhood, they develop back stresses with respect to the applied tensile load. Vice versa, grains 2 and 4 are expected to reside in a weaker (or more compliant) surrounding which yields earlier (or elastically deforms more) than the two grains under consideration. Grains 2 and 4 will hence develop additional forward stresses, i.e. residual tensile stresses, to keep a compatible deformation.

A linear dependence of the peak position q_{mean} on the macroscopic applied stress becomes evident from figure 6b. Considering that the average elastic strain $\varepsilon_{el} = (q_0/q_{mean})-1$ of each grain is related to the mean peak position, this is expected from elastic behaviour. An apparent elastic modulus $E^* = |\sigma|/|\varepsilon_{el}|$ can be determined for each of the grains. The resulting values specified in figure 6b are all above the elastic modulus $E_{[100]} = 63.5$ GPa of aluminium along its crystallographic [100] direction [24]. The increased values are partially caused by the misalignment between the tensile axis and the [100] direction along which the elastic strain is determined. Due to elastic cross contraction, any such misalignment leads to a reduction of the measured elastic strain along the 400 diffraction vector and hence to an overestimation of the elastic modulus. The variations between the apparent moduli of the four grains are, however, too large to be caused by their slightly different orientation (which can be assessed from the azimuthal position of the diffraction peaks and differ by less than 5°). The different values for the apparent moduli must hence result from the neighbouring grains affecting the stress state. The effect of the neighbourhood will not only cause load transfer between grains of different stiffness and strength, but also to possible deviations from a uniaxial stress state. According to figure 6b, a correlation exists between the apparent moduli and the residual stresses in a grain: Grains 1 and 3 which developed compressive residual stresses during tensile pre-deformation have a lower apparent modulus than grains 2 and 4 which developed tensile residual stresses. This correlation can be rationalized by the same reasoning about the neighbourhood of the different grains as outlined above. The strong and stiff neighbourhood of grains 1 and 3 experiences higher local stresses $|\sigma_{loc,high}|$ and higher elastic strains $|\varepsilon_{el,high}|$ than the weak or compliant surrounding of grains 2 and 4 ($|\sigma_{loc,low}|<|\sigma_{loc,high}|$ and $|\varepsilon_{el,low}|<|\varepsilon_{el,high}|$). Hence, the apparent Young's moduli $E^* = |\sigma|/|\varepsilon_{el,high}|$ of grains 1 and 3 become smaller than the apparent moduli $E^* = |\sigma|/|\varepsilon_{el,low}|$ of grains 2 and 4.

4.2. Shape of radial profiles

Radial profiles of plastically deformed metals are in general asymmetric [25-27] as rationalized by the composite model of deformation structures [28,29]. The refined version of the composite model [5] takes into account individual, dislocation-free subgrains and dislocation-rich dislocation walls. During loading, subgrains yield before the walls due to their lower dislocation density. In order to maintain compatible deformation, load is transferred from the weaker subgrains to the stronger walls leading to back stresses (with respect to the applied stress) in the subgrains and forward stresses in the walls. This difference in stresses causes differences in the elastic strains between subgrains and walls which are revealed by diffraction as both components will diffract at different diffraction vectors $q = q_0/(1+\varepsilon_{el})$. Furthermore, the subprofile arising from dislocation walls is significantly broadened due to their high dislocation density. As rationalized in the refined composite model [5], each dislocation-free subgrain will contribute with its own sharp profile at a slightly different diffraction vector. The summed subprofile from all subgrains will be broadened slightly due to the Gaussian distribution of the elastic strains of individual subgrains [2,13-15,17]. The combination of the shift of the two subprofiles with their different width and volume fraction leads to asymmetric profiles. For instance, in tension, subgrains come under compressive back stresses leading to compressive strains along the loading direction and a higher diffraction vector than that of the walls which are under forward stresses and hence experience tensile elastic strains. Due to the rather sharp subprofile contributed by all subgrains and their slightly different stresses, the maximum intensity of the diffraction peak at q_{max} is close to the average position of the subgrains, whereas the contribution of the walls occurs at lower diffraction vectors. Consequently, the average position q_{mean} of the entire peak has a lower value than q_{max} and the radial diffraction profile becomes asymmetric with larger width for lower q, hence, leading to a positive (absolute) asymmetry κ.

This asymmetry of diffraction peaks has been discovered first in tensile deformed copper [23,24] and rationalized by the original composite model [25-27]. Occurrence of the corresponding asymmetry in aluminium after tensile deformation has been confirmed much later [28]. During cyclic deformation in tension and compression, the situation becomes reversed between tension and compression and the asymmetry is expected to change sign from being positive under tension to becoming negative in compression. This has been observed on copper after unloading from tension or compression [22].

The present investigation confirms the general trend of reversing the sign of asymmetry between tension and compression as seen in figure 7, but even more importantly it reveals that the asymmetry is much more pronounced in the unloaded state than under maximum tensile or compressive load: The absolute asymmetry shows the proper sign at maximum tension and compression, but with quite small absolute values only, whereas during unloading a significant increase of the absolute amount of the asymmetry is observed. The reason for this behaviour roots in the different elastic strains of the individual subgrains and their succession of yielding during strain reversal. This idea has to be investigated further and may potentially rationalize the butter-fly behaviour of the peak width as well.

4.3. Increase in strain amplitude

An increase in strain amplitude is connected with an increase in the amplitude of the applied stresses due to work-hardening of the material mainly by dislocation accumulation in the walls. The increased dislocation density in the walls leads – from statistical reasons – even during cyclic deformation to an increase of the small orientation differences between the subgrains [31]. The increasing mutual disorientations cause an increasing orientation spread in the grains which explains both, the observed larger spread of the intensity distribution in the azimuthal maps (figure 3) and the larger width of the rocking curves (figure 10) after an increase in strain amplitude. The increased dislocation density in the walls causes a larger difference between the elastic strains of subgrains and walls and consequently a larger peak width and a larger amount of the absolute asymmetry after the increase in strain rate.

5. Conclusion

Cyclic deformation in tension and compression of macroscopic, commercially pure, polycrystalline aluminium samples was performed with a custom-made load frame. High-resolution reciprocal space mapping was used to investigate the effect of changing the strain amplitude during cycling. Independent of the strain amplitude, the acquired azimuthal maps show only minor changes along a load cycle. A peak shift of the radial profiles following the macroscopically applied stress was observed as well as indications for a characteristic behaviour of the profile width and asymmetry along a tension-compression load cycle. On the other hand, a clear difference in both shape and intensity distribution in the azimuthal maps is observed for different strain amplitudes. An increase in strain amplitude results in broadening, which can be quantified from both radial profiles and azimuthal maps and linked to structural reorganization occurring in the monitored grains. Further insight into the deformation-induced processes is of fundamental importance to understand and predict materials behaviour during cyclic deformation in order to facilitate future materials design.

Acknowledgements

This research used resources of the Advanced Photon Source, a U.S. Department of Energy (DOE) Office of Science User Facility operated for the DOE Office of Science by Argonne National Laboratory under Contract No. DE-AC02-06CH11357. The authors gratefully acknowledge the financial support from the Danish Council for Independent Research through DANSCATT and like to thank Jun-Sang Park for assistance in using beamline 1-ID.

References

[1] Jakobsen B, Poulsen H F, Lienert U, Almer J, Shastri S D, Sørensen H O, Gundlach C and Pantleon W 2006 *Science* **312** 889-92
[2] Jakobsen B, Poulsen H F, Lienert U and Pantleon W 2007 *Acta mater.* **55** 3421-30
[3] Jakobsen B, Poulsen H F, Lienert U, Huang X and Pantleon W 2007 *Scripta mater.* **56** 769-72
[4] Jakobsen B, Poulsen H F, Lienert U and Pantleon W 2008 *Mater. Sci. Engng. A* **483-484** 641-3
[5] Pantleon W, Wejdemann C, Jakobsen B, Lienert U and Poulsen H F 2010 Advances in characterization of deformation structures by high resolution reciprocal space mapping *Proc. 31st Risø Intern. Symp. Mater. Sci.* eds N Hansen et al. (Roskilde: Risø DTU) pp 79-100
[6] Pantleon W, Wejdemann C, Jakobsen B, Poulsen H F and Lienert U 2014 High-resolution reciprocal space mapping for characterizing deformation structures *Strain and dislocation gradients from diffraction* eds R Barabash and G Ice (London: Imperial College Press) chapter 9 pp 322-57
[7] Pantleon W, Wejdemann C, Jakobsen B, Lienert U and Poulsen H F 2009 *Mater. Sci. Engng. A* **524** 55-63
[8] Jakobsen B, Poulsen H F, Lienert U, Bernier J, Gundlach C and Pantleon W 2009, *phys. stat. sol. (a)* **206** 21-30
[9] Wejdemann C, Poulsen H F, Lienert U and Pantleon W 2009 *IOP Conf. Series: Mater. Sci. Engng.* **3** 012003:1-6
[10] Wejdemann C, Lienert U, Nielsen H B and Pantleon W 2010 Identifying individual subgrains in evolving deformation structures by high angular resolution x-ray diffraction *Proc. 31st Risø Intern. Symp. Mater. Sci.* eds N Hansen et al. (Roskilde: Risø DTU) pp 477-87
[11] Wejdemann C, Lienert U and Pantleon W 2010 *Scripta mater.* **62** 794-7
[12] Wejdemann C, Lienert U and Pantleon W 2010 *J. Phys.: Conf. Series* **240** 012160:1-4
[13] Wejdemann C, Poulsen H F, Lienert U and Pantleon W 2013 *JOM* **65** 35-43
[14] Lienert U, Ribárik G, Ungar T, Wejdemann C, Pantleon W 2017 *Synchr. Rad. News* **30** 35-40
[15] Diederichs A M, Thiel F, Fischer T, Lienert U and Pantleon W 2017 *IOP Conf. Series: J. Phys.: Conf. Ser.* **843** 012031-1:9
[16] Diederichs A M, Thiel F, Lienert U and Pantleon W 2017 *Procedia Struct. Integr.* **7** 268-74
[17] Diederichs A M, Thiel F, Lienert U and Pantleon W 2018 *Int. J. Fatigue* **117** 206-16
[18] Huang X and Hansen N 1997 *Scripta mater.* **37** 1-7
[19] Vidum M and Ryum N 1996 *Mater. Sci. Engng. A* **219** 1-10
[20] Gundlach C, Pantleon W, Lauridsen E M, Margulies L, Doherty R D and Poulsen H F 2004 *Scripta mater.* **50** 477-81
[21] Holste C and Burmeister H J 1979 *phys. stat. sol. (a)* **57** 269-80
[22] Biermann H, Ungar T, Pfannenmüller T, Hoffmann G, Borbely A and Mughrabi H 1993 *Acta metall. mater.* **41** 2743-53
[23] Diederichs A M 2019 *Structural reorganziation during cyclic deformation* (Kongens Lyngby: DTU) Ph.D. thesis 252 pp
[24] Tallon J L and Wolfenden A 1979 *J. Phys. Chem. Solids* **40** 831-7
[25] Ungar T, Mughrabi H, Rönnpagel D and Wilkens M 1984 *Acta metall.* **32** 333-42
[26] Mughrabi H, Ungar T, Kienle W and Wilkens M 1986 *Philos. Mag. A* **53** 793-813
[27] Mughrabi H and Ungar T 2002 *Long-range internal stresses in deformed single-phase materials: the composite model and its consequences* (*Dislocations in solids* vol 11) (Amsterdam: Elsevier) chapter 60 pp 343-411
[28] Mughrabi H 1978 *Mater. Sci. Engng.* **33** 207-23
[29] Mughrabi H 1983 *Acta metall.* **31** 1367-79
[30] Pantleon W, Poulsen H F, Almer J and Lienert U 2004 *Mater. Sci. Engng. A* **387-389** 339-42
[31] Pantleon W 1999 Dislocation boundaries: formation, otrientation and implications *Proc. 20th Risø Intern. Symp. Mater. Sci.* eds J B Bilde Sørensen et al. (Roskilde: Risø National Laboratory) pp 123-46

IOP Conf. Series: Materials Science and Engineering **580** (2019) 012005 doi:10.1088/1757-899X/580/1/012005

Roles of transformation interfaces in the design of advanced high strength steels

Tadashi Furuhara, Yongjie Zhang and Goro Miyamoto

Institute for Materials Research, Tohoku University
2-1-1 Katahira, Aoba-ku, Sendai, 980-8577, Japan

furuhara@imr.tohoku.ac.jp

Abstract. Alloy partitioning and segregation or precipitation at transformation interfaces have been important in designing the mechanical properties of final products. Some representative examples in high-strength low-alloy steels are briefly discussed, with an emphasis on the transformation interface. For achieving better strengthening-ductility balance in TRIP-aided low alloy steels, it is essential to control carbon enrichment in austenite during ferrite and bainite transformations and the stability of retained austenite. Recent studies on alloy partitioning including carbon enrichment into austenite, provides a deeper insight into transformation mechanisms. We show that energy dissipation during interface migration and strain energy accumulation are important factors to control carbon enrichment into untransformed austenite during the ferrite/bainite transformation. As a counterpart of TRIP steels, interphase precipitation of alloy carbides during ferrite transformation leads to high yield strength and good elongation in low-carbon ferritic steels. A key for further strengthening is clarification of the conditions for refining the carbide dispersion. It is shown that incoherent ferrite/austenite interfaces are superior nucleation sites for carbides.

1. Introduction

A variety of phase transformations products, i.e., ferrite, pearlite, bainite and martensite, in low-alloy steels containing carbon, are used in steel products. In development of modern high strength steels, the importance of strength and the ductility/toughness balance is increasing and more sophisticated control of the microstructure formed during phase transformations is utilized. In order to control the microstructure and properties of steels, it is inevitable to use alloying elements in Fe-C alloys.

In the recent development of low-carbon high strength sheet steels, two distinctive categories have been extentively investigated [1]. One is mutiphase steels, represented by dual-phase and TRIP steels produced by hot-rolling and subsequent interupted ferrite/bainite transformation followed by rapid cooling or by intercritical annealing of cold rolled sheet accompanying reverse transformation to austenite and subsequent forward transformation. During forward or reverse transformation in the (ferrite+austenite) dual-phase condition, fast diffusing carbon enriches rapidly the austenite. After final cooling, carbon-rich martensite (for DP steel) or metastable high-carbon retained austenite (for TRIP steel) is obtained. For TRIP steels, the mechanical properties of the carbon-enriched second phase (retained austenite) are significantlly influenced by the amount of carbon in solution [2, 3]. In addition, slow diffusing substitutional alloying elements may be also partitioned between the two phase in complex manners. Recent systematic efforts studying on alloy partitioning during ferrite or bainite transformation reveal the importance of transformation interface [4]. Also substantial efforts are being made to investigate the partitioning behavior of carbon and other elements in Q&P steels for exploring higher strength and ductility balance [5,6].

Another is ferritic steels with nanoscale precipitation. In this case, ferrite transformation occurs during cooling after hot rolling. Addition of microalloying elements such as Ti, Nb and V, results in alloy carbide precipitation of which nucleation is promoted at growth interface of ferrite. This phenomena denoted as 'interphase precipitation' was extensively studied in 70's as summarised by Honeycombe [7,8]. Recently, a group of microalloyed low-carbon steels is commercialized for automobile application as high strength, high ductility sheet steels produced by hot rolling and cooling [9]. It was shown that combined addition of alloying elements was effective for strengthening due to refined dispersion of carbide. This kind of steels shows superior ductility in terms of local elongation whereas high strength multiphase steels exhibit superior uniform elongation achieved by large work hardening.

It is obvious that the natures of transformation interface have direct impact on the behaviors of the alloy partitioning and interphase precipitation and resultant mechanical properties. We have been investigating on the roles of crystallography and interfacial structure of transformation products on microstructure development in steels [10]. In this paper, some representative examples in high-strength low-alloy steels will be discussed in this context.

2. Non-equilibrium alloy partitioning during transformation in low-alloy steels

2.1. Partitioning during ferrite/bainite transformation

Proeutectoid ferrite shows a variety of morphologies [11, 12]. Grain boundary ferrite (GBF), which is formed under low supersaturation, shows a characteristic morphology described as allotriomorphs elongated along austenite grain boundaries with smoothly curved intrefaces. It has a near Kurdjumov-Sachs orientation relationship (K-S OR) with respect to one of the adjacent austente grains and tends to grow toward the another austenite grain by motion of an 'incoherent' interface due to a non K-S and irrational OR. As temperature decreases, plate- or needle-shaped Widmanstätten ferrite (WF) is formed at austenite grain boundaries or on top of pre-existing allotriomorph and grows into the austenite grains by holding a more rigid OR and with partly coherent interfaces. As the transformation temperature further decreases typically below 600°C, bainitic ferrite (BF) with high interfacial coherency, similar to WF, starts to form both at austenite grain boundaries but also by more frequent intragranular nucleation.

The growth of GBF in Fe-C alloys is contolled by volume diffusion of carbon in austenite. In the Fe-C system, ferrite growth with incoherent boundary is well explained by local equilibrium in a decarburization study [13]. The thickening of GBF is also C diffusion-controlled but is slower than the expected from calculation assuming the migration of incoherent boundaries. The reason for this was considered to be effects of the boundary structures mixed by partly coherent and incoherent ones [14].

In alloyed steel, GBF grows more slowly than the Fe-C case [15] with carbon partitioning. Alloy partitioning accompanied results where variations of phase stability and transformation kinetics are more complex. There are three distinctive local equilibrium growth modes in Fe-C-M (M: substitutional alloying element) system. These are para-equilibrium (PE), negligible-partitioning local equilibrium (NPLE) and partitioning local equilibrium (PLE). PE and NPLE are essentially C diffusion-controlled modes without macroscopic partitioning of M. The former is a local equilibrium in the (Fe, M)-C pseudo-binary system whereas the latter satisfies local equilibrium at the interface for all elements with a sharp spike of M near the interface. Recent efforts in decarburized experiments in the ternary systems reveal that the formation limit of ferrite in the Fe-Mn-C and Fe-Ni-C systems is below the PE lines and close to the NPLE/PLE limit [4]. In the beginning of transformation, para-equilibrium (PE) growth seems to occur and make a transition to NPLE growth mode and accordingly a model to treat the PE-NPLE/PLE transition has been proposed [16].

It has been recognized that there are various 'frictions' on the interphase boundary migration exerted from various components of the microstructure as represented in figure 1. Atomic displacement at the interface causes intrinsic friction. Solute atoms segregated at the interface provide another dissipation known as 'solute drag'. Dislocations or fine second-phase particles may add further barriers for interface migration.

Hillert and his co-workers discussed the energy dissipation by such friction extensively [17]. Figure 2 shows the effect of the energy dissipation in boundary migration on partitioning of carbon in the Fe-C binary system. The same situation can be imagined for the (Fe, M)-C pseudo-binary case. (a) is for

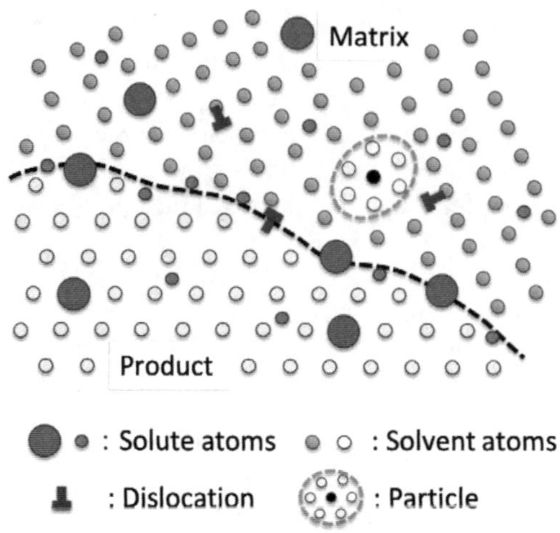

Figure 1. Schematic illustration of microstructure components causing energy dissipation in interface migration.

forward transformation to ferrite whereas (b) represents the case of reverse transformation to austenite.

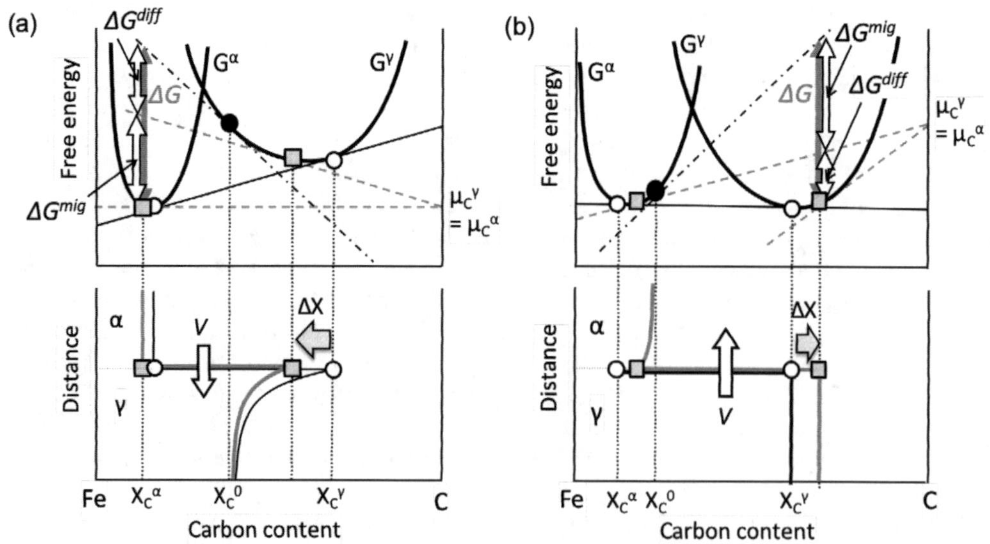

Figure 2. Schematic diagrams of deviation of carbon partitioning from local equilibrium due to energy dissipation at a migrating interface; (a) austenite → ferrite forward transformation and (b) ferrite → austenite reverse transformation to austenite. White circles: carbon contents at local equilibrium, gray squares: carbon contents achieved with energy dissipation, Black circle: initial matrix composition.

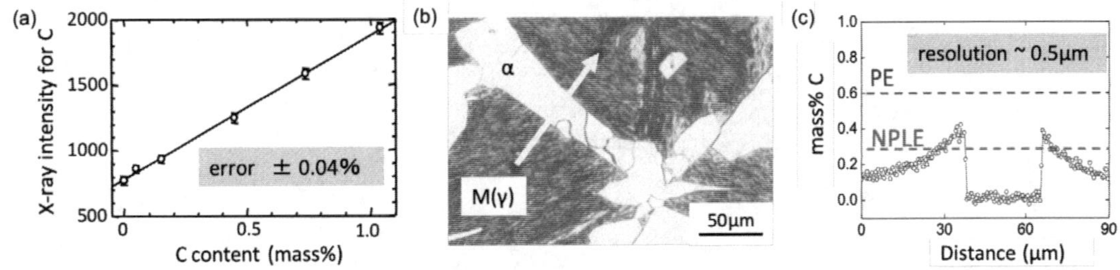

Figure 3. FE-EPMA measurement of carbon concentration at ferrite/austenite interface in Fe-0.15C-2Mn (mass%) alloy [18].

Local equilibrium at the interface is achieved at carbon concentrations represented by white circles given by drawing the common tangent to free energy curves of ferrite and austenite phases. At this condition, there is no force exerted at the interphase boundary. When some friction accompanies boundary migration, an extra driving force to make the boundary move forward is necessary and a deviation from local equilibrium occurs, described as shift of the boundary concentration from the open circles to squares in the figure. Under partial equilibrium of carbon as a fast diffuser, the extra driving force or the energy dissipation by boundary migration is given by ΔG^{mig}. The total driving force for transformation, ΔG, is the sum of ΔG^{mig} and ΔG^{diff} which is consumed by long-range carbon diffusion. As seen in figure 2(a), such energy dissipation leads to lower carbon content in austenite for forward transformation and higher carbon content in austenite for reverse transformation (figure 2(b)). Furthermore, additional partitioning of substitutional alloying elements gives more complex effects on the stability of austenite.

By measuring the chemical composition, in particular the carbon content across the interface, it is possible to estimate the amount of such dissipation. We recently developed quantitative measurement method using field-emission electron probe micro-analysis [18]. Figure 3 shows an example applied to GBF in Fe-0.15C-2Mn. After we constructed calibration line (figure 3(a)) using as-quenched martensite specimens with a variety of carbon contents, measurement was made. In the analysis of a ferrite in (b), a clear concentration profile is obtained along a linescan with a spatial resolution of ca. 0.5μm in (c).

Figure 4 summarizes calculated phase boundaries and growth limits with measured carbon content in austenite at the interface in various Fe-C-Mn alloys [10]. Increase of Mn content stabilizes austenite and also shifts the NPLE limit toward the lower temperature region. The NPLE limit becomes almost coincident with the T_0 line, which is the growth limit for composition-invariant diffusional growth in massive transformation and diffusionless growth in martensitic transformation, at 2mass%Mn and finally surpasses the T_0 line at 3mass%Mn. The growth of GBF seems to correspond well to the NPLE mode for all the Mn-contained alloys, except at one temperature in the 1.5Mn alloy. On the other hand, it is interesting to see that the growth of WF at 923K deviates from NPLE in the 0.75Mn-0.1C alloy whereas GBF in the 1.5Mn-0.15C alloy grows in the NPLE mode at the same temperature. This difference might be affected by interfacial coherency. As the temperature is lowered, carbon enrichment at austenite/GBF deviates to a lower side than the NPLE case. In 2mass%Mn, those two limits both can explain the ferrite growth successfully. However, in 3mass%Mn alloy, carbon enrichment exceeds the NPLE limit and is coincident with T_0' for undercooling of 400 J/mol from T_0 [19].

The solute drag effect [20] is another mechanism causing energy dissipation when alloying with substitutional atoms. It is considered that Mn has some tendency to segregate at the ferrite/austenite boundary and provides a dragging force. On the other hand, Si is thought to have weak interaction to ferrite/austenite boundaries. Thus, Fe-C-Si is a good candidate system to study dissipation mainly due to interfacial coherency.

Figure 5(a) shows measured carbon contents in austenite at the interface in a Fe-0.4C-3Si alloy [21, 22]. Similar to the Fe-C-Mn cases, carbon contents in austenite for GBF correspond well to the NPLE

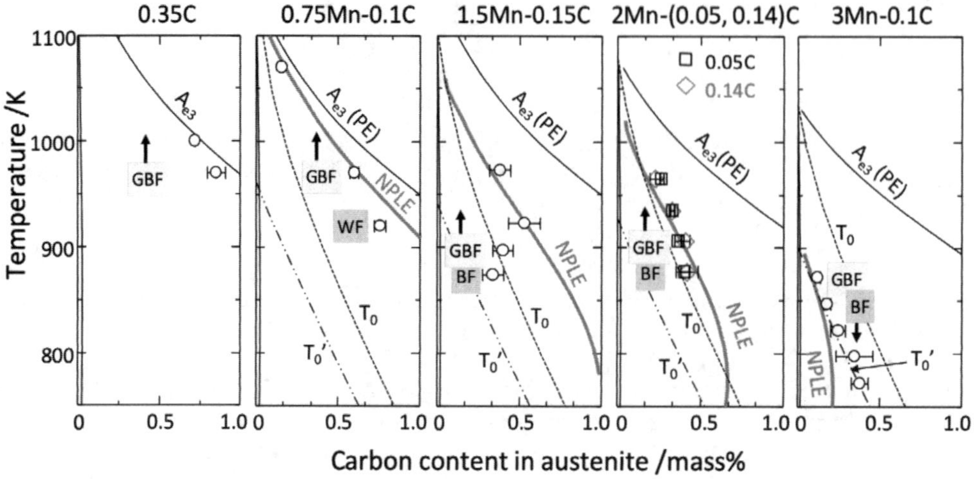

Figure 4. Carbon concentration at the interphase boundary in austenite matrix determined by FE-EPMA analysis in Fe-Mn-C ternary alloys [10]. GBF: Grain boundary allotriomorphic ferrite, WF: Widmanstätten ferrite, BF: Bainitic ferrite. NPLE represents the upper limit of carbon concentration for the NPLE mode and PE represents the phase boundary for para-equilibrium. T_0' is for undercooling of 400J/mol from T_0 as proposed for bainite transformation.

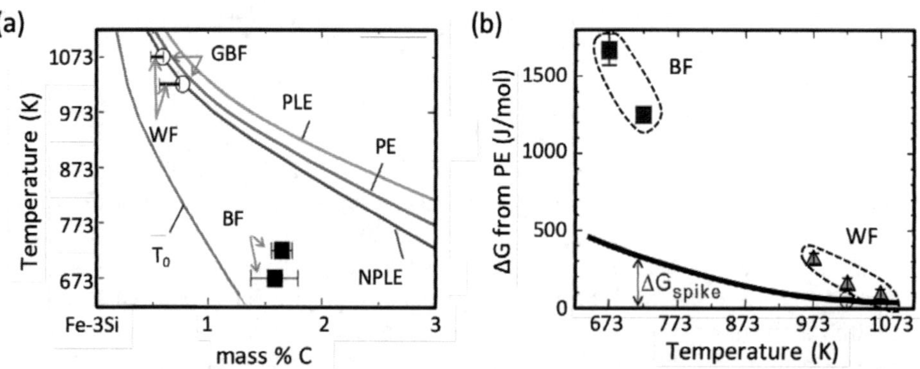

Figure 5. (a) Carbon enrichment in austenite and (b) energy dissipation for various types of ferrite in Fe-0.4C-3Si (mass%) [20,21]. ΔG_{spike} is dissipation due to a Si concentration spike in the NPLE mode.

line, whereas those for WF are lower than this. For BF, the deviation is much larger than for GBF and WF but carbon enrichment is still higher than the T_0 line. The estimated energy dissipation from para equilibrium (PE) exceeds over 300J/mol for WF and 1500J/mol for BF (figure 5(b)). This result clearly indicates that interface coherency causes the friction in boundary migration.

As already stated, both WF and BF hold a near K-S OR so that they should grow by migration of partially coherent boundaries. Figure 6 shows that a periodic array of lattice dislocations exist at the BF/austenite boundary in a Fe-C-Si-Mn alloy [10]. Those dislocations at the boundary create a pinning force. Also, it is seen that plastic accommodation takes place during bainite transformation. Strong interaction between the boundary and dislocations in austenite exerts another dissipation in boundary migration. Furthermore, elastic strain energy associated with the transformation needs to be accommodated. Dislocation slip can accommodate shear strain but diffusion must play a major role in

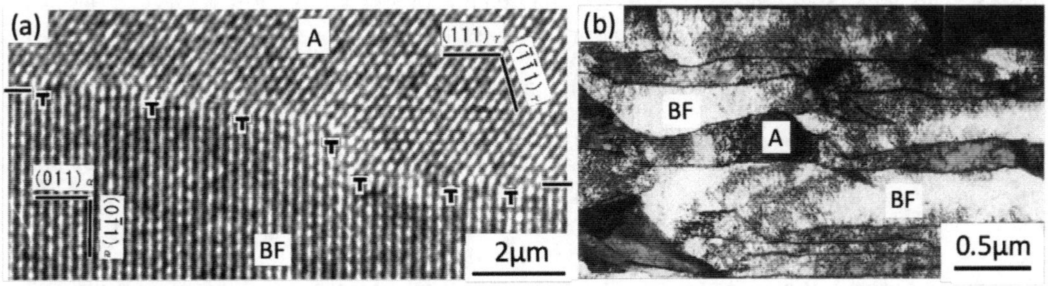

Figure 6. (a) HREM image of the BF/ austenite(A) boundary with interfacial dislocations. (b) TEM image of BF and austenite with high densities of dislocations in Fe-0.6C-2Si-1Mn [10].

accommodation of volumetric strain. It is concluded that bainite transformation accompanies a larger energy dissipation caused by the difficulty of strain accommodation for both components at lower transformation temperatures.

2.2. Effects of solute drag in energy dissipation during ferrite/bainite transformation

Addition of carbide-forming elements often retards the ferrite transformation. It has been proposed that segregation of alloying elements with strong affinity for carbon, such as Mo, at ferrite/austenite boundaries causes incomplete transformation [23]. Also, Nb is considered to be effective to exert a solute drag effect on the ferrite transformation [24]. However, the effects of solute drag are still controversial for bainite transformation since less potency of elemental segregation is expected and also reported [25, 26]. We have recently examined the energy dissipation for ferrite and bainite transformation in an Fe-C-Mo alloy [27]. Figures 7(a) and (b) show typical Mo and C concentration profiles across the interface between grain boundary allotriomorphic ferrite (AF) and austenite (γ) (transformed to martensite (M) by quenching). Extensive Mo segregation is seen at the non K-S interface in (a) but segregation at the near K-S interface is limited in (b), which indicates that Mo segregation is difficult at partly coherent interfaces. Figure 7(c) summarizes relations between the interfacial excess

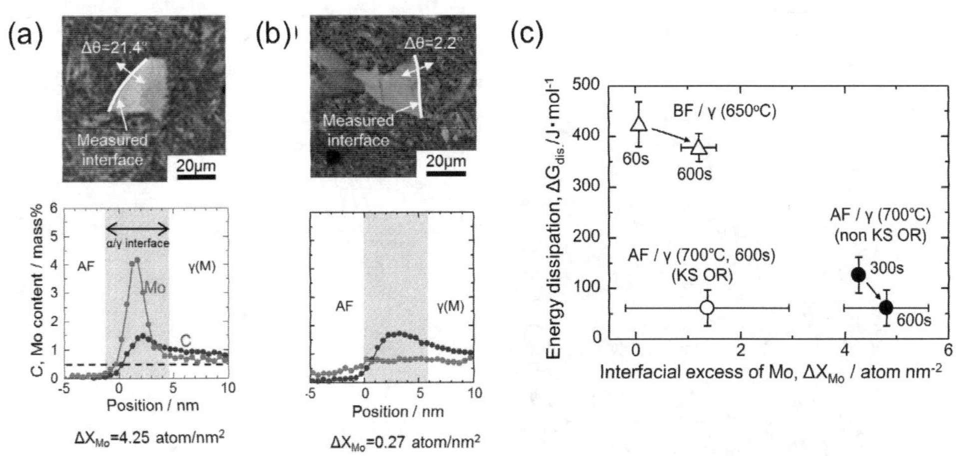

Figure 7. Ferrite orientation maps and Mo and C profiles across grain boundary allotriomorphic ferrite (AF) /austenite (γ) interface obtained by 3DAP analysis in the Fe-0.4C-0.5Mo (mass%) alloy transformed at 700 °C. (a) and (b) were taken from a non K-S interface and a near K-S interface, respectively. (c) Relation between amount of Mo segregation and energy dissipation at AF or bainitic ferrite (BF) / γ interfaces [27].

of Mo atoms and energy dissipation at the AF/γ interface with and without a K-S OR and the bainitic ferrite (BF)/γ interface at 650 °C. Mo segregation increases at longer holding time and energy dissipation becomes small mainly due to decrease in growth rate. It is noted that the amount of Mo segregation at BF/γ interfaces is much lower than that at AF/γ interface without a K-S OR. On the other hand, BF experiences much larger dissipation even at the same (but small) amount of Mo segregation as near K-S AF/γ interfaces. This indicates that the solute drag effect is small for bainite transformation, as described in the previous sub-section.

2.3. Austenite reversion behavior dictated by crystallography

The interfacial coherency also plays an important role during reverse transformation from ambient temperature microstructure to austenite. During reverse transformation, it is well known that two major morphologies of reverted austenite are formed; granular (γ_G) and acicular (γ_A) austenite [28]. Growth of γ_G takes place by migration of incoherent interface but γ_A grows with partly coherent interface [29]. Since the interfacial mobility is different between these two kinds of interface, as in the case of forward transformation, alloy partitioning is strongly affected by the crystallography at the interface. Energy dissipation and higher carbon content in austenite at the interface, described in figure 2(b) was detected in Fe-C and Fe-C-Mn systems [30]. However, since ferrite, in which diffusion is faster than in austenite, is the matrix phase, carbon partitioning reaches local equilibrium easily. Additionally, partitioning of substitutional alloying elements starts to take place even at the early stages of reverse transformation.

Figure 8 shows the partitioning behavior of substitutional elements in Fe-C-Mn-Si during intercritical annealling [31]. γ_G grows fast via the NPLE mode in the early stage of annealing, followed by slower growth kinetics with Mn and Si partioning at the later stage. For γ_A, on the other hand, Mn and Si partioning is already seen in the early stage of growth. Such a difference in partitioning kinetics also contributes to interface coherency of reverted austenite. The overall transformation kinetics were found to be much slower than para-equilibrium, and diffusional growth under low interfacial mobility was suggested from DICTRA calculation.

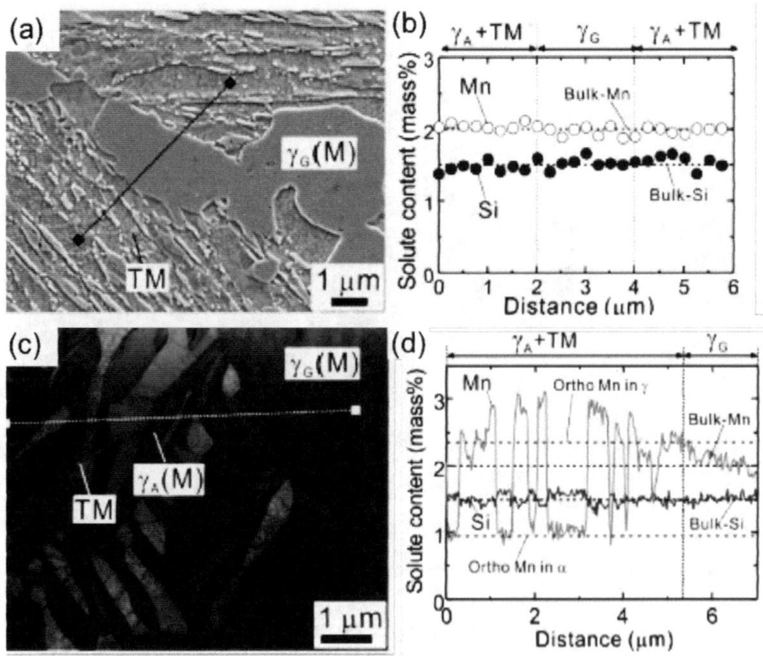

Figure 8. Microstructure of reverted austenite and partitioning in Fe-0.3C-2Mn-1.5Si annealed at 1048K; (a) SEM image and (b) line profile obtained by EDS for γ_G. (c) STEM-BF image and (d) line profile obtained by EDS for γ_A [31].

3. Strengthening by nanoscale precipitation at migrating interfaces

Interphase precipitation of alloy carbides during ferrite transformation establishes high yield strength and good elongation in low-carbon ferritic steels. A key for further strengthening is clarification of the conditions for refining the carbide dispersion, because high strength can be achieved via the Ashby-Orowan strengthening mechanism with moderate work hardening by introducing a high density of dislocations uniform [9, 32]. It was confirmed that a nano-sized carbide dispersion increases yield strength and work hardening in the early stage of deformation but later uniformly introduced dislocations interact each other and dynamic recovery is enhanced, resulting in suppression of inhomogeneous deformation.

The classical theory of interphase precipitation was reviewed by Honeycombe [33]. There are two distinctive mechanisms in terms of ferrite/austenite interface characteristics: (1) a ledge mechanism and (2) a bowing/quasi-ledge mechanism. In the ledge mechanism, the ferrite/austenite interface, across which the K-S OR is held, consists of an immobile coherent plane (terrace) and a mobile incoherent plane (ledge or riser), and then carbides nucleate at the immobile terrace plane. In the quasi-ledge mechanism, bulging of an interface with irrational OR takes place between particles with relatively large interparticle spacing after the incoherent and mobile interface is immobilized by carbide pinning on the interface. Subsequent nucleation of carbides on the bulged interface takes place and produces the next sheet.

We have demonstrated recently that a larger driving force for precipitation results in a finer carbide dispersion with incoherent ferrite/austenite interfaces [34-36]. Figure 9 is statistical data for orientation relationships at ferrite formed at austenite grain boundaries [34]. At higher temperatures, grain boundary

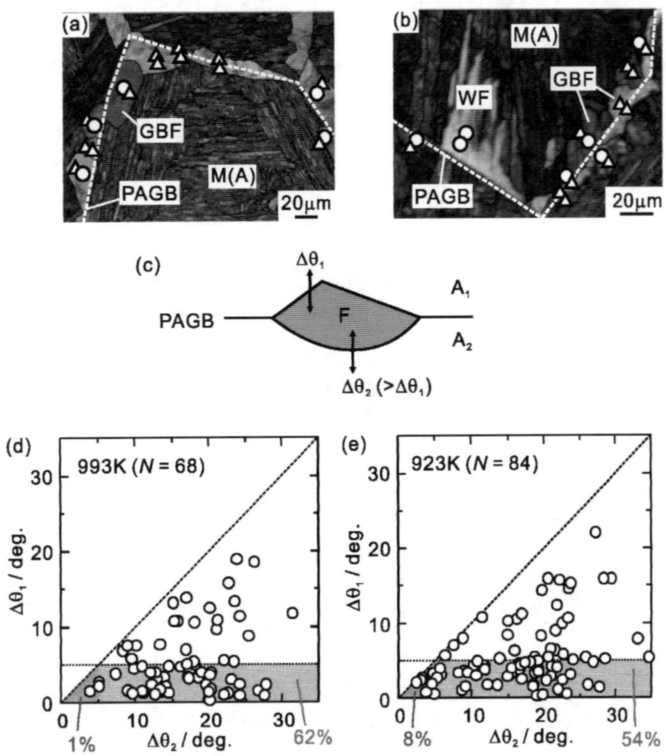

Figure 9. Ferrite orientation maps of the Fe–0.1C–1.5Mn–0.4V alloy transformed at (a) 993 K and (b) 923 K. Circles represent near K-S interfaces and triangles indicate non K-S interfaces. (c) Schematic for angular deviations from K-S OR; (d, e) distribution of $\Delta\theta_1$ and $\Delta\theta_2$ at 993K and 923K, respectively [34].

allotriomorphic ferrite (GBF) is dominant and grows toward to the matrix grain with non K-S OR (figures 9(a), (d)). As temperature decreases, Widmanstätten ferrite starts to form at grain boundaries and near K-S interfaces are increasingly formed (figures 9(b), (e)). Figure 10 shows effects of incoherency at the interface (here represented as deviation angle from the K-S OR) on the VC particle dispersion and corresponding nanohardness [35]. As the deviation angle becomes larger, up to 5-10 degrees, the number density of VC particles increases drastically with slight increase in particle size, as shown in (a) and (b), respectively. Such an increase of particle density results in higher hardness through the dispersion hardening as confirmed previously. The reason why nucleation of VC is promoted at non K-S interfaces is as follows. V is thought to have an attractive interaction to ferrite/austenite interface and exert solute drag effect as in the case of Nb addition. Fig. 11 shows the result of 3DAP analysis on near K-S and non K-S interfaces [36]. It is clear that more V atoms segregate at non K-S interfaces and provide a large advantage for carbide precipitation thermodynamically and kinetically.

It is rather challenging to make incoherent ferrite/austenite interfaces to obtain fine carbide dispersion through interphase precipitation. One possible way is use of ferrite transformation from deformed austenite. Fine-grained austenite structure inhibits formation of Widmanstätten ferrite so that recrystallization of austenite can be effective to obtain granular ferrite structure. During ferrite transformation in unrecrystallized austenite can be promising since local orientation fluctuation in the deformed austenite will accomplish loss of interfacial coherency during growth even though initial nucleus had near K-S OR with austenite. Use of nucleation at particles is also a good way. We demonstrated that nucleation at grain boundary precipitate or intragranular inclusion effectively enhances the formation of granular ferrite grains with non K-S OR, rather than Widmanstätten or bainitic

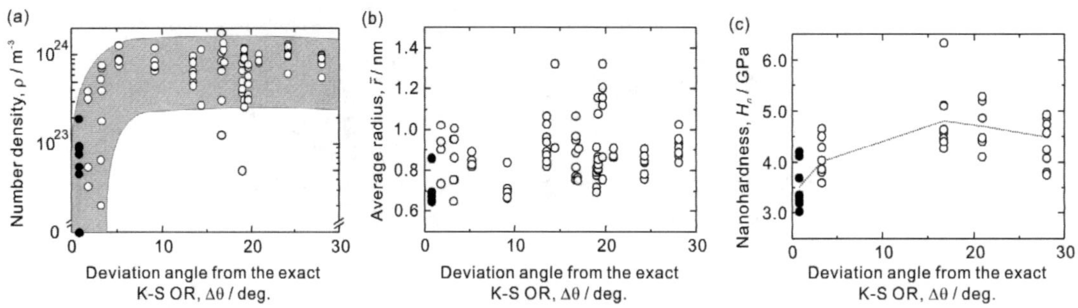

Figure 10. Variations in (a) number density and (b) average radius of VC precipitates, (c) nanohardness of α grains with Δθ in Fe-0.1C-1.5Mn-0.4V transformed at 923K for 60s. Solid and open symbols indicate WF and GBF, respectively [35].

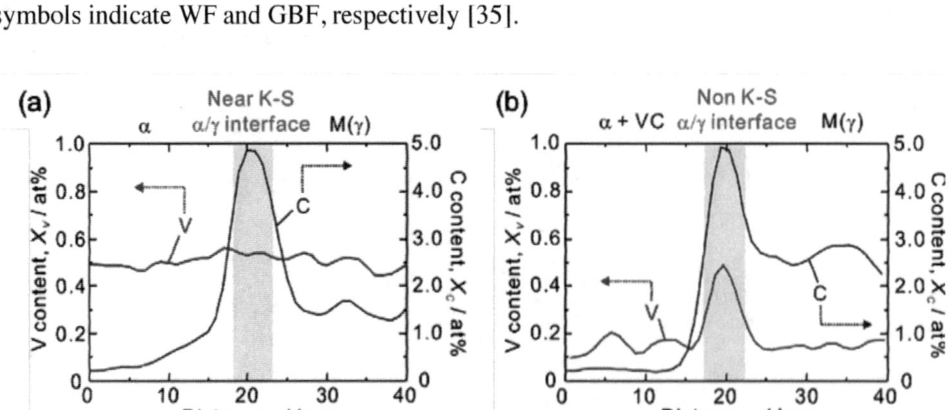

Figure 11. 1D concentration profiles of V and C obtained by 3DAP analyses of (a) near K-S and (b) non K-S ferrite (α)/ austenite (γ) interfaces in Fe-0.1C-1.5Mn-0.1V alloy transformed at 923K [36].

ferrite without nanosized VC precipitation in as-transformed condition [34, 37]. This ferrite morphology change results in the significant increase of overall hardness in V-microalloyed carbon steels.

4. Concluding remarks

In this paper, the effects of transformation interfaces on two important phenomena, namely, alloy partitioning/segregation and nanoscale precipitation at migrating interfaces, in design of advanced high strength steels with good ductility have been summarized. Interfacial coherency plays major roles in the control of microstructure and mechanical properties. Since it is important to control crystallographic constraints during nucleation and growth in phase transformations, a variety of alloy design and thermomechanical processing is expected to contribute to further advances in the development of novel high strength steels.

Acknowledgements

This work was financially supported by the Innovative Structural Materials Project funded by the New Energy and Industrial Technology Development Organization and CREST Basic Research Program "Creation of Innovative Functions of Intelligent Materials on the Basis of Element Strategy" funded by Japan Science and Technology Agency, Japan. Partial support from JSPS Grant-in-Aid for Scientific Research (B) No. 23360316 (2011–2013) and (A) No. 17H01330 (2017-2019) funded by Japan Society for the Promotion of Science are also gratefully acknowledged.

References

[1] Takahashi M 2015 *ISIJ Int.*, **55** 79
[2] Kawano O, Wakita J and Takahashi M 1996 *Tetsu-to-Hagané*, **82** 333
[3] Sakuma Y, Matsumura O and Akisue O 1991 *ISIJ Int.*, **31** 1348
[4] Purdy G, Ågren J, Borgenstam A, Bréchet Y, Enomoto M, Furuhara T, Gamsjäger E, Gouné M, Hillert M, Hutchinson C, Militzer M and Zurob H 2011 *Metall. Mater. Trans. A*, **42A** 3703
[5] Toji Y, Matsuda H, Herbig M, Choi P P and Raabe D 2014 *Acta Mater.* **65** 215
[6] Pierce D T, Coughlin D R, Clarke K D, De Moor E, Poplawsky J, Williamson D M, Mazumder B, Speer J G, Hood A and Clarke A J 2018 *Acta Mater.* **151** 454
[7] Honeycombe R W K 1976 *Metall. Trans. A*, **7A** 915
[8] Honeycombe R W K 1980 *Met. Sci.* **14** 201
[9] Funakawa Y, Shiozaki T, Tomita K, Yamamoto T and Maeda E 2004 *ISIJ Int.* **14** 1945
[10] Furuhara T, Chiba T, Kaneshita T, Wu H D and Miyamoto G 2017 *Metall. Mater. Trans. A*, **48** 2739
[11] Dubé C A, Aaronson H I and Mehl R F 1958 *Rev. de Met.*, **55** 201
[12] Aaronson H I *Decomposition of Austenite by Diffusional Processes* 1962 (USA: Interscience) eds V F Zackay and H I Aaronson, pp. 387-548
[13] Purdy G R and Kirlaldy J S 1963 *Trans. TMS-AIME* **227** 1255
[14] Bradley J R, Rigsbee J M and Aaronson H I 1977 *Metall. Trans. A* **7A** 327
[15] Bradley J R and Aaronson H I 1981 *Metall. Trans. A* **12A** 1729
[16] Zurob H S, Hutchinson C R, Bréchet Y, Seyedrezai H and Purdy G R 2009 *Acta Mater.* **57** 2781
[17] Hillert M and Rettenmayr M: 2003 *Acta Mater.* **51** 2803
[18] Liu Z Q, Miyamoto G, Yang Z G and Furuhara T 2013 *Acta Mater.* **61** 3120
[19] Liu Z Q, Miyamoto G, Yang Z G, Zhang C and Furuhara T 2015 *Metall. Mater. Trans. A*, **46A**1544.
[20] Aaronson H I, Reynolds Jr. W T and Purdy G R 2004 *Metall. Mater. Trans. A* **35A** 1187
[21] Wu H D, Miyamoto G, Yang Z G, Zhang C and Furuhara T 2018 *Acta Mater.* **149** 68
[22] Wu H D, Miyamoto G, Yang Z G, Zhang C and Furuhara T 2017 *Acta Mater.* **133** 1
[23] Reynolds Jr. W T, Liu S K, Li F Z, Shui C K and Aaronson H I 1990 *Metall. Trans. A* **21A** 1433
[24] Suehiro M, Liu Z K and Ågren J 1996 *Acta Mater.* **44** 4241

[25] Furuhara T, Tsuzumi K, Miyamoto G, Amino T and Shigesato G 2014 *Metall. Mater. Trans. A* **45A** 5990
[26] Xia Y, Miyamoto G, Yang Z G, Zhang C and Furuhara T 2015 *Acta Mater.* **91** 10
[27] Miyamoto G, Yokoyama K and Furuhara T 2019 *Acta Mater.* submitted
[28] Watanabe S and Kunitake T 1975 *Tetsu-To-Hagané* **61** 96
[29] Zhang X G, Miyamoto G, Toji Y, Nambu S, Koseki T and Furuhara T 2018 *Acta Mater.* **144** 601
[30] Liu Z Q, Miyamoto G, Yang Z G and Furuhara T 2013 *Metall. Mater. Trans. A* **44A** 4872
[31] Zhang X G, Miyamoto G, Kaneshita T, Yoshida Y, Toji Y and Furuhara T 2018 *Acta Mater.* **154** 1
[32] Kamikawa N, Sato K, Miyamoto G, Murayama M, Sekido N, Tsuzaki K and Furuhara T 2015 *Acta Mater.* **83** 383
[33] Honeycombe R W K *Proc. Int. Conf. Phase Transformations in Ferrous Alloys* 1984 (USA: TMS-AIME) eds A R Marder and J I Goldstein, pp. 259-291
[34] Zhang Y J, Miyamoto G, Shinbo K, Furuhara T, Ohmura T, Suzuki T and Tsuzaki K 2015 *Acta Mater.* **84** 375
[35] Zhang Y J, Shinbo K, Ohmura T, Suzuki T, Tsuzaki K, Miyamoto G and Furuhara T 2018 *ISIJ Int.* **58** 542
[36] Zhang Y J Miyamoto G, Shinbo K and Furuhara T 2017 *Acta Mater.* **128** 166
[37] Miyamoto G, Hori R, Poorganji B and Furuhara T 2013 *Metall. Mater. Trans. A* **44A** 3436

IOP Conf. Series: Materials Science and Engineering **580** (2019) 012006 doi:10.1088/1757-899X/580/1/012006

Additive manufacturing of metal components – process-material interaction in different process chains.

H N Hansen, J H Hattel, D B Pedersen, S Mohanty, S A Andersen, V K Nadimpalli, C G Klingaa, T Dahmen, Y Zhang

Department of Mechanical Engineering, Technical University of Denmark, Building 425, DK-2800 Kgs. Lyngby, Denmark

hnha@mek.dtu.dk

Abstract. In this paper we describe metal additive manufacturing (AM) processes in general, and laser bed fusion processes in particular. The process characteristics are described, and the various modelling approaches demonstrated. The open AM architecture developed at DTU Mechanical Engineering is introduced, as well as an alternative process chain based on AM and subsequent injection moulding.

1. Introduction

Additive manufacturing (AM) [1,2], is a hypernym that covers a family of technologies that conversationally is known as 3D printing. AM is regarded as a digitally born family of processes, that embodies the digital and physical aspects of layered manufacturing into a cyber-physical unit, thus nourishing by its nature a wider trend, namely the digitization and automation of the manufacturing industry [3,4]. For this reason, the increasing adoption of AM within the manufacturing industry is pushing companies to research new ways of adapting and optimizing their manufacturing strategies by integrating these digital technologies of tomorrow into existing production, and by bolstering their strategies towards a physical to digital conversion.

To illustrate the link between the physical and digital process of AM, figure 1 serves as an overview of the gross elements for a generic AM process. All elements pertaining to the physical and digital process chain are listed in white slanted boxes.

The subdivision of technologies within AM that pertains to metal is provided in figure 2 below. Figure 2 offers an overview of which technologies exist for the additive manufacture of metal components, abstracted in terms of the state of fusion, feedstock condition, and how the feedstock is distributed and fused to solid matter. Additionally the technologies are named according to their generic names as indicated by the ASTM/ISO 52900:2015 standard [1].

Process 1 – A filament arc welding process by which a metal wire is fed through a nozzle that also serves as the anode in the weld circuit. The wire is advanced under the cover of an inert gas towards the work-piece, which is the cathode of the circuit. As the wire is advanced, a plasma arc is formed between the anode and cathode, melting and depositing the wire onto the work-piece. This process has in recent years seen adaption for niche big-area additive manufacturing applications.

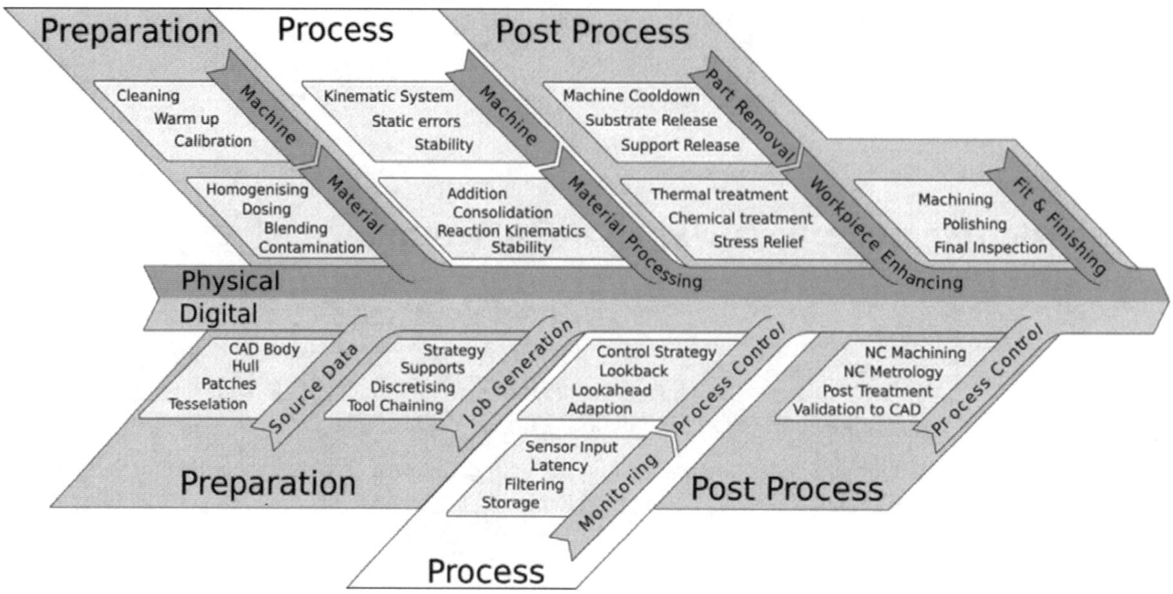

Figure 1. Overview of the physical and digital links of an AM process chain [5].

Process 2 – A cladding processes by which a metal powder is fluidised and jetted onto the work-piece while being intersected by a laser beam to melt the powder to a liquefied spray. As the high-velocity spray hits the work-piece the droplets adhere to the surface forming a thin layer. This process is known as cladding and seen in niches metal additive manufacturing where wear/resistant coatings are sprayed onto the work-piece as a finishing procedure or in hybrid manufacturing where each added layer is subsequently machined.

Process 3 – A powder-bed process where a thin layer of metal powder is coated over the work-piece in a confined build-chamber, known as a powder-bed. After deposition of a layer, the layer is consolidated by means of a high powered electron-beam emitted under vacuum by a magnetic field deflected electron beam gun. This process is renowned for allowing for high-speed beam modulation as no mechanical subsystems are involved in deflecting the electron beam and for allowing for high build-chamber temperatures, whereas the process is limited by the availability of process materials and from cumbersome maintenance of a high vacuum process environment.

Process 4 – Laser induced powder-bed fusion. Very similar to the e-beam technology, this process relies on a high-powered laser for melting of the powder. and has therefore been highlighted in the matrix. The process will be described in detail in the following sections.

Process 5 – A variation of Process 4, employing slightly less laser power while fusing the powder. Through this process the individual powder grains are now fully melted through the formation of a melt-pool. The powder is heated to some 10-20 degrees below its melting point allowing for the powder to become sticky such that the selectively heated grains stick and adhere to neighbouring grains. This will form a considerably weaker but porous metal structure that is often sought for metal components for medical implants. The porous structure allows the implants to form a mechanical inter-lock with organic tissue inside the human body, thereby to a certain extent eliminating the need for implant screws and adhesives.

Process 6 – The final process in figure 2 is the only process by which a work-piece is built from metal sheets. This process involves cutting thin metal sheets to contour by means of a high-powered laser or by a punch. Subsequently the sheets are stacked and welded by use of ultrasonic welding to form the component. This process is at the time of writing rarely used in industry, but is included here for legacy purposes.

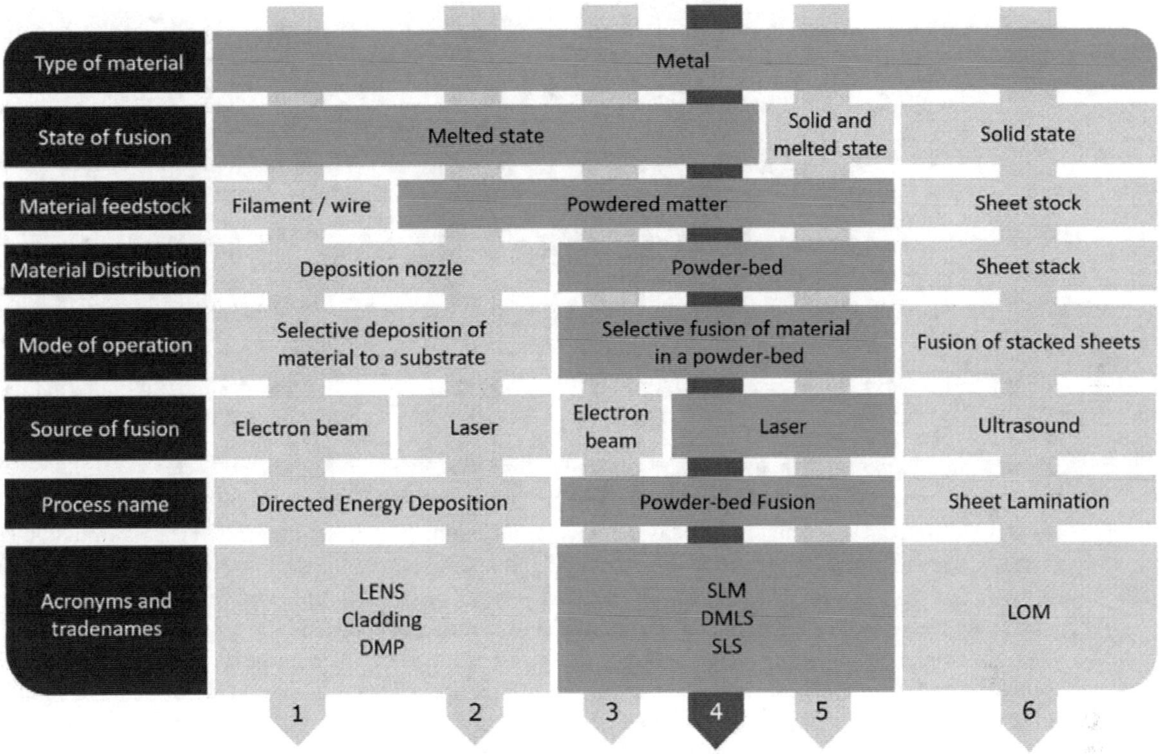

Figure 2. The realm of metal based AM processes, and their characteristics [5].

2. Basic process steps of powder bed fusion (PBF)

The metal powder bed fusion (PBF) process can be construed to have five major aspects which have considerable impact on the overall process:

1) Powder delivery and powder bed formation
2) Generation and delivery of energy source
3) Energy-material interaction
4) Local material consolidation
5) Global thermal processing

The basic process chain, however, follows up the PBF with an equally important heat treatment post-process aimed at relieving stresses and/or controlling the microstructure. Each of these five steps are described in greater details in subsequent sections.

2.1. Powder delivery and powder bed formation

The metal powder bed fusion process, as the name indicates, begins with the formation of the powder bed. A powder delivery system e.g. a hopper, deposits a heap of fine powder particles over the build-plate, which is then spread using the sweeping motion of a recoater blade (or a roller or scrapper). The system is also responsible for relocating the build plate along the vertical-direction, and therefore defines the accuracy and resolution of the build in the vertical direction. This relocation of the build plate creates a small gap between the tip of the recoater blade and the build plate to allow the deposition of the required powder layer. At subsequent recurring stages of the PBF process, the powder is spread on top of the previously manufactured layers and thus the interaction of the recoater blade with the surface of the previous layer becomes significant as well. See figure 3.

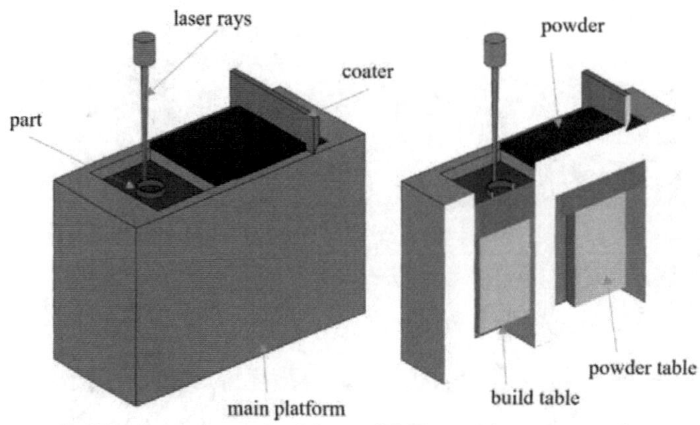

Figure 3. Schematic 3D view of a typical laser-PBF machine along with the components [6].

2.2. Generation and delivery of energy source

The typical energy source used in the case of metal PBF is either a focused laser beam or an electron beam with high energy density. The laser is essentially light (i.e. electromagnetic waves), created by a phenomenon denoted stimulated emission, with high spatial and temporal coherence properties, which allow it to be focused in a very small spot and achieve a narrow spectrum, respectively. The generated laser beam is passed through several lenses that regulate the optical characteristics of the beam, and is then made incident on a mirror that controls the incident location of the laser beam on the powder bed. Following the wave-particle duality of light, the interaction of the laser beam with the lenses and the mirrors can typically be described by the concepts of ray optics while effects such as absorption of the laser beam in the optical elements (mirrors and lenses), polarization, or thermal lensing, invoke the electromagnetic wave characteristics of the laser. During the additive manufacturing process, the laser beam is selectively manoeuvred across the powder layer to consolidate the metal powder particles and to create the desired geometry.

2.3. Energy-material interaction

The core of the metal AM process begins when the laser beam (or electron beam) is incident upon the powder bed. In broad terms, a portion of the laser beam is directly reflected away, a second portion undergoes scattering, a third portion is absorbed by the powder/bulk material and a fourth portion is reflected into the powder bed and undergoes multiple cycles of reflection, absorption & scattering. The overall absorption of the laser beam depends upon the wave characteristics of the laser (wavelength, monochromaticity, etc.), the properties of the material (albedo, refractive index, emissivity, etc.) as well as the characteristics of the powder bed (powder packing, powder morphology, powder composition, etc.). The actual laser-material interaction is often quite complex and can only accurately be defined through the radiative transfer equations.

2.4. Local material consolidation

The energy-material interaction can lead to two distinct types of material consolidation, namely sintering and melting/fusion.

2.4.1. Sintering: Sintering bonds powder particles at their interface through diffusion mechanisms, as shown in figure 4, and thus forms a solid (porous) structure through heating without complete melting.

Neck growth occurs between formerly separate metal particles followed by shrinkage of the particles into a solid [7]. The bonded bulk material exhibits different global properties (such as strength, density, ductility, and thermal/electrical conductivity) as compared to the powder bed. The nature of the sintering process makes it prone to the formation of pores medially between the surrounding particles. Material consolidation through sintering always occurs in a PBF process.

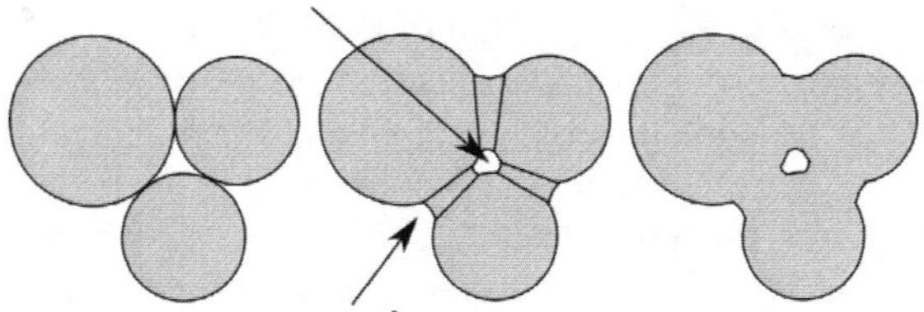

Figure 4. The principle of solid state sintering.

2.4.2. Melting/fusion: In regions with sufficiently high energy density, the consolidation of the powder particles can instead proceed by complete melting as shown in figure 5. The rapid melting and solidification enabled by the process, allows for parts with tailored properties, distinct from those obtained from traditionally processed parts [8]. This facilitates the exploitation of localized material properties, obtained from refined microstructures, formation of non-equilibrium phases, and supersaturated solutions. This mode of material consolidation is more sensitive to process instabilities, potentially leading to formation of melt spatters, localized secondary-phase particles (such as oxide and carbide inclusions) or extremely fine refined microstructures.

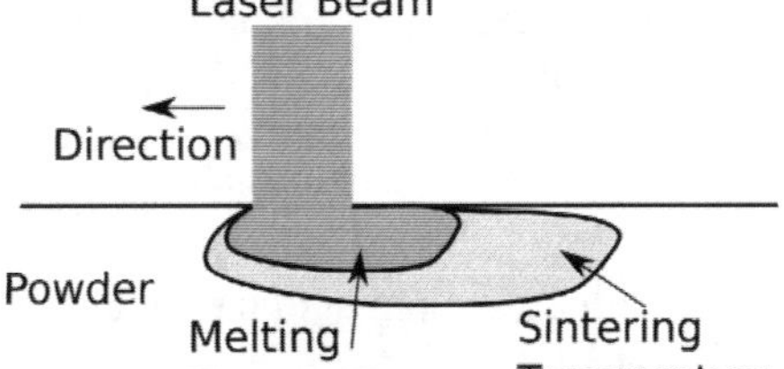

Figure 5. The principle of full melting/fusion of a workpiece.

2.5. Global thermal processing

Neglecting the small region with the melt-pool (and/or being sintered), the metal AM process globally resembles a non-homogenous non-isothermal heat treatment processes. This in-situ heat treatment can generate considerable transient thermal stresses, initiate local plastic yielding and hardening, and finally cause large residual stresses within the part. Furthermore, the rapid cyclic thermal conditions result in non-equilibrium microstructures with varied mechanical properties (as compared to the powder material). Thus, as mentioned earlier, a typical PBF process chain always involves a secondary heat treatment and/or stress relief post-process.

Following the PBF process, several post-processes can also typically be required before a functional part within dimensional and surface tolerances is achieved, e.g. multiple heat/solution treatments, machining, grinding, and polishing. A removal process of the part from the build plate must also be carried out by, e.g., EDM (electro-discharge machining) or band-sawing, and support structures (if present) must also be removed during the post-processing stages. A post-process step of hot-isostatic pressing is also common in parts produced with densities below 99% (compared to bulk material).

3. Modelling of metal AM

The PBF process chain results in a large set of significant control parameters. Consequently, determining stable and optimum process windows is an arduous activity for metal PBF. Empirical models relating the most common parameters (e.g. power and scan speed) with typical properties (tensile strength, fatigue life, etc.) are difficult to obtain, and instead the current research across the literature aims to identify process signatures that can indicate resultant properties and defects.

In metal AM (including PBF), the distinct process signatures and defects arise due to various phenomena occurring across several length and time scales. From the process viewpoint, the smallest length scale typically corresponds to the wavelength of the laser beam being used (e.g. 1.06 μm for a Nd YAG laser system). However, from a material science perspective, even smaller features such as nano-precipitates and micro-/nano-porosities become relevant for the metal AM process, thus making nanometer length scales significant. On the other hand, warpage effects (i.e. localized distortion/bending) on metal AM components only become practically relevant in parts which are several centimeters large (in each dimension). Thus, the range of relevant length scales for metal AM can vary from the nanometer (10^{-9} m) to centimeter (10^{-2} m) scale, corresponding to at least seven orders of magnitude. Similarly the relevant time scales for the process can vary from microseconds (time for the laser beam to pass over a powder particle) to several hours (time for a complete build), corresponding to 10-11 orders of magnitude.

Consequently, a goal-oriented approach (typically motivated by the type of process defect/signature of interest) of grouping the various phenomena into broader categories of micro-, meso- and macro-scale is followed, wherein the multiphysics involved at the relevant length and time scales are handled with greater accuracy, while the contributions from phenomena occurring at larger or smaller scales are typically approximated

3.1. Empirical models

As mentioned earlier empirical models for additive manufacturing are difficult to generate due to the extensive set of significant parameters and the corresponding volume of experiments that would be required. However, the ease of application of empirical models during decision-making (e.g., during the design and/or process planning phases) has nonetheless promoted experimental campaigns aimed at such model development. As a case in point, simple prediction models for predicting the expected roughness at a specific location within a channel produced by AM are being generated that can be used during the component design phase – with a potential application in conformal channel design.

The benefits of conformal cooling channels over conventional cooling channels have been known for some time [9], with the performance of such channels known to be determined by the actual geometry of the channel, the cross-sectional area and the internal surface texture. However, build features of PBF components have been shown to be dependent on the angle of the component with respect to the build plate, and the roughness on the surface has been found to increase with decreasing inclination angle [10-11]. Investigations further indicate that the roughness of PBF channels are dependent on two orientations, namely the global orientation (α) and the local orientation (β). The investigated components are shown in figure 6 and the definitions of global and local orientations may also be seen in figure 6.

The test samples seen in figure 6 were manufactured using the laser PBF process in a 17-4PH stainless steel. The analysis of these test samples was conducted by X-ray CT (computed tomography) scanning seven channels with an equal diameter of two millimeters in seven different global orientations: 0°, 15°, 30°, 45°, 60°, 75° and 90°. The roughness of the channels was analyzed using an in-house Python image analysis code. The results of the Python code analysis are visualized in figure 7 and can be summarized as: "the larger the α angle, the smaller the induced roughness". More quantitatively, the surface roughness follows a Student's t-distribution with respect to the α angle – thus enabling development of an empirical model relating design feature characteristics with surface roughness.

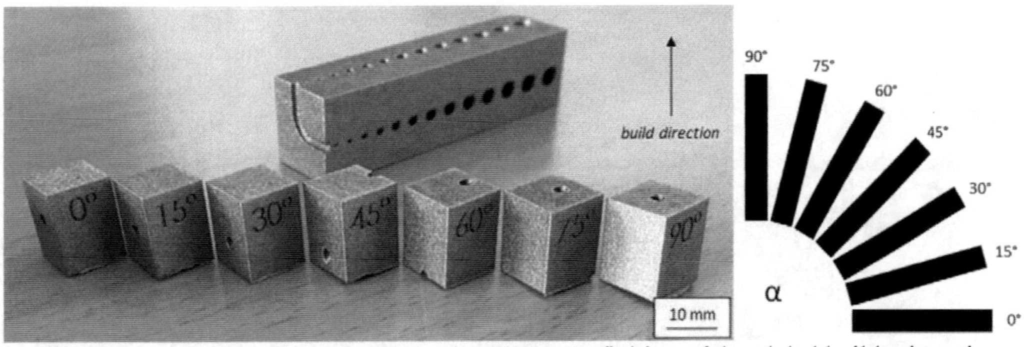

Figure 6. Left: Investigated test samples. Right: Definition of the global build orientation α.

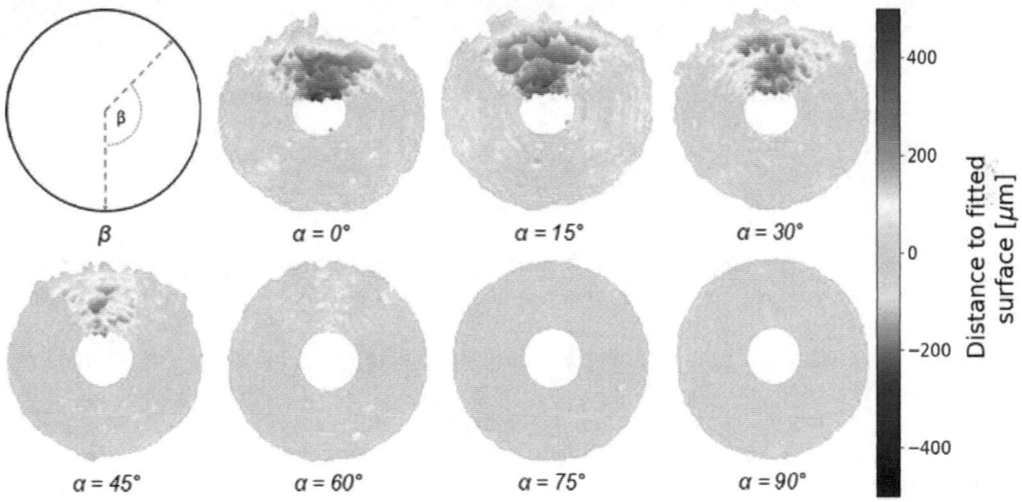

Figure 7. Through view of the channels, visualizing the roughness dependence on the location inside a channel (local orientation β) and build direction (global orientation α).

3.2. Meso-scale model

Meso-scale models of the PBF process are focused on simulating the powder bed formation, the laser-material interaction and the local material consolidation aspects. Powder bed formation is typically modelled using either the Raindrop packing algorithms (which randomly deposit powder particles) or using DEM (discrete element method), which also considers interaction forces among powder particles and from the roller/scrapper/blade. Models for laser-material interaction at this scale range from complex ray-tracing methods tracking specular reflection, to computationally-intensive radiative transfer equations that consider diffuse reflection and wave optical characteristics of the laser. The local material consolidation is also simulated with similar high fidelity by involving thermo-fluid dynamic models that can consider temperature-dependent material properties, solid-liquid-vapor phase changes, surface tension gradient driven Marangoni effects and thermo-capillary effects of the powder bed. Figure 8 shows a simulation of a single track formed during PBF through such a complex meso-scale model involving DEM for powder bed formation, ray tracing algorithms for laser material interactions, and the above mentioned thermo-fluid model [6]. Through such modelling, it is possible to predict the irregularity and stability of the individual tracks, the formation and evolution of porosities, and the resultant metrological characteristics of the surface. The simulated domains, however, are small in size due to the large computational requirements.

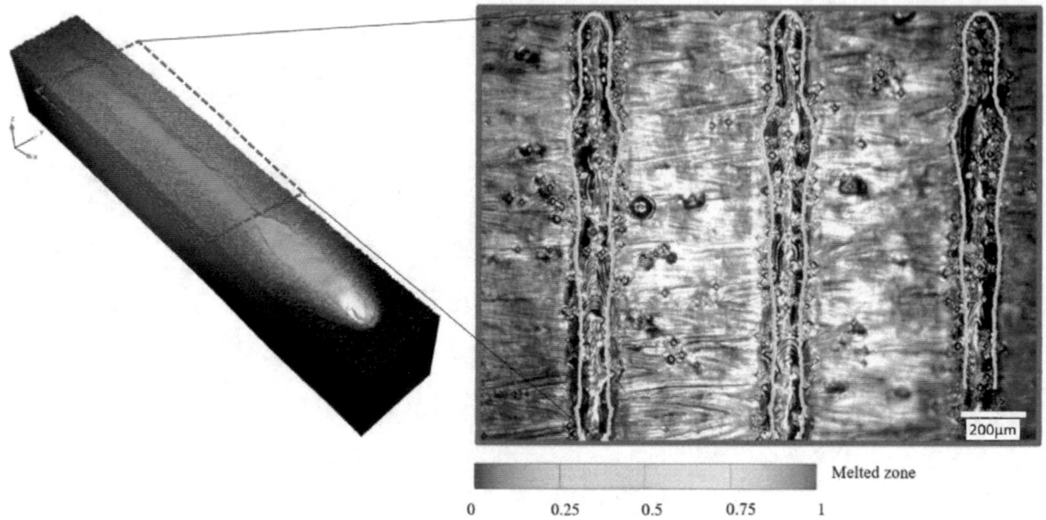

Figure 8. Left: Thermo-fluid dynamic simulation of melt track formation. Right: Comparison between the predicted top melt pool indicated by the colored borders and experiments. [6]

3.3. Micro-scale model

Micro-scale models of the PBF process have a focus on simulating the laser-material interaction, the local material consolidation and the global thermal processing aspects. The typical goal of these models is to determine the thermal (and mechanical) conditions and therefrom predict the emergent metallurgical and material properties. The geometric and optical/electromagnetic complexities of meso-scale models are often not of primary interest here and hence the powder bed is typically modelled using either an equivalent porous continuum assumption or using the Raindrop algorithms. Similarly, laser-material interaction models range from simplified assumptions of surface fluxes/volumetric sources calculated through empirical models, to ray-tracing methods tracking specular reflection. The local material consolidation can be simulated with a similar high fidelity as for meso-scale models, but use of simpler equivalent conductivity-based thermal models is also common. The micro-scale simulations also include models to predict the metallurgical microstructure, either directly through cellular automata or phase field techniques, or indirectly through the use of empirical models for phase fraction predictions. Figure 9 shows temperature results from such an implementation for simulating PBF of two different tensile bars of Ti6Al4V [12]. The difference in the thermal history pattern at different locations and due to orientation resulted in different microstructures (phase composition as well as morphology) at the middle and bottom parts of the samples.

Figure 10 shows the microstructure predicted during a parametric study of laser scan speed and laser beam size. The solid red and black lines show the borders of the region corresponding to equiaxed and columnar grain formation, while the individual dots correspond to different (fixed) locations at which the solidification and thermal parameters are calculated. The figure displays the relative abundance of columnar grains in the domain for all investigated parameters – corresponding to the typical experimental observation for metal PBF. Thus, through such modelling, it is possible to predict the local microstructure and mechanical properties of a part produced by PBF. The size of the simulated domains for micro-scale models is typically correlated to the computational requirements of the chosen metallurgical models.

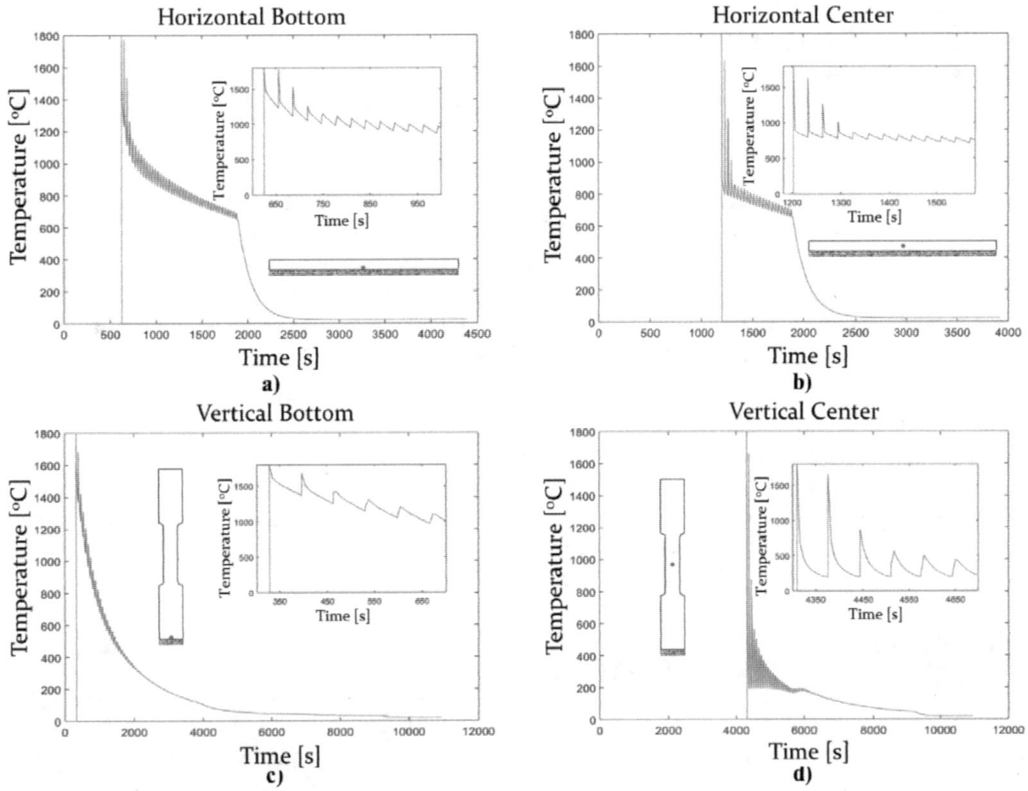

Figure 9. Thermal history modelling results. a) Bottom and b) center of a horizontal tensile bar specimen. Vertical tensile bar specimen: c) bottom and d) center. [12]

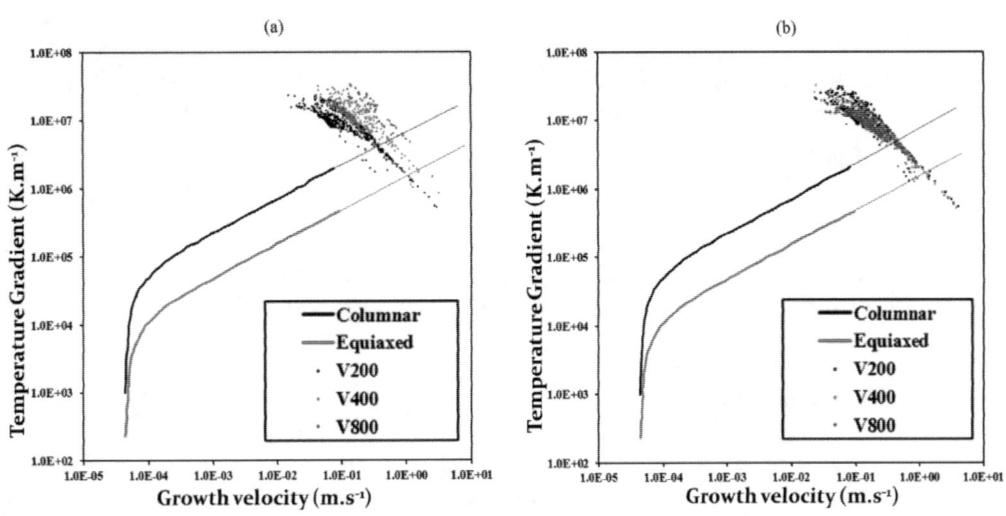

Figure 10. Thermal gradient-vs-growth velocity maps extracted at melt pool boundaries during single track PBF simulations by varying (a) scan speed and (b) beam size [13].

3.4. Macro-scale model

Macro-scale models of the PBF process are focused on simulating the global thermal processing aspect. Powder bed formation at this scale is typically modelled using an equivalent porous continuum assumption. Laser-material interaction models apply equivalent volumetric sources calculated through either empirical models or analytical solutions based on simplified assumptions. The local material consolidation is typically simulated through simpler equivalent conductivity-based thermal models. At this scale, the thermo-mechanical effects become significant and so prediction of residual stresses and distortions are among the primary goals. Constitutive models for elasticity, plasticity and visco-plasticity are used along with models for thermal-softening and work-hardening to predict both the local material properties, and the resultant behavior of parts produced by PBF under the large thermal (stress) loads. Figure 11 shows a macro-scale thermo-mechanical model implementation where two cantilevers have been produced using PBF. The cantilever corresponding to the top of the image is then detached from the build plate (leading to a certain warpage). Subsequently, both cantilevers are subjected to the same stress-relief and heat treatment cycle, and then the second cantilever is also detached from the build plate. The figure clearly shows the much larger warpage in the component detached prior to stress relief as compared to the cantilever detached at the end. Through such modelling, it is possible to predict the warpage, residual stresses, and stress-driven delamination, as well as the fatigue/damage behaviour. The size of simulated domains in such models are typically in the order of centimetres i.e. at the scale of the parts.

Figure 11. Different distortions observed in two identical cantilevers produced by PBF subjected to different sequences of the same chosen post-process steps.

4. Open metal AM architectures

Although academic research within the field is growing exponentially, with 2000 publications in 2010 and 10000 publications in 2018, the vast majority of these publications are on research peripherally related to AM, such as design for AM, applications of AM, simulations of AM processes, characterization of input and output material, and post-manufacturing quality assurance. Research that pertains directly to the principle of operation of the AM system to improve the capabilities of the process is virtually non-existent in said statistics. This is due to the complexity of the principle of operation of AM processes as shown in figure 1, and, to a large extent, due to the very closed and proprietary nature of industrial AM systems. It is necessary therefore to institute a liberation of this restricted area of development by stringent advocation and practice of open research on the philosophy of open architectures and open source platforms, so that academia may assist in driving research in AM beyond the current state of art. By this philosophy, the Technical University of Denmark has developed and is continuously improving on, two metal laser powder bed systems that challenge the established industrial standard by introducing capabilities such multiple powder hoppers for multi-material manufacturing, as well as a fully open laser and galvanometer modulation system that allows for high speed laser pulse shaping and an open hardware melt pool monitoring system that is currently under development.

4.1. Multi-material capable Laser PBF system

Since Laser PBF involves continuous laser welding in a small cubic volume, it presents a unique materials challenge. Suitable material alloys need to be able to withstand very high cooling rates while simultaneously retaining a good microstructure and low distortion from residual stress. Hence, many of the alloys developed for casting and welding are not exactly ideal for laser PBF. However, due to the wide availability of these alloys they are predominantly used with some degree of success. Working with open-architecture AM systems enables tailored materials development at a significantly faster pace. So far, directed energy deposition processes (DED) are mainly associated with multi-material additive manufacturing in metals, due to their intrinsic system-related flexibility of changing to different feedstock during operation. However, PBF based additive manufacturing processes exhibit specific advantages over DED, such as the possibility to generate more intricate geometries and to achieve smaller feature sizes. At the same time, most commercially available PBF systems are limited to a single powder feedstock restricting the deployment of different materials during the build process. Aurora Labs S-titanium pro system utilizes a unique powder hopper delivery system as shown in figure 12. The open architecture allows for a precise layer-by-layer compositional control by mixing up to three different powders. The power source is a set of two 150 W CO2-lasers fired simultaneously. The two beams pass separately through a set of optics before entering into the focus lens resulting in a single spot of approx. 150 μm in diameter in the processing plane. Thereby, the focus lens is mounted onto a moving print head in an X-Y gantry system. The average maximum power output of the focused beam is 255 W at the processing plane. In the Aurora labs system, we successfully made discrete and continuously changing interfaces between stainless steel (SS 316L) and maraging steel (MS1). The etched microstructure in figure 12 shows the ability to gradually transition from one material to the other. Vickers micro hardness and elemental maps showed a smooth as designed transition, without stress concentrations [14]. The ability to make functionally graded multi-material components provides an expanded materials design paradigm

4.2. Fully customizable open-architecture laser PBF system

The main constituents of the Technical University of Denmark (DTU) open architecture laser PBF system are the laser scanning module, the powder-bed, the recoater/scraper, and the control system of the machine. The optical module is comprised of a 250W laser, SPI-250C, that can be modulated at 100kHz, and, a galvanometer system (Thorlabs GVS012). The powder-bed and recoater module have been tailored towards research purposes [15]. The hardware modularity allows for experiments comparing scan strategies, feedstock and in-situ sensor integration. It also allows the study and causality of every physical component within the system on part quality, thus enabling an iterative and quick redesign capability.

This section presents the chosen approach to gain full control over the process flow in laser PBF. The control system enveloping the open architecture platform includes a laser modulation and galvanometer unit that can allow for high-speed modulation of both the beam trajectory and laser output power [15]. The control system is comprised of a hardware controller, its embedded firmware and an interface to a host-side job-planner, all of which have been tailored to experimental powder-bed fusion, allowing the research scientist to freely explore scan strategy, pulse modulation, speed, laser power, and beam shaping. The system will also be capable of cross-talk to industrial PLCs and controllers for NC stages, such that the powder-handling system allows for complete low-level hardware control.

Powder-bed fusion relies heavily on automation throughout the entire process flow. This starts with an approximation of the computer model into a triangularly tessellated surface, described by vertices and direction. Next, the tessellated surface approximation is interpreted through a job planner. Aside from geometrical interpretation, the job planner is also responsible for introducing the process parameters applied throughout the build. Thus, a custom and fully open job planner has been developed to take a 3D geometry and process it into a layer-wise pattern, then to convert it into a job-file that can be processed by the laser and galvanometer controller unit. To increase usability of the planner a graphical user interface allows adjustment of process parameters such as hatch line spacing, laser-power, and scan speed amongst others.

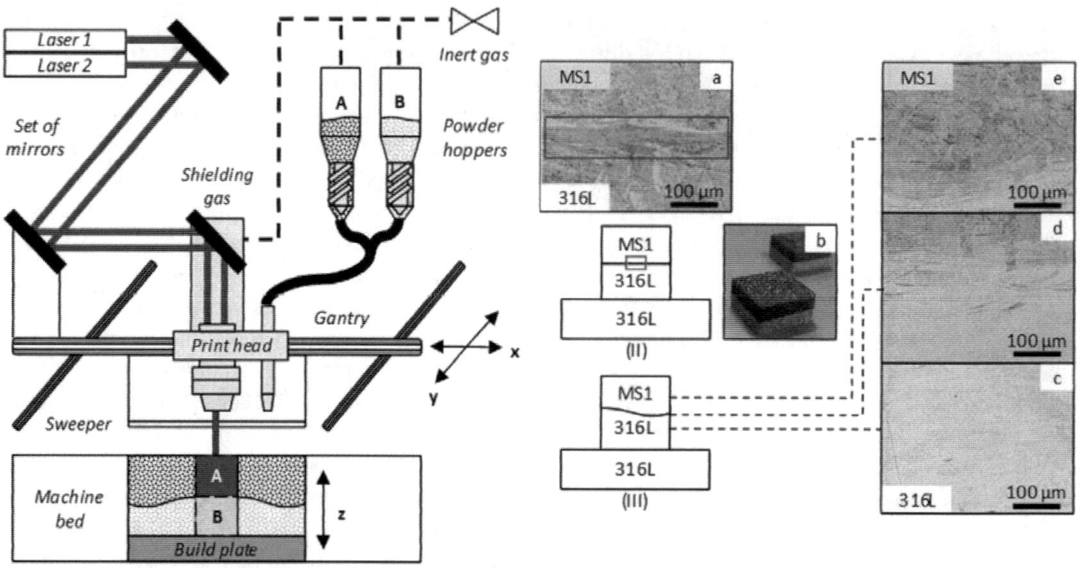

Figure 12. Left: schematic of the Aurora Labs system. Right: microstructure of discrete vs functionally graded interface between stainless (316L) and maraging (MS1) steels [14]

The laser and galvanometer controller unit has been designed such that it receives angular mirror positions and laser modulation parameters from a host computer over a high-speed USB interface and from this generates the appropriate control signals to the interfaces available on the laser and galvanometer subsystems. Most industrial fibre lasers can be modulated from an analogue 10V control signal, and most analogue galvanometers can be interfaced with an analogue ±10 V control signal. At the heart of the Galvanometer Laser and Modulation System, a high performance 32-bit ARM® Atmel SAM3X8E Cortex® microcontroller is implemented. In addition to the on-board capabilities, 16 bit digital-to-analogue converter and operational amplifiers have been implemented in order to achieve the necessary interfacing capabilities. In addition to the interfacing enabled by the modulation and control unit, the Atmel SAM3X8E Cortex® allows interfacing to an additional 47 digital I/O streams and 12 analogue inputs.

5. Alternative process chain using AM for making metal components

Powder injection molding (PIM) is a well-established process that allows mass production of metal and ceramic components with fairly complex geometries [16]. Depending on the complexity of the produced parts and materials choice, the process usually become more economical above 1000 to 10000 parts. The high cost of the production of the molds makes this technology non-remunerative for low rate manufacturing. Moreover, another constraint of this process is the need for a two plate mold, to allow removal of the manufactured piece after injection moulding. This generates a certain degree of restriction when complex pieces have to be produced. A common problem for injection molding parts, for example, is the one caused by the undercuts. Due to these problems, when highly customized parts are needed, or in the case of prototyping and pilot production, machining or additive manufacturing has to be chosen instead of PIM [17].

The process so-called "3D-Printing Integrated Manufacturing System (3DIMS)" is under investigation as an innovative process for powder injection molding, making possible the use of this technology particularly when low-volume production and high geometrical complexity of the part are required. In this process, AM is used to fabricate a sacrificial mold for the use in PIM. The process chain is illustrated in figure 13. The introduction of AM in the process chain leads to necessary adjustments in the feedstock for PIM, as well as in the process parameters. The mold is fabricated by a photocurable polymer (or photopolymer) using a DLP (Direct Light Processing) or SLA (stereolithography) process. In this process, hardening of the liquid resin occurs as a result of cross-linking of the polymer when it is exposed to UV (ultraviolet) light. The strength of the material is sufficient to withstand a certain level of injection pressure. The mold has a cavity with the desired geometry. However, the printed part is often only part of a mold, the so called "insert", instead of an entire mold plate. The design of the AM parts and its material properties are crucial for the success of the process. Most of the commercially available photopolymers are rather durable with regard to exposure to all kinds of chemicals. In 3DIMS, a special grade of photopolymer is used. After UV curing, the resin is able to be dissolved in aqueous solution.

The printed mold insert (after cleaning and a series of post processing steps) is mounted in an injection molding machine. The feedstock which contains metal power is injected into the mold by a screw or a piston. Typically the metal content is more than 60% in order to assure the strength during debinding and sintering. A modified Polyethylene-Paraffin Wax binder system can be used in the feedstock. The green body forms inside the mold. Compared with a conventional mold that is made in metal, the strength of photopolymer resin can be up to four orders of magnitude lower than steel. The chosen binding system must lead to feedstock with low viscosity during injection molding in order to reduce injection pressure as much as possible. At the same time, the geometry must be well maintained during all other steps (demolding, debinding and sintering).

In the ejection step of injection molding, the printed mold is ejected with the green body inside. This means that the actual demolding is executed outside the injection molding machine in an additional step. The mold should be removed by dissolution in an aqueous solution. Due to the non-water-soluble binding system in the feedstock, the geometry of the cavity will be well replicated by the feedstock material and not damaged during the dissolution process. The green body is then debinded and sintered in a furnace in a protective atmosphere. The selection of binder system and a suitable debinding procedure is believed to play a key role in the entire process chain. Nowadays there are different lost-forming technologies for PIM. For example Hein et al. [18] proposed a process chain with a sacrificial mold for PIM, where in this process the DLP printed mold is removed by thermal degradation after PIM. Similar to 3DIMS, the general feasibility could be demonstrated, but further optimization of the process is required before it can be transferred to industrial production.

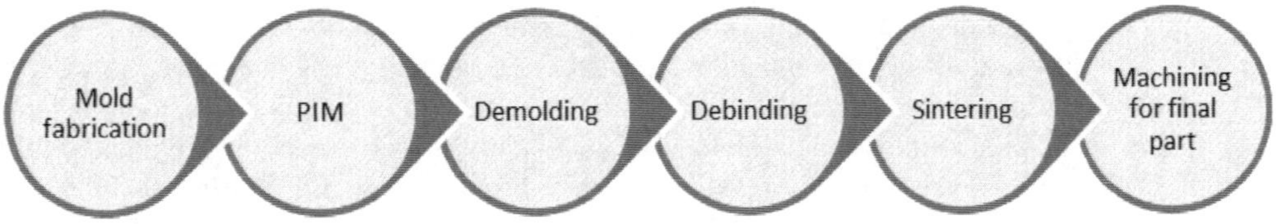

Figure 13. Process chain of the 3D-Printing Integrated Manufacturing System.

6. Summary

In this paper we have discussed a classification of AM processes and more specifically described metal AM processes. Details of the PBF process have been described, alongside with the various aspects of modelling. The approach of an open AM platform has been introduced as a key enabler for process development and optimization. Finally, an alternative process chain for realizing metal parts based on AM has been introduced.

Certification and qualification of AM components is the primary driver necessary for wide-scale implementation of metal AM technologies in industry. The components that are typically suited for metal AM also require high geometrical and mechanical performance tolerances. In a dynamic process such as PBF, there are hundreds of process variables that could affect part quality and performance. Hence, the best certification methodologies use in-situ sensors to gather process data that are otherwise unavailable ex-situ. The open AM architecture philosophy at the authors' university helps us work in synergy with process data and multi-physics models to ensure that realistic and industrially applicable quality management (QM) techniques are developed for AM processes and component certification.

References

[1] ISO/ASTM 52900:2015 (ASTM F2792) Additive manufacturing — General principles — Terminology

[2] ISO 17296-1 2015 Additive manufacturing — General principles part 1: Terminology

[3] Wohlers Associates Inc. 2017 Wohlers Report 2017, 3D Printing and Additive Manufacturing State of the Industry – Annual Worldwide Progress Report ISBN 978-0-9913332-3-3

[4] Program for the 29th Annual International Solid Freeform Symposium 2018, Austin TX

[5] Pedersen D B, Andersen S A and Hansen H N 2019 Measurements in Additive Manufacturing In: Gao W. (eds) *Metrology. Precision Manufacturing*. Springer, Singapore, https://doi-org.proxy.findit.dtu.dk/10.1007/978-981-10-4912-5_13-1

[6] Bayat M, Thanki A, Mohanty S, Witvrouw A, Yang S, Thorborg J, Tiedje N S and Hattel J H Keyhole-induced porosities in Laser-based Powder Bed Fusion (L-PBF) of Ti6Al4V: High-fidelity modelling and experimental validation *Additive Manufacturing (submtted)*

[7] Kalpakjian S and Schmid S R 2010 *Manufacturing Engineering and Technology*, Pearson, Ch. 17. ISBN: 9810681445.

[8] Gibson I, Rosen D W and Stucker B 2010 *Additive Manufacturing Technologies: 3D Printing, Rapid Prototyping, and Direct Digital Manufacturing*, Springer, section 5.6.2. ISBN: 1493921134

[9] Thompson M K, Moroni G, Vaneker T, Fadel G, Campbell R I, Gibson I, Bernard A, Schulz J, Graf P, Ahuja B and Martina F 2016 Design for Additive Manufacturing: Trends, opportunities, considerations, and constraints *CIRP Annals* **65/2** 737-760

[10] Chen H, Gu D, Xiong J and Xia M 2017 Improving additive manufacturing processability of hard-to-process overhanging structure by selective later melting *J. Mater. Process. Technol.* **250** 99-108

[11] Fox J C, Moylan S P and Lane B M 2016 Effect of process parameters on the surface roughness of overhanging structures in laser powder bed fusion additive manufacturing *Proc. CIRP* **45** 131-134

[12] Funch C V, Somlo K, Poulios K, Mohanty S, Somers M A J and Christiansen T L 2019 The influence of microstructure on mechanical properties of SLM 3D printed Ti6Al4V *Titanium (submitted)*

[13] Bayat M, Mohanty S and Hattel J H A 2019 Systematic investigation of the effects of process parameters on heat and fluid flow and metallurgical conditions during Laser-based powder bed fusion of Ti6Al4V alloy *Int. J. of Heat and Mass Transfer* (accepted).

[14] Nadimpalli V K, Dahmen T, Valente E H, Mohanty S and Pedersen D B 2019 Multi-material additive manufacturing of steels using laser powder bed fusion *Proc. Euspen Int. Conf. & Exh.*

[15] Andersen S A, Nielsen K E, Pedersen D B and Nielsen J S 2017 Considerations on the construction of a Powder Bed Fusion platform for Additive Manufacturing *Physics Procedia* **89** 3-10

[16] Rak Z S 1999 New trends powder injection moulding *Powder Metallurgy and Metal Ceramics*

38(3-4) 26–132
[17] Frazier W E 2014 Metal additive manufacturing: A review *J. of Mat. Eng. and Perf.* **23(6)** 1917–1928
[18] Hein S B and Otte A L 2018 Lost-form powder injection moulding - combining additive manufacturing and PIM *European Powder Metallurgy Association -EPMA-: Euro PM2018 Proceedings*

IOP Conf. Series: Materials Science and Engineering **580** (2019) 012007 doi:10.1088/1757-899X/580/1/012007

The ESRF dark-field x-ray microscope at ID06

M Kutsal[1,2], P Bernard[1], G Berruyer[1], P K Cook[1], R Hino[1], A C Jakobsen[1], W Ludwig[1,3], J Ormstrup[2], T Roth[1], H Simons[2], K Smets[4], J X Sierra[5], J Wade[1,2], P Wattecamps[1], C Yildirim[1,6], H F Poulsen[2] and C Detlefs[1]

[1] European Synchrotron Radiation Facility, 38043 Grenoble, France
[2] Department of Physics, Technical University of Denmark, 2800 Kgs. Lyngby, Denmark
[3] MATEIS, INSA-Lyon, 69621 Villeurbanne Cedex, France
[4] LAB Motion Systems, 3001 Leuven, Belgium
[5] Department of Energy, Technical University of Denmark, 4000 Roskilde, Denmark
[6] OCAS, Pres. J.F. Kennedylaan 3, BE-9060, Zelzate, Belgium

E-mail: `detlefs@esrf.fr`

Abstract. We present an instrument for dark-field x-ray microscopy installed on beamline ID06 of the ESRF — the first of its kind. Dark-field x-ray microscopy uses full field illumination of the sample and provides three-dimensional (3D) mapping of micro-structure and lattice strain in crystalline matter. It is analogous to dark-field electron microscopy in that an objective lens magnifies diffracting features of the sample. The use of high-energy synchrotron x-rays, however, means that these features can be large and deeply embedded. 3D movies can be acquired with a time resolution of seconds to minutes. The field of view and spatial resolution can be adapted by simple reconfiguration of the x-ray objective lens, reaching spatial and angular resolution of 30-100 nm and 0.001°, respectively. The instrument furthermore allows pre-characterization of samples at larger length scales using 3DXRD or DCT, such that a region of interest (e.g. a single grain) can be selected for high-resolution studies without the need to dismount the sample. As examples of applications we show work on mapping the subgrains in plastically deformed iron and aluminum alloys, mapping domains and strain fields in ferroelectric crystals, and studies of biominerals. This ability to directly characterize complex, multi-scale phenomena *in-situ* is a key step towards formulating and validating multi-scale models that account for the entire heterogeneity of materials. As an outlook, we discuss future prospects for such multi-scale characterization by combining DFXM with 3DXRD/DCT, and coherent x-ray methods for coarser and finer length-scales, respectively.

1. Introduction

Many, if not most, technological materials are composed of crystalline elements that are hierarchically organized over length scales ranging from millimeters to nanometers, spanning up to 6 orders of magnitude. The same is true for biominerals, ice, sand and geological materials in general. Crystalline elements and substructures such as grains, domains and atomic-scale defect networks determine many of the macroscopic physical and mechanical properties of these materials. Understanding the interplay of physical phenomena and structural dynamics at, and between, these different lengths scales is therefore a critical and persistent issue across materials and geological sciences. Our understanding, however, is still limited by the lack of a non-destructive three-dimensional (3D) probe of the local crystal lattice (structure, strain, and orientation) that can be rapidly switched between different length scales and that enables acquisition of movies during processing.

The need to probe the local crystallography favours a diffraction-based approach. Existing 3D techniques, however, have shortcomings with respect to multi-scale characterization. Electron-based methods can provide very high spatial resolution, but are either limited to thin foils [1, 2] or involve serial sectioning [3, 4]. Scanning [5, 6, 7, 8, 9, 10] and coherent [11, 12, 13, 14, 15, 16, 17, 18] x-ray methods are rapidly progressing towards 20 nm resolution, but are limited to small sampling volumes. Furthermore, all methods face the challenge that the illuminated part of a bulk sample may comprise millions or even billions of structural elements whose diffracted signals overlap, rendering data analysis and interpretation complicated and in many cases impossible.

Aiming to overcome these limitations, dark-field x-ray microscopy (DFXM) is a full-field imaging technique for non-destructive mapping of the structure, orientation, and strain of deeply embedded crystalline elements in 3D [19, 20, 21, 22, 23, 24]. By placing an x-ray objective lens in the diffracted beam, direct space resolutions of 30–100 nm can be reached while maintaining a comfortable working distance of 10 cm or more between the sample and x-ray lens. The spatial resolution and field of view can be adapted by changing the focal length of the lens and thus "zooming" in or out. Furthermore, through its narrow angular and real space field of view, the objective also filters stray diffraction signals, suppressing unwanted overlap and isolating the individual structural element of interest.

As a first implementation of the method a dedicated dark-field x-ray microscope was recently installed on beamline ID06 of the European Synchrotron Radiation Facility (ESRF, Grenoble, France). The instrument is designed to combine DFXM with coarse scale 3D grain mapping techniques such as 3D X-Ray Diffraction (3DXRD) [25, 26, 27, 28, 29] and diffraction contrast tomography (DCT) [30, 31], as well as classical tomography and diffraction topography. This combination enables the user to rapidly progress from fast overviews of the entire specimen to detailed studies of local phenomena in a single experimental setting, without the need to dismount the sample.

Following an introduction to DFXM, in this article we present the **instrumentation** of the microscope at ID06, providing details on all relevant hardware components. In particular we present the refractive optics used for condensing and imaging. Compound refractive lenses (CRLs) were until recently the only choice, providing a reasonably large field of view but being limited in terms of spatial resolution by lens aberrations. As a newly developed alternative, multilayer Laue lenses (MLLs) offer larger numerical apertures (NAs) and improved spatial resolution, at the expense of a smaller field of view and reduced working distance. Next, we demonstrate the performance of the instrument on three selected examples in materials science: the processing of plastically deformed metals, the distribution of strain and orientation gradients in ferroelectrics, and hierarchical organization in bio-minerals.

1.1. The geometry and principle of DFXM

The geometry and operational principle of dark-field x-ray microscopy (see figure 1) is conceptually similar to dark-field transmission electron microscopy (TEM): the diffracted beam passes through an x-ray objective lens, creating a magnified image of a specific region of interest with contrast from local variations in lattice symmetry, orientation and strain [21]. The sample-to-detector distance $d_1 + NT + d_2$ is 2–6 m, enabling magnification ratios of up to 50 while still maintaining sufficient space around the sample for complex sample environments.

A defining feature of the dark-field x-ray microscope is the x-ray objective. Like a visible light microscope, the x-ray objective can be reconfigured to adjust the magnification, field-of-view, and numerical aperture, hence allowing adjustment of the spatial resolution according to specific experimental requirements. So far primarily compound refractive x-ray lenses made of

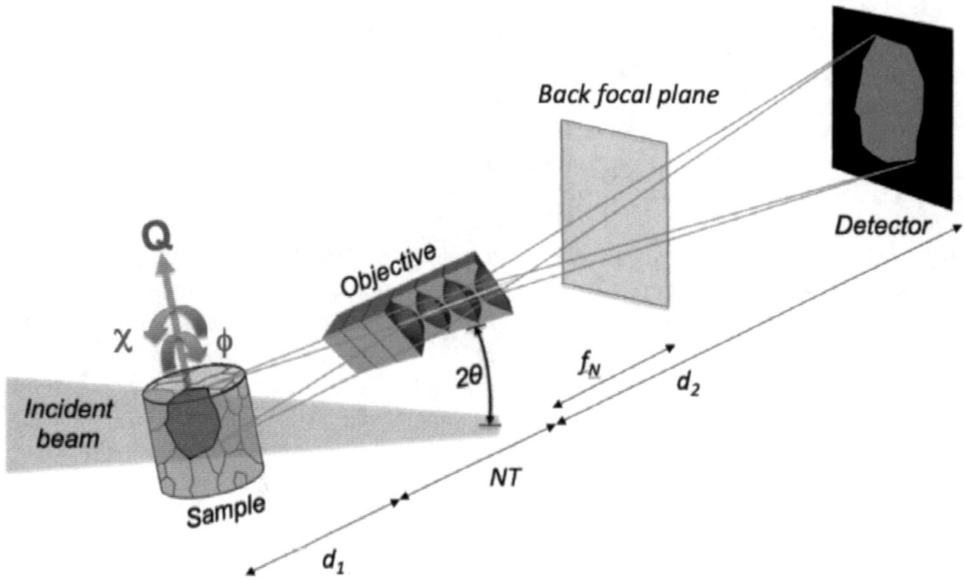

Figure 1. Principle of dark-field x-ray microscopy. Monochromatic x-rays illuminate a crystalline element of interest, and the diffracted radiation is imaged by means of an x-ray objective and a 2D detector. The objective is here a compound refractive lens, comprising N lenslets each with a thickness of T. d_1, d_2, f_N are the sample plane-entry of CRL distance, exit of CRL – image plane distance and the focal distance, respectively. Scanning the sample tilt (χ, ϕ), and scattering (2θ) angles facilitate mapping of orientation and axial strain respectively, while different projections can be obtained by rotating the sample about its scattering vector, $\vec{Q}$.

Beryllium [32] or SU-8 polymer [33] have been used. Their performance is discussed below along with a presentation of the first results using multilayer Laue lenses as objectives — with the aim to improve the spatial resolution. Irrespective of the choice, the imaging system is associated with a Fourier plane, the back focal plane downstream of the objective, see figure 1. Similar to dark field microscopy operation in a TEM, the direct space (imaging) information in the image plane and the Fourier space (diffraction) information in the back focal plane may be combined in a variety of ways [22].

Like the TEM, the dark-field x-ray microscope can be operated in a variety of modes. Most experiments so far have used a one-dimensionally-focusing condenser to create a narrow line-beam [34] that illuminates a 'layer of the material (typically $200\,\mu m \times 200\,\mu m \times 200\,nm$), which is then imaged at an oblique angle. Experiments typically involve the use of a succession of modalities including

- *Rocking curve imaging in section topography:* The dependence of the intensity on the Bragg angle (rocking scan) is analyzed pixel by pixel. It is possible to combine rocking curve imaging in magnified and non-magnified (using a near-field camera, see below) modes without unmounting the sample.

- *Mosaicity scans:* By systematically varying the sample tilts, χ and ϕ, a spatially resolved local pole figure can be acquired [21].
- *Strain scans:* By scanning longitudinally (θ–2θ-scan) the strain component along the scattering vector is imaged [21]. Typically, this is combined with a rocking scan or with integration over the rocking profile at each 2θ setting. As an alternative, strain mapping may be performed in the back focal plane [22].
- *Reciprocal space maps:* A high resolution reciprocal space map of the illuminated volume is available either in the back focal plane [22] or in the far-field regime (see below).

3D mapping can be performed in two ways. Firstly, by stacking layers of the kind described above. This is performed by translating the sample through the planar beam in small increments. A second, faster method is *magnified topo-tomography.* Here, projections of the sample are acquired while the sample is rotated about the scattering vector [35, 21], and a 3D representation is reconstructed using tomographic principles. Again, data can be taken in magnified and non-magnified mode. Experimental protocols and reconstruction codes for magnified topo-tomography are currently under development.

By setting 2θ to zero, magnified bright field imaging is obtainable [36, 37]. Translating the sample along the optical axis makes it possible to acquire pure absorption contrast images or phase contrast images at any Fresnel number. Moreover, by placing a phase plate in the back focal plane, Zernike type phase contrast images are created [38].

When based on the use of CRLs, the entire imaging system is accurately described by geometrical optics. However, the length of the CRL, NT, tends to be of the same order as the focal length f_N, implying that the thin lens approximation is not valid. Analytical expressions for the NA, field of view, depth of field, and spatial resolution in bright field mode are provided in [39] . For dark field microscopy, operating in the image plane, expressions for the resolution in both direct space and reciprocal space spatial resolution are provided in [21]. Correspondingly, in [22] equations are provided for the reciprocal space resolution in the back focal plane and the corresponding strain resolution.

1.2. Complementary imaging on longer length scales

It is necessary to combine DFXM with more traditional imaging in order to identify local regions of interest and to provide overviews during processing studies. The microscope at ID06 enables this is several ways:

- A *near-field camera* with high resolution placed close to the sample can provide classical absorption or phase contrast tomography, non-magnified diffraction topographies, and DCT measurements.
- A *diffraction camera* with large pixels can provide 3DXRD type mapping as well as classical (sample averaged) diffraction information.
- *Far-field imaging* can provide high resolution reciprocal space maps [27]. Conveniently this modality can be obtained with the same detector as shown in figure 1 by simply translating the objective out of the diffracted beam.

More details about these cameras are provided below.

2. Description of beamline components

2.1. Overall beamline layout and optical scheme

The beamline is located at the high-beta straight section ID06 of the ESRF in Grenoble, France. It consists of one white beam optics hutch, OH, and two monochromatic beam experimental

Table 1. Undulator parameters obtained from fits of the observed gap with maximum flux as function of photon energy to the theoretical curve.

		U18 CPMU	U27
Min. gap	[mm]	6.05	11
Max. gap	[mm]	30	300
Period λ_u	[mm]	18.25(2)	27.19(2)
B_0	[T]	2.44(2)	1.93(2)
α		1.10(1)	1.00(1)

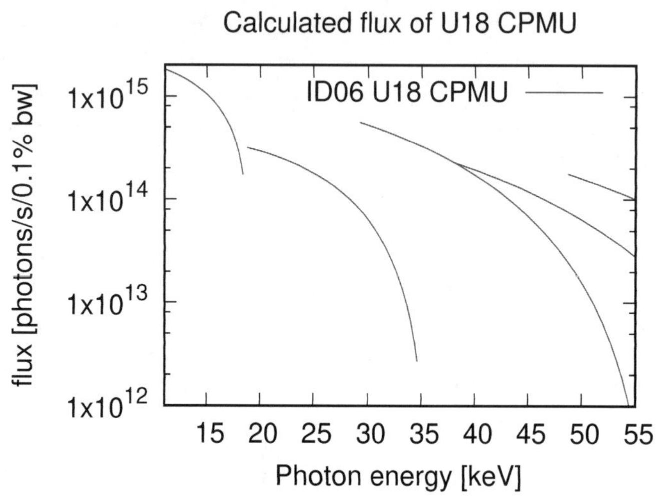

Figure 2. Calculated flux of the ID06 U18 CPMU undulator as function of photon energy, in photons per second into a $1 \times 2\,\text{mm}^2$ aperture at 28 m from the source, and for a 0.1% band width. This calculation is for the pre-upgrade ESRF, without taking into account absorption in diamond and Beryllium windows.

hutches, EH1 and EH2. OH houses standard ESRF high power primary slits, a UHV section where temporary white beam experiments can be set up [40, 41, 37], the monochromator, secondary slits and intensity monitor, a Transfocator, and, finally, the safety shutter for EH1. The dark-field x-ray microscope is implemented in EH1, whereas EH2 houses the Large Volume Press (LVP) that is described elsewhere [42].

The coordinate system used at ID06 has x along the beam axis, y horizontal and away from the storage ring, and z upwards.

2.2. Source

Two source devices are installed in the straight section (see Table 1): A U18 cryogenic permanent magnet in-vacuum undulator (CPMU) with minimum gap 6 mm [43] covers the energy ranges from 11–20 keV and 33–100 keV, see figure 2. Additionally a U27 out-of-vacuum permanent magnet undulator with minimum gap 11 mm covers the energy ranges 7.5–11 keV below the fundamental, and 20–33 keV in between the fundamental and third harmonic of the U18.

2.3. Monochromator

The ID06 double crystal monochromator, located in OH at 35.8 m downstream of the nominal source point, is the first optical element of the beamline apart from a 300 μm diamond window that separates the beamline and storage ring vacuum systems, and the white beam primary slits located 27.8 m downstream of the source. The monochromator was designed and built by Cinel Strumenti Scientifici (Vigonza, Italy) to ESRF specifications. It has been in continuous operation on ID06 since 2008. It houses two pairs of flat crystals in Bragg-Bragg geometry mounted side by side. By horizontally translating the entire monochromator vessel, either symmetric Si(111) or Si(311) reflections can be selected. The Bragg angle θ can be adjusted between 2.05° to 53.5°, covering 2.46–55 keV using the Si(111) pair of crystals, and 4.72–105 keV using the Si(311) pair. In practice, the vast majority of experiments uses the Si(111) pair. The monochromator is operated in fixed exit geometry with an offset of $z = +15$ mm between the white and monochromatic beams.

The peak monochromatic flux at the sample position is $\approx 10^{13}$ photons/s at 11 keV, with an unfocused spot size of $\approx 1\,\mathrm{mm}(h) \times 0.5\,\mathrm{mm}(v)$.

2.4. Transfocator

The UHV Transfocator [44] is located in OH at a distance of 38.7 m from the source point. The mechanical design is based on in-vacuum piezo-actuators for groups of lens elements. The vacuum chamber can be moved in y, z, pitch and yaw to align the optical axis of the Transfocator to the beam (ESRF in-house design, which is based on a similar device at Petra III [45]). It contains sets of 1D (vertically focusing) and 2D (bi-dimensionally focusing) compound refractive lenses made of Be (RXOptics, Juelich, Germany) that can be moved into and out of the beam in groups. The Transfocator can be used as a collimator or pre-focusing device. Due to the ratio of the distances towards the source and the sample, however, the demagnification of 1:3.9–1:4.8 (depending on the position of the main block of the instrument) remains fairly moderate. For larger demagnification, i.e. smaller spot size or line height at the sample position, condenser optics can be employed close to the sample, see section 2.5.3. The installed lenses allow collimation from 7.5–30 keV and focusing from 7.5–30 keV.

2.5. Microscopy bench

The hard x-ray microscope was implemented in the existing experiments hutch EH1 of ID06. The design was therefore restricted by the existing infrastructure and beamline optics as described above.

A Be exit window at 52.4 m from the source ends the UHV section of the beam path. The remainder of EH1 is utilized by the microscopy bench and far-field detector frame, see figure 3. The beam travels in air, without flight paths.

The design goal of the hard x-ray microscope was to achieve a resolution better than 50 nm. For this, optimum mechanical stability is required on short (vibrations) and long (thermal drifts) time scales. Due to its high density, high damping factor and low thermal expansion coefficient, granite is very well suited as the construction material.

In particular, the relative position of the sample and objective lens is critical, as any differential movement of these elements will be amplified in the far-field detector by the geometrical magnification. Common-mode vibrations, however, are not magnified. We have therefore chosen to implement the critical elements on a common granite support (the "main block"), which can be moved on air pads on a support bench, also made from granite. In choosing the position of the main block a compromise has to be made between the maximum scattering angle (limited by the maximum height of the far-field detector, favoring position further downstream) and maximum geometrical magnification (limited by the focal length of the objective lens and the distance to the far-field detector, favoring a position further upstream).

IOP Conf. Series: Materials Science and Engineering **580** (2019) 012007 doi:10.1088/1757-899X/580/1/012007

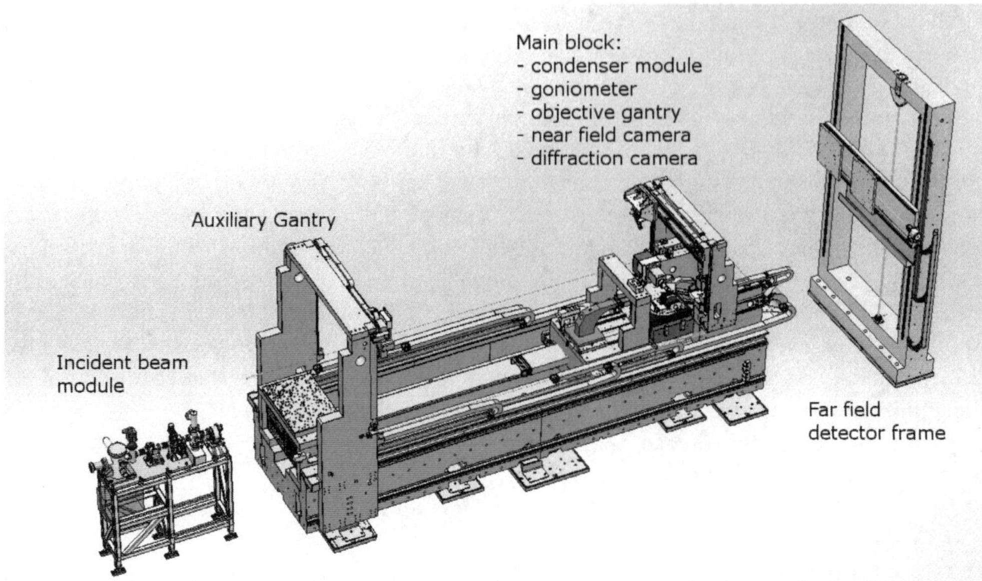

Figure 3. The microscopy bench. Along the beam (left to right), elements are: Incident beam module, main block with condenser module, goniometer, near-field camera, objective gantry and diffraction camera. The auxiliary gantry can be moved along the entire length of the bench. The far-field detector frame is located at the downstream end of the hutch.

The support bench, main block, and alignment mechanics were designed and built by LAB motion systems (Leuven, Belgium) to ESRF specifications. An overview is shown in figure 3, further details on the different components are given below.

The bench is assembled from several granite blocks, whose weight was limited by the capacity of the available overhead crane (6 t). Its total weight is about 22 t. It is mounted on 13 adjustable Wedgmount levelers (AirLoc Schrepfer, Oetwil am See, Switzerland) – this lucky number is the result of the blocks being individually supported to aid with their relative alignment before they are rigidly bolted together. Steel plates have been set in the concrete floor to spread the load, provide a flat and level mounting surface, and to anchor the tension bars of the levelers. The complete bench can be disassembled, e.g. to remount it on a different beamline. It is, however, also possible to inject epoxy resin under the bench to permanently bond it to the concrete floor. The bench is 6 m long and 1.4 m wide. The air bearing surface is 0.5 m above the floor.

2.5.1. Incident beam module The incident beam module consists of a 0.6 m long optical bread board mounted in fixed position at the upstream end of the bench, 53.7 m from the source. The optical elements on it are (upstream to downstream):

- Beam position monitor based on a quadrupole ionization chamber and ESRF Quadera error amplifier. The vertical beam position detected here is used for feedback on the monochromator's second crystal pitch.
- Secondary slits (AT-F7-AIR, JJ X-ray, Hoersholm, Denmark) at 53.9 m from the source.
- Decoherer (optional) a spinning disk of graphite (Goodfellow 613 136 00) to produce a fluctuating wave front that will average out any speckle from structure in the incident beam's wave front [46].

- Ionization chamber intensity monitor (ESRF design). The ionization chamber can be moved out of the beam by a motorized translation to insert a beam profile monitor (ESRF design using a Basler ace acA1300-30gm CCD camera and Infinity InfiniStix 2× objective) or visible light mirror reflecting a pre-aligned laser beam as an alignment aid.
- Fast shutter, synchronized with the exposure signal of the different cameras.

2.5.2. Main block The main block serves as common support for the sample goniometer and for optical elements in close vicinity of the sample, i.e. condenser, near-field camera, diffraction camera, and objective. It can be moved on air pads along the length of the bench. For experiments, its position is typically fixed and the air is turned off. The main block is 2.3 m long and can be translated by 2.8 m.

2.5.3. Condenser The condenser unit is the first element on the main block. A translation stage with 340 mm stroke along x can be equipped with a small hexapod (H-824, Physik Instrumente, Karlsruhe, Germany) and compact Transfocator (ESRF in-house development) with Be lenses for moderate focusing, or a granite arch with y, z, pitch and yaw alignment mechanics for Si, polymer or multilayer Laue lenses for more aggressive focusing. Downstream of the lenses a set of 4-blade motorized slits (AT-F7-AIR, JJ X-ray, Hoersholm, Denmark) and a compact ionization chamber (ESRF design) for intensity monitoring can be mounted.

In recent practice, the most frequently used condenser is a set of 58 vertically focusing 1D Be lenses with apex radius of curvature 100 μm (RXOptics, Juelich, Germany) mounted in a fixed enclosure on the hexapod. This is used in conjunction with 2D lenses in the OH Transfocator to optimize the flux on the entrance pupil of the condenser. A set of 1D SU-8 polymer lenses (KIT, Karlsruhe, Germany) for use at higher photon energies is under commission.

2.5.4. Sample goniometer The sample goniometer is mounted on the main block, downstream of the condenser. By moving the main block, the sample can be positioned between 53.6 and 56.4 m from the source and, correspondingly, between 2.2 and 5.0 m from the plane of the far-field detector frame (see section 2.6).

The sample goniometer is assembled from several components. The goniometer base consists of two orthogonal air-bearing rotation stages rotating about the z (outside) and y (inside) axes, and an air-bearing y-stage. It was designed and manufactured by LAB motion systems (Leuven, Belgium) to INSA (Lyon, France) and ESRF specifications. The goniometer provides high quality rotational movement about the horizontal and vertical axes of each respective air bearing rotation stage – the large rotation stage about the z axis has a maximum rotational error of 36 nm and axial error of 23 nm, and the smaller rotation stage about the y axis has a maximum rotation error of 80 nm and axial error of 56 nm. The center of rotation can be aligned to the beam axis by the goniometers y-stage and a stepper motor driven z-stage (AXMO, Brétigny/Orge, France) underneath the goniometer.

Shown in figure 4 is the part of the goniometer that is closest to the sample. Mounted on the horizontal ω rotation stage are a motorized crossed swivel stage (SA04B-RS, Kohzu, Japan) for the "sample tilts" ϕ and χ, and a piezo-based xyz micropositioner for sample centering (SmarAct, Oldenburg, Germany).

The goniometer pre-dates the dark field microscopy project. Its configuration allows for rotations about a horizontal scattering vector ("topo-tomo" or azimuthal scan) [30, 31] when the Bragg angle is set by the outsize z rotation stage. As, due to the asymmetry of the source, the preferred scattering geometry is in the vertical plane, the goniometer should be replaced by a new instrument optimized for topo-tomo in the vertical scattering plane, i.e. with a y-rotation stage outside and z-rotation stage inside [21].

IOP Conf. Series: Materials Science and Engineering **580** (2019) 012007 doi:10.1088/1757-899X/580/1/012007

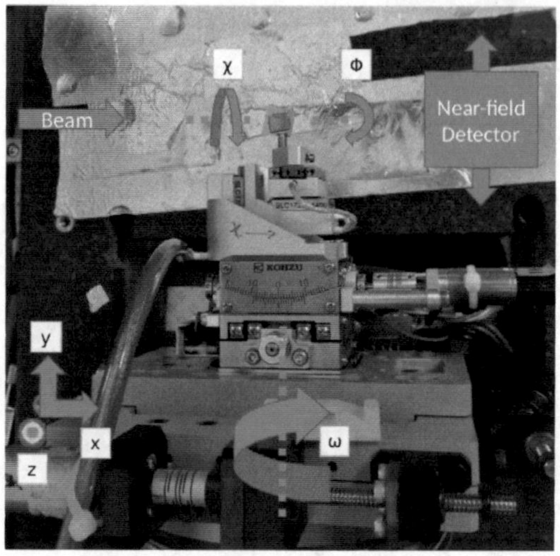

Figure 4. View of goniometer: LAB goniometer, Kohzu tilts and SmarAct xyz stage. Showing beam direction in blue, xyz coordinate system, and goniometer axes of rotation.

2.5.5. Near-field camera The near-field camera is an indirect detector, consisting of a thin scintillator, x-ray transparent mirror, and an optical microscope coupled to a CCD camera. The optics were designed and built by Optique Peter (Lentilly, France) to ESRF specifications. The semi-transparent scintillator can be positioned within millimeters of the sample, e.g. for classical tomography or for DCT-type experiments [30, 31]. Free-standing gadolinium gallium garnet (GGG) single crystals of $25\,\mu$m and $50\,\mu$m thickness are available, as well as thinner scintillator films on a substrate. The objective lens (Mitutoyo M-Plan Apo 10x/0.28 or 5x/0.14) is protected from radiation damage by a lead glass window. Optical parameters of the near-field camera are listed in table 2. The CCD camera is a ESRF FreLoN 2K camera [47].

In addition to xyz linear alignment axes, the near-field camera is mounted on a rotation table to swing it out of the way when other optical elements, such as the objective lens, need to be positioned in close proximity to the sample, or when a full 360° rotation about the goniometers z axis is required for alignment purposes.

Table 2. Optical parameters of the different cameras of the dark field microscope.

CCD	**pixel size** [μm^2]	**Magnification**			**Effective pixel size** [μm^2]	**Field of view** [mm^2]
Near-field camera		Objective	Tube lens	Eye piece		
Atmel TH7899M	14×14	5x	0.9x	2.5x	1.24×1.24	2.49×2.49
		10x			0.62×0.62	1.24×1.24
Diffraction camera		Fiber taper				
Atmel TH7899M	14×14	(0.296x)			47.3×47.3	94.6×94.6
Wide-field far field camera		Objective (12 mm f/1.8)				
Sony ICX445	3.75×3.75	≈0.08x			48×48	62×45
High resolution far field camera		Objective	Tube lens			
Atmel TH7899M	14×14	10x	0.9x		1.56×1.56	3.1×3.1

2.5.6. Diffraction camera The diffraction camera is a fiber-taper coupled FreLoN camera [48, 47] with effective pixel size (after distortion correction) of 47.3 μm × 47.3 μm, and field of view 94.6 mm × 94.6 mm (see table 2). The camera can be inserted at a distance of 180 to 560 mm downstream of the sample, or be removed completely from the beam path to make room for the objective lens or auxiliary optics and detectors.

It is equipped with a movable beam stop which in turn can be equipped with a photodiode to monitor the intensity of the direct beam. The purpose of the diffraction camera is to record complete diffraction rings at elevated photon energies in order to reconstruct 3D grain maps of the sample using 3DXRD techniques [25, 26, 27, 28]. Grains thus determined can then be studied with ultimate resolution using dark field microscopy or, eventually, coherent techniques.

2.5.7. Objective The objective stage allows for xyz, pitch and yaw adjustment of an objective lens. The pitch (model 5202.60) and yaw (model 410) stages were supplied by Huber Diffraktionstechnik (Rimsting, Germany). The linear and angular travel ranges are large enough to keep the lens aligned with the axis of a diffracted beam up to scattering angles (2θ) of 30°.

The centers of rotation of the pitch and yaw stages intersect in the center of the lens assembly. For rotations about the sample position, e.g. for varying the scattering angle 2θ during strain scans, the rotations therefore have to be combined with the appropriate translations. The objective stages center of rotation can be positioned between 0 and 380 mm downstream of the sample. The finite length of the lens assemblies imposes further practical limits. In the y-z plane the objective stage can be moved up to 500 mm away from the beam axis.

Mechanically, the objective stage is implemented as a gantry with two lead screw drives for the x- and z- motions, and one lead screw drive for the y-motion. The x and z motions need to be driven simultaneously to avoid skew motions and eventual mechanical damage to the system. This is handled through the ESRF Icepap's "linked axis" firmware option. The x, y and z-stages are equipped with pneumatic clamps to further improve the mechanical stability when at fixed position.

During the conception of the objective gantry particular attention was paid to minimizing thermal and vibration effects, as the relation of these movements relative to the sample are significantly magnified on the far-field camera based upon the ratios shown in figure 1. Thermal

Table 3. Parameters of the different objective lenses. E is the working energy, NA the numerical aperture, N and T the number and distance between lens elements, and f is the focal length.

	E [keV]	NA [mrad]	Resolution [nm]	$N \times T$ [mm]	f [mm]
Be	17	0.55	≈ 100	$68 \times 1.6 + 20 \times 2.0$	277
SU-8	33	0.18	≈ 140	$(65(h) + 65(v)) \times 0.77$	330
MLL	17.3	2(h), 1.6(v)	≈ 30	–	9.5(h), 13.7(v)
MLL	34.5	0.4(h), 0.7(v)	≈ 30	–	37(h), 40(v)

stability was ensured by the choice of granite as the base material linked with the stabilization of the climate in EH1, and reducing wherever possible the thickness of materials with higher coefficients of expansion. To create a rigid structure to reduce the vibration effects, this led to several developments in the placement of rails and sliding carriages to ensure the maximum possible rigidity based on space and translation constraints. This can be seen particularly in the gantry legs whereby two rails have been mounted each side at 90 degrees to each other, which provides a better average stiffness in the z and y axes. The rails are also spaced apart and the lower rail provides a strong lever arm against the vibrations in the tops of the structure.

After measuring the frequency response of the objective gantry was found to have its first harmonic at 40 Hz. Given the size of the gantry, this is encouraging and correlates with the original computer simulations completed by both the ESRF and LAB Motion Systems.

At present, three different types of objective lens are used for microscopy experiments:

Beryllium CRL Low absorption and high ratio of refractive index decrement over absorption (δ/β) make Beryllium in theory the best CRL material at all photon energies [49]. For working energies of $\approx$ 15–20 keV we therefore use a CRL of 88 2D Be lenses with apex radius of curvature of 50 μm (RXOptics, Juelich, Germany). In practice the CRLs are limited by aberration to a resolution around 100 nm.

Optical parameters of the Be CRL are summarized in table 3.

SU-8 polymer CRL The chromaticity of the CRL (focal length is approximately proportional to the square of the working energy) limits the practical use of Be in objective lenses to working energies below $\approx$ 20 keV.

While Si-based lenses can potentially achieve very high resolution [50, 51, 39], strong absorption limits their apertures to very small values, which in turn dictates very short working distances – thus the resolution obtained under the practical working conditions of a materials science x-ray microscope (working distance $\approx$ 100–300 mm) is much worse than the ultimate limit of $\approx$ 50 nm that has been demonstrated in optics tests.

SU-8 polymer lenses as developed by the IMT group at KIT present an interesting solution for working energies above 20 keV, combining manufacturing precision, homogeneous material and cost-effectiveness [33].

Optical parameters of the SU-8 CRL are summarized in table 3.

Multilayer Laue lenses Recently we successfully tested Multilayer Laue Lenses (MLLs) as objectives for both phase and diffraction contrast imaging [52]. MLLs can have a NA 10 times better than compound refractive lenses, and virtually no manufacturing errors [53]. MLLs are therefore a candidate for achieving a spatial resolution as good as 10 nm or alternatively 10–100 times faster data acquisition. Similar to Si refractive lenses, however, the MLLs available today have small physical apertures and thus require very short working distances [54].

Optical parameters of the two MLL objective lenses that have been tested on the instrument are summarized in table 3. Further optimization in terms of both manufacturing (a larger physical aperture) and alignment is required, but the work is seen as encouraging.

2.5.8. Auxiliary gantry The auxiliary gantry is a xyz alignment stage that can travel along the entire length of the bench. It allows the mounting of additional optical elements for future experiments beyond simple microscopy. Possible uses include the insertion of slits, phase shifters or cameras upstream of the condenser or between the objective lens and far-field detector, e.g. in the back focal plane.

2.6. Far-field detectors

The far-field detector frame is located at the downstream end of EH1, separate from the support bench. It allows yz positioning of two detectors, a wide-field detector and a high resolution detector. It consists of a granite frame with y and z translation stages (AXMO, Brétigny/Orge, France), a cradle that can be inclined to orient the cameras perpendicular to the diffracted beam (Design et Mechanique, Montaud, France), and the detectors.

The far-field detector frame is located 60.64 m downstream of the source, i.e. between 2.2 and 5.0 m downstream of the sample, depending on the position of the main block. Its travel range (using the high resolution detector and direct beam axis as reference points) is −0.8 to +0.6 m in the horizontal (y) and −0.5 to +2.3 m in the vertical (z) direction.

2.6.1. Wide-field far camera This is a simple indirect detector based on a plastic scintillator screen, Linos MeVis-C 12mm f/1.8 machine vision lens, and a C-mount CCD camera (Basler acA1300-30gm). Details are given in table 2.

2.6.2. High resolution detector The high resolution detector is an indirect detector composed of a scintillator screen (typically 10 μm LuAG:Eu on a 170 μm YAG substrate) [55], optical microscope with Olympus UIS2 UPlanSApo 10x/0.40 objective and Olympus U-TLU-1-2 tube lens. The optical microscope is operated without an eye piece, projecting the intermediate image directly onto an ESRF FreLoN camera [47] with 2048 × 2048 pixels. This combination yields and effective pixel size of $1.56 \times 1.56\,\mu m^2$ and a field of view of $3.1 \times 3.1\,mm^2$. The thickness of the scintillator screen is a compromise between quantum efficiency (x-ray absorption) and resolution (due to the limited depth of field of the objective) [56].

The narrow field of view of the high-resolution camera can make finding the diffracted beam somewhat challenging. To facilitate this, both cameras are rigidly mounted on the same support. It is thus possible to first locate and center the diffracted beam in the wide-field camera, and to then switch to the high resolution camera by simply translating the camera support by the predetermined spacing between the cameras.

For dark field microscopy in the horizontal scattering plane (e.g. topo-tomo experiments with the present goniometer), an additional far-field detector can be mounted on a table equipped with y and z translations and a rotation about the z axis. The detector configuration is identical to the high-resolution far-field detector described above.

3. Examples

Below we provide three examples of DFXM applications. Other lines of work performed so far include imaging of dislocations within semiconductors and oxides [23], and processing studies of fuel cells [57].

IOP Conf. Series: Materials Science and Engineering **580** (2019) 012007 doi:10.1088/1757-899X/580/1/012007

3.1. Deformed metals and alloys

When a pure metal or a metallic alloy is plastically deformed, dislocations self organize into walls surrounding defect-free regions known as cells or subgrains with a size of $\approx 1\,\mu$m. Moreover, these cells organize into intricate patterns inside existing grains. During subsequent annealing the stored energy is released by the processes of recovery and recrystallisation. In the former the cells refine and strain is released. In the latter new strain-free grains are nucleated and grow at the expense of the deformed matrix. Recovery and recrystallisation co-exist within the sample. Electron microscopy has been extensively used for post mortem studies of both the plastic deformation and the annealing phenomena. DFXM provides a unique opportunity to study these multi-scale phenomena *in situ* from deeply embedded grains within $\approx$ mm-sized specimens [19, 58, 59, 24].

In a first example, DFXM was used to study a deeply embedded alpha iron grain within a 330 μm thick sample of Fe with 850 ppm Nitrogen [24]. Prior to the experiment, the sample was solutionized at 580 C for 15 hours. Using a 33 keV full field beam and the SU-8 polymer lens as objective, projections were acquired. The exposure time was 1 s. The high angular resolution of DFXM enables us to resolve the mosaic spread in the rocking (θ) and rolling (χ) directions, within the grain. Despite the annealing the mosaicity map of the (110) reflection shows a substantial angular spread (figure 5(a)). The orientation gradient in this figure can be resolved with EBSD. The local misorientation, however, falls below the angular resolution of conventional EBSD.

In a second example, DFXM was used to study a deeply embedded recrystallized grain in a 50% recrystallized Al 1050 alloy [19]. A cross-sectional layer was illuminated with a condensed line beam and imaged with the Be CRL as objective at 17 keV with exposure time of 1 s. The mosaicity map presented in figure 5(b) clearly shows the subgrain structures separated by low-angle boundaries within the recrystallized grain.

In a recent experiment, we have generalised DFXM to allow observation of the microstructure in highly deformed materials of key industrial interest. Specifically the microstructural changes during *in-situ* annealing of a steel sample deformed to a true strain of $\epsilon \approx 2$ were studied (to be presented elsewhere [60]).

3.2. Strain and domain mapping in ferroelectrics

Ferroelectric, ferroelastic and multiferroic materials are typified by the presence of domains: distinct regions of similar lattice orientation, separated by atomically-coherent domain walls (e.g. inversion about a zone axis) [61]. In response to stresses, electric fields, or other perturbations, these domain structures self-organize via the collective motion of domain walls. Given the strong anisotropy of many of these materials, the modulation of domain structures is often accompanied by significant macroscopic responses (e.g. changes in the dielectric and piezoelectric coefficients). These large responses are widely utilized in a range of applications including shape-memory alloys, digital memory and solid-state capacitors. Furthermore, the discovery of emergent phenomena at the domain walls (e.g. photoelectricity) has renewed interest in the local structure at and around such walls [62]. The ability to quantitatively probe the lattice strain and misorientation in the vicinity of the domain walls is expected to yield important insights into what drives these emergent phenomena.

Here, DFXM was used to reveal the domain structure embedded within a bulk single crystal of barium titanate ($BaTiO_3$). Figure 6 shows the image intensity acquired in a single exposure, which shows clear contrast from the individual domains. Given the 001-orientation of the crystal, it is evident that the domain walls lie on $\langle 110 \rangle$ lattice planes, and that the domain walls are not completely straight. The curvature of domain walls in this manner indicates the presence of underlying electric or elastic heterogeneities, such as vacancies or dislocations, that cause the walls to distort from their lowest-energy configuration. The key advantage of investigating such

IOP Conf. Series: Materials Science and Engineering **580** (2019) 012007 doi:10.1088/1757-899X/580/1/012007

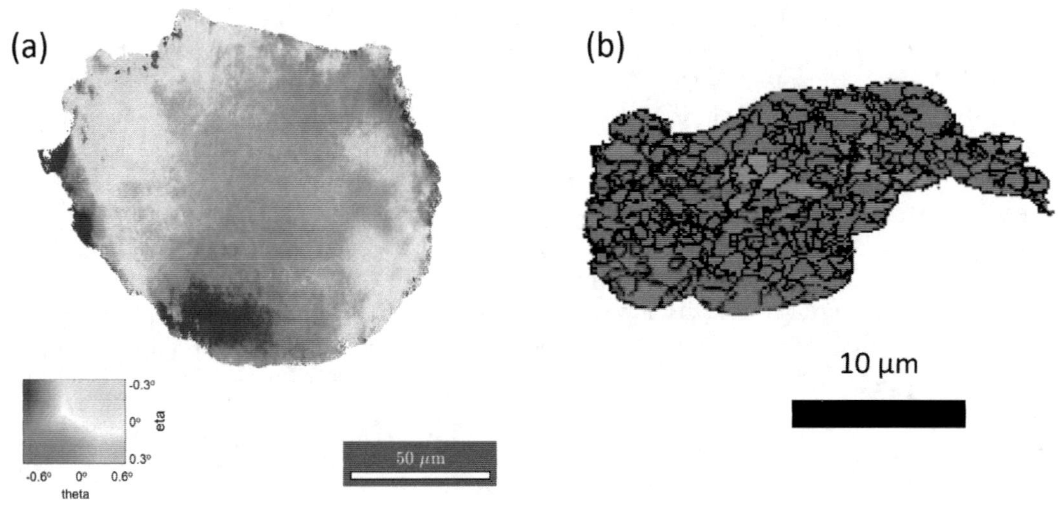

Figure 5. (a) Mosaicity map of a Fe-850 ppmN grain after annealing at 580 C for 15 h. This is a projection image with the color code indicating angular variation within parts of a (110) pole figure. (b) Mosaicity map of a layer within a recrystallized grain in 50% recrystallized Al 1050. The colors indicates angular variation within parts of a (200) pole figure.

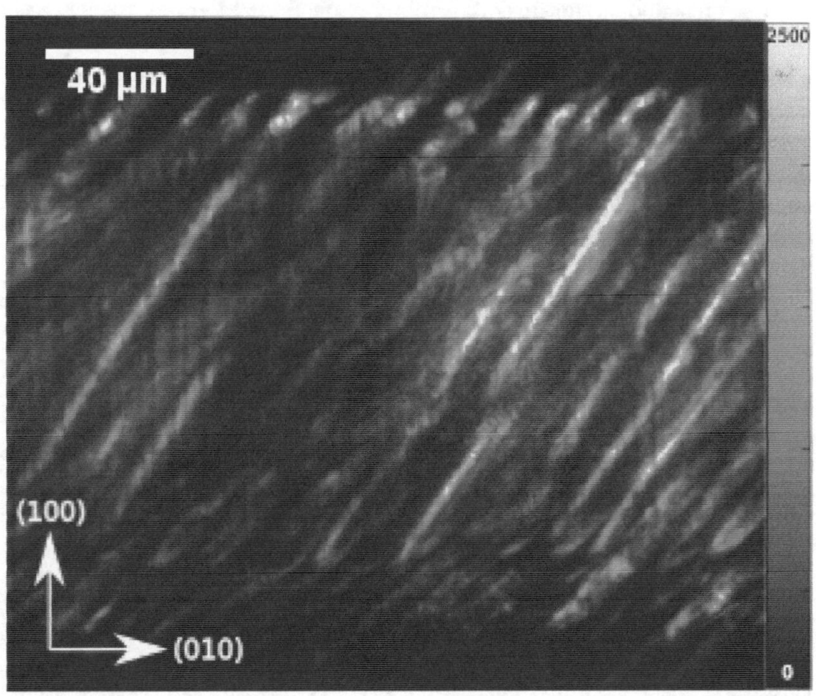

Figure 6. Dark field x-ray microscopy image of domains embedded in a single crystal of $BaTiO_3$. The domain walls are clearly curved, indicating the presence of underlying elastic or electric heterogeneities. The color code to the right indicates intensity in arbitrary units.

domain structures using DFXM is that it is a full-field imaging technique. As such, full images of embedded domain structures may be captured in a single exposure within a second or less. This allows for dynamic studies of multiscale structures under applied electric fields or at elevated temperatures [63]. These measurements are key to understanding and accurate modelling of ferroelectricity and ferroelasticity, and are directly comparable to multi-scale models, such as phase-field and Monte-Carlo simulations.

Shown in figure 7 are the first DFXM results using the MLL as the objective. The figure shows a map of the average lattice misorientation from a projection through the $BaTiO_3$ crystal. The ferroelastic domain walls are clearly visible as diagonal lines aligned with the 110 direction of the sample. In this map, the transition region between neighbouring domain orientations at the domain wall is in the order of $1\,\mu$m thick. The map also shows a clear misorientation between the top and bottom parts of the image, which is likely to be attributable to an out-of-plane domain wall along 101-type planes. This allows the local lattice distortions at the intersection of the 101- and 110-oriented domain walls to be investigated directly and in detail.

3.3. Microstructure of biominerals

Biominerals are hierarchical materials found throughout nature [64]. Examples are widespread, including bone, tooth, corals, shells, and otoliths. Biominerals form as a combination of inorganic mineral and organic template under strict biological control by the organism, yielding an exceptional control of final structure in both crystallography and habit [64]. This control has evolved to yield materials whose properties are optimized for particular functions, such as wear resistance for teeth, or strength and crack resistance for protective shells [64]. Like nature, materials science aims to produce function-optimized materials. Bio-inspired structures are becoming increasingly common as scientists seek to understand and replicate natural growth processes in vitro. Nonetheless, there remain many aspects of biomineralization that remain to be fully understood.

As examples, we present two types of biominerals that have been examined with DFXM [65]. The first relates to the nacreous layer in a mussel shell. This grows in a brick-and-mortar-like assembly, composed of calcium carbonate platelets surrounded by proteinaceous organic matrix. A piece of shell from the edge was broken off and mounted on the sample support. Examination with the dark-field microscope permitted visualization of a multitude of platelets with a large range in orientation (figure 8) [65]. Reflections were observed across the full scan range of 0.6° on both θ and in the azimuth. This unique view of the platelets *in-situ* illustrates their highly controlled assembly and inter-platelet relationships.

The second biomineral examined was fish otolith [65]. Otoliths are calcium carbonate accretions that grow in the inner ear of vertebrate fish. They are composed of aragonite fibres which extend from the nucleus to the outer surface [66]. A hierarchical structure is present, comprising the fibre, bundles of prismatic crystalline units, and nanogranules. DFXM reveals evidence of all of these levels of organization (figure 9). The visible otolith fibre is $\approx 100\,\mu$m long and individual bundles of prismatic crystalline units are visible along its length. The interruptions between bundles along the growth axis likely correspond to daily growth layers governed by the individuals circadian rhythm. Single crystalline units can be seen stacked side by side across the fibre. Individual nanogranules cannot be resolved, but small clusters can be seen. The prismatic crystalline units are closely oriented with respect to the growth axis, with the crystal c axes coherent to within $\pm 2°$. The orientation of the a and b axes, however, appears to be isotropic, with prisms occupying a wide range of orientations around the c axis.

Figure 7. Dark-field x-ray microscopy images using the MLL as objective. Projection image of lattice misorientations around ferroelastic/ferroelectric domains in a $BaTiO_3$ single crystal. The misorientation is represented by colour according to the part of a pole figure below

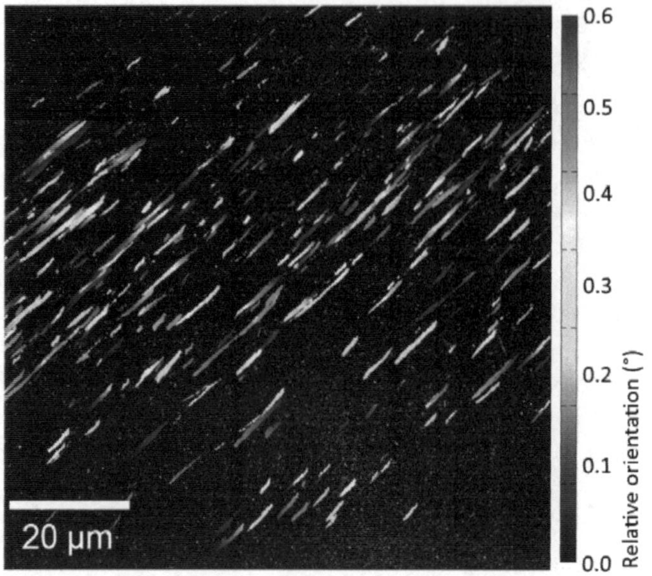

Figure 8. Orientations of nacre platelets in a mussel shell imaged by dark-field X-ray microscopy. The colours indicate orientation differences according to the scale to the right. Figure adapted from [65].

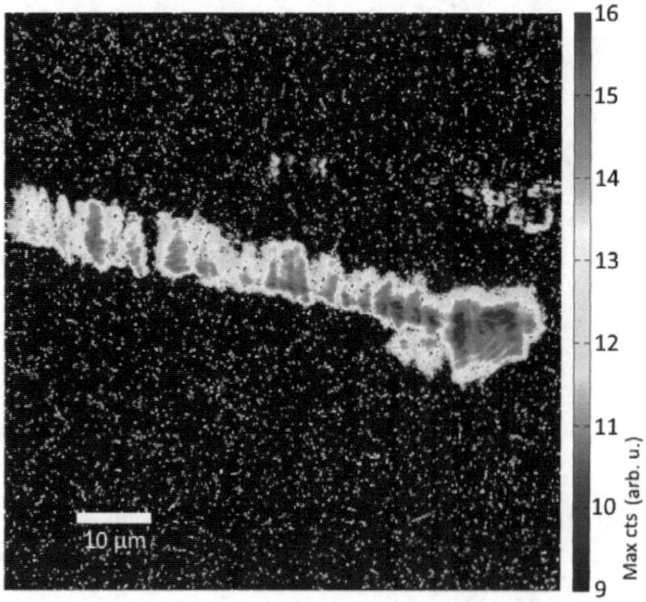

Figure 9. A dark-field image of an aragonite fibre from a fish otolith shows its growth from left to right in sequential layers, each composed of multiple crystalline units. The colour scale indicates intensity. Figure taken from [65].

Table 4. Present and future source parameters of ID06 [71]

		Present source	EBS Upgrade
Horizontal rms source size	[μm]	387.8	27.2
Horizontal rms source divergence	[μrad]	10.3	5.2
Vertical rms source size	[μm]	3.5	3.4
Vertical rms source divergence	[μrad]	1.2	1.4

4. Outlook

4.1. A multi-scale view of microstructure

Dark-field x-ray microscopy is a promising technique for multi-scale characterization of hierarchical microstructures in materials. Yet, the field of view and spatial resolution of DFXM cover only a portion of this range. Moreover, the field of view in reciprocal space is very restricted, such that only one Bragg reflection of one grain can be studied at any given time. In order to obtain a complete multi-scale view, DFXM has to be combined with complementary diffraction and imaging techniques that can provide the missing information. In particular, DCT or 3DXRD can be used to obtain three-dimensional maps of the position and orientation of all grains within a sample, with relatively lower real space and reciprocal space resolution. Acquisition of such maps would enable users to differentiate and select the relevant regions of interest (ROI) in the microstructure. Furthermore, we have ongoing work to develop a high-resolution variant of 3DXRD to bridge the gap between conventional far-field 3DXRD and DFXM.

The spatial resolution of DFXM is ultimately limited by the numerical aperture of the objective lens. In the need of higher resolution, coherent x-ray techniques such as (Bragg) coherent diffraction imaging [11, 15, 16, 13, 17, 18, 67] or Fourier ptychography [68, 69, 70] can be employed to extend DFXM towards higher spatial and reciprocal space resolution.

Common to all these techniques is that they employ hard x-rays. Therefore, it is possible, with some technical development, to perform all of the aforementioned techniques on the same instrument without the need to re-mount the sample.

The instrument at ID06 is designed with this need in mind. Efforts for on-line data analysis and integration of DCT and 3DXRD results into the DFXM data acquisition strategies are ongoing. The integration of all of the mentioned techniques is the main focus of the upcoming EBS upgrade beamline.

4.2. EBS upgrade beamline

The HXRM project has been selected for implementation as an Upgrade beamline for the ESRF EBS (Extremely Bright Source) upgrade. The microscope will be installed at a dedicated beamline with optics and infrastructure fully optimized for x-ray microscopy. We expect that the project will be implemented in 2022, and that the new beamline will be open to external users via the ESRF general user proposal system in 2023.

With the EBS upgrade [71], the source parameters will be improved significantly, as shown in table 4. In combination with a new, fully optimized undulator this will lead to a 10–30 fold increase in the maximum photon flux at the sample, but more importantly to a 200–600 fold increase in coherent flux.

The dedicated beamline will also allow pink beam operation. For bright field imaging the chromatic aberration associated with both refractive and diffractive optics can be overcome by focusing the incoming beam on the back focal plane [41], and pink beam operation can therefore lead to an increase in useful flux by two orders of magnitude. For dark field operation a bandwidth of 1 % is not optimal, but numerical simulations indicate that it may be of relevance [72].

Acknowledgments
The authors thank Élise Dufour (UMR 7209 CNRS/Muséum National d'Histoire Naturelle) for providing otolith samples for study. We acknowledge the Karlsruhe Nano Micro Facility (KNMF) at Karlsruhe Institute of Technology (KIT) for the provision of SU-8 polymer lenses and DTU Nanolab for provision of a Si based condenser. We acknowledge the S. Bajt at the CFEL for the provision of the multilayer Laue lenses. The authors thank Tao Zhou for assistance during Fe-N experiment and the loan of SU-8 polymer lenses.

References

[1] Liu H H, Schmidt S, Poulsen H F, Godfrey A, Liu Z Q, Sharon J A and Huang X 2011 *Science* **332** 833–834
[2] Midgley P A and Dunin-Borkowski R E 2009 *Nature Materials* **8** 271–280
[3] Zaafarani N, Raabe D, Singh R, Roters F and Zaefferer S 2006 *Acta Mater.* **54** 1863–1876
[4] Uchic M D, Groeber M A, Dimiduk D M and Simmons J 2006 *Scripta Mater.* **55** 23–28
[5] Larson B C, Yang W, Ice G E, Budai J D and Tischler J Z 2002 *Nature* **415** 887–890
[6] Hofmann F, Abbey B, Liu W, Xu R, Usher B F, Balaur E and Liu Y 2013 *Nat. Commun.* **4** 2774
[7] Schroer C G, Kurapova O, Patommel J, Boye P, Feldkamp J, Lengeler B, Burghammer M, Riekel C, Vince L, van der Hart A and Küchler M 2005 *Appl. Phys. Lett.* **87** 124103
[8] Mimura H, Handa S, Kimura T, Yumoto H, Yamakawa D, Yokoyama H, Matsuyama S, Inagaki K, Yamamura K, Sano Y, Tamasaku K, Nishino Y, Yabashi M, Ishikawa T and Yamauchi K 2009 *Nat. Phys.* **6** 122–125
[9] Ice G E, Budai J D and Pang J W L 2011 *Science* **334** 1234
[10] Xu C, Zhang Y, Godfrey A, Wu G, Liu W, Tischler J Z, Liu Q and Juul Jensen D 2017 *Scientific Reports* **7**
[11] Miao J, Charalambous P, Kirz J and Sayre D 1999 *Nature* **400** 342–344
[12] Dierolf M, Menzel A, Thibault P, Schneider P, Kewish C M, Wepf R, Bunk O and Pfeiffer F 2010 *Nature* **467** 436–439
[13] Pfeifer M A, Williams G J, Vartanyants I A, Harder R and Robinson I K 2006 *Nature* **442** 63–66
[14] Shapiro D, Thibault P, Beetz T, Elser V, Howells M, Jacobsen C, Kirz J, Lima E, Miao H, Neiman A M and Sayre D 2005 *Proc. Nat. Acad. Sci. USA* **102** 15343–15346
[15] Chapman H and Nugent K 2010 *Nature Phot.* **4** 833–839
[16] Miao J, Ishikawa T, Robinson I K and Murnane M 2015 *Science* **348** 530–535
[17] Chamard V, Stangl J, Carbone G, Diaz A, Chen G, Alfonso C, Mocuta C and Metzger T 2010 *Phys. Rev. Lett.* **104** 165501
[18] Yau A, Cha W, Kanan M, Stephenson G and Ulvestad A 2017 *Science* **356** 739–742
[19] Simons H, King A, Ludwig W, Detlefs C, Pantleon W, Schmidt S, Stöhr F, Snigireva I, Snigirev A and Poulsen H F 2015 *Nat. Commun.* **6** 6098
[20] Simons H, Jakobsen A C, Ahl S R, Detlefs C and Poulsen H F 2016 *MRS Bulletin* **41** 454
[21] Poulsen H F, Jakobsen A C, Simons H, Ahl S R, Cook P K and Detlefs C 2017 *J. Appl. Cryst.* **50** 1441
[22] Poulsen H F, Cook P K, Leemreize H, Pedersen A F, Yildirim C, Kutsal M, Jakobsen A C, Trujillo J X, Ormstrup J and Detlefs C 2018 *J. Appl. Cryst.* **51** 1428–1436
[23] Jakobsen A C, Simons H, Ludwig W, Yildirim C, Leemreize H, Porz L, Detlefs C and Poulsen H F 2019 *J. Appl. Cryst.* **52** 122
[24] Mavrikakis N, Detlefs C, Cook P, Kutsal M, Campos A, Gauvin M, Calvillo P, Saikaly W, Hubert R, Poulsen H, Vaugeois A, Zapolsky H, Mangelinck D, Dumont M and Yildirim C 2019 *Acta Materialia* **174** 92–104
[25] Poulsen H F, Nielsen S F, Lauridsen E M, Schmidt S, Suter R M, Lienert U, Margulies L, Lorentzen T and Juul Jensen D 2001 *J. Appl. Cryst.* **34** 751
[26] Schmidt S, Nielsen S F, Gundlach C, Margulies L, Huang X and Juul Jensen D 2004 *Science* **305** 229–232
[27] Jakobsen B 2006 *Science* **312** 889–892
[28] Hefferan C M, Lind J, Li S F, Lienert U, Rollett A D and Suter R M 2012 *Acta Mater.* **60** 4311–4318
[29] Pokharel R 2018 *Overview of High-Energy X-Ray Diffraction Microscopy (HEDM) for Mesoscale Material Characterization in Three-Dimensions* (OSTI, 2018)
[30] King A, Johnson G, Engelberg D, Ludwig W and Marrow J 2008 *Science* **321** 382–385
[31] Ludwig W, Reischig P, King A, Herbig M, Lauridsen E M, Johnson G, Marrow T J and Buffiere J Y 2009 *Rev. Sci. Instrum.* **80** 033905
[32] Snigirev A, Kohn V G, Snigireva I I and Lengeler B 1996 *Nature* **384** 49

[33] Marschall F, Last A, Simon M, Kluge M, Nazmov V, Vogt H, Ogurreck M, Greving I and Mohr J 2014 *Journal of Physics: Conference Series* **499** 012007
[34] Stöhr F, Wright J, Simons H, Michael-Lindhard J, Hübner J, Jensen F, Hansen O and Poulsen H F 2015 *J. Micromech. Microeng.* **25** 125013
[35] Ludwig W, Cloetens P, Härtwig J, Baruchel J, Hamelin B and Bastie P 2001 *J. Appl. Cryst.* **34** 602–607
[36] Lengeler B, Schroer C, Tümmler J, Benner B, Richwin M, Snigirev A, Snigireva I and Drakopoulos M 1999 *J. Synchrotron Rad.* **6** 1153–1167
[37] Falch K V, Casari D, Di Michiel M, Detlefs C, Snigirev A, Snigireva I, Honkimäki V and Mathiesen R H 2016 *J. Mater. Sci.* **52** 3497–3507
[38] Falch K V, Lyubomirskij M, Casari D, Detlefs C, Snigirev A, Snigireva I, Detlefs C, Di Michiel M, Lyatun I and Mathiesen R H 2018 *Ultramicroscopy* **184** 267
[39] Simons H, Ahl S R, Poulsen H F and Detlefs C 2017 *J. Synchrotron Rad.* **24** 392
[40] Van Vaerenbergh P, Detlefs C, Härtwig J, Lafford T A, Masiello F, Roth T, Schmid W, Wattecamps P and Zhang L 2010 *AIP Conf. Proc.* **1234** 229
[41] Falch K V, Detlefs C, Di Michiel M, Snigireva I, Snigirev A and Mathiesen R H 2016 *Appl. Phys. Lett.* **109** 054103
[42] Guignard J and Crichton W A 2015 *Rev. Sci. Instrum.* **86** 085112
[43] Chavanne J, Le Bec G and Penel C 2011 *Synchrotron Rad. News* **24** 10–13
[44] Snigirev A, Snigireva I I, Vaughan G B M, Wright J P, Rossat M, Bytchkov A and Curfs C 2009 *J. Phys. Conf. Ser.* **186** 012073
[45] Zozulya A V, Bondarenko S, Schavkan A, Westermeier F, Grübel G and Sprung M 2012 *Opt. Express* **20** 18967
[46] Falch K V, Detlefs C, Paganin D, Christensen M S and Mathiesen R 2019 *Opt. Express* In press
[47] Labiche J C, Mathon O, Pascarelli S, Newton M A, Guilera Ferre G, Curfs C, Vaughan G, Homs A and Fernandez Carreiras D 2007 *Rev. Sci. Instrum.* **78** 091301
[48] Coan P, Peterzol A, Fiedler S, Ponchut C, Labiche J C and Bravin A 2006 *J. Synchrotron Rad.* **13** 260–270
[49] Serebrennikov D, Clementyev E, Semenov A and Snigirev A 2016 *J. Synchrotron Rad.* **23** 1315–1322
[50] Schroer C G and Lengeler B 2005 *Phys. Rev. Lett.* **94**(5) 054802
[51] Simons H, Stohr F, Michael-Lindhart J, Jensen F, Hansen O, Detlefs C and Poulsen H F 2015 *Opt. Commun.* **359** 460
[52] Murray K T, Pedersen A F, Mohacsi I, Detlefs C, Morgan A J, Prasciolu M, Yildirim C, Simons H, Jakobsen A C, Chapman H N, Poulsen H F and Bajt S 2019 *Opt. Express* **27** 7120
[53] Bajt S, Prasciolu M, Fleckenstein H, Domaracký M, Chapman H N, Morgan A J, Yefanov O, Messerschmidt M, Du Y, Murray K T, Mariani V, Kuhn M, Aplin S, Pande K, Villanueva-Perez P, Stachnik K, Chen J P, Andrejczuk A, Meents A, Burkhardt A, Pennicard D, Huang X, Yan H, Nazaretski E, Chu Y S and Hamm C E 2017 *Light: Science & Applications* **7** 17162–17162
[54] Morgan A J, Prasciolu M, Andrejczuk A, Krzywinski J, Meents A, Pennicard D, Graafsma H, Barty A, Bean R J, Barthelmess M, Oberthuer D, Yefanov O, Aquila A, Chapman H N and Bajt S 2015 *Sci. Rep.* **5** 9892
[55] Martin T and Koch A 2006 *Journal of Synchrotron Radiation* **13** 180–194
[56] Koch A, Raven C, Spanne P and Snigirev A 1998 *J. Opt. Soc. Am. A* **15** 1940–1951
[57] Sierra J, Poulsen H, Jørgensen P, Detlefs C, Cook P, Simons H, Jakobsen A and Bowen J 2019 *Journal of Power Sources* **413** 351–359
[58] Ahl S, Simons H, Zhang Y, Detlefs C, Stöhr F, Jakobsen A, Juul Jensen D and Poulsen H 2017 *Scripta Materialia* **139** 87–91
[59] Ahl S, Simons H, Detlefs C, Juul Jensen D and Poulsen H F 2019 Subgrain dynamics during recovery of partly recrystallized aluminium to be published
[60] Yildirim C *et al.* 2019 Dark field x-ray microscopy study of in-situ annealing of a highly deformed steel sample in preparation
[61] Tagantsev A K, Cross L E and Fousek J 2010 *Domains in ferroic crystals and thin films* vol 13 (Springer)
[62] Catalan G, Seidel J, Ramesh R and Scott J F 2012 *Reviews of Modern Physics* **84** 119
[63] Simons H, Haugen A B, Jakobsen A C, Schmidt S, Stöhr F, Majkut M, Detlefs C, Daniels J E, Damjanovic D and Poulsen H F 2018 *Nature Materials* **17** 814
[64] Gilbert P U P A, Abrecht M and Frazer B H 2005 *Rev. Mineral. Geochem.* **59** 157–185
[65] Cook P K, Simons H, Jakobsen A C, Yildirim C, Poulsen H F and Detlefs C 2018 *Microscopy and Microanalysis* **24** 8889
[66] Dauphin Y and Dufour E 2008 *Micron* **39** 891 – 896 ISSN 0968-4328
[67] Pedersen A F, Chamard V, Detlefs C, Zhou T, Carbone D and Poulsen H F 2018 X-ray coherent diffraction imaging with an objective lens: towards 3d mapping of thick polycrystals (*Preprint* `arxiv:1810.04268`)
[68] Simons H, Poulsen H F, Guigay J P and Detlefs C 2016 X-ray fourier ptychographic microscopy (*Preprint*

arxiv:1609.037513)

[69] Wakonig K, Diaz A, Bonnin A, Stampanoni M, Bergamaschi A, Ihli J, Guizar-Sicairos M and l A M 2019 *Sci. Adv.* **5** eaav0282

[70] Detlefs C, Beltran M A, Guigay J P and Simons H 2019 Translative lens-based full field coherent x-ray imaging (*Preprint* arxiv:1905.01916)

[71] Dimper R, Reichert H, Raimondi P, Sanchez Ortiz L, Sette F and Susini J 2014 *The Orange book: ESRF Upgrade Programme Phase II (2015-2022)* (ESRF, Grenoble (France))

[72] Pedersen A F, Chamard V and Poulsen H F 2018 *Optics Express* **26** 23411

Cold rolling behaviour of Cu with highly oriented nanotwins: the importance of local shear strain

Q H Lu, J S Bai, Z Cheng, L Lu

Shenyang National Laboratory for Materials Science, Institute of Metal Research, Chinese Academy of Sciences, Shenyang 110016, P. R. China

Email: LLu@imr.ac.cn

Abstract. To clarify the deformation mechanism of nanotwinned materials under a complex stress state and large strain level, the microstructure evolution of pure Cu samples with high density of preferentially oriented nanotwins cold rolled to a strain up to 65% was systematically characterized by means of electron backscatter diffraction and transmission electron microscopy in this study. Heterogeneous deformation behavior was observed in nanotwinned Cu at different rolling strains. At a rolling strain of 15%, uniform deformation carried by the interactions between dislocations and twin boundaries dominate. At the same time, a small amount of strain localization in the direction ±45° respective to the rolling direction takes place, carried by detwinning and shear bands. As the rolling strain increased to 50%, detwinning becomes a dominant deformation mechanism and produces a large amount of coarse twin/matrix lamellae. A unique laminated structure with high angle lamellar boundaries (larger than 50° misorientation angle) and an average lamellar thickness of about 150 nm evolves from the coarse twin/matrix lamellae. As the rolling strain is increased up to 65%, an extensive laminated structures embedded with a few preserved twin blocks are prevalently formed. It is found that the local shear strain plays a critical role in the microstructure evolution of the nanotwinned Cu under cold rolling. When the local shear strains are so small as to be negligible, uniform deformation is sustained by interactions of dislocation/twin boundaries and the nanoscale twins survive. As the local shear strain increases up to 1, detwinning prevails and results in coarse twin/matrix lamellae. When the local shear strain is larger than 1, formation of shear bands composed of dislocation cells, sub-grains and fragments of twin lamellae instead of the detwinning process takes place, finally evolving into a laminated structure.

1. Introduction

Nanotwinned (nt) metals have attracted extensive interest for their superior mechanical properties, such as high strength, considerable ductility and good work hardening [1-5]. The intrinsic deformation mechanisms of nt-metals under uniaxial deformation have been extensively investigated by experimental characterization [6-8], crystal plasticity modeling [9] and by molecular dynamics simulations [10-12]. It is well established that the plastic deformation of nt-metals is dominated by the interactions between dislocations and twin boundaries (TBs). The deformation behavior and the deformation mechanisms of nt-metals not only strongly depend on the inherent microstructural parameters, such as twin length and twin thickness [13, 14], but also closely depend on the extrinsic deformation conditions, such as the strain rate [15, 16], temperature [6, 17], and loading direction [7].

For example, under uniaxial compression, the plastic deformation of nanotwin structures can be switched among three specific dislocation modes, namely, dislocations transferring across and piling up at twin boundaries (TBs) [18], dislocations gliding in-between the confined twin/matrix layers [3] and dislocation-mediated TBs migration (or detwinning) [19], only by changing the uniaxial loading orientation with respect to the twin plane [7]. Correspondingly, different dislocation-based deformation mechanisms result in quite different mechanical properties, including yield strength, ductility and work hardening [7].

Compared to uniaxial deformation, the plastic deformations of nt-metal under complex axial stress states, such as cold rolling [20], torsion [21], and nano-indentation [22] are much more complicated. Detwinning, the general deformation mechanism by coarsening of nanoscale twins, is predominantly observed under complex axial deformations [19, 21, 23-26]. Detwinning caused by the migration of TBs is driven by gliding of Shockley partials along TBs, which can be active from incoherent twin boundaries (ITBs) and twin steps on the coherent twin boundaries (CTBs) or nucleated from grain boundaries [19, 26, 27]. In-situ transmission electron microscopy (TEM) investigations have indicated that Shockley partial dislocations can be multiplied from CTBs via the dissociation of a perfect dislocation crossing a CTB under certain stress states, leading to the migration of CTBs [28]. Multiaxial stress states and high shear stresses under non-uniform deformation facilitate transmission of dislocations across TBs and promote the glide of Shockley partial dislocations along the TBs, enhancing the detwinning process in nt-metals.

In addition, the formation of shear bands due to severe local deformations also typically occurs in nt-metals deformed under high strain and complex stress conditions [29-32]. Hong et al. [26] studied the evolution of shear bands in Cu-Al under the dynamic plastic deformation (DPD) process and found that only when the activation of dislocations is heavily suppressed in the nanoscale twin structures, are shear bands nucleated from detwinning area, with further increase of the local shear strains. Specifically, the formation and evolution of shear bands are closely related to the deformation strain and the local shear strain.

Obviously, both detwinning and shear bands are kinds of unstable, non-uniform plastic deformation, and frequently appear in multiaxial stress states with large deformation strains and high shear stresses or strains [19, 21, 26]. As such, a quantitative understanding the deformation mechanisms related to detwinning and shear bands of nt-metals under complex stress states and large strain conditions is imperative.

In this work, the microstructure evolution of bulk nt-Cu with preferentially oriented nanotwins subjected to cold rolling with different strains of 15%, 50% and 65% is systematically investigated. Deformation microstructures, such as dislocation morphology, twin thickness distribution, and boundary misorientation have been characterized by electron backscatter diffraction (EBSD) and TEM to reveal the deformation mechanisms of nt-Cu under complex axial deformation. Local shear strains in detwinning regions and shear bands are measured using the TBs as a reference. Our work reveals that the local shear strain plays an important role in the formation of detwinning and shear bands under cold rolling. Eventually, both detwinning and shear bands lead to the evolution of a stable laminated structure.

2. Experimental

High-purity Cu sheets with nanoscale growth twins were synthesized by means of direct-current electrodeposition using an electrolyte of $CuSO_4$. The substrate was a pure Ti sheet. By carefully controlling the deposition parameters, a nt-Cu sheet with preferentially oriented nanoscale twins was deposited to a thickness of ~1.5 mm. More details of the procedure can be found in Ref. [7].

Specimens with an area of 5 × 3 mm^2 were cut from the as-deposited Cu sheet. Before cold rolling, the specimens were mechanically ground to a final thickness of ~1 mm after mechanical polishing and chemical polishing. Cold rolling was carried out on a twin-roller apparatus with roll diameter of 40 mm at room temperature. During cold rolling, the specimens were placed with the normal direction (ND) of the rolling plane parallel to the growth direction (GD) (figure 1d). As a consequence, both the rolling direction (RD) and transverse direction (TD) are perpendicular to GD. The rolling strain (ε) is defined as the thickness reduction,

$$\varepsilon = (h_0 - h)/h_0 \tag{1}$$

where h_0 is the initial sample thickness, and h is the final thickness after rolling. The strain rate of the cold rolling is estimated to be 10^{-2} ~ 10^{-1} s^{-1}. Three rolling strains of 15%, 50% and 65% were selected to evaluate the microstructure evolution of the cold-rolled nt-Cu.

The microstructures of the as-deposited Cu specimens were examined using a field emission gun scanning electron microscope (SEM) in a FEI Nova NanoSEM 430 with backscattered electron (BSE) imaging using a VCD detector. The electron backscattered diffraction (EBSD) measurements and analysis were performed on the longitudinal section (perpendicular to TD) of the as-rolled samples using the HKL channel 5^{TM} software suite. The step sizes were 200 nm and 50 nm for each sample. Samples for EBSD measurements were first mechanically ground, followed by electro-polishing in a solution of 25% phosphoric acid, 25% alcohol and 50% deionized water at room temperature.

The microstructures of both the as-deposited and the as-rolled nt-Cu samples taken from the longitudinal plane were further characterized using a JEOL 2010 TEM at an accelerating voltage of 200 kV. The TEM foils fixed to 3 mm diameter Cu rings were thinned by twin-jet polishing in an electrolyte of 33% phosphoric acid, 33% alcohol and 33% deionized water at about -10 °C. This microscope was also equipped with a semi-automatic Kikuchi diffraction analysis technique [33] to allow measurement of the crystallographic orientation of individual crystallites.

3. Results

3.1. Microstructure characteristics of the as-deposited nt-Cu sample

The microstructure of the as-deposited nt-Cu was characterized from both the plane view and the cross-section view. The plane-view SEM image in figure 1a reveals that the as-deposited Cu consists of homogenously distributed equiaxed polycrystalline grains with an average grain size of 3 μm (the grain size distributes from 1 to 6 μm). Most grains are separated by sharp and clear grain boundaries. Only in a portion of grains are straight planar substructures clearly seen.

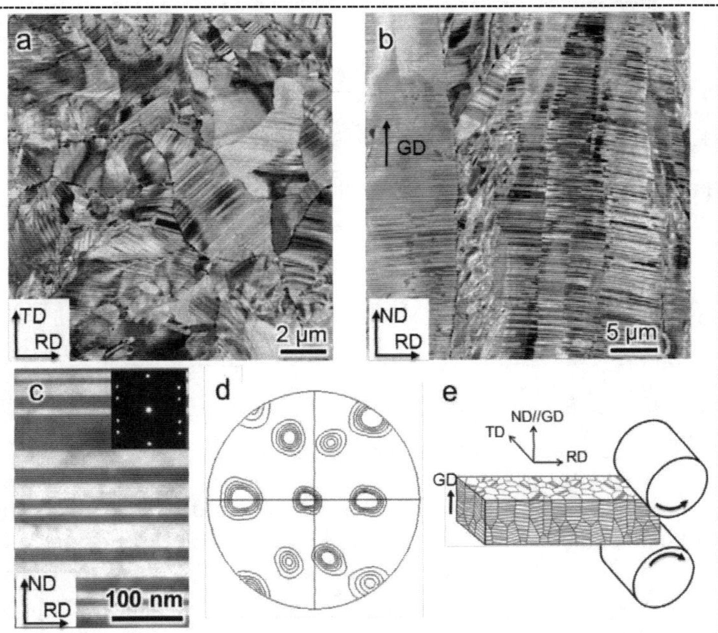

Figure 1. SEM images of the as-deposited columnar nt-Cu sample in plane view (a) and cross-section view (b). GD represents the grown direction of the deposited samples. (c) Cross-sectional TEM observation. The inset shows the corresponding selected area electron diffraction (SAED) pattern. (d) Pole figure taken from plane view data showing a strong <111> out of plane texture. (e) The rolling geometry with respect to the as-deposited specimen. Here ND, TD and RD are the normal direction, transverse direction and rolling direction of the rolled sample, respectively.

Cross-sectional SEM observations (figure 1b) indicate that most grains in the as-deposited Cu are columnar in shape with their longitudinal axis parallel to the GD. Each columnar grain is further divided by a high density of TBs, which are roughly parallel to the deposition plane. Cross-sectional TEM observation (figure 1c) indicates that the TBs are clear {111} coherent TBs with few dislocations. Analysis of a statistically representative number of edge-on TEM images reveals an average twin thickness (λ) of about 30 nm. The as-deposited sample shows a strong <111> out-of-plane texture in GD, as displayed in the {111} pole figure derived from the EBSD data (figure 1d). A three-dimensional (3D) sketch showing the microstructure of the as deposited Cu is illustrated in figure 1e. The ND during cold rolling is parallel to the GD of the as-deposited nt-Cu sample (ND//GD). As a consequence, the rolling plane (containing RD and TD) is parallel to the TBs.

3.2. Microstructure characteristics of the as-rolled nt-Cu with a rolling strain of 15%

Figure 2 shows the deformation microstructure of nt-Cu sample with a rolling strain of 15%. As displayed in the inverse pole figure (IPF) map colored according to ND (figure 2a), most grains (volume fraction of more than 80%) have a ND close to <111>, which suggests a strong <111> texture in the GD, coinciding with that of the as-deposited nt-Cu. There is no apparent change in the shape of the columnar grains, compared to the as-deposited structure (figure 1b). The grain boundaries are still clear and sharp. Some local shear deformation takes place, however, inside the columnar grains (as indicated by the white arrows in figure 2a), along the direction at ±45° respective to RD, and resulting in obvious deviation from the original <111> orientation.

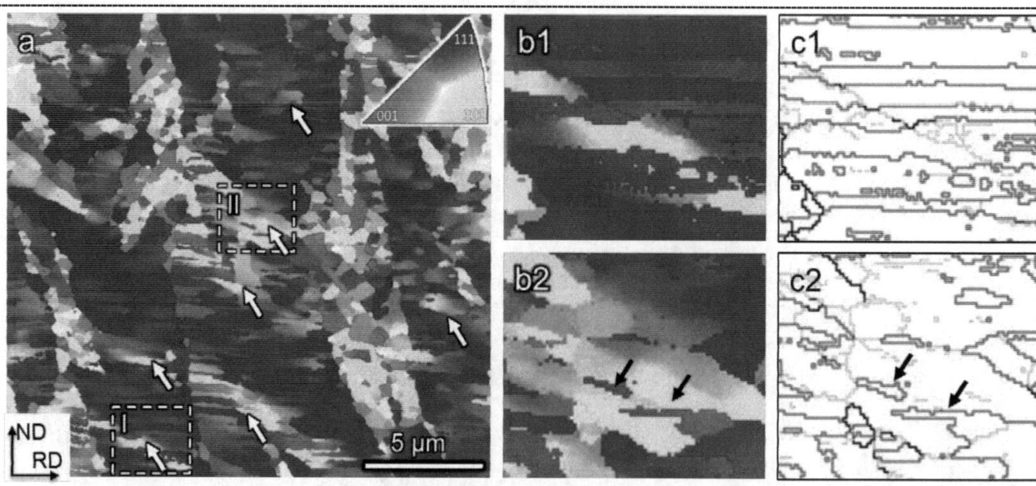

Figure 2. (a) IPF map of cold rolled nt-Cu sample after a strain of 15%, colored according to the ND. Local shear regions are indicated by white arrows. (b1-b2) Enlarged IPF maps of the white squares labeled as "I" and "II" in (a), respectively. (c1-c2) Boundary maps corresponding to (b1-b2). Twin fragments are indicated by black arrows. In the boundary maps, green and black lines represent boundaries with misorientation angles, θ, of $\theta \leq 15°$ and $\theta > 15°$, respectively; red lines represent the Σ3 TBs.

Two sorts of local shear deformation structures with different morphologies can be distinguished in the nt-Cu sample after a rolling strain of 15%. Typical examples for these two deformation structures, white squares I and II, are further revealed by the enlarged IPF maps (figure 2b1 and 2b2) and the boundary maps (figure 2c1 and 2c2). The enlarged IPF map of the local shear region I (figure 2b1) is composed of coarse twin/matrix (T/M) lamellae of more than 100 nm thick, which still have coherent TBs, confirmed by the boundary map in figure 2c1. Low angle grain boundaries with boundary misorientations of 8~10° (measured from figure 2b1) are formed along the edge of this local shear region I (green lines in figure 2c1), leading to deviations from the original <111> direction. This observation suggests extensive slip and accumulation of perfect dislocations in this region.

The local shear region II (figure 2b2) shows a typical morphology of shear bands composed of elongated sub-grains or dislocations cells inclined to the original TBs along ±45° direction with respect to RD. The misorientations of deformed structures in the shear bands is about 8~15° deviation to the original <111> orientation. The corresponding boundary map (figure 2c2) indicates that most sub-grains in the shear band are associated with low angle dislocation boundaries. The narrow sub-grains surrounded by TBs (black arrows in figure 2b2 and 2c2) in the shear bands are preserved twin lamellae fragments. Clearly, the formation of shear bands destroys the original twin lamellae, and then dislocation cells and boundaries are formed along the shear direction.

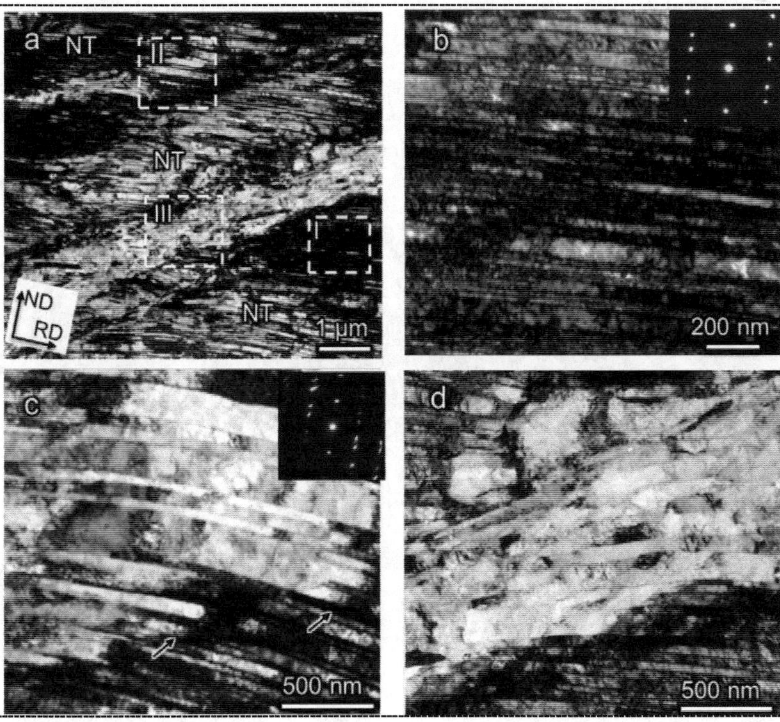

Figure 3. (a) TEM image of cold rolled nt-Cu sample after a strain of 15%. (b-d) High magnification images of the corresponding regions labeled as "I", "II" and "III" in (b), showing the nanoscale twin structure, the coarse twin structure and a shear band in detail, respectively. The insets in (b) and (c) are the corresponding SAED patterns of the nanotwins and the coarse twin structure. The interface between the detwinning region and preserved nanotwins is indicated by black arrows in (c).

More detailed TEM observations were carried out to characterize the detailed microstructure of nt-Cu samples after a rolling strain of 15%. The low magnification TEM observation of the nt-Cu in figure 3a shows the appearance of a few parallel locally sheared regions along a direction about 35° with respect to RD, cutting through the original nanoscale twin lamellae (marked by "NT"). Three typical microstructures are selected for further detailed analysis, namely preserved nanoscale twin structure, coarse T/M lamellae, and shear band, as indicated by the white squares "I", "II" and "III", respectively. An enlarged image of the preserved nanoscale T/M lamellae in white square "I" (figure 3a) shows that the TBs are well preserved, with a strict T/M relationship, as detected by SAED. A high density of dislocations appears inside each T/M lamellae and is accumulated at the TBs, which indicates that the plastic deformation in this region is mainly carried by extensive activity of perfect dislocations.

In comparison with the well preserved fine twin lamellae, in a local detwinning (region "II" , figure 3c), most of the fine twin lamellae of width less than 100 nm are annihilated and only coarse lamellae with lamellar thicknesses of 100 nm ~200 nm are detected. The SAED pattern (the inset in figure 3c) indicates that the coarse lamellae are still separated by TBs but that the zone axis deviates slightly from the [011] axis of the well preserved nanoscale twins (see inset in figure 3b). A high density of dislocations are accumulated at the interface between the detwinning region and the preserved nanoscale twins, lead to an obvious misorientation (arrows in figure 3c).

In contrast to the detwinning region (figure 3c), much more severe deformation takes place in the shear band, as shown in figure 3d (white square "III" in figure 3a), where the nanoscale twins are completely destroyed. Elongated sub-grains or dislocation cells are formed inside the shear band and lie along the direction of the shear bands, which are inclined to the preserved TBs.

To quantify the local plastic deformation associated with a given shear band or detwinning region, local shear strains in detwinning and shear bands are measured by using the curved TBs as markers [32]. Figure 4a and 4b are typical TEM images of a detwinning region and a shear band of 15% cold rolled Cu sample, in which the preserved TBs (red dotted lines in figure 4a and 4b) are used as a reference to calculate the local shear strains. A fiducial line is drawn following the TB and is divided into a series of segments of equal width, ***d***, in the direction perpendicular to the shear banding direction (figure 4c). The average shear strain associated with each segment is then calculated according to:

$$\gamma = d/\delta = ctg\theta - ctg\theta_0, \quad (2)$$

where the parameters d, δ, θ, θ_0 are defined as shown in figure 4c. θ_0 is the angle of the shear direction (*AA'*) to the direction of the original TB (*BB'*); θ is the angle of the shear direction (*AA'*) to the direction of curved the TB (*CC'*); δ is the width (transverse to the shear direction) of a unit segment, and d is the shear displacement of a unit segment.

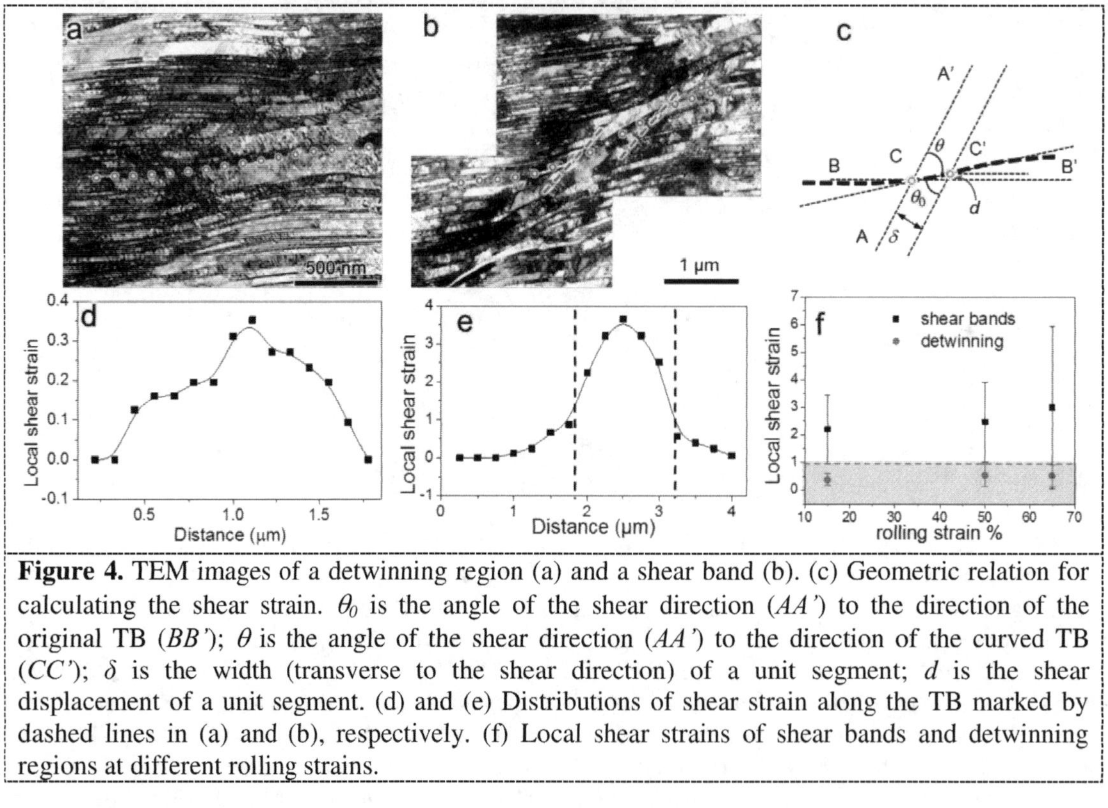

Figure 4. TEM images of a detwinning region (a) and a shear band (b). (c) Geometric relation for calculating the shear strain. θ_0 is the angle of the shear direction (*AA'*) to the direction of the original TB (*BB'*); θ is the angle of the shear direction (*AA'*) to the direction of the curved TB (*CC'*); δ is the width (transverse to the shear direction) of a unit segment; d is the shear displacement of a unit segment. (d) and (e) Distributions of shear strain along the TB marked by dashed lines in (a) and (b), respectively. (f) Local shear strains of shear bands and detwinning regions at different rolling strains.

Figures 4d and 4e indicate the distribution of the local shear strains in the detwinning region and the shear band, respectively. A single saddle-shape distribution of the local shear strains is observed in both the detwinning region and in the shear bands, i.e., the local shear strains rise up from almost zero on the edge to a peak in the core. The maximum shear strain in the center of the detwinning region approaches $\gamma_{max} = 0.4$ (figure 4d), whereas this value can be as large as 3.6 for the shear band (figure 4e).

Statistic results of local shear strains measurements in the detwinning region and in shear bands in the nt-Cu samples under different strains are summarized in figure 4f. It is found that the local shear strains in detwinning regions are less than 1, and are insensitive to the rolling strain. In contrast, in the shear bands, the local shear strains are larger than 1 and increase slightly with increasing rolling strain. This means that it is the local shear strain, rather than the nominal rolling strain that induces detwinning or the formation of shear bands during cold rolling of nt-Cu.

Based on the above SEM and TEM observations, it is found that both homogeneous and inhomogeneous deformations are observed in nt-Cu after a rolling strain of 15%. More than 80% of the nanotwins are preserved. Local shear deformation microstructures including detwinning and shear bands are already present at this strain level.

3.3. Microstructure characteristics of the as-rolled nt-Cu with a rolling strain of 50%

As the rolling strain is increased to 50%, non-uniform deformation is prevalent in the nt-Cu. As shown in the IPF map (figure 5a), a large amount of detwinning (white arrows) and shear bands (black arrows) form along ±45° to RD, which can be easily recognized by orientations deviated from the original <111> direction. As a result, the deformation structure presents a net-like morphology, where the original columnar grains are hard to identify. Nanoscaled twins with the original <111> orientation are cut into blocks (labeled as "NT"). The volume fraction of preserved nanoscaled twins is reduced to less than 30%. Compared to the nt-Cu sample after a rolling strain of 15%, shear bands run across many grains, and are wider and longer, and tend to lie parallel to RD direction. Moreover, the orientations within shear bands deviate 20-25° from the original <111> orientation, a much larger deviation than that in 15% cold rolled samples.

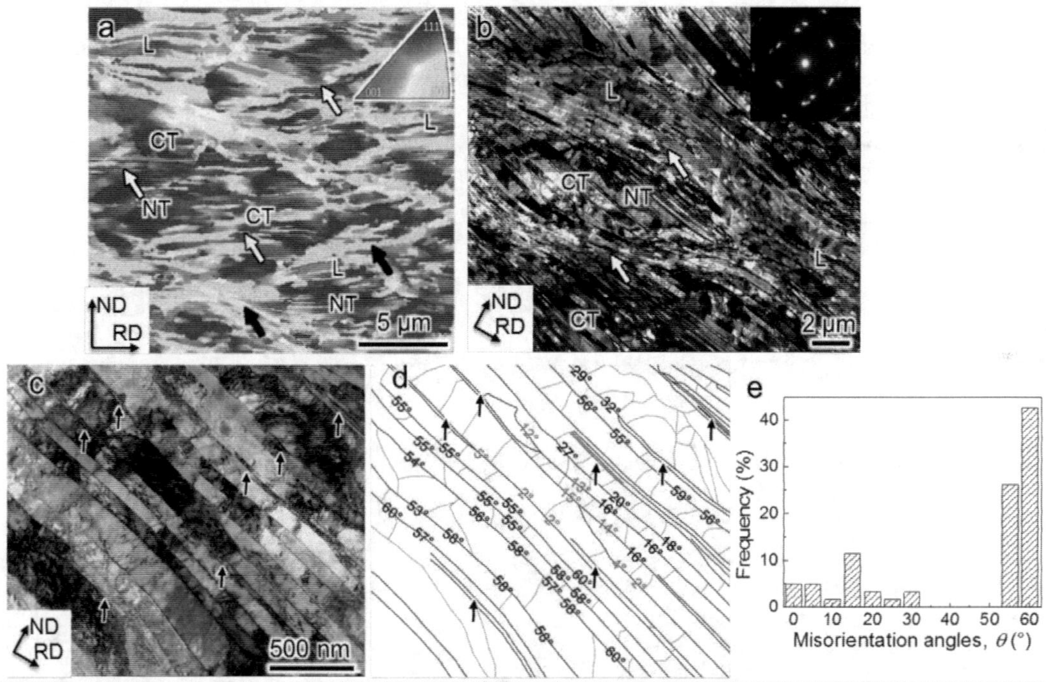

Figure 5. IPF map (a) and TEM image (b) of cold rolled nt-Cu sample after a strain of 50%. The inset in (b) shows the corresponding SAED pattern of the NT bundle. Nanoscaled twins, coarse T/M lamellae and laminated structure are labeled as "NT", "CT" and "L", respectively. The detwinning regions and shear bands are indicated by white and black arrows, respectively. (c) High magnification TEM image of the laminated structure. (d) Sketch corresponding to (c), where misorientation angles are marked across the lamellar boundaries. Fine twin lamellae fragments are indicated by black arrows in (c) and (d). (e) Misorientation angle distribution of the lamellar boundaries.

As shown in figure 5a, a large amount of coarse twins (CT), with a volume fraction of ~50%, are formed by extensive detwinning. The orientations of the detwinning regions are deviated by 10-12° from the original <111> orientation, which is almost the same as for the detwinning regions in the 15% rolled samples. In addition, a kind of newly formed laminated structure ("L"), with similar morphology and comparable lamellar thicknesses to that of the coarse T/M lamellae but without any

T/M relationship, is observed. The orientations in laminated structure deviate 20-25° from the original <111> orientation, larger than that of the coarse twin regions.

The TEM image of nt-Cu after a rolling strain of 50% in figure 5b indicates typical laminated structures along RD, with some preserved twin blocks, including nanoscale twins and coarse T/M lamellae surrounded by shear bands. The SAED pattern shows that some lamellar boundaries are no longer TBs but high angle boundaries (HABs). Sub-grains in shear bands are stretched and curved along the edge of twin blocks, presenting an elongated morphology similar to the laminated structures but finer.

The higher magnification image of the laminated structures in figure 5c reveals that the extended lamellar boundaries are sharp and flat with thicknesses of more than 100 nm. Numerous dislocations are present inside the lamellar channels, forming dislocation cells and interconnecting boundaries. The misorientation angles across the lamellar boundaries analyzed by Kikuchi diffraction (figure 5d) show a representative character of very high angle misorientations (larger than 50°) across the laminated boundaries. This is much higher than for lamellar boundaries produced by cold rolling of conventional coarse grained metals, but close to 60° for a coherent TB. A few fine twin lamellae fragments with twin thickness less than 50 nm (black arrows) can be observed. Two fragments of fine twin lamellae are connected by low angle boundaries (LABs), indicating the annihilation of fine twin lamellae. Extensive statistical analysis (figure 5e) indicates that the laminated structures are composed ~80% of HABs with more than 50° misorientation, and ~10% of LABs. The above observations suggest that these laminated structures are evolved from coarse T/M lamellae formed by detwinning.

Clearly, when the nt-Cu sample is rolled to a strain of 50%, detwinning and shear bands are extensively activated, cutting the original nanoscale twins into blocks. As a result a typical laminated structure, separated by HABs of more than 50°, begins to form.

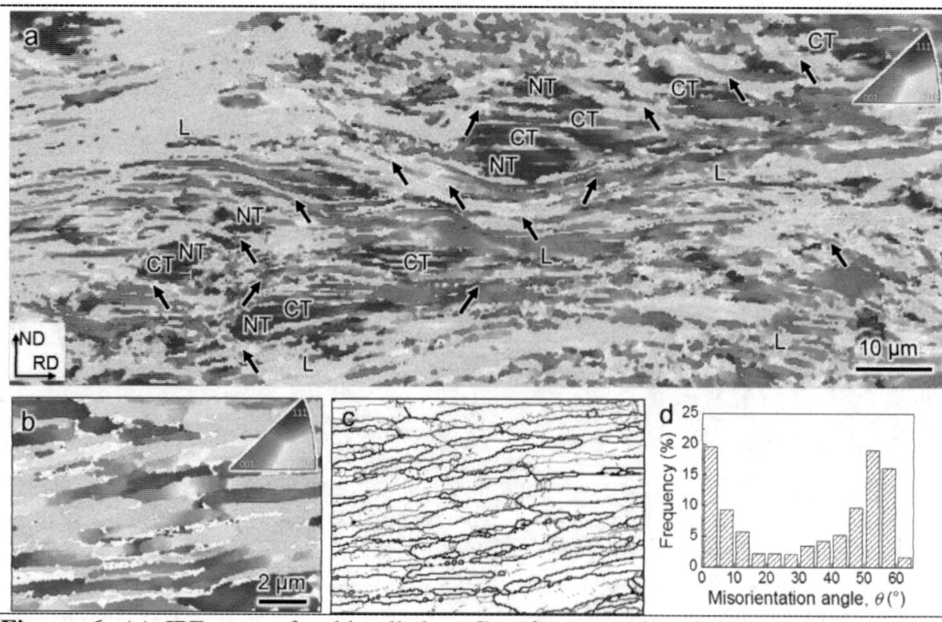

Figure 6. (a) IPF map of cold rolled nt-Cu after a strain of 65%. Shear bands are indicated by black arrows. (b) High magnification IPF map of the laminated structure. Shear bands are indicated by black arrows. (c) Boundary map corresponding to (b). In the boundary maps, green and black lines represent boundaries with misorientation angles θ, $\theta \leq 15°$, and $\theta > 15°$, respectively; red lines represent the $\Sigma 3$ TBs. (d) Misorientation angle distribution of the lamellar boundaries. The nanoscaled twins, coarse T/M lamellae and laminated structures are labeled as "NT", "CT' and "L", respectively.

3.4. Microstructure characteristics of the as-rolled nt-Cu with a rolling strain of 65%

As shown in the IPF map (figure 6a), as the rolling strain increases to 65%, the microstructure presents a wavy lamellar morphology. Most of the original nanoscale twins are replaced by laminated structures ("L") with a volume fraction of more than 70%. The original nanoscale twins are cut into twin blocks including most of the coarse T/M lamellae ("CT") and very few preserved nanoscale twins ("NT") by shear bands (black arrows in figure 6a) and laminated structures. The coarse T/M lamellae ("CT") deviate by 8-12° from the original <111> direction, whereas the nanoscale twins still maintain their original <111> direction. Sub-grains in the shear bands are further elongated and rotate, then evolve into laminated structures, slightly inclined to the RD.

A high magnification IPF map (figure 6b) and corresponding grain boundary map (figure 6c) reveal that the laminated structures are separated by longitudinal boundaries composed of many HABs of angle in the range $50° < \theta < 60°$ (blue lines in figure 6c), a few Σ3 coherent TBs (red lines in figure 6c) and few LABs of angle $\theta < 15°$ (green lines along RD in figure 6c) roughly parallel to RD. As a result, a statistical analysis of boundary misorientations (figure 6d) shows a bimodal distribution with peaks at 5° and 55°. Here the HABs corresponding to the peak at 55° come from the longitudinal boundaries evolving from the TBs, in accordance with the observations on the 50% rolled sample. The large number of LABs should, however, be attributed to dislocations cells and boundaries formed inside the lamellae (short green lines in figure 6c).

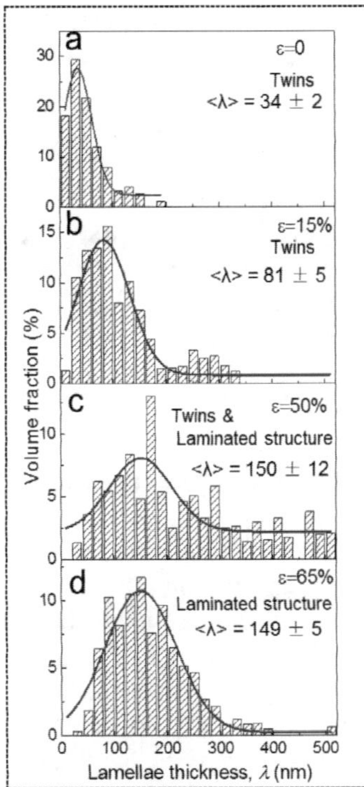

Figure 7. Lamellae thickness distributions of as deposited (a) and cold rolled nt-Cu samples at different strains of 15% (b) , 50% (c) and 65% (d).

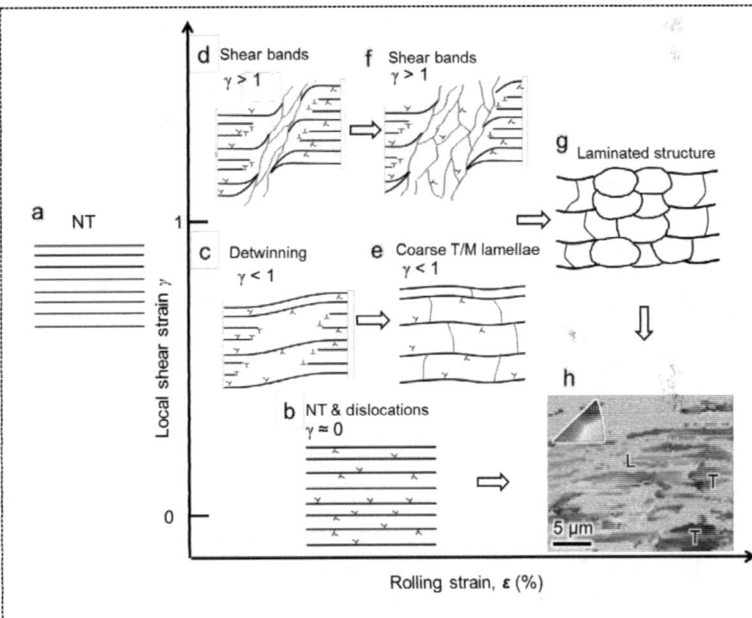

Figure 8. Schematic illustration of the deformation structure evolution during cold rolling of nt-Cu. The twin blocks and laminated structures are labeled as "T" and "L", respectively.

The distribution of lamellae thickness values for the laminated structure, as measured from TEM data in samples at different rolling strains is shown in figure 7. When detwinning begins to happen at 15% rolling strain, coarse T/M lamellae with an average thickness of ~80 nm and a wide distribution extending to more than 300 nm are formed. After universal detwinning, the average thickness of coarse T/M lamellae reaches 150 nm (figure 7c). The average lamellar thicknesses of the laminated structure, as well as of the coarse T/M lamellae, reaches a saturated value about 150 nm (figure 7c and 7d) during microstructure evolution to larger rolling strains. This also strongly supports the idea that the laminated structure evolves from the coarse T/M lamellae. It should be noted that this saturated lamellar thickness of the as-rolled nt-Cu is finer than that of lamellar boundaries (230 nm) formed in conventional coarse Cu cold rolled to a sufficiently high strain [34].

For the sample rolled to a strain of 65% the results show that a laminated structure containing HABs with misorientation angles of more than 50° and average thicknesses of ~150 nm embedded with a few preserved twin blocks are eventually formed.

4. Discussion

The structural evolution and deformation mechanism of polycrystalline metals have been well studied and reported in the literature [34-40]. It has been found that the microstructure evolution is one of continuous subdivision by dislocation boundaries, including incidental dislocation boundaries (IDBs) and geometrically necessary boundaries (GNBs), arising from the extensive forest dislocation activity with increasing strain [35-37]. The boundary alignment is inclined to the rolling direction at low strains in a manner closely related to the grain orientation [35]. At a strain of about 65% and above, this structure evolves into a saturated structure of lamellar dislocation boundaries almost parallel to the rolling plane [34, 35]. With increasing cold rolling strain, the boundary spacing decreases and can reach 50–300 nm, as exemplified in cold rolled Al, Ni, Cu and Fe [38-40]. At the same time, the misorientation angle across the boundaries increases substantially and reaches a saturated average value of 40° as the dislocation boundaries finally develop into HABs.

Different from plastic deformation mechanisms dominated by the interactions of forest dislocations in coarse grained face-centered cubic metals, the plastic deformation of nt-Cu is dominated by interactions between dislocations and TBs [7, 19]. The stress state of nt-Cu in this rolling deformation can be assumed as a combination of both compression in the ND (perpendicular to the TBs) and tension in the RD (parallel to the TBs). The uniform deformation will be accommodated by perfect dislocations transmitting across, and piling up at, the TBs or threading inside the T/M lamellae [41]. The uniform plastic strain is, however, very limited (~1%) for nt-Cu with fine grain sizes of less than 3 μm [3]. Both of these two kinds of dislocation behavior are heavily suppressed by the TBs with an average spacing in the nanoscale. Meanwhile, partial dislocations slipping along the TBs can hardly be activeted, due to the limited shear stress on the TBs. Consequently, the uniform deformation of nt-Cu is limited, and strain localization takes place ±45° with respect to the RD direction, where there is an obvious stress concentration during cold rolling.

Strain localization leads to numerous perfect dislocations transmitting and interacting with TBs by dislocation reactions, producing many Shockley partials [19, 41]. Both transmission of perfect dislocations, or pile ups on TBs and partial dislocations gliding along the TBs sustain the local shear strains. The motion of Shockley partials on the TBs leads to detwinning of fine twin lamellae, similar to the detwinning process observed in in-situ TEM experiments [19]. Detwinning contributes a shear strain of $b\langle 112\rangle/d\{111\}=0.707$ along the detwinning direction [42], which has a component of $0.707\cos\theta$ (figure 6c) along the shearing direction (inclined to the TBs). Combined with the contribution of perfect dislocations, the maximum local shear strain taken by detwinning reaches a value of about 1, as illustrated in the former observation.

After detwinning, further rolling strain is taken by the activity and interaction of perfect dislocations from different slip systems, instead of the activity of a dominant single slip system in the nanoscaled twins. Dislocations continuously accumulate on the TBs, breaking the T/B relationship. The coarse T/M lamellae finally evolve into a laminated structure with HABs of more than 50° misorientation angle and a comparable lamellar thickness to the coarse T/M lamellae. The HABs of the laminated structure originate from the initial TBs, in contrast to HABs formed in cold rolled coarse-grained metals, which instead originate from dislocation boundaries with increasing boundary misorientations as the rolling strain increases [40].

In the region with severe local shear strains (>1), accumulation of perfect dislocations and transmission across TBs dramatically happens under higher stress concentrations, leading to the destruction of the fine twins. The local shear strains are taken by continuous slip and interaction of perfect dislocations. As a result, shear bands containing dislocation cells, and sub-grains as well as a few broken twin lamellae are formed. Upon further rolling, the shear bands grow longer and wider under a high local shear strain. Further deformation in these shear bands is taken by the activation and interaction of dislocations. The dislocation cells, sub-grains and fragments of twin lamellae inside the shear bands eventually evolve into laminated structures, which are slightly inclined to RD. Similar to laminated structures evolving from coarse T/M lamellar, the laminated structures from shear bands are also composed of HABs resulting from the twin boundaries and of LABs coming from dislocation cells.

The evolution of the deformation microstructure of nt-Cu (figure 8a) under cold rolling is summarized in figure 8. At a very small rolling strains, uniform deformation assisted by dislocation slip inside the nanoscale twins and accumulation at TBs (figure 8b) dominates the plastic defamation. Strain localization takes place in some grains, taken by detwinnning (figure 8c) and shear bands (figure 8d), closely related to the local shear strain with a critical value about 1, regardless of local crystal orientation. Detwinning is substantially promoted in the grains with local shear strains <1, produced extensive coarse T/M lamellae (figure 8e). Additionally, the shear bands grow wider and longer (figure 8f) under a large local shear strain (>1). With further increase of rolling strain, strain localization intensifies. Finally, both the coarse T/M lamellae and the shear bands evolve into a saturated laminated structure containing HABs (misorientation angle larger than 50°) (figure 8g). The eventual deformation structure of the nanotwinned Cu sample under cold rolling consist of the massive laminated structures embedded with a few blocks of preserved twins (figure 8h).

5. Conclusions

We have systematically studied the microstructure evolution of highly-oriented nt Cu samples under cold rolling to strains of 15%, 50% and 65%. At a rolling strain of 15%, uniform deformation, carried by the interactions between dislocations and twin boundaries, dominates. At the same time, a small amount of strain localization, carried by detwinning and shear bands in a direction 45° respective to the rolling direction, take place. As the rolling strain is increased to 50%, detwinning becomes a dominant deformation mechanism and produces a large amount of coarse twin structure. As the rolling strain is further increased up to 65%, an extensive stable laminated structure, embedded with a few preserved twin blocks, is formed.

It is found that the local shear strain plays a critical role in the microstructure evolution of the nanotwinned Cu under cold rolling. When the local shear strains are so small as to be negligible, uniform deformation is accommodated by the interactions of dislocations with twin boundaries, and the nanoscale twins survive. As the local shear strains reach up to 1 (<1), detwinning prevails and results in coarse twin/matrix lamellae. When the local shear strain is larger than 1, the formation of shear bands, composed of dislocation cells, sub-grains and fragments of twin lamellae, takes place instead of detwinning, eventually evolving to produce an extensive laminated structure.

This study sheds lights on the understanding on the deformation physics of nt-metals under complex stress states and large strain, in particular regarding the heterogeneous deformation behavior including detwinning and formation of shear bands in nt-metals.

Acknowledgments

L. Lu acknowledges the financial support by the National Science Foundation of China (Grant Nos. 51471172, 51420105001 and U1608257) and the Key Research Program of Frontier Science, CAS. We thank Mr. S. Jin for his assistance in sample preparation.

6. References

[1] Lu L, Shen YF, Chen XH, Qian LH, Lu K 2004 *Science*, **304** 422-26.
[2] Zhang X, Wang H, Chen XH, Lu L, Lu K, Hoagland RG, Misra A 2006 *Appl. Phys. Lett.*, **88** 173116.
[3] You ZS, Lu L, Lu K 2011 *Acta Mater.*, **59** 6927-37.
[4] Pan Q, Zhou H, Lu Q, Gao H, Lu L 2017 *Nature*, **551** 214-17.

[5] Lu L, Chen X, Huang X, Lu K 2009 *Science*, **323** 607-10.
[6] Hodge AM, Wang YM, Barbee TW Jr 2008 *Scr. Mater.*, **59** 163-66.
[7] You ZS, Li XY, Gui L J, Lu QH, Zhu T, Gao HJ, Lu L 2013 *Acta Mater.*, **61** 217-27.
[8] Lu Q, You Z, Huang X, Hansen N, Lu L 2017 *Acta Mater.*, **127** 85-97.
[9] Dao M, Lu L, Shen YF, Suresh S 2006 *Acta Mater.*, **54** 5421-32.
[10] Li X, Wei Y, Lu L, Lu K, Gao H 2010 *Nature*, **464** 877-80.
[11] Jin ZH, Gumbsch P, Albe K, Ma E, Lu K, Gleiter H, Hahn H 2008 *Acta Mater.*, **56** 1126-35.
[12] Jin ZH, Gumbsch P, Ma E, Albe K, Lu K, Hahn H, Gleiter H 2006 *Scr. Mater.*, **54** 1163-68.
[13] Chen XH, Lu L, Lu K 2011 *Scr. Mater.*, **64** 311-14.
[14] Shen YF, Lu L, Lu QH, Jin ZH, Lu K 2005 *Scr. Mater.*, **52** 989-94.
[15] Jia D, Ramesh K T, Ma E, Lu L, Lu K 2001 *Scr. Mater.*, **45** 613-20.
[16] Shen YF, Lu L, Dao M, Suresh S 2006 *Scr. Mater.*, **55** 319-22.
[17] Wang HT, Tao N R, Lu K 2012 *Acta Mater.*, **60** 4027-40.
[18] Zhang X, Misra A, Wang H, Nastasi M, Embury JD, Mitchell TE, Hoagland RG, Hirth J P 2004 *Appl. Phys. Lett.*, **84** 1096-98.
[19] Wang J, Li N, Anderoglu O, Zhang X, Misra A, Huang JY, Hirth JP 2010 *Acta Mater.*, **58** 2262-70.
[20] Lu QH, Sui ML, Huang XX, Li DX, Hansen N 2014 *Philos. Mag.*, **94** 2262-80.
[21] Hodge AM, Furnish TA, Shute CJ, Liao Y, Huang X, Hong CS, Zhu YT, Barbee T W, Weertman J R 2012 *Scr. Mater.* **66** 872-77.
[22] Liu Y, Jian J, Chen Y, Wang H, Zhang X 2014 *Appl. Phys. Lett.*, **104** 231910.
[23] Cao Y, Wang YB, Chen ZB, Liao XZ, Kawasaki M, Ringer SP, Langdon TG, Zhu YT 2013 *Mater. Sci. Eng. A*, **578** 110-14.
[24] Lee S, Im J, Yoo Y, Bitzek E, Kiener D, Richter G, Kim B, Oh SH 2014 *Nat. Commun.*, **5** 3033.
[25] Cheng G, Yin S, Chang TH, Richter G, Gao H, Zhu Y 2017 *Phys. Rev. Lett.*, **119** 256101.
[26] Hong CS, Tao NR, Huang X, Lu K 2010 *Acta Mater.*, **58** 3103-16.
[27] Wang J, Misra A, Hirth JP 2011 *Phys. Rev. B*, **83** 064106.
[28] Li N, Wang J, Zhang X, Misra A 2011 *JOM*, **63** 62.
[29] Hodge AM, Furnish TA, Navid AA, Barbee TW Jr. 2011 *Scr. Mater.*, **65** 1006-09.
[30] Duggan BJ, Hatherly M, Hutchinson WB, Wakefield PT 1978 *Met. Sci.*, **12** 343-51.
[31] Xiao GH, Tao NR, Lu K 2009 *Mater. Sci. Eng. A*, **513-514** 13-21.
[32] Hong CS, Tao NR, Huang X, Lu K 2010 *Acta Mater.*, **58** 3103-16.
[33] Liu Q 1994 *J. Appl. Crystallogr.*, **27** 755-61.
[34] Takata N, Yamada K, Ikeda K-i, Yoshida F, Nakashima H, Tsuji N 2007 *Mater. Trans.*, **48** 2043-48.
[35] Huang X, Winther G 2007 *Philos. Mag.*, **87** 5189-14.
[36] Winther G, Huang X 2007 *Philos. Mag.*, **87** 5215-35.
[37] Bay B, Hansen N, Hughes DA, Kuhlmannwilsdorf D 1992 *Acta Metall. Mater.*, **40** 205-19.
[38] Liu Q, Huang X, Lloyd DJ, Hansen N 2002 *Acta Mater.*, **50** 3789-02.
[39] Malin AS, Hatherly M 1979 *Met. Sci.*, **13** 463-72.
[40] Hughes DA, Hansen N 2000 *Acta Mater.*, **48** 2985-3004.
[41] Zhu YT, Wu XL, Liao XZ, Narayan J, Kecskes LJ, Mathaudhu SN 2011 *Acta Mater*, **59** 812-21.
[42] De Cooman BC, Estrin Y, Kim SK 2018 *Acta Mater.*, **142** 283-362.

IOP Conf. Series: Materials Science and Engineering **580** (2019) 012009 doi:10.1088/1757-899X/580/1/012009

Understanding the evolving state of deforming polycrystals using synchrotron x-rays

Matt Miller[1,2], Chris Budrow[1,2], Tim Long[1] and Mark Obstalecki[3]

[1] Sibley School of Mechanical and Aerospace Engineering, Upson Hall, Cornell University, Ithaca, NY 14853

[2] Cornell High Energy Synchrotron Source, Wilson Laboratory, Synchrotron Drive, Ithaca, NY 14853

[3] Air Force Research Laboratory, Wright-Patterson AFB, OH 45433

E-mail: mpm4@cornell.edu

Abstract. The multiscale nature of irreversible processes like plasticity make them particularly challenging. One must be looking everywhere, all the time to gain a predictive understanding of plasticity–related processes such as fatigue crack initiation. High energy (HE) synchrotron x-ray diffraction is a unique tool for characterizing the state of an entire aggregate of metallic crystals with sub-crystal resolution. Combined with *in situ* loading, HE x-ray diffraction provides experimental hope for characterizing the conditions leading to the initiation of a crack. In this work, we conducted a HE x-ray diffraction experiment on a commercially pure copper sample subjected to *in situ* cyclic loading at the F2 Station at the Cornell High Energy Synchrotron Source (CHESS). We observed that the evolution of lattice orientation distributions within each crystal evolved as the specimen was loaded through complete fatigue cycles. We defined a "size" of these orientation clouds, which we called $\bar{\Theta}$. Changes in $\bar{\Theta}$ are related to the plasticity-induced state of each crystal. We saw significant evolution of the $\bar{\Theta}$ distributions as the sample went through cycle 2. By cycle 256, the evolution of the $\bar{\Theta}$ distribution was significantly reduced - consistent with the saturation of cyclic hardening that we see on the macroscale. We also saw that the 40 crystals with the largest values of $\bar{\Theta}$ in the first half cycle had a larger amount of $\bar{\Theta}$ accumulation on average over the 256 cycles than the rest of the crystals in the aggregate.

1. Introduction

High energy (HE) synchrotron x-ray methods have significantly impacted the scope of the scientific questions we are able to explore. Working together with x-ray scientists and the manufacturers of sophisticated specimen interrogation tools and ever faster x-ray detectors, the structural metals community has transformed x-ray science *experiments* into a new generation of materials characterization *measurements*. HE x-ray diffraction experiments employing *in situ* loading are particularly well–suited for studying loading history–dependent processes such as plasticity. Indicators of material state provided by traditional "before and after" 2D snapshots of material microstructure from a micrograph or an electron backscatter diffraction (EBSD) map are greatly augmented by the real–time measures of the distributions of lattice spacings (strains) and lattice orientations at the size scale of a polycrystalline aggregate and below that can be obtained using HE x-ray diffraction. This is particularly true for processes like fatigue crack initiation, which are among the most challenging to understand, model and ultimately predict

IOP Conf. Series: Materials Science and Engineering **580** (2019) 012009 doi:10.1088/1757-899X/580/1/012009

because of their seemingly random nature. The ability to "track" the plasticity of each crystal during each cycle could lead to the understanding necessary to predict the initiation of a fatigue crack. A state feature (state variable) with subgrain scale resolution that would provide coverage over an aggregate is necessary; something that would mark grains that are more susceptible to crack initiation by virtue of the fact that they have incurred "more plasticity". This paper describes a study of the evolution of an aggregate of commercially pure copper crystals subjected to cyclic loading conditions consistent with low cycle fatigue (LCF); each fatigue cycle contains macroscopically elastic-plastic deformation. The uniaxial test specimen initially loads elastically and yields in tension, then unloads elastically and re–yields in compression. The macroscopic cyclic stress-strain curve forms characteristic hysteresis loops that change in size (typically increase) with each cycle reflecting the plasticity-induced material evolution that is taking place within each crystal. It has been hypothesized that at some point, the to-and-fro plastic slip within one crystal or a group of crystals initiates a fatigue crack. Using HE x-ray diffraction we studied the cyclic plasticity within every deforming crystal by measuring how the lattice *misorientation* changes as the sample is loaded around the hysteresis loop.

2. Experiments and Results

X-ray diffraction experiments were performed at the F2 beamline at the Cornell High Energy Synchrotron Source (CHESS). The sample was electrical discharge machined from 6.35 mm diameter commercially pure copper rod. The square gauge region had a cross-section measuring 1 mm by 1 mm. After manufacturing, the sample was heat treated in a nitrogen environment at 750° F for two hours. A GE area detector was utilized with an x-ray energy of 61.332 keV and a sample to detector distance of about 850 mm. During this experiment, a total of five, 125 micron tall layers were scanned at each load step, such that the previously mapped volume was contained within the far-field data. The sample was deformed for a total of 256 tension-compression cycles at a fixed strain amplitude of 0.3%. Cycling of the sample was performed under displacement control and the macroscopic strain was monitored using digital image correlation (DIC). The displacement endpoints at the hysteresis loop tips were periodically adjusted to maintain a consistent strain amplitude based on the DIC data. Figure 1 shows the hysteresis loops at cycles 2 and 256 along with the load steps where diffraction measurements were made. The differences between the size of the loops is indicative of the cyclic loading–induced evolution (cyclic hardening) that is a characteristic of commercially pure copper.

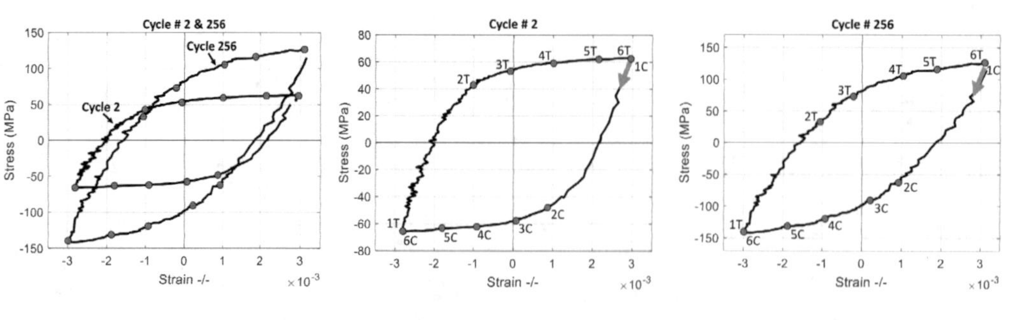

Figure 1. Macroscopic stress-strain data from the commercially pure copper sample deformed *in situ* cyclically in uniaxial tension/compression. Cycles 2 and 256 are depicted. Load steps where HE x-ray diffraction experiments were performed are also shown as red circles. (Note: load step 1T is the same as load step 6C but 1C and 6T are different.)

2.1. Diffraction data

Diffraction experiments (far field) were performed at 11 points (load steps) around the stress-strain hysteresis loops. The first load step (1C) is at the hysteresis loop tip corresponding to the maximum tensile strain in the sample, 0.3%. Load step 2C roughly corresponds to the specimen yield point, then 3 load steps approximately evenly spaced in strain followed by load step 6C, which was taken at the minimum strain point. Load step 1T is the same as 6C and the process is repeated on the tensile-going half of the loop. Note that point 6T is different than 1C. At each load step, deformation of the sample was halted and the load on the specimen was reduced by approximately 10%. Data from the approximately 350 crystals within the aggregate were gathered by rotating the sample around the loading axis. Diffraction data were collected continuously on a GE area detector and separated into 0.25^o bins. Full 360° rotations, which enable collection of all peaks associated with a crystal, were made possible using the RAMS loading system, which is resident at CHESS F2 [1].

We are interested in understanding the evolution of plastic straining in each deforming crystal by examining the evolution of intragrain lattice misorientations. In previous work, we have connected the azimuthal spread of each diffraction peak associated with a crystal as an indicator of misorientation [2, 3, 4, 5]. This work employs a different methodology. Instead of fitting each diffraction peak, the orientation spread within each grain is estimated though a forward projection approach by employing a virtual diffractometer [6]. The set of orientations associated with each grain can then be examined. The orientations within one crystal at the tensile and compressive ends of hysteresis loops 2 and 256 are shown in Figure 2. Rodrigues orientation space - parameterized with the Rodrigues vector, $\mathbf{r} = \mathbf{n} \tan \frac{\phi}{2}$, where $\mathbf{n}$ is the rotation axis and ϕ is the rotation angle - is discretized in Figure 2 and each subvolume containing mapped detector intensity is shaded. Predicted locations of intensity on the detector for each discretized point (defining a unique orientation) in Rodrigues space are compared to the experimental detector images [7]. Changes between tension and compression in cycle 2 - the cloud actually shrinks when the sample plastically deforms in compression - and the increase in the size of the regions (orientation clouds) with little difference in tension as compared to compression in cycle 256 are visually evident. A simple measure to quantify the "size" of the cloud for each crystal at each load step can be defined as

$$\bar{\Theta} = \frac{2}{N}\left(\sum_{i=1}^{N} \tan^{-1} |\mathbf{w}_i|\Big|_{\text{load step}} - \sum_{i=1}^{N} \tan^{-1} |\mathbf{w}_i|\Big|_{\text{initial}}\right) \qquad (1)$$

where $\mathbf{w}_i = \mathbf{r}_i - \bar{\mathbf{r}}$ is the misorientation of each subvolume relative to the current average orientation of the grain, $\bar{\mathbf{r}}$. Figure 3 and Figure 4 depict the $\bar{\theta}$ histograms at each load step in cycles 2 and 256.

3. Discussion

We associate $\bar{\Theta}$ with the "amount" of plasticity–induced lattice misorientation present within a particular crystal - $\bar{\Theta}$ is an indicator of the plastic *state* of the crystal and changes in $\bar{\Theta}$ from one load step to the next are indicative of the changes in that state due to plasticity–induced orientation evolution within a crystal. In the data, we see $\bar{\Theta}$ values getting larger and smaller - both are indicative of plastic deformation. By definition, the $\bar{\Theta}$ *distributions* shown in Figure 3 and Figure 4 are snapshots of the state of each and every crystal at each load step; the attributes of these histograms are related to the *state* of the aggregate. As can be seen, during cycle 2 there are significant changes in the histograms as the loading proceeds around the hysteresis loop. Beginning at the tensile loop tip (2-1C) the large positive tail on the distribution is seen to contract until by 2-5C it looks nearly Gaussian. It then spreads before the stress-strain curve hits the compressive loop tip (2-6C=2-1T). Evolution during the tensile-going part of the loop

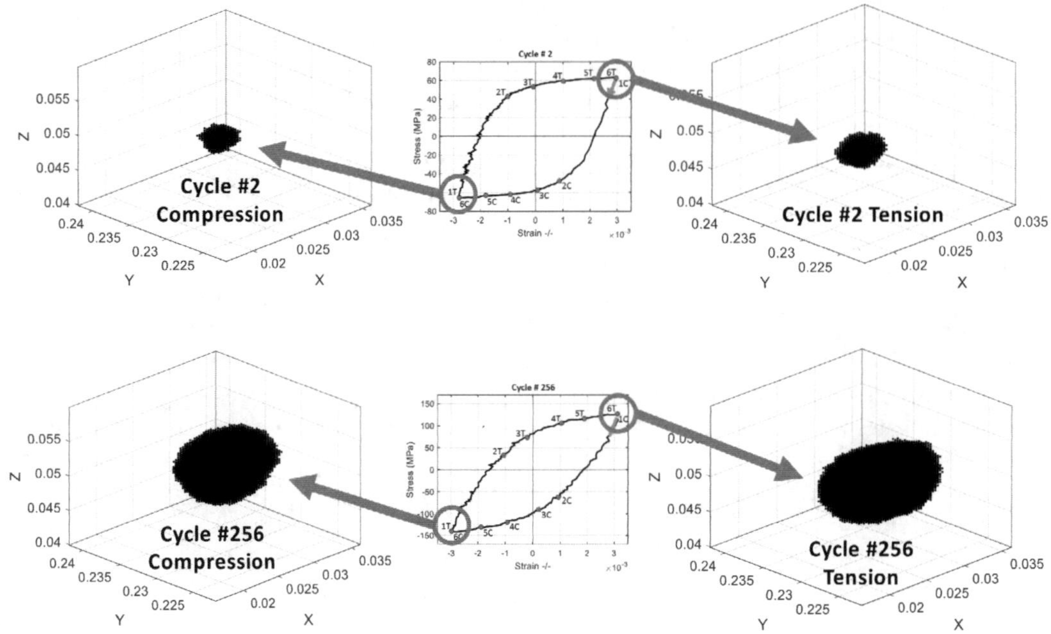

Figure 2. Orientation clouds within one crystal inside the deforming copper aggregate at the hysteresis loop tips during cycles 2 and 256. The clouds are depicted in Rodrigues orientation space.

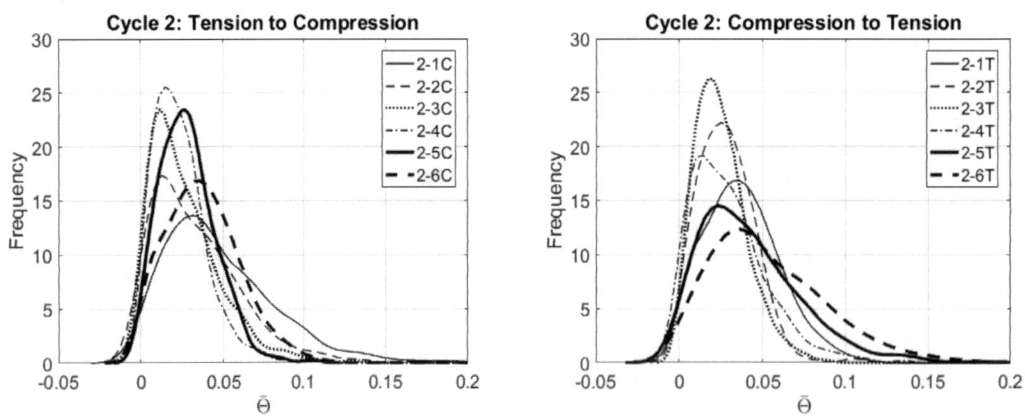

Figure 3. Histograms of the misorientation metric, $\bar{\Theta}$, for the crystals within the copper aggregate for cycle 2. (See Figure 1 for locations of the load steps.)

is somewhat similar, with the distribution actually narrowing from 2-1T to 2-2T and to 2-3T as the specimen re–yields in tension, then broadens again by 2-4T and eventually taking on a shape at 2-6T that is similar to but different than 1C, where the cycle started. The histograms in Figure 4 are much broader by cycle 256 at every point around the loop and the evolution of the histograms from one load step to the next is much more muted.

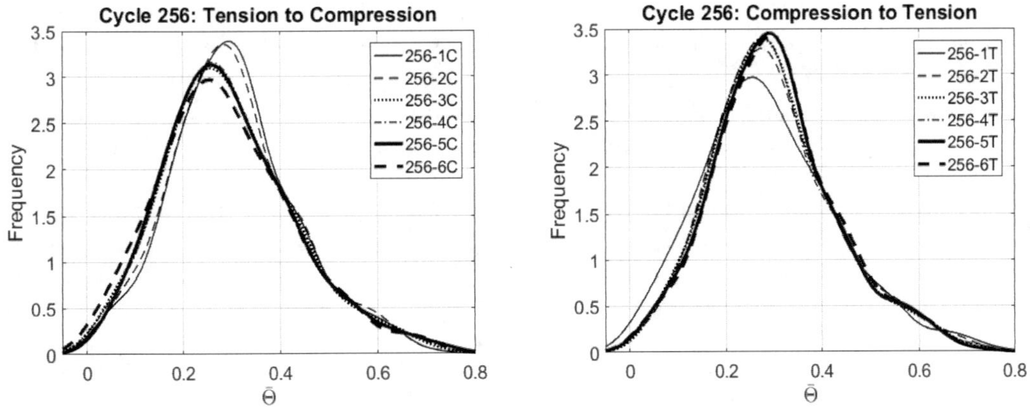

Figure 4. Histograms of the misorientation metric, $\bar{\Theta}$, for the crystals within the copper aggregate for cycle 256. (See Figure 1 for locations of the load steps.)

3.1. Upper Tail of the Initial $\bar{\Theta}$ *Distribution*

One of the most important developments from the recent HE x-ray diffraction research on polycrystalline metals is the ability to quantify how the crystal-scale stresses and strains – and underlying crystal states – vary from one crystal to another and how those crystal states evolve with plastic deformation. One crystal or set of crystals within the copper aggregate will eventually accumulate plastic slip, become more prone to crack initiation and a crack will form. The overarching goal of this project is predicting that process. With the hypothesis that "more slip" will lead to crack initiation, the first crystals to examine as likely crack initiation sites are those experiencing the largest inelastic strains. An interesting set of crystals to track, therefore, are those that yield first and accumulate the most plastic strain during the first half cycle. The $\bar{\Theta}$ distributions at the hysteresis loop tips of cycle 1, (1-1C and 1-1T), are shown in Figure 5 along with the first loop tip of cycle 2 (2-1C). The $\bar{\Theta}$ distribution at the 1-1C load step is due to the initial loading; the crystals in the upper tail were those most susceptible to slip in the initial state of the specimen. This seems a likely set of crystals to track in terms of $\bar{\Theta}$, so we begin our search for the most *fatigue vulnerable* crystals by examining the 40 crystals with the largest $\bar{\Theta}$ values at point 1-1C. We call those crystals the *Top 40*. The rest of the crystals we label as *Bulk*. Figure 5 also depicts how the characteristic "long upper tail" in the 1C $\bar{\Theta}$ distributions (from both cycles 1 and 2) could actually be the superposition of 2 Gaussian distributions, the Top 40 and the Bulk.

At load step 1-1C, we identify the Top 40 and Bulk crystals then compute the mean $\bar{\Theta}$, $\bar{\Theta}|_{\text{avg}}$, for both groups of crytals at the available load steps in cycles 1, 2 and 256. Since we are actually interested in how plastic slip increases in each crystal during a cycle, changes in $\bar{\Theta}|_{\text{avg}}$ from one load step to the next and how those changes accumulate during a cycle are most interesting. We therefore define the sum of the absolute value of each *increment* of $\bar{\Theta}|_{\text{avg}}$ from one load step to the next within a cycle by creating a quantity, $\bar{\Sigma}_L$, for load step L:

$$\bar{\Sigma}_L = \bar{\Theta}|^1_{\text{avg}} + \sum_{i=2}^{L} \left|\bar{\Theta}|^i_{\text{avg}} - \bar{\Theta}|^{i-1}_{\text{avg}}\right| \quad (2)$$

Values for $\bar{\Theta}|_{\text{avg}}$ and $\bar{\Sigma}$ at each load step are given in Table 1 and $\bar{\Sigma}$ is plotted vs. load step in Figure 6.

From Figure 6 we see a much larger increase in $\bar{\Sigma}$ in the Top 40 crystals compared to the Bulk. The $\bar{\Sigma}$ curves become roughly parallel until a slope change in the Top 40 curve at 2-4T, which corresponds to a reversal in the $\bar{\Theta}|_{avg}$ data trend, which can be seen in Table 1. In terms of the underlying cloud data, this reversal means that, on average, the clouds stopped getting smaller and began getting larger again. There was also a reversal in $\bar{\Theta}|_{avg}$ in the compressive-going data between 2-5C and 2-6C but this didn't produce any noticeable slope change in the $\bar{\Sigma}$ data. During cycle 256, the $\bar{\Sigma}$ data remain roughly parallel - separated by nearly the same amount as at load step 2-6T taken 254 cycles earlier. Note, however, that this may not mean that the differences in the $\bar{\Sigma}$ values that we see in cycle 256 represents all the changes in $\bar{\Theta}|_{avg}$ that accumulated between cycles 2 and 256. Reversals of the change in $\bar{\Theta}|_{avg}$ similar to what we saw in cycle 2 are most likely present in other cycles - especially early cycles - these reversals could produce additional increases in $\bar{\Sigma}$.

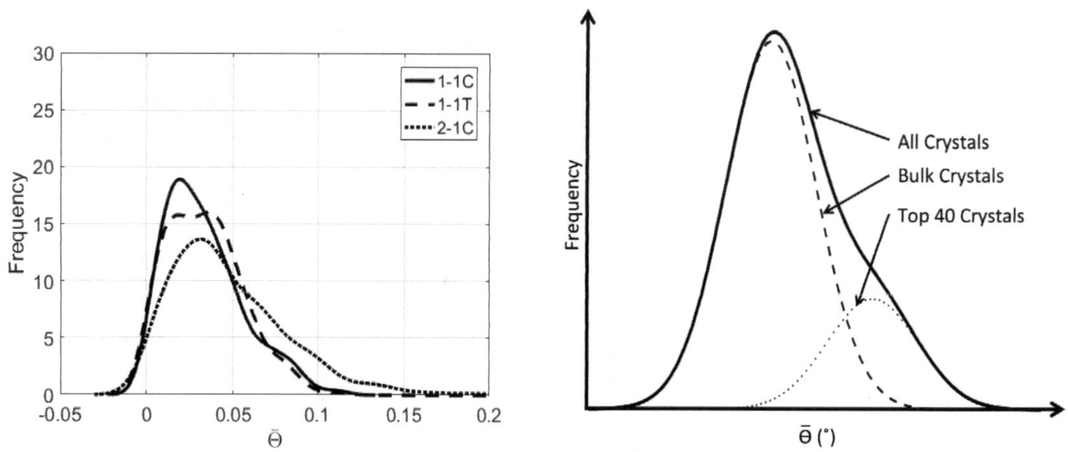

Figure 5. (Left) Cycle 1 $\bar{\Theta}$ distributions at the hysteresis loop tips. (Right) Schematic depicting how the 1C $\bar{\Theta}$ histogram could be decomposed into 2 Gaussian distributions: the Top 40 crystals and the Bulk crystals.

Table 1. $\bar{\Theta}|_{avg}$ and $\bar{\Sigma}$ in degrees for the Top 40 and Bulk sets of crystals.

Load	Cycle 1				Cycle 2				Cycle 256			
Step	Top 40		Bulk		Top 40		Bulk		Top 40		Bulk	
	$\bar{\Theta}\|_{avg}$	$\bar{\Sigma}$	$\bar{\Theta}\|_{avg}$	$\bar{\Sigma}$	$\bar{\Theta}\|_{avg}$	$\bar{\Sigma}$	$\bar{\Theta}\|_{avg}$	$\bar{\Sigma}$	$\bar{\Theta}\|_{avg}$	$\bar{\Sigma}$	$\bar{\Theta}\|_{avg}$	$\bar{\Sigma}$
1C	0.0787	0.0787	0.0272	0.0272	0.1097	0.1894	0.0395	0.0395	0.3775	0.6070	0.2937	0.3603
2C	-	-	-	-	0.0826	0.2165	0.0269	0.0520	0.3832	0.6126	0.2916	0.3624
3C	-	-	-	-	0.0610	0.2381	0.0221	0.0568	0.3799	0.6159	0.2828	0.3712
4C	-	-	-	-	0.0492	0.2499	0.0207	0.0582	0.3762	0.6196	0.2836	0.3720
5C	-	-	-	-	0.0375	0.2616	0.0271	0.0645	0.3718	0.6240	0.2862	0.3745
6C = 1T	0.0389	0.1185	0.0342	0.0342	0.0445	0.2687	0.0376	0.0750	0.3686	0.6272	0.2774	0.3832
2T	-	-	-	-	0.0419	0.2714	0.0271	0.0854	0.3732	0.6318	0.2910	0.3968
3T	-	-	-	-	0.0459	0.2754	0.0229	0.0896	0.3717	0.6333	0.2887	0.3990
4T	-	-	-	-	0.0698	0.2993	0.0240	0.0906	0.3716	0.6334	0.2870	0.4007
5T	-	-	-	-	0.0993	0.3288	0.0352	0.1018	0.3745	0.6363	0.2910	0.4047
6T	0.1097	0.1894	0.0395	0.0395	0.1234	0.3529	0.0452	0.1119	0.3783	0.6401	0.2936	0.4073

3.2. Crystal Stresses, Strain Hardening and Crystal-Based Modeling

Of course plastic straining, in general, and the heterogeneity of plastic slip on the crystal scale, in particular, are intimately coupled to the stress experienced by the crystal as the sample is cycled.

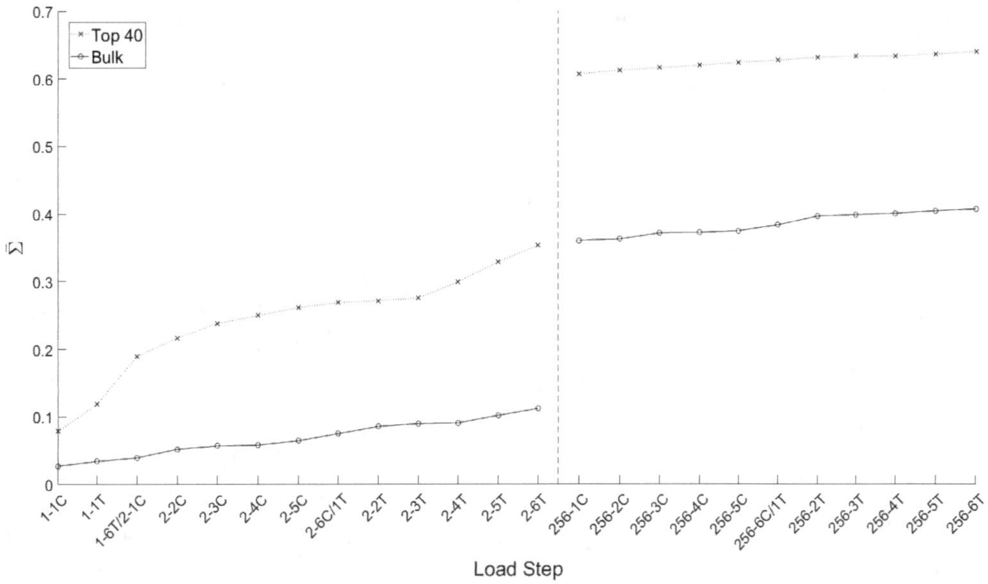

Figure 6. $\bar{\Sigma}$, the accumulated average $\bar{\Theta}$ within the Top 40 and Bulk crystals, at each load step around the hysteresis loops.

In pure copper - due to its high degree of elastic anisotropy and complex dislocation interaction potential - the fully three-dimensional stress states at the crystal scale, even under uniaxial loading conditions, can be quite complicated and can evolve significantly during deformation. The high degree of cyclic strain hardening manifests on the macroscale as the differences in the cycle 2 and cycle 256 hysteresis loops presented in Figure 1. For a truly comprehensive understanding of the factors leading to fatigue crack initiation, lattice strains (and resulting crystal stress tensors) should be collected along with the orientation clouds. Unfortunately, due to the low strength of the pure copper, the lattice strains acquired during our experiment were very near the resolution level of the HE x-ray diffraction method - especially during cycle 2 - so no lattice strain or crystal stress data are presented. The question of how to bring the various aspects of the diffraction data - orientation clouds and crystals stresses, for instance - remains, however, The "best", most rational use of the diffraction data for truly predicting fatigue crack initiation is in conjunction with a crystal-based deformation model. Comparisons between measured and simulated clouds and stresses can be made most readily. In simulations conducted in parallel with our experiments, the organization of crystals into persistent slip networks is proposed as an initiation mechanism [8]. We will explore this hypothesis in future work.

4. Summary, Conclusions and Future Work

We conducted HE x-ray diffraction experiments *in situ* during cyclic loading experiments on a polycrystalline sample containing over 300 commercially pure copper crystals. The focus of this work is on the plasticity–induced fatigue crack initiation that occurs in ductile metals such as copper; the to and fro plastic slip that occurs within a copper crystal during cycling has long been associated with crack initiation. The main goal of this particular study was to use the diffraction data and a virtual diffractometer employing a forward projection algorithm to approximate the evolution of the cloud of orientations within each crystal at load steps around the hysteresis loops. Changes in the size and shape of the orientation cloud associated with a crystal indicate that plastic straining has occurred and, to a first approximation, the magnitude of the cloud change is related to the amount of accumulated plastic strain. We created a scalar measure of cloud size, $\bar{\Theta}$, to track the plastic state of the aggregate as loading proceeded around hysteresis loops of a $\pm$ 0.3% uniaxial cyclic test. Histograms of $\bar{\Theta}$ early in the life of the sample (cycles 1 and 2) and later (cycle 256) were presented and the average $\bar{\Theta}$ value, $\bar{\Theta}|_{avg}$, from the 40 crystals in the upper tail of the histogram from the first cycle were tracked over cycles 2 and 256. We saw significant evolution of the $\bar{\Theta}$ distributions in cycle 2 and almost no evolution in cycle 256; evolution on the crystal scale mimics the cyclic hardening we see on the macroscale in the commercially pure copper. The magnitude of the changes in $\bar{\Theta}|_{avg}$ from one load step to the next, $\bar{\Sigma}$, was computed and plotted. The $\bar{\Sigma}$ values for the Top 40 crystals began the experiment larger than the Bulk crystals and were larger at cycle 256 indicating that on average, the crystals that yielded and plastically deformed the most at the start of the experiment, ended the experiment having accumulated "more" plasticity than the other crystals in the aggregate.

We introduced $\bar{\Theta}$ and the Top 40 as a first step to studying crystals we feel might be most susceptible to accumulating plastic strain. The HE x-ray dataset is incredibly rich and we have the details of the cloud distributions, as well as the average orientation of every crystal and their locations. We are currently examining the individual crystal data to understand vulnerability to fatigue crack initiation on a crystal by crystal basis. We are examining orientation dependence on $\bar{\Theta}$ and the role that crystal topology plays. Are the Top 40 crystals somehow organized so that persistent slip networks form [8]? Also, since fatigue crack initiation is a surface phenomena, how many vulnerable grains are on the surface of the sample? We will build on the concept of the orientation clouds and especially $\bar{\Theta}$ in the future - along with incorporating simulation results - as we continue to create new ways to use HE x-ray data to create understanding of complex phenomena like fatigue crack initiation.

5. Acknowledgments

Funding for this work has been provided by the US Department of Energy, Materials Sciences and Engineering Division, Basic Energy Sciences under Grant No. DE-SC0018901 (Dr. John Vetrano, Program Manager). This work is based upon research conducted at the Cornell High Energy Synchrotron Source (CHESS) which is supported by the National Science Foundation under award DMR-1332208.

References

[1] Shade P et al 2015 *Rev. Sci. Inst.* **86**(9):093902
[2] Obstalecki M, Wong S, Dawson P, and Miller M 2014 *Acta Mater* **75** 259
[3] Wong S, Obstalecki M, Miller M and Dawson P 2015 *J. Mech. Phys Sols* **79** 157
[4] Carson R, Obstalecki M, Miller M and Dawson P 2017 *Mod. Sim. Mats. Sci. Eng.* **25** (5):055008
[5] Obstalecki M, Carson R, Dawson P, and Miller, M 2019 *Mat. Char.* Submitted
[6] Wong S, Park J, Miller M and Dawson P 2013 *Comp. Mat. Sci.* **77** 456
[7] Obstalecki M 2017 Ph.D. Dissertation
[8] Carson R and Dawson P 2019 *J. Mech. Phys Sols* **126** 1

IOP Conf. Series: Materials Science and Engineering **580** (2019) 012010 doi:10.1088/1757-899X/580/1/012010

Following Microstructures during Deformation: In situ X-ray/Neutron Diffraction and HRDIC

E Polatidis[1,2], K Sofinowski[1,3], W-N Hsu[1,3], H Van Swygenhoven[1,3]

[1] Photons for Engineering and Manufacturing Group, Swiss Light Source, Paul Scherrer Institute, Villigen-PSI 5232, Switzerland
[2] Currently: Laboratory for Neutron Scattering and Imaging (LNS), Paul Scherrer Institute, Villigen-PSI 5232, Switzerland
[3] Neutrons and X-rays for Mechanics of Materials, IMX, Ecole Polytechnique Federale de Lausanne, CH-1012 Lausanne, Switzerland

helena.vanswygenhoven@psi.ch

Abstract. The mechanical behavior of three engineering materials is studied employing in situ deformation methods. The study covers metastable austenitic steels with different stacking fault energies during multiaxial loading, a Ti-6Al-4V alloy processed by electron beam melting during uniaxial deformation and a commercial nanocrystalline NiTi alloy during multiaxial deformation. The experimental results obtained by in situ X-ray or neutron diffraction elucidate the load transfer and phase transformation mechanisms, information that is averaged over a relatively large volume containing a statistically representative number of grains. Complementary in situ high resolution digital image correlation allows details to be revealed regarding the localized strain accommodation and slip activity with a sub-grain spatial resolution. It is demonstrated that the synergy of the different length-scale investigations provides a better understanding of the complex relationship between microstructure and deformation behavior in these materials.

1. Introduction

In situ neutron and synchrotron x-ray diffraction are well established techniques for studying internal stress and microstructure evolution [1–3]. The evolution of diffraction peak positions, width and intensity can provide insight on the average intergranular and intragranular strains, and texture evolution within different grain families [4–8], where these grain families are classified according to their crystallographic orientation with respect to the diffraction vector. The average lattice strain of a grain family represents the fraction of applied load, i.e. type-I or macroscopic stresses, shared by that grain family. Elastic anisotropy, plastic slip, grain neighborhood interactions and the direction of loading significantly influence this evolution. In situ diffraction is also particularly useful for studying multi-phase materials during deformation, as it reveals the load transfer between the phases and provides information on the role of each phase for the strain hardening behavior of the composite material. The technique has the advantage of providing average information over relative large volumes that are representative of the studied microstructure. It misses, however, direct spatial microstructure correlations, i.e. observations in individual grains, grain interactions or strain concentrations in particular places.

To have a better link with the microstructure, often electron backscatter diffraction (EBSD) measurements are carried out before and after an in situ deformation and diffraction experiment. Such studies provide additional information on the change of the microstructure in terms of deformation texture, grain refinement or the formation of sub-grain boundaries. However, these observations only reveal the microstructure in stationary states before and after the deformation, and do not provide information on where and how strain is accommodated.

Digital image correlation (DIC) is an optical method that employs tracking and image-registration techniques and is used to follow changes in images. The method is employed in the engineering community for measuring full-field displacement and strains. When applied at higher magnifications, DIC of images obtained during mechanical testing makes it possible to study the localization of plastic deformation on the surface of specimens. Coupling of DIC with scanning electron microscopy (SEM) enables significantly more precise data to be obtained down to the submicron scale [9]. The spatial resolution that can be obtained depends to a great extent on the surface quality of the sample and on the pattern that is used, for instance by using nano-sized microstructural independent patterns such as gold speckle patterns [9] or particles from a colloidal silica suspension [10].

In situ high resolution digital image correlation (HRDIC) can offer a direct link between the microstructure and the strain produced by dislocation slip, twinning and martensitic phase transformations when the evolution during deformation is monitored in the same material portion. Such localized information complements the average information obtained from diffraction techniques. Apart from complimenting macroscopic experimental results, the strain localization obtained by HRDIC can be coupled with crystal plasticity simulations [11].

Here we show that by using a miniaturized multiaxial deformation rig, implementable in an SEM, in situ HRDIC can be carried out to capture the details of the strain hardening mechanisms that are missing to explain the averaged load sharing and the phase transformation behavior obtained by in situ synchrotron X-ray/neutron diffraction techniques. The cases illustrated are the transformation induced plasticity (TRIP) effect in a 304 austenitic steel, the extraordinary hardening behavior in a Ti-6Al-4V alloy processed by electron beam melting (EBM) and the multiaxial mechanical loading behavior of a commercial superelastic NiTi alloy.

2. Methods

2.1. In situ multiaxial deformation

For the in situ multiaxial loading and diffraction experiments, cruciform-shaped samples have been designed with their geometry optimized with the aid of finite element analysis [12]. In situ synchrotron X-ray diffraction experiments have been carried out at the material science beamline (MS) of the Swiss synchrotron (SLS). The miniaturized multiaxial deformation rig [13] or a uniaxial tensile rig [14] are mounted on the goniometer of the powder diffraction station at the MS beamline. The in situ measurements are conducted in transmission mode using a Mythen II microstrip detector with counting time of 30-60 seconds, depending on the material. More details about the experimental setups at the MS beamline and the cruciform sample geometry are given in [14,15].

The multiaxial load rig available at POLDI is used for in situ neutron diffraction studies [16]. The biaxial deformation system is equipped with a 2-camera digital image correlation (DIC) system (GOM, Aramis 5M) for measuring the in-plane strain at the center of the cruciform-shaped sample. Uniaxial loading and equibiaxial loading are performed with a loading rate of 80 N/s. The neutron diffraction measurements are carried out in predefined force intervals after interrupting the loading and holding the displacement. Typically the counting times range from approximately 30 minutes to 120 minutes depending on the required accuracy of the diffraction information. The neutron diffraction data is reduced and fitted using the open source software Mantid [17].

2.2. HRDIC in SEM

For the in situ HRDIC measurements, the miniaturized multiaxial deformation rig is installed inside the chamber of a FEG SEM Zeiss ULTRA 55 [13]. The electron beam with acceleration voltage of 3 kV in combination with a 20 μm aperture is used for the image acquisition. The images are acquired with an in-lens detector at a working distance of approximately 7.5 mm in order to minimize the topographic contrast by gathering low energy electrons, providing a good signal/noise ratio. SEM images are taken at various magnifications depending on the size of the deformation features to be tracked with a resolution of 3072 × 2304 pixels at the center of the cruciform/dogbone sample after reaching certain strains determined based on the aim of the study. The HRDIC data is analyzed with Ncorr [18], using typically a subset radius of 15 pixels, 0 subset spacing and a strain radius of 3 pixels, determining the circle in which the strain is calculated. Pseudo-strains due to thermal drifts and lens distortion are negligible compared to the strain magnitude observed in e.g. slip bands and therefore, no correction is applied.

2.3. Patterns for DIC

The remodeling of a sputtered gold film can produce nanoscale aggregates of gold particles that can serve as patterns for HRDIC analysis, resulting in strain mapping with sub-micron resolution [9,19]. As a flat and well-polished surface is required in order to achieve good speckle patterns, the NiTi and Ti-6V-4Al samples are ground and polished until a mirror-like surface is achieved. For the 304 steel cruciforms, the novel fabrication process incorporating electrochemical micromachining results in a high-quality surface, ready for direct application of the gold speckle pattern [20]. By adapting the method proposed in [9], fine gold speckle patterns are applied to commercial superelastic NiTi and 304 steel to study the strain accommodation in commercial NiTi and the slip activity in 304 steel under multiaxial deformation. An image of these patterns is shown in figure 1a and figure 1b.

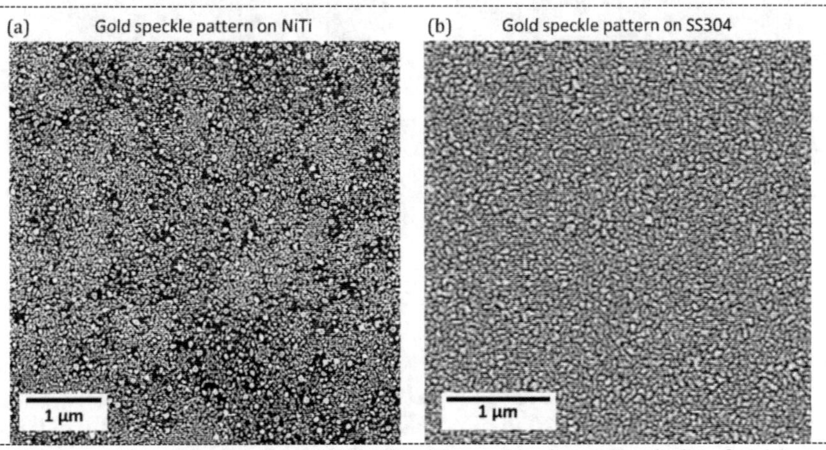

Figure 1. Fine gold speckle patterns allow HRDIC with a sub-micron resolution for (a) commercial superelastic NiTi and (b) austenitic stainless steel 304.

3. Metastable austenitic steels under multiaxial loading

The deformation mechanisms under uniaxial and biaxial load paths were studied in two types of metastable austenitic stainless steels, namely a low staking fault energy (SFE) 201 steel and a medium SFE 304 steel. Both materials are fully austenitic, i.e., have a face-centred cubic crystal structure in the unloaded state and exhibit martensitic phase transformation upon deformation. The evolution of the phases was followed during in situ uniaxial and equibiaxial deformation by neutron diffraction. More details on the experiments are given in [21,22].

In situ neutron diffraction performed on the low SFE 201 steel showed that for the same von Mises strain, more martensite is formed under uniaxial tensile loading than under biaxial tensile deformation. Following the evolution of the neutron diffraction patterns with increasing strain, it is clear that ε-martensite, which has a hexagonal closed packed crystal structure, forms as an intermediate phase prior to the body centered α′-martensite [21]. Post-mortem EBSD investigations show that the majority of the grains under uniaxial deformation favour the formation of ε-martensite, and that the intercepts of the ε-martensite plates are preferred nucleation sites for α′-martensite. The combined effect of the crystallographic orientation and the load path on the strain-induced martensitic transformation is rationalized as described in the following. The leading and trailing partial dislocations have different Burgers vectors. In low SFE steels for grains with crystallographic orientation such that the leading partial dislocation experiences a higher resolved shear stress than the trailing partial dislocation, the partials separate and form an extensive stacking fault (SF), which allows the local formation of ε-martensite, [23,24].

The opposite behavior is seen for 304 steel, i.e. more martensite is formed under biaxial than under uniaxial deformation [22,25], as apparent by the diffraction peaks corresponding to martensite which appear in figure 2-b, after approximately 15% von Mises strain. Under uniaxial deformation, no diffraction peaks corresponding to martensite are seen until the end of the deformation, at approximately 27% strain, as shown in figure 2-a. Instead, a typical deformation behavior by slip is seen, where a strong <111>-texture forms, parallel to the loading direction, accompanied by significant peak broadening, due to dislocation activity, as seen in figure 2-a. Post mortem EBSD investigations show that the majority of the grains under uniaxial deformation favor the formation of SFs (similarly to 201 steel) which in the case of the higher SFE material, leads to the formation of deformation twinning. A significant number of twins was observed by EBSD in favorably orientated grains under uniaxial deformation [22].

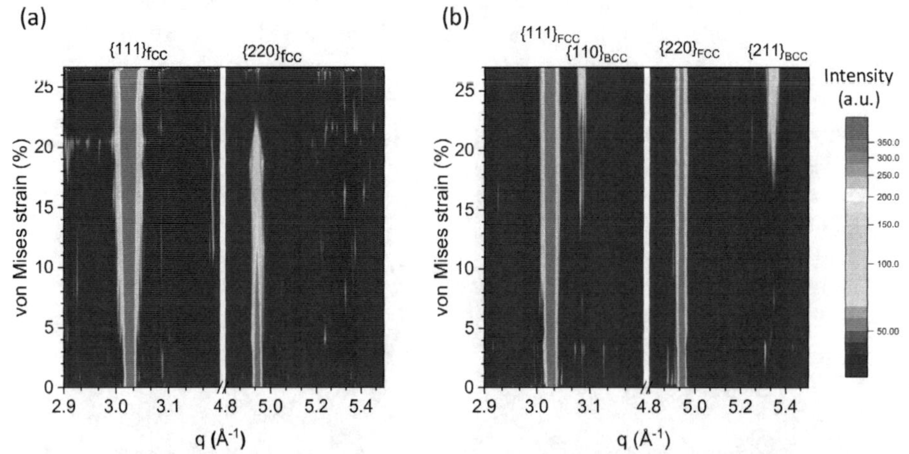

Figure 2. The evolution of the neutron diffracted intensities with increasing strain for a) uniaxial b) equibiaxial tensile deformation of 304 steel. The neutron scattering vector is parallel to the loading direction under uniaxial deformation and parallel with one of the loading axes under equibiaxial deformation.

The interpretation of the different amount of strain-induced α′-martensite under uniaxial and equibiaxial tensile deformation in 201 steel is evident by the suppression or facilitation of the formation of the intermediate phase, i.e. ε-martensite. However, in the medium SFE 304 steel, the formation of deformation twinning and its interaction with the martensite formation is not sufficiently understood to explain the opposite behavior, i.e., why less transformation under uniaxial deformation than under biaxial deformation is observed. Here HRDIC is a particularly useful tool for understanding the interplay between deformation by slip, twinning and martensite formation in these materials, by following the slip activity at the grain level.

Figure 3-a shows an image of the miniaturized cruciform-shaped 304 steel sample and its surface quality. For this study, a novel design for miniaturized cruciform samples was developed where the center of the cruciform is thinned by electrochemical machining (figure 3-a), as described in [20]. The sample surface can be readily used for electron backscatter diffraction (EBSD) and after applying a gold speckle pattern, high resolution digital image correlation (HRDIC) can be carried out.

Thanks to the gold speckle pattern, the evolution of the slip activity within each grain can be followed at different strain levels, as shown in figure 3-b and figure 3-c). Careful examination of 21 grains in each loading state shows that equibiaxial deformation promotes the activation of slip on multiple slip planes in the {111} lattice plane family within each grain, as shown in figure 3-e. Fewer grains with multiple activated slip planes are observed under uniaxial loading (figure 3-d). It is thus concluded that the enhanced twinning under uniaxial tensile deformation retards the activation of slip in secondary slip planes, which in turn results in less martensite nucleation (as a consequence of the reduction in the number of shear band intercepts, which are favorable nucleation sites for martensite formation). In contrast, under equibiaxial tensile deformation, less twinning results in more martensite formation via the activation of slip on more than one slip plane in each grain.

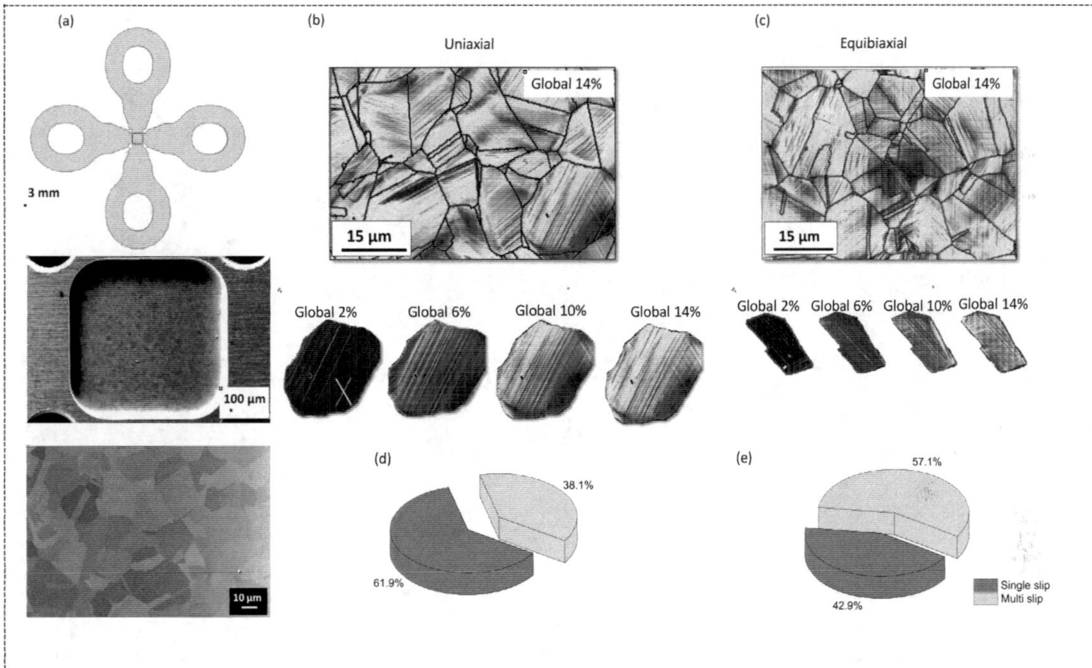

Figure 3. (a) A miniaturized cruciform-shaped 304 steel sample with reduced thickness in the centre, ready to be used for EBSD and HRDIC; (b-c) HRDIC showing slip traces at the grain level during deformation under uniaxial and equibiaxial tensile deformation. The isolated grains (indicated with red squares), show examples where a second slip system is activated. The statistical analysis (percent of grains) shows that less grains exhibit multiple slip under (d) uniaxial than (e) equibiaxial deformation [20].

4. Ti-6Al-4V

The source of high-work hardening behavior was investigated for a Ti-6Al-4V alloy processed by EBM and a novel post heat treatment. The samples were hot isostatic pressed (HIPed), annealed for two hours at sub-β transus temperatures and then water quenched. The resulting microstructure consisted of large α lamellae surrounded by regions of acicular α'-martensite, with the volume fraction of the martensite regions depending on the annealing temperature. A full description of the preparation techniques is provided in [26,27]. The post-treated material exhibits a higher ultimate tensile strength and, in some cases, a higher ductility than the standard dual- phase α+β material obtained in the as-HIPed (no post treatment) condition. This was found to be due to a high work hardening in the post-treated alloy, 2-3 times higher than in the as-HIPed samples [27]. The load sharing between the phases was investigated with in situ XRD at the MS beamline (Swiss Light Source) and the results were corroborated with transmission electron microscopy (TEM), transmission Kikuchi diffraction (TKD), and in situ HRDIC [28].

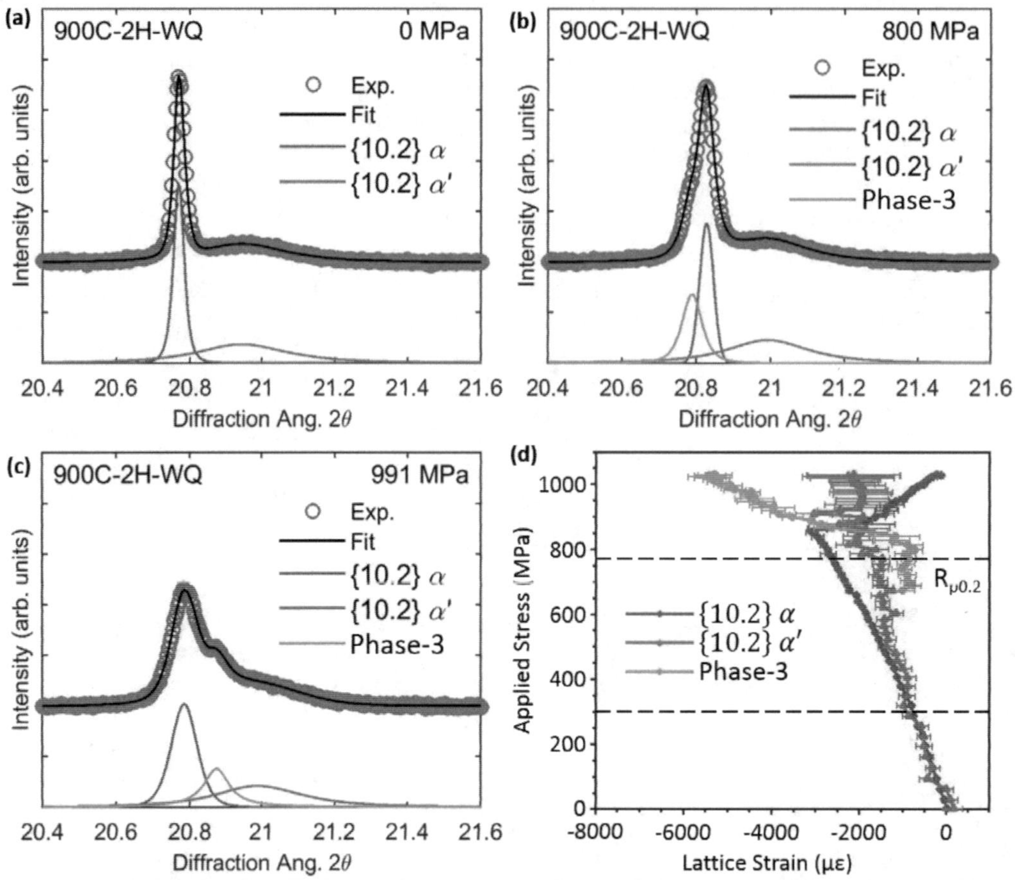

Figure 4. X-ray diffraction of the $\{10.2\}\ \alpha$, $\{10.2\}\ \alpha'$, and a Phase-3 peaks for Ti-6Al-4V at (a) 0 MPa (initial, unloaded state), (b) 800 MPa, and (c) 991 MPa,(d) the elastic lattice strain as function of the applied stress(diffraction vector $\boldsymbol{Q}_{\perp}$ perpendicular to the loading direction)

In situ uniaxial tensile loading experiments with XRD revealed the existence of a third phase in the martensitic regions, mechanically distinct from the α and α' phases (figure 4). The XRD peaks of this phase are visible as shoulders at several HCP reflections before loading, and from additional peaks that initially are obscured by nearby α reflections but which appear during loading as the phases take load differently. This is shown in figure 4-a to c. The diffracted peaks could be fit as either a secondary hexagonal closed packed (hcp) phase (secondary α) or an orthorhombic phase (α''-martensite) and could not be identified with full confidence. Therefore, the phase will be henceforth referred to as "Phase-3." The martensitic regions therefore consist of small laths of both α' and Phase-3. They are a few 100 nm long, approximately one order of magnitude smaller than the α lamellae. The deformation is the result of the load-sharing between the three phases, shown in figure 4-d for the peaks near the $\{10.2\}_{hcp}$ reflection of a sample heat treated at 900°C. Note that the diffraction vector **Q** is perpendicular to the loading direction, i.e. the calculated elastic lattice strains measure the Poisson contraction of the grain families. Thus, in figure 4-d, a more negative value of the lattice strain implies that the sample takes more load along the loading direction.

Initially, the sample is in the elastic regime and the elastic lattice strains for all phases evolve linearly. At around 300 MPa, the sample enters a microplastic regime where the elastic lattice strains for α' and Phase-3 deviate from linearity and begin to take less load. The α' peak broadens, suggesting it is deforming plastically, while the peak width of the observable Phase-3 peaks does not change. Load is transferred to the α phase. The onset of microplastic yield corresponds to the onset of plastic deformation in several soft α grain families, which rapidly shed load. The α' grains continue to deform plastically during this regime, and the load is transferred to the harder α grain families and Phase-3, which behaves as the hardest phase. The presence of an additional hard phase allows the α and α' phases to deform plastically beyond what they could in a dual-phase material, resulting in the enhanced work hardening and ductility.

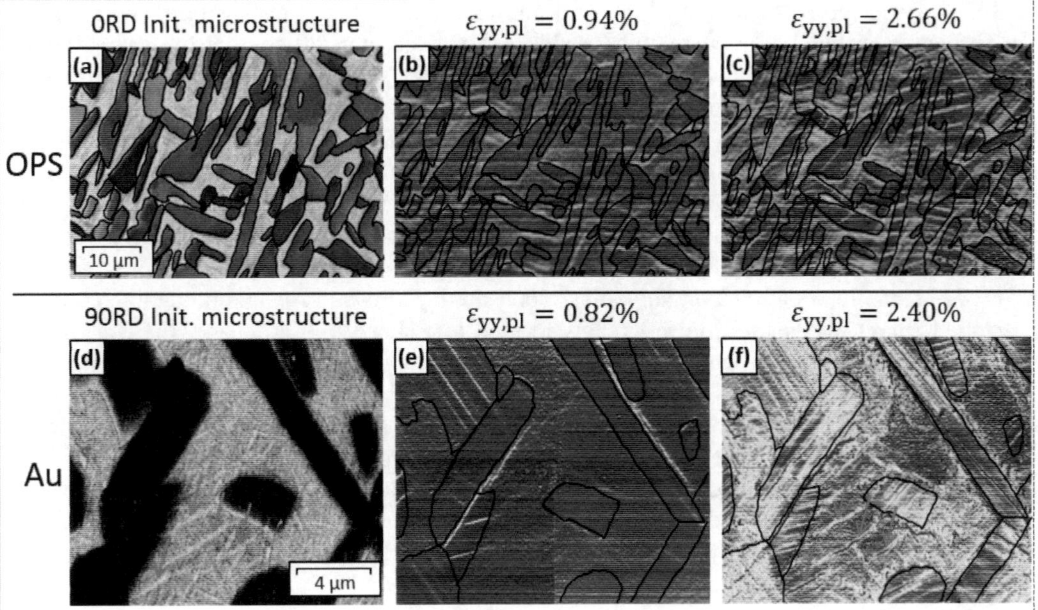

Figure 5. In situ HRDIC of (a,b,c) OPS and (d,e,f) gold speckle pattern on Ti-6Al-4V (heat treated at 900°C). Initial microstructures are backscatter electron images. Dark grains are α lamellae, light gray regions are martensitic regions. Grain boundaries are drawn manually as a guide for the eye.

To better understand the composite behavior of the martensitic regions, in situ HRDIC was performed during uniaxial tensile loading. Dogbone-shaped samples were deformed inside of an SEM and paused at various points during loading to collect highly magnified images of the sample surface. Two types of samples, one coated with OPS and one coated with gold speckles, were tested to examine the strain partitioning at two length scales. The strain maps are shown in figure 5. HRDIC confirmed that, as expected from the XRD, strain was accommodated first within the martensite regions. Additionally, the images show that the strain is not evenly distributed within the martensite regions. Rather, strong strain concentrations form along the interfaces between the α lamellae and martensite regions (figure 5-a) and within the martensite regions themselves (figure 5-b). The strain concentrations between the martensite regions and α lamellae form very early in the deformation and imply that the α lamellae are delaminating due to the mechanical differences between them. Delamination explains how Phase-3 could shed load without deforming plastically in the microplastic regime. After microplastic yield, strain concentrations are observed at the interfaces of the larger laths within the martensite regions themselves (figure 5-b). This confirms the composite behavior of the martensite regions, which, prior to these experiments, were thought to be homogeneous. Strain concentrations at the interfaces grow and form microcracks, which ultimately lead to fracture of the samples. The in situ mechanical results imply that varying the post-treatment annealing temperature can tune the volume fraction of the martensite regions and the size of the laths within them to optimize work hardening and ductility.

5. Strain accommodation in commercial superelastic NiTi

Superelasticity of NiTi allows it to sustain large strain without introducing permanent deformation owing to the reversible martensitic transformation. The B2 austenite phase transforms into the B19' martensite phase upon loading. Once the external load is removed, martensite can transform back to austenite. The transformation behavior in the commercial superelastic nanocrystalline NiTi (Memry GmbH) was studied by combining cruciform multiaxial deformation tests with in situ synchrotron X-ray diffraction. Details about the sample geometries and experimental setups are given in reference [15].

Figure 6 shows the evolution of the diffraction reflections in a selected 2θ range during a "square" load path change. The initial microstructure is full austenite phase, as evidenced by the presence of reflections only corresponding to the B2 phase before loading. During uniaxial loading along the F1 direction, austenite transforms into martensite when a critical transformation force (~24 N) is reached. The intensity of $\{110\}_{B2}$ drops drastically, while the martensite reflections $\{002\}_{B19'}$ and $\{012\}_{B19'}$ appear. Subsequent loading along the F2 direction does not introduce new martensite reflections but shows a slight decrease in the B19' volume fraction, suggesting that a reverse transformation is induced by changing the load path. Upon unloading along the F1 direction (changing loading condition from F1F2 to F2max), the martensite reflections formed at F1max, $\{002\}_{B19'}$ and $\{012\}_{B19'}$, decrease in intensity and eventually disappear. Meanwhile new martensite reflections, $\{110)_{B19'}$ and $\{020\}_{B19'}$, appear and grow in intensity. Different sets of martensite reflections corresponding to different loading directions suggest a dependency of the martensitic variant selection on the load path, which is also reported in [29], where uniaxial and equibiaxial loads result in different martensite footprints in diffraction spectra. The final unloading of F2 induces a complete reverse transformation from martensite to austenite. Upon unloading, the strong $\{110\}_{B2}$ appears again, and no martensite reflection is retained. In situ X-ray diffraction clearly demonstrates the path dependency of the martensite variants during the load path change in the commercial NiTi.

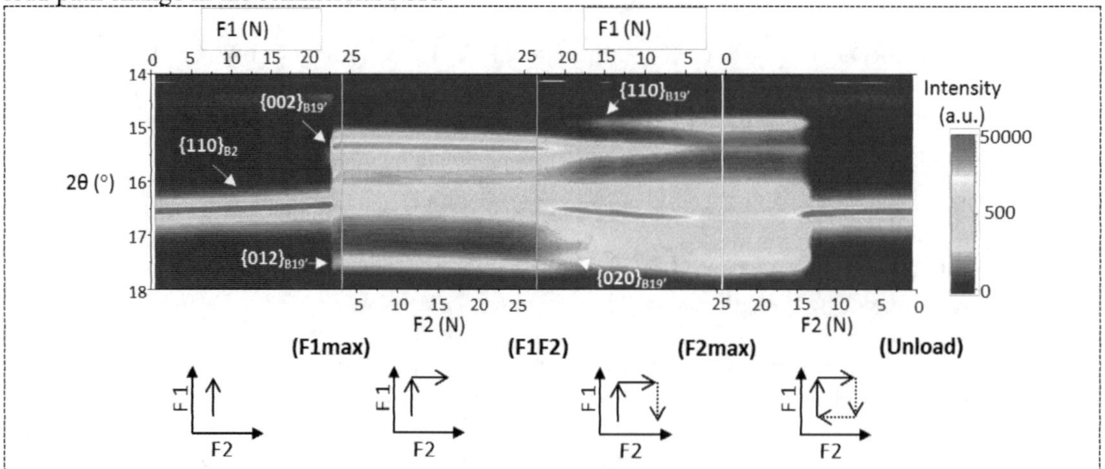

Figure 6. Evolution of the diffraction intensity of different reflections during a square load path change in NiTi alloy.

As characterized by EBSD, the initial austenite microstructure is complex (see figure 7-a). The large austenite grains delineated with dashed line in figure 7-a contain bands varying in width, direction and crystallographic orientation. High-magnification EBSD and TEM reveal that these bands consist of agglomerations of subgrains (average crystallite size: 40 nm) having similar crystallographic orientations. In order to correlate the transformation behavior observed by in situ X-ray diffraction with the complex microstructure, in situ HRDIC using gold speckle pattern applied on the surface of the cruciform-shape sample is performed. More experimental details are given in [15].

Figure 7-b shows the in-plane strain maps at F1max and F2max during the square load path. Note that the field of view for HRDIC is the same as the EBSD map shown in figure 7-a. HRDIC shows a non-uniform strain distribution inside the large austenite grains, suggesting that the strain accommodated by the phase transformation is microscopically heterogeneous. Depending on the loading direction, the high-strain traces appear in different places in the same large austenite grains. For instance, in Grain A, different inclinations and locations of the high-strain traces can be observed under F1max and F2max. The high-strain traces mostly orient similarly to the band structure shown in the EBSD crystallographic orientation map, underlying the importance of the initial austenite microstructure on the transformation behavior. A careful inspection of the behavior in different grains confirms the important role of the austenite subgrain structure [15]. Together with the in situ diffraction data shown in figure 6, the link between the austenite microstructure and the transformation behavior during the square load path change is uncovered. The martensitic transformation occurs in the nanoscaled austenite subgrains. Depending on the loading direction, different groups of subgrains transform collectively into different sets of martensitic variants, hence producing the heterogeneous strain distributions observed in the HRDIC maps.

After one cycle of the square load path change, diffraction data shows a full austenite phase and slight peak broadening in austenite reflections, which implies the formation of dislocations. The HRDIC strain map taken after unloading (figure 7-c) shows residual strain traces, which correspond to the sharp contrast features in electron channeling contrast image (ECCI). These results affirm the formation of dislocations during a single load path change cycle.

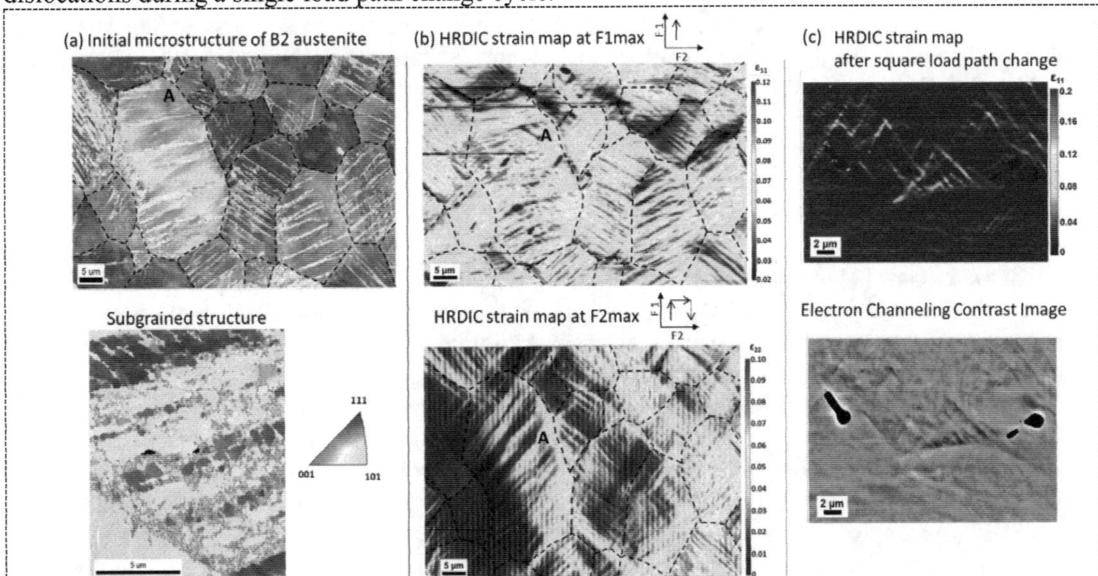

Figure 7. (a) The initial microstructure of the commercial NiTi. A full B2 austenite phase is indexed with EBSD. The subgrains are revealed in a higher magnification map. (b) The HRDIC strain map taken at F1max and F2max during the square load path change. (c) The HRDIC strain map and the electron channeling contrast image taken at the same area, after fully unloading.

6. Summary

Three examples have been presented to highlight the complementary aspects of HRDIC and in situ diffraction methods for investigating the mechanical behavior of engineering materials. Neutron diffraction showed that in a 304 austenitic steel the martensitic transformation is enhanced during equibiaxial deformation. HRDIC provided the additional understanding that under biaxial loading more grains exhibit multiple slip systems, and hence more nucleation sites for martensite are provided. For the Ti-6Al-4V alloy processed by EBM, in situ XRD revealed a rather complex load sharing between three mechanically distinct phases, whereas HRDIC demonstrated an inhomogeneous strain accommodation and delamination between very fine laths. Together the results of the two techniques explain why this composite microstructure exhibits a high work hardening with good ductility. Finally, synchrotron XRD showed that in a commercial NiTi alloy different martensite variants appear during a strain path change, while HRDIC revealed how this strain is accommodated and underlined the role of the nano-sized subgrain structure on the mechanical behavior of these material.

Acknowledgements
The authors thank the European Research Council for the ERC advanced Grant MULTIAX (339245).

References
[1] Clausen B 1997 *Characterisation of polycrystal deformation by numerical modelling and neutron diffraction measurements* (Roskilde: Risø National Laboratory)
[2] Oliver E C 2002 The generation of internal stresses in single and two phase materials *PhD Thesis*
[3] Oliver E C, Daymond M R and Withers P J 2004 Interphase and intergranular stress generation in carbon steels *Acta Mater.* **52** 1937–51
[4] Hutchings M T, Withers P J, Holden T M and Lorentzen T 2005 *Introduction to the Characterization of Residual Stress by Neutron Diffraction* (CRC Press)
[5] Dawson P, Boyce D, MacEwen S and Rogge R 2001 On the influence of crystal elastic moduli on computed lattice strains in AA-5182 following plastic straining *Mater. Sci. Eng. A* **313** 123–44
[6] Wong S L and Dawson P R 2010 Influence of directional strength-to-stiffness on the elastic–plastic transition of fcc polycrystals under uniaxial tensile loading *Acta Mater.* **58** 1658–78
[7] Van Swygenhoven H and Van Petegem S 2013 In-situ mechanical testing during X-ray diffraction *Mater. Charact.* **78** 47–59
[8] Upadhyay M V, Capek J, Van Petegem S, Lebensohn R A and Van Swygenhoven H 2017 Intergranular Strain Evolution During Biaxial Loading: A Multiscale FE-FFT Approach *JOM* **69** 839–47
[9] Di Gioacchino F and da Fonseca J Q 2013 Plastic strain mapping with sub-micron resolution using digital image correlation *Exp. Mech.* **53** 743–754
[10] Yan D, Tasan C C and Raabe D 2015 High resolution in situ mapping of microstrain and microstructure evolution reveals damage resistance criteria in dual phase steels *Acta Mater.* **96** 399–409
[11] Tasan C C, Hoefnagels J P M, Diehl M, Yan D, Roters F and Raabe D 2014 Strain localization and damage in dual phase steels investigated by coupled in-situ deformation experiments and crystal plasticity simulations *Int. J. Plast.* **63** 198–210
[12] Upadhyay M V, Panzner T, Van Petegem S and Van Swygenhoven H 2017 Stresses and Strains in Cruciform Samples Deformed in Tension *Exp. Mech.* **57** 905–20
[13] Van Petegem S, Guitton A, Dupraz M, Bollhalder A, Sofinowski K, Upadhyay M V and Van Swygenhoven H 2017 A Miniaturized Biaxial Deformation Rig for in Situ Mechanical Testing *Exp. Mech.* **57** 569–80
[14] Van Swygenhoven H, Schmitt B, Derlet P M, Van Petegem S, Cervellino A, Budrovic Z, Brandstetter S, Bollhalder A and Schild M 2006 Following peak profiles during elastic and plastic deformation: A synchrotron-based technique *Rev. Sci. Instrum.* **77** 013902
[15] Hsu W-N, Polatidis E, Šmíd M, Casati N, Van Petegem S and Van Swygenhoven H 2018 Load path change on superelastic NiTi alloys: In situ synchrotron XRD and SEM DIC *Acta Mater.* **144** 874–83
[16] Van Petegem S, Wagner J, Panzner T, Upadhyay M V, Trang T T T and Van Swygenhoven H 2016 In-situ neutron diffraction during biaxial deformation *Acta Mater.* **105** 404–16

[17] Arnold O, Bilheux J C, Borreguero J M, Buts A, Campbell S I, Chapon L, Doucet M, Draper N, Ferraz Leal R, Gigg M A, Lynch V E, Markvardsen A, Mikkelson D J, Mikkelson R L, Miller R, Palmen K, Parker P, Passos G, Perring T G, Peterson P F, Ren S, Reuter M A, Savici A T, Taylor J W, Taylor R J, Tolchenov R, Zhou W and Zikovsky J 2014 Mantid—Data analysis and visualization package for neutron scattering and μ SR experiments *Nucl. Instrum. Methods Phys. Res. Sect. Accel. Spectrometers Detect. Assoc. Equip.* **764** 156–66

[18] Blaber J, Adair B and Antoniou A 2015 Ncorr: Open-Source 2D Digital Image Correlation Matlab Software *Exp. Mech.* **55** 1105–22

[19] Scrivens W A, Luo Y, Sutton M A, Collette S A, Myrick M L, Miney P, Colavita P E, Reynolds A P and Li X 2007 Development of Patterns for Digital Image Correlation Measurements at Reduced Length Scales *Exp. Mech.* **47** 63–77

[20] Polatidis E, Hsu W-N, Šmíd M and Van Swygenhoven H 2019 A High Resolution Digital Image Correlation Study under Multiaxial Loading *Exp. Mech.* **59** 309-317

[21] Polatidis E, Hsu W-N, Šmíd M, Panzner T, Chakrabarty S, Pant P and Van Swygenhoven H 2018 Suppressed martensitic transformation under biaxial loading in low stacking fault energy metastable austenitic steels *Scr. Mater.* **147** 27–32

[22] E. Polatidis, M. Smid, W.-N. Hsu, M. Kubenova, J. Capek, T. Panzner and H. Van Swygenhoven The interplay between deformation mechanisms in austenitic 304 steel during uniaxial and equibiaxial loading, under review, Materials Science and Engineering A.

[23] Byun T S 2003 On the stress dependence of partial dislocation separation and deformation microstructure in austenitic stainless steels *Acta Mater.* **51** 3063–71

[24] Martin S, Wolf S, Martin U, Krüger L and Rafaja D 2016 Deformation Mechanisms in Austenitic TRIP/TWIP Steel as a Function of Temperature *Metall. Mater. Trans. A* **47** 49–58

[25] Zecevic M, Upadhyay M V, Polatidis E, Panzner T, Van Swygenhoven H and Knezevic M 2019 A crystallographic extension to the Olson-Cohen model for predicting strain path dependence of martensitic transformation *Acta Mater.* **166** 386–401

[26] de Formanoir C, Michotte S, Rigo O, Germain L and Godet S 2016 Electron beam melted Ti–6Al–4V: Microstructure, texture and mechanical behavior of the as-built and heat-treated material *Mater. Sci. Eng. A* **652** 105–19

[27] de Formanoir C, Brulard A, Vivès S, Martin G, Prima F, Michotte S, Rivière E, Dolimont A and Godet S 2017 A strategy to improve the work-hardening behavior of Ti–6Al–4V parts produced by additive manufacturing *Mater. Res. Lett.* **5** 201–8

[28] Karl Sofinowski, Miroslav Šmíd, Ivo Kuběna, Solange and Vivès, Nicola Casati, Stéphane Godet, Helena Van Swygenhoven In Situ Characterization of a High Work Hardening Ti-6Al-4V prepared by Electron Beam Melting *Under review, Acta Materialia*

[29] Hsu W-N, Polatidis E, Šmíd M, Van Petegem S, Casati N and Van Swygenhoven H 2019 Deformation and degradation of superelastic NiTi under multiaxial loading *Acta Mater.* **167** 149–58

Morphological characterization of bicontinuous microstructures using two-point statistics

Y Sun, K L M Elder and P W Voorhees

Department of Materials Science and Engineering, Northwestern University, Evanston IL 60208

E-mail: p-voorhees@northwestern.edu

Abstract. Bicontinuous structures are found in a wide range of multiphase materials. We employ the prototypical bicontinuous structure that is found following spinodal decomposition to probe the morphology and evolution of bicontinuous two-phase mixtures during coarsening using two-point statistics of the mean, total curvature, and the interfacial velocity. We find that the relative spatial variations of the interfacial velocity are larger than those of the mean curvature. In addition, interfaces with high positive and high negative values of the mean curvature, total curvature and interfacial velocity are spatially localized. Finally, the spatial distribution of interfaces with near zero interfacial velocity largely resembles that of near zero mean curvature, and have correlations only over very short distances.

1. Introduction

Bicontinuous microstructures usually form as a result of phase ordering or phase decomposition [1, 2] such as spinodal decomposition [3, 4]. These microstructures have complex morphologies and topologies, and have been observed in various types of materials in nature, including metals [1], polymers [3, 4], and ceramics [5]. These bicontinuous structures are a result of phase decomposition which occurs when a homogeneous single-phase mixture is cooled into a two-phase region and the alloy phase separates. After the initial phase decomposition, the system may continue to evolve towards a lower total energy by reducing the total interfacial area, in other words, the system will undergo coarsening. During the coarsening processes, the amount of each phase remains constant throughout the evolution process. In this case, the system evolves under *conserved dynamics.* In other cases, such as spin ordering in magnetic materials, a region in the system may transition from one phase to the other without being compensated by an opposite transition somewhere else in the system. This type evolution is said to follow *non-conserved dynamics.*

Although it has been shown that dendritic microstructures do not show self-similarity during coarsening [6, 7], studies have shown that bicontinuous microstructures do evolve in a self-similar manner during coarsening. Jinnai *et. al.* [4] examined the time-evolution of the probability distributions of the mean and Gaussian curvatures in a binary polymer mixture formed via spinodal decomposition. They showed that the probability distributions of the mean and Gaussian curvatures, when scaled by $S_v(t)$ and $S_v^2(t)$, respectively, where S_v is the surface area per volume and t is time, become independent of time in the evolution. This implies that the microstructure evolves in a self-similar manner. In other investigations [8, 9, 10] the morphological evolution of simulated bicontinuous structures where the morphology is characterized using interfacial shape distributions (ISDs) was studied. The ISD is the probability of finding a patch of interface with a given pair of principle curvatures κ_1, κ_2. It was demonstrated that, for both bicontinuous structures simulated via Cahn-Hilliard (conserved) [11] and Allen-Cahn (non-conserved) [12] dynamics, the scaled morphology remains constant during the late stage of coarsening. This again supports the self-similarity of the coarsening process in bicontinuous microstructures. During coarsening of these structures under conserved dynamics, the characteristic length scale, given by S_v^{-1}, follows the expected $t^{1/3}$ power law [13]. This temporal scaling law of S_v^{-1} applies to bicontinuous structures regardless of the phase volume fractions of the phases.

While S_v^{-1} can be used as a descriptor of the characteristic length scale of the bicontinuous microstructure during coarsening which follows a temporal power law, using S_v^{-1} alone is far from sufficient to describe the complicated morphology of the bicontinuous microstructure during coarsening. Although the ISD, similar to a probability distribution of a polydisperse array of spheres, is an important measure of the interfacial morphology, it does not give any insights into how these patches of curvature are distributed in the structure. To address the spatial distribution of curvature in the structure and further characterize these complex structures, we shall investigate two-point statistics of the interfacial curvature.

2. Two-point statistics introduction

We split the interface, Ω, into small patches at spatial locations denoted by $\boldsymbol{x}$. Interfacial properties (interfacial curvatures H, K, d, and normal velocity V, where the mean curvature $H = (\kappa_1 + \kappa_2)/2$, the Gaussian curvature $K = \kappa_1\kappa_2$, and the total curvature $d = \sqrt{\kappa_1^2 + \kappa_2^2}$) are calculated at these locations, which constitute a set of microstructure functions as a function of $\boldsymbol{x}$ denoted by $m(u, \boldsymbol{x})$ with u being the interfacial property considered. In practice, we will have a set of microstructure functions $m(H, \boldsymbol{x})$, $m(K, \boldsymbol{x})$, $m(V, \boldsymbol{x})$, etc. defined at each interface patch at $\boldsymbol{x} \in \Omega$. We then calculate the two-point Pearson correlation of these microstructure functions. The two-point Pearson correlation of microstructure functions associated with interfacial properties u and u' is defined as

$$g(u, u' \,|\, \boldsymbol{r}) = \frac{1}{|\Omega(\boldsymbol{r})|} \frac{1}{\sigma(\boldsymbol{r})\sigma'(\boldsymbol{r})} \int_{\boldsymbol{x}\in\Omega(\boldsymbol{r})} [m(u, \boldsymbol{x}) - \mu] \left[m(u', \boldsymbol{x} + \boldsymbol{r}) - \mu'\right] \mathrm{d}\boldsymbol{x}, \tag{1}$$

where μ and μ' are the mean values of $m(u, \boldsymbol{x})$ and $m(u', \boldsymbol{x})$ on the interface Ω, $\sigma(\boldsymbol{r})$ and $\sigma'(\boldsymbol{r})$ are the standard deviations, given by

$$\sigma(\boldsymbol{r}) = \left\{ \frac{1}{|\Omega(\boldsymbol{r})|} \int_{\boldsymbol{x}\in\Omega(\boldsymbol{r})} [m(u, \boldsymbol{x}) - \mu]^2 \, \mathrm{d}\boldsymbol{x} \right\}^{\frac{1}{2}}, \tag{2}$$

$$\sigma'(\boldsymbol{r}) = \left\{ \frac{1}{|\Omega(\boldsymbol{r})|} \int_{\boldsymbol{x}\in\Omega(\boldsymbol{r})} \left[m(u', \boldsymbol{x} + \boldsymbol{r}) - \mu'\right]^2 \mathrm{d}\boldsymbol{x} \right\}^{\frac{1}{2}}, \tag{3}$$

and $\Omega(\boldsymbol{r})$ is the set of all interfacial patches (locations) where both $m(u, \boldsymbol{x})$ and $m(u', \boldsymbol{x} + \boldsymbol{r})$ are defined, and $|\Omega(\boldsymbol{r})|$ is the total interfacial area represented by $\Omega(\boldsymbol{r})$.

Although this function is fully capable of mapping over continuous spaces, any practical implementation on either experimental or simulated datasets demands the adoption of one of the many discretized variants of this function. It is also noted that the discretized versions are better suited for computer programming. We examine next the details of these discretized variants.

It is assumed that the microstructure volume of interest occupies a rectangular parallelepiped space $\mathbb{X}$ that can be uniformly tessellated into spatial bins (i.e., voxels) $X_s \in \mathbb{X}$, where

$s = 1, 2, \ldots$ is the integer index enumerating the spatial bins. Note that s may be a vector in a high-dimensional space. In other words, for a three-dimensional volume, one might use a vector $s = (s_1, s_2, s_3)$ to index each spatial bin X_s in the microstructure volume, where s_1, s_2, and s_3 take integer values. In a similar way, we could tessellate the local state space $\mathbb{U}$ in some suitable manner to identify discretized bins in the local state space. In other words, we can discretize $\mathbb{U}$ into bins $U_n \in \mathbb{U}$, with $n = 1, 2, \ldots$ (could also be a vector index if needed) enumerating these local state bins. Building on the discretized representations, we can now invoke a discretized representation for the microstructure function as $m[n, s]$, which denotes the volume fraction in the spatial bin X_s occupied by local states identified by U_n. Specifically,

$$m[n, s] = \int_{u \in U_n} \int_{x \in X_s} m(u, x)\, \mathrm{d}x \mathrm{d}u. \tag{4}$$

With this definition, it is easy to see that, at each spatial bin X_s, the volume fractions of all local states U_n have values between 0 and 1, and they sum to 1,

$$\sum_n m[n, s] = 1, \qquad 0 \leq m[n, s] \leq 1 \;\forall\; n. \tag{5}$$

When the discretization in space is fine enough, each spatial bin X_s will only contain local states belonging to one local state bin U_n. In that case, the microstructure function $m[n, s]$ will be a binary function taking values of 0 or 1,

$$m[n, s] = \begin{cases} 1, & u(x) \in U_n \;\forall\; x \in X_s \\ 0, & \text{otherwise} \end{cases} \tag{6}$$

It should be noted that the binning of the local state space occurs quite naturally in some situations. For example, if one is dealing with a multiphase material system, then each phase can be simply assigned a different value of n. In particular, in a two-phase system, n could take on values of 0 and 1.

Following the approaches described earlier, it is relatively straightforward to write discrete versions of both two-point Pearson correlations introduced above. In this work, because the local states of interest exhibit values on a continuous range, we are primarily interested in the application of the two-point Pearson correlations based directly on the values of the local states (e.g., interface curvature and interface velocity). Therefore, the corresponding discretized versions are

$$\tilde{g}[u, u' \,|\, t] = \frac{1}{|\mathbb{S}[t]|} \frac{1}{\tilde{\sigma}[t]\tilde{\sigma}'[t]} \sum_{s \in \tilde{\mathbb{S}}} \left(u[s] - \tilde{\mu}\right) c[s] \cdot \left(u'[s + t] - \tilde{\mu}'\right) c[s + t], \tag{7}$$

$$\tilde{\mu} = \frac{1}{|\mathbb{S}[0]|} \sum_{s \in \tilde{\mathbb{S}}} u[s]\, c[s], \tag{8}$$

$$\tilde{\mu}' = \frac{1}{|\mathbb{S}[0]|} \sum_{s \in \tilde{\mathbb{S}}} u'[s]\, c[s], \tag{9}$$

$$\tilde{\sigma}[t] = \left\{ \frac{1}{|\mathbb{S}[t]|} \sum_{s \in \tilde{\mathbb{S}}} \left| u[s] - \tilde{\mu} \right|^2 c[s]\, c[s + t] \right\}^{\frac{1}{2}}, \tag{10}$$

$$\tilde{\sigma}'[t] = \left\{ \frac{1}{|\mathbb{S}[t]|} \sum_{s \in \bar{\mathbb{S}}} \left| u'[s+t] - \tilde{\mu}' \right|^2 c[s]\, c[s+t] \right\}^{\frac{1}{2}}, \tag{11}$$

where

$$|\mathbb{S}[t]| = \sum_{s \in \bar{\mathbb{S}}} c[s]\, c[s+t]. \tag{12}$$

However, we will also use the discretized local states to aid our analysis, in which case the discretized two-point Pearson correlations computed on the microstructure functions (as given in equation 1) are also needed. These can be computed as

$$g[n, n' \,|\, t] = \frac{1}{|\mathbb{S}[t]|} \frac{1}{\sigma[t]\sigma'[t]} \sum_{s \in \bar{\mathbb{S}}} \Big(m[n, s] - \mu \Big) c[s] \Big(m[n', s+t] - \mu' \Big) c[s+t] \tag{13}$$

where

$$\mu = \frac{1}{|\mathbb{S}[0]|} \sum_{s \in \bar{\mathbb{S}}} m[n, t]\, c[s], \tag{14}$$

$$\mu' = \frac{1}{|\mathbb{S}[0]|} \sum_{s \in \bar{\mathbb{S}}} m[n', t]\, c[s], \tag{15}$$

$$\sigma[t] = \left\{ \frac{1}{|\mathbb{S}[t]|} \sum_{s \in \bar{\mathbb{S}}} \left| m[n, t] - \mu \right|^2 c[s]\, c[s+t] \right\}^{\frac{1}{2}}, \tag{16}$$

$$\sigma'[t] = \left\{ \frac{1}{|\mathbb{S}[t]|} \sum_{s \in \bar{\mathbb{S}}} \left| m[n', s+t] - \mu' \right|^2 c[s]\, c[s+t] \right\}^{\frac{1}{2}}. \tag{17}$$

3. Preparation of microstructural data

3.1. Data generation by phase field simulation

The microstructure studied is a bicontinuous two-phase mixture with equal volume fractions (50 : 50 mixture) generated by phase field simulations using the conserved Cahn-Hilliard dynamics. Phase field method is a computational technique for simulating phase transformations [14, 15, 16]. It evolves the values of an order parameter ϕ and other spatially varying field quantities of interest by some dynamical equations governing the evolution process. The order parameter has different values in different phases, and smoothly varies across a finite-thickness interface between adjacent phases. The diffuse-interface approach used in the phase field method allows one to apply correct dynamics in different phases without explicitly tracking the locations of the interface. The governing equation that was used to generate the data used in this study is the Cahn-Hilliard equation [11, 17],

$$\frac{\partial \phi}{\partial t} = L_\phi \nabla^2 \left(\frac{\partial f}{\partial \phi} - \epsilon^2 \nabla^2 \phi \right) \tag{18}$$

where L_ϕ and ϵ are the mobility and the gradient energy coefficients, respectively. The bulk free energy, $f(\phi)$, is assumed to be a double-well potential,

$$f(\phi) = \frac{W}{4} \phi^2 (1 - \phi)^2 \tag{19}$$

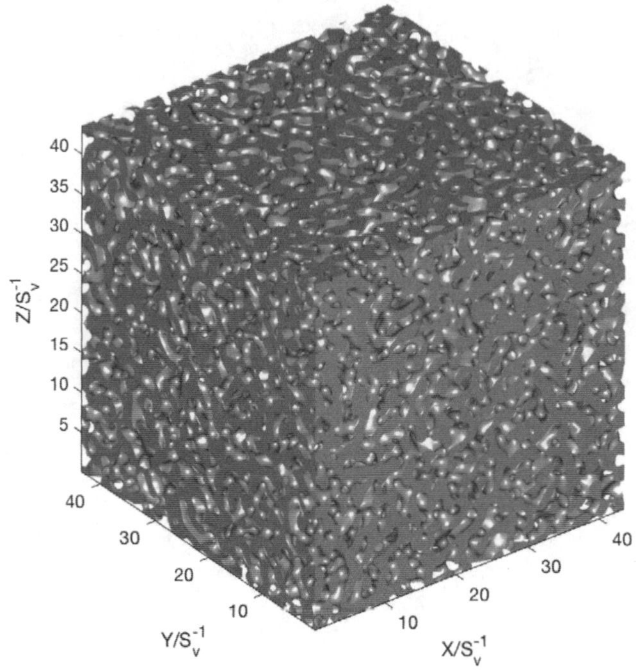

Figure 1: Three-dimensional plot of the simulated bicontinuous microstructure showing the whole simulation box.

where W is the well-height parameter controlling the height of the double-well potential. The double-well potential given above has two minima at $\phi = 0$ and $\phi = 1$, which represent the equilibrium values of the two phases.

The simulation is performed using a dimensionless form of the Cahn-Hilliard equation, where the variables are scaled by constant scaling factor so that the scaled quantities no longer carry any physical units. Specifically, we define the spatial variable $x = \tilde{x}/l$, where l is a characteristic length, and $t = \tilde{t}/\tau$, where τ is the associated time scale and the tilde indicates the original dimensional variables. We choose l and τ such that the grid spacings and the mobility coefficient are one: $\Delta x = \Delta y = \Delta z = 1.0$, and $L_\phi = \tilde{L}_\phi \tau / l^2 \tilde{W} = 1.0$. The time step is chosen to be $\Delta t = 0.05$. To generate sufficient statistics for accurate analyses of interfacial morphologies, a large computational domain size of $1024 \times 1024 \times 1024$ on a Cartesian grid was used in the simulation. In the phase field method, the interfacial thickness δ is determined by the values of ϵ^2 and W in the Cahn-Hilliard equation (equations 18 and 19). In order to ensure numerical stability while maintaining sufficient interfacial resolution, we use $\epsilon^2 = 0.2$ and $W = 0.4$ in the simulation, which gives an interfacial thickness of $\delta = 4.0$. This results in the interfacial region (defined by ϕ in the range of 0.1 to 0.9) to be approximately four grid points wide. Random noise of amplitude 0.1 is used as the initial condition for the order parameter ϕ in the simulation. The initial input has an average value of $\phi = 0.5$, which results in the $50 : 50$ volume fraction formed after spinodal decomposition. Since the Cahn-Hilliard equation is conserved, the volume fractions of the phases remain constant throughout their evolution in the simulation.

The microstructure is evolved for a dimensionless simulation time of $t = 65000$, which is sufficient to produce self-similar structures [18]. The resultant microstructure has a characteristic

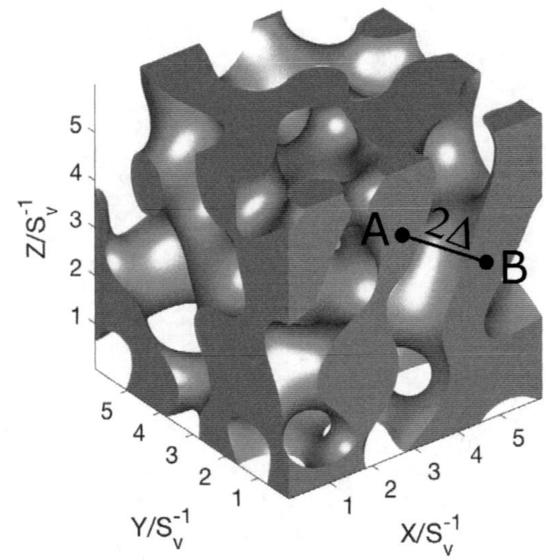

Figure 2: A subregion of the simulated bicontinuous microstructure shown in a cubic domain with side lengths six times the characteristic length scale S_v^{-1}. The parameter, 2Δ, denotes the spacing between regions of the same phase.

length of $S_v^{-1} = 23.5$, and characteristic time given by $S_v^{-1}/(dS_v^{-1}/dt) = 1.11 \times 10^{-4}$, where S_v^{-1} is the inverse of interfacial area per unit volume. A three-dimensional illustration of the microstructure is given in figure 1, in which the entire periodic simulation box is shown. In this plot, the phase corresponding to order parameter $\phi = 0$ is shown in solid color, whereas the other phase with $\phi = 1$ is shown as transparent. From now on, we will refer to these two phases as the α and β phases, respectively. In figure 2, a subregion of the microstructure is shown in a cubic domain. The cubic domain has side lengths equal to six times the characteristic length scale S_v^{-1}. From the microstructure plots, we see that the two phases in the microstructure are interconnected throughout the simulation domain, and they also penetrate into each other frequently.

3.2. Calculation of interfacial quantities

In order to capture the interfacial morphology and interfacial dynamics of the microstructure during coarsening, we use the interfacial curvatures and interfacial normal velocity to quantify the microstructure and its evolution. The interfacial curvatures are calculated using a level-set method [19, 20, 21], in which a signed distance function (SDF) is used to implicitly define the interface locations. The SDF gives the distance to the nearest interface at a specific location in space. It has opposite signs on the two sides of the interface, therefore the interface is defined as where the SDF crosses zero. The SDF can be constructed from a binary representation of the two phases, in which the order parameter has only two values – each representing one of the two phases. The construction of the SDF follows an iterative reinitialization routine described in reference [21]. With the SDF, the interfacial mean curvature and Gaussian curvature can be calculated from the spatial derivatives of the SDF [21], and other forms of curvatures can be subsequently calculated from the two. To reduce noise in higher-order derivatives of the SDF, the SDF is smoothed by evolving it in a diffusion process [19]. By comparing the interfacial locations between adjacent time steps, the interfacial normal velocity may be obtained.

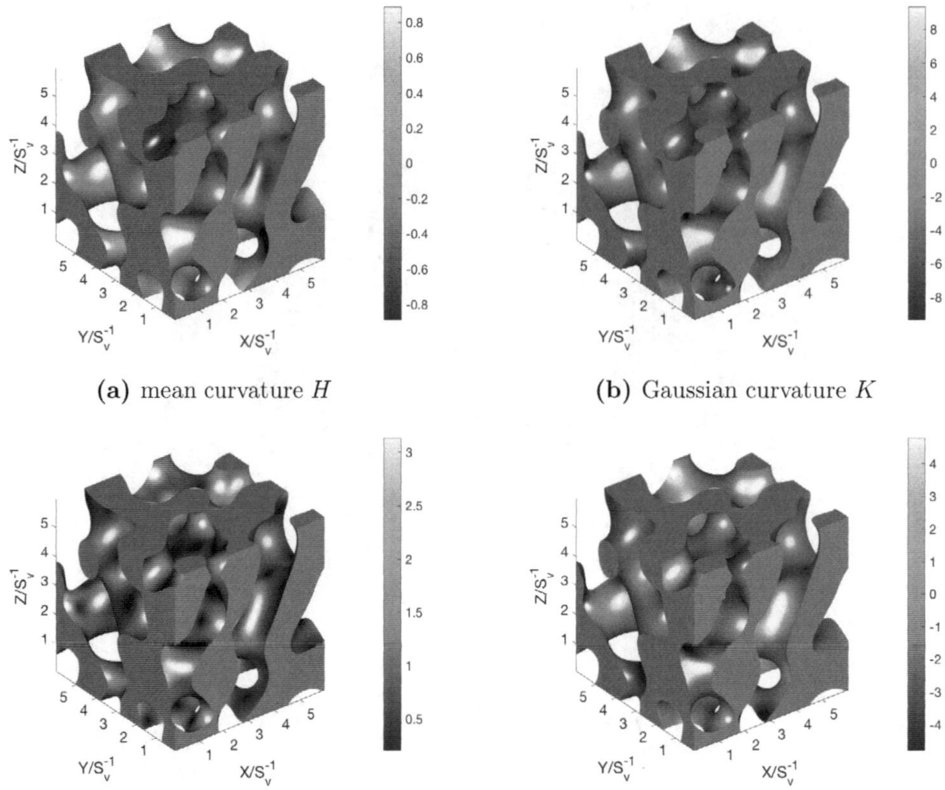

(a) mean curvature H **(b)** Gaussian curvature K

(c) net curvature d **(d)** interfacial normal velocity V

Figure 3: Interfacial curvatures and interfacial normal velocities of the interface within a sub-domain in the microstructure showing (a) mean curvature, (b) Gaussian curvature, (c) net curvature, and (d) interfacial normal velocity.

In figure 3, the interfacial mean curvature H, Gaussian curvature K, net curvature d, and interfacial normal velocity V are shown for the interface within a subregion of the microstructure. All interfacial quantities are calculated assuming the interfacial normal points from the α-phase to the β-phase, which holds for the rest of this paper.

3.3. Local state selection for two-point statistics calculation

Since the evolution of the interface during coarsening is primarily controlled by the local interfacial mean curvature H, we choose the interfacial mean curvature H as the main descriptor for the interfacial morphologies, and the interfacial normal velocity V as the descriptor for interfacial dynamics. In addition, we will use the interfacial net curvature d as an auxiliary to the interfacial mean curvature in describing interfacial morphologies. We will use the interfacial mean curvature H, the interfacial net curvature d, and the interfacial normal velocity V as the spatially varying interfacial quantities, and build local states from these quantities in the calculation of two-point statistics.

We first use the values of H, d and V as local states with continuous range of values, and calculate their spatial correlations on the interface. Then, we discretize these local states into bins with equal amounts of interfacial data points (which approximately results in equal amounts of interfacial area) according to their values, and use these discretized local states to build microstructure functions for calculating two-point statistics. Since the two-point Pearson correlation is applicable to both continuous and discrete local states, we will use the two-point Pearson correlation as the form of two-point statistics. We thus use equation 7 to calculate the two-point statistics of continuous local states, and equation 13 for discrete local states. In all of these calculations, periodic boundary conditions are assumed since the simulation is conducted under such boundary conditions. In addition, as these interfacial quantities are only defined on the interface, we use a mask array which allows only statistics on the interface. The mask is constructed such that interfacial voxels have values of 1 and non-interfacial voxels have values of 0.

For comparison, we also use the binary order parameter as the local state in the calculation of two-point statistics. The binary order parameter is defined as

$$\phi_b = \begin{cases} 0 & \text{if in phase } \alpha, \\ 1 & \text{if in phase } \beta. \end{cases} \tag{20}$$

The binary order parameter is a representation of the two phases in the microstructure in which the interfaces between phases are infinitely thin. Since it is defined in the bulk phases all over the microstructure, we will use the whole simulation box as the mask in the calculation of the two-point statistics, which effectively means that no mask is applied as the microstructure is periodic.

4. Results

We investigate the bicontinuous microstructure using statistical analysis. First, we examine the statistics of the local interfacial quantities (H, d, and V) by considering their univariate distributions. Then, we calculate the spatial distributions of these interfacial quantities using the two-point statistics method.

4.1. Statistics of local interfacial quantities

The probability densities of the three interfacial quantities of interest (H, d, and V) are plotted in figure 4. The probability densities are calculated from a triangular mesh along the isosurface where the order parameter crosses zero. Each of the triangular patches on the mesh is associated with a value in one of the interfacial quantities (H, d, and V), and the probability densities of the interfacial quantities are weighted by the areas of the patches. Therefore, the probability densities plotted in figure 4 show the interfacial area weighted probability distributions of the interfacial quantities.

From figure 4 we see that the distributions of H and V are perfectly symmetric about zero, which indicates that the microstructure is symmetric with respect to the two phases. This is because the two phases in the microstructure have identical bulk free energies and volume fractions. In other words, if we find an interface with a specific morphology in the microstructure, there should be an equal probability of finding another interface with the same morphology but the two phases on switched sides. On the other hand, the net curvature d only resides on the positive side of the probability distribution due to the way d is defined – it only accounts for the unsigned magnitude of the interfacial curvature.

In order to investigate the relations between the interfacial quantities, we calculate the joint probability distributions of the interfacial quantities. Figure 5a shows the joint probability density of the interfacial principal curvatures κ_1, κ_2. This is the interfacial shape distribution (ISD) commonly used in determining the overall morphologies of the microstructure. On the ISD, the principal curvatures are primarily concentrated in the second quadrant and along the $\kappa_1 = -\kappa_2$ line. This indicates that the interfaces in the microstructure largely constitute of saddle-like shapes with principal curvatures of equal magnitude but opposite signs. In figure 5b

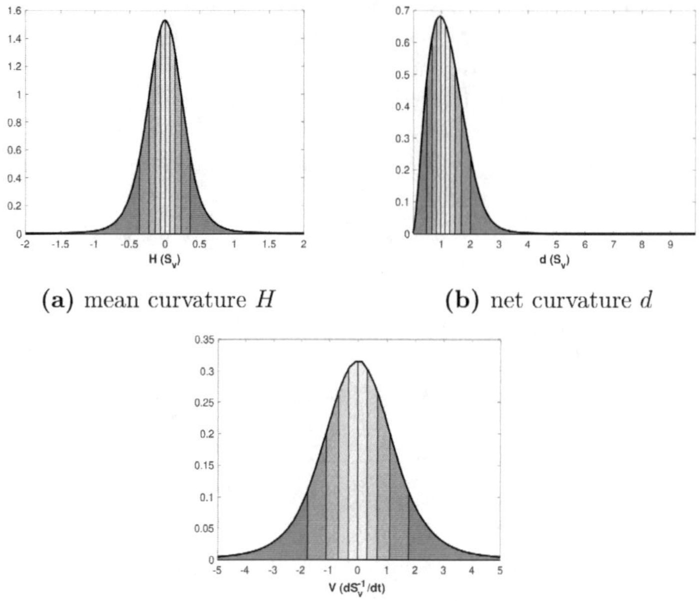

(a) mean curvature H

(b) net curvature d

(c) interfacial normal velocity V

Figure 4: Probability density of (a) interfacial mean curvature H, (b) interfacial net curvature d, and (c) interfacial velocity V. All values are in units of S_v^{-1} (curvature) and dS_v^{-1}/dt (velocity).

the ISD is converted to the joint probability density in the $H - d$ space. In this distribution plot, the probability is clustered along the $H = 0$ line. This means that the distribution in H is much narrower than that in d. Moreover, the distribution of d is high at finite values but low at near $d = 0$. This indicates that the majority of the interfaces have near zero mean curvature but finite net curvature. In other words, the microstructure mostly consists of interfaces with saddle shapes rather than flat interfaces.

Next we investigate the local correlation between the interfacial curvatures and the interfacial normal velocity. Figure 5c shows the joint probability density of H and V. The plot shows a narrow distribution along a straight line across the second and forth quadrants, which indicates a strong linear relation between H and V which correlates the local mean curvature with the negative of the local interfacial normal velocity. This suggests that the evolution of the interfaces in the bicontinuous microstructure is indeed governed by the Gibbs-Thomson effect. However, the correlation is not perfect as would be expected for Allen-Cahn dynamics [12]. On the other hand, the joint probability density of d and V shown in figure 5d displays a distribution with a much larger spread. This implies a weak correlation between the local values of d and V.

4.2. Two-point Statistics of Continuous Interfacial Quantities

After examining the local statistics of the interfacial quantities, we now investigate the spatial distributions of these quantities. We do this through calculating the two-point spatial autocorrelations of the interfacial quantities using the two-point statistics method.

First, we calculate the two-point Pearson auto-correlations of the interfacial mean curvature H, interfacial net curvature d, and interfacial normal velocity V. The results are shown in figures 6b, 6c, and 6d in the form of two-dimensional cross-sections across the origin of the three-dimensional correlation maps. For comparison, we also calculate the auto-correlation of the phases in the microstructure (i.e. the auto-correlation of the binary order parameter ϕ_b),

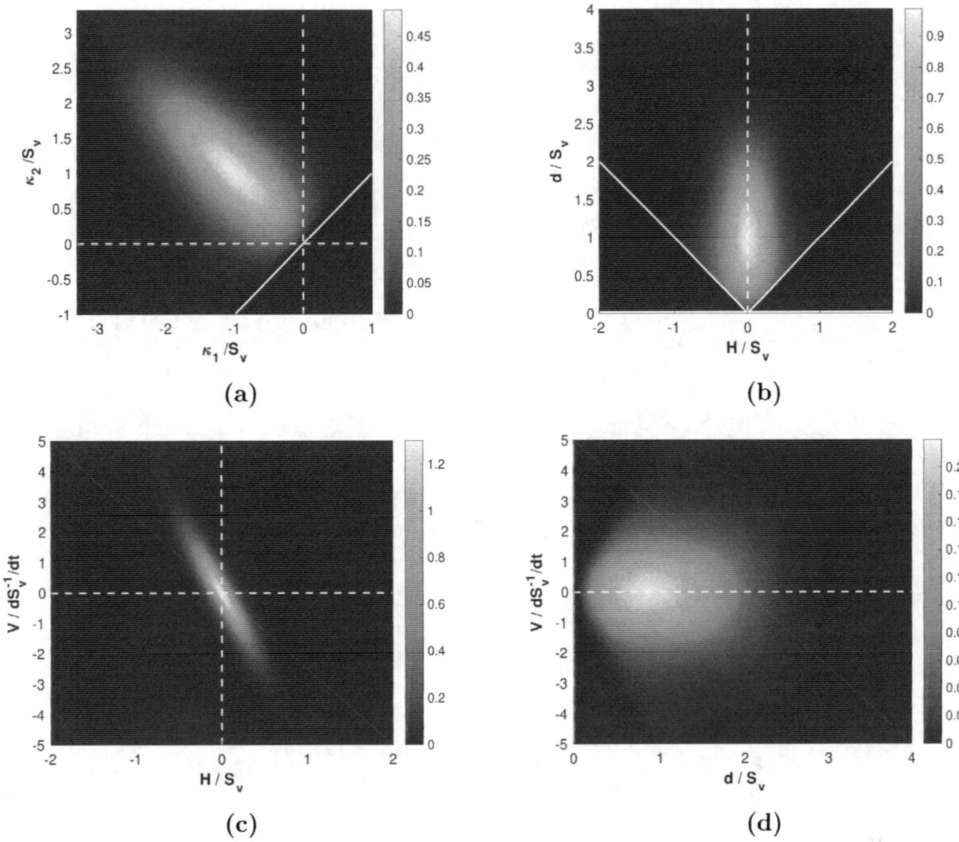

Figure 5: Probability density in (a) $\kappa_1 - \kappa_2$, (b) $H - d$, (c) $H - V$ and (d) $d - V$ spaces. All values are in units of S_v^{-1} (curvature) and dS_v^{-1}/dt (velocity).

which is shown in figure 6a. For clear visualization, the values in these correlation plots are capped at ± 0.02. From figure 6 we can see that the auto-correlation of the phases shows a spherically isotropic correlation pattern within a radius of 5 S_v^{-1}. Outside of this region, the correlation exhibits random patterns. In a slightly larger range, the auto-correlations of H and V show a similar isotropic correlation pattern, outside of which the correlation appears random as well. As for the auto-correlation of d, the isotropic correlation pattern only shows within $r < 3$ S_v^{-1}. Outside of that range, the correlation becomes random. The isotropy of these correlations reflects that the bicontinuous microstructure is isotropic in a statistical sense. This means that the morphological features in the microstructure are randomly oriented all over the microstructure domain.

Due to the isotropy of the correlations, we radially average the two-point Pearson correlations shown in figure 6. This will collapse the three-dimensional correlations into one-dimensional plots. In figure 7a, the radially averaged two-point Pearson auto-correlations of the phases, H, d and V are shown. The correlations are truncated at a radius of about 8 S_v^{-1}. All of the auto-correlations show a large peak with maximum equal to 1 at the origin, which is always the case for Pearson auto-correlations. The correlations then become negative after the first peak, and then they oscillate about zero with diminishing amplitudes as r increases.

In particular, the auto-correlation of the binary order parameter ϕ_b shows nearly periodic oscillations about zero with exponentially decaying amplitudes. The positive correlation peaks

IOP Conf. Series: Materials Science and Engineering **580** (2019) 012011 doi:10.1088/1757-899X/580/1/012011

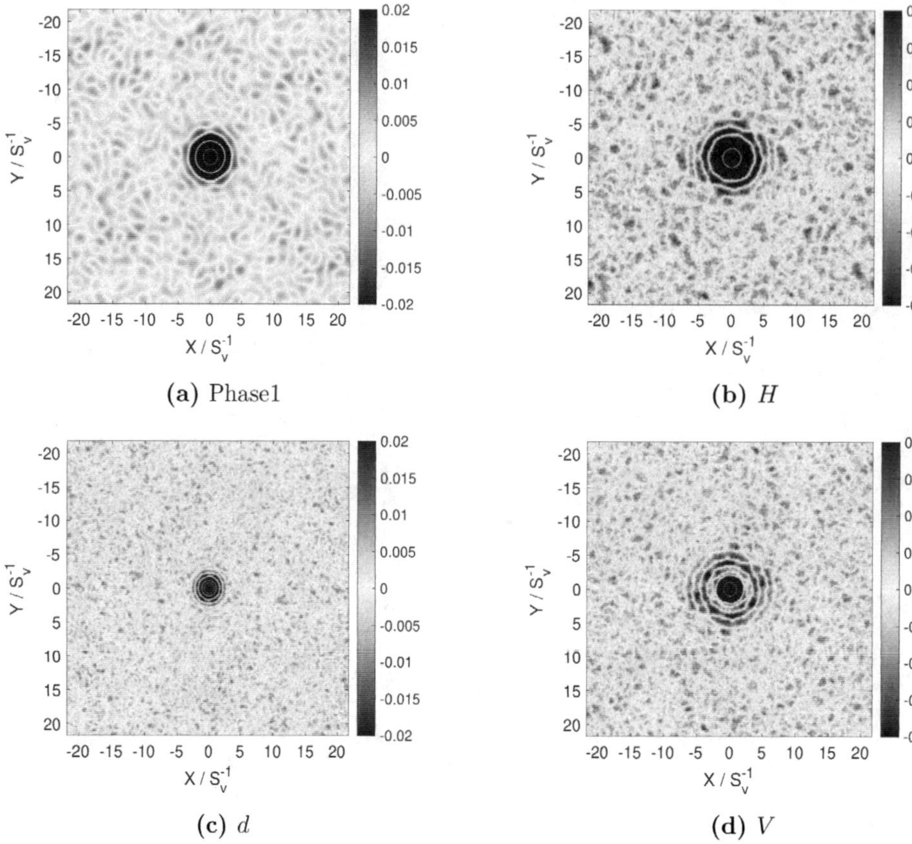

Figure 6: Two-dimensional cross-sections of the two-point Pearson auto-correlations of (a) binary order parameter (labeled as "Phase1") (b) interfacial mean curvature H (c) interfacial net curvature d, and (d) interfacial normal velocity V.

can be explained as the correlations between different portions of the same phase, and the negative local minima are results from the anti-correlations between opposite phases. One representation of the structure is that it is, on average, a series of channels where the distance between neighboring channels of the same phase is given by the location of the first local maximum, i.e. $r = 2.8\ S_v^{-1}$. Following this logic, the distance between neighboring channels of opposite phases, or the inter-phase distance Δ, is given by half of the first local maximum location – i.e. $\Delta = 1.4\ S_v^{-1}$. On the other hand, the first local minimum appears at $r = 1.6\ S_v^{-1}$, which is smaller than the inter-phase distance Δ obtained above. This is because the wide self-correlation peak at $r = 0$ shifts the first local minimum to the right. Since the three-dimensional microstructure consists of channel structures randomly oriented in space, the spatial correlation along the cross-axial direction of one channel may be along the axial direction of some other channel with a different orientation. Therefore, the primary correlation peak also includes the self-correlation along the axial direction of the channels, which widens the peak.

At larger r, the spacing between the local minima and maxima become smaller, which seems to be contradictory to the periodicity of the structure. This again can be explained by the three-dimensionality of the microstructure. In two or more dimensions, the distance between second nearest neighbors may be smaller than twice of the first nearest neighbor distance, because the second and first nearest neighbors and the reference point may not be on a straight line. For

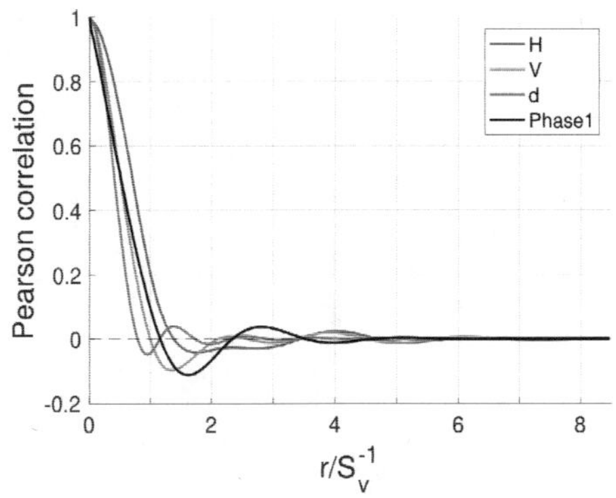

(a)

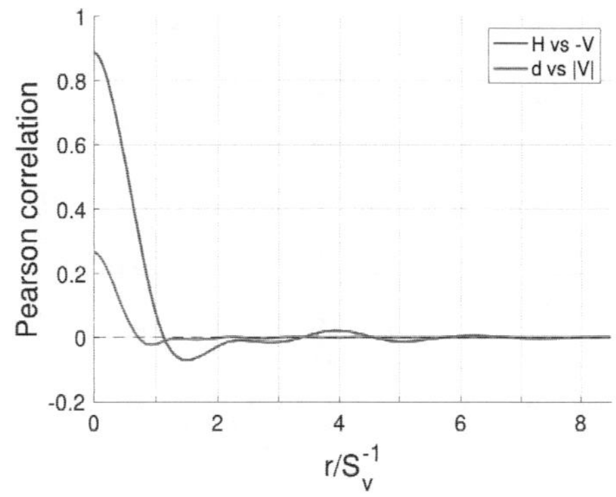

(b)

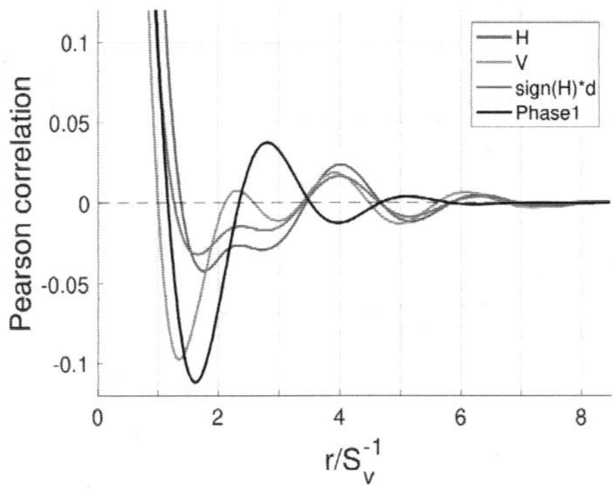

(c)

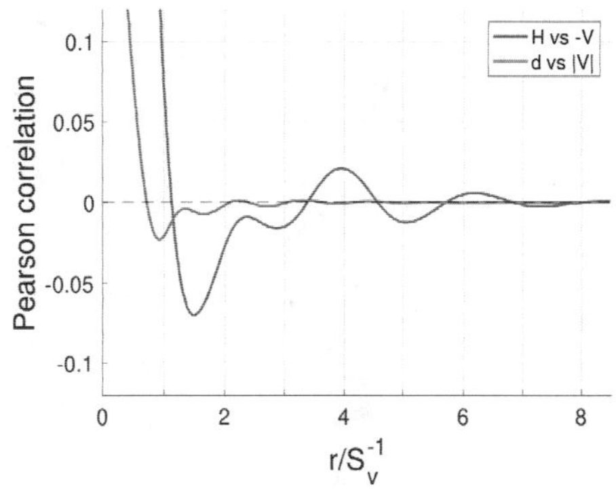

(d)

Figure 7: Radially averaged two-point Pearson correlations of binary order parameter, interfacial curvatures, and interfacial normal velocity. (a) Auto-correlations of the binary order parameter (labeled as "Phase1"), interfacial mean curvature H, interfacial net curvature d, and interfacial normal velocity V. (b) Correlations of the interfacial mean curvature H vs. the negative of interfacial normal velocity $-V$ (blue), and the interfacial net curvature d vs. the absolute value of interfacial normal velocity $|V|$ (red). (c-d) Zoomed-in views of (a-b).

example, in a three-dimensional cubic lattice with lattice constant a, the distance between first nearest neighbors is a, second nearest neighbors $\sqrt{2}a$, and third nearest neighbors $\sqrt{3}a$.

In contrast to the binary order parameter, the spatial auto-correlation of the interfacial mean curvature H shows a less periodic pattern. In particular, the auto-correlation of H shows a wide region of anti-correlation from 2.2 S_v^{-1} to 3.5 S_v^{-1}. At $r > 3.5$ S_v^{-1}, the auto-correlation of H is perfectly out of phase with the auto-correlation of ϕ_b. In other words, the two correlations have minima and maxima at the same locations but with opposite signs, and they also share the same nodes. The first local maximum appears at $r = 4$ S_v^{-1}. From the auto-correlation of ϕ_b we know that this distance is about three times the inter-phase distance, or 3Δ. Since the interfacial mean curvature is only defined on the interface, the two-point spatial auto-correlation of H will only include interfacial locations. Therefore, in this context, the distance of 3Δ would be the distance between the interfaces on the opposite sides of two neighboring channels of the same phase and the auto-correlation pattern of H could come from a microstructure with two channels of the same phase bending towards opposite directions. In this case, the high positive curvatures on the outer interfaces will correlate with each other, resulting in the peak at $r = 3\Delta$; whereas the

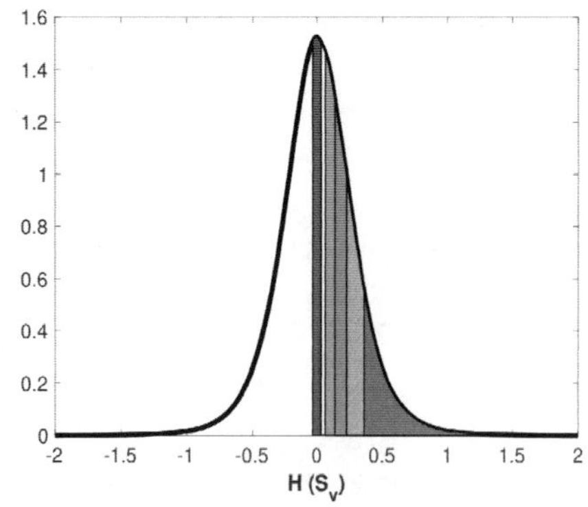

(a) mean curvature H

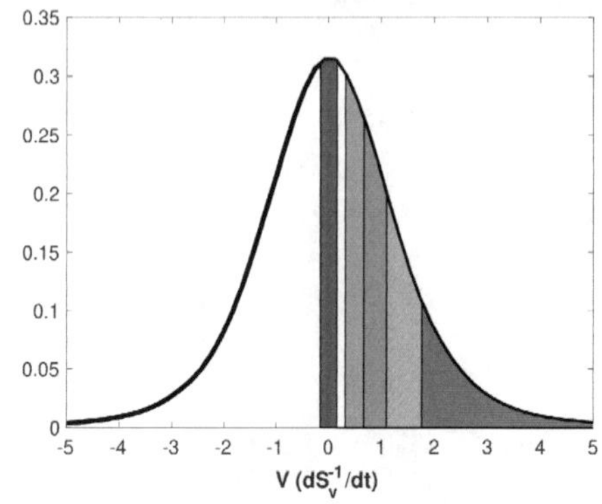

(b) negative interfacial normal velocity $-V$

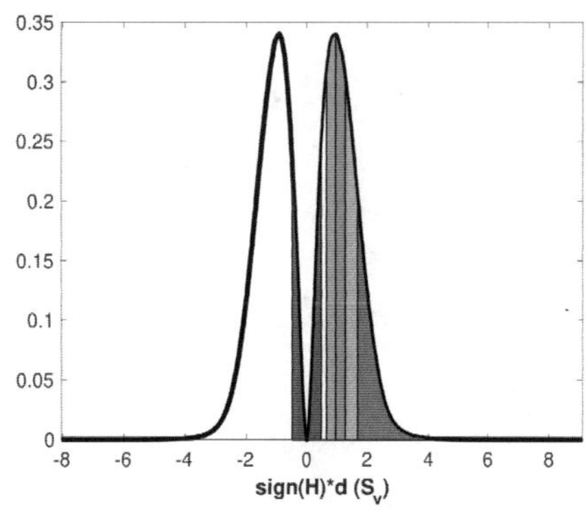

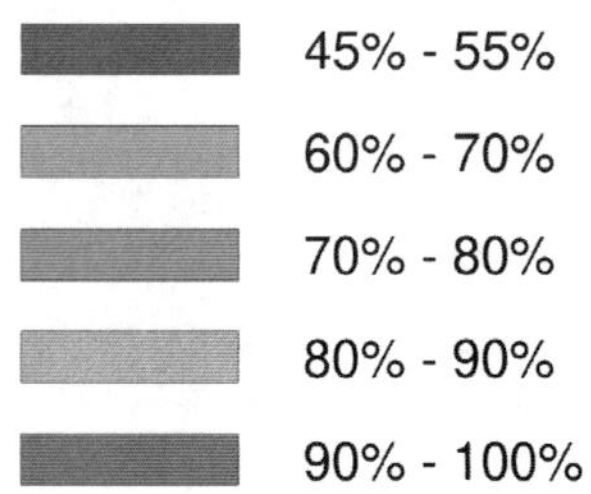

(c) signed net curvature sign$(H)d$

Figure 8: Illustration of the 10% bins on the histograms of (a) interfacial mean curvature H, (b) negative interfacial normal velocity V, and (c) signed interfacial net curvature sign$(H)d$, used in the calculation of two-point statistics. The 10% bins shown are the 45%−55% (blue), 60%−70% (light blue), 70% − 80% (green), 80% − 90% (yellow), and 90% − 100% (red) percentiles of H, V, and sign$(H)d$. They represent the 10% interfaces with near zero (blue), high positive (red), and intermediate positive (light blue, green, yellow) values in the three interfacial quantities.

negative curvatures on the inner interfaces will anti-correlate with the outer interfaces, resulting in the two negative local minima at around $r = \Delta$ and $r = 2\Delta$. Again the first local minimum is right-shifted due to the wide primary peak caused by the self-correlation effect. After the first positive local maximum, the auto-correlation of H oscillates with the periodicity of the microstructure, and quickly decays to zero.

The auto-correlation of the interfacial normal velocity V exhibits the same overall trend as that of H. Specifically, it also has negative local minima at about $r = \Delta$ and $r = 2\Delta$, and positive local maximum at around $r = 3\Delta$, with smaller oscillatory peaks to follow. The similarity between the auto-correlations of H and V directly follows from the fact that H and V are strongly (anti-)correlated locally. Nevertheless, the auto-correlation of V also shows some distinct features. In particular, the first local minimum is larger in magnitude and appears at a shorter r, and the local maximum at $\sim 2.2\ S_v^{-1}$ becomes positive. This indicates that the relative spatial variations of V is larger than that of H, and V changes sign more frequently than H. The positive correlation at $\sim 2.2\ S_v^{-1}$ implies that the local correlation between V and H may be reversed at certain locations on the interface. This is possible, because the competition among neighboring interfaces during interfacial evolution may cause the movement of the interfaces near an interface with a large curvature to be suppressed or even reversed.

Different from all of the three auto-correlations presented above, the auto-correlation of d shows an oscillatory pattern with a period about half of the other three. This is because the interfacial net curvature d is an unsigned value, and hence interfaces with high positive and high negative mean curvatures will positively correlate with each other in d even though they have opposite signs in H (and also V). In fact, the first correlation peak after then primary peak appears at $r = 1.4\ S_v^{-1}$, which is exactly the inter-phase distance Δ. This implies that this positive correlation comes from the correlation between the two walls of the channel, which both have high curvatures but with opposite signs. In addition, the auto-correlation of d decays quickly to zero after $r = 4\ S_v^{-1}$, which indicates that the interfacial net curvature has a much shorter spatial correlation compared to the phases, the interfacial mean curvature and the interfacial normal velocity.

4.3. Two-point statistics of discretized interfacial quantities

While the two-point statistics presented in the previous section show spatial correlations of the values of the interfacial quantities in general, they do not provide information on the contributions of the subranges in these values to the correlations. Specifically, the two-point statistics of the continuous interfacial quantities does not tell us how the interfaces with high, intermediate, and near zero values of these quantities correlate in space. We investigate such correlations in the current section.

We select 10% bins from the probability distributions of the interfacial curvatures and normal velocity, each of which contain 10% of the total interface, but with different values of the interfacial quantities. Since the two phases in the bicontinuous microstructure are equivalent, the two sides of the interface are completely symmetric, in terms of both morphology and dynamics. Therefore, we can examine only the interfaces with positive interfacial mean curvature H and negative interfacial normal velocity V, and the results should also apply to the other half of the interface. Nevertheless, the interfacial net curvature d is defined in an unsigned form. In order to distinguish between positive and negative curvatures while using d, we multiply the sign of the interfacial mean curvature H to the value of d, which gives the signed interfacial net curvature $\text{sign}(H)d$.

Based on these definitions, five 10% bins from the probability distributions of H, $-V$ and $\text{sign}(H)d$ are selected. As illustrated in figure 8, these bins are the 45% – 55% (blue), 60% – 70% (light blue), 70% – 80% (green), 80% – 90% (yellow), and 90% – 100% (red) percentiles of H, V, and $\text{sign}(H)d$, which represent the 10% interfaces with near zero (blue), high positive (red), and intermediate positive (light blue, green, yellow) values in the three interfacial quantities. We then construct discretized local states using these 10% bins of the interfacial quantities, and build binary microstructure functions as spatial indicators of the discrete local states, as described earlier. These binary microstructure functions are used in the calculation of the two-point statistics of the 10% bins, in which equation 13 is used.

Figure 9 shows the radially averaged results of the two-point Pearson auto-correlations of the five 10% bins in H, $-V$ and $\text{sign}(H)d$. In this figure, the auto-correlations of H and $-V$ both show that the auto-correlations of the largest values of H and V are the most spatially correlated among all values of H and V. Moreover, the spatial auto-correlations of the largest values of H and $-V$ show approximately the same correlation patterns as the correlation of H and V when continuous values of them are used. For the near zero bins of H and $-V$, the auto-correlation decays monotonously to zero within $r \leq 2\ S_v^{-1}$. The auto-correlations of the other bins with intermediate values of H and $-V$ follow the same trend as the largest bins. Specifically, they have local minima and maxima at the same locations but with decreasing amplitudes as we go to smaller value bins, and they also share the same nodes. This indicates that most of the

IOP Conf. Series: Materials Science and Engineering **580** (2019) 012011 doi:10.1088/1757-899X/580/1/012011

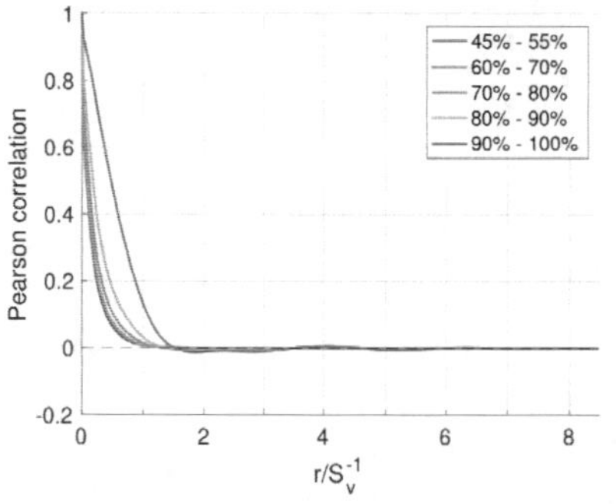

(a) H

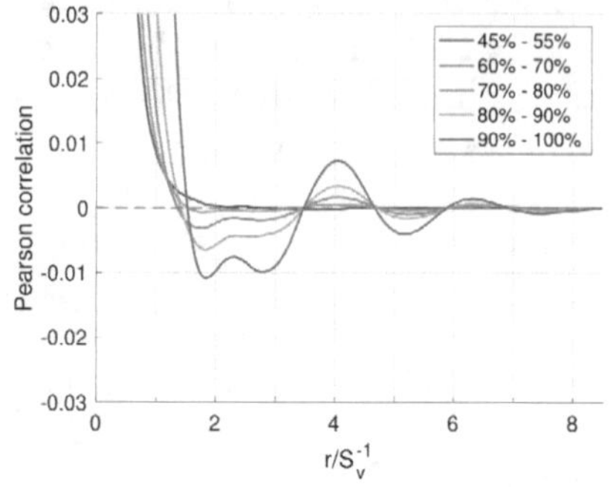

(b) H

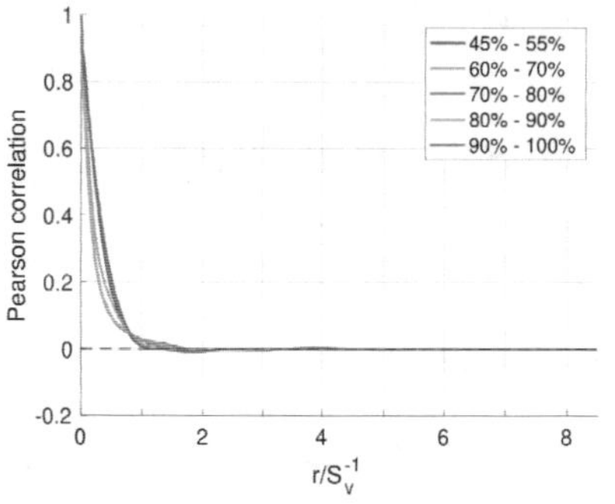

(c) $\text{sign}(H)d$

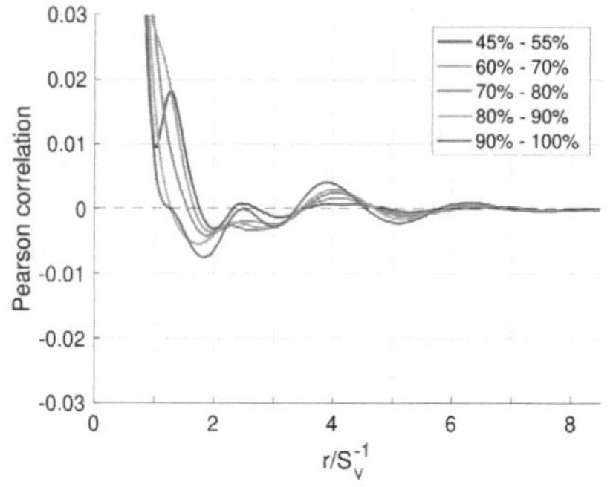

(d) $\text{sign}(H)d$

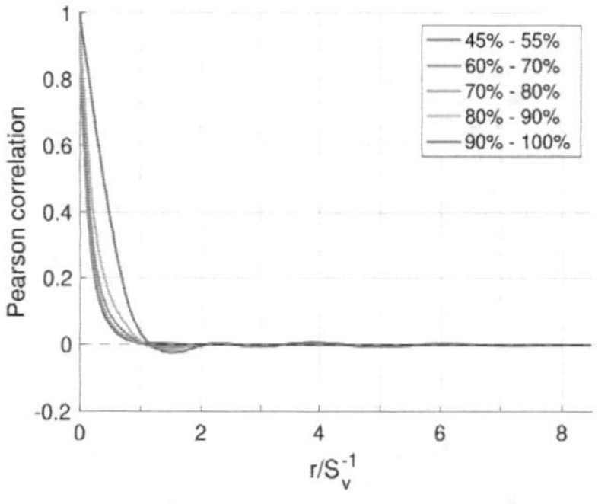

(e) $-V$

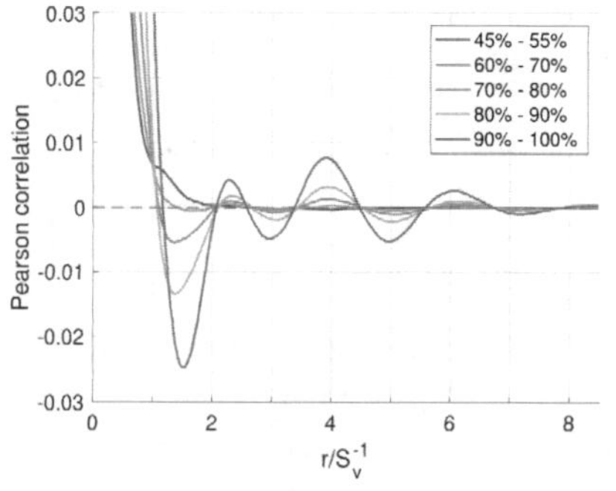

(f) $-V$

Figure 9: Radially averaged two-point Pearson auto-correlations of interfaces with values of (a,b) interfacial mean curvature H (c,d) signed interfacial net curvature $\text{sign}(H)d$, and (e,f) negative interfacial normal velocity $-V$, in the $45\% - 55\%$ (blue), $60\% - 70\%$ (light blue), $70\% - 80\%$ (green), $80\% - 90\%$ (yellow), and $90\% - 100\%$ (red) percentiles.

spatial correlation in H and V come from the extreme values. Interfaces with large (positive or negative) values of H and V form clusters in the microstructure. The average distance between these clusters is given by the first significant local maximum on the correlation curve, i.e. 4 S_v^{-1}, or $\sim 3\Delta$. In between these interfaces with large H and V, the interface smoothly varies in H and V, which results in a similar correlation pattern of the intermediate values in H and V. For the interfaces with near zero values of H and V, it shows that they are randomly distributed in space.

In contrast to the auto-correlations of the bins in H and $-V$, the auto-correlations of the five bins in sign$(H)d$ do not share the same patterns. In particular, the high positive values of sign$(H)d$ shows a similar auto-correlation pattern as high positive H. This is because all interfaces with large H will also have large d, although the inverse statement is not necessarily true. However, as is shown in figure 9, the interfaces with large d but smaller or near zero H are either correlated in the same way as the largest H, or not correlated in the long range at all. Therefore, the net result is that the auto-correlation of large sign$(H)d$ shows a similar correlation pattern as that of large H. For the near zero bin of sign$(H)d$, a positive correlation peak appears at about $r = 1.4\ S_v^{-1}$, or 1Δ, on the auto-correlation curve. This indicates that interfaces within this bin, i.e. flat interfaces, are clustered in the microstructure with an average inter-cluster distance equal to the average inter-phase distance. Unlike the corresponding bins in H and $-V$, the bins of sign$(H)d$ with intermediate values do not have the same local minima, local maxima, or nodes. In addition, in the range of $r = 2.2 - 3\ S_v^{-1}$, the auto-correlation of the intermediate bins of sign$(H)d$ show negative correlations, while the near zero and high positive bins are not correlated in this range.

5. Conclusions

The spatial distribution of the interfacial quantities and the bulk phases is determined using two-point statistics. From the spatial correlation of the order parameter or bulk phases, we find that the average distance between neighboring regions in opposite phases is $\Delta = 1.4\ S_v^{-1}$. Using two-point statistics of interfacial quantities, we demonstrate that the relative spatial variations of V are larger than those of H, and V changes signs more frequently than H with distance from a point in space. This large spatial variation in interfacial velocity is likely caused by the excess or depletion of solute induced by the large velocity of interfaces with high curvatures. This also leads to large velocity of flat interfaces nearby. In addition, for interfaces with high positive and high negative values of H, the signed total curvature sign$(H)d$ and V tend to be spatially localized. The average spacing between regions with the same sign of curvature or velocity is $\sim 4\ S_v^{-1}$ or $\sim 3\Delta$. Interfaces with near zero values of H are only correlated over very short distances interfaces with near zero d are clustered in space, with an average inter-cluster spacing of 1.4 S_v^{-1}, or 1Δ. The spatial distribution of interfaces with near zero V largely resembles that of near zero H interfaces.

Acknowledgments

This work is supported by U.S. Department of Energy grant DE-FG02-99ER45782 and the Natural Sciences and Engineering Research Council of Canada (NSERC), PGSD3-516809-2018.

References

[1] Hellwig O, Berger A, and Fullerton E E 2003 *Phys. Rev. Lett.* **91** 197203
[2] Fialkowski M and Holyst R 2002 *Phys. Rev. E* **66** 046121
[3] Jinnai H, Nishikawa Y, Koga T, and Hashimoto T 1995 *Macromolecules* **28** 4782-84
[4] Jinnai H, Nishikawa Y, Morimoto H, Koga, T and Hashimoto T 2000 *Langmuir* **16** 4380-93
[5] Chan V Z, Lee V Y, Iatrou H, Avgeropoulos A, Hadjichristidis N, Miller R D, and Thomas E L 1999 *Science* **286** 1716-19
[6] Marsh S P and Glicksman M E 1996 *Metall. Mater. Trans. A* **27** 557-67

[7] Kammer D and Voorhees P W 2006 *Acta Mater.* **54** 1549-58
[8] Kwon Y, Thornton K, and Voorhees P W 2007 *Phys. Rev. E* **75** 021120
[9] Park C L, Voorhees P W, and Thornton K 2015 *Acta Mater.* **90** 182-93
[10] Park C L, Gibbs J W, Voorhees P W, and Thornton K 2017 *Acta Mater.* **132** 13-24
[11] Cahn J W 1961 *Acta Metall.* **9** 795-801
[12] Allen S M and Cahn J W 1979 *Acta Metall.* **27** 1085-95
[13] Lifshitz I M, and Slyozov V V 1961 *J. Phys. Chem. Solids* **19** 35-50
[14] Chen L Q 2002 *Annu. Rev. Mater. Res.* **32** 113-40
[15] Karma A and Rappel W J 1996 *Phys. Rev. E* **53** R3017
[16] Warren J A and Boettinger W J 1995 *Acta Metall. et Mater.* **43** 689-703
[17] Cahn J W and Hilliard J E 1958 *J. Chem. Phys.* **28** 258-67
[18] Kwon Y, Thornton K and Voorhees P W 2010 *Philos. Mag.* **90** 317-35
[19] Park C L, Voorhees P W, and Thornton K 2014 *Comput. Mater. Sci.* **85** 46-58
[20] Xu J J and Zhao H K 2003 *J. Sci. Comput.* **19** 573-94
[21] Seithian J A 1999 *Level Set Methods and Fast Marching Methods: Evolving Interfaces in Computational Geometry, Fluid Mechanics, Computer Vision, and Materials Science* Cambridge University Press **3**

Linking local and heterogeneous deformation behavior to global deformation of materials by *in-situ* experimental techniques

Nobuhiro Tsuji[1,2], Myeong-heom Park[1,2], Si Gao[1], Yu Bai[1], Yan Chong[1,3]

[1] Department of Materials Science and Engineering, Kyoto University, 606-8501 Kyoto, Japan
[2] Elements Strategy Initiative for Structural Materials, Kyoto University, 606-8501 Kyoto, Japan
[3] Department of Materials Science and Engineering, University of California Berkeley, 94720, Berkeley, CA, U.S.A.

Email: nobuhiro-tsuji@mtl.kyoto-u.ac.jp

Abstract. Plastic deformation of metallic materials is essentially heterogeneous and localized. In the case of dislocation slip, for example, slip of dislocations occurs on particular crystallographic planes and in particular directions, and dislocations are multiplied from dislocation sources located on limited slip planes. Heterogeneous deformation is also induced by the constraint of adjacent grains in polycrystalline materials and appears in different length scales in the form of different slip patterns within each grain, deformation bands, and so on. To achieve higher strengths, the microstructures of structural metallic materials have become more and more complicated and are often composed of different phases (or domains) having different strength levels. In such complicated microstructures, deformation becomes even more heterogeneous. Therefore, it is important, especially in high strength metals, to link local and heterogeneous deformation aspects to the global deformation behavior, in order to fundamentally understand the nature of deformation and then to design advanced materials for performing desired mechanical properties. Conventional material testing, such as tensile tests can determine the global deformation behavior of materials, but the obtained stress-strain curves show only the average material response, in which local information is hidden. Local and heterogeneous characteristics of deformation have been observed by various kinds of microscopy, but it is typically difficult to correlate such local and heterogeneous features with the global and average materials response, especially during deformation. Recently, significant progress has been made in various kinds of *in-situ* experimental techniques, and a breakthrough in deepening the fundamental understanding of deformation of materials is expected. Such techniques include *in-situ* neutron/synchrotron diffraction during deformation, *in-situ* deformation observations in scanning and transmission electron microscopy, nano-indentation, and digital image correlation (DIC) during deformation. In the current paper, some recent results obtained in the author's group linking local and heterogeneous characteristics to the global deformation response for several metallic materials are introduced, illustrating the use of state-of-the-art techniques.

1. Introduction

Plastic deformation of crystalline materials is essentially heterogeneous. In the case of metallic materials, dislocation slip is the most common mechanism for plastic deformation. Dislocation slip takes place on particular crystallographic planes corresponding to the crystal structure, and a relative displacement of crystalline lattices on both sides of the slip plane occurs along a particular crystallographic direction, expressed by the Burgers vector, ***b*** [1]. For face-centered cubic (FCC) crystals, for example, the slip plane is {111} and ***b*** is parallel to the <110> direction. Figure 1 schematically illustrates plastic deformation by dislocation slip in a FCC single crystal having (111)[-110] orientation under a shear stress, τ, applied on the upper and lower surfaces of the crystal. Two types of dislocation slip are illustrated in the figure. If a dislocation moves on every (111) plane and the operated dislocations exit the material at the side surfaces, the shape change illustrated in figure 1(b) will result.

On the other hand, if many dislocations operate on a particular single (111) slip plane, the shape change of the crystal would be as shown in figure 1(c). The resultant macroscopic plastic deformation, expressed by parallelograms superimposed on figures 1(b) and 1(c), is the same for the both cases, assuming that the total number of operated dislocations (or the total slip length) is equivalent. One can understand from figure 1, however, that the plastic deformation shown in figure 1(b) is more homogeneous than that shown in figure 1(c). In most cases, dislocation slip in metallic crystal takes place in a heterogeneous manner like that illustrated in figure 1(c), since dislocation sources (such as the Frank-Read source [1], shown in figure 2) multiply dislocations on a particular slip plane.

The heterogeneity of plastic deformation is more obvious in other deformation mechanisms. In deformation twinning and martensitic transformation, a particular volume of matrix crystal transforms to a twin or to martensite resulting in a localized shear deformation intrinsic to the deformation twinning or the martensitic transformation. In grain boundary sliding, deformation localizes on grain boundaries.

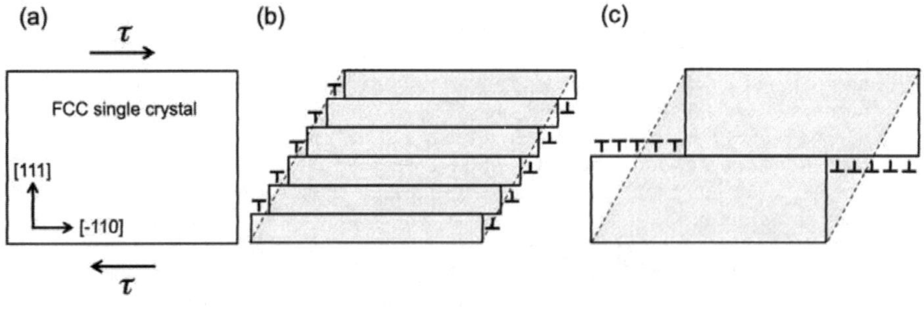

Figure 1: Schematic illustrations showing shape changes of a (111)[-110] single crystal under a shear stress, τ, on (111) plane in the [-110] direction. (a) Initial geometry of the single crystal. (b) After dislocation slip on successive (111) planes (more homogeneous deformation). (c) After dislocation slip on a particular (111) plane (more heterogeneous deformation). Parallelograms in (b) and (c) indicate the macroscopic shape change after the plastic deformation.

In polycrystalline metals composed of many grains having different crystallographic orientations, the shape changes of neighboring grains (like those illustrated in figure 1(b) and 1(c)) must be compatible to each other. In order to satisfy this compatibility, other slip system(s) should operate in local areas within each grain, resulting in another kind of heterogeneity in deformation. Such a difference in slip patterns within one grain leads to the formation of different domains having misorientations to each other within the grain (*grain subdivision* [2]). Grain boundaries generally act as a strong obstacles for dislocation slip due to the discontinuity of crystallographic orientation, so that dislocations are stored near grain boundaries. Dislocations are also trapped by various obstacles (such as other dislocations, and precipitates) within each grain. Stored dislocations have their own elastic fields that can induce other heterogeneities in subsequent plastic deformation. Heterogeneity in deformation also appears in the form of deformation bands and shear bands in which different amounts or/and combinations of slip systems from those in the matrix operate. When the microstructure involves different phases, each having a different hardness (strength), the compatibility situation at interphase boundaries is still more complex than that at grain boundaries in single-phase materials, resulting in a higher degree of heterogeneity.

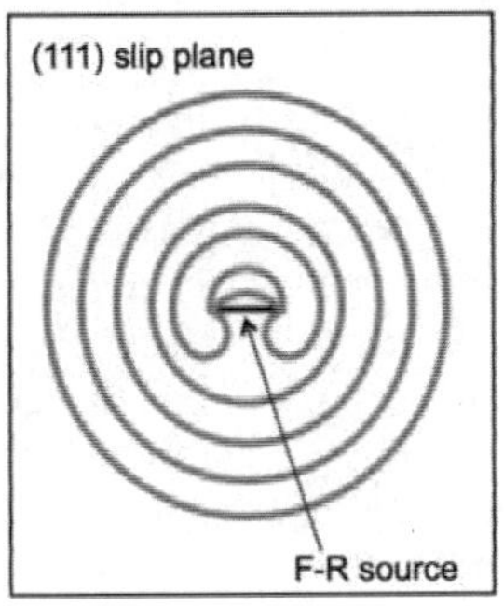

Figure 2: Schematic illustration showing multiplication of dislocations from a Frank-Read (F-R) source on a particular slip plane. The deformation shown in figure 1(c) can be considered as a result of such a multiplication process on a single slip plane.

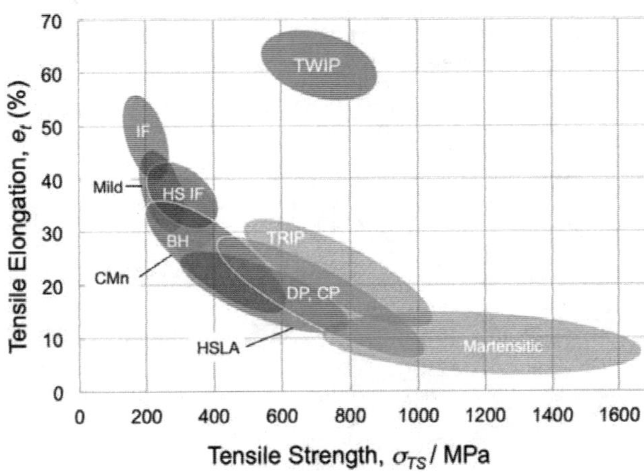

Figure 3: Strength-ductility balance for various kinds of sheet steel: IF (interstitial free) steels; HS IF (high strength IF) steels, BH (bake hardened) steels; CMn (carbon-manganese) steels; HSLA (high strength low-alloy) steels; TRIP (transformation induced plasticity) steels; DP (dual phase) steels; CP (complex phase) steels; and TWIP (twinning induced plasticity) steels.

2. Usefulness of nano-indentation for determining strength of different phases/domains

Nano-indentation technique is useful to determine the strength (and elastic modulus) of different phases or domains involved in the given material microstructure. Here one interesting

IOP Conf. Series: Materials Science and Engineering **580** (2019) 012012 doi:10.1088/1757-899X/580/1/012012

result revealed by nano-indentation mapping is shown. Ti-6Al-4V (mass%) is one of the most popular α+β titanium alloys, showing microstructures composed of the HCP α phase and the BCC β phase [5]. If such alloys having lamellar or martensitic structures of α are hot-deformed and then annealed at high temperatures in the α+β two phase region, α plates are broken and equiaxed α+β microstructures are formed. If such equiaxed α+β microstructures are then cooled at relatively high cooling rates, β grains transform to finer α plates having various elongated directions and crystallographic orientations during cooling. The resultant microstructure, composed of equiaxed primary α grains and transformed β regions, are called "bimodal" microstructures in the field of α+β titanium alloys [5].

Figure 4 (a) shows an EBSD orientation map of a bimodal microstructure in Ti-6Al-4V, where primary α grains with equiaxed morphology, and transformed β regions composed of fine α plates with various elongated directions and crystallographic orientations are clearly recognized. Nano-indentation mapping was carried out on the bimodal microstructure; some triangular indentations are seen in the scanning electron microscope (SEM) image of figure 4(b). From the nano-indentation mapping, it was found that the transformed β regions having much finer microstructures were basically softer than the primary α grains [6]. Actually in figure 4(b), indentations in equiaxed α grains are smaller than those in transformed β regions. Figure 4(c) shows average hardness values of the primary α grains and transformed β regions in bimodal microstructures obtained after annealing at different temperatures in the α+β two-phase region. The difference in hardness between two domains decreases with increasing intercritical annealing temperature. Chemical analysis by energy dispersive X-ray spectroscopy in the SEM have clarified that the difference in hardness of the two domains is due to a difference in solid solution strengthening by Al [6].

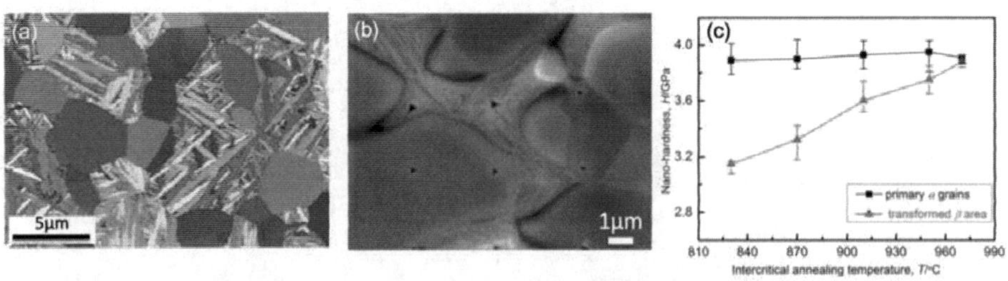

Figure 4: (a) EBSD orientation map of a bimodal microstructure in a Ti-6Al-4V alloy. (b) SEM image showing nano-indentations on a bimodal microstructure in Ti-6Al-4V. (c) Average nano-hardness values of primary α grains and transformed β regions in bimodal microstructures obtained by hot-deformation and intercritical annealing at different temperatures in α+β two-phase region of Ti-6Al-4V. [6]

The Ti-6Al-4V alloy with the bimodal microstructures prepared by hot-deformation and intercritical annealing at different temperatures showed different stress-strain behaviors at room temperature, as shown in figure 5(a). The shape of each stress-strain curves in figure 5(a) can be classified into two groups, depending on the range of intercritical annealing temperature. For annealing temperatures in the range from 830°C to 910°C, the specimens have a relatively low yield strength but show good strain hardening, subsequently resulting in large uniform elongation and total elongation. In contrast, the bimodal specimens annealed at higher temperatures (950°C and 970°C) showed high yield strength but limited uniform elongation. It is known that the uniform elongation during tensile testing is determined by the plastic instability condition, expressed as,

$$\sigma \geq \left(\frac{d\sigma}{d\varepsilon}\right) \tag{1}$$

where σ is the true flow stress and (dσ/dε) is the strain-hardening rate. Strain-hardening curves of the bimodal specimens are shown together with the corresponding true stress-strain curves in figure 5(b). It is clearly seen that the bimodal specimens fabricated through intercritical annealing at the lower temperature range maintain higher strain-hardening rate at the later stages of tensile deformation, resulting in delayed occurrence of plastic instability, i.e., a larger uniform elongation. Compared with the nano-hardness result shown in figure 4(c), it can be concluded that a better balance of strength and ductility (or better strain-hardening ability) in the global mechanical response of Ti-6Al-4V with bimodal microstructures is obtained when two domains (primary α grains and transformed β regions) have different hardness (strength), compared to the case where the microstructure has more uniform hardness distributions.

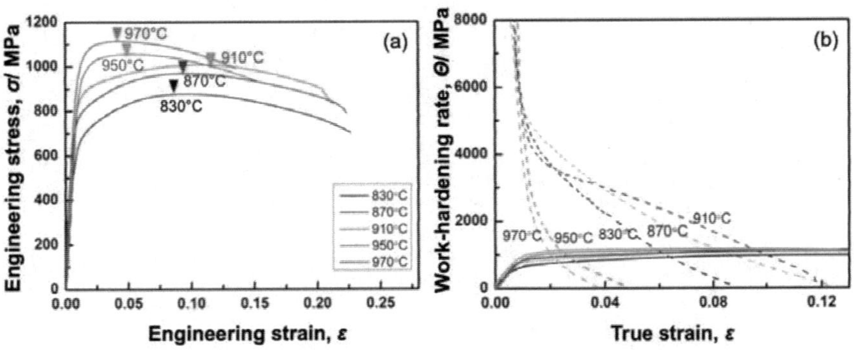

Figure 5: (a) Engineering stress-strain curves of the Ti-6Al-4V specimens (corresponding to figure 4(c)) with bimodal microstructures obtained by hot-deformation and intercritical annealing at different temperatures in the α+β two-phase region. Tensile test were conducted at RT. (b) True stress-strain curves and work-hardening rate curves of Ti-6Al-4V with bimodal microstructures [6].

3. Digital Image Correlation (DIC) analysis in specimen-scale

The digital image correlation (DIC) technique is an optical method to measure the displacement of local points by comparing digital images of identical areas in an undeformed and deformed material [7,8]. From the obtained displacement data in the measured area, distributions of local strains can be reconstructed. Here an example of DIC analysis for heterogeneous deformation at a relatively macroscopic scale (namely the specimen scale) is introduced.

Lüders deformation is a well-known form of localized and heterogeneous deformation typically observed in carbon steels. Recently, huge Lüders deformation in which the Lüders strains reach to 10 ~ 20% has been reported for some advanced high strength steels having an FCC austenite matrix or austenite retained partly in the microstructure [9-11]. As those steels exhibit both high strength and large tensile ductility, it is considered that the huge Lüders deformation plays a key role in determining the superior mechanical properties. We have recently succeeded in fabricating a SUS304 austenitic stainless steel having fully-recrystallized ultrafine grained (UFG) microstructure with sub-micrometer grain size. The stress-strain curves of the both coarse grained (CG) and UFG specimens at RT are shown in figure 6 [12]. The CG specimen showed a typical stress-strain curve of this steel, with continuous yielding at a 0.2% proof stress of 340 MPa, subsequent strain hardening, and macroscopic necking (plastic instability) to failure. In contrast, the yield strength of the UFG specimen greatly increased to 950 MPa (0.2% proof stress) and a clear yield-drop was observed. After the yield-drop, huge Lüders-type deformation was observed, where the Lüders strain surprisingly reached 30% strain.

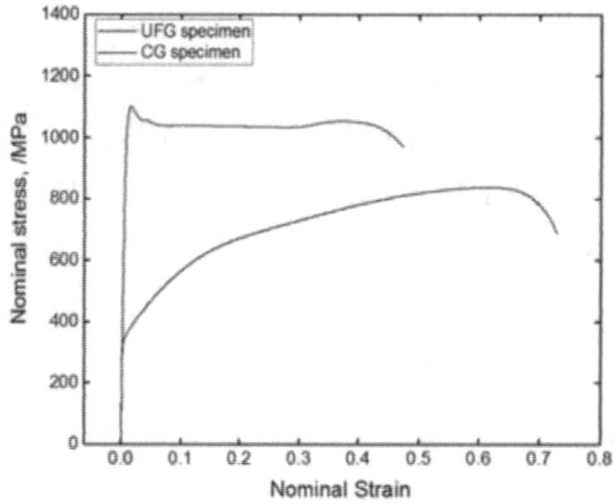

Figure 6: Engineering stress-strain curves of SUS304 austenitic steel specimens having a coarse grained (CG) microstructure with an average grain size of 17 μm and a UFG microstructure [12].

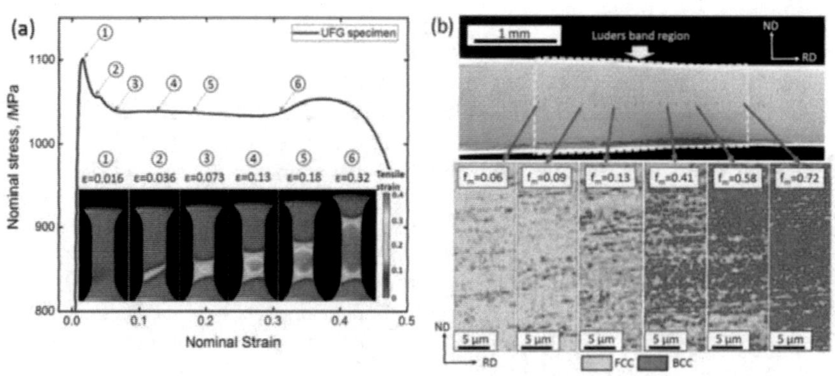

Figure 7: (a) Engineering stress-strain curve of the UFG specimen and local strain maps obtained from DIC analysis. Six strain maps are shown in the figure, corresponding to different tensile strains indicated by the numbers and arrows on the stress-strain curve. (b) SEM image (upper half) observed through the propagating Lüders band in the tensile specimen, and EBSD phase maps (lower half) obtained at different positions of the Lüders band region along the tensile direction (parallel to RD) on the section perpendicular to TD. Austenite (FCC) and martensite (BCC) are indicated by green and red colors, respectively. [12]

In order to understand the nature of the huge Lüders-type deformation in the UFG SUS304 specimen, DIC analysis on the specimen scale was carried out. The surface of the sheet-type tensile specimen was painted in white first, and then black dots were sprayed on the surface. The dots played the role of markers to track the displacements of the material at different positions. A video of the surface was captured during tensile testing at RT, and DIC analysis was conducted by the use of the VIC2D software. The DIC result for the UFG specimen is shown in figure 7(a) [12]. Around the stress peak of the yield-drop, a band of localized plastic strain appeared near the lower left shoulder of the tensile specimen. Plastic strains furthermore localized in the band, but macroscopic necking stopped, and the front of the strain localized band propagated along the gage part of the specimen, characteristic of typical Lüders deformation. Tensile testing of another UFG specimen was halted during the propagation of a Lüders band, after which the microstructure was analyzed by EBSD (figure 7(b)). The EBSD maps clearly indicated that deformation induced martensitic transformation took place within the Lüders band, and the fraction of martensite reached up to 72%. Another tensile test of the UFG specimen combined with DIC analysis was also conducted at 200°C. The results are shown in figure 8 [12]. The stress-strain curve at 200°C was significantly different from that at RT. A yield-drop was also observed and strain-localized band appeared in the specimen. However, plastic strain continuously localized within the band, leading to macroscopic necking to failure. EBSD phase analysis showed that deformation induced martensitic transformation within the band was greatly suppressed at 200°C, at which temperature the austenite becomes more stable against martensitic transformation. The results clearly indicate that a sufficient amount of martensite is necessary to allow propagation of the Lüders band resulting in good tensile elongation. That is, the propagation of the Lüders band in the UFG specimen could be understood in terms of overcoming the plastic instability condition (equation 1) by extra hardening (strain-hardening) brought about by deformation induced martensite. In other words, the yield-drop observed in this UFG material is essentially the same as the initiation of macroscopic necking.

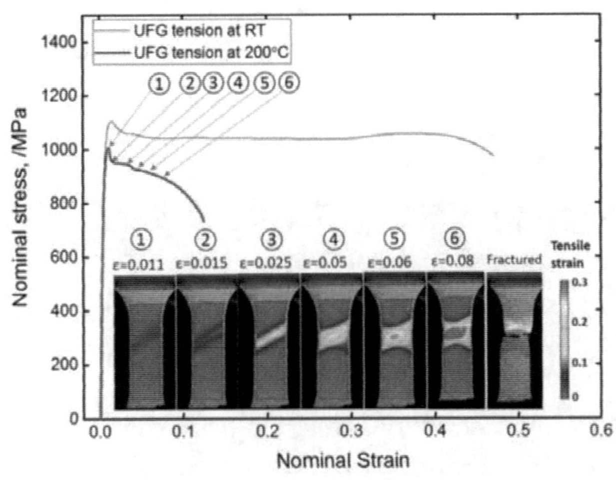

Figure 8: Nominal stress-strain curve of the UFG specimen tensile tested at 200℃ and local strain maps corresponding to different deformation stages, obtained by DIC analysis. The stress-strain curve obtained at room temperature is also shown for comparison. [12]

IOP Conf. Series: Materials Science and Engineering **580** (2019) 012012 doi:10.1088/1757-899X/580/1/012012

4. Digital Image Correlation (DIC) analysis in microstructure-scale

DIC analysis can be applied to micrometer or even nanometer scales, by making the microstructure itself act as the markers for tracking. Results of the a micro-scale DIC analysis on the deformation of a dual-phase (DP) steel composed of soft ferrite phase and hard martensite phase are shown here. A low-carbon steel (Fe-0.087C-0.79Si-1.77Mn: mass%) was used. Two different kinds of DP structures, having different distribution morphologies of martensite were prepared. In the first case, the steel was austenitized at 950°C for 1.8 ks and then air-cooled down to RT to obtaining a ferrite + pearlite structure. The steel was then intercritically annealed at different temperatures from 760°C to 840°C, where the ferrite (BCC) and austenite (FCC) two phases are stable. After intercritical annealing for 1.8 ks, the specimens were water-quenched, during which thet austenite phase transformed to martensite, to make a DP structure. The volume fraction of martensite in the obtained DP structures varied depending on the different intercritical annealing temperature. In the second case, the steel was also autenitized at 950°C for 1.8 ks, but cooled in this case directly to an intercritical temperature in the range from 730°C to 820°C to achieve a austenite + ferrite microstructure. After intercritical annealing for 1.8 ks the samples were water-quenched to obtain DP structures composed of ferrite and martensite.

In the first heat treatment route, the austenite phase preferentially appeared along grain boundaries of ferrite during intercritical annealing, resulting in network-like morphologies of the austenite phase. As a result, DP structures having network distributions of martensite were obtained in this route. In contrast, the ferrite phase precipitated from austenite during intercritical annealing in the second heat treatment route, and the remaining austenite transformed into martensite during water-quenching. As a result, the DP structures obtained in the second route contained martensite with an isolated morphology. Two types of DP structures obtained through the former and latter routes are hereafter called networked-DP and isolated-DP, respectively.

SEM microstructures of two types of DP structures are shown in figure 9. The volume fraction of martensite (V_m) was kept at 28% for both cases, by adjusting the intercritical annealing temperatures in the different heat treatment routes. In the SEM images, black areas correspond to the soft ferrite phase, while light regions having substructure correspond to the hard martensite phase. The average grain size of ferrite (D_F) and martensite (D_M) are also indicated in figure 9. It was confirmed by nano-indentation mapping that average hardness values of ferrite and martensite were nearly the same for the networked-DP and isolated-DP structures having equivalent martensite fractions, V_m.

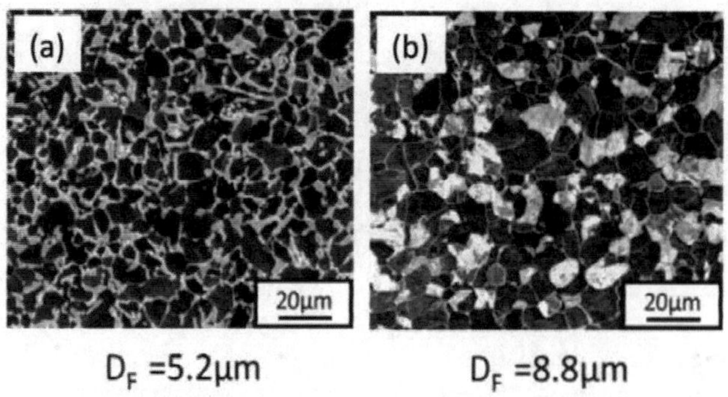

Figure 9: SEM images of two different types of DP structure having martensite volume fraction (V_m) of 28%. (a) Networked-DP. (b) Isolated DP. Average grain sizes of ferrite (D_F) and martensite (D_M) are also indicated.

Figure 10 shows true stress-strain curves and strain-hardening rate curves of the networked-DP and isolated-DP specimens with $V_m = 28\%$. The two specimens showed similar strength levels, but interestingly the networked-DP showed larger tensile elongation compared to the isolated-DP. In fact, the decreasing strain-hardening rate ($d\sigma/d\varepsilon$) with increasing tensile strain was suppressed in the networked-DP, compared with that in the isolated-DP. For understanding the reason for the differences in the global strain hardening and the tensile elongation between two types of DP structures, microscopic DIC analysis was conducted on identical areas in the DP specimens. The DP specimens were tensile tested at RT, with the tensile tests interrupted at different tensile strains to allow recording of the microstructure of a fixed area in the SEM. Local strain maps showing distributions of plastic strains in the microstructures obtained by the DIC analysis are shown in figure 11. The distribution of local equivalent plastic strain (ε_{eq}) for each macroscopic strain (e) is indicated by the different color scales below each map. The strain maps are superimposed on SEM images of the microstructure, where dark regions and light regions correspond to soft ferrite and hard martensite, respectively. In both specimens it was observed that the plastic strain localized in the soft ferrite grains (dark regions). However, the strain distributions in the networked-DP structure were more homogeneous than those in the isolated-DP structure. The plastic strain was more localized in soft ferrite in the isolated-DP, and the strain localized regions detoured the hard martensite. That is, partitioning of plastic strain between soft ferrite and hard martensite was more homogeneous in the networked-DP structure than in the isolated-DP structure, which interestingly resulted in better global strain-hardening. The more localized strain in the isolated-DP can be expected to lead to macroscopic localization, which might correspond to the smaller post-uniform elongation and the earlier fracture of the isolated-DP specimen compared to the networked-DP specimen. It can be concluded from the DIC results that local heterogeneity in deformation certainly controls the global deformation behavior in DP steels.

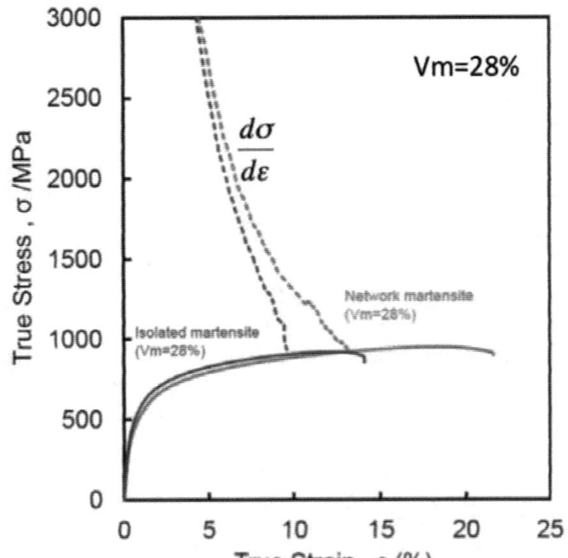

Figure 10: True stress-strain curves and corresponding strain-hardening rate curves of specimens with the networked-DP and isolated-DP structures shown in figure 9. The volume fraction of martensite (V_m) was 28% in both specimens.

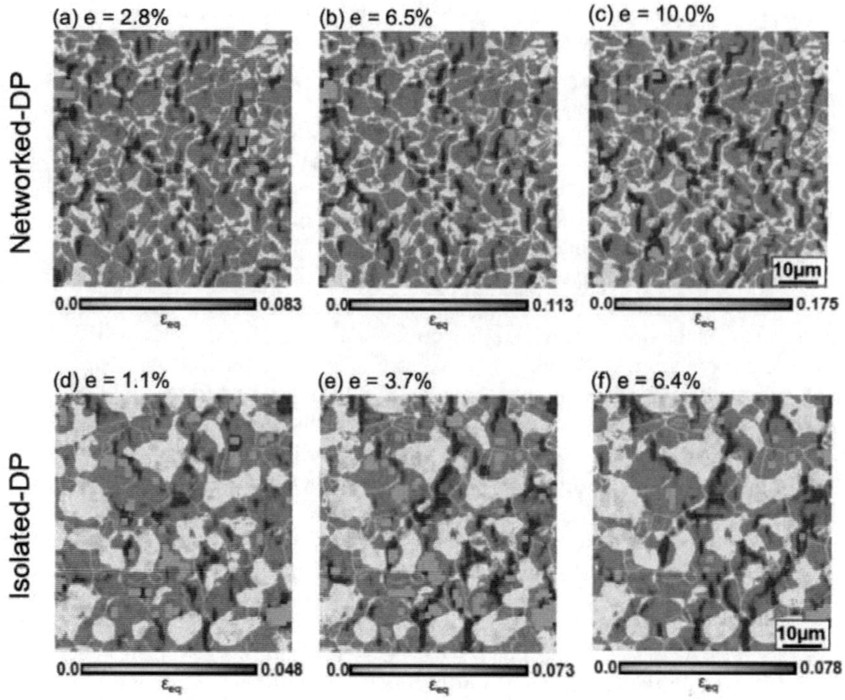

Figure 11: Local strain maps for two types of DP structures at different tensile strain (*e*) levels. (a, b, c) Networked-DP. (d, e, f) Isolated-DP. Distributions of equivalent plastic strains (ε_{eq}) are expressed in different color scales indicated below each maps. Strain maps are superimposed on SEM microstructures, where dark regions and light regions correspond to soft ferrite and hard martensite, respectively. The tensile direction is parallel to the horizontal direction.

5. Summary

Several experimental results from our efforts to link local and heterogeneous deformation behavior to the global mechanical response in metallic materials with different microstructures have been introduced. The results confirmed the importance of local heterogeneities in deformation for understanding the global mechanical properties of bulk metals. We may also conduct another experimental techniques to quantitatively analyze deformation parameters of materials. For example, *in-situ* synchrotron diffraction during tensile deformation can provide quantitative and precise values of parameters such as the dislocation density, the fraction of deformation twins or deformation induced martensite, and the internal stresses in different phases, during deformation dynamically. In-situ neutron diffraction will allow for the possibilty to get such parameters from large bulk specimens. We now have many tools for deepening our understanding of the deformation of materials.

Acknowledgments
This work was financially supported by the Elements Strategy Initiative for Structural Materials (ESISM), and the Grant-in-Aid for Scientific Research (S) (No.15H05767), both

through the Ministry of Education, Culture, Sports, Science and Technology (MEXT), Japan. The supports are gratefully appreciated.

References

[1] Weertman J and Weertman J R 1992 *Elementary Dislocation Theory* (Oxford: Oxford University Press)
[2] Hansen N 2001 *Metall. Metall. Mater. Trans. A* **32A** 2917-2935
[3] Demeri M Y 2016 *Advanced High-Strength Steels: Science, Technology and Applications* (Materials Park, Ohio: ASM International)
[4] http://www.autosteel.org/
[5] Froes F H 2015 *Titanium: Physical Metallurgy, Processing, and Applications* (Materials Park, Ohio: ASM International)
[6] Chong Y, Bhattacharjee T, Park M-H, Shibata A and Tsuji N 2018 *Mater. Sci. Eng. A* **730** 217-222
[7] Chu T C, Ranson W F and Sutton M A 1985 *Exp. Mechanics* **25** 232-244
[8] Sutton M A A, Orteu J J and Schreier H 2009 *Image Correlation for Shape, Motion and Deformation Measurements: Basic Concepts, Theory and Applications* (Neew York: Springer)
[9] He B B, Hu B, Yen H W, Cheng G J, Wang Z K, Luo H W, and Huang M X 2017 *Science* **357** 1029-1032
[10] Furuta T, Kuramoto S, Ohsuna T, Oh-ishi K and Horibuchi K 2015 *Scripta Mater.* **101** 87-90
[11] Zhang Y, Wang L, Findley K O and Speer J G 2017 *Metall. Mater. Trans. A* **48** 2140-2149
[12] Gao S, Bai Y, Zheng R, Tian Y, Mao W, Shibata A and Tsuji N 2019 *Scripta Mater.* **159** 28-32.

IOP Conf. Series: Materials Science and Engineering **580** (2019) 012013 doi:10.1088/1757-899X/580/1/012013

Multi-dimensional characterization and controlling of microstructure evolution during solidification of metallic alloys

Tongmin Wang

Key Laboratory of Solidification Control and Digital Preparation Technology (Liaoning Province),School of Material Science and Engineering, Dalian University of Technology, Dalian 116024, China

Email: tmwang@dlut.edu.cn

Abstract. Metallic alloys are important structural and functional materials and have been widely applied in industry. The performance of metallic alloys strongly depends on the microstructures after solidification. Many effective methods are introduced to control the microstructure evolution during solidification, including grain refining, adding rare earth (RE) elements, applying electric current and/or magnetic field, etc. However the inherent mechanisms affecting the microstructure formation are still unclear due to the limitation of real time and multi-dimensional characterization technique. With the development of the third generation synchrotron radiation light source, a real time imaging technology in 2D (two dimensional) or 3D (three dimensional) view with strong penetrability, high spatial and temporal resolution, nondestructive features has been developed to break those limitations. In this paper, some application cases during solidification performed in our group are reported in different alloys and using different solidification controlling techniques. Different solidification behaviors in those cases are revealed and discussed.

Keywords. Metallic alloys; solidification control; in-situ observation; microstructure evolution; synchrotron radiation

1. Introduction

Microstructure evolution during solidification can be divided into two stages: grain nucleation and growth, which determine the final solidification structure. Finding an effective way to better control the grain nucleation and growth is the goal of solidification researchers. The key is to accurately understand the working mechanism on the grain nucleation and growth with different solidification controlling ways.

The convection in a solidifying melt can significantly affect the solute distribution directly in metallic alloys and hence the solidified microstructures [1, 2]. Many studies have shown that the application of external fields on the solidifying melts is an effective method to achieve high quality castings, either directly by stirring [3, 4] or indirectly by applying an electromagnetic field [5-7] or ultrasonic field [8, 9]. Among all the above mentioned methods, electromagnetic field offers an efficient way for a contactless influence on the solidifying melts, resulting in the convective transport of solute and the destabilization of growth conditions which lead to grain refinement and morphological change [10-14].

IOP Conf. Series: Materials Science and Engineering **580** (2019) 012013 doi:10.1088/1757-899X/580/1/012013

Besides, the addition of a grain refiner master alloy to liquid melt before pouring, such as adding Al-Ti-B into aluminum alloys, has been well-documented to be highly effective in grain refinement [15, 16]. Adding trace metal additives, such as Ni, Zn and rare-earth (RE) elements into liquid melt is also a common way to improve the microstructure and thus the mechanical performance [17, 18].

However, traditional experimental investigations are mainly restricted to a post-mortem analysis of the solidified microstructures [3, 8, 10-14] and fail to reveal the dynamics about the interactive relationships between the solute distribution and the microstructure formation under the electromagnetic field. Though experiments of transparent organic alloys [9, 19-21] delivered valuable information for verification of solidification theories, these experiments do not really represent the situation in a solidifying metal alloys due to the differences in the thermophysical parameters. Now, synchrotron X-ray radiography with satisfying spatial and temporal resolution has become an important diagnostic tool for *in situ* and real time observation of solidification process in opaque metal alloys [22-27].

The main research in our group is to characterize and control the microstructure evolution during solidification of metallic alloys. Since 2006, the solidification behavior of different alloys (e.g. Sn-, Al-, Zn-, Cu-) under different controlling ways (e.g. grain refiner, RE, electric current, magnetic field) have been studied in our laboratory to reveal the controlling mechanisms on microstructure evolution. Some typical cases are selected to be reported in this paper.

2. Case I: 3D microstructure of Al-Bi immiscible alloy with different grain refiners

In this case, the 3D microstructures of Al-Bi alloys with different grain refiners (Al-5Ti-B, Al-3B and Al-3Ti) were studied using synchrotron X-ray microtomography. Samples were prepared by high purity Al and Bi (both purity ⩾ 99.99%). The investigated Al-20wt. %Bi alloys with or without 1wt. % Al-3B, Al-5Ti-B and Al-3Ti grain refiners were melted in a graphite crucible and then cast into a water-cooled copper mold. Small cylindrical samples of 1 mm in diameter and 20 mm in height were machined at a distance of 10 mm from the bottom of the ingots (Φ45 mm×65 mm) along the transverse direction and used for the synchrotron X-ray microtomography experiments. The tomography experimental set up is illustrated in Figure 1. The experiments were carried out on the beam line BL13W1 at Shanghai Synchrotron Radiation Facility (SSRF), a 3rd generation synchrotron radiation facility in China. During the experiment, the sample was fixed on a stainless steel rod mounted on a 6D assembled Kohzu rotating stage.

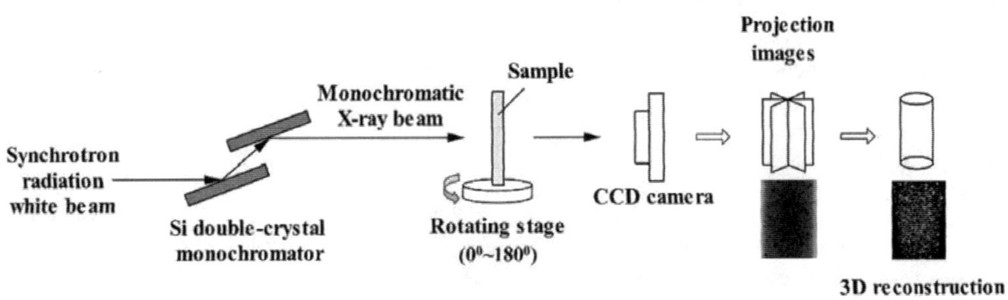

Figure 1. Schematic of synchrotron X-ray microtomography experimental setup[27].

The pixel size was 0.74 μm and the sample-to-detector distance was set at 3 cm. A set of 1260 projections was obtained during the 180°C rotation with an exposure time of 2.7 s per projection. The 2D projections were reconstructed in 3D using a filtered-back-projection algorithm.

Figure 2 shows the projection images of Al-20wt. %Bi alloys with different grain refiners. It can be seen that the images show good contrast for Bi-rich particles (black) in the Al matrix (gray), which is beneficial to the subsequent 3D reconstruction of microstructures. The reference sample without grain refiners shows largely oversized particles with a non-uniform distribution. Samples with grain refiners are different from the reference sample. Slight reduction of the exceptional coarse Bi-rich particles by the grain refiners can be identified. Besides, the size and distribution of Bi-rich particles in the Al matrix inoculated with different grain refiners have been improved to some extent.

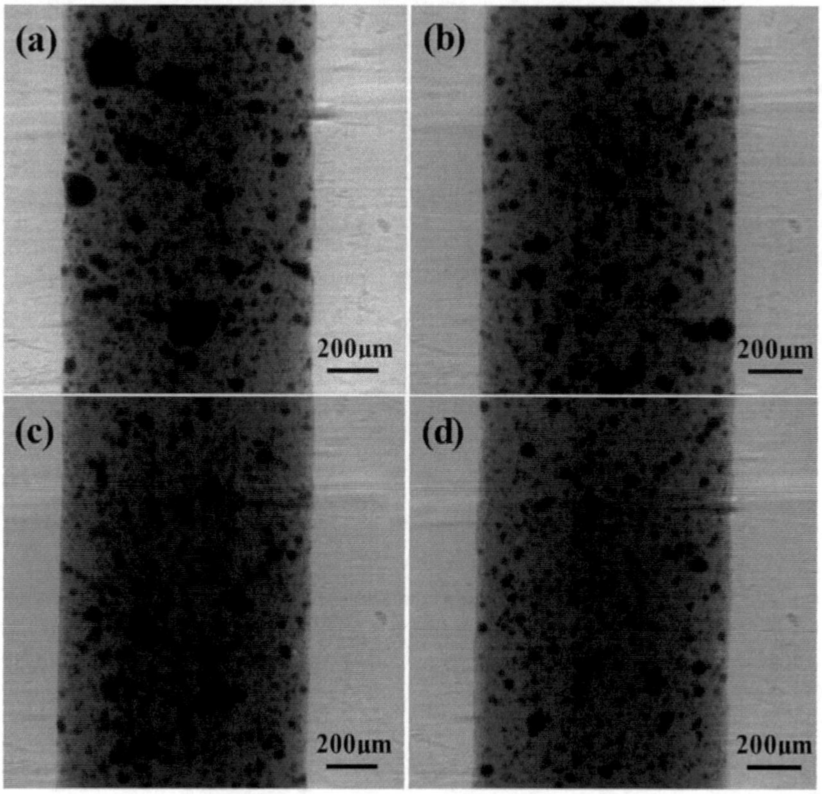

Figure 2. Projection images of cylindrical Al-20wt. %Bi alloy with different grain refiners: (a) reference sample without grain refiners; (b) with 1 wt. % Al-3B; (c) with 1 wt. % Al-5Ti-B; (d) with 1 wt. % Al-3Ti[27].

Three dimensional microstructures can clearly reveal the size, shape and distribution of the phases, providing a novel way to investigate the refining efficiency of different grain refiners. Figure 3 shows the 3D X-ray microtomography images of the Al-20wt. %Bi alloys with different grain refiners and the size distribution obtained by applying image analysis on the 3D X-ray microtomography images. Without any grain refiner inoculated in Al-Bi alloy, the Bi-rich particles are very large with irregular morphology and are non-uniformly distributed in the Al matrix (Figure 3a). This microstructure essentially derives from the decomposition of homogeneous hypermonotectic liquid, $L \rightarrow L_1 + L_2$, where Bi-rich L_2 droplets nucleate and grow by diffusion in the Al-rich L_1 matrix melt until the monotectic temperature is reached. The former decomposes at the monotectic temperature by the reaction $L_1 \rightarrow S_1 + L_2$, and the latter then solidifies at the eutectic temperature through the reaction $L_2 \rightarrow S_1 + S_2$, where S_1 is solid phase of nearly pure Al and S_2 is Bi-rich particle. During solidification, the Bi-rich droplets may be applied by several forces, such as gravity, Marangoni force due to temperature difference and interfacial tension. As a result of the collision and coalescence, some of those droplets grow faster than others before eutectic reaction (solidification of L_2) and inhomogeneous microstructures are formed consequently.

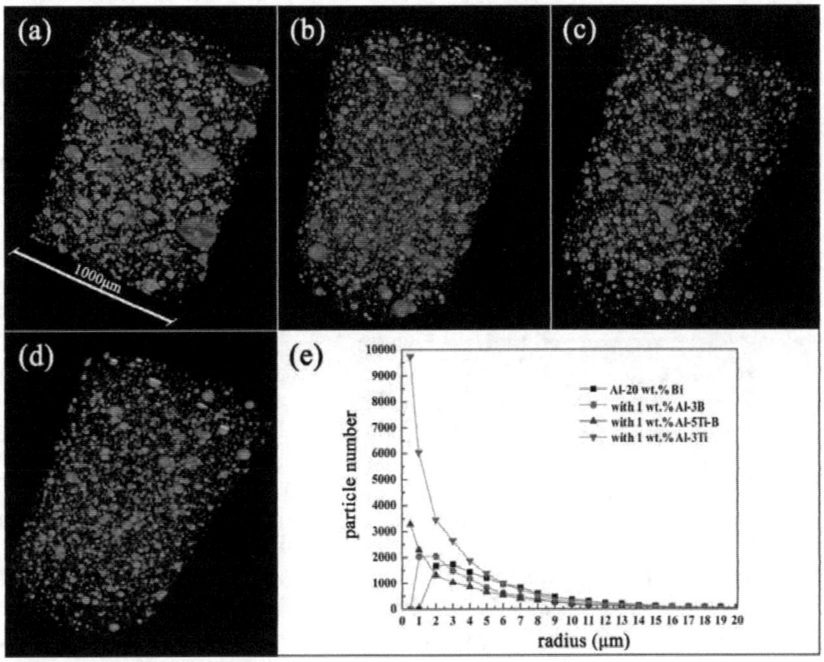

Figure 3. Three dimensional X-ray microtomography images of Al-20wt.%Bi alloys with different grain refiners, showing the size and spatial distribution of Bi-rich particles in Al matrix: (a) reference sample without grain refiners; (b) with 1 wt.% Al-3B; (c) with 1 wt.% Al-5Ti-B; (d) with 1 wt.% Al-3Ti; (e) Size distribution of Bi-rich particles. The Al matrix has been made transparent for better presentation of the Bi-rich particles[27].

When Al-3B grain refiner is added, however, the morphology and distribution of the Bi-rich particles in Al matrix are slightly improved (Figures 3b and e). The most prominent change, which was brought to the Al-20wt. %Bi alloy refined by Al-3B, is the elimination of exceptional coarse Bi-rich particles. Considering the good nucleating potency of AlB_2 substrate to Al, this change may take place in the stage $L_1 \rightarrow S_1 + L_2$. The refined Al reduces the possibility of collision and coalescence of Bi-rich L_2 droplets, but the Bi-rich L_2 droplets might be coagulated with each other by the growing α-Al grains. Therefore, similar particle distributions are found in the alloys with or without addition of Al-3B.

Unlike Al-3B grain refiner, the Bi-rich particles are refined to a significant extent by adding Al-5Ti-B grain refiner, although the distribution of particles is still not very uniform (Figures 3c and e). In the Al-5Ti-B mater alloy, $TiAl_3$ and TiB_2 are the dominant intermetallic phases. Again the good nucleating potency of $TiAl_3$ and TiB_2 helps to eliminate the exceptional coarse Bi-rich particles and increase the number of Bi-rich L_2 droplets. It is inferred that TiB_2 has an analogous effect in treating the Al-Bi melt. Hence the refined Bi-rich particles in the Al-20wt. %Bi alloy by adding Al-5Ti-B are attributed to the $TiAl_3$ and TiB_2 particles in the refiner.

As shown in Figures 3(d) and (e), the size and spatial distribution of Bi-rich particles are further improved when Al-3Ti refiner is inoculated and the radius size of over 60% Bi-rich particles is less than 1 um. The Bi-rich particles are not only more spherical in morphology with smaller size but also much more uniform in distribution after adding Al-3Ti grain refiner. Needless to say, this improvement is ascribed to the presence of $TiAl_3$ introduced by Al-3Ti master alloy. The refinement mechanism of the Al-20wt. %Bi alloy by adding Al-3Ti is analogous to that by adding Al-5Ti-B. The relatively weaker refining efficiency of Al-5Ti-B with respect to Al-3Ti in treating Al-Bi alloy may be due to the agglomeration of TiB_2 in the former.

3. Case II: Growth behaviour of Cu_6Sn_5 during solidification in RE-doped Sn–Cu solder alloys

In this case, the growth behaviour of Cu_6Sn_5 intermetallic compounds (IMCs) in rare earth (RE, La)-doped Sn–Cu solder alloys was investigated using *in situ* synchrotron radiation imaging technology. The raw materials were button-like Sn ingot (99.99 wt. % purity), Cu–La master alloy (9.33 wt. % La purity) and Cu foil (99.99 wt. % purity). The Sn–6.5 Cu and Sn–6.5 Cu–0.06 La alloys (all compositions in wt. %) were prepared by those raw materials molten in a resistance furnace. After being held in the furnace for 30 min at certain temperatures (300±10 °C, 520±10 °C and 550±10 °C respectively), the melt was chill cast as an ingot in a copper mold for the preparation of synchrotron radiation experiment. The imaging method was coaxial phase-contrasting imaging method. The prepared specimens were completely melted with a superheat of 20 K and held for 2 min to homogenize the composition, then cooled down at the same cooling rate of 3 °C/min as that of as-cast specimens mentioned above. The view size of CCD camera was 2048 × 2048 pixel^2 and the pixel size was down to 3.7 μm. The energy of the X-ray beam and the exposure time were 25 keV and 1 s respectively, which were quite adequate to capture the growth behaviour of IMCs.

Figure 4 shows the time-sequence images, where the growth behaviour of Cu_6Sn_5 in various compositions of Sn–Cu alloys were obtained. The time when the specimen began to cool down was set as initial time (0 s). With the time increasing, the Cu_6Sn_5 of each shape began to nucleate at different instants of time, and then grew. The centres of these shapes were Cu_6Sn_5 nucleation sites. Obviously, the I-like structure had just two growth directions with an angle of 180°; the Y-like structure had three growth directions with two obtuse angles (nearly 150°) and one acute angle (nearly 60°); the bird-like structure was small and irregular with several uncertain angles. In general, the morphology of Cu_6Sn_5 is closely related to crystal structure [28]. Cu_6Sn_5 with angles of > 186 °C possesses a hexagonal structure, whose orientation indexes are <0001> = [0001], [000$\bar{1}$] with angles of 60°, 150° or 180° [29]. Energy densities along these three directions are lower than those along other directions; therefore crystals grow preferentially along those directions [30].

La has a significant effect on the regularity and refinement of Cu_6Sn_5. In general, the number of bird-like Cu_6Sn_5 in Sn–6.5 Cu–0.06 La alloy was smaller than that in Sn–6.5 Cu alloy, providing an evidence that doping La had led to the reduction of bird-like Cu_6Sn_5 that was beneficial to the homogenization of microstructures, as shown in Fig. 4. Comparing Fig. 4a with Fig. 4b, it was clearly seen that Cu_6Sn_5 began to nucleate at 50 s and then grew to the final sizes at about 850 s in Sn–6.5 Cu alloy. However, in Sn–6.5 Cu–0.06 La alloy, they emerged to nucleate at 15 s and grew to the final sizes at 815 s, meaning that adding La lowered the nucleation energy, and thus the critical nucleus can be formed more easily. Here, a thermodynamic analysis is proposed to demonstrate the effect of La on the growth of Cu_6Sn_5. According to the total free energy change during heterogeneous nucleation [31], the equation is expressed as follows:

$$\Delta G_{het} = \left(\frac{4}{3}\pi r^3 \Delta G_V + 4\pi r^2 \sigma_{\alpha L}\right) f(\theta) = \Delta G_{hom} f(\theta) \qquad (1)$$

Where ΔG_{het} is the total free energy change of heterogeneous nucleation, ΔG_{hom} is the total free energy change of homogeneous nucleation, θ is the contact angle, which is considered to be constant. An assumption is $\frac{dG}{dr}$, thus the critical nucleus radius is deduced as follows:

$$r^* = -\frac{2\sigma_{\alpha L}}{\Delta G_V} \qquad (2)$$

Combined equation (a) and (2), the nucleation energy of heterogeneous nucleation is generated as follows:

$$\Delta G_{het}^* = \Delta G_{hom}^* f(\theta) < \Delta G_{hom}^* \qquad (3)$$

Equation (3) shows that the nucleation energy of heterogeneous nucleation is less than that of homogeneous nucleation. Actually, all of the real solidification cases are heterogeneous nucleation. The contact angle θ is responsible for the facilitation of substrate on the heterogeneous nucleation. That is, the smaller θ is the better facilitation of substrate on the heterogeneous nucleation will be. Since La element is very active, there is a larger possibility for La and Sn elements to form a compound [32] in the liquid melt as a nucleation site. Therefore, La element plays a role of nucleation agents, which significantly decreases the contact angle θ.

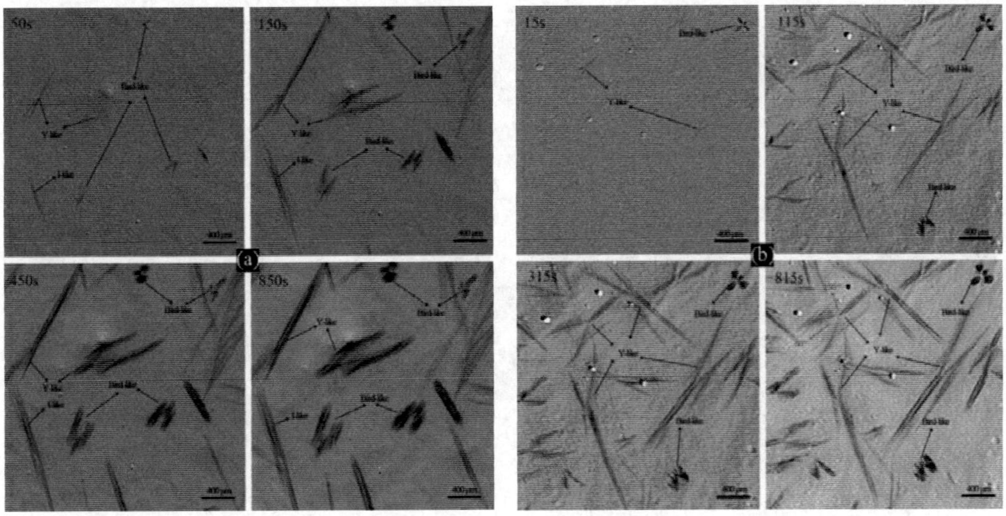

Figure 4. Sequences of *in situ* radiographs showing the growth behavior of Cu_6Sn_5 in solidifying Sn–Cu solder alloys. The cooling rate is 3°C/min: (a) Sn–6.5 Cu alloy; (b) Sn–6.5 Cu–0.06 La alloy[33].

4. Case III: Evolution of dendrite morphology of Sn-Bi alloy under an applied electric current

In this case, we focused on observing the evolution of dendrite morphology of a solidifying Sn-Bi alloy under an applied direct current (DC) field. A binary alloy with a composition of Sn-12wt%Bi was prepared. The behaviour of dendrite growth can be directly observed due to the X-ray absorption difference between Sn and Bi. The high temporal and spatial resolutions for clear and fast imaging can be achieved by using synchrotron X-rays. The DC was imposed on the alloy by connecting the sample to a DC generator. Two pieces of aluminium foils acted as the positive electrode (upside) and negative one (downside), respectively. The alloy was melted completely with a superheat of 8 K, and then cooled at a cooling rate of 1.5 K/min. A small gradient might exist in the vertical direction of sample due to the internal construction of furnace, and it is hard to be measured quantitatively. Three experiments were performed: one case is the solidification without DC; others are with various electric current densities of 7 A/cm^2, 19 A/cm^2, respectively. DC was applied to the entire solidification process.

Figure 5 shows the evolution of grains' morphology of different experiments. Without DC, the grains grew in a columnar manner due to a small temperature gradient in the vertical direction, as shown in Fig. 5(a). Very thin primary, secondary, tertiary dendrite arms and sharp dendrite tips were also found. The columnar dendritic grains grew fast and finally reached a large size of around 2 mm.

Once a DC density of 7 A/cm^2 was imposed, the dendrite branching and arm growth were significantly restrained as shown in Fig. 5(b). The primary arms grew to be thick, and the secondary arms were suppressed to be degenerate. Both the primary and secondary arm tips tended to be round or even flat. The sharp tips and tertiary dendrites were hardly found. The grains tended to be an equiaxed cellular morphology.

With increasing DC density (~19 A/cm^2), as shown in Fig. 5(c), some primary arm tips tended to split and form concave pits. Several secondary dendrites started to branch obliquely, whose directions are no longer perpendicular to their normal parent-primary arms.

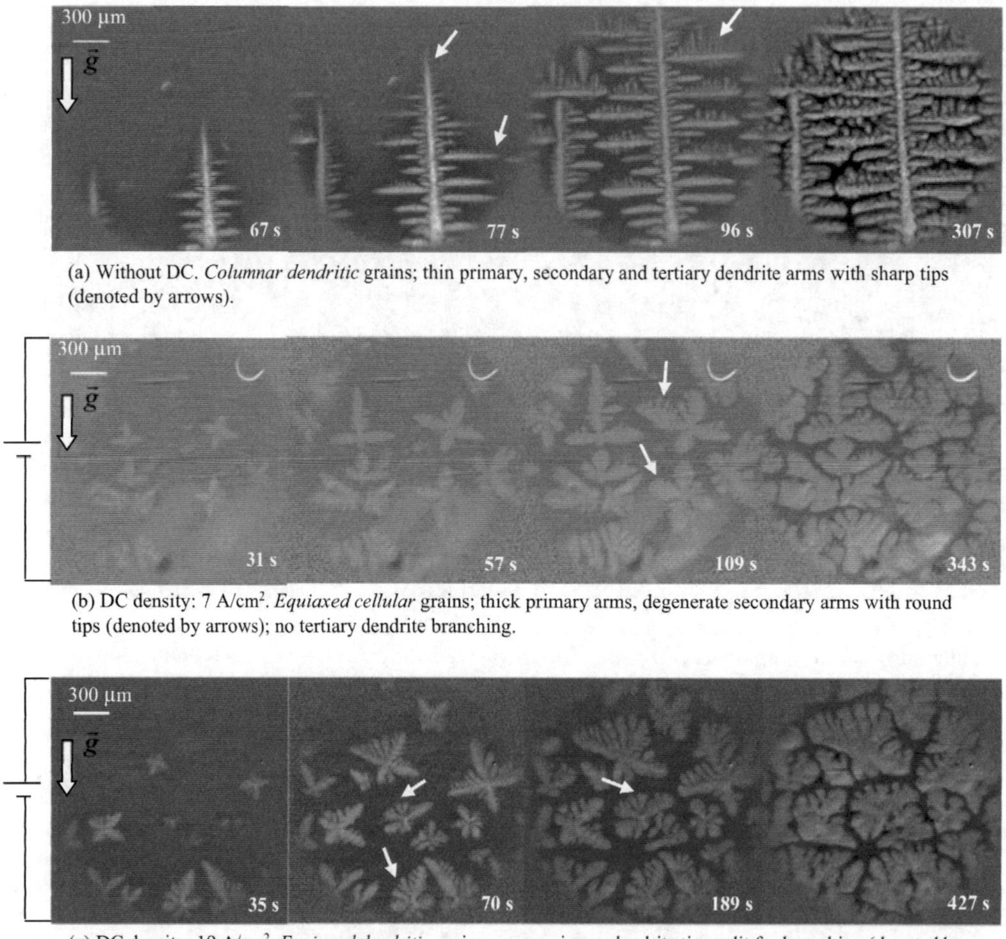

(a) Without DC. *Columnar dendritic* grains; thin primary, secondary and tertiary dendrite arms with sharp tips (denoted by arrows).

(b) DC density: 7 A/cm^2. *Equiaxed cellular* grains; thick primary arms, degenerate secondary arms with round tips (denoted by arrows); no tertiary dendrite branching.

(c) DC density: 19 A/cm^2. *Equiaxed dendritic* grains; some primary dendrite tips split for branching (denoted by arrows); no tertiary dendrite branching.

Figure 5. *In situ* observations on the dendritic growth of a solidifying Sn-12wt%Bi alloy under various DC conditions. The cooling rate is 1.5 K/min[34].

An explanation to the influence mechanism of DC on the evolution of dendrite morphology is schematically illustrated in Fig. 6 based on our *in situ* observations and the previous work by other researchers [35, 36]. Since the electrical conductivity of solid Sn-12wt%Bi alloy is much bigger than that of liquid alloy, the electric streamlines prefer to pass through the solid to liquid. They easily gather in a sharp tip of dendrite to generate much more heat there due to both the Joule heating effect and the current crowding effect. The heat causes a rising temperature around the sharp dendrite tip by which the tip growth can be significantly suppressed. Accordingly, the sharp tip tends to be round or even flat in a self-adaptive way to scatter the electric streamlines for reducing Joule heat in order to keep growing. However, with increasing electric current density, a special tip-split scenario occurred to the primary arm resulting in a 'branching' of the primary dendrite. As illustrated in Figure. 6, the solute is rejected gradually to the front of a growing dendrite tip during solidification. It is more difficult for the rejected solute to transversely diffuse from the tip center towards its lateral sides in case of a round or flat (higher DC density case) tip rather than a sharp one. The pileup solute becomes an obstacle opposite to the tip growing by changing the constitutional undercooling, thus a concave pit is formed. The pit is expended with the continuous growing of tip and the segregation of solute. Finally, a tip is split into two tips.

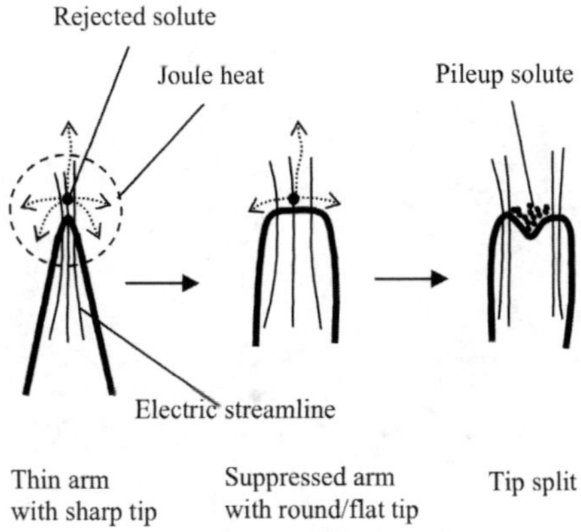

Figure 6. Schematic explanation to the influence of DC on the evolution of dendrite morphology.

5. Case IV: Effect of traveling magnetic field (TMF) on solute distribution and dendritic growth in unidirectionally solidifying Sn-Pb alloy

In this case, the solute distribution and the dendritic growth during the bottom-up solidification of Sn-50wt. %Pb alloy were studied by using synchrotron X-ray radiography. The TMF is generated by means of applying out-of-phase alternating-current (40 A) with the normal frequency of 50 Hz to the TMF generator which was placed at the bottom of the lower heaters. As a result, an electromagnetic force (EMF) from left to right is induced into the melt.

A series of solidification experiments with and without TMF were carried out as follows. Firstly, both the upper and lower plate heaters were set to a temperature above the liquidus temperature of Sn-50wt. %Pb alloy. The lower heater temperature was reduced to trigger the solidification when the sample was completely melted and homogenously distributed. The temperature along the sample cell was measured by two embedded K-type thermocouples placed 1 cm away from each other. The cooling rate of the sample was $R = 5\ K\ min^{-1}$. The mean temperature gradient over the sample was $G = 6.3\ K\ cm^{-1}$ for all experiments. As a consequence, the dendrite front grew from bottom to top (antiparallel g) unidirectionally. The TMF was applied after $t = 248$ s when the dendrite front had reached about one third of the field of view, $t = 0$ s is assigned to the onset of cooling.

Figure 7 shows a sequence of images recorded during the solidification of Sn-50wt. %Pb alloy with and without a TMF. Figure 7(a)-(d) shows the solidification of Sn-50wt. %Pb alloy without a TMF. The dark dendritic network denoted by the arrows corresponds to the primary Pb-rich dendrites growing from bottom to top. The Sn-rich zones can be identified in the interdendritic regions and in the vicinity of the dendrite front as a consequence of solute rejection. The Sn-rich melt is less dense as compared to the melt with initial composition. As a consequence, the melt density difference in different areas provokes the formation of ascending plumes containing Sn-rich solute. The number, size and position of the plumes are varied during the solidification, as shown in Figure 7(a)-(d). Many small plumes arise at the early stage of the solidification, and then the plumes are well developed. The plumes span a distance of several millimeters and produce a long range diffusion of solute towards the melt region.

Finally, those small crystals are blocked by the overgrowing dendrite side arms. Figure 7(e)-(h) shows the development of solute concentration field and the evolution of dendritic structure under the influence of TMF. A forced melt flow from left to right is induced when the TMF is applied. As a consequence, the well-developed upright ascending plumes become unstable and tend to incline to the right denoted by the red arrow in Figure 7(e)-(g). Finally, the plumes on the right side of the solidification front gradually disappeared. In the process of solidification, the growth of Pb-rich dendrites causes an accumulation of rejected Sn in the interdendritic zone. The solute pile-up reduces the local undercooling and promotes a partly remelting of the dendritic structures labeled by the red ellipses in Figure. 7(h).

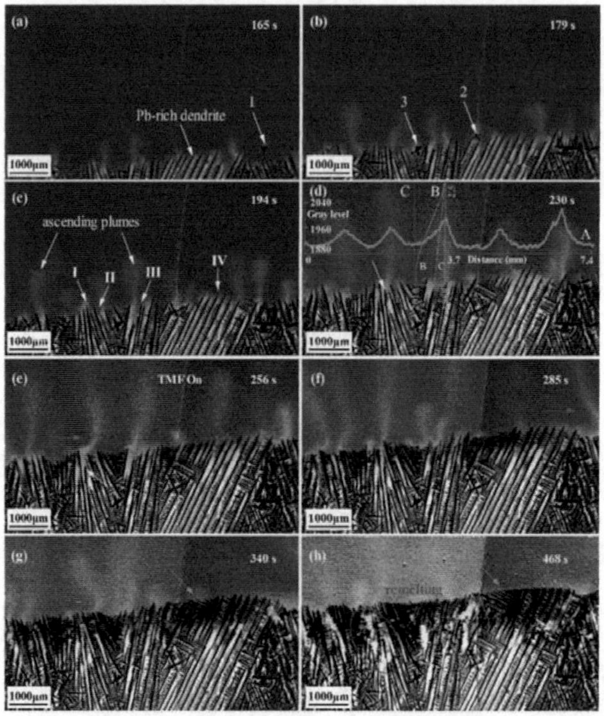

Figure 7. Sequence of *in situ* radiographs showing the solute distribution and the dendritic growth during the solidification of Sn-50wt. %Pb alloy: (a)-(d) solidification without TMF; (e)-(h) solidification under the influence of TMF ($t = 0$ s is assigned to the onset of cooling)[37].

The concentration variation of Sn can be reflected by the evolution of the gray level extracted from the sequenced radiographs. Since the X-ray transmittance in Sn is higher than that in Pb, the gray level is proportional to the Sn concentration in Sn-Pb system. The gray level curves are obtained by smoothing the measured points. The blue curve in Figure 7(d) shows the gray level profile along the red horizontal line A. A wavy distribution of the Sn-rich solute along the solidification front is presented corresponding to the plume pattern. The orange curves in Figure 7(d) show the gray level distribution in the melt ahead of the solid-liquid interface measured at positions of vertical lines B and C, respectively. Higher solute concentration at the interface is found for the position of line B where plume occurs.

Figure 8(a) displays the evolution of the gray level at the solid-liquid interface in horizontal direction (X = 0, starting from the left). A stable wavy distribution of the Sn-rich solute at the solidification front is disturbed when the TMF was applied. The significant fluctuation of the solute profile on the right side disappears gradually as a consequence of forced convection induced by the TMF. However, the fluctuation of the solute profile on the left side still exists due to the weak EMF at this area. Figure 8(b) shows the gray value profile ahead of the interface (Z=0) front recorded along the vertical line B (X = 3622.3 μm) at different time.

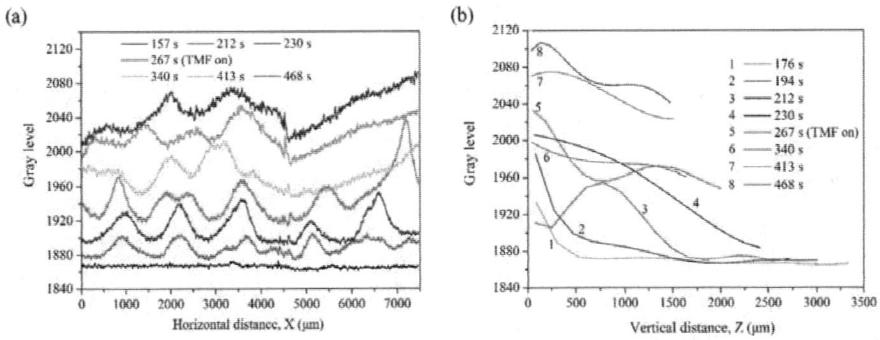

Figure 8. Evolution of the gray level ahead of the solid-liquid interface. (a) Gray value profile in horizontal direction at the interface; (b) gray value profile ahead of the interface front recorded along the vertical line B[37].

Figure 9 shows the schematic of the forced melt flow that interact with the dendrites at different inclination angles to illustrate the mechanism of the growth behaviour of the secondary arms under TMF. As shown in Figure 9, the Sn-rich solute at the left side of dendrite II and the right side of dendrite IV is not easy to be transported into the melt when the melt flows from left to right. As a consequence, the growth of the secondary arms at these areas is suppressed due to the solute enrichment. While the Sn-rich solute transportation at the red elliptic areas shown in Figure. 9 is enhanced by the forced melt flow, which facilitates the growth of secondary arms.

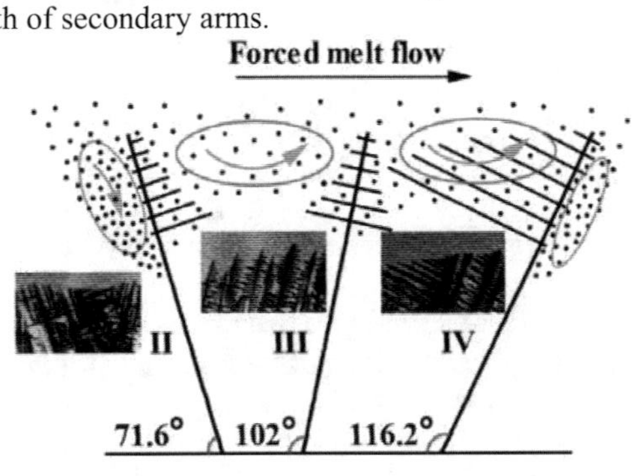

Figure 9. Schematic of the forced melt flow interacting with the dendrites at different angles[37].

6. Conclusions

(1) Synchrotron radiation real-time imaging technology provides an *in situ*, nondestructive way to investigate the solidification behaviour of alloys.

(2) The Bi-rich particles in Al-20wt. %Bi alloy are more uniformly distributed and spherical with finer size with the addition of Al-3Ti grain refiner.

(3) In RE(La)-doped Sn–6.5wt.%Cu alloy, the morphology of Cu_6Sn_5 tended to be more uniform owing to the adsorption effect of RE. In addition, La played a role of nucleating agent, which significantly increased the nucleation rate due to the decrease of nucleation energy.

(4) When DC density is below a critical value, the microstructure of Sn-12wt. %Bi alloy can be refined prominently which is ascribed to the enhanced nucleation and depressed growth rate.

(5) A forced melt flow in Sn-50wt.%Pb alloy melt from left to right is induced with the application of TMF. The forced melt flow ahead of the solidification front facilitates the solute transportation and reduces the local fluctuations of solute concentration considerably.

Acknowledgments
The authors gratefully acknowledge the supports of National Key Research and Development Program of China (No. 2017YFA0403803), the National Natural Science Foundation of China (Nos. 51525401, 51774065, 51601028, 51690163).

References

[1] R. Trivedi, P. Mazumder and S.N. Tewari 2002 The effect of convection on disorder in primary cellular and dendritic arrays *Metall. Mater. Trans. A* **33** 3763-75
[2] H. Nguyen Thi, Y. Dabo, B. Drevet, M.D. Dupouy, D. Camel, B. Billia, J.D. Hunt and A. Chilton 2005 Directional solidification of Al-1.5wt% Ni alloys under diffusion transport in space and fluid-flow localisation on earth *J. Cryst. Growth* **281** 654-668
[3] D. Brabazon, D.J. Browne and A.J. Carr 2002 Mechanical stir casting of aluminium alloys from the mushy state: process, microstructure and mechanical properties *Mater. Sci. Eng. A* **326** 370-381
[4] Z. Fan 2002 Semisolid metal processing *Int. Mater. Rev.* **47** 49-85
[5] S. Asai 2000 Recent development and prospect of electromagnetic processing of materials *Sci. Technol. Adv. Mater.* **1** 191-200
[6] H.K. Moffatt 1991 Electromagnetic stirring *Phys. Fluids A* **3** 1336-43
[7] P. Davidson 1999 Magnetohydrodynamics in materials processing *Annu. Rev. Fluid. Mech.* **31** 273-300
[8] R. Haghayeghi and P. Kapranos 2014 Solidification of Pb-Sn alloys under ultrasonic field *Mater. Lett.* **126** 244-248
[9] D. Shu, B. Sun, J. Mi and P. Grant 2012 A High-Speed Imaging and Modeling Study of Dendrite Fragmentation Caused by Ultrasonic Cavitation *Metall. Mater. Trans. A* **43** 3755-66
[10] L. Wang, J. Shen, L. Qin, Z. Feng, L. Wang and H. Fu 2012 The effect of the flow driven by a travelling magnetic field on solidification structure of Sn-Cd peritectic alloys *J. Cryst. Growth* **356** 26-32
[11] Y.J. Li, W.Z. Tao and Y.S. 2012 Yang Grain refinement of Al-Cu alloy in low voltage pulsed magnetic field *J. Mater. Process. Technol.* **212** 903-909
[12] J.C. Jie, Q.C. Zou, J.L. Sun, Y.P. Lu, T.M. Wang and T.J. Li 2014 Separation mechanism of the primary Si phase from the hypereutectic Al–Si alloy using a rotating magnetic field during solidification, *Acta Mater.* **72** 57-66
[13] L. Hachani, K. Zaidat and Y. Fautrelle 2015 Experimental study of the solidification of Sn-10wt.%Pb alloy under different forced convection in benchmark experiment *Int. J. Heat Mass. Tran.* **85** 438-454
[14] E.J. Zoqui, M. Paes and E. Es-Sadiqi 2002 Macro- and microstructure analysis of SSM A356 produced by electromagnetic stirring *J. Mater. Process. Technol.* **120** 365-373
[15] Z.N. Chen, T.M. Wang, L. Gao, H.W. Fu and T.J. Li 2012 Grain refinement and tensile properties improvement of aluminum foundry alloys by inoculation with Al–B master alloy *Mater. Sci. Eng. A.* **553** 32-36
[16] A.G. Murphy, W.U. Mirihanage, D.J. Browne, and R.H. Mathiesen 2015 Equiaxed dendritic solidification and grain refiner potency characterised through in situ X-radiography *Acta Mater.* **95** 83-89
[17] A.A. El-Daly, A.E. Hammad, G.S. Al-Ganainy and M. Ragab 2014 Influence of Zn addition on the microstructure, melt properties and creep behavior of low Ag-content Sn–Ag–Cu lead-free solders, *Mater. Sci. Eng. A* **608** 130-138

[18] F.J. Wang, X. Ma and Y.Y. Qian 2005 Improvement of microstructure and interface structure of eutectic Sn–0.7Cu solder with small amount of Zn addition *Scr. Mater.* **53** 699-702
[19] E. Çadirli, İ. Karaca, H. Kaya and N. Maraşli 2003 Effect of growth rate and composition on the primary spacing, the dendrite tip radius and mushy zone depth in the directionally solidified succinonitrile–Salol alloys *J. Cryst. Growth* **255** 190-203
[20] J. Teng, S. Liu and R. Trivedi 2009 Onset of sidewise instability and cell-dendrite transition in directional solidification *Acta Mater.* **57** 3497-3508
[21] S.S.L. Peppin, H.E. Huppert and M.G. Worster 2008 Steady-state solidification of aqueous ammonium chloride *J. Fluid. Mech.* **599** 465-476
[22] R. Mathiesen, L. Arnberg, F. Mo, T. Weitkamp and A. Snigirev 1999 Time resolved x-ray imaging of dendritic growth in binary alloys *Phys. Rev. Lett.* **83** 5062
[23] G. Reinhart, C.A. Gandin, N. Mangelinck-Noël, H. Nguyen-Thi, J.E. Spinelli, J. Baruchel and B. Billia 2013 Influence of natural convection during upward directional solidification: A comparison between in situ X-ray radiography and direct simulation of the grain structure *Acta Mater.* **61** 4765-77
[24] T.M. Wang, F. Cao, P. Zhou, H.J. Kang, Z.N. Chen, Y.N. Fu, T.Q. Xiao, W.X. Huang, Q.X. Yuan 2014 Study on diffusion behavior and microstructural evolution of Al/Cu bimetal interface by synchrotron X-ray radiography *J. Alloys Compd.* **616** 550-555
[25] H. Yasuda, I. Ohnaka, K. Kawasaki, A. Sugiyama, T. Ohmichi, J. Iwane and K. Umetani 2004 Direct observation of stray crystal formation in unidirectional solidification of Sn–Bi alloy by X-ray imaging *J. Cryst. Growth* **262** 645-652
[26] G. Reinhart, A. Buffet, H. Nguyen-Thi, B. Billia, H. Jung, N. Mangelinck-Noel, N. Bergeon, T. Schenk, J. Hartwig and J. Baruchel 2008 In-Situ and real-time analysis of the formation of strains and microstructure defects during solidification of Al-3.5 wt pct Ni alloys *Metall. Mater. Trans. A* **39A** 865-874
[27] T.M. Wang, F. Cao, Z.N. Chen, H.J. Kang, J. Zhu, Y.N. Fu, T.Q. Xiao and T.J. Li 2015 Three dimensional microstructures and wear resistance of Al-Bi immiscible alloys with different grain refiners *Sci. China Tech. Sci.* **58** 870-875
[28] K. Nogita 2010 Stabilisation of Cu6Sn5 by Ni in Sn-0.7Cu-0.05Ni lead-free solder alloys *Intermetallics.* **18** 145-149
[29] W. Liu, Y.H. Tian, C.Q. Wang, X.L. Wang and R.Y. Liu 2012 Morphologies and grain orientations of Cu-Sn intermetallic compounds in Sn3.0Ag0.5Cu/Cu solder joints *Mater. Lett.* **86** 157-160
[30] J.W. Mullin 1971 *Crystallization 2nd edn.* (London: Butterworth)
[31] G.X. Hu, X. Cai and Y.H. Rong 2006 *Fundamentals of Materials Science* (*Shanghai: Shanghai Jiaotong University Press*), 227-241
[32] X. Ma, Y.Y. Qian, F. Yoshida 2002 Effect of La on the Cu-Sn intermetallic compound (IMC) growth and solder joint reliability *J. Alloys Comp.* **334** 224-227
[33] T. M. Wang, P. Zhou, F. Cao, H. J. Kang, Z. N. Chen, Y. N. Fu, T. Q. Xiao, W. X. Huang, Q. X. Yuan 2015 Growth behavior of Cu6Sn5 in Sn-6.5 Cu solders under DC considering trace Al: In situ observation *Intermetallics* **58** 84-90
[34] T. M. Wang, J. Zhu, H. J. Kang, Z. N. Chen, Y. N. Fu, W. X. Huang, T. Q. 2014 In situ synchrotron X-ray imaging on morphological evolution of dendrites in Sn-Bi hypoeutectic alloy under electric currents *Appl Phys A-Mater Sci Process* **117** 1059-66
[35] G. W. Chang, J. P. Yuan, Z. D. Wang, C. J. Wu, X. H. Wang and H. Q. Hu 2000 Mechanism of reduction of columnar dendrite spacing in unidirectional solidification caused by electric current passing through solid-liquid interface *Trans. Nonferrous Met. Soc. China.* **10** 610-613
[36] S. R. Coriell, G. B. McFadden, A. A. Wheeler and D. T. J. Hurle 1989 The effect of an electric field on the morphological stability of the crystal-melt interface of a binary alloy: II. Joule heating and thermoelectric effects *J. Cryst. Growth* **94** 334-346

[37] F. Cao, F. F. Yang, H. J. Kang, Z. N. Chen, Y. N. Fu, T. Q. Xiao, W. X. Huang, Q. X. Yuan 2016 Effect of traveling magnetic field on solute distribution and dendritic growth in unidirectionally solidifying Sn-50 wt%Pb alloy: An in situ observation *J. Cryst. Growth* **450** 91-95

IOP Conf. Series: Materials Science and Engineering **580** (2019) 012014 doi:10.1088/1757-899X/580/1/012014

Rich multi-dimensional correlative imaging

Philip J Withers and Timothy L Burnett

The Henry Royce Institute, School of Materials, University of Manchester, M13 9PL, United Kingdom
Corresponding author: p.j.withers@manchester.ac.uk

Abstract. The use of microstructural design to tailor materials properties has increased sharply in recent years. In parallel the number and the capability of techniques able to characterise materials microstructures has increased sharply too providing structural, chemical and crystallographic information. Here we examine how correlated 3D, 4D (3D+time) and multi-dimensional imaging enable a much richer picture to be built up of a materials microstructure. We look at how a data-centric approach can support the use of materials informatics, digital twinning and machine learning to accelerate the design of new materials systems and to optimise the manufacturing of established ones. However for this to happen we need to develop ways to digitally fingerprint the microstructural images and maps we collect such that they can be incorporated into machine learning schemes. Through the use of case studies (multimodal imaging) we look at correlative imaging across scales, across time (the dilation of electrode materials in lithium batteries during discharging and fast corrosion of magnesium), as well as across multiple modalities (butterfly defects in bearings steels and the sintering and recrystallization of powders). These demonstrate how different techniques can come together to provide complementary aspects of the bigger picture.

1. Introduction

The number of techniques available to interrogate materials microstructures has increased sharply over recent years. In particular we have seen measurements extended from two dimensions to three by X-ray and transmission electron computed tomography (CT), atom probe and serial section scanning electron tomography. By exploiting the non-destructive nature of X-rays and visible light this has been extended to four dimensions through temporal studies. With the ability to map chemical and crystallographic information the number of dimensions has been extended yet further. These can be combined to obtain a rich, multifaceted picture of the structure and microstructure of materials and devices.

At the same time manufacturing methods are now able to rapidly build up and assemble materials and devices engineered across multiple hierarchical scales. This is alongside a concurrent development of advanced multiscale models for their design and simulation. In order to exploit this ability to rapidly manufacture an almost infinite variety of materials and structures, a step jump in materials imaging and characterization will be required.

This article examines the trajectory of this development which has led to the recent advancements of correlative microscopy and looks forward towards a wider, and more integrated, approach to characterization through correlative frameworks where it is possible to exploit the capability of individual pieces of equipment whilst keeping them contextually connected to other analyses. We consider how we can connect information efficiently and effectively across scales and modalities and the benefits this brings for characterization and modelling. We assess the current status of correlative strategies and look at opportunities for application in materials science looking first at manufacturing and then at materials degradation.

2. Multiscale 2/3D imaging

As illustrated in figure 1, almost all materials are hierarchical in some sense. In each case it is important to identify the key length scales that control behaviour and to develop the tools needed to probe these scales.

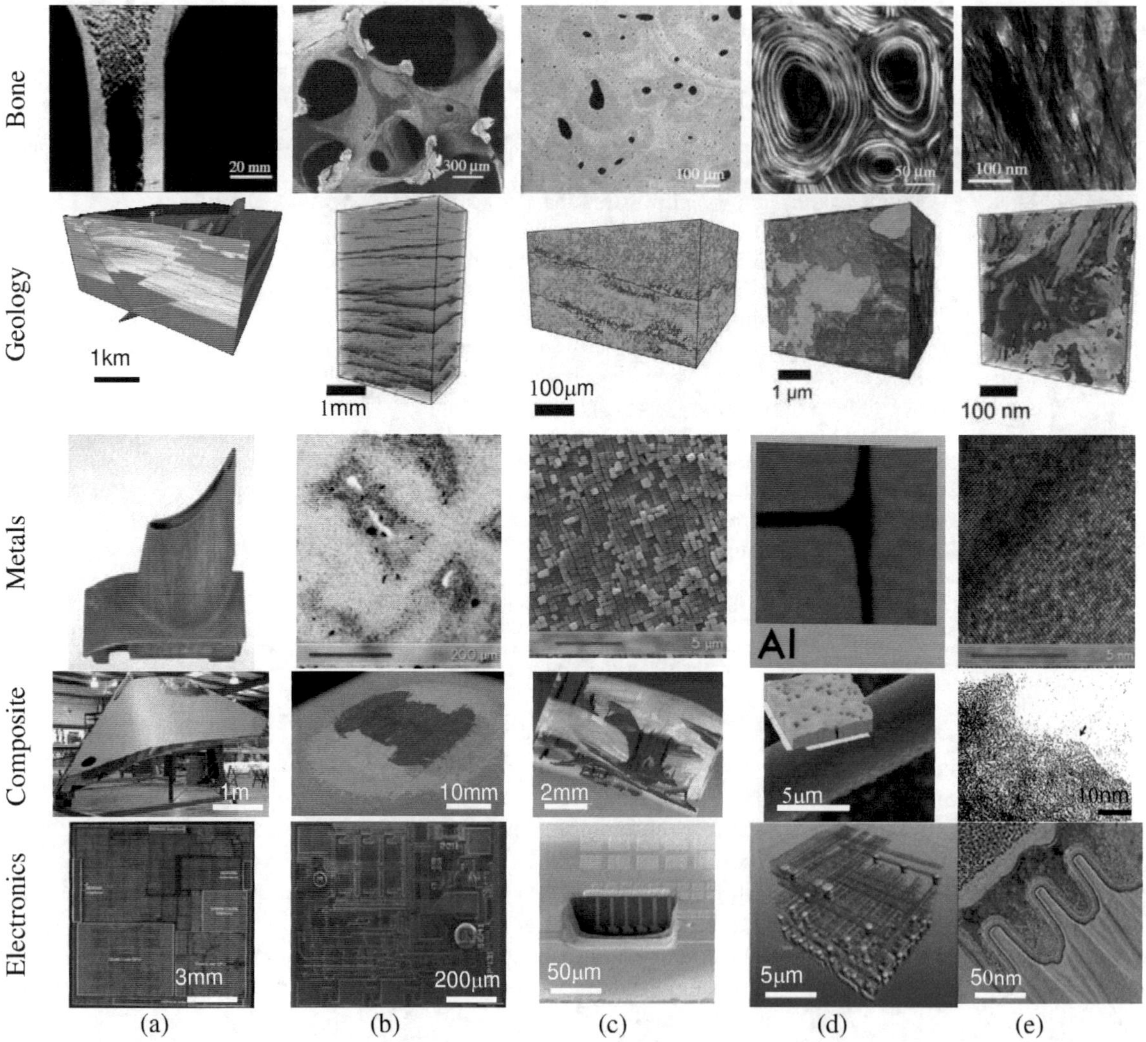

Figure 1. Row 1: Bone at different hierarchical levels; a) an optical micrograph showing the cortical and trabecular bone, b) SEM of the trabecular network, c) micro CT of the Haversian system, d) Lamellar structure in polarized light, e) TEM of collagen fibrils [1], Row 2: Shale rock at a) reservoir scale, b) fractures (MacroCT), c) laminations and microstructures (MicroCT), d) Organics & clay (FIB-SEM), e) nanopores (TEM tomography) [2], Row 3: a) Ni superalloy single crystal turbine blade, b) dendrites, c) γ/γ' arrangement, d) EDX map of Al segregation to γ', e) atomic structure across boundary (courtesy of [3]), Row 4: C fibre composite a) composite wing, b) CT image of impact damage, c) CT of woven tow structure, d) SEM image of C fibre, e) TEM image of c-fibre, Row 5: Microchip at increasing levels of magnification, a) overview, b) SEM of integrated circuit, c) FIB-SEM tomograph of 3D circuit architecture, d) 3D serial section PFIB-SEM of a through Si via, e) TEM of 14nm chip (images courtesy of TESCAN).

Of course, linking together microscopes to examine different length scales in 2D is not new. However, linking them together in 3D is more challenging because locating a potentially sub-merged region (volume) of interest identified at one scale for examination in greater detail at another is difficult. This has been done for a number of applications in metallurgy, particularly with respect to studies of degradation [4, 5]. To date, this has mostly been through the registration of sample features with manual transfer of the sample coordinates from one machine to another. Going forwards, routes for automatically transferring the sample coordinates, with very high accuracy, from one machine to another are required if multiscale 3D imaging is to become routine. A number of instrument manufacturers are aiming to do this predominantly through the use of standard stages and fiducials [6].

A critical aspect in multiscale imaging is managing the volume of the acquired dataset. If the whole region is collected at increasingly high spatial resolutions (termed the 'Googlemaps' approach in [7]) then the data requirements soon become the limiting factor (see figure 2). Alternatively targeted trajectories [7] have to be identified that track certain features at increasingly high spatial resolutions, but over concomitantly smaller volumes.

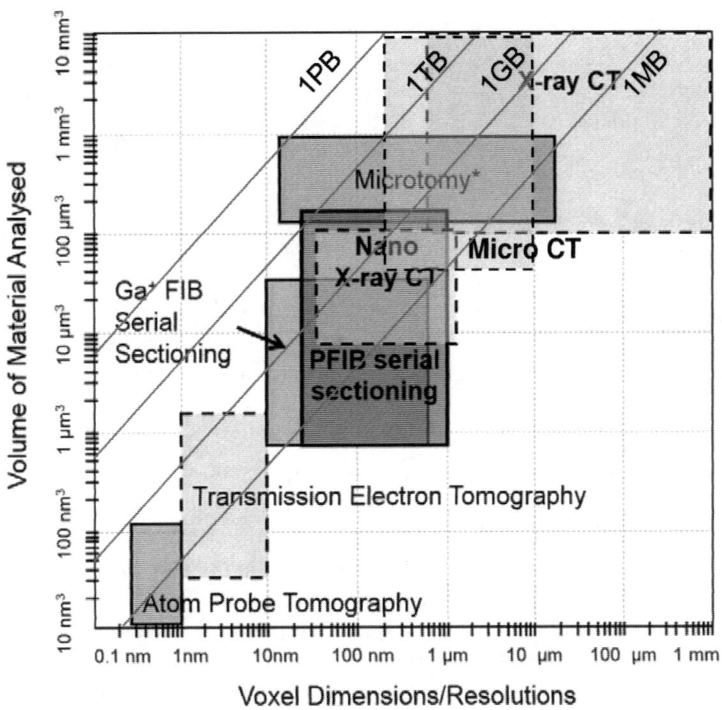

Figure 2. The relationship between spatial resolution, sample size and typical data storage requirements (non-destructive techniques denoted by dashed lines).

3. Temporal 4D (3D+time) imaging

The non-destructive nature of X-ray imaging opens the way for acquiring a sequence of images over time. For phenomena that occur quickly relative to the acquisition times needed to collect sufficient X-ray projections (radiographs) to reconstruct a tomograph (typically 200-3000 projections), such as ductile fracture [8], then continuous streaming of the data during the process is common [9]. For longer timespans, e.g. corrosion [10], or cases where the sample must be incrementally withdrawn from a particular environment for X-ray scanning, such as during sintering [11], time-lapse intervals are often chosen [9]. Image sequences can be correlated over time using digital image, or volume, correlation to extract deformation fields, for example to map the dilatation of lithium batteries during charging [12] as illustrated in figure 3. In addition to conventional CT, 3D XRD has been developed to obtain very high spatial resolution images of grains in 3D over time [13, 14].

Figure 3. A representative region of a coin cell electrode material showing the relative dilation of the manganese oxide grains during discharge. The grains closest to the electrode-terminal interface experience large dilation whereas grains further away experience much smaller dilation [12].

4. Multidimensional (>4D) imaging

Structural or microstructural images are useful, but often only provide part of the picture. Greater understanding can be gleaned by combining structural images with maps of chemistry or crystallography for example. This is now routine in 2D where scanning electron microscopy is combined with energy dispersive X-ray (EDX) mapping and electron backscattered diffraction (EBSD) to provide insights into the elemental distribution and the grain textures respectively. This technique has been extended to 3D using serial sectioning. The volume that can be mapped extends up to $(50\ \mu m)^3$ using traditional Ga^+ focused ion beam, to ~$(300\ \mu m)^3$ using a plasma FIB [15] and beyond with ultramicrotomy for relevant materials [16], broad ion beam sectioning [17] or laser sectioning [18].

Non-destructive multimodal imaging can be undertaken to provide a richer picture of behaviour that achievable with single modalities. For example with regard to the time-lapse sintering of metallic powders, diffraction contrast tomography (DCT) has provided detailed maps of the 3D grain structure while attenuation CT provided complementary information about the elimination of porosity during consolidation [11]. Similarly in medical imaging (positron emission tomography (PET) and CT scanning are combined – the former identifies specific metabolic functions using radiopharmaceuticals while the CT provides anatomical information and can also be used to help reconstruct the PET scans too.

Given that we have many destructive and non-destructive tools with which to probe a rich multi-dimensional space there are inevitably many choices to be made in terms of the trajectory by which we explore it (See figure 4a).

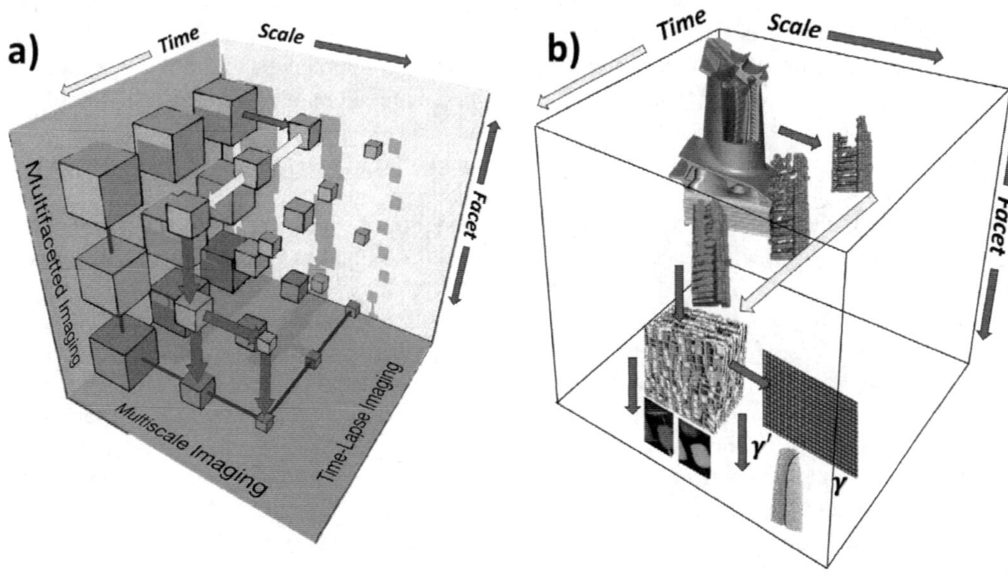

Figure 4. a) Correlative frameworks can illuminate a multidimensional space via many potential trajectories. Consequently the user needs to consider data gathered to date in determining the most useful pathways b) an illustrative workflow aimed at understanding the microstructure of a superalloy turbine blade [19-23].

Such workflows demand significant planning, both to take into account information gathered at one stage when considering the next step, but also because experimental considerations restrict the order of certain pathways (e.g. selecting when to excise certain regions or to employ certain destructive methods) etc). Figure 4b outlines a study aimed at understanding the manufactured microstructure in a single crystal nickel superalloy jet engine turbine blade. Here lab. X-ray CT reveals the overall 3D architecture of the turbine blade while at a finer scale time-lapse synchrotron X-ray CT is able to capture the morphology and formation of the dendrites during the single crystal solidification process. Subsequently serial section FIB-SEM is needed to characterize the finer γ matrix/γ' precipitate microstructure and map their chemistry by EDX. To examine the atomic structure and local chemistry transmission electron microscopy (TEM) and atom probe tomography (APT) are applied respectively.

5. Exploiting rich multidimensional datasets: Microstructural Informatics

Computational materials science has been proposed as a means to shorten the time to design and manufacture new materials systems [24]. Initiatives such as the Materials Genome Imitative (MGI) and Integrated Computational Materials Engineering (ICME) have focused considerable effort into the use of computational materials science to discover new materials systems, as well as to optimise the manufacturing processes needed to deliver high performance components.

Perhaps the greatest successes to date have been in areas where the materials properties depend little on microstructure. In such cases the desired properties (stiffness, thermal expansion, electrical or thermal or basic coupled properties) can be predicted on the basis of the crystal unit cell. In such circumstances thousands of materials systems can be trialled *in silico* and the most promising ones selected and means investigated to make them. Unfortunately, the majority of performance characteristics of materials depend heavily on many aspects of microstructure and defect structure not captured by the unit cell, from the atomic defects right up to the macroscale.

Reflective of the fact that there are key length scales that can control specific phenomena, materials models have particular scales over which they are valid, from ab initio methods to continuum finite element models. Multiscale modelling sets out to address a number of challenges to effectively represent hierarchical systems. Further multiscale, multi-physics models provide an opportunity to make engineering predictions rigorously based on physical principles. In this context it should be remembered that properties are scale-specific and that the microstructure at each scale can play a key role in determining these properties with microstructural variation being a key aspect in terms of understanding scatter in performance. Weinan [25] identifies 4 challenges to multiscale modelling:

1) We need to understand the mechanisms, model parameters, boundary conditions, microstructural varaiation, etc. operative at each scale .
2) Ideally, we should be able to derive coarse-grained models from more detailed ones in the appropriate limit, providing a link from scale to scale.
3) We need to understand how to couple together models at different levels together smoothly, without creating artifacts. It is common for large errors to occur at the interface where two models join.
4) We need to understand how to formulate models at intermediate levels where physics based models do not exist. Although our understanding of some scales may be mature it is common for certain domains to be poorly understood, e.g. the mesoscale gap relating discrete dislocations, dislocation networks and grain boundaries in metallic systems which mean that experiments play a critical role in filling this gap.

Rich multidimensional microstructure datasets provide an opportunity to tackle some of these challenges and thereby to improve the accuracy of such models. Ideally, the models at each length scale should be tied to the microstructures they represent (see Figure 5). In this way each model can be separately tested and validated. Furthermore in cases where materials models do not exist or where the couplings between models at different scales are poorly defined it may be possible to design experiments or workflows which deliberately address these gaps to allow the bridging of scales. In practice the link between materials models and microstructure are not so well defined and many semi-empirical models use parameters which are not simply related to the microstructures they represent (e.g. continuum damage parameters). At the present time, one of the main obstacles is the lack of well characterised, and widely available, multi-dimensional microstructural datasets.

In this respect, one of the principle issues is that at present we have no agreed way of quantifying microstructures beyond a few simple parameters (grain size, phase fractions, etc). To make progress we need to configure our microstructural models in terms of agreed microstructural parameters that can be unequivocally measured and develop machine learning methods that can relate aspects of the microstructural features to material processing or performance. As Wodo et al. point out [27] the goal of microstructural imaging is to provide detailed, high-resolution maps, and hence, imaging techniques inevitably produce high-dimensional data sets, while the goal of establishing quantitative process-microstructure-property relations (PMPRs) is to derive the smallest set of variables that explains most of the variability of the data. Consequently, microstructural data need to be reduced to a small number meaningful descriptors, a digital fingerprint (or gene sequences), to enable them to be used in developing property microstructure processing relationships. Discovering this reduced descriptor space as applied to microstructure is the basis of what has been termed microstructural informatics. At the same time such fingerprints would also enable archives of large 2D or 3D microstructural images or volumes to be stored efficiently.

ICME approaches exploit a digital twin (see figure 5) that represents *in silico* the attributes of the manufactured component as it undergoes the manufacturing process. In essence the microstructural data and the associated microstructurally cognisant materials models are like the two strands of DNA providing the underlying characteristics that control behaviour. Gaps in one strand can be compensated for by information from the other. Further machine learning approaches can then help to identify microstructural traits (or genes sequences) that give rise to specific performance characteristics or that can be used to assess the degree to which two microstructures have similar characteristics helping to direct microstructural control for performance optimisation.

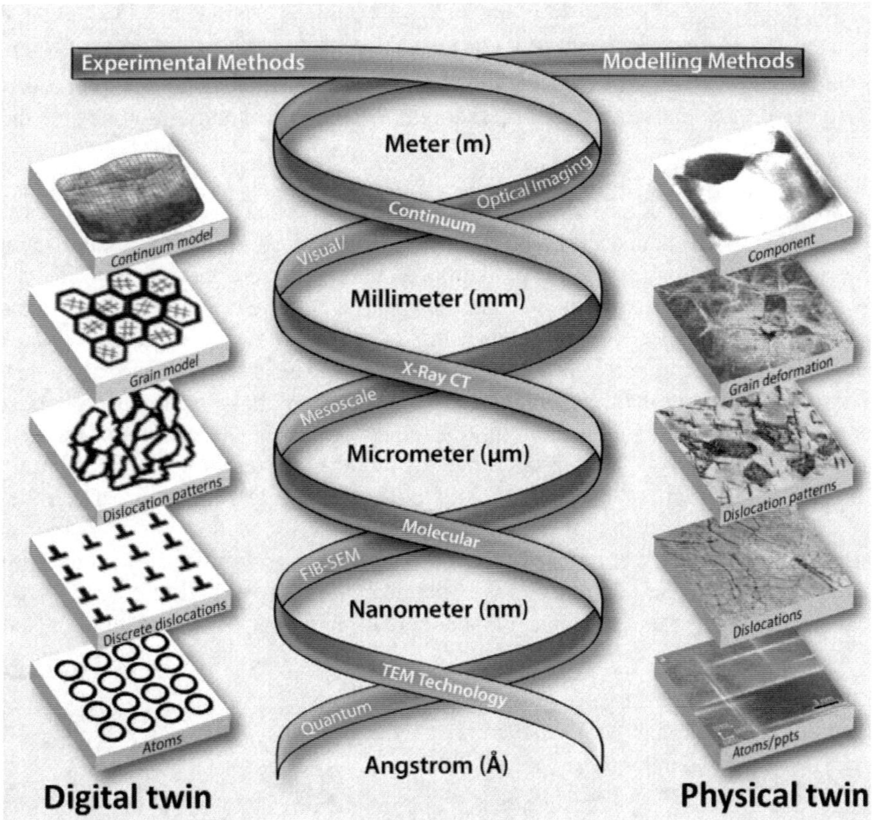

Figure 5. Multiscale experiments and multiscale models need to be developed synergistically and in parallel both to improve our understanding of materials across the key length scales, but also to enable the multilevel design of new materials. The important microstructural scales are illustrated for the deep drawing of an aluminium alloy [28-32]. Model schematics after [33] and layout after [7], FE model courtesy Dirke Raabe.

6. 4D imaging: Corrosion

Magnesium alloys are attracting a lot of interest across a broad spectrum of applications, where light weighting is a key attribute. Conventional processing to sheet involves a number of steps and for magnesium, being a hexagonally closed packed material, this are hindered by its limited formability. Twin roll casting (TRC) is a viable, cost effective alternative, which involves steep cooling rates. This affects the distribution of microstructural features across the section thickness [34], in addition to significantly influencing the near surface texture [35]. This is a critical factor because many magnesium alloys show a strong link between crystallographic texture and corrosion behaviour.

For TRC AZ31 Mg alloy (2.9 wt% Al, 0.88% Zn, 0.34 % Mn, 0.02 % Si, 0.007 % Fe, 0.009 % Cu) under saline conditions, corrosion initiates via micro-galvanic coupling between cathodic intermetallic phases including sub-micron sized, rosette-shaped Al_8Mn_5 particles [36], β-$Mg_{17}(Al,Zn)_{12}$ phase and the adjacent Mg matrix, once the hydroxide surface layer has been compromised [37]. Propagation of filiform-like corrosion fronts are then observed running along the dendrite arms whilst being constrained by the Al- and Zn-rich interdendritic boundaries [35]. As a result, corrosion can be highly microstructure dependent and very non uniform.

There is still a real need to better understand the fundamentals associated with corrosion of Mg and Mg-based alloys as the low corrosion resistance of Mg is its "Achilles heel" hindering uptake in many engineering applications. Previous time-lapse nano-CT imaging conducted on TRC AZ31 Mg alloys showed that the corrosion fronts were restricted within the interdendritic regions [38]. However, the exposed samples were intermittently removed from the saline environments between consecutive CT scans, essentially rendering it an "ex-situ" characterization study. The CT images of the corroded specimens showed shrinkage cracks within the dried corrosion products; where the possibility of their influence on the corrosion propagation during the successive exposures cannot be ruled out. Secondly, only a relatively small sample surface could be accommodated within the field of view of the nano-CT scans.

In situ studies of the evolution of the interdendritic corrosion fronts in TRC AZ31 Mg alloy have avoided these issues [39]. Here X-ray imaging was able to resolve the Al- and Zn- rich phases thereby delineating the interdendritic boundaries (see figure 6). These show a range of dendrite orientations relative to the surface of the sheet with a gradation of microstructure from the surface towards the central plane of the sheet. The experiment benefited from a relatively larger field of view, and the samples were continuously exposed to the saline electrolyte, whereby the sequential corrosion events could be captured, preventing any changes upon removal and exposure to air, in particular preventing the drying and cracking of the hydrated hydroxide surface layer that forms during exposure to the saline environment. However, this adds an additional challenge as the attenuation of the metal and the water are similar using relatively hard X-rays making contrast generation more difficult. Normally sub-micron resolution scans (on a Zeiss Versa 520) would take too long to acquire given the rapid rate of corrosion. Here image quality has to some extent been sacrificed to achieve shorter acquisition times. An acquisition time of 4 h per scan was adopted, which was followed by an additional 8 h of electrolytic exposure, before the next scan was conducted. As shown in figure 6, the images are of sufficient quality to observe the advance of the corrosion fronts as a function of the dendrite orientations.

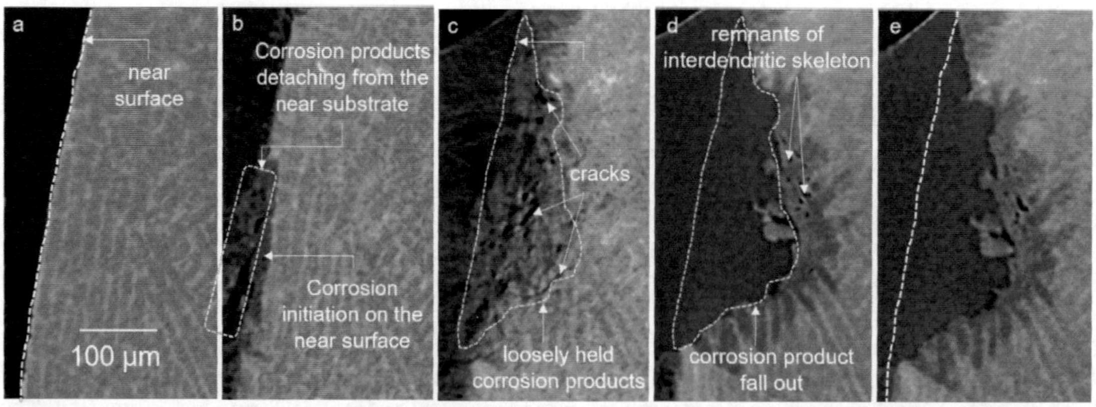

Figure 6. X-ray time-lapse sequence showing a region of interest from a virtual slice located at the surface of TRC AZ31 sheet subjected to saline corrosion in situ, a) as received, b) after 10.8 h, c) after 51.6 h, d) 59.5 h, and e) 78.3 h of exposure [39]. In situ imaging was necessary to prevent drying out and the associated phase changes upon exposure to air.

The X-ray micro-CT results show the manner in which the corrosion preferentially attacks the Mg matrix along the dendrite arms, with the Al- and Zn-rich interdendritic and grain boundaries showing increased corrosion resistance and acting as a local barrier (Figure 6b). At the same time, evidence of detachment of the corrosion products was also captured within the CT scans, as highlighted in image (b). The images show gradual increase in the local corrosion depths with respect to time. The results also confirm that the hydrated, near-surface oxide/hydroxide layer of corrosion products are relatively loosely held and render minimal protection to underlying microstructures. More importantly, after 51.6 h of exposure, the alloy microstructure showed a significant number of cracks within the corrosion products, as evident in image (c), and also observed in the previous studies [38] and the fall out of bulky chunks of corrosion products as confirmed in image (d) after 59.5 h. The formation of cracks and the falling out of corrosion products can be attributed to the weight of the loosely held interdendritic skeleton along with hydrated corrosion products. The 3D structure of Mg-dendrites in TRC AZ31 Mg alloy shows a non-continuous distribution of eutectic along the interdendritic regions

[35]. Therefore, it can be suggested that the interdendritic regions without any eutectic are the locations where the separation/fallout of the bulkier portions of corrosion products occurs. This observation obtained from the micro-CT results is very important from a corrosion propagation perspective. Whilst we have studied the initiation at the surface elsewhere by SEM, [34, 37, 39] the time-lapse CT allows us to track the propagation of corrosion across the surface and then critically when it starts to penetrate deeper into the material at a later stage. It is this later stage that is most damaging to the structural integrity of the material. .

7. Correlative multimodal 3D mapping: Butterfly defects

Under rolling contact fatigue bearings tend to produce a population of defects subsurface that initiate from inclusions at around 10^5 cycles. These arise well before failure normally after around 10^8 cycles [40]. The defects that grow from these inclusions typically have a 'butterfly' morphology and they are associated with intense plastic deformation and dissolution of the metal carbide particles. These regions appear as white etching matter (WEM) and include the butterfly wings as well as regions which decorate cracks. Traditionally these defects have been observed in 2D in metallographic sections such as that in figure 7a. Limited 3D information of butterfly defects has been obtained by coarse serial sectioning [41].

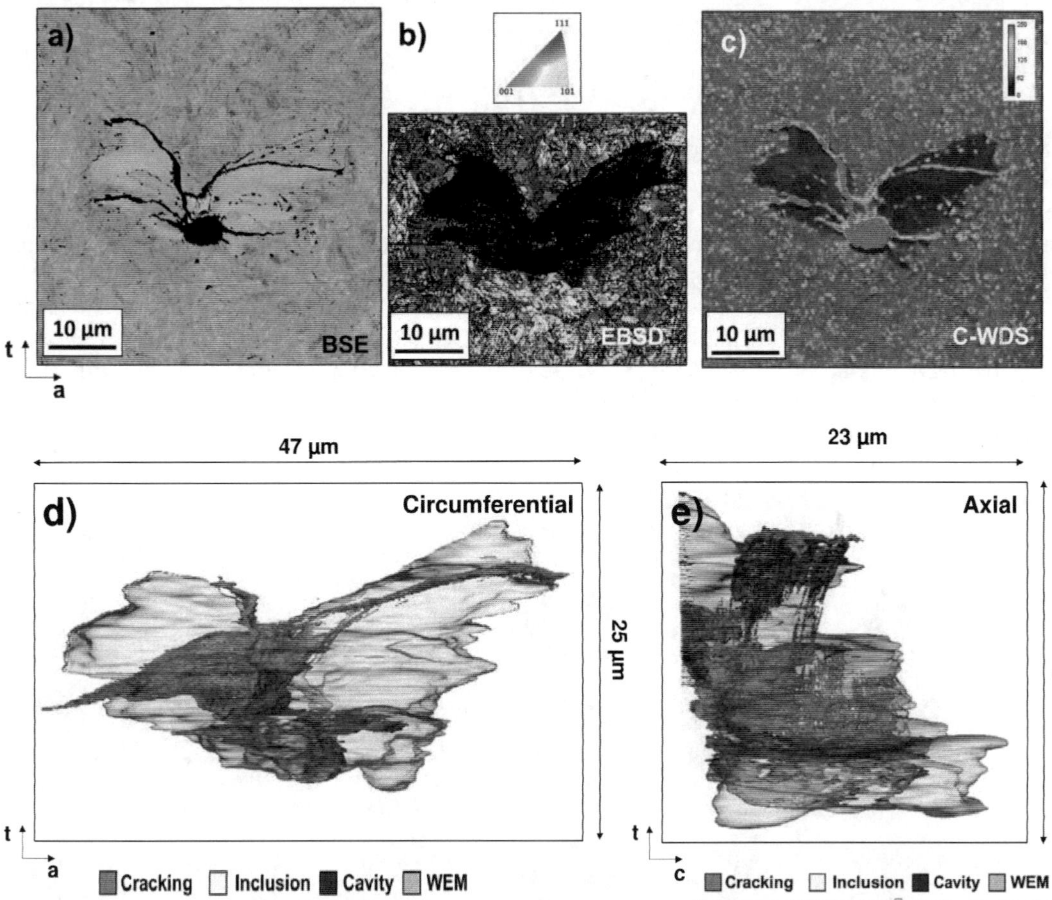

Figure 7. Complementary 2D images of the same butterfly defect, a) secondary electron imge showing the cracks, b) inverse pole figure from EBSD delineating the WEM as a black (non-indexed) region as well as showing the orientations of the parent material surrounding the butterfly c) EPMA C wavelength dispersive map showing the C solute level in the parent and WEM and the dissolution of carbides in the WEM, d) 3D rendering of the butturefly viewed in the circumferential direction, e) the

same butterfy viewed from the axial direction. [26]

In order or build up a more detailed and complete picture of the nature of butterfly defects correlative imaging is required because no single technique is capable of distinguishing all the features:

- secondary electron (SE) imaging can delineate the cracks (see figure 7a),
- electron back scattered diffraction (EBSD) imaging can distinguish the WEM as a non-indexing feature (shown black in figure 7b),
- EPMA is needed to detect changes in the level of carbon in solid solution (figure 7c).

The EPMA maps in figure 7c clearly show that, for the alloys they have studied at least, the WEM has less C in solution than the parent alloy. This is surprising given the dissolution of the carbides in this region and runs counter to previous inferences [40-42] .

Curd et al. [26] have brought these techniques together with serial sectioning by broad ion beam to build up a 3D detailed picture of the spatial distribution of these features (figure 7d and e). This shows the butterfly is extended in the circumferential direction in accordance with the sketched representation drawn by Becker [41]. Furthermore there is cracking at the interface between the WEM and the parent as well as cracking through the WEM itself. In this case two dominant cracks are seen either side of the inclusion giving a bi-plane appearance. The WEM tends to extend further behind the particle than the cracks.

8. Multimodal 4D imaging: Sintering of Copper

Pressureless sintering of loose or compacted granular bodies at elevated temperature occurs by a combination of particle rearrangement, rotation, local deformation, diffusion, and grain growth. In order to optimise the performance of the sintered body and eliminate defects, it is important to be able to understand the relative importance of these mechanisms as well as the extent of grain growth that occurs during densification. Conventional attenuation based X-ray CT can provide 4D information about the evolution of the general shape and volume as well as tracking voids, while X-ray diffraction can be exploited to learn about the rotations of the particles as well as information about the grain growth. Diffraction contrast tomography [43] enables the combination of attenuation and diffraction based CT to provide all this information. Originally developed for application at synchrotrons it has recently be developed to run on laboratory X-ray sources[44].

Here the technique is used in a time-lapse manner to follow the sintering process in pure copper. It is evident from figure 8 that as the sample consolidates some grains grow and others shrink. A clear example of the recrystallized gran growth is given by the abnormally large grain (rendered purple) in the centre of the compact.

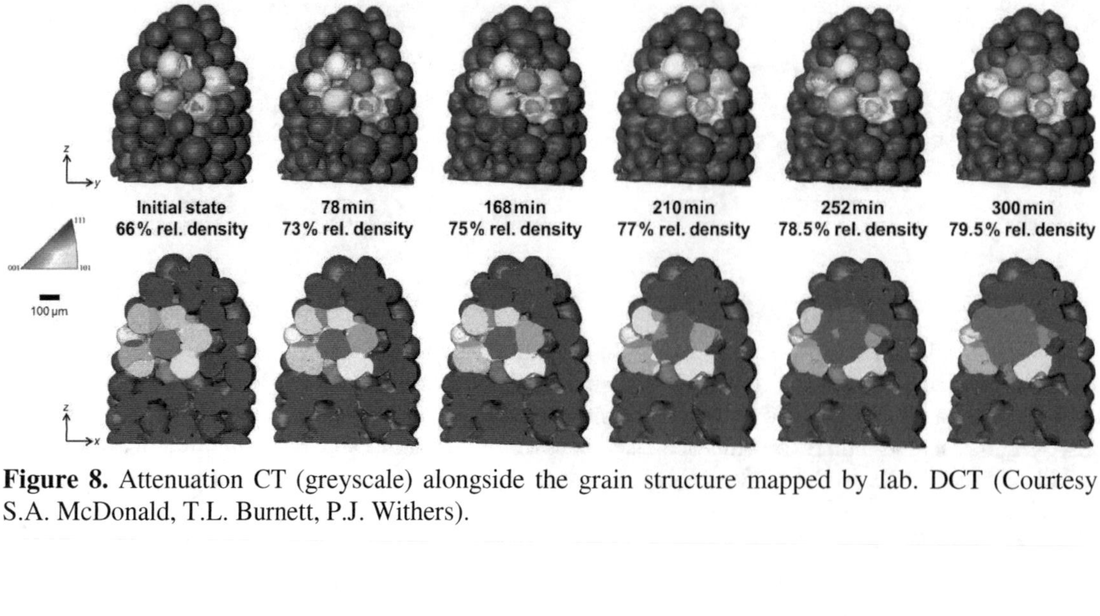

Figure 8. Attenuation CT (greyscale) alongside the grain structure mapped by lab. DCT (Courtesy S.A. McDonald, T.L. Burnett, P.J. Withers).

The densification was recorded by the attenuation CT and is shown in figure 9a showing that initially densification is quick but that this slows as 80% density is approached. It can be seen in figure 9b by combining the DCT and attenuation CT that there is a degree of correlation between the number of touching particle neighbours and the extent of particle rotation (quantified by the rotations of the constituent grains in each particle). These rotations become smaller as the sintering process develops.

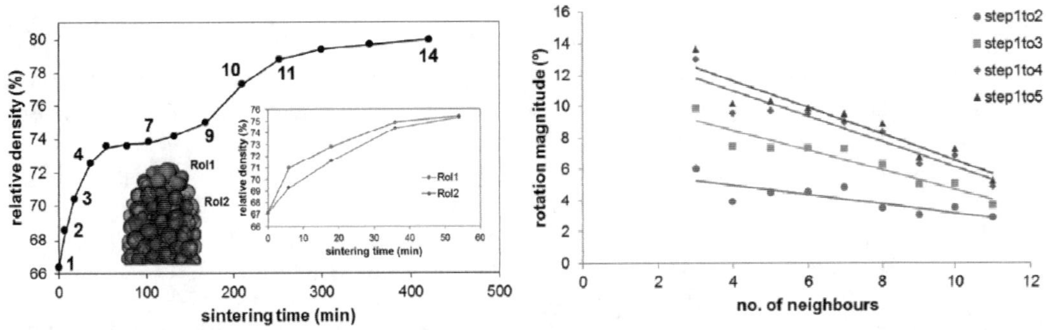

Figure 9. a) Densification of the powder body across the sintering steps at 1050 °C showing the increase in density for the full sample and, inset, comparing two regions of interest (RoI) for the first 5 steps, RoI2 corresponding to an interior region and RoI1 to an exterior facing ragion, b) Quantification of particle rotations and effect of particle coordination Showing the mean particle rotations relative to the original arrangement as a function of the number of contacting neighbours, for the whole sample studied [11]

9. Conclusions

While in many cases a great deal can be learnt from 2D images using a single instrument there are many other instances where a more complete picture is required. In particular the multiscale models needed to drive the computational design of new materials and manufacturing processes are highly dependent on rich, multiscale and multidimensional information for validation and to ensure the applicability across the required range of scales. X-ray imaging can provide 3D information across a range of length scales and can be combined with 3D serial sectioning techniques within the electron microscope to extend such data to finer length scales while TEM tomography and atom probe methods extend this to finer scales still.

The non-destructive nature of X-ray imaging opens up 4D studies, using either time-lapse, or fast streaming high frame rate imaging, to cover a wide range of time scales. Here an example was described where the crystallographically dependent rate of corrosion can be followed over time for twin roll cast AZ31 Mg alloy.

The ability to bring together complementary insights from multiple imaging or mapping modalities enables the superposition of data to form multi-layered (multifaceted) 3D datasets. Here we describe an example where secondary electron imaging, electron backscattered imaging, energy dispersive mapping and electron microprobe analysis are all brought together first in 2D to understand the relationship between cracking, crystallography and white etch matter, inclusion chemistry and carbide and carbon solute distributions associated with a butterfly defect. Using a coupled broad ion beam – scanning electron microscope system (BIB-SEM) the morphology of the WEM and cracks that make up the butterfly could be exposed in three dimensions. Bringing together information from different instruments is more complex in terms of the co-registration of data, as well as the transfer of samples between instruments.

In particular complementary X-ray techniques open the way for multi-modal 4D datasets. Here we presented data combining attenuation CT to record in a time-lapse manner the consolidation of copper particles during sintering alongside information about particle rotations as well as the consumption of certain grains by the growth of others. This is important if the sintering conditions are to be optimised to control grain growth during sintering yet achieve high levels of densification.

10. Acknowledgements
The authors are grateful to the EPSRC for access to facilities in the Multidisciplinary Characterisation Facility (MCF) within the Henry Royce Institute (EP/P025021/1). We are grateful to S. Pawar and V.C. Gudla for providing the time lapse study of Mg alloy. In particular we acknowledge the support for the Henry Moseley X-ray Imaging Facility through grants (EP/F007906/1, EP/F001452/1, EP/I02249X, EP/M010619/1, EP/F028431/1, and EP/M022498/1). PJW acknowledges support from the European Research Council grant No. 695638 CORREL-CT.

11. References
[1] Georgiadis M, Müller R and Schneider P 2016 Techniques to assess bone ultrastructure organization: orientation and arrangement of mineralized collagen fibrils *Journal of the Royal Society Interface* **13** 10.1098/rsif.2016.0088
[2] Ma L, Dowey P J, Rutter E, Taylor K G and Lee P D 2019 A novel upscaling procedure for characterising heterogeneous shale porosity from nanometre- to millimetre-scale in 3D *Energy & Environmental Science (Submitted)*
[3] Eggeler G 2015 From atoms to turbine blades – a scale bridging journey into the nanocosmos of a Ni-base superalloy. In: *https://www.youtube.com/watch?v=wYHch5QIWTQ,*
[4] Daly M, Burnett T L, Pickering E J, Tuck O C G, Léonard F, Kelley R, Withers P J and Sherry A H 2017 A multi-scale correlative investigation of ductile fracture *Acta Mater.* **130** 56-68
[5] Slater T J A, Bradley R S, Bertali G, Geurts R, Northover S M, Burke M G, Haigh S J, Burnett T L and Withers P J 2017 Multiscale correlative tomography: an investigation of creep cavitation in 316 stainless steel *Scientific Reports* **7** 7332
[6] Merkle A P, Lechner L, Steinbach A, Gelb J, Kienle M, Phaneuf M W, Unrau D, Singh S S and Chawla N 2014 Automated correlative tomography using XRM and FIB-SEM to span length scales and modalities in 3D materials *Microsc. Anal.* **28** S10-S3
[7] Burnett T L and Withers P J 2019 Completing the picture through correlative characterisation *Nature Materials* DOI: 10.1038/s41563-019-0402-8
[8] Maire E, Le Bourlot C, Adrien J, Mortensen A and Mokso R 2016 20 Hz X-ray tomography during an in situ tensile test *International Journal of Fracture* **200** 3-12
[9] Garcea S C, Wang Y and Withers P J 2018 X-ray computed tomography of polymer composites *Composites Science and Technology* **156** 305-19
[10] Bradley R S, Liu Y, Burnett T L, Zhou X, Lyon S B, Withers P J, Gholinia A, Hashimoto T, Graham D, Gibbon S R and Hornberger B 2017 Time-lapse lab-based x-ray nano-CT study of corrosion damage *Journal of Microscopy* **267** 98-106
[11] McDonald S A, Holzner C, Lauridsen E M, Reischig P, Merkle A P and Withers P J 2017 Microstructural evolution during sintering of copper particles studied by laboratory diffraction contrast tomography (LabDCT) *Scientific Reports* **7** 5251
[12] Eastwood D S, Yufit V, Gelb J, Gu A, Bradley R S, Harris S J, Brett D J L, Brandon N P, D.Lee P, Withers P J and Shearing P R 2014 Lithiation induced dilation mapping in a Li-ion battery electrode by 3D X-ray Microscopy and Digital Volume Correlation *Advanced Energy Materials* **4** 1300506
[13] Simons H, Jakobsen A C, Ahl S R, Detlefs C and Poulsen H F 2016 Multiscale 3D characterization with dark-field x-ray microscopy *MRS Bulletin* **41** 454-9
[14] Poulsen H F, Nielsen S F, Lauridsen E M, Schmidt S, Suter R M, Lienert U, Margulies L, Lorentzen T and Juul Jensen D 2001 Three-dimensional maps of grain boundaries and the stress state of individual grains in polycrystals and powders *Journal of Applied Crystallography* **34** 751-6
[15] Burnett T L, Kelley R, Winiarski B, Contreras L, Daly M, Gholinia A, Burke M G and Withers P J 2016 Large volume Serial Sectioning Tomography by Xe Plasma FIB Dual Beam Microscopy *Ultramicroscopy* **161** 119-29
[16] Hashimoto T, Thompson G E, Zhou X and Withers P J 2016 3D imaging by serial block face scanning electron microscopy for materials science using ultramicrotomy *Ultramicroscopy* **163** 6-18

[17] Winiarski B, Gholinia A, Mingard K, Gee M, Thompson G E and Withers P J 2017 Broad ion beam serial section tomography *Ultramicroscopy* **172** 52-64
[18] Echlin M P, Mottura A, Torbet C J and Pollock T M 2012 A new TriBeam system for three-dimensional multimodal materials analysis *Rev. Sci. Instr.* **83** 023701
[19] Azeem M A, Lee P D, Phillion A B, Karagadde S, Rockett P, Atwood R C, Courtois L, Rahman K M and Dye D 2017 Revealing dendritic pattern formation in Ni, Fe and Co alloys using synchrotron tomography *Acta Mater.* **128** 241-8
[20] Parsa A B, Wollgramm P, Buck H, Somsen C, Kostka A, Povstugar I, Choi P P, Raabe D, Dlouhy A, Muller J, Spiecker E, Demtroder K, Schreuer J, Neuking K and Eggeler G 2015 Advanced Scale Bridging Microstructure Analysis of Single Crystal Ni-Base Superalloys *Advanced Engineering Materials* **17** 216-30
[21] Ziętara M, Kruk A, Gruszczyński A and Czyrska-Filemonowicz A 2014 FIB–SEM tomography of 4th generation PWA 1497 superalloy *Materials Characterization* **87** 143–8
[22] Kulawik K, Buffat P A, Kruk A, A.M.Wusatowska-Sarnek and Czyrska-Filemonowicz A 2015 Imaging and characterization of γ' and γ'' nanoparticles in Inconel 718 by EDX elemental mapping and FIB–SEM tomography *Materials Characterization* **100** 74-8
[23] Mignanelli P M, Jones N G, Pickering E J, Messé O M D M, Rae C M F, Hardy M C and Stone H J 2017 Gamma-gamma prime-gamma double prime dual-superlattice superalloys *Scr. Mater.* **136** 136–40
[24] Agrawal A and Choudhary A 2016 Perspective: Materials informatics and big data: Realization of the "fourth paradigm" of science in materials science *APL Materials* **4** 053208
[25] Weinan E 2011 *Principles of Multiscale Modeling* (Cambridge: Cambirdge University Press)
[26] Curd M E, Burnett T L, Donoghue J, Yan P and Withers P J 2019 The 3D morphology of subsurface butterfly defects in bearing steels *In preparation*
[27] Wodo O, Broderick S and Rajan K 2016 Microstructural informatics for accelerating the discovery of processing–microstructure–property relationships *MRS Bulletin* **41** 603-9
[28] Rodriguez-Veiga A, Bellon B, Papadimitriou I, Esteban-Manzanares G, Sabirov I and Llorca J 2018 A multidisciplinary approach to study precipitation kinetics and hardening in an Al-4Cu (wt. %) alloy *Journal of Alloys and Compounds* **757** 504-19
[29] Sehitoglu H, Foglesong T and Maier H J 2005 Precipitate effects on the mechanical behavior of aluminum copper alloys: Part I. Experiments *Metallurgical and Materials Transactions a-Physical Metallurgy and Materials Science* **36A** 749-61
[30] Han W Z, Vinogradov A and Hutchinson C R 2011 On the reversibility of dislocation slip during cyclic deformation of Al alloys containing shear-resistant particles *Acta Mater.* **59** 3720-36
[31] Qi M L, Yao Y, Bie B X, Ran X X, Ye W, Fan D and Li P 2015 Nucleation and growth of damage in polycrystalline aluminum under dynamic tensile loading *Aip Advances* **5**
[32] Wilson D V 1966 Plastic anisotropy in sheet metals *Journal of the Institute of Metals* **94** 84
[33] McDowell D L 2018 *Computational Materials System Design,* ed D Shin and J Saal (Cham: Springer International Publishing) pp 1-25
[34] Bayandorian I, Huang Y, Fan Z, Pawar S, Zhou X and Thompson G E 2012 The Impact of Melt-Conditioned Twin-Roll Casting on the Downstream Processing of an AZ31 Magnesium Alloy *Metallurgical and Materials Transactions a-Physical Metallurgy and Materials Science* **43A** 1035-47
[35] Pawar S, Slater T J A, Burnett T L, Zhou X, Scamans G M, Fan Z, Thompson G E and Withers P J 2017 Crystallographic effects on the corrosion of twin roll cast AZ31 Mg alloy sheet *Acta Mater.* **133** 90-9
[36] Pawar S, Zhou X, Hashimoto T, Thompson G E, Scamans G and Fan Z 2015 Investigation of the microstructure and the influence of iron on the formation of Al8Mn5 particles in twin roll cast AZ31 magnesium alloy *Journal of Alloys and Compounds* **628** 195-8
[37] Pawar S, Zhou X, Thompson G E, Scamans G and Fan Z 2015 The Role of Intermetallics on the Corrosion Initiation of Twin Roll Cast AZ31 Mg Alloy *Journal of the Electrochemical Society* **162** C442-C8

[38] Burnett T L, Holroyd N J H, Lewandowski J J, Ogurreck M, Rau C, Kelley R, Pickering E J, Daly M, Sherry A H, Pawar S, Slater T J A and Withers P J 2017 Degradation of metallic materials studied by correlative tomography *IOP Conference Series: Materials Science and Engineering* **219** 012001

[39] Pawar S, Gudla V.C, Kefallinou Z, Thompson G E, Zhou X, Burnett T L and Withers P J 2019 Time-lapse X-ray CT imaging of saline corrosion in twin roll cast AZ31 Mg alloy sheet *Chemistry of Materials In preparation*

[40] Bhadeshia H K D H and Solano-Alvarez W 2015 Critical Assessment 13: Elimination of white etching matter in bearing steels *Materials Science and Technology* **31** 1011-5

[41] Becker P C 1981 Microstructural changes around non-metallic inclusions caused by rolling-contact fatigue of ball-bearing steels *Metals Technology* **8** 234-43

[42] Grabulov A, Petrov R and Zandbergen H W 2010 EBSD investigation of the crack initiation and TEM/FIB analyses of the microstructural changes around the cracks formed under Rolling Contact Fatigue (RCF) *IJ Fatig.* **32** 576-83

[43] Johnson G, King A, Honnicke M G, Marrow J and Ludwig W 2008 X-ray diffraction contrast tomography: a novel technique for three-dimensional grain mapping of polycrystals. II. The combined case *Journal of Applied Crystallography* **41** 310-8

[44] McDonald S A, Reischig P, Holzner C, Lauridsen E M, Withers P J, Merkle A P and Feser M 2015 Non-destructive mapping of grain orientations in 3D by laboratory X-ray microscopy *Scientific Reports* **5** 14665

IOP Conf. Series: Materials Science and Engineering **580** (2019) 012015 doi:10.1088/1757-899X/580/1/012015

Quantification of local boundary migration in 2D/3D

Yubin Zhang

Department of Mechanical Engineering, Technical University of Denmark, Kgs. Lyngby, 2800, Denmark

Email: yubz@mek.dtu.dk

Abstract. With the development of advanced electron and X-ray microscopy techniques, the local boundary migration during recrystallization and grain growth can be followed in 2D at a sample surface and/or in 3D inside bulk samples during in/ex situ annealing. The results show that locally boundaries migrate in a much more complex way than commonly imagined, for example by the development of local protrusions and retrusions and by migrating in a stop-go type of fashion. A quantitative analysis of the local boundary migration is essential for understanding this heterogeneous process. In this paper, methods for quantifying local boundary migration are summarized, including methods for quantification of local boundary migration velocities, misorientation (angle/axis pair) and plane normal of migrating/non-migrating boundaries, driving force (from both the energy stored in the deformation matrix and local boundary curvature), boundary mobility and activation energy. Through a quantitative analysis of both experimental and simulation data, local boundary migration is rationalized in several material systems during recrystallization and grain growth.

1. Introduction

In polycrystalline materials, grains with different crystallographic orientations are separated by grain boundaries. Boundary migration is involved when a grain grows into, or is consumed by, neighboring grains. The boundary migration rate, v, is generally expressed as:

$$v = MF, \tag{1}$$

where M is the boundary mobility and F is the driving force for migration. F is typically considered as the driving force from the energy stored in the deformed matrix for recrystallization and as the boundary curvature induced driving force for grain growth [1]. New studies focusing on the local boundary migration during recrystallization have shown that the curvature-based driving force, F_σ, on recrystallizing boundaries from local protrusions/retrusions (developed when boundary segments migrate ahead/behind neighboring segments) [2,3] can be comparable in magnitude to that from the stored energy in the deformed matrix, F_s. Therefore equation (1) is modified as [4,5]:

$$v=M\cdot(F_S + F_\sigma), \tag{2}$$

for quantifying local recrystallization boundary migration.

Despite the simple equations, local boundary migration is not understood. This may be related to the fact that both boundary mobility and energy depend on temperature in complex ways, and are affected by material purity, and depend on the misorientation and plane normal of the boundary, which covers a 5D parameter space (3D for the misorientation plus 2D for the boundary plane normal) [1].

The migration of recrystallization boundaries is even more complex, where the local variation in the deformed microstructure has an additional impact. Experimentally, only equation (1) has been validated for boundary migration during the grain growth process using idealized samples, e.g. bicrystals [6–11]. However, the results cover only a very small proportion of the possible 5D parameter space, and the validity of such results on recrystallization boundaries and on boundaries in engineering materials have not yet been studied. Computer modeling using molecular dynamics has been conducted to determine both the boundary energy and mobility for a relatively large number of boundaries in different material systems [12–17]. Experimental validations on such results are, however, missing. As such new, sufficiently fast, experimental methods for quantification of boundary migration are required.

In the last 2-3 decades, several new experimental characterization techniques have been developed and matured, such as in-situ/ex-situ measurements using electron backscattered diffraction (EBSD) [18,19] and transmission electron microscopy (TEM) [20–22], 3D X-ray diffraction (3DXRD) [23–25], synchrotron and laboratory-based diffraction contract tomography (DCT) [26–30], and 3D X-ray Laue micro-diffraction [31–34]. With these techniques *local* migration of individual boundaries during recrystallization and grain growth can be followed in 2D or 3D in *polycrystalline* materials. By correlating the migrating boundary with the microstructure ‘swept’ by the boundary, the local boundary migration velocity, misorientation and plane normal (for 3D data), and driving force for individual boundary segments can be *quantified*. With such quantitative analyses, the physical mechanisms behind the local boundary migration behavior may be rationalized, which is essential for developing advanced models that can predict local migration behavior more precisely.

Nevertheless, quantitative analysis of local boundary migration based on crystallographic orientation data in polycrystalline is still relatively new. The analysis procedure is also rather complicated, involving sophisticated image processing steps, such as data registration and segmentation, morphological operations, and in some cases surface meshing. So far no commercial software is available to perform all these steps. The present paper aims to summarize the required analysis procedure. The results obtained using such quantitative analysis methods on several material systems are then be summarized for a better understanding of local boundary migration during recrystallization and grain growth.

2. Quantification of procedure

The procedures described in this section are those required for analysing local boundary migration based on 2D/3D crystallographic orientation data collected from roughly the same sample area/volume over a certain time period during in situ or ex situ annealing.

2.1. Data registration

Data registration is typically the first task for the quantitative analysis. This is because data collected at different annealing time steps may not be aligned, especially for data collected during ex-situ annealing, where the sample has been demounted and remounted from the measurement stage for annealing treatment. Also in practice, different characterization techniques can be combined to optimize the spatial, temporal and angular resolutions for comprehensive characterization of the microstructural evolution [35]. In such a case, data registration between multiple data sources is also required. Moreover, if the time spent on collecting data at each individual time step is long, quite often the data can be spatially distorted due to thermal and/or mechanical drift of the sample and/or drift of the electron/X-ray beam [36]. In that case, data correction (registration) according to some distortion-free data has to be carried out for each time step.

For the registration, markers or control points (CPs) selected according to features in the microstructure or on the sample are typically required to determine the transformation function between data collected from neighboring time series. An example of how a spatially distorted EBSD map can be registered is given here to demonstrate the registration procedure (see figure 1). The SEM image collected after EBSD mapping is shown in figure 1a. The non-rectangular darker shaded region shows clearly the drifting that has taken place during the mapping. In this case an electron channeling contrast (ECC) image (figure 1c) is used as a distortion-free reference image for the registration, as such images can typically be collected within a relative short time (few minutes). A small horizontal twin marked by the white arrow in figure 1c appears inclined in the EBSD map (figure 1b), demonstrating the local spatial distortion in the EBSD map. To register the EBSD map, 60 CPs were selected manually from the microstructure (figure 1d), based on which a transformation function was computed. Many transformation functions can be used. For example, rotation, translation, similarity and affine transformation are well-known functions, all of which transform an image rigidly, i.e. line features in the image are reserved after transformation. More sophisticated non-rigid transformation functions include polynomial and thin plate spline (TPS) functions. Registered EBSD maps using TPS and polynomial functions are shown in figures 1e and 1f, respectively. The shape of the registered EBSD map using a TPS function matches well the darker shaded region in figure 1a, and locally the twin marked by the white arrow has also been correctly restored (see figure 1e), demonstrating the accuracy of the data registration on both the global and local scales [36]. Registration with the more commonly used polynomial function [37] is less satisfactory (see figure 1f), as polynomial functions are more suitable for correcting global distortions while TPS is suitable also for correcting locally varying distortions [36].

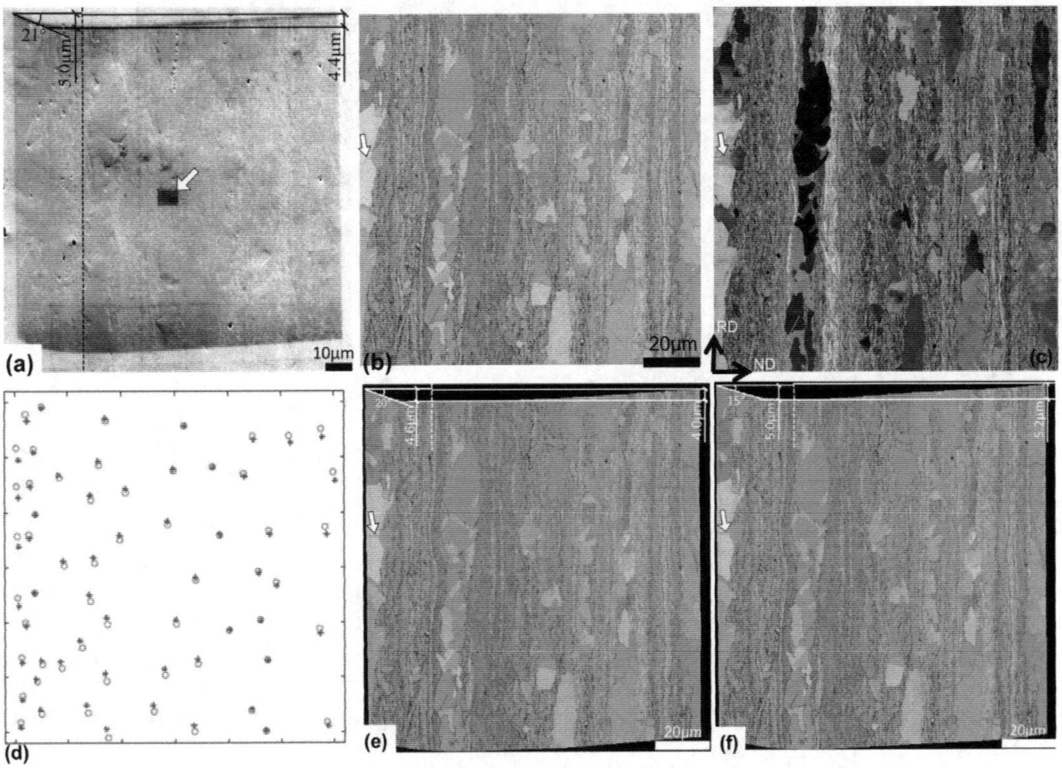

Figure 1. Example showing the procedure for data registration of an EBSD map. (a) secondary electron image showing carbon contaminated regions (darker shaded areas) after EBSD mapping. (b) and (c) are EBSD map and ECC image, respectively, collected at the same surface region in a partially recrystallized Ni sample. (d) Control point pairs collected manually based on the microstructure in (b) and (c). (e) and (f) are registered EBSD maps using a TPS and a polynomial function, respectively. The white arrows in (b), (c), (e) and (f) mark the same twin. Reproduced from [36].

The registration procedure for 3D time series data sets is similar to that described above for 2D data, except that the transformation functions are 3D functions determined based on 3D CP coordinates. For 3D data sets, the overall 3D sample shape and fiducial markers on the sample surface can be used as markers [38]. Crystallographic orientations of the same grain measured at different time steps can also be used to determine any rigid rotation of the sample. For systems containing many grains, the center of mass coordinates of the grains can be used as CPs [39]. However, the center of mass coordinate of a grain may change significantly during annealing if the grain grows or shrinks anisotropically. Transformation results based on this measure typically therefore need some crosscheck using other types of markers.

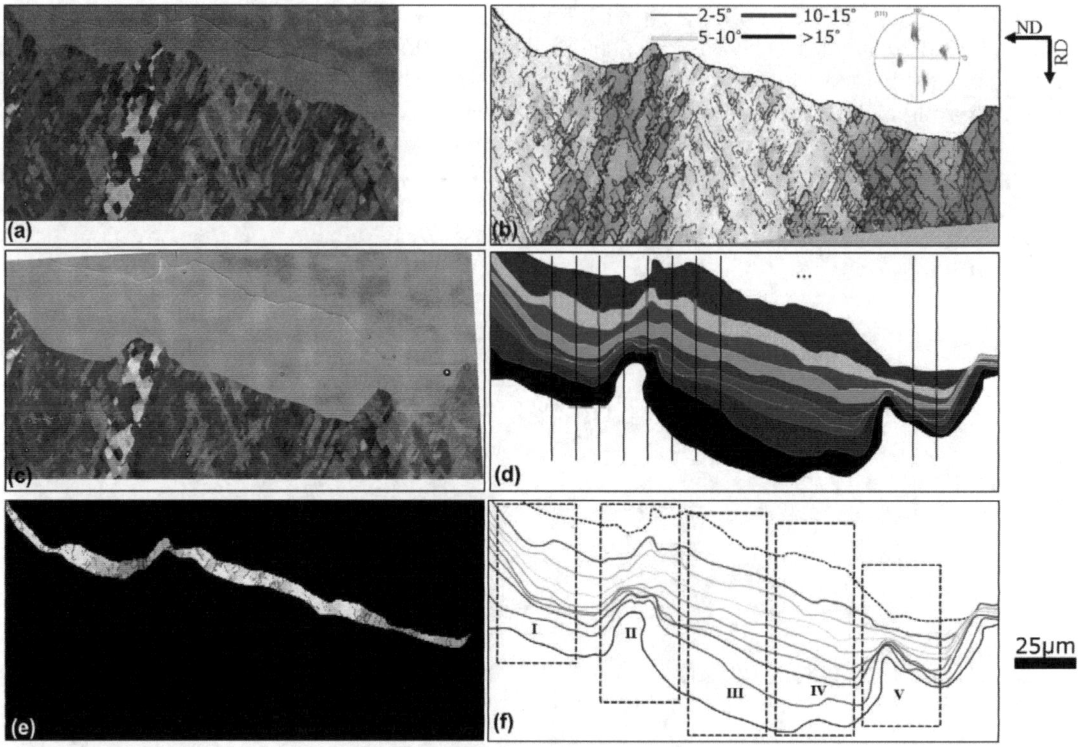

Figure 2. Examples showing data segmentation and methods for calculation of migration velocity based on aligned ex-situ data. (a) and (c) are aligned ECC images showing the microstructure of partially recrystallized pure Al at the starting state (annealed at 250 °C for 10+15 min) and after 6 steps of annealing at 250°C for 15 min each. (b) EBSD map showing the same microstructure as that in (a). The colors in (b) represent the crystallographic direction parallel to sample ND (see the inset pole figure). (d) Traces of the recrystallization boundary after each annealing step; 7 steps at 250 °C and the last 2 steps at 260 °C for 15 min each. Areas of different color/shading show the regions consumed during different annealing steps. Vertical lines are used for determining local migration velocity using a line intercept method. (e) Segmented EBSD data using the area masks in (d) for the first annealing step. (f) Segmentation to local regions marked by the five numbered boxes. The dashed lines in (d) and (f) represent the boundary grooves seen in (a) and (c), developed during electropolishing before the ex-situ annealing experiments. Reproduced from [33].

2.1.1. Data segmentation. After data registration, time series of data can then be segmented into subsets for further quantification. An example of a registered 2D data set is shown in figure 2, where the migration of a recrystallizing boundary into a deformed matrix in pure Al (50% cold rolled) has been followed using ECC after ex situ annealing at 250 °C and 260 °C for various time intervals [40]. An EBSD map (figure 2b) has also been collected before the ex-situ study and registered according to the corresponding ECC image. By tracking the boundary position at each annealing step, the relative

IOP Conf. Series: Materials Science and Engineering **580** (2019) 012015 doi:10.1088/1757-899X/580/1/012015

position of the boundary of interest can be determined as a function of annealing time (see the coloured boundary traces in figures 2d and f). From the aligned boundary traces, the area 'swept' by the recrystallizing boundary during each annealing step can be segmented (see the areas of different colour in figure 2d). The segmented areas can then be used as masks to segment the EBSD data for local quantification of temporal changes (see e.g. figure 2e). In addition, the boundary can also be segmented spatially along the boundary into local segments for quantification of spatial variations (see e.g. the five boxes in figure 2f).

2.2. *Quantification of local boundary migration velocity*

Three commonly used methods [35,41] for quantification of local boundary migration velocity are summarized in the following:

i) The migration velocity, v, of the boundary can be determined using equation (3)

$$v = A_{Cons}/lt, \tag{3}$$

where A_{Cons} is the area through which a migrating boundary of a length, l, migrates over a given annealing time interval, t. Both A_{Cons} and l can be quantified from the aligned data (see for example the colored areas in figure 2d). For 3D boundaries, A_{Cons} and l are replaced with volume, V_{Cons}, and area, A, of a migrating boundary, respectively. The measured velocity is an average migration velocity for local boundary segments, with length/area of l/A in 2D/3D.

ii) The velocity can also be measured by a line-intersect method, i.e. by drawing a line across the migrating boundary and measuring the migration distance during each time interval (see the red intersect lines in figure 2d). Using this method, the measured velocity values are valid for specific points on the boundary along a specific direction. The values depend therefore on the line direction, which sometimes can be difficult to define when the migration pattern is complex and the migration direction is difficult to define.

iii) To overcome the line direction problem, a Euclidean distance transformation method can be used to determine the migration distance for specific points on the boundary. For each pixel/voxel on the initial boundary, the Euclidean distance transformation determines the migration distance as the minimum distance from the pixel/voxel of interest to all the pixels/voxels on the new boundary. An example is shown in figure 3, where the migration distance for a 3D recrystallizing boundary in pure Al is colored according to the migration distance in figure 3e [41]. The migration direction determined with this method is different for each pixel/voxel. This method is well suited for boundary migration along multiple directions for convex grains.

2.3. *Boundary characteristics for migrating and non-migrating boundary segments*

A boundary can be separated into migrating and non-migrating parts, based on the quantified local migration velocities. The non-migrating parts are defined as those parts that migrate less than a certain amount (e.g. the step size of the measurement), while the remaining parts are defined as the migrating parts. In practice all pixels/voxels within consumed regions (see e.g. figure 2e) are considered for calculation of misorientations (angle/axis pairs) for migrating boundary segments, as at a certain time during annealing the migrating boundary has reached this position.

For determining the boundary plane normal based on 3D data set, segmentation is carried out only for the starting boundary for each annealing step. An example of the segmentation for the 3D boundary in figure 3 is shown in figure 4a and 4b. The voxelized data is then meshed into triangles for calculation of the boundary plane normal distribution (see figure 4c and 4d).

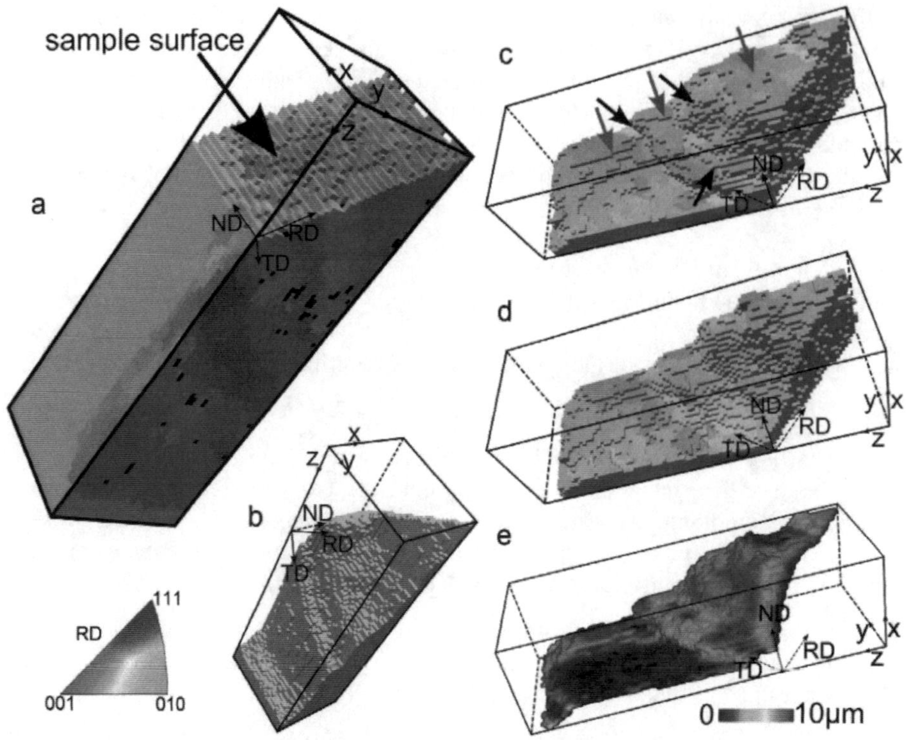

Figure 3. An example showing the calculation of the local migration velocity for a 3D migrating boundary. (a) 3D view of the gauge volume in partially recrystallized Al mapped by micro-diffraction. (b) and (c) are subsets of the volume in (a) visualizing the recrystallized grain and the deformed matrix, respectively. (d) the deformed part of the gauge volume after two annealing steps. The colors in (a)-(d) represent crystallographic orientations along the sample rolling direction (RD) (see the color code below (a)). (e) the initial recrystallizing boundary, colored by the migration distance. The black voxels in (a) represent non-indexed voxels. The box size along x, y, z, is 28, 32, 95μm, respectively. The gray and black arrows in (c) mark faceted and curved parts of the boundary, respectively. [41]

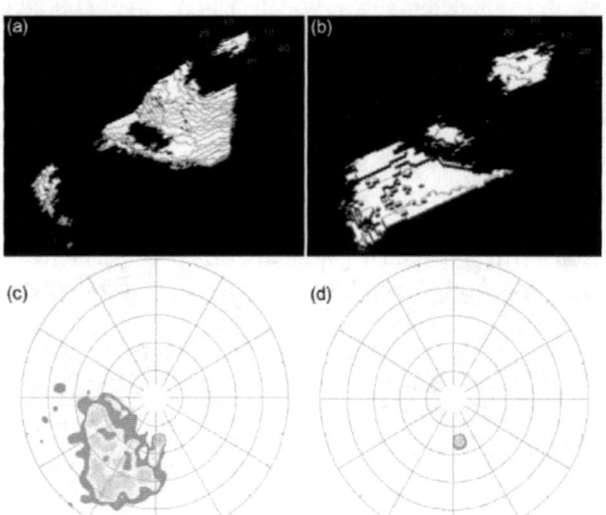

Figure 4. Partitioning of the 3D recrystallization boundary in figure 3a, based on the calculated migration distance at each position on the boundary, into (a) migrating parts and (b) non-migrating parts (defined here as parts migrating by less than 1 voxel) [41]. (c) and (d) show the plane normal distributions in the crystal system for the boundary segments in (a) and (b), respectively.

2.4. Quantification of driving force

The driving force for recrystallization is considered as both the driving force from the energy stored in the deformed grains and that induced by boundary curvature (see equation (2)), while only the latter is typically considered for grain growth. For both cases, a knowledge of grain boundary energies is required, which is typically determined using the Read-Shockley equation to describe the energy as a function of the misorientation of each boundary. A recent 3D study using LabDCT has shown that this equation describes in general well the boundary energies in real engineering materials [42].

For recrystallization, the stored energy can be calculated using the methods described in [43]. In some cases, it may be required to perform orientation filtering to reduce orientation noise [40,44]. Similar to the case for quantification of boundary misorientations, the analysis can also be carried out for local areas/volumes consumed by the growing grains, which represents the 'true' stored energy that leads to the migration, and for areas/volumes of a certain thickness in front of non-migrating boundary segments. This type of quantification is more accurate for understanding the local boundary migration than the averaged stored energy determined from whole samples [19,45,46].

For both recrystallization and grain growth, the curvature-based driving force, $\boldsymbol{F}_{\sigma}$, is calculated numerically based on the boundary shape. The driving force, $\boldsymbol{F}_{\sigma,\mathrm{i}}$, for a position, i, at a boundary is determined using equation $\boldsymbol{F}_{\sigma,\mathrm{i}} = 2\sigma/r_{\mathrm{i}}$, where σ is the boundary energy and r_{i} is the radius of the curvature at that position. $\boldsymbol{F}_{\sigma}$ is then calculated by integrating $\boldsymbol{F}_{\sigma,\mathrm{i}}$ over the length, Δl, of boundary segments within the area investigated, i.e. $\boldsymbol{F}_{\sigma} = \int_{\Delta l} \boldsymbol{F}_{\sigma,\mathrm{i}}$. So far, this kind of quantification has been carried out mainly for 2D data sets. For 3D data sets, the integration has to be performed over the whole area of a boundary.

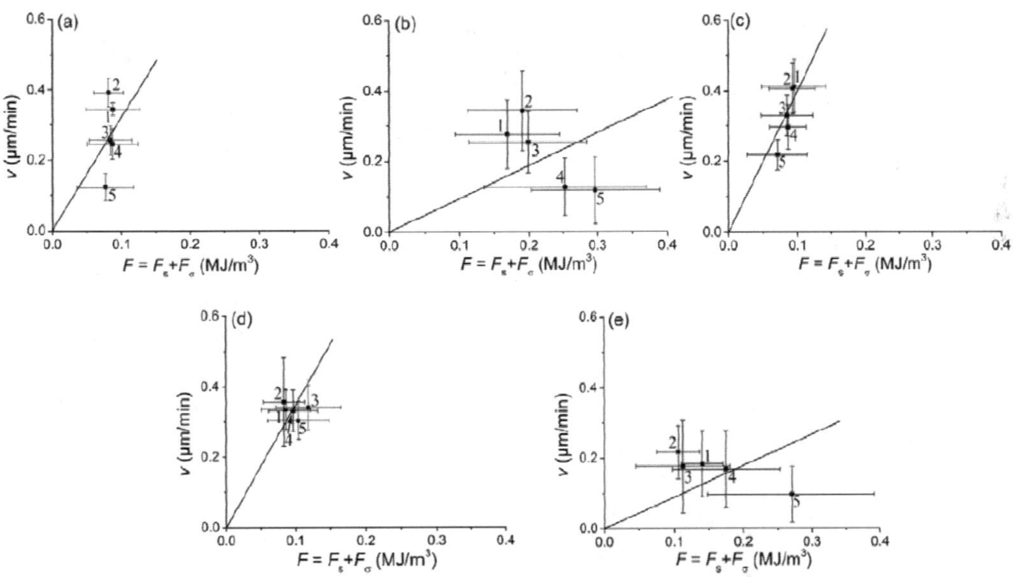

Figure 5. Migration velocity v for the boundary segments in the five boxes marked in figure 2f as a function of total driving force, $F = F_S + F_\sigma$ for recrystallization at 250 °C. (a) – (e) are for boundary segments in the boxes I–V, respectively. The error bars represent standard deviations. Numbers are used to identify annealing intervals. [40]

2.5. Quantification of local mobilities

With both v and F quantified for local boundaries segments, the boundary mobility can be determined for cases when a linear function fits the v and F data well. For example, the v-F plots for the boundary segments in the five boxes in figure 2f for recrystallization at 250 °C are shown in figure 5, based on which the apparent mobility for the boundary segments within boxes III and IV is determined to be

$6.2 \pm 0.4 \times 10^{-14} m^4 s^{-1} J^{-1}$. It is not possible to determine the apparent mobilities for the boundary segments in other boxes due to the large scatter in the *v*-*F* plot (i.e. poor linear fitting), which matches a large orientation spread in the deformed matrix in front of the those local boundary segments [40].

2.6. Quantification of activation energy

Boundary mobility is a thermally activated parameter, and can be expressed by:

$$M = M_0 \exp(-Q/RT), \tag{4}$$

where Q is activation energy, M_0 a pre-exponential factor, R the gas constant, and T temperature. Much work has been carried out on grain growth using bicrystal experiments (e.g. [7]). In contrast very little effort has been given to local analysis of individual recrystallizing boundaries, due to the complexity of the boundaries. Here a new method is proposed, which is based on the local analysis of in-situ/ex-situ data sets.

By measuring the boundary apparent mobilities at different temperatures using the method described in section 2.5, the activation energy can be obtained. For example, if the boundary apparent mobilities measured at temperatures T_1 and T_2 are M_1 and M_2, respectively, the activation energy can be determined as:

$$Q = \ln(M_2/M_1) R T_1 T_2/(T_2 - T_1). \tag{5}$$

For example, for the boundary segments within box III in figure 2, the activation energy determined using equation (5) is ~280 kJ/mol. This value is well within the range determined based on grain growth experiments for pure Al [7,9].

Note that one has to be careful when determining activation energy for local recrystallizing boundaries. The characteristics (misorientation and plane normal) of the boundary should not change during the investigated annealing period. For example, the boundary segments within box IV in figure 2 migrates into a region with new misorientation during the last annealing steps at 260 °C with a relatively low mobility [40]. If the activation energy for boundary segments is determined using equation (5) assuming that the boundary characteristics have not changed, an activation energy value of ~100 kJ/mol is obtained. This value is only about 1/3 of that for the boundary segments with the same misorientation in box III, and this is clearly not correct.

3. Local boundary migration

The local boundary migration has been followed and *quantified* using the methods described in section 2 in several material systems, including recrystallization boundary migration in heavily deformed as well as medium strained metals, and boundary migration during grain growth. Several examples are presented in the following. Quantitative studies of local boundary migration based on phase field simulations are also exemplified.

3.1. Boundary migration in heavily deformed metals

3.1.1. Triple junction motion. After rolling deformation to high strains (ε_{vM}>3.5), the microstructure contains typically lamellar boundaries almost parallel to the sample rolling direction (RD), connected by incidental dislocation boundaries (IDBs) within the lamellae (see e.g., figure 6a). During annealing, boundary migration in connection to triple junction (TJ) motion prior recrystallization nucleation has been seen in Al and Ni [21,47–49]. Figure 6 shows an example of TJ motion in heavily deformed pure Ni followed during in situ annealing using TEM. The migrating TJs and the moving direction is marked by the red arrows in figure 6c. It is seen that TJ motion leads to microstructural coarsening without much changing of the morphology of the microstructure (figures 6b and 6d).

By combining in situ TEM with transmission Kikuchi diffraction (TKD), the boundary characteristics of the migrating TJs can be studied. It has been shown that both low angle boundaries (LABs) and high angle boundaries (HABs) can migrate in connection with TJ motion; see the thin and bold lines, respectively, in figures 6c and 6d. The misorientation distribution for the moving boundaries have been quantified based on ex-situ EBSD studies (see figure 7). The results show that the misorientations for the migrating boundaries cover a large range, 3-63°. The relative frequency of moving LABs is much lower than that of HABs, considering the fact that the fraction of LABs is about 50% in the deformed matrix in this material [50,51]. Among all the migrating boundaries, boundaries with rotation axes close to ⟨111⟩ have a relatively high frequency (see figure 7).

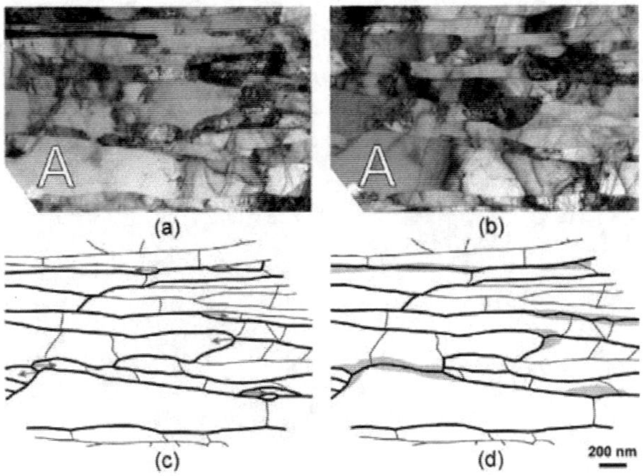

Figure 6. An example showing TJ motion in heavily deformed pure Ni prepared by accumulative rolling bonding to a von Mises strain of 6.4. (a) and (b) are TEM images showing the microstructure before and after in situ annealing. Letter A marks the same position. (c) and (d) are drawings showing boundary networks and TJs. Bold and thin lines indicate HABs and LABs, respectively. In (c) arrows indicate the migration direction of Y-junctions. Changes caused by TJ motion are shown in grey in (d). The LABs and HABs are defined as boundaries with misorientation <15° and ≥15°, respectively, determined based on TKD results. [48]

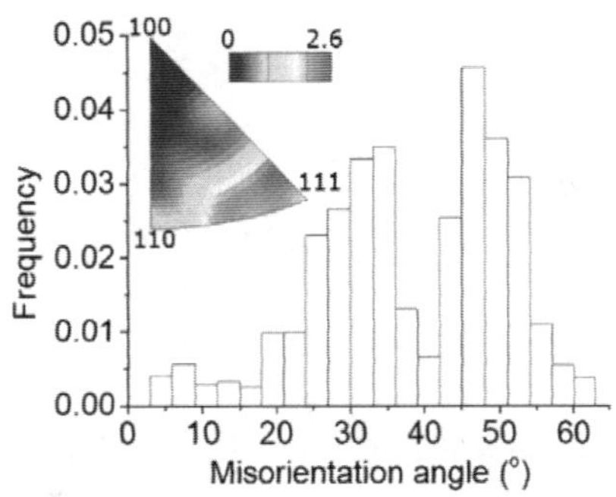

Figure 7. Misorientation (angle/axis pairs) distribution for migrating boundaries in connection with TJ motion during annealing at 200 °C for 1h of heavily deformed Ni prepared by ARB to a von Mises strain of 4.8. The result is obtained from an ex-situ EBSD study.

3.1.2. Recrystallization boundary migration. Boundary migration during recrystallization has been followed by ex-situ EBSD in heavily deformed Ni, Cu and steel. It is found that the recrystallizing boundaries migrate relatively faster along RD than along ND [52,53]. A large misorientation range is seen for the migrating and non-migrating boundaries. A relatively high frequency of HABs with rotation axes close to ⟨111⟩ is seen for both FCC and BCC metals (see figure 8) [53–55]. A comparison between the misorientations for migrating and non-migrating boundaries has shown no significant differences, except that there is a relatively high fraction of LABs and boundaries with misorientation close to 60°⟨111⟩ for the non-migrating boundaries (see figure 8a,b) [52]. The local stored energies quantified based on EBSD maps within the consumed regions and the regions in front of non-migrating boundaries are also similar [53,54]. The results have also shown that the boundary segments along RD are relatively flat and typically migrate slowly [52,53].

The results have further shown that some parts of the non-migrating boundary segments at any given annealing step can migrate upon further annealing, i.e. the boundary segments migrate in a stop-go fashion. Quantification of local stored energy in front of the non-migrating and migrating boundary segments after different annealing steps shows that concurrent recovery of the deformation matrix is not the reason for the stop-go motion [53].

3.2. Boundary migration during recrystallization of pure Al deformed to medium strain

3.2.1. 3D boundary migration. Local boundary migration during recrystallization in 50% cold rolled pure Al has been studied using 3D synchrotron Laue micro-diffraction (see figure 3). The 3D boundary has a staircase shape, consisting of both curved and facet parts (see figure 3c and d). Quantitative analysis shows that the curved parts migrate faster than the facets (see figure 3e). The boundary migrates by movement of the curved parts laterally along the facets, thereby maintaining the staircase shape of the boundary. As a result, the facet edges are either extended or shortened during boundary migration [41].

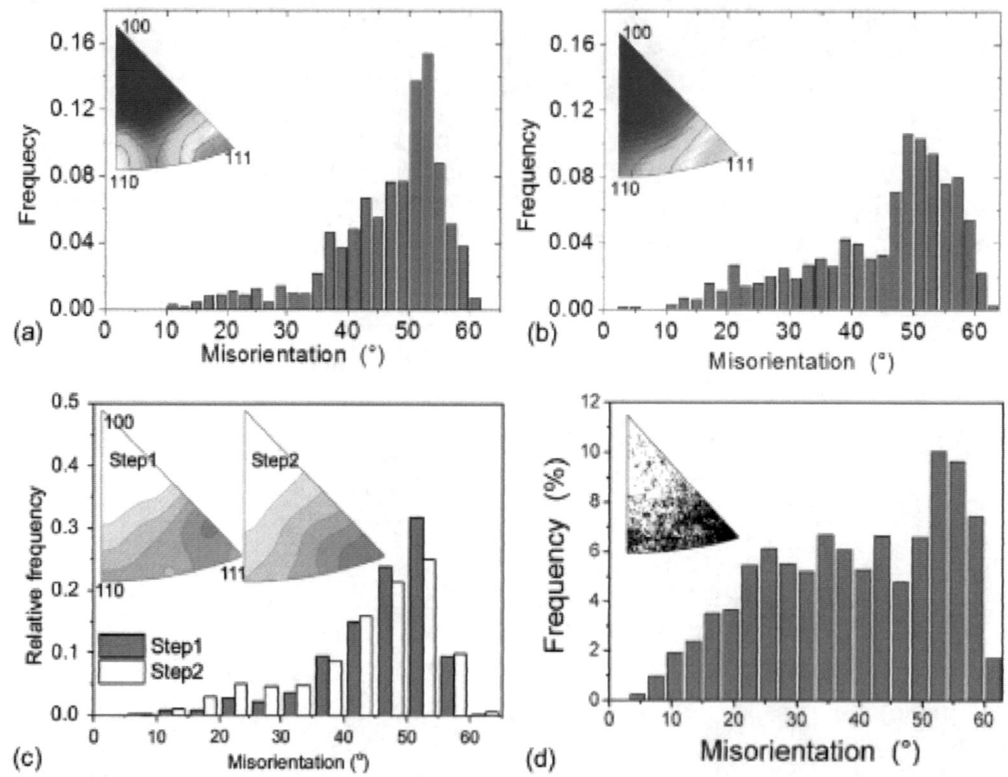

Figure 8. Migration distribution for migrating and non-migrating boundaries during recrystallization. (a) and (b) are for migrating and non-migrating boundaries, respectively, for cold rolled pure Ni [52]. (c) and (d) are for migrating boundaries in cold rolled pure Cu [53] and steel processed by dynamic plastic deformation [55].

Misorientation analysis shows that there is no significant difference between the misorientation distributions for curved parts and facets, i.e. migrating and non-migrating boundary segments, respectively. The stored energies quantified based on the misorientations for the regions consumed by the curved migrating boundary are similar to those for the regions in front of non-migrating facets (see figure 9). The curvature-based driving force is relatively small and is not the main reason for the observed difference between the migrating and non-migrating boundary segments [41].

An analysis of boundary plane normals from the 3D studies reveals a significant difference for the curved migrating and facet non-migrating boundary segments (see figure 4). However, MD simulations reveal that the mobility difference for the curved and facets parts is small, suggesting that the mobility difference between the migrating and non-migrating parts can be small, i.e. the boundary plane normal is also not the main reason for the observed difference for this particular 3D boundary.

By correlating the deformed microstructure and the migration pattern of the recrystallizing boundary, it is found that the migration of the curved parts follows the (-111) set of dislocation boundaries (see figure 9e). This result suggests that the curved part is likely to be promoted by these favorably aligned dislocation boundaries, which supports a previous suggestion based on destructive 3D [56] and in-situ 2D investigations [35,57].

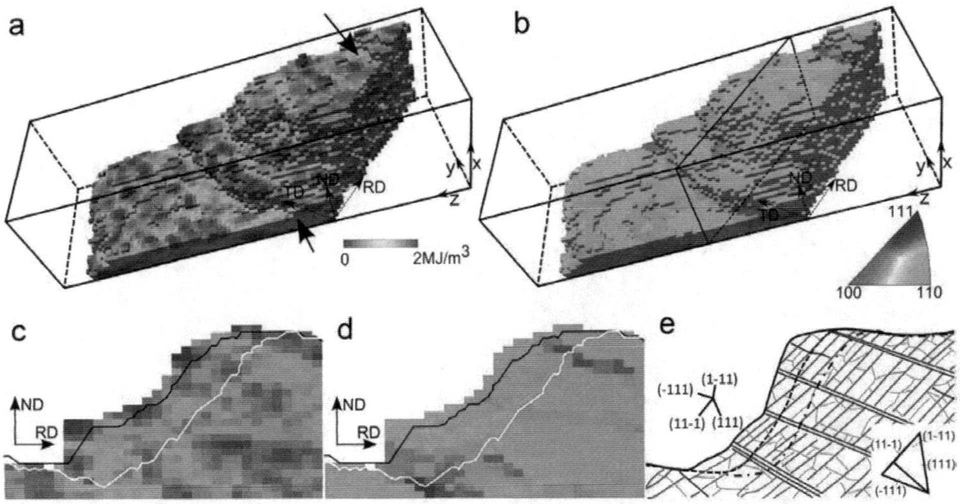

Figure 9. Effects of the deformation microstructure on local boundary migration. The deformation matrix at the initial stage is colored according to the local stored energy (a) and local misorientation rotation axis (b). (c) and (d) show a cross-section view of the microstructure at the position marked in (b). Black and white lines mark the boundary positions after the two annealing steps. (e) Sketch drawn based on the 3D X-ray data and 2D data from electron microscopy showing the microstructure and the boundary traces. The blue and red lines represent the two sets of intersecting dislocation boundaries, while the light blue lines represent incidental dislocation boundaries. The black solid, dashed and dotted-dashed lines represent three recrystallization boundary positions. The traces of the four {111} planes and the projection of a {111} tetrahedron for the deformed grain on the RD-ND section are also shown. [41]

3.2.2. Boundary mobility. The local boundary migration during recrystallization in the same pure Al sample discussed above in section 3.2.1 has been followed during ex-situ annealing at 250 °C using ECC and EBSD (see figure 2). Based on the local analysis in the regions marked by the five boxes in figure 2, the local migration behaviour has been analysed in detail, from which the local mobilities have been quantified (see section 2.5).

By correlating the deformed microstructure and the boundary traces (see figure 10), some further information can be obtained. It is found that most of the pinning points (e.g. A–D in figure 10) coincide with ⟨110⟩ rotation axes (green in figure 10b), while the misorientation angles are quite scattered. The local stored energies at these points are relatively high (figure 10c). It is therefore very likely that the boundary segments with misorientation axes close to ⟨110⟩ have comparatively lower apparent mobilities than those with rotation axes close to ⟨111⟩.

3.3. Boundary migration during grain growth

3.3.1. Grain growth of cube textured Ni. Grain growth of a cube textured pure Ni has been followed during ex-situ annealing at 600 °C using EBSD. It is found that the misorientation angles for the migrating boundaries cover a wide range (see figure 11). Compared to the misorientation distribution for all the boundaries in the sample (figure 11a, twin boundaries are not included), the results show that except for boundaries with misorientations lower than 5° the two distributions are similar [58].

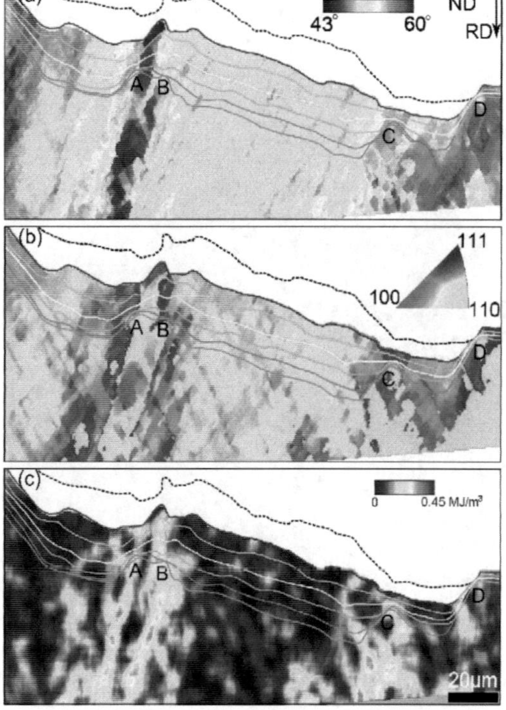

Figure 10. Boundary traces overlaid on an EBSD map for the corresponding area. In the EBSD map, each pixel in the deformed matrix is colored according to the misorientation angle (a), the misorientation rotation axis (b) between the pixel and the recrystallized grain, and the local stored energy (c). A–D mark pinning points. The local stored energy at each pixel is calculated from a subset of 11 × 11 pixels centered at each selected pixel using the method described in section 2.4. [40]

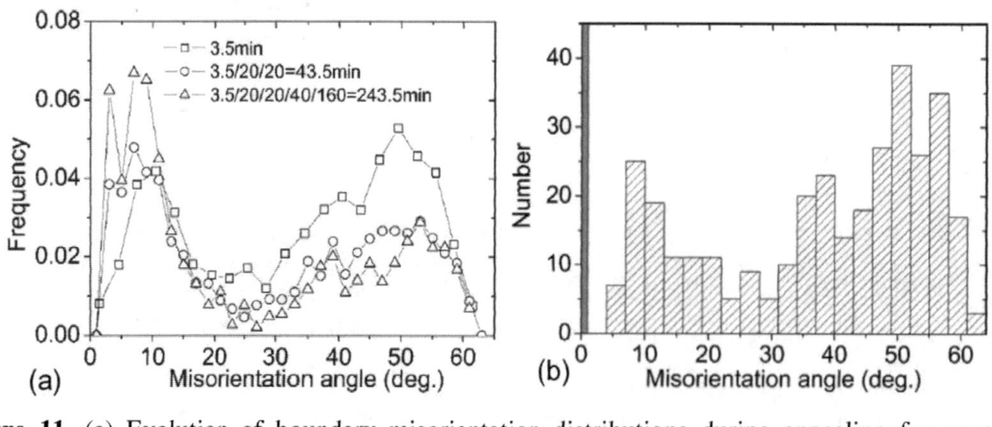

Figure 11. (a) Evolution of boundary misorientation distributions during annealing for annealing times of 3.5, 43.5 and 243.5 min in pure Ni. (b) Misorientation distribution for all boundaries observed to migrate, summarized over all annealing steps. Reproduced from [58].

A close-up examination of the individual migrating boundaries in this sample shows that some boundaries with misorientation close to Σ3 boundaries (following the Brandon criteria) are highly mobile. These boundaries typically have a large deviation angle (>4°) to the ideal twin relationship, i.e. 60°⟨111⟩, and therefore are not twin boundaries (see figure 12a-d). A coherent or incoherent twin boundary can be elongated by the migration of their connected boundaries (see figure 12e and f); similar results are seen in pure Al [59]. More surprisingly, it is also observed that a boundary can migrate even though it has a negative curvature (see the black boundary around grain C in figure 12c and d). This underlines the importance of the grain boundary network connectivity.

IOP Conf. Series: Materials Science and Engineering **580** (2019) 012015 doi:10.1088/1757-899X/580/1/012015

3.3.2. 3D grain growth of pure iron. The grain growth of pure iron has been followed in 3D during ex-situ annealing using LabDCT and synchrotron DCT (see e.g., figure 13). Similar to pure Ni, a large misorientation spread is seen for the migrating boundaries, and the misorientation distribution for these boundaries is similar to that for all the boundaries in the characterized volume [60]. The 3D result shows that in this sample locally some boundaries having negative curvature also migrate. Quantification of the growth rate for individual grains as a function of mean integrated curvature of the grains shows that grain growth slows down and the overall boundary mobility decreases during grain growth [39].

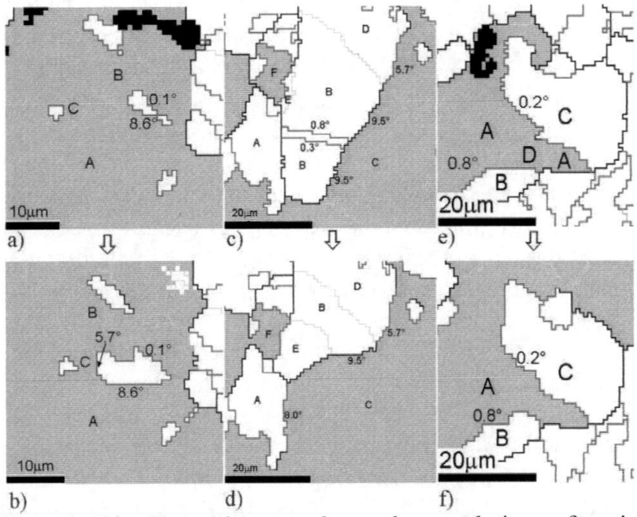

Figure 12. Examples to show the evolution of twin boundaries during grain growth: numbers in each figure indicate deviation angle from ideal 60°⟨111⟩. [58]

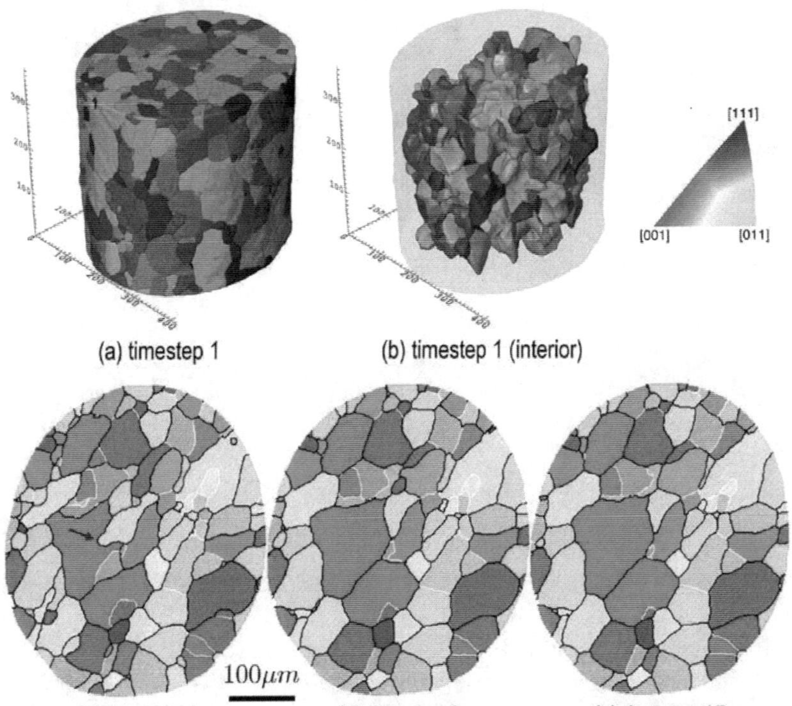

Figure 13. Experimental results from a 3D grain growth experiment in pure iron annealed at 800 °C. Above: 3D grain maps for time-step 1, displaying (a) all grains and (b) only interior grains. Below: one section of the 3D grain map for time steps 1 (c), 8 (d), and 15 (e). The colours represent the grain orientation along the sample RD (see the insert triangle), while black and white lines in (c)-(e) represent boundaries with misorientations above and below 15°, respectively. [39]

3.4. Modelling of local boundary migration during recrystallization

A phase-field model and a numerical model have been developed to simulate local boundary migration into spatially varying deformation energy fields during recrystallization [61–64]. An example of phase-field modelling is shown in figure 14, where the movement of an initially planar boundary into energy fields with 2D variations representing two sets of dislocation boundaries lying at equal, but opposite angles to the boundary are considered. When the boundary migrates from bottom to top of the energy field, parts of the boundary, indicated with the red oval, show a stop-and-go type of movement with pinch-off: boundary parts near minima in the energy density field remain for a long period at the same position, while the rest of the boundary continues to advance. At a certain point, the stagnated parts are pinched off, leaving behind a small part of deformed material (see figure 14b). As the boundary migrates from left to right, asymmetrical protrusions and retrusions develop along the migrating recrystallization front with a considerably higher curvature at the retrusions than at the protrusions, and no boundary segment stagnates (see figure 14c). The average migration velocity is therefore different for the two cases. The results suggest that the geometrical alignment of the dislocation boundaries has a strong impact on the formation of local protrusions and retrusions as well as on the local boundary migration. More work has to be done to quantify the effect of protrusions/retrusions on such non-uniform migration behaviour.

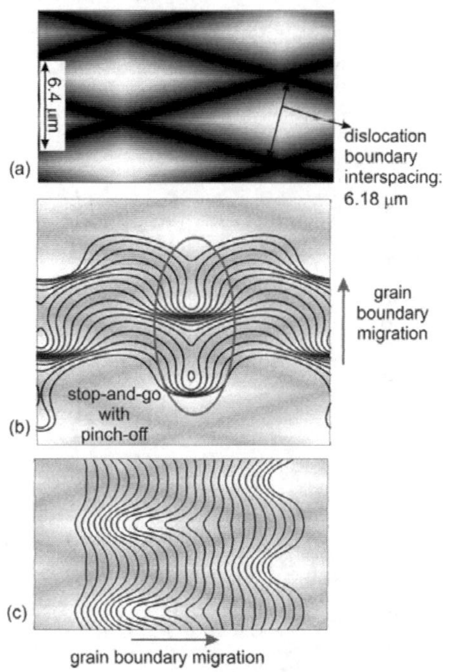

Figure 14. Simulation results obtained for a recrystallization boundary migrating in a deformation energy field with variations. (a) Representation of the considered energy field using a color scaling with black representing the highest energy density and white an energy density equal to 0. (b), (c) Grain boundary positions as a function of time obtained in the simulations, plotted every 2000 time steps starting from time step 60000. (b) Boundary moving from bottom to top. (c) Boundary moving from left to right. [61]

4. Conclusions and outlook

Based on 3D(x, y, time) and 4D(x, y, z, time) data, several methods have been developed for quantifying local boundary migration, with which the local boundary migration in several material systems during both recrystallization and grain growth has been quantified in detail. Based on such analyses, the physical mechanisms underlying the migration process have been quantified for the selected materials. The conclusions are:

i) The misorientations of migrating boundaries cover a wide range for *both* recrystallization and grain growth in *both* FCC and BCC metals. Boundaries with misorientation angle < 5° are in all cases rarely mobile, which agrees with the common knowledge that LABs have lower mobilities and with the orientation pinning mechanism [65]. High angle boundaries with misorientation rotation axis close to ⟨111⟩ have relatively high mobilities compared to other high angle boundaries for both FCC and BCC metals. The results further show that the misorientation distributions for the migrating and non-migrating boundaries are quite similar for recrystallization boundaries. This is partly because that the boundaries are connected in a network: the non-migrating parts can eventually be driven by the connected migrating parts through the development of local retrusions, which provide an additional curvature based driving force.

ii) For the migration of recrystallizing boundaries, the geometrical alignment of the dislocation and lamellar boundaries in the deformed microstructure is another important parameter defining the boundary migration. It affects the overall recrystallization boundary shape, e.g. staircase shaped consisting of curved parts and facets, and the formation of local protrusions and retrusions, which can in turn accelerate the local boundary migration speed. Boundaries migrate differently when they meet the dislocation/lamellar boundaries at different angles, even if the misorientations of the boundaries are very similar.

iii) The local apparent mobility and the activation energy for a recrystallization boundary with misorientation around 50° ⟨221⟩ has been quantified to be $6.2 \pm 0.4 \times 10^{-14} m^4 s^{-1} J^{-1}$ and 280 kJ/mol, respectively, for pure Al annealed at 250 °C. The results show also that the boundary segments with misorientation axes close to ⟨111⟩ have comparatively higher apparent mobilities than those with rotation axes close to ⟨110⟩.

Acknowledgements

The author thanks financial support by the European Research Council (ERC) under the European Union's Horizon 2020 research and innovation programme (M4D – grant agreement No. 788567), and support by the Danish Council for Independent Research Fund (8022-00340B).

References

[1] Humphreys F J and Hatherly M 2004 *Recrystallization and related annealing phenomena.*(Oxford: Elsevier)
[2] Van Boxel S, Schmidt S, Ludwig W, Zhang Y B, Juul Jensen D and Pantleon W 2014 *Mater. Trans.* **55** 128–36
[3] Zhang Y B and Juul Jensen D 2015 *IOP Conf. Ser. Mater. Sci. Eng.* **89** 012015
[4] Zhang Y B, Godfrey A and Jensen D J 2009 *Comput. Mater. Contin.* **14** 195–205
[5] Zhang Y, Godfrey A and Jensen D J 2011 *Scr. Mater.* **64** 331–4
[6] Gorkaya T, Molodov K D, Molodov D A and Gottstein G 2011 *Acta Mater.* **59** 5674–80
[7] Molodov D A, Swiderski J, Gottstein G, Lojkowski W and Shvindlerman L S 1994 *Acta Metall. Mater.* **42** 3397–407
[8] Winning M and Rollett A D 2005 *Acta Mater.* **53** 2901–7
[9] Molodov D A, Czubayko U, Gottstein G and Shvindlerman L S 1995 *Scr. Metall. Mater.* **32** 529–34
[10] Winning M, Rollett A D, Gottstein G, Srolovitz D J, Lim A and Shvindlerman L S 2010 *Philos. Mag.* **90** 3107–28
[11] Winning M, Gottstein G and Shvindlerman L S 2002 *Acta Mater.* **50** 353–63
[12] Olmsted D L, Foiles S M and Holm E A 2009 *Acta Mater.* **57** 3694–703
[13] Holm E A, Rohrer G S, Foiles S M, Rollett A D, Miller H M and Olmsted D L 2011 *Acta Mater.* **59** 5250–6
[14] Olmsted D L, Holm E A and Foiles S M 2009 *Acta Mater.* **57** 3704–13
[15] Priedeman J L, Rosenbrock C W, Johnson O K and Homer E R 2018 *Acta Mater.* **161** 431–43
[16] Han J, Vitek V and Srolovitz D J 2015 *IOP Conf. Ser. Mater. Sci. Eng.* **89** 012004
[17] Ratanaphan S, Olmsted D L, Bulatov V V, Holm E A, Rollett A D and Rohrer G S 2015 *Acta Mater.* **88** 346–54
[18] Nakamichi H, Humphreys F J and Brough I 2008 *J. Microsc.* **230** 464–71
[19] Lens A, Maurice C and Driver J H 2005 *Mater. Sci. Eng. A* **403** 144–53
[20] Yu T, Hughes D A, Hansen N and Huang X 2015 *Acta Mater.* **86** 269–78
[21] Yu T and Hughes D A 2019 *Philos. Mag.* **99** 869–86

[22] Mompiou F, Caillard D and Legros M 2009 *Acta Mater.* **57** 2198–209
[23] Poulsen H F 2004 *Three-dimensional X-ray diffraction microscopy : mapping polycrystals and their dynamics.* (Berlin:Springer)
[24] Poulsen H F 2012 *J. Appl. Crystallogr.* **45** 1084–97
[25] Li S F and Suter R M 2013 *J. Appl. Crystallogr.* **46** 512–24
[26] Viganò N, Tanguy A, Hallais S, Dimanov A, Bornert M, Batenburg K J and Ludwig W 2016 *Sci. Rep.* **6** 20618
[27] King A, Reischig P, Adrien J and Ludwig W 2013 *J. Appl. Crystallogr.* **46** 1734–40
[28] Viganò N, Ludwig W and Batenburg K J 2014 *J. Appl. Crystallogr.* **47** 1826–40
[29] McDonald S A, Holzner C, Lauridsen E M, Reischig P, Merkle A P and Withers P J 2017 *Sci. Rep.* **7** 5251
[30] McDonald S A, Reischig P, Holzner C, Lauridsen E M, Withers P J, Merkle A P and Feser M 2015 *Sci. Rep.* **5** 14665
[31] Larson B C and Levine L E 2013 *J. Appl. Crystallogr.* **46** 153–64
[32] Levine L E, Larson B C, Yang W, Kassner M E, Tischler J Z, Delos-Reyes M A, Fields R J and Liu W 2006 *Nat. Mater.* **5** 619–22
[33] Barabash R I, Ice G E, Liu W and Barabash O M 2009 *Micron* **40** 28–36
[34] Zhang Y B, Andriollo T, Fæster S, Liu W, Hattel J and Barabash R I 2016 *Acta Mater.* **121** 173–80
[35] Zhang Y, Godfrey A and Juul Jensen D 2014 *Metall. Mater. Trans. A* **45** 2899–90
[36] Zhang Y B, Elbrønd A and Lin F X 2014 *Mater. Charact.* **96** 158–65
[37] Huang Y, Laflen B, Deal A, Spinelli I, Barbuto A and Hanlon T 2013 *1st Int. Conf. 3D Mater. Sci.* 165–70
[38] Xu C, Zhang Y, Godfrey A, Wu G, Liu W, Tischler J Z, Liu Q and Juul Jensen D 2017 *Sci. Rep.* **7** 42508
[39] Zhang J, Zhang Y, Ludwig W, Rowenhorst D, Voorhees P W and Poulsen H F 2018 *Acta Mater.* **156** 76–85
[40] Zhang Y B 2018 *Scr. Mater.* **146** 286–9
[41] Zhang Y B, Budai J D, Tischler J Z, Liu W, Xu R, Homer E R, Godfrey A and Juul Jensen D 2017 *Sci. Rep.* **7** 4423
[42] Sun J, Zhang Y, Lyckegaard A, Bachmann F, Lauridsen E M and Juul Jensen D 2019 *Scr. Mater.* **163** 77–81
[43] Godfrey A, Mishin O V and Yu T 2015 *IOP Conf. Ser. Mater. Sci. Eng.* **89** 012003
[44] Seret A, Moussa C, Bernacki M, Signorelli R and Bozzolo N 2019 *J. Appl. Crystallogr.* **52** 1–16
[45] Vandermeer R A, Juul Jensen D and Woldt E 1997 *Metall. Mater. Trans. A* **28** 749–754.
[46] Basu I, Chen M, Loeck M, Al-Samman T and Molodov D A, 2016 *Mater. Charact.* **117** 99–112.
[47] Yu T, Hansen N and Huang X 2011 *Proc. R. Soc. A Math. Phys. Eng. Sci.* **467** 3039–65
[48] Zhang Y B, Yu T and Mishin O V. 2017 *Mater. Lett.* **186** 102–4
[49] Mishin O V, Zhang Y B and Godfrey A 2015 *IOP Conf. Ser. Mater. Sci. Eng* **89** 012035
[50] Zhang Y B, Mishin O V, Kamikawa N, Godfrey A, Liu W and Liu Q 2013 *Mater. Sci. Eng. A* **576** 160–6
[51] Zhang Y B, Mishin O V and Godfrey A 2014 *J. Mater. Sci.* **49** 287–93
[52] Zhang Y B, Juul Jensen D and Godfrey A 2012 *Mater. Sci. Forum* **715–716** 329–32
[53] Lin F, Zhang Y, Pantleon W and Juul Jensen D 2018 *Metall. Mater. Trans. A Phys. Metall. Mater. Sci.* **49A** 5246–58
[54] Zhang Y B, Godfrey A, Liu Q, Liu W and Juul Jensen D 2009 *Acta Mater.* **57** 2631–9
[55] Zhang Z, Zhang Y, Mishin O V, Tao N, Pantleon W and Juul Jensen D 2016 *Metall. Mater. Trans. A Phys. Metall. Mater. Sci.* **47** 4682–93
[56] Fan G H, Zhang Y B, Juul Jensen D and Driver J H 2014 *Scr. Mater.* **72–73** 9–12
[57] Zhang Y B, Godfrey A and Juul Jensen D 2010 *Proc. of the 31st Risø Int. Sym. on Mater. Sci.*

IOP Conf. Series: Materials Science and Engineering **580** (2019) 012015 doi:10.1088/1757-899X/580/1/012015

(Roskilde: Risø - DTU) p 497–503
[58] Zhang Y B, Godfrey A, Liu W and Liu Q 2010 *Mater. Sci. Technol.* **26** 197–202
[59] He Q, Huang T, Shuai L, Zhang Y, Wu G, Huang X and Jensen D J 2018 *Scr. Mater.* **153** 68–72
[60] Sun J, Lyckegaard A, Zhang Y B, Catherine S A, Patterson B R, Bachmann F, Gueninchault N, Bale H, Holzner C, Lauridsen E and Juul Jensen D 2017 *IOP Conf. Ser. Mater. Sci. Eng.* **219** 012039
[61] Moelans N, Zhang Y B, Godfrey A and Juul Jensen D 2015 *IOP Conf. Ser. Mater. Sci. Eng.* **89** 012037
[62] Moelans N, Godfrey A, Zhang Y and Juul Jensen D 2013 *Phys. Rev. B* **88** 054103
[63] Martorano M A, Fortes M A and Padilha A F, 2006 *Acta Mater.* **54** 2769.
[64] Martorano M A, Sandim H R Z, Fortes M A and Padilha A F, 2007 *Scripta Mater.* **56** 903.
[65] Juul Jensen D 1995 *Acta Metall. Mater.* **43** 4117–29

Thermochemical surface engineering: a playground for science and innovation

T L Christiansen and M A J Somers

Technical University of Denmark, Department of Mechanical Engineering, Produktionstorvet b.425 DK 2800 Kongens Lyngby, Denmark

Abstract. This article illustrates the plethora of different possibilities in thermochemical surface engineering (TSE) and the close interlink between science and innovation. Three widely different examples of gaseous TSE are presented: surface engineering of austenitic stainless steel by high- and low- temperature surface treatment, surface hardening of titanium by incorporation of nitrogen, oxygen and carbon, and direct synthesis of magnetic iron-nitride α''-$Fe_{16}N_2$. It is shown how 3D printed materials made of stainless steel and titanium can benefit from TSE.

1. Introduction

Thermochemical surface engineering (TSE) is widely applied for improvement of the surface performance of metallic parts, subjected to wear, fatigue or corrosion. Some of the well-known classical methods are gaseous carburising (or "case hardening") and "ferritic nitriding" of steels. TSE is the *intentional change* of the composition (of the surface) of a material at elevated temperature with the purpose to *improve materials performance*. In contrast to deposition and coating techniques, such as physical vapour deposition, thermochemical processing entails the development of a solid solution, a dispersed phase or a reaction layer within (and involving) the original material, rather than on top of it.

In the present contribution a few examples of the current research and innovation activities in our group are used to illustrate how TSE can bring about new (surface) features and thus enhanced performance in different types of materials. The examples given include:

- TSE of stainless steel by dissolution of nitrogen and carbon at high- and low temperature;
- TSE of titanium alloys by formation of mixed interstitial phases based on oxygen, nitrogen and carbon;
- TSE for low temperature *synthesis* of magnetic α''-$Fe_{16}N_2$ in (bulk) iron.

The last example is - strictly speaking - not surface engineering, but "classical" surface engineering methods are applied for the purpose of bulk engineering.

Pertinent details of the various treatments and materials are provided in the figure captions. Otherwise the reader is referred to patent literature for a full description of the technologies noting that some processes entail proprietary steps, which are not disclosed. The scope of this article is not a detailed treatise of the different topics, but rather an illustration of what is possible for the different materials systems using TSE.

2. TSE of stainless steel: high temperature solution nitriding and expanded austenite

Stainless steels are used in numerous applications owing to their favourable corrosion properties in addition to deformability. However, this steel class typically suffers from insufficient wear resistance, especially for austenitic stainless steels. A remedy to the wear problem was originally found in gaseous nitriding [1] by formation of a case containing small Cr-nitrides resulting in an increase in hardness, albeit at the expense of corrosion resistance. In the 1980s it was discovered that low temperature nitriding and carburising could offer both increased surface hardness and unaltered or even improved corrosion resistance [2,3]. In the 1990s another surface hardening concept was introduced by Berns

[4,5], so-called high temperature solution nitriding (HTSN). Here nitrogen was introduced to the surface of stainless steel at very high temperatures. The concepts of low- and high temperature surface hardening of (austenitic) stainless steel can be illustrated by the use of a hypothetical time-temperature-transformation (TTT) diagram as given in figure 1.

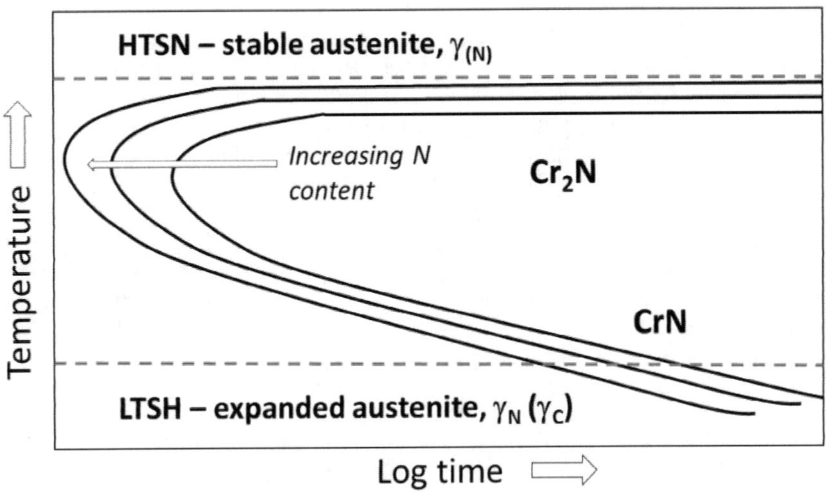

Figure 1. Time-temperature-transformation diagram of hypothetical austenitic stainless steel containing different amounts of nitrogen.

At high temperatures, say, above 1050°C there is a significant solubility of nitrogen in austenite, which can be exploited as a surface engineering method. The nitrogen solubility in austenite can be controlled by adjusting the partial pressure of molecular nitrogen, N_2. The N_2 can dissociate on metallic surfaces above, say, 800°C and enter into the solid state as atomic N. The presence of, in particular, Cr in stainless steels enhances the solubility of nitrogen and thus enables nitriding at high temperatures in N_2. Hence, a case of nitrogen-rich austenite can be realized in the surface of stainless steel, and case depths of more than 1 mm can be achieved due to the high temperatures involved. From the TTT diagram in figure 1 it is evident that cooling from the stable austenite phase field is critical: formation of Cr_2N will occur very quickly at intermediate temperatures around 800-900°C (in a matter of seconds). By applying a high pressure N_2 gas quench, chromium nitride formation can be avoided during cooling and a supersaturated nitrogen-rich austenite case in the treated part can be obtained. Typical nitrogen contents are in the range 0.3 to 0.8 wt% depending on the specific grade. One problem associated with this (conventional) quenching method is that the high N_2 pressure applied in the cooling stage (up to 20 bars) can result in formation of Cr_2N at the surface and at the grain boundaries in the surface adjacent zone; most pronounced if the nitrogen content in the austenite is high. The formation of surface nitrides will partially impair the otherwise improved corrosion properties resulting from having nitrogen in solid solution in austenite. A recently patented method that fully circumvents this problem is the use of inert gas for quenching in the HTSN process [6].

The other option for surface engineering of stainless steel is to apply low temperatures – lower than those conventionally used for ferritic nitriding/nitrocarburising of steels. At temperatures below, say, 500°C for carburising and 450°C for nitriding a highly supersaturated solid solution of N/C in austenite can be obtained, so-called expanded austenite. The interstitial contents that can be realized are several weight percent, which results in very high surface hardness, albeit with a case depth limited to typically less than 35 μm. The formation of expanded austenite exploits the fact that interstitial diffusion still takes place in the low temperature range, whereas substitutional diffusion is largely non-existent. In addition, the presence of Cr in stainless steel, which has a high affinity to interstitial atoms, results in a high solubility. At the treatment temperature the equilibrium N/C solubility in austenite is low, as nitrides/carbides are stable, but the formation of these phases is kinetically hindered due to the limited diffusivity of substitutionally dissolved atoms, leading to a supersaturated solid solution of interstitials in austenite (expanded austenite). The formation of an expanded austenite case results in a surface hardness of more than 1200 HV. Moreover, beneficial compressive residual stresses in the surface occur, which significantly improves fatigue resistance. One major problem with low temperature surface hardening, which is particularly relevant for nitrogen, is that most technical stainless steel parts contain deformation-induced martensite from production, machining etc. Deformation-induced martensite is often desired to provide additional strength to a component and additional bearing capacity for a hard case. However, strain-induced (or otherwise) martensite is undesirable for low temperature surface hardening as it promotes the formation of CrN due to a favourable orientation relations. Stable or stabilized austenitic stainless steels are therefore be preferred. A patented remedy to this problem is the combination of HTSN followed by low temperature surface hardening (LTSH) [7]. HTSN effectively removes cold work and transforms martensite to austenite while, at the same time providing solid solution strengthening of the austenite due to the dissolved nitrogen. In-between the high- and low-temperature treatments an additional deformation of the stabilized austenite can be performed [8]. In this case the combined solid solution strengthening and cold work can bring about unique properties in the substrate.

The high/low- temperature process is particularly relevant for 3D-printed austenitic stainless steel. The hierarchical microstructure in 3D printed stainless steel contains a cellular microstructure with strong compositional segregation [9], which promotes (local) sensitization. In order to “restore” the stainless character of a 3D-printed stainless steel heat treatment is required. However, this will give rise to a reduction in strength due to recovery and/or recrystallization of the printed microstructure (containing the strengthening features just mentioned). The straightforward solution is HTSN, with the optional combination with LTSH. The HTSN treatment will homogenize the “chaotic” as-printed microstructure and, at the same time, introduce more nitrogen in solid solution, which imparts strength without a reduction of the ductility. As a result, the corrosion performance is drastically improved. An example of a HTSN and a low temperature carburised 3D-printed austenitic stainless steel part is shown in figure 2.

As shown in figure 2 the “footprint” of the SLM printing process is largely nullified and the microstructure is homogenized, i.e. cell structures and melt pool boundaries have disappeared. The nitrogen enriched zone from the HTSN treatment is not observable with light optical microscopy (LOM), but is evident from the hardness depth profile (not shown here). The hardness can be increased up to 260-270 HV resulting from the dissolution of nitrogen and this imparts good load bearing capacity to the expanded austenite case. The low temperature carburising treatment conducted after HTSN results in a carbon expanded austenite case with a hardness of approximately 1200 HV. Hence, the HTSN and low temperature carburising combination imparts the 3D printed austenitic stainless steel with highly favourable properties, such as excellent corrosion and wear performance, together with high fatigue resistance.

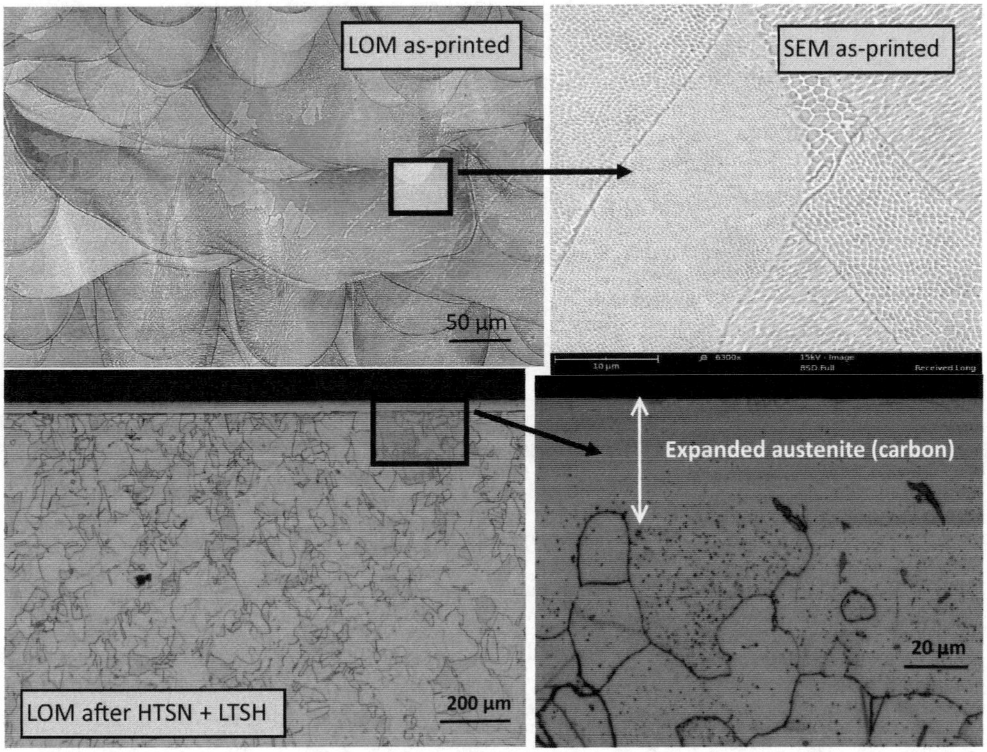

Figure 2. Microstructure of as-printed and surface hardened 3D-printed (selective laser melting, SLM) austenitic stainless steel AISI 316. The as-printed microstructure contains a clear trademark of the printing process, namely melt pool boundaries; at higher magnification a characteristic cellular structure is observed (SEM image). HTSN was performed at 1100°C for two hours followed by low temperature carburising at 470°C for 40 hours. HTSN has a significant impact on the bulk microstructure and results in a deep zone with an augmented nitrogen content (cannot be distinguished in the micrograph). The outermost case is carbon expanded austenite. From work in preparation [10].

3. TSE of titanium; mixed interstitial phases

Titanium is in many ways an excellent engineering material, as it has a high specific strength, is light weight, is extremely corrosion resistant and is fully biocompatible. However, until now it has been considered a relatively costly material. The advent of additive manufacturing (AM) and 3D metal printing can change the niche applicability of titanium, as it is one of the materials of choice for 3D metal printing due to favourable properties and the low buy-to-fly ratio inherent to AM. One obvious benefit is that printed parts of titanium have excellent corrosion performance, which is not true for printed stainless steel (see the previous section in this respect). The world production of titanium powder for AM is rapidly increasing, which implies a large scope for reduction in price. It is anticipated, therefore, that titanium can replicate the history of aluminium as a low cost technical material. Hence, titanium is about to experience a renaissance. However, titanium still suffers from poor tribological properties, but these can be improved by TSE. Recently, methods for surface hardening were developed by the present authors and rely on the use of so-called mixed interstitial phases (cf. patent application [11]). Titanium is unique in many ways, but one specific feature is the very high interstitial solubility of oxygen (and to a lesser extent nitrogen). Interestingly, also carbon and nitrogen can be dissolved *together* with oxygen resulting in mixed interstitial phases. These mixed interstitial systems are not fully investigated but synergistic effects in terms of enhanced mixed interstitial solubility have been observed [12,13]. The rock-salt type compounds based on oxygen, carbon and nitrogen in the titanium system, i.e. titanium oxide, carbide and nitride, show almost full intersolubility. This is schematically shown in figure 3 (left). The mixed interstitial compounds are claimed to have other properties than the binary compounds [14].

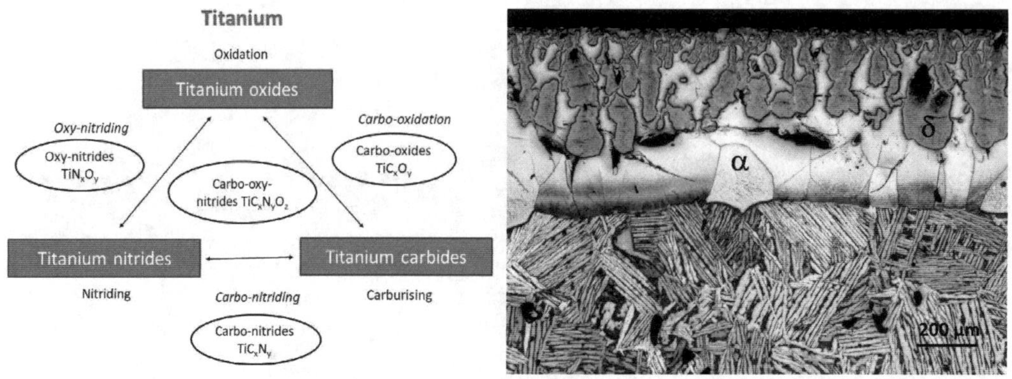

Figure 3. Left) Interstitial compounds in the titanium system containing nitrogen, carbon and oxygen. Right) carbo-oxidizing at 1050°C followed by nitriding at 1100°C of 3D-printed (SLM) titanium grade 2. The resulting case consists of carbo-oxi-nitrides (δ) and expanded h.c.p. titanium (α); the core has a Widmanstätten structure.

Mixed interstitial phases can be exploited for surface engineering of titanium and this is shown in figure 3 (right). Here a 3D-printed CP titanium grade 2 part has been carbo-oxidized in a gaseous atmosphere with a chemically controlled low partial pressure of oxygen and high carbon activity, and subsequently nitrided in an atmosphere of N_2. The deep hardened case consists of mixed interstitial compounds and a solid solution. The carbo-oxy-nitride (δ) phase will typically have a hardness in the range 2500 to 3000 HV depending on the specific composition (of N, C and O). The solid solution of nitrogen, carbon and oxygen in the (metallic) titanium (α) can reach hardness values in excess of 1500 HV, which should be compared with the hardness of the untreated titanium of 200-250 HV. It is also possible to apply low temperature gaseous surface hardening process on titanium alloys, which results exclusively in a diffusion zone of interstitials in solid solution in h.c.p. α-Ti.

4. Bulk synthesis of magnetic α''-$Fe_{16}N_2$ by TSE

It has long been anticipated and proved by theoretical calculations that α''-$Fe_{16}N_2$ nitride has attractive magnetic properties, including a so-called giant magnetic moment [15]. In this sense this nitride can be considered the "holy grail" in magnetic materials. However, a major challenge is the synthesis. Since the work of K.H. Jack it is known that α''-$Fe_{16}N_2$ develops from a supersaturated solid solution of nitrogen in ferrite or in martensite at annealing temperatures below, say, 200°C [16]. Also, decomposition of ε-Fe_2N_{1-z} with low interstitial nitrogen content develops α''-$Fe_{16}N_2$ [17,18]. Recent work has demonstrated that nano-crystalline iron-oxide powder, reduced in hydrogen to nano-crystalline iron can be converted into α''-$Fe_{16}N_2$ in NH_3 [19]. This synthesis route does not lend itself for upscaling because of the requirement for handling of hazardous nano-powders, where nano-size of the particles may facilitate a release of the strain energy induced by conversion of iron into α''-$Fe_{16}N_2$. In order to verify this hypothesis, we used gaseous nitriding of iron to synthesize extremely porous iron-nitride and, after de-nitriding, porous iron from pure iron foils or powders, such that the wall thickness of solid iron is reduced to (much) less than a micron. Accordingly, the diffusion distance for nitrogen in the solid iron bridge connecting adjacent pores is short and strains developing during the conversion of iron can relax at the pervasive surface. The reason that nitriding can be used for the synthesis of porous iron is the occurrence of two competing reactions, the nitriding reaction:

$$NH_3 = [N] + 3/2\ H_2 \tag{1}$$

and the de-nitriding reaction

$$[N] + [N] = N_2 \tag{2}$$

where [N] represents nitrogen in solid solution. Reaction (1) is the intended reaction on nitriding, but reaction (2) is the thermodynamically preferred reaction and becomes important for temperatures above approx. 450 °C, where N_2 can desorb. It is reaction (2) that leads to porosity in iron nitrides [20]. The synthesis of porous iron can be achieved by multi-step nitriding as schematically depicted for thermogravimetry in figure 4. Pure iron (foil, strip or powder) is nitrided at successively lower temperatures; at each temperature step the nitrogen uptake shows a fast increase and an overshoot, whereafter the incorporated nitrogen content tends to decrease to approach a plateau of constant N-uptake. The reduction in nitrogen content after each overshoot is explained from the development of porosity induced by N_2 formation, which leads to nitrogen loss as the pores become connected with the atmosphere. The plateau reflects the stationary state between nitrogen uptake from the dissociation of NH_3 and N_2 desorption from the (enlarged) surface area. Subsequently, the temperature is reduced leading to another increase in nitrogen content, because at a lower temperature the N_2 formation becomes less dominant and thus a higher stationary nitrogen content can be attained. Again, a nitrogen overshoot occurs followed by a weight loss to approach the next plateau. At a lower temperature the distance that nitrogen atoms can diffuse from the interior to a free (pore) surface is shorter than at higher temperature.

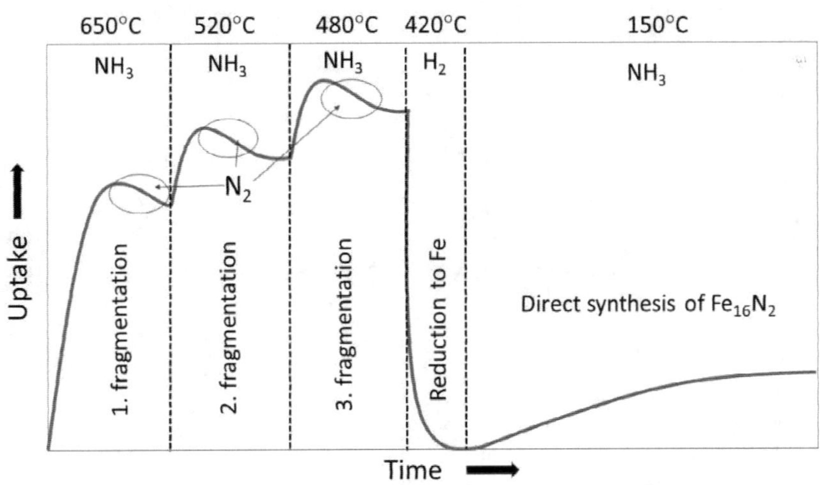

Figure 4. Schematic illustration of the concept for synthesis of α''-$Fe_{16}N_2$ from bulk iron using thermogravimetry. The temperatures stated are merely examples of typical temperatures, which will result in a tri-modal distribution of porosity. A multi-model distribution can be obtained by adding more nitriding steps. N_2 development (mass loss) is indicated for each nitriding temperature. The loss of mass is related to release of N_2 gas from the specimens (through surface connected pores); N_2 trapped within the specimen will not result in a mass loss.

Accordingly, new pores develop at a shorter separation than those formed at higher temperature. This stepwise decrease of the nitriding temperature after each overshoot in nitrogen content can be repeated to obtain a progressively higher nitrogen content in the stationary state and progressively more porous iron, with progressively shorter iron-bridges in-between the pores.

After the sequence of nitriding steps, the sample is exposed to pure hydrogen and all nitrogen is removed by the reverse reaction (1). Thereafter, the temperature is lowered well below 200°C and exposed to ammonia again. The nitrogen uptake now leads to the direct development of α"-$Fe_{16}N_2$. Clearly, the kinetics is slow in the final nitriding stage. This is mainly a consequence of slow diffusion kinetics and slow dissociation of ammonia. SEM micrographs of different steps in the development of multi-model porosity and of the final microstructure/morphology of the synthesized α"-$Fe_{16}N_2$ are shown in figure 5. The specific processing routes are described in more detail in a patent application [21].

The porous iron foils synthesized by the step-wise lowering of the nitriding temperature in NH_3 followed by de-nitriding in H_2 can be seen as a precursor for the synthesis of iron-based phases, for example (metastable) iron-nitrides or carbides, allowing scientific investigation of the fundamental properties of such compounds. Also, with the aim of surface engineering, the porous region can be confined to the surface region of an iron-based material. Such a deliberately porous region could be used as a scaffold for the deposition of other materials, as for example catalysts.

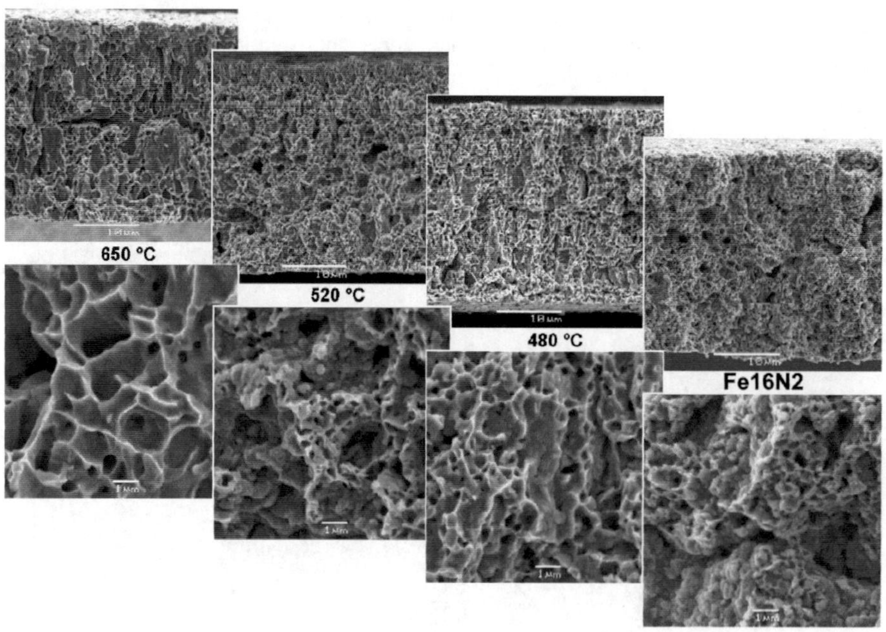

Figure 5. Conversion of iron foil to progressively more porous iron nitride after step-wise lowering of the nitriding temperature (top row shows entire foil thickness; bottom row shows enlargements at the indicated temperatures). See also the conceptual thermogravimetric curve in figure 4. The foil denoted $Fe_{16}N_2$ was denitrided and converted to α"-$Fe_{16}N_2$ by low temperature nitriding in flowing NH_3.

5. Summary and outlook

Widely different aspects of thermochemical surface engineering have been illustrated in the presented examples. TSE can be used to improve the performance of self-passivating stainless steels and titanium alloys by addition of interstitial elements. Both stainless steel and titanium parts obtained by 3D printing can benefit from TSE. Additionally TSE can also be used to synthesize magnetic α"-$Fe_{16}N_2$ from pure iron. The examples shown in this paper illustrate how science and innovation go hand-in-hand and how

TSE can be used to dramatically change the properties of technical materials and push the limit of materials performance.

References

[1] Malcolm V T 1949 *Product Engineering*, Jan. 84.
[2] Kolster B H 1983 *VDI-Berichte* **506** 107.
[3] Zhang Z L and Bell T 1985 *Surf. Eng.* **1(2)** 131.
[4] Berns H 1996 *ISIJ Int. Manufacture and application of high nitrogen steels* **36** 909.
[5] Berns H 2000 *Conference proceedings: Thermochemical Surface Engineering of Stainless Steel* Osaka, Japan 5 - 8 Nov. 111.
[6] Christiansen T L, Hummelshøj T S and Somers M A J 2013 Patent, PCT WO2013/159781-A1.
[7] Christiansen T L, Hummelshøj T S and Somers M A J 2012 Patent, PCT WO2012146254-A1.
[8] Bottoli F, Winther G, Christiansen T L, Dahl K V and Somers M A J 2016 *Metall. Mater. Trans. A* **47A(6)** 4146.
[9] Wang Y M, Voisin T, McKeown J T et al. 2017 *Nat. Mater.* **17** 63.
[10] Siyun G 2019 M.Sc Thesis The Technical University of Denmark.
[11] Christiansen T L, Jellesen M J, Somers M A J, Gammeltoft-Hansen N B 2016 EP patent application 16172699.7.
[12] Stone L and Margolin H 1953 *Transactions AIME, Journal of Metals* 1498.
[13] Gammeltoft-Hansen N B, Munch S, Jellesen M S, Somers M A J and Christiansen T L 2017 *Materials Performance and Characterization* **6(3)** 298.
[14] Fedirko V M, Pohrelyuk I M and Yas'kiv O I 2006 *Materials Sciences* **42(3)** 299.
[15] Kim T K and Takahashi T 1972 *Appl. Phys. Letters* **20** 492.
[16] Jack K H 1951 *Proc. R. Soc.* **208A** 216.
[17] El Hajjaji M, Foos M, Michel H and Gantois M 1983 *Scripta Metall.* **17** 879.
[18] Malinov S, Böttger A J, Mittemeijer E J, Pekelharing M and Somers M A J 2001 *Metall. Mater. Trans. A* **32A** 59.
[19] Nagatomi A, Kikkawa S, Hinomura T, Nasu S and Kanamaru F 1999 *J. Jpn. Powder Powder Metal.* **46** 151.
[20] Somers M A J, Kooi B J, Małdzinski L, Mittemeijer E J, van der Horst A A, van der Kraan A M and van der Pers N M 1997 *Acta Mater.* **45** 2013.
[21] Christiansen T L, Somers M A J and Frandsen R B 2015 Patent Application WO 2015 193295 A1.

IOP Conf. Series: Materials Science and Engineering **580** (2019) 012017 doi:10.1088/1757-899X/580/1/012017

Using synchrotron radiation to improve understanding of deformation of polycrystalline metals by measuring, modelling and publishing 4D information

D Greeley[1], M Yaghoobi[1], D Pagan[2], V Sundararaghavan[1] and J Allison[1]

[1]University of Michigan, Ann Arbor, MI

[2]Cornell High Energy Syncrotron Source, Cornell University, Ithaca, NY

E-mail: johnea@umich.edu

Abstract. The evolution of deformation in a Mg-Nd alloy has been investigated using 3D HEDM and crystal plasticity simulation using the *PRISMS-Plasticity* CPFE Code. Using a simplified representative volume element (RVE), the average stress-strain response is well predicted by the *PRISMS-Plasticity* simulation. The distribution of the resolved shear stresses corresponding to the basal mode for all grains is obtained using both HEDM and the CPFE. simulation. The results show that the CPFE predicted substantially narrower distributions, thus indicating the need for improved digital representation of the RVE. The results of these HEDM experiments and *PRISMS-Plasticity* simulations have been stored in the Materials Commons in a straight-forward manner and are being made available as a published dataset. Both *PRISMS-Plasticity* and the Materials Commons are open-source and available for use by the global materials community.

1. Introduction

The Center for PRedictive Integrated Structural Materials Science (PRISMS) is a major software innovation center that is a component of the US Materials Genome Initiative. The PRISMS Center is developing a framework comprised of advanced open-source multi-scale software integrated with experiments and a data repository, the Materials Commons, to accelerate the predictive science for structural metals [1]. The open-source software includes real space density functional software, a statistical mechanics code, a phase field code and crystal plasticity software. In this paper we present an overview of one project within the Center, the use of high energy x-ray diffraction (HEXD) to map the evolution of local strains with time (the fourth dimension) measured in 3D coupled with simulation of this strain evolution using the *PRISMS-Plasticity* open source full field crystal plasticity finite element code. The results of both are stored and being published as a public dataset using the Materials Commons. This research project is investigating grain size effects on monotonic and cyclic deformation in a Mg-Nd alloy. This paper describes initial results of simulations and experimental results on tensile behaviour of this alloy.

2. Experimental

2.1 Materials

The material used in this experiment was extruded Mg-2.4 weight percent Nd (Mg-2.4Nd), provided by CanmetMATERIALS. Samples were solution heat treated at 525°C for 30 minutes, with an average grain size of 145μm. Mechanical test specimens were wire electrical discharge machined with the loading direction aligned parallel to the bar extrusion direction. Test specimens were machined with a gage cross section of 1mm x 1mm, and a gage length of 1.52mm.

2.2 Experimental Setup

High energy X-ray diffraction microscopy (HEDM) measurements were performed at Sector ID1-A3 beamline at the Cornell High Energy Synchrotron Source (CHESS). A solution heat treated (SHT) Mg-2.4Nd specimen was imaged through a single displacement-controlled, uniaxial tensile loading to track grain scale deformation evolution. Mechanical loading was performed on the Rotational and Axial Motion System, 2nd Build (RAMS2) load frame [2] with the experimental test setup shown in Figure 1. The validity of this experimental framework to measure the average grain strain has been demonstrated by Pagan et al. [3-5].

Samples were displaced uniaxially in the RAMS2 load frame with macroscopic stress measured through a load cell and macroscopic strain measured through surface digital image correlation (DIC) measurements. Far-field high energy X-ray diffraction microscopy (ff-HEDM) measurements were obtained at load steps corresponding to 0.2 and 0.5% macroscopic strain in tension. Six 150μm-height layers were imaged and stacked to illuminate a 1mm x 1mm x 0.9mm volume of the gage section. The samples were box beam scanned with a beam energy of 41.991 keV, and the illuminated volume location was tracked with a surface fiducial wire marker. At each HEDM measurement load step, the specimens were unloaded by 15% to avoid stress relaxation during diffraction imaging. Far-field diffraction patterns were captured at 0.25° increments over a 360° specimen rotation on two Dexela 2923 NDT panel detectors located 878mm from the test specimen.

High energy x-ray diffraction data was processed and reconstructed using the HEXRD software package [6]. Grain center of mass, average crystallographic orientation, and elastic strain tensor were tracked at each load step for 376 grains in the SHT condition through spot analysis of far-field diffraction peak position.

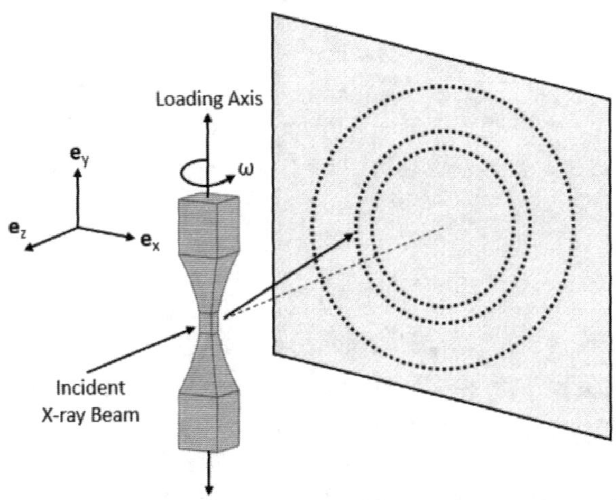

Figure 1. Test geometry in the far-field diffraction condition. Laboratory coordinates are listed, with the beam direction along z and tensile axis along y.

3. *PRISMS-Plasticity* Crystal plasticity model and model set-up

The open-source parallel 3-D crystal plasticity finite element (CPFE) software package *PRISMS-Plasticity* [7] was used to capture the sample response and evolution of slip system strengths. This code is available on Github at prisms-center/plasticity. The constitutive model of each grain is modelled using the rate-independent single crystal plasticity theory developed by Anand and Kothari [8]. A multiplicative decomposition of the deformation gradient tensor **F** into its elastic and plastic components, i.e., $\mathbf{F}^{\mathrm{e}}$ and $\mathbf{F}^{\mathrm{p}}$, is considered here as follows:

$$\mathbf{F} = \mathbf{F}^{\mathrm{e}}\mathbf{F}^{\mathrm{p}} \tag{1}$$

Two independent deformation mechanisms of elastic distortion of the crystal lattice and plastic slip accommodate the applied deformation. The macroscopic velocity gradient tensor **L** can be additively decomposed into the elastic and plastic components, i.e., $\mathbf{L}^e$ and $\mathbf{L}^{\mathrm{p}}$, respectively, as follows:

$$\mathbf{L} = \mathbf{L}^e + \mathbf{L}^p \tag{2}$$

The key idea of crystal plasticity is to link the plastic part of the velocity gradient to the superposition of shear deformation induced by crystallographic slip on multiple slip as follows:

$$\mathbf{L}^{\mathrm{p}} = \sum_{\alpha} \dot{\gamma}^{\alpha}\mathbf{S}^{\alpha}\,\mathrm{sign}(\tau^{\alpha}) \tag{3}$$

where $\dot{\gamma}^{\alpha}$ is the shearing rate on slip system α, τ^{α} is the resolved shear stress acting on the slip system α, and $\mathbf{S}^{\alpha}$ is the Schmid tensor for the slip system α, which can be defined as follows:

$$\mathbf{S}^{\alpha} = \mathbf{m}^{\alpha}\otimes\mathbf{n}^{\alpha} \tag{4}$$

where unit vectors $\mathbf{m}^{\alpha}$ and $\mathbf{n}^{\alpha}$ denote the slip direction and slip plane normal, respectively, in the deformed configuration.

The resolved shear stress on the slip system α can be obtained as follows:

$$\tau^{\alpha} = \boldsymbol{\sigma}:\mathbf{S}^{\alpha} \tag{4}$$

where $\boldsymbol{\sigma}$ is the Cauchy stress tensor.

The rate-independent yield surface of slip system α can be defined as follows:

$$f^{\alpha} = |\tau^{\alpha}| - s^{\alpha} \tag{5}$$

where s^{α} is the slip resistance on slip system α. Finally, the slip resistance evolution for slip system α can be obtained as follows:

$$\dot{s}^{\alpha} = \sum_{\beta} h^{\alpha\beta}\dot{\gamma}^{\beta} \tag{6}$$

where $h^{\alpha\beta}$ describes the variation of slip resistance for slip system α due to the slip rate on slip system β. The hardening moduli $h^{\alpha\beta}$ can be described as a power-law relationship considering the combined effect of work hardening and recovery as follows:

$$h^{\alpha\beta} = h_0^{\beta} q \left[1 - \frac{s^{\beta}}{s_s^{\beta}}\right]^{a^{\beta}} \tag{7}$$

where q is the latent hardening ratio ($q = 1$ in the case of $\alpha = \beta$), h_0^{β} denotes the hardening parameter for slip system β, s_s^{β} is the slip resistance at hardening saturation for slip system β, and a^{β} is a material constant for slip system β which governs the sensitivity of the hardening moduli to the slip resistance.

For this investigation, using the *PRISMS-Plasticity* code, a polycrystalline model of the Mg-2.4Nd alloy was generated as a stacked aggregate of single crystal eight-node cubic elements in a 6 x 7 x 9 = 378 finite element mesh. Each element is assigned a separate lattice orientation from the 376 far-field HEDM (ff-HEDM) identified initial grain orientations, with two orientations duplicated and element spatial arrangement randomly seeded. For the initial calibration simulation, the orientation distribution function of the sample is captured but local neighborhood crystallographic orientations and grain morphology are not. Future work will include use of the local neighborhood and grain morphologies.

Tensile loading and unloading displacements were applied at a fixed strain rate of $1\mathrm{x}10^{-4}\ \mathrm{s}^{-1}$, and the degree of load drop at each measurement point set by global strain measurements from surface digital image correlation (DIC). Deformation in the HCP crystal structure was modelled with four slip mechanisms including Basal ⟨a⟩ ({0001}⟨11$\bar{2}$0⟩), Prismatic ⟨a⟩ ({10$\bar{1}$0}⟨11$\bar{2}$0⟩), Pyramidal ⟨a⟩ ({10$\bar{1}$1}⟨11$\bar{2}$0⟩), and Pyramidal ⟨c + a⟩ ({11$\bar{2}$2}⟨11$\bar{2}$3⟩), with a latent hardening parameter of q = 1. The extruded basal texture of the Mg-2.4Nd alloy in this study results in basal poles aligned perpendicular to the extrusion direction, favourably orienting grains for twinning upon uniaxial compressive loading [9-13]. Deformation twinning at low tensile strains is not expected and were not included in this initial simulation. *PRISMS-Plasticity* has the capability for modelling twinning deformation [7]. The single crystal elastic constants used in the simulations are listed in Table 1. This *PRISMS-Plasticity* CPFE simulation was performed using 32 processors on the PRISMS FLUX cluster at the University of Michigan with a simulation time (wall time) of 2700 seconds. *PRISMS-Plasticity* has the capability of automatically uploading simulation results and analysis meta-data to the Materials Commons and this was accomplished.

Table 1. Mg elastic constants used in this study (GPa) [14]

C11	C12	C13	C33	C44
59.40	25.61	21.44	61.60	16.40

4. Results

Grain-level elastic strain tensors were tracked through the stacked volume through far-field spot diffraction analysis. Grain average stress tensors were calculated with the elastic stiffness constants listed in Table 1 and grain average orientation from ff-HEDM. The resolved shear stresses of Basal ⟨a⟩, Prismatic ⟨a⟩, Pyramidal ⟨a⟩, and Pyramidal ⟨a+c⟩ slip modes were calculated in each grain using its average stress. Next, in each grain and for each slip mode, the maximum resolved shear stress is determined. The values obtained are averaged for all grains and reported in Table 2 for each slip mode along with the corresponding standard deviation. The measured values given in Table 2 are at the 15% load drop point; the strains (load steps) listed are the initial strains prior to the 15% load drop. The average values of resolved shear stresses for all slip modes increase as the strain increases from 0.2 % to 0.5 % due to work hardening. This increase is also observed using the CPFE simulation as shown in Figure 2.

Since the basal slip mode is the dominant slip mechanism for the range of strains investigated, the data obtained for the basal mode is investigated further. Accordingly, the distribution of the resolved shear stresses corresponding to the basal mode for all grains is obtained using both HEDM and CPFE (at the 15% load drop point) at the strain load steps of 0.2% and 0.5%, as shown in Figure 2. At the 0.2% load step, the distribution of the resolved shear stresses ranges from 0 MPa to 30 MPa. At the 0.5% load step, the distribution of the resolved shear stresses spans from 0 MPa to 45 MPa. The wide range of basal resolved shear stresses observed for both strain values can be attributed the wide grain size distribution, magnesium extrusion texture, and the hcp crystal structure's extensive plastic anisotropy [15]. Similar spread in basal resolved shear stress distribution has been observed in other hcp materials such as Ti-7Al [3, 5].

In the case of the CPFE simulation, all grains are modelled to have a similar size. Accordingly, the distribution of basal resolved shear stress is sharp. Future plans for addition of grain size, grain morphology, and spatial distribution from nf-HEDM will integrate local grain neighborhood characteristics, more accurately accounting for local strain accommodation.

Figure 3 (a) presents the response of the sample during uniaxial tensile loading. It includes the stresses obtained from the DIC measurement, i.e., load divided by the cross-section area, the CPFE

IOP Conf. Series: Materials Science and Engineering **580** (2019) 012017 doi:10.1088/1757-899X/580/1/012017

simulation results, and the stress obtained from the HEDM results (after the 15% load drop). In the case of HEDM, one can obtain the average stress of each grain using its average strain. Here, due to the lack of experimental data regarding to the size of each grain corresponding to each strain, it is assumed that all grains have equal size. Accordingly, the sample stress can be obtained by averaging the stresses for all grains. The stresses obtained from HEDM for 0.2 % and 0.5% (initial) strains are 20% and 13% lower than those of the DIC experiment, respectively. The observed differences can be attributed to the grain size distribution and elastic moduli used in the calculations. First, it is assumed here that the grains have a similar size, which is not precise. Second, the elastic modulus incorporated here is for pure Mg, which is slightly different from the Mg-2.4Nd alloy. The difference is not considered to be related to the accuracy of the experimental framework. Future work will integrate weighting of grain size from near field HEDM and simulated Mg-2.4Nd moduli [3-5].

Fig. 3. (a) shows that the stress-strain response of the sample can be successfully captured using CPFE. Furthermore, the relative slip activity of the sample obtained by CPFE simulation is presented in Figure. 3 (b). As expected, the basal mode is dominant at low strains with minor activation of prismatic slip and pyramidal accommodation of c-axis deformation [15-23].

Table 2. Average ff-HEDM resolved shear stress for 0.2% and 0.5% strain load steps, solution heat treated Mg-2.4Nd

	Average Resolved Shear Stress (MPa)			
Load Step	Basal	Prismatic	Pyramidal <a>	Pyramidal <c+a>
0.2% Strain	10.7 ± 4.7	9.1 ± 5.6	12.0 ± 5.0	14.6 ± 6.9
0.5% Strain	17.4 ± 10.3	19.6 ± 12.9	23.7 ± 12.3	26.1 ± 11.7

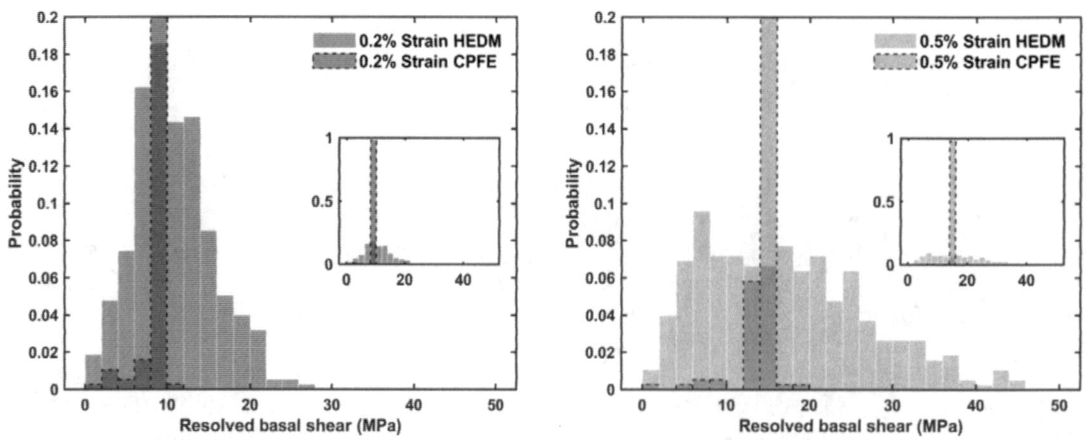

Figure 2. Distribution of the resolved shear stress for the basal slip mode in every grain tracked in solution heat treated Mg-2.4Nd. CPFE simulated distribution overlaid in dark blue (0.2% strain - left) and light blue (0.5% strain - right). (For interpretation of color, the reader is directed to the web version of the article). Note: HEDM and CPFE results are at 15% load drop condition for each strain step.

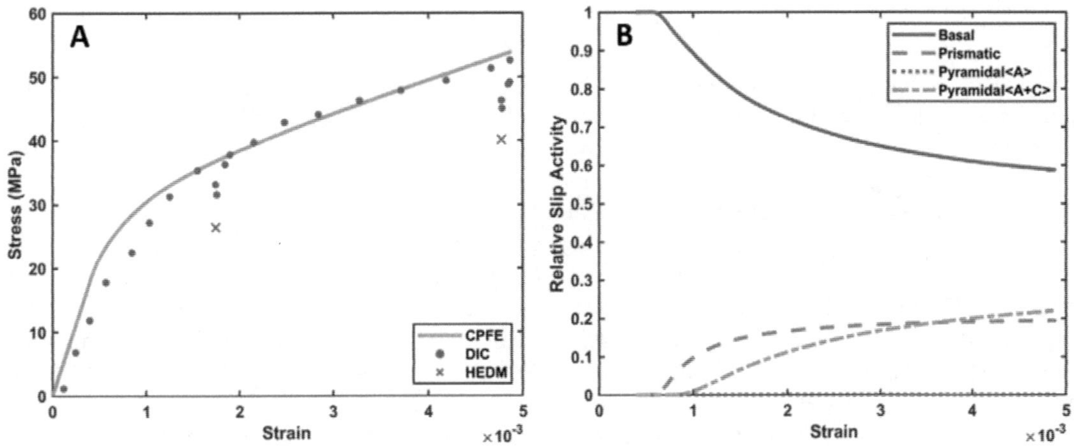

Figure 3. Uniaxial response of the solution heat treated Mg-2.4Nd sample during uniaxial tensile loading: (a) The stress-strain response is obtained using DIC, CPFE simulation, and (after the 15% load drop) the far-field HEDM; (b) Relative slip activity obtained by CPFE.

5. Publishing 4D Information Using the Materials Commons

The Materials Commons is an information repository and virtual collaboration space for curating, archiving and disseminating information from experiments and computations [24]. It is an integral part of the PRISMS framework [1] and is readily available to the global materials community at https://materialscommon.org. It provides solutions to the data problems facing the 4D materials researcher. Synchrotron-based 4D HEDM experiments require proposals, scheduling, sample preparation and complex data analysis making these datasets relatively difficult and costly to secure. The data contained in these datasets is also rich, requires large storage and is often incompletely analysed – thus reuse of these datasets is important for continuously improving our understanding of the phenomena uncovered by these important measurements. Similarly, when these experiments are analysed using advanced computational methods it is instructive to understand all of the modelling parameters used in these simulations. The Materials Commons is aimed at assisting the 4D community in secure storage and re-use of these valuable datasets.

The Materials Commons is designed to capture meta-data seamlessly either during an experiment or after an experiment or simulation is completed. It has a straight forward web-based user-interface that guides the user through data and meta-data upload. Data can be automatically uploaded from spreadsheets or by using the Materials Commons web user interface or using the Materials Commons Application Program Interface (API). The results of *PRISMS-Plasticity* simulations can be automatically stored in the Materials Commons in a consistent format using the Materials Commons API. For large datasets typical of HEDM information, the Materials Commons uses the Globus file transfer service for fast, secure data transfers [25]. Custom templates have been designed for use with HEDM from CHESS and APS beamlines. These templates are readily adaptable and customizable for other synchrotron facilities. When ready for publication, datasets can be easily formed and published with the push of a virtual button. With over 400Tb of storage, the Materials Commons is capable of storing and publishing large datasets. Future embodiments of the Materials Commons will provide access to large datasets stored on "cloud services".

6. Summary

The evolution of deformation in a Mg-Nd alloy has been investigated using 4D HEDM at CHESS and full field crystal plasticity simulation using the *PRISMS-Plasticity* CPFE Code. The base microstructural RVE for the CPFE simulation was a simple grid microstructure which captures the overall texture of the experimental sample but not the local grain neighborhoods and grain morphologies. The average stress-strain response is well predicted by the CPFE simulation. At global strains up to 0.5%, deformation is dominated by basal slip. In addition to the average stress-strain response, resolved basal stress distributions have been characterized experimentally and CPFE for each grain in the solution treated sample. The CPFE simulated stress distribution is substantially narrower than that experimentally determined by HEDM. This indicates the need for more precise RVE in the CPFE to include local grain neighborhoods and morphologies and intergranular misorientation developments [26]. The results of these HEDM experiments and *PRISMS-Plasticity* simulations can be readily stored in the Materials Commons and provided to the wider materials community. Both the *PRISMS-Plasticity* code and the Materials Commons are available as open-source software and key components of the PRISMS framework for use in accelerating the prediction of material responses.

Acknowledgements
This work was supported by the U.S. Department of Energy, Office of Basic Energy Sciences, Division of Materals Sciences and Engineering under Award #DE-SC0008637 as part of the Center for PRedictive Integrated Structural Materials Science (PRISMS Center) at University of Michigan. This work is based upon research conducted at the Cornell High Energy Synchrotron Source (CHESS) which is supported by the National Science Foundation under award DMR-1332208. We acknowledge with appreciation CANMETMaterials who provided the materials used in this study.

References

[1] Aagesen L K, Adams J F, Allison J E, Andrews W B, Araullo-Peters V, Berman T, Chen Z, Daly S, Das S, DeWitt S, et al 2018 JOM 70 2298–314
[2] Shade P A, Blank B, Schuren J C, Turner T J, Kenesei P, Goetze K, Suter R M, Bernier J V, Fai Li S, Lind J, et al 2015 Rev. Sci. Instrum. 86 93902-1-93902-8.
[3] Pagan D C, Shade P A, Barton N R, Park J S, Kenesei P, Menasche D B and Bernier J V. 2017 Acta Mater. 128 406–17.
[4] Guillen D P, Pagan D C, Getto E M and Wharry J P 2018 *Mater. Sci. Eng. A* **738** 380–8.
[5] Pagan D C, Bernier J V, Dale D, Ko J Y P, Turner T J, Blank B and Shade P A 2018 *Scr. Mater.* **142** 96–100.
[6] Bernier J V, Barton N R, Lienert L J and Miller M P, 2011 *Journal of Strain Analysis* **46** 527-47.
[7] Yaghoobi M, Ganesan S, Sundar S, Lakshmanan A, Rudraraju S, Allison J and Sundararaghavan V 2019, in press, *Computational Materials Science, https://doi.org/10.1016/j.commatsci.2019.109078.*
[8] Anand L and Kothari M 1996 *J. Mech. Phys. Solids* **44** 525-58.
[9] Wu L, Jain A, Brown D W, Stoica G M, Agnew S R, Clausen B, Fielden D E and Liaw P K 2008 *Acta Mater.* **56** 688–95.
[10] Murphy-Leonard A D, Pagan D C, Beaudoin A, Miller M P and Allison J E 2019 *Int. J. Fatigue* **125** 314–23.
[11] Mirza F A, Chen D L, Li D J and Zeng X Q 2013 *Mater. Sci. Eng. A* **575** 65–73.
[12] Brown D W, Jain A, Agnew S R and Clausen B 2007 *Mater. Sci. Forum* **539–543** 3407–13.
[13] Kumar AM, Beyerlein I J and Tome C N 2017 J. Alloys Compd. 695 1488–97.
[14] Zhang J and Joshi S P 2012 *J. Mech. Phys. Solids* **60** 945–72.
[15] Agnew S R and Duygulu Ö 2005 *Int. J. Plast.* **21** 1161–93.

[16] Wang H, Boehlert C J, Wang Q D, Yin D D and Ding W J 2016 *Mater. Charact.* **116** 8–17.
[17] Agnew S R, Tomé C N, Brown D W, Holden T M and Vogel S C 2003 *Scr. Mater.* **48** 1003–8
[18] Koike J, Kobayashi T, Mukai T, Watanabe H, Suzuki M, Maruyama K and Higashi K 2003 *Acta Mater.* **51** 2055–65.
[19] Jain A and Agnew S R 2007 *Mater. Sci. Eng. A* **462** 29–36.
[20] Beyerlein I J, Capolungo L, Marshall P E, McCabe R J and Tome C N 2010 *Philos. Mag.* **90** 2161–90.
[21] Cheng J and Ghosh S 2015 *Int. J. Plast.* **67** 148–70.
[22] Cheng J, Shen J, Mishra R K and Ghosh S 2018 *Acta Mater.* **149** 142–53.
[23] Kadiri H El, Kapil J, Oppedal A L, Hector L G, Agnew S R, Cherkaoui M and Vogel S 2013 *Acta Mater.* **61** 3549–63.
[24] Puchala B, Tarcea G, Marquis E A, Hedstrom M, Jagadish H V, Allison, J E 2016 *JOM* **68** 2035-44.
[25] Kettimuthu R, Liu Z, Wheeler D, Foster I, Heitmann K and Cappello F 2018 Future Generation Computer Systems 88 191–8.
[26] Acar P, Ramazani A and Sundararaghavan V 2017 *Metals* **7(11)** 459.

IOP Conf. Series: Materials Science and Engineering **580** (2019) 012018 doi:10.1088/1757-899X/580/1/012018

Sub-grain orientation resolution during continuous loading using only far-field HEDM

K E Nygren[1], D C Pagan[1], and M P Miller[1,2]

[1] Cornell High Energy Synchrotron Source, Cornell University, 161 Synchrotron Drive, Ithaca, NY 14853

[2] Sibley School of Mechanical and Aerospace Engineering, Cornell University, 313 Campus Rd, Ithaca, NY 14853

E-mail: `k.nygren@cornell.edu`

Abstract. A far-field data reduction algorithm for high-energy X-ray diffraction microscopy (HEDM) has been adapted to extract sub-grain orientation descriptors in a polycrystalline material **during** continuous *in situ* loading experiments. Previously, the standard data reduction algorithm would only extract grain-averaged strains, orientation, and the centroid position. A new descriptor, the grain orientation envelope (GOE), is introduced as a measure of the intragranular microstructure extracted from every grain during a continuous loading experiment. Initial results showing the evolution of a GOE from one grain in the titanium alloy, Ti-7Al, during a tension test is presented as a demonstration of the nature of the data.

1. Introduction

Establishing the relationships between microstructure and the thermomechanical response of complex engineering alloys is critical for alloy selection, development, and adoption. Historically, these relationships have been based on decades of empirically derived behaviors, rather than fundamental mechanistic understandings. Even with modern advances in characterization techniques and computational tools, there are still hurdles to reaching this mechanistic understanding such as: bridging characterization techniques between multiple length-scales, simultaneously assessing macroscopic response and microstructural evolution, and identifying significant features that should directly inform predictive models.

Titanium alloys are a favorite among a diverse group of industries for their excellent mechanical properties, relatively low density, and outstanding corrosion resistance even at moderate temperatures. In service, these alloys are often subjected to extreme conditions, such as the complex thermomechanical cycles within a gas turbine engine. These complexities, in addition to those imposed intrinsically by the material response - including the elastic and plastic anisotropies associated with hexagonal metals - make titanium a particularly challenging system for investigating the underlying relationship between microstructures and thermomechanical properties. One response that notably proves challenging for both material performance and characterization is the susceptibility of many titanium alloys to creep and stress relaxation at room temperature. This phenomenon is thought to be one of the primary sources leading to a significant failure mode known as cold dwell fatigue, where cracks nucleate and grow in low temperature environments. In order to understand the fundamental connections between failure modes like cold dwell fatigue and microstructure, the kinetics of deformation and microstructural response must be understood, ideally by obtaining both sets of information simultaneously. Unfortunately, room temperature creep (and stress relaxation) not only affects the performance of the material, but the ability to measure microstructure under an applied load. Traditional microstructure characterization methods, which encompass spatial mapping techniques and analysis of defects and chemistries, either require the loading to be paused during data acquisition or are destructive in nature. The collection times of these traditional techniques are on the order of minutes to days. For non-destructive techniques, the acquisition time constraint combined with the room temperature creep of titanium alloys leaves two options for *in situ* experiments: (i) hold the applied load during the measurement, or (ii) reduce the load away from the yield surface to take the measurement. Neither option is ideal considering that inelastic deformation proceeds at any non-zero stress level. High energy X-ray diffraction during continuous loading overcomes these constraints and offers an opportunity for the simultaneous extraction of stress state and microstructural evolution.

High-energy X-ray diffraction microscopy (HEDM) performed at a synchrotron light source enables *in situ* bulk analysis of crystal-scale stresses and orientations for all crystals in a diffraction volume (hundreds to thousands of grains) [1, 2]. HEDM has two primary experimental modes: far-field, which is sensitive to orientation and strain, and near-field, which is sensitive to position [3]. As their names suggest, a detector is placed far from the sample at $\approx$ 1 m downstream for the far-field technique, and a detector is placed near to the sample at $\approx$ 5 mm downstream for the near-field technique. While the near-field technique produces 3D spatially resolved grain maps, collection times are on the order of hours for relevant volumes and therefore this technique cannot be performed *during* continuous loading. The shorter acquisition times for far-field HEDM experiments, however, enable sufficient data collection for some continuous loading experiments, admittedly at relatively slow strain rates. From these experiments, the real-time grain-scale stress evolution is readily extracted using data processing algorithms, which for far-field HEDM are optimized to extract grain-averaged data, with the primary output being an elastic strain tensor, the grain centroid, and the grain orientation. This strain tensor can then be used to calculate a grain-averaged stress, capturing the stress evolution during continuous loading without ever pausing the experiment, albeit with slight change in load.

In this manuscript, the standard far-field HEDM data reduction algorithm [1, 4] is modified to capture the evolution of intragranular orientations. A quantity we call the grain orientation envelope (GOE) is introduced. Results for a Ti-7Al alloy are used as the example throughout the paper, but the GOE is discussed broadly for application in any polycrystalline metal.

2. Far-field HEDM: Experiments

General aspects of a far-field diffraction experiment are presented below, followed by details of the Ti-7Al alloy experiment used for analysis. As can be seen from the schematic in Figure 1, an area detector is placed $\approx$ 1 m from the sample while a relatively large box-beam (achieved through orthogonal beam-defining slits) is transmitted through a large volume of the material. We note that it is imperative that diffraction peaks be *distinct* and not overlap one another for this technique. To capture the behavior of each crystal in the volume, diffraction events from multiple families of lattice planes ({hkl}s) are required. This is achieved through rotation of the crystal, with each revolution of 360 degrees capturing the average response over the duration of the revolution. To avoid halting the experiment for data collection, the sample is continuously loaded and rotated while the detector collects diffraction images.

2.1. Ti-7Al Experiment

The Ti-7Al alloy was machined into the geometry shown in Figure 1(a). A schematic of the experiment is presented in Figure 1(b). Two Dexela 2923 detectors (74.8 um effective pixel pitch, 3888 pixels x 3072 pixels) were positioned 884 mm downstream of the sample. An example image from one of the detectors, Figure 1(b), shows the necessary discrete diffraction peaks. The sample was continuously loaded in uniaxial tension at a constant cross-head displacement velocity of $1x10^{-5}$ mm/s. The experiment was stopped at a final macroscopic strain of 0.03. Throughout loading, the sample was continuously rotated, with diffraction images taken every 0.25 degrees for a total of 1440 frames per 360 degree revolution. Each revolution, or measurement point, took 5 minutes, averaging over an $\approx$ 0.0005 spread in strain. A total of 54 total measurements were collected throughout the duration of the experiment. The macroscopic strain was measured using digital image correlation in the gauge region and the load was monitored using a load cell placed above the specimen.

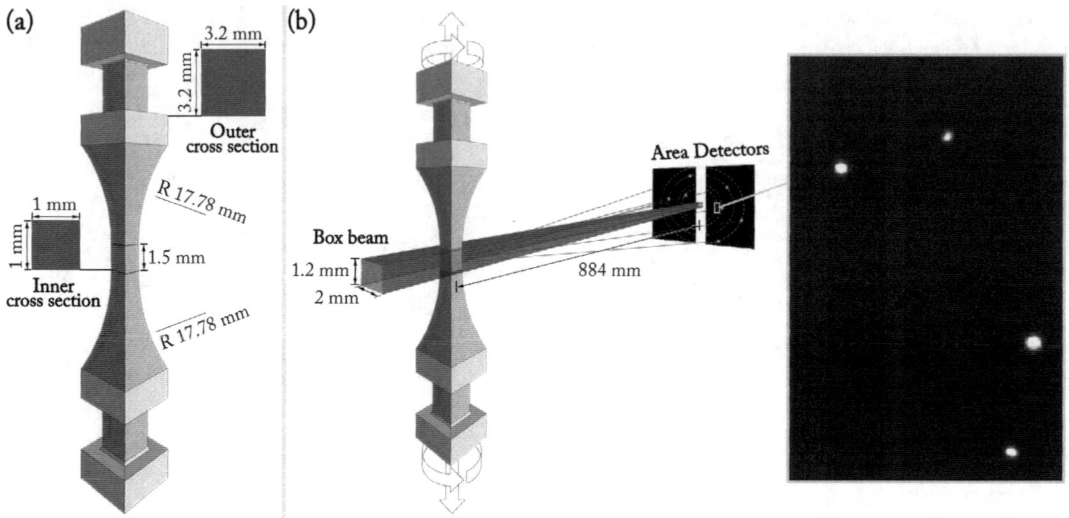

Figure 1. Schematic of continuous-loading far-field X-ray diffraction experiment.

3. Far-field HEDM: Data Reduction

The HEXRD data reduction package was used in this study, details of which can be found in [4]. This section will first discuss the standard data reduction scheme for obtaining the grain-averaged data, followed by the presentation of our new intragranular orientation extraction method.

3.1. Grain Averaged Orientation, Position, and Stress

After calibration, the HEXRD processing algorithm begins with an indexing routine, followed by a peak fitting routine. In the indexing step, the alloy crystal structure, experimental geometry, and X-ray energy are used to simulate diffraction events expected on the detector; for a given lattice orientation, the positions and angles of individual diffraction peaks are simulated. A large number ($\approx$ 1 million) of possible discrete orientations, seeded from crystallographic fibers of orientation determined from peaks on the detector, are generated. The peaks simulated using the seeded orientations are compared to the experimental data. A 'completeness', defined by the ratio of how many expected peaks from the simulation were measured in the experimental data, is assigned to each orientation. Orientations that are closely clustered together are considered one orientation. In this study, a completeness of 0.98 indicated a high likelihood that the crystal orientation was present in the diffraction volume. After all of the orientations within the diffraction volume have been identified, a peak fitting algorithm identified and optimized the peak centroids to refine the grain averaged orientation, grain centroid, and elastic strain tensor.

This procedure is repeated for every load step, with the refined values of grain orientation, grain centroid, and elastic strain tensor used as the starting values for refinement of each subsequent step.

The data reduction has been limited to extracting *grain-averaged* data since the peak-fitting algorithm is designed to track peak centroids. To illustrate this point, Figure 2 shows a schematic of two diffraction peaks on a dashed Debye-Sherrer ring. Figure 2(a) represents a specific peak for a set of lattice planes which have almost no misorientation, such as would be the case for a well annealed grain before plastic deformation. As metal alloys plastically deform, misorientations develop within the grain. This spread in orientation manifests as an azimuthal spread in the diffraction spot of the same peak in Figure 2(b). Of course, although these spots are spreading, the peak centroid is still what is measured in the peak fitting routine, indicated by purple cross-hairs in Figure 2. Therefore, any information about the morphology of the peak and intragranular spread in misorientation is being ignored. It is precisely this peak spread and identifying the contributing additional orientations, whose peak centroids are indicated by white cross-hairs in Figure 2, that is addressed in the following section.

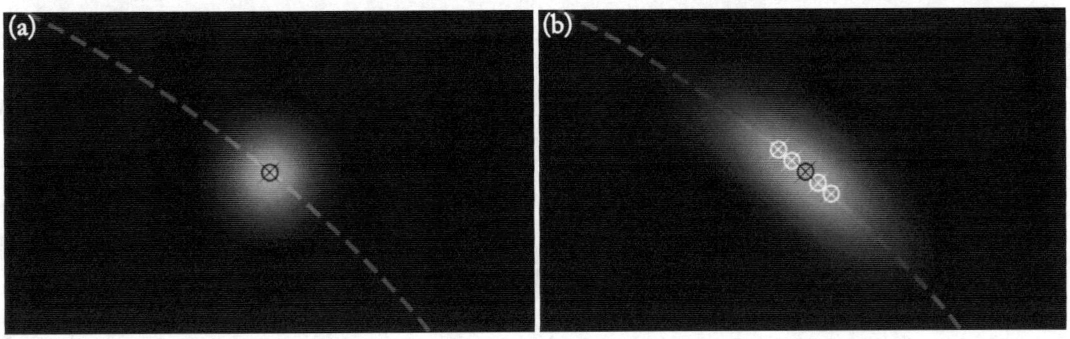

Figure 2. Schematic of diffraction peak for (a) undeformed and (b) plastically deformed metal. Dashed line indicates Debye-Sherrer ring, dark cross-hairs indicate centroid of overall peak, white cross-hairs indicate centroids of slightly offset (hkl) planes, such as would arise from misorientations within a grain.

3.2. Intragranular Orientations

To quantify the misorientation developing within a grain during plastic deformation, an additional indexing routine was developed to find all orientations present within a grain, not just the average. The method begins by first running the grain-averaged indexing and fitting routines to identify all unique grains in the diffraction volume. Then, for each grain, a series of new test orientations is generated by discretizing continuous points in orientation space (angle-axis parameterization) around the average orientation. The bounds are defined by the orientation spread in the material at the final load of the experiment, in this case 5 degrees. This new list of possible orientations for each grain is simulated in the indexing algorithm, and once again a 'completeness' score assigned based on the comparison to experimental data. Unlike the previous algorithm, no clustering is performed, and 'completeness' scores for trial orientations are recorded. We assume the orientations have a probability of existing in the grain commensurate with their completeness value. Similar to the grain-averaged dataset, a threshold completeness is decided upon to generate a list of orientations most likely to exist within the grain. A threshold of 0.98 was used in the Ti-7Al experiment. It is important to mention here that this analysis does not take into account the intensity of the peaks, and therefore only produces a list of unique orientations. In other words, no frequency of occurrence for each orientation is obtained.

IOP Conf. Series: Materials Science and Engineering **580** (2019) 012018 doi:10.1088/1757-899X/580/1/012018

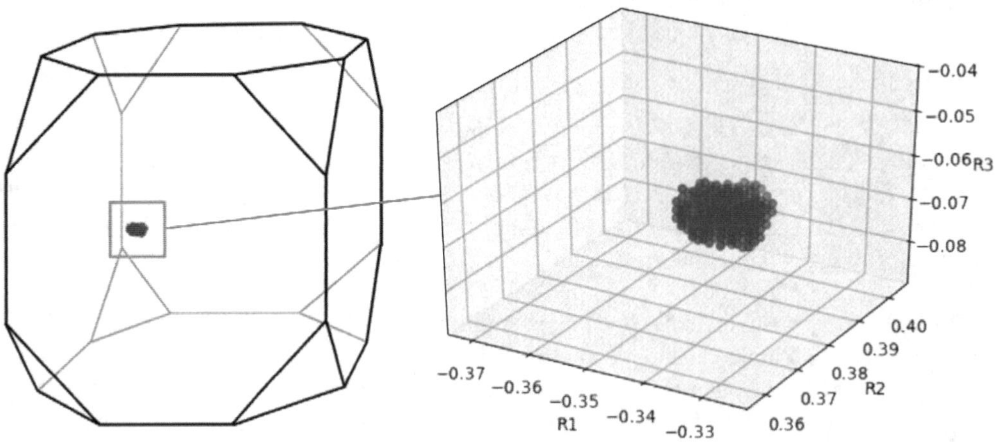

Figure 3. Grain orientation envelope (GOE) in Rodrigues space with all of fundamental region (left) and magnified region (right) for Grain No. 143 at a macroscopic strain of 0.015.

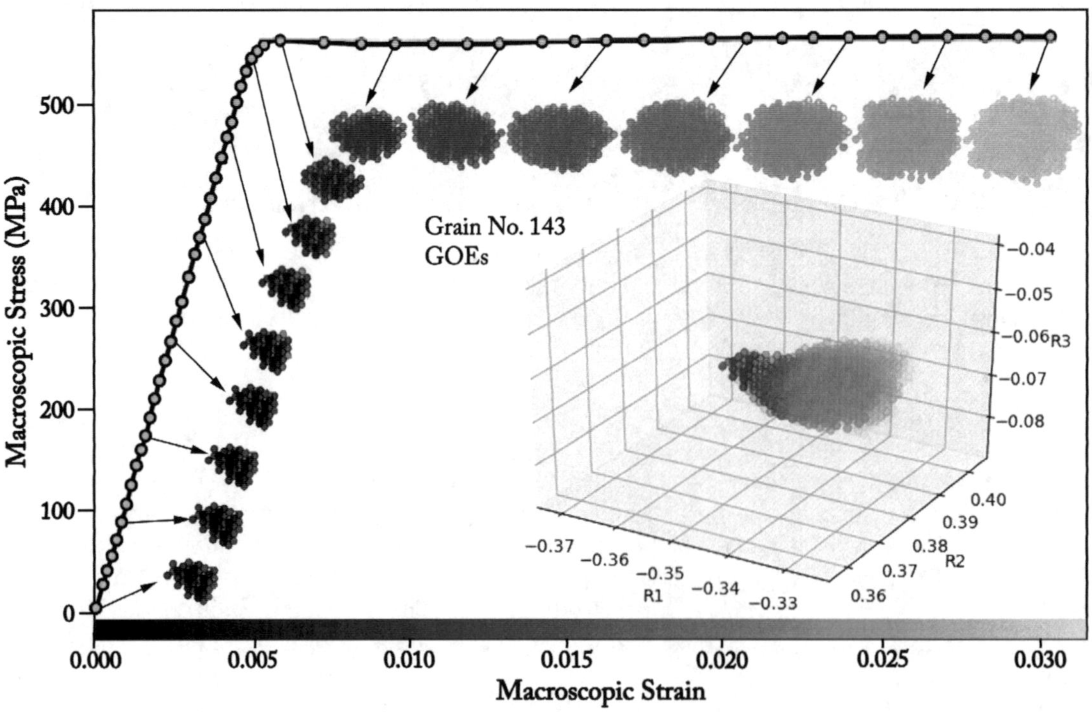

Figure 4. Macroscopic stress strain curve with circles corresponding to start of far-field diffraction measurements. GOEs of grain No. 143 plotted for incremental measurements, colored according to their corresponding macroscopic strain. Inset: GOEs for grain No. 143 are plotted in Rodrigues space.

IOP Conf. Series: Materials Science and Engineering **580** (2019) 012018 doi:10.1088/1757-899X/580/1/012018

3.3. Grain Orientation Envelope

The list of orientations identified as belonging to a grain is defined as the grain orientation envelope (GOE). While related to a grain-scale orientation distribution function (ODF), the GOE contains only unique orientations that exist within the grain and therefore is the envelope of the distribution function. An important caveat of the GOE is that it is obtained entirely from far-field data and therefore is not spatially resolved - there is only knowledge of which grain the orientations belong to, but not where the orientations reside within the grain.

An illustration of the GOE in orientation space for grain No. 143 at 0.015 macroscopic strain in the Ti-7Al diffraction volume is plotted in Figure 3. The left side of this figure shows the GOE in the cubic fundamental region in Rodrigues space; the right side of the figure shows a magnified view of the GOE region. Each orientation is plotted as a point with the same color and value to emphasize that this is a single measurement as compared to a montage of measurements depicting the evolution of the misorientation which is shown next.

This GOE is extracted for every measurement point taken during the continuous loading experiment. Figure 4 shows the evolution of the GOE for grain No. 143. In this Figure, the macroscopic stress-strain curve is plotted, with open circles indicating the start of far-field scans. For a subset of these measurement points, the GOE is shown next to the stress-strain curve to demonstrate the size and morphological changes with increasing deformation. Each GOE is colored based on its macroscopic strain state; the colorbar is shown along the strain axis. A GOE is obtained for every measurement point, but only 15 of the 54 are presented for illustrative purposes. The size and shape of the GOE is shown to be relatively constant throughout the elastic region, followed by immediate expansion at yield, and then by growth and shape changes as the plastic deformation proceeds. The inset of Figure 4 shows an overlay of the GOE for each macroscopic strain measurement plotted in Rodrigues space. The centroid of the GOE is shifting over time, corresponding to an average rotation of the grain. These GOE shape changes and translations are measures of the microstructural evolution, not only providing a snapshot of orientation content at each time step, but providing a means to measure rotation *rate changes*.

The enormous utility of this method is that *every* grain in the diffraction volume has a corresponding GOE evolution history. For this Ti-7Al alloy, 792 grains were illuminated in the diffraction volume. As shown in Figure 5, during a continuous loading experiment every grain in the diffraction volume now can be probed for the evolution of its grain-scale stress tensor and orientation envelope. The far-field technique can be used to simultaneously examine deformation kinetics and to resolve microstructural evolution without the need to pause the experiment.

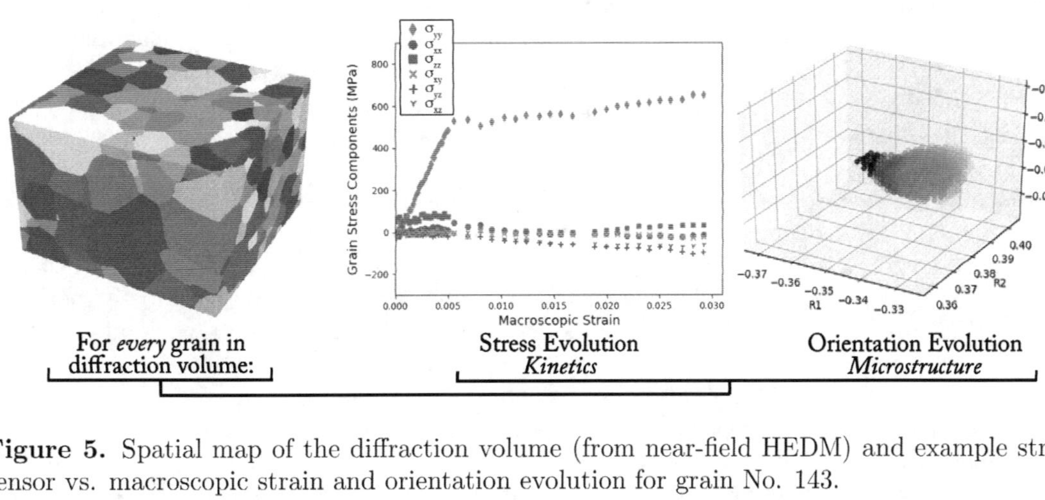

Figure 5. Spatial map of the diffraction volume (from near-field HEDM) and example stress tensor vs. macroscopic strain and orientation evolution for grain No. 143.

IOP Conf. Series: Materials Science and Engineering **580** (2019) 012018 doi:10.1088/1757-899X/580/1/012018

4. Discussion

A far-field data reduction algorithm was introduced which extracts a measure of the microstructure in the form of intragranular orientation spread. This modification is an expansion of existing data reduction routines already used to extract grain-scale strains and requires no modification of the HEDM experiment. The significance of the GOE measurements extracted from these data are now discussed.

4.1. Identifying important behaviors and regions of interest and informing models

Identifying *which* microstructural features are most important for understanding mechanical response and informing constitutive models is of great importance. With the increasing interest in utilizing big data sets and machine learning to independently identify behaviors and trends, it is important to have robust and representative data. Unfortunately, many characterization techniques can only capture small regions of interest at a time, with the added problem that identification of these regions is highly subjective. The GOEs provide a complete history of evolution for not only an isolated grain, but subsets of grains within a polycrystal including all grain neighborhoods. Because of the simultaneous extraction of stress and intragranular orientation evolution, the combined kinetic and microstructural response can be investigated directly for the hundreds of grains interrogated in a diffraction volume. This type of analysis can be used to identify regions of interest by qualifying which behaviors are representative, anomalous, or linked to particular material responses. These features can then be incorporated into modeling efforts.

HEDM datasets are already frequently paired with crystal plasticity models [5, 6] and the GOE can be readily computed from a simulated deforming polycrystal. This connection provides a directly comparable experimental and simulated result that can be used both to interpret the experimental GOE and validate the crystal plasticity model for a given system. Augmenting the GOE by including the intensity information to create the grain-scale ODF is under development, but the evolution of the GOE from *every* grain in real time provides unprecedented opportunities for the crystal scale modeling community.

4.2. Integrating spatially resolved data and electron microscopy

There are many useful measures of microstructure that, in practice, do not make use of a spatial component (e.g. dislocation densities, precipitate densities, grain sizes, textures, etc.). Like these quantities, the GOE has merit in its own right, but a full mechanistic understanding of microscale deformation also requires information about the microstructure which cannot be fully represented with non-spatial orientation information. The GOE can, however, be supplemented with other spatial techniques in a rather straight forward manner. Firstly, segmented grains from near-field HEDM maps and 3D EBSD serial sectioning can be, with some effort, correlated with far-field HEDM data [7, 6, 8, 9, 10, 11, 12]. These datasets provide spatial resolved orientation maps which can be of great importance for understanding orientation gradients and local features of interest in the material. Because it is non-destructive, the near-field mapping in particular has the potential to be taken both before and after the experiment, and at increments during an experiment should pausing the experiment be appropriate. Having the spatial map of the same diffraction volume not only provides context to the GOEs by identifying the full neighborhood of each grain, but also provides a road map for subsequent characterization techniques such as electron microscopy.

The microstructure characterizations mentioned thus far only provide information about the orientation, which is insufficient to understanding other important aspects of deformation, such as dislocation structures. Imaging of these structures often requires the use of scanning / transmission electron microscopy, which can be particularly difficult to pair with bulk studies due to the small nature of the sample extracted. Therefore, determining regions of interest and understanding extraction geometries provided by the GOE histories and spatial map pairings can bolster these complementary techniques by providing context to a powerful, but small-scale sampling tool.

5. Summary

The grain orientation envelope (GOE) has been presented as a measure of the microstructure that can be extracted from far-field HEDM data during continuous loading experiments. An example of the GOE was shown for one grain in Ti-7Al, capturing the non-spatially resolved evolution of orientations up to a strain of 0.03. This measure of microstructure is extracted in addition to the average elastic strain tensor and centroid of a grain. Since a GOE is extracted for every grain in the diffraction volume, the GOE provides a means to discover regions of interest and identify regions for subsequent high-spatial resolution characterizations. In the future, incorporation of the intensity of the signal may be used to populate an orientation distribution function (ODF), rather than simply the envelope. The method presented shows promise for new micromechnical studies to directly link local stress states to microstructural evolution.

6. Acknowledgments

This work is supported financially by the Office of Naval Research Department of Defence contract N00014-16-1-2982, Dr. William Mullins, program manager. The Ti7Al material was provided by Adam Pilchak, Air Force Research Laboratory. This work is based upon research conducted at the Cornell High Energy Synchrotron Source (CHESS) which is supported by the National Science Foundation under award DMR-1332208.

References

[1] Bernier J, Barton N, Lienert U and Miller M 2011 *J. Strain. Anal. Eng.* **46** 527–547
[2] Poulsen H F 2004 *Three-dimensional X-ray diffraction microscopy: mapping polycrystals and their dynamics* vol 205 (Springer Science & Business Media)
[3] Pagan D C and Miller M P 2014 *J. Appl. Crystallogr.* **47** 887–898
[4] Boyce D and Bernier J 2013 hexrd: Modular, open source software for the analysis of high energy x-ray diffraction data Tech. rep. Lawrence Livermore National Lab.(LLNL), Livermore, CA (United States)
[5] Miller M and Dawson P 2014 *Curr. Opin. Solid State Mater. Sci.* **18** 286–299
[6] Turner T J, Shade P A, Bernier J V, Li S F, Schuren J C, Kenesei P, Suter R M and Almer J 2017 *Metall. Mater. Trans. A* **48** 627–647
[7] Lenthe W C, Echlin M P, Trenkle A, Syha M, Gumbsch P and Pollock T M 2015 *J. Appl. Crystallogr.* **48** 1034–1046
[8] McDonald S, Reischig P, Holzner C, Lauridsen E, Withers P, Merkle A and Feser M 2015 *Sci Rep.* **5** 14665
[9] Turner T J, Shade P A, Bernier J V, Li S F, Schuren J C, Lind J, Lienert U, Kenesei P, Suter R M, Blank B and Almer J 2016 *Integr. Mater. Manuf. Innov.* **5** 5
[10] Juul N Y, Winther G, Dale D, Koker M K, Shade P and Oddershede J 2016 *Scr. Mater* **120** 1 – 4
[11] Pagan D C, Bernier J V, Dale D, Ko J P, Turner T J, Blank B and Shade P A 2018 *Scr. Mater* **142** 96 – 100
[12] Tayon W A, Nygren K E, Crooks R E and Pagan D C 2019 *Acta Mater* **173** 231 – 241

IOP Conf. Series: Materials Science and Engineering **580** (2019) 012019 doi:10.1088/1757-899X/580/1/012019

Advantages of architectured harmonic structure in structural performance

Dmytro Orlov[1*], Jinming Zhou[1], Stephen Hall[2], Mie Ota-Kawabata[3] and Kei Ameyama[3]

[1]Department of Mechanical Engineering, LTH, Lund University, P.O. Box 118, Lund 22100, Sweden;
[2]Department of Construction Sciences, LTH, Lund University, P.O. Box 118, Lund 22100, Sweden;
[3]Department of Mechanical Engineering, College of Science and Engineering, Ritsumeikan University, 1-1-1 Noji-higashi, Kusatsu, Shiga 525-8577, Japan;

*e-mail: dmytro.orlov@material.lth.se; tel +46 46 - 222 9095

Abstract. The concept of heterogenous materials with architectured harmonic structure (HS) has emerged recently with the aim of enhancing the structural performance of metals. This work presents critical microstructure characteristics contributing to the superior structural performance of harmonic-structured materials using as an example of commercially-pure nickel. Optical and scanning electron microscopy, along with nano-indentation and tensile testing, as well as digital image correlation analysis, were used for this purpose. A brief overview of the structural performance of several representative systems with HS is provided and interesting directions for future studies are discussed.

1. Introduction

For many years, the main goal in fabricating structural polycrystalline materials was to produce *homogeneous* microstructures [1]. These typically possess either high ductility – low strength in the case of coarse-grained (CG, crystallite or grain size $d \geq 10\mu$m) or low ductility – high strength in the case of ultrafine-/nano-grain structures (UFG, $d \leq 1.0$ μm / $d \leq 0.1\mu$m) [2-6]. On the other hand, all natural materials have *heterogeneous* structures architectured at the micro-/nano-scale, which leads to higher structural efficiency [7-10]. Therefore, fabrication of *heterogeneous* structures having a bimodal grain size distribution was proposed recently [11-14]. In such structures, the UFG-scale grains provide high strength, while the coarser grains store dislocations therefore enhancing the ductility [6, 15].

Metallic materials with a bimodal grain size distribution can be fabricated by a number of recently developed techniques based either on traditional thermo-mechanical processing [16, 17] or on more novel severe plastic deformation principles [14, 18, 19], as well as through a powder metallurgy approach [20, 21]. Although these techniques are proven to be effective in controlling the majority of bimodal microstructure parameters [14, 16-21], they all lack the control of one essential characteristic, namely the spatial *topological distribution of coarse and fine grains*. Poor control of the topology of grain distribution can result in the fabrication of simply heterogeneous structures that are known for poor performance, for instance in fatigue [22], and modest reproducibility of results. By contrast, appropriate control of topology can ensure truly high-performance of bimodal structured material [10,

23-25]. The most recent developments in this direction include fabrication of micro-architectured materials with 'gradient-structure', 'heterogeneous-lamella' and 'harmonic' structures [23-26].

The harmonic structure (HS) has bimodal grain size distribution and fully controllable 3D topology, which is heterogeneous on the micro-scale but homogeneous on macro-scales. Materials with such a structure demonstrate superior structural performance in comparison with CG and UFG counterparts, as well as with bimodal structured materials having a non-controlled topology of the grains [26]. Furthermore, materials with a HS have a reduced variation of properties, therefore improved reproducibility of results [27]. The HS design concept is particularly useful when alloying and where other means of microstructure control are restricted by environmental or other requirements. The concept can also be extended to a variety of materials.

In this paper, we explain the concept of harmonic-structured materials, report recent insights in their stress-strain behaviour using as an example pure nickel, and provide an overview of structural performance in several representative systems with such microstructures.

2. Experimental materials and methods

The following procedures were used for the fabrication of samples, specimen preparation and testing.

2.1. Material fabrication

Commercially-pure (at least 99.2%) Ni powders with a mean particle size of 150 μm were prepared by plasma rotating electrode processing (PREP) in an Ar gas atmosphere. Some fraction of the powder was preserved in this condition and considered 'initial powder' (IP) while the rest was processed by mechanical milling (MM) at ambient temperature to produce gradient structure in the particles, as illustrated schematically in the inset of figure 1. A planetary ball mill (Fritch P-5) with a SKD11 vials and SUJ2 steel balls, mixed with the powders at a mass ratio of 1.8:1, was used for this purpose. The rotation speed was 150 rpm, and milling times were up to 150h. Afterwards, the powders were consolidated by spark-plasma sintering (SPS) at 1073 K for 30 min in vacuum and applied pressure of 100 MPa. As-sintered specimens were 15 mm in diameter and up to 2 mm in thickness.

Based on previous studies [28, 29], the samples with nominal UFG fraction of 40% only were fabricated in this work. For the fabrication of bimodal-HS sample with 40% UFG fraction (bHS40), MM was carried out for 100h. In order to produce a material having a bimodal structure with random distribution of 40% UFG (bR40), the powder was milled for 150h, then mixed with initial powder in a proportion allowing production of a UFG fraction similar to that in the bHS40 material.

2.2. Characterisation of microstructure

The microstructures were characterised using light-optical microscopy (LOM) and scanning-electron microscopy (SEM). A Leica DM RME optical microscope equipped with polarized light and digital camera was used for the former, while a field-emission gun SEM Hitachi SU6600, fitted with HKL system for electron backscatter diffraction (EBSD) analysis was used for the latter.

The specimens for both types of characterisation were cut by electric-discharge machining (EDM), ground on SiC paper and polished in diamond suspensions and colloidal silica as a last step. These were sufficient for SEM and EBSD analysis. Specimens for LOM were further etched in a sulphuric-nitric acid-based solution for revealing the material microstructures.

2.3. Mechanical testing

Mechanical testing was carried out by means of tensile testing and nano-indentation. Due to the limited dimensions of the fabricated samples, tensile tests were carried out on miniaturised dog-bone shaped specimens with nominal gage dimensions (length x width x thickness) of 5.0 x 1.5 x 1.0 mm^3. The initial strain rate was 3.6 x 10^{-3} s^{-1}. At least three specimens were tested for each material condition. An in-house built miniaturized tensile rig was used for the majority of the tensile testing, while an Instron ElectroPulse-10000 instrument was used for high-accuracy testing combined with digital image correlation (DIC) analysis. Nano-hardness testing was carried out using Micro Materials NanoTest Vantage instrument at 50 mN loads with Berkovich-type indenter.

The specimens for both types of testing were cut from the samples by EDM. Afterwards, they were ground on SiC paper and then polished in diamond suspensions and colloidal silica as a last step. Finally, they were etched in a sulphuric-nitric acid-based solution for revealing the material microstructure in optical microscopy. The resulting contrast was sufficient for DIC analysis in tensile testing and for the positioning of indentation areas in nano-hardness testing.

3. Results and analysis

3.1. Evolution of microstructure

The results of the characterization and analysis of microstructure evolution, from the initial PREP powders to final samples are summarised in figures 1 and 2. The upper part of figure 1 illustrates the evolution of shape and microstructure in Ni powder particles before (initial powder, i.e. as-PREP processed before MM) and after mechanical milling for 100h. The processing variables in MM were tailored for producing the gradient structure shown schematically in the inset between the particle micrographs. Such processing should lead to a minimum distortion of the net shape of powder particles with significant reduction of grain sizes on their periphery. The results of this study are in good agreement with earlier reports already demonstrated for several metals and alloys, confirming that when the variables are optimised, an increase of milling time leads to an increase in the peripheral area consisting of UFG microstructure.

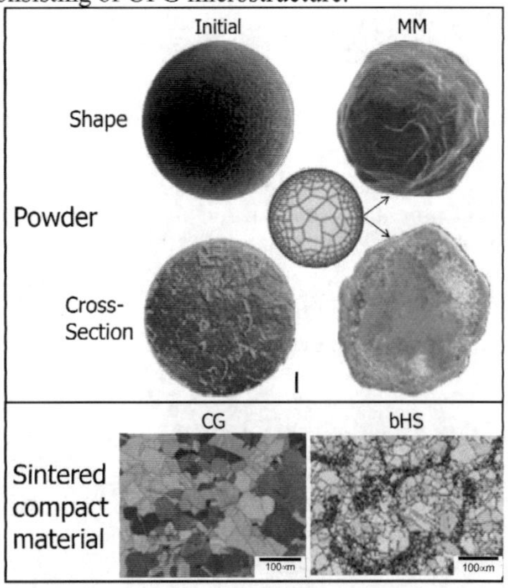

Figure 1. SEM micrographs showing the external shape and cross-section in the initial and mechanically-milled Ni powder particles (above) along with EBSD grain boundary maps of compact nickel sintered from these powders. The inset between the particle images illustrates the schematic microstructure of a powder particle with gradient structure after MM.

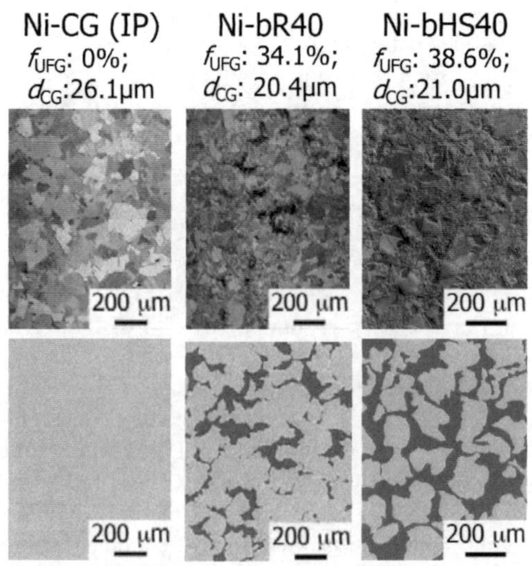

Figure 2. Original (top row) and thresholded (bottom row) optical micrographs of sintered compacts from pure Ni illustrating the microstructure and the distributions of UFG (green) vs CG (pink) areas, respectively. The three columns correspond to homogeneous 'coarse-grained' (CG), and heterogeneous material with bimodal-random (bR40) and bimodal-harmonic (bHS40) grain size distributions. Numbers in the latter indicate nominal fractions of UFG grains. Actual fraction f_{UFG} and mean diameter of coarse grains d_{CG} are shown at respective column tops.

The lower part of figure 1 shows EBSD maps of as-sintered compacts from the powders described above. The SPS process was, again, optimised to achieve maximal consolidation of the material at minimal UFG grain growth.

Optical micrographs and their thresholded counterparts representing microstructures in the materials used in this study are shown in figure 2. The thresholding of the UFG grains was done with tools available in Adobe Photoshop, for the visualization and quantification of the UFG volume fractions. The as-fabricated materials with bimodal grain size distributions have approximately the same fractions of UFG grains and grain sizes of the CG component just slightly exceeding 20 μm. The latter is close enough to the same characteristic of the CG material to conclude that the selected parameters of MM had little effect on grain sizes in the particle cores. Hence, the only difference between these two samples is the topological distribution of the UFG structure, forming a continuous 'skeleton' in the bHS40 material while it appears as disconnected patches in the bR40 material.

3.2. Evolution of mechanical properties.

The outcomes from tensile and nano-hardness testing along with respective analysis are presented in figures 3-5. Tensile testing reveals that producing a bimodal-random distribution of UFG phase in pure Ni leads to an increase of strength along with a decrease in ductility when compared to the initial CG material. This is typical for deformation microstructures widely reported in the literature. At the same time, the fabrication of a harmonic structure leads to a further increase in strength while preserving the ductility characteristics at the level of the initial CG counterpart. This is a unique behaviour, which can be attributed to the specific UFG distribution control in harmonic-structured materials. The presence of a long strain-hardening stage in Ni-bHS40 material also demonstrates its advantage over 100%-UFG counterparts.

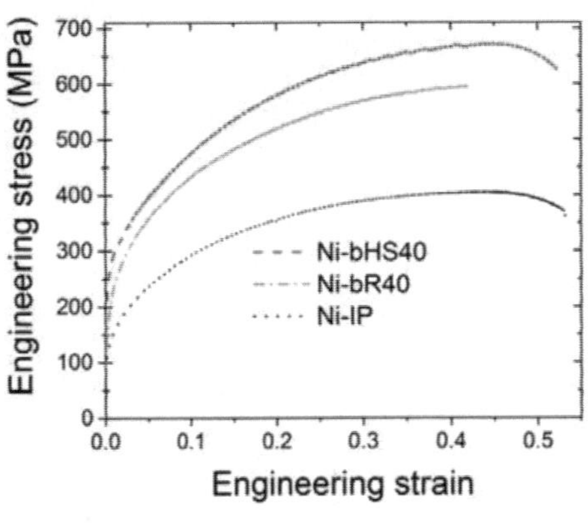

Figure 3. Stress-strain curves from the tensile testing of homogeneous 'coarse-grained' (IP) and heterogeneous Ni with nominal UFG fraction of 40% having bimodal-random (bR40) and bimodal-harmonic (bHS40) grain size distributions.

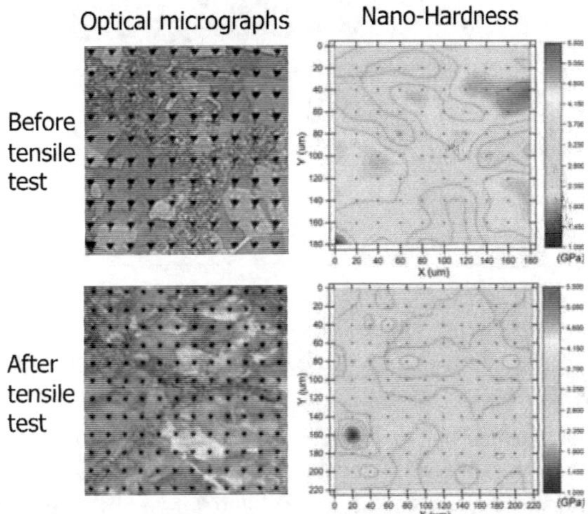

Figure 4. Results of nano-hardness measurements of harmonic-structured Ni-bHS40 before and after tensile testing. Optical micrographs from the indentation area in the left column correspond to the distributions of hardness in the right column.

Figure 4 shows the results of nano-indentation testing in the area of a triple-junction of original MM-processed powder particles in the Ni-bHS40 sample. For the sake of clarity, optical micrographs with the traces of the indents are presented alongside the respective nano-hardness distribution maps. In the as-fabricated material before tensile testing, areas corresponding to the CG fraction have the lowest hardness, at the level of 2 GPa, while the UFG regions demonstrate rather localised higher hardness up to the level of approximately 4.6 GPa. The post mortem analysis of specimens after tensile testing reveals an increase of the lowest hardness to approximately 4 GPa on the background of broadening of the hardness peaks, with virtually no increase in their maximum level. One indent with a hardness significantly lower than all the others on the post mortem map can be considered as an outlier due to a local effect from a residual pore or similar defect. Other important observations that can be discerned from the optical micrographs are the distortion of UFG area and the coarse grains along with the appearance of alternating contrast, i.e. the traces of dislocation lines, in the latter.

The main findings from tensile testing with DIC analysis are presented in figure 5. In the acquisition of images for DIC analysis, an optical stereo-microscope was used with polarized light mounted perpendicular to the loading axis within the instrument for tensile testing. The 'native' contrast generated on the surface of etched specimens by microstructure was sufficient for DIC analysis. The photograph of a specimen gage area is presented as an inset within figure 5a. Rubber bands within the field of view were used for mounting a mechanical extensometer as an additional way to accurately measure the specimen strain. The open specimen area between the bands indicated in figure 5a by a blue orthogonal mesh was used for DIC analysis. The maximum level of tensile deformation for reliable strain quantification from image correlation was limited by distortions arising from plastic flow at the surface leading to (i) excessive surface reflectivity making the algorithm fail, and (ii) excessive out-of-plane strain components generating significant discrepancies with the presumed model of a uniaxial strain mode.

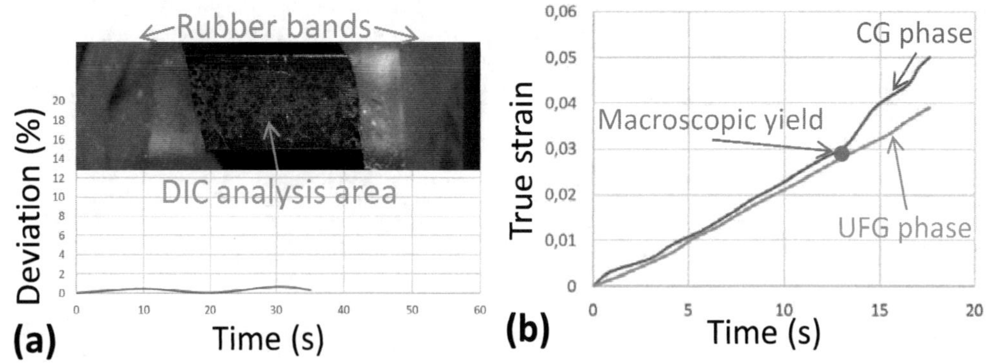

Figure 5. (a) Optical photograph of Ni-bHS40 specimen area used for DIC analysis along with a graph showing the difference in true strain values calculated from the extensometer and from DIC data, and (b) dependence of accumulated true strain on time during tensile testing, shown for UFG and CG areas separately.

It can be seen in figure 5a that the deviation of strain calculated as an average of the entire DIC field from that measured by the mechanical extensometer is less than 1% for the entire range of strains presented in figure 5b. Accumulated true strain values presented in figure 5b were calculated as an average of five correlation areas within the CG and UFG phases, respectively, for each time step. A point corresponding to the macroscopic yield at the yield stress of engineering stress-strain curve is also presented in the figure. It can be seen that until that point true strain accumulates in both the phases at virtually the same rate. Upon reaching macroscopic yield, the rate of true strain accumulation starts increasing rapidly in the CG phase while remaining the same in the UFG phase. Such a behaviour suggests strain partitioning between the phases and agrees well with the findings of Park et al in [30].

This result is very consistent with other observations in this study. In particular, strain partitioning and enhanced plastic flow in CG regions lead to extensive strain hardening in these areas, see figure 4, and increased strain hardening rate in bimodal-grained materials, see figure 3. These seem to be more intensive in the HS material, most likely due to the continuous UFG skeleton encompassing isolated CG islands. The combined effect of elevated yield and ultimate tensile strengths, along with extended high-rate strain hardening suggest superior performance of harmonic-structured nickel.

3.3. Effect of harmonic structure in various materials.
At the stage of writing this paper, the concept of fabricating harmonic structures has been applied to a number of materials including pure metals, steels and some alloys for permanent bio-implants. The results on their structural performance are summarised in figure 6. They are presented as the ratios of total elongation, ultimate tensile strength (UTS) and toughness in HS material compared to the CG counterparts. Toughness was calculated as the total area under a stress-strain curve.

It can be found from figure 6 that although harmonic structure formation leads in general to improved properties, see average values, specific results depend on the material system. For example, pure Fe and Al, as well as alloy Ti64 fabricated by jet-milling, and stainless steel SUS304L, all demonstrate reduced total elongation, while the elongation remains essentially unchanged in pure Ti and Ni, and even increases in pure Cu, and Ti fabricated by jet-milling, in the bio-medical alloy CCM and in stainless steel SUS316L. In contrast, the UTS and especially toughness increase significantly in all the HS metals. The exact reasons for such dependence are still to be discovered in future studies.

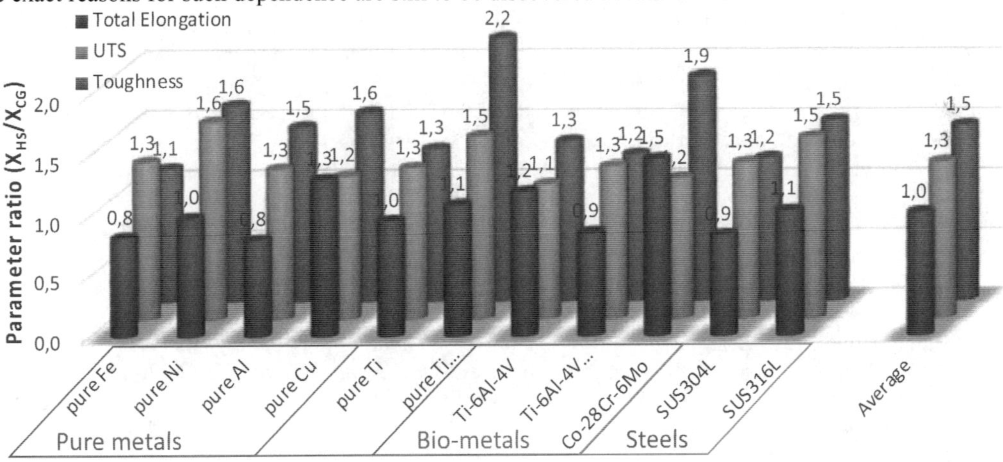

Figure 6. Summary and examples of principal structural characteristics for various kinds of HS materials fabricated thus far. The use of '…' indicates that jet-milling was used for processing the metal powders, while mechanical-milling was used in all other cases.

Conclusions
In this study, the critical microstructure characteristics contributing to the superior structural performance of harmonic-structured materials have been presented using as an example commercially-pure nickel. Optical and scanning electron microscopy along with nano-indentation and tensile testing as well as digital image correlation analysis were used for this purpose. Further insights explaining at least some unique features of stress-strain behaviour in HS metals have also been reported. Finally, an overview of the structural performance of several representative systems with such microstructures has been provided and interesting directions for future studies discussed.

Acknowledgements
Swedish Research Council (Vetenskapsrådet) is acknowledged for the financial support of this project through a research grant # 2016-03811. The authors also deeply appreciate technical support from graduate students Naoki Horikawa of the Graduate School of Science and Engineering at Ritsumeikan University in Japan and Jérôme Hobier of Ecole Nationale Superieure de Mecanique et D'Aerotechnique (ENSMA) in France for the fabrication of samples, majority of experimental work and DIC analysis during their internship projects in Lund University.

References

[1] Hansen N, Huang X and Huges D A 2004 Deformation Microstructures. In: *2004 TMS Annual Meeting and Exhibition,* ed T G L Y.T. Zhu, R.Z. Valiev, S.L. Semiatin, D.H. Shin, T.C. Lowe (Charlotte, North Carolina, USA: TMS (The Minerals, Metals & Materials Society)) pp 67-74
[2] Gleiter H 1989 Nanocrystalline materials *Progress in Materials Science* **33** 223-315
[3] Valiev R 2002 Materials science: nanomaterial advantage *Nature* **419** 887-9
[4] Meyers M A, Mishra A and Benson D J 2006 Mechanical properties of nanocrystalline materials *Progress in Materials Science* **51** 427-556
[5] Langdon T G 2013 Twenty-five years of ultrafine-grained materials: Achieving exceptional properties through grain refinement *Acta Materialia* **61** 7035-59
[6] Estrin Y and Vinogradov A 2013 Extreme grain refinement by severe plastic deformation: A wealth of challenging science *Acta Materialia* **61** 782–817
[7] Mandelbrot B B 1983 *The Fractal Geometry of Nature*: Henry Holt and Company)
[8] Wegst U G K and Ashby M F 2004 The mechanical efficiency of natural materials *Philosophical Magazine* **84** 2167-86
[9] Meyers M A, Chen P-Y, Lin A Y-M and Seki Y 2008 Biological materials: Structure and mechanical properties *Progress in Materials Science* **53** 1-206
[10] Meyers M A, McKittrick J and Chen P-Y 2013 Structural Biological Materials: Critical Mechanics-Materials Connections *Science* **339** 773-9
[11] Wang Y, Chen M, Zhou F and Ma E 2002 High tensile ductility in a nanostructured metal *Nature* **419** 912-5
[12] Gil Sevillano J and Aldazabal J 2004 Ductilization of nanocrystalline materials for structural applications *Scripta Materialia* **51** 795-800
[13] Zhao Y-H, Liao X-Z, Cheng S, Ma E and Zhu Y T 2006 Simultaneously Increasing the Ductility and Strength of Nanostructured Alloys *Advanced Materials* **18** 2280-3
[14] Höppel H W, Korn M, Lapovok R and Mughrabi H 2010 Bimodal grain size distributions in UFG materials produced by SPD: Their evolution and effect on mechanical properties *Journal of Physics: Conference Series* **240** 012147
[15] Huang X, Hansen N and Tsuji N 2006 Hardening by Annealing and Softening by Deformation in Nanostructured Metals *Science* **312** 249-51
[16] Azizi-Alizamini H, Militzer M and Poole W J 2007 A novel technique for developing bimodal grain size distributions in low carbon steels *Scripta Materialia* **57** 1065-8
[17] Ravi Kumar B and Raabe D 2012 Tensile deformation characteristics of bulk ultrafine-grained austenitic stainless steel produced by thermal cycling *Scripta Materialia* **66** 634-7
[18] Orlov D, Todaka Y, Umemoto M and Tsuji N 2011 Formation of bimodal grain structures in high purity Al by reversal high pressure torsion *Scripta Materialia* **64** 498-501
[19] Zherebtsov S, Kudryavtsev E, Kostjuchenko S, Malysheva S and Salishchev G 2012 Strength and ductility-related properties of ultrafine grained two-phase titanium alloy produced by warm multiaxial forging *Materials Science and Engineering: A* **536** 190-6
[20] Srinivasarao B, Oh-ishi K, Ohkubo T, Mukai T and Hono K 2008 Synthesis of high-strength bimodally grained iron by mechanical alloying and spark plasma sintering *Scripta Materialia* **58** 759-62
[21] Dirras G, Gubicza J, Ramtani S, Bui Q H and Szilágyi T 2010 Microstructure and mechanical characteristics of bulk polycrystalline Ni consolidated from blends of powders with different particle size *Materials Science and Engineering: A* **527** 1206-14
[22] Estrin Y and Vinogradov A 2010 Fatigue behaviour of light alloys with ultrafine grain structure produced by severe plastic deformation: An overview *International Journal of Fatigue* **32** 898-907
[23] Zhou M 2013 Exceptional Properties by Design *Science* **339** 1161-2

[24] Ma E and Zhu T 2017 Towards strength–ductility synergy through the design of heterogeneous nanostructures in metals *Materials Today* **20** 323-31
[25] Ovid'ko I A, Valiev R Z and Zhu Y T 2018 Review on superior strength and enhanced ductility of metallic nanomaterials *Progress in Materials Science* **94** 462-540
[26] Vajpai S K, Ota M, Zhang Z and Ameyama K 2016 Three-dimensionally gradient harmonic structure design: an integrated approach for high performance structural materials *Materials Research Letters* **4** 191-7
[27] Zhang Z, Orlov D, Vajpai S K, Tong B and Ameyama K 2015 Importance of Bimodal Structure Topology in the Control of Mechanical Properties of a Stainless Steel *Advanced Engineering Materials* **17** 791-5
[28] Sawangrat C, Kato S, Orlov D and Ameyama K 2014 Harmonic-structured copper: performance and proof of fabrication concept based on severe plastic deformation of powders *Journal of Materials Science* **49** 6579-85
[29] Zhang Z, Vajpai S K, Orlov D and Ameyama K 2014 Improvement of mechanical properties in SUS304L steel through the control of bimodal microstructure characteristics *Materials Science and Engineering A* **598** 106-13
[30] Park H K, Ameyama K, Yoo J, Hwang H and Kim H S 2018 Additional hardening in harmonic structured materials by strain partitioning and back stress *Materials Research Letters* **6** 261-7

Integral mean curvature analysis of 3D grain growth: Linearity of dV/dt and grain volume

BR Patterson[1], RT DeHoff[1], CA Sahi[1,2], J Sun[3], J Oddershede[3], F Bachmann[3], E Lauridsen[3] and D Juul Jensen[4]

[1] University of Florida, Gainesville, FL 32611, USA
[2] Pratt & Whitney, 400 Main Street, East Hartford, CT 06118, USA
[3] Xnovo Technology Aps, Køge, Denmark
[4] Technical University of Denmark, 2800 Kgs. Lyngby, Denmark

Email: patters@mse.ufl.edu

Abstract The volumetric growth rate of individual grains has been found to be directly proportional to the individual grain volume V as $dV/dt = \beta(V_0\text{-}V)$. This simple result is explained through the DeHoff relationship $dV/dt = -kM_S$ between growth rate and the integral mean curvature of individual grains, M_S, combined with the experimentally observed relationship between M_S and grain volume, $M_S = \alpha(V_0\text{-}V)$. These relationships have now been observed consistently in both 3D grain growth simulations and experiments. This paper describes the relationships among these kinetic and geometric grain characteristics that provide this simple description of 3D grain growth.

1. Introduction

Grain growth models have historically focused on the change in the average linear grain size $\bar{D}$, perhaps because that was essentially the only parameter previously available for describing the process, i.e., through the line intercept count. Unfortunately, this viewpoint of studying the average behavior of growing and shrinking grains masks the important geometric details of what drives and controls the processes that occur. Recent experimental and theoretical developments have led to a new and surprisingly simple picture of grain growth. The advent of 3D methods such as DCT and computer modelling enable imaging and following the change in volume of grains that describes their growth and shrinkage rates, not previously possible. New theoretical developments have particularly involved the application of integral mean curvature of grains, M_S, to describe these volume changes resulting from curvature driven boundary motion. This descriptor, which was previously unattainable for 3D grains, can now be easily computed from the images produced by the above 3D experimental methods.

1.1 Integral mean curvature, M_S

The integral mean curvature M_S [1-3] of a grain boundary is the integral of its mean local curvature H over the surface area of a grain boundary, i.e., the product of the average mean curvature of the boundary $\bar{H}$ and the total surface area,

$$M_S = \iint H\, dS = \bar{H}S \tag{1}$$

and is defined through the sign of H as negative for concave faces, typical of relatively large, many-faced grains, and positive for smaller convex few-faced grains. Large grains with relatively large surface and sharply concave faces have large negative values of M_S, while intermediate sized grains have relatively flat faces with $M_S \cong 0$ and small concave grains have smaller, positive values of M_S.

2. Experimental methods

The examples throughout this paper of the relationships that have been experimentally found between growth rate, grain volume and M_S, have been obtained through both 3D computer simulations and LabDCT studies [4, 5]. These methods are described below.

2.1 3D Monte Carlo grain growth studies

3D kinetic Monte Carlo simulations were performed on the SPPARKS [6] platform using a simulation temperature of $kT = 0.5$ and periodic boundary conditions. Large 500^3 voxel arrays were used to produce large numbers of grains for good statistical averaging of the measured grain characteristics. These large initial numbers are necessary to allow for the large fraction of grains that are lost in the early transient stage of grain growth preceding the stable self-similar stage in which the analyses are performed.

The starting simulation grain structures were synthesized using DREAM3D [7] with initial grain volume distribution widths of $ln\sigma = 0.3$. This width, typical of grain structures in the self-similar state, enabled achieving stable grain growth early in the process while many grains were still present. Simulations typically ran for approximately 100 Monte Carlo equivalent seconds since this amount of time captured the necessary information during stable grain growth, starting at 10-20 seconds.

2.2 Laboratory DCT studies

Laboratory grain growth studies were also performed employing DCT imaging of grains in a real material, Armco iron, to test the validity of the relationships found in the prior simulations. These studies proved exciting in that they confirmed the exact trends and relationships found previously. The excellent comparison of such disparate methods validates both techniques for detailed geometric and kinetic studies such as these.

2.2.1 Material preparation. Low carbon Armco iron with a single-phase ferrite grain structure was employed as the experimental material for these studies. The grain sizes of the as-obtained hot-rolled strips were refined by cold working either 50 or 75%, for different portions of the study, respectively, and giving recrystallization and grain growth anneals for several hours at 880°C to produce an initial mean grain intercept of approximately 75μm, the minimum allowable for the following laboratory X-ray Diffraction Contrast Tomography (LabDCT [5]) analysis. Strips of the prepared Armco iron were electro-discharge machined to form cylinders with a 1mm diameter without surface damage.

2.2.2 DCT analysis. The DCT scans of the Armco iron specimens were performed using a ZEISS Xradia 520 Versa with LabDCT. 3D grain reconstruction was performed using GrainMapper3D and volume and face measurements were performed. Multiple scans, each covering an area of 0.3mm length, 0.3mm width, and 1mm height were taken at adjacent positions and stitched together, to increase the number of grains included in each analysis [8]. For both the 50 and 75% cold worked specimens a total of 14 volumes were scanned covering five different annealing states. Figure 1 shows a typical scanned image of the specimen.

Subsequent 880°C anneals of ~8hr were given to produce incremental growth steps, after which DCT scans were performed to image individual grains and obtain their volumes, dV/dt, and M_S.

2.3 Determination of M_S

Several methods have been developed in recent years for determining M_S of individual grains and grain size or face classes from 3D images. Although the methods vary greatly in application, they show surprisingly good comparison. These methods are described below.

2.3.1 Rowenhorst method. Rowenhorst, et al. [9] performed the first measure of M_S of individual grains using a serial sectioned titanium alloy. After reconstruction, meshing and smoothing values of M_S were computed from the triangulated structure using equation (2a),

$$M_S = 1/2 \sum \chi \, dl \quad (2a)$$

$$= \frac{\pi}{4}(O - I) \quad (2b)$$

in which χ is the angle between face normals of pairs of contacting triangles and dl is the length of their shared edge. Equations (1) and (2) are equivalent except that in (1) M_S is computed for a smooth face while (2) sums the integral mean curvature contributions of the edges making up a non-smooth face.

2.3.2 DeHoff and I-O methods. Other methods for computing M_S have since been developed, including (a) the stereological method of DeHoff [10], employing measures of the areas and numbers of edges on 2D grain sections for grains identified by their 3D volume or face class, and (b) the Innie-Outtie *(I/O)* method [11], applicable to voxelated structures and with M_S computed from Eq.(2b). This latter technique is based on the same principles as Rowenhorst's, treating a surface as a network of flat facets bounded by edges, but requires no smoothing or meshing of the surfaces.

Here, Innies *(I)* are the numbers of edges between pairs of inward angled (concave) voxel faces and Outties *(O)* are the numbers of voxel edges between outward angled (convex) faces. These numbers can be computed via computer code on either an individual grain face or over an entire grain. Convex faces have a majority of *O*-edges, giving positive M_S, concave faces have a majority of *I*-edges and negative M_S and flat faces have similar *I* and *O* values with $M_S \cong 0$. Equation (2b) derives from (2a) for the special case of all edges having a length of unity and all edges having dihedral angles χ of either $+\pi/2$ for *O*-edges or $-\pi/2$ for *I*-edges, as occurs with voxelated structures from simulations or reconstructed DCT images.

The current studies have employed both the DeHoff and *I-O* methods. Figure 2 shows the close comparison between the Rowenhorst and the I-O methods after normalizing M_S by size to make it unitless, which enables comparison of grains of different relative scales such as these. Note that the plot shows $M_S = 0$ for grains with ~15 faces, which are often assumed to be fairly flat-faced and is essentially linear as will be discussed later.

$$M_S norm = M_S / V^{-2/3} \quad (3)$$

3. Integral mean curvature driven growth rate of individual grains

3.1 DeHoff model of dV/dt vs. M_S

The recent DeHoff model [10] of integral mean curvature driven grain growth has provided a simple assumption-free description of the rate of volumetric growth or shrinkage of individual grains or grain classes of similar size or number of faces:

Figure 1. Image of LabDCT scanned Armco iron specimen reconstructed from stitching several individual scans to maximize the analysis volume [7,8].

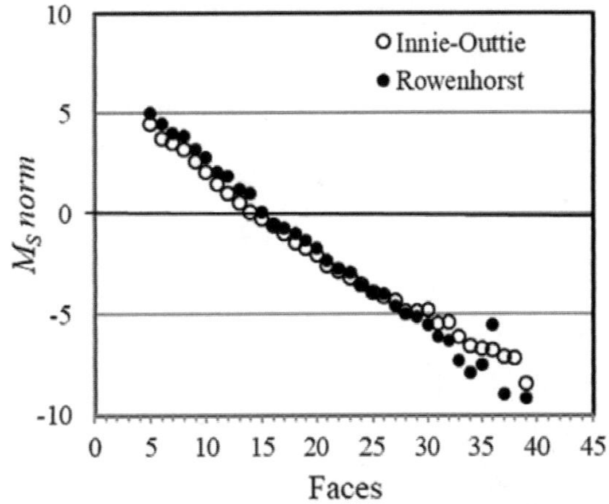

Figure 2. Comparison of the Rowenhorst and I-O methods for determining M_S. Data are for individual grains grouped by face class. The M_S values have been normalized by size to make unitless for comparing grains of different scales.

$$dV/dt = -k\,M_S \tag{4}$$

where k is the rate constant. Note that $dV/dt = 0$ when $M_S=0$, as for flat-faced grains. This linear relationship, shown in figure 3 for (a) 3DMC simulated grain growth and (b) LabDCT characterized grain growth in Armco iron, has now been confirmed in numerous simulations and experiments, and provides the basis for the later analyses in this paper. In (b) the data include an abnormal grain in the specimen that received heavy cold work prior to recrystallization. Note that this grain has an extremely high dV/dt and a high negative value of M_S but fits the same trend as the normal grains, implying that its linear growth rate v is dependent on the same proportionality to H, through k.

$$v = -k\,H \tag{5}$$

Combining equations (1) and (5) with the kinematic description of dV/dt as the product of boundary velocity and surface area shows the origin of equation (4), describing growth rate as only a function of the value and sign of M_S for a boundary or grain.

$$dV/dt = \iint v\,dS = \iint -kH\,dS = -k\,M_S \tag{6}$$

3.2 Simple relationships between dV/dt, M_S and V of individual grains
The DeHoff relationship forms the basis for the following investigations of the relationships among those grain characteristics that are known to control growth and shrinkage of individual grains, such as face class, curvature and size. The goal of this study is to find and understand the relationships of these characteristics to M_S which is the final element controlling growth kinetics.

3.2.1 Relationship between M_S and individual grain volume V. Through these studies where the properties of individual grains, M_S, V and dV/dt are now available in simulations or experiments several novel and

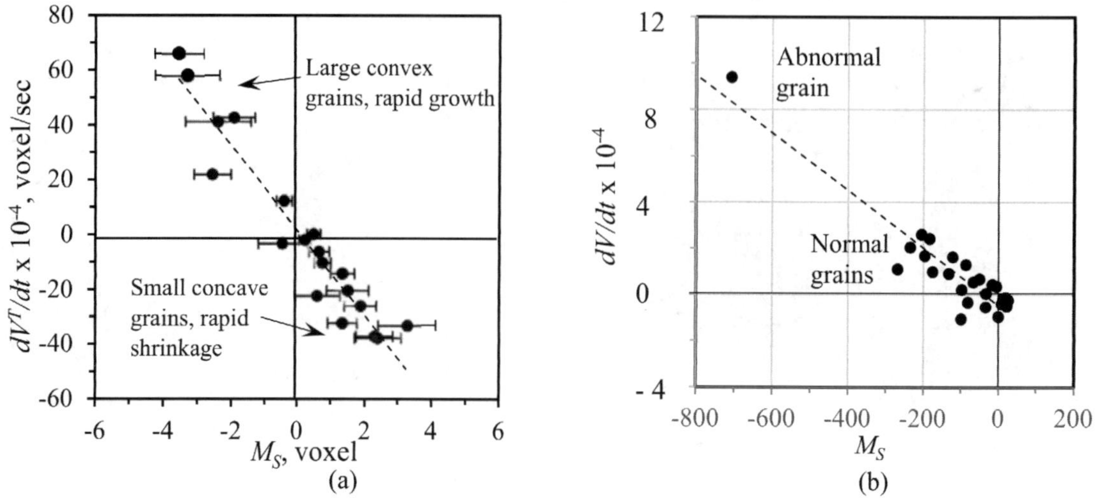

Figure 3. Linear relationship between dV/dt and M_S of individual grains averaged by size (a), or face class (b), passing through the origin and supporting equation (4). In (a) dV/dt was computed from 3DMC simulation data and M_S found by the DeHoff method. In (b) DCT analysis was used to compute dV/dt and the automated I-O method used for obtaining M_S.

consistent relationships have been found, including a linear relationship between M_S and V, described by equation (7) and shown in figure 4,

$$M_S = \alpha(V_0 - V) \tag{7}$$

The individual lines of data points in figure 4 represent different simulation-equivalent times of grain growth, i.e., 10, 20 and 40 seconds, and the slopes α decrease in magnitude with increasing grain growth time, i.e., with increasing scale of the system. The grain volume at the zero-growth rate x-intercept, V_0, represents the volume of the grains that at that time neither grow nor shrink. V_0 increases with time and the scale of the system and has been found to equal the mean grain volume, $\langle V \rangle$ at each time.

Since M_S has units of length, α in equation (7) and the slopes of the lines in figure 4 must have units of $V^{-2/3}$. In addition to α connecting M_S to grain size, equation (1) shows that slope must also carry information about grain shape (number of faces, through H), which also varies with grain size. This grain size information exists both through the overall scale of the structure, where each separate line has a different mean volume $\langle V \rangle$ and through the variation in grain volume along each line at fixed times.

Figure 5 shows the effect of *overall structural scale* on the slope, plotting α vs. $\langle V \rangle^{-2/3}$. This linear path throughout grain growth passes through the origin with the relationship of

$$a = \gamma \bar{V}^{-2/3} \tag{8}$$

In this plot of α vs. $\langle V \rangle^{-2/3}$, the slope $\gamma \cong -2.7$. Since α has units of $\langle V \rangle^{-2/3}$, this proportionality constant is unitless and describes a self-similar condition throughout the process. The significance of this plot is that although a contains both shape and size components of M_S, it still scales in a steady-state manner with *overall structural scale*, expressed as a function of mean grain volume. Thus, α still carries information about the variation in grain size and shape at any *single* state, expressed through the slope γ.

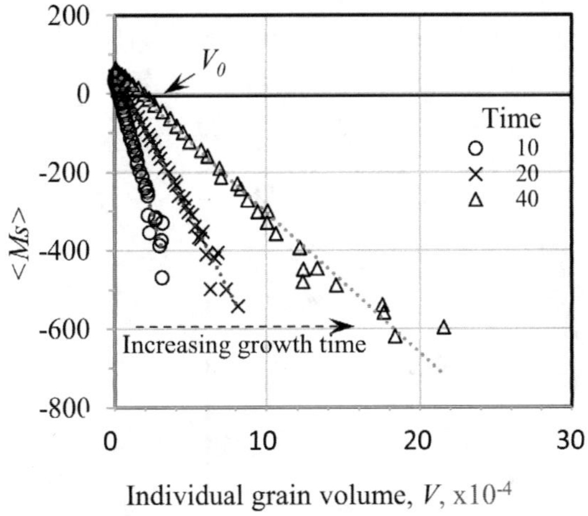

Figure 4. Linear relationship between M_S and individual grain volume, here grouped by face class, as described by equation (7). Data from 3DMC simulation.

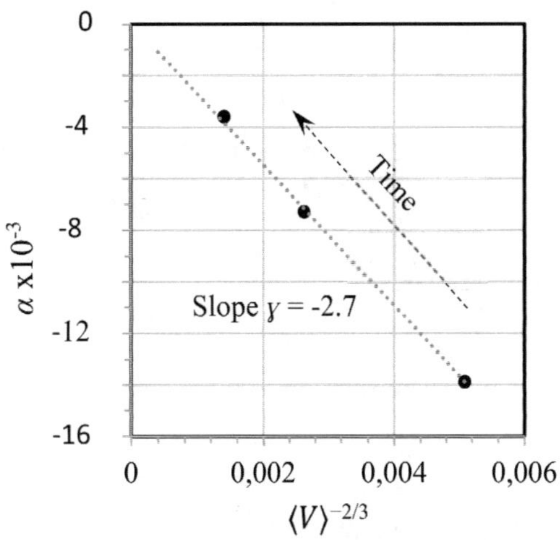

Figure 5. Unitless scaling relationship between the term α in equation (7) and overall scale of the system, expressed as $\langle V\rangle^{-2/3}$.

Combining the above relationships,

$$M_S = \alpha(V_0 - V) = \gamma\langle V\rangle^{-2/3}(V_0 - V) \tag{9}$$

it is apparent that a plot of

$$M_S/\ \langle V\rangle^{-2/3}\ vs. (V_0 - V) \tag{10}$$

still with slope γ, should provide a linear scaling relationship over all times and grain structure scales, as shown to occur in figure 6.

3.2.2 Relationship between dV/dt and individual grain volume V. An equally simple though fundamental relationship has also been observed for the first time between *dV/dt* and individual grain volume *V*, described by equation (11) and shown in figure 7.

$$dV/dt = \beta(V_0 - V) \tag{11}$$

The lines of data in figure 7 represent different times of grain growth, with the slopes, β, decreasing with increased growth time. The data points on each line represent the average properties for the grains grouped by face class, with the grains with larger numbers of faces showing the highest growth rate and those with the fewest faces shrinking. For the plots of each growth time, the face class on the x-axis with zero-growth rate is typically ~ 15 faces. These grains have volume V_0 in equation (11) and represent the volume the grains that at that time neither grow nor shrink. V_0 is essentially identical to the mean grain volume $\langle V\rangle$ at all times throughout grain growth, with the identity:

$$V_0 \cong \langle V\rangle \tag{12}$$

IOP Conf. Series: Materials Science and Engineering **580** (2019) 012020 doi:10.1088/1757-899X/580/1/012020

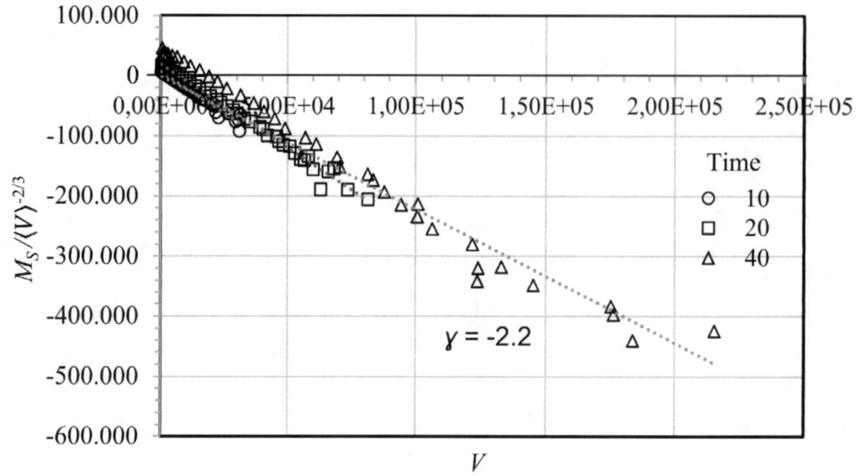

Figure 6. Unitless scaling relationship between $M_S/\langle V\rangle^{-2/3}$, i.e., M_S normalized by the *scale of the grain structure* and individual grain volumes, *V*, here, averaged by face class.

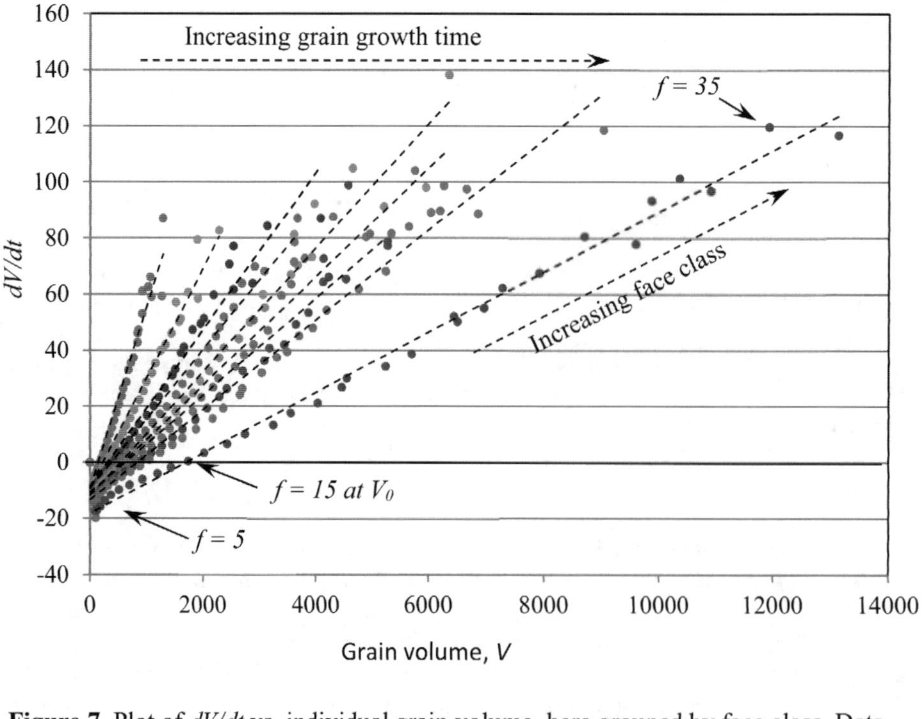

Figure 7. Plot of *dV/dt* vs. individual grain volume, here grouped by face class. Data obtained through 3DMC simulation.

Summary
The volumetric growth rate of individual grains in an aggregate has been consistently observed to be directly proportional to the individual grain volume V as $dV/dt = \beta(V_0-V)$. This relationship is explainable through the DeHoff relationship $dV/dt = -kM_S$ between growth rate and the integral mean curvature of individual grains, M_S, combined with the experimentally observed linear relationship between M_S and individual grain volume, $M_S = \alpha(V_0-V)$.

Acknowledgments
The overall program and BRP, RTD and CAS gratefully acknowledge financial support from NSF Grants DMR-1035188 and 1307665. Dr. Juul Jensen acknowledges funding from the European Research Council, ERC MYD Grant No. 788567. They also thank Dr. Veena Tikare at Sandia National Laboratories, New Mexico for guidance in the computer modelling portion of these studies.

References
[1] DeHoff R and Rhines F 1968 Quantitative Microscoppy, McGraw Hill
[2] DeHoff R 1967 *Trans. Met. Soc. AIME* **239** 617
[3] Cahn J 1967 *Trans. Met. Soc.* AIME **239** 610
[4] Sahi C 2018 Ph D Dissertation, Univesity of Florida
[5] Bachmann F, Bale H, Gueninchault N, Holzner C and Lauridsen E *J. Appl. Crystalloger.* in press.
[6] http://spparks.sandia.gov - Sandia National Laboratories
[7] http://dream3d.bluequartz.net/
[8] Keinan R, Bale H, Gueninchault N, Lauridsen E and Shahani A 2018 *Acta Mater.* **148** 225
[9] Rowenhorst D, Lewis A and Spanos G 2010 *Acta Mater.* **58** 5511
[10] DeHoff R, Patterson B, Sahi C and Chiu S 2015 *Acta Mater.* **100** 240
[11] Sun Z, Tikare V, Patterson B and Sprague A 2012 *Comp. Mater. Sci.* **55** 329

IOP Conf. Series: Materials Science and Engineering **580** (2019) 012021 doi:10.1088/1757-899X/580/1/012021

Understanding structural evolution of nanostructures during deformation from 2D (3D) experiments

O Renk[1*], M W Kapp[2] and R Pippan[2]

[1] Montanuniversität Leoben, Department of Materials Science, Chair of Materials Physics, Jahnstrasse 12, 8700 Leoben, Austria
[2] Erich Schmid Institute of Materials Science, Austrian Academy of Sciences, Jahnstrasse 12, 8700 Leoben, Austria

* corresponding author email: oliver.renk@oeaw.ac.at

Abstract. Understanding how microstructures evolve during deformation and/or annealing treatments is of fundamental interest, as the underlying structure dictates the resulting property spectrum. This issue is even more important for nanomaterials, where interfaces and defect populations have a stronger effect on mechanical properties. In addition, for nanometals the grain architecture can also be an important parameter for improvement of mechanical properties. However, synthesis of bulk high strength materials by severe deformation with tailored and desired properties requires a thorough understanding about the processes governing their microstructural evolution at large and severe strains. Here, we summarize findings from 2D (3D) experiments uncovering these restoration processes. Interestingly, the main mechanisms seem to change with deformation temperature, causing formation of extremely elongated grain structures at intermediate deformation temperature regimes. Apart from discussing the reasons for this finding, we also present recent results focusing on the mechanisms triggering these restoration processes at low temperatures, highlighting the crucial role of plastic strain to facilitate these processes.

1. Introduction

Mechanical properties of metals are inherently determined by the underlying defect and microstructure. Accordingly, any changes occurring during deformation and/or annealing will greatly influence their performance. This dependency seems to become even more important for ultra-strong nanomaterials. In such confined volumes, properties become extremely sensitive to defect concentrations, as well as the interfaces themselves, causing phenomena not present in conventional grain size regimes. For instance, deformed or distorted boundaries possess enhanced diffusivity [1] and when required, can emit defects much more easily compared to boundaries in the relaxed state, which can lead to further strengthening if mobile defects and boundaries are recovered [2–4]. Apart from this, the nanograin architecture may also play an important role for mechanical properties. As cracks preferentially grow along interfaces or grain boundaries, experiments suggest that preferentially aligned grains or phases could provide a promising strategy to unite strength and exceptional damage tolerance in one or two loading directions [5–7]. Accordingly, to make use of this knowledge, in-depth understanding of how specific interfaces, defect structures or grain architecture can be produced or how they result from a synthesis process is required. Nevertheless, although bulk nanostructures have been successfully fabricated by severe plastic deformation (SPD) [8] for several decades, knowledge of how desired nanostructures can be generated is extremely limited. So far, the resulting nanostructures have been processed by trial and error methods, rather than based on an understanding what deformation and restoration processes determine the resulting nanostructure after severe strains. However, such knowledge is necessary to deliberately process nanomaterials with desired and exceptional property combinations.

In recent years, research has tackled these questions and provided a considerable understanding of the processes that govern microstructural evolution after severe strains, already allowing a qualitative understanding of these mechanisms. It is thus the aim of this work to summarize these results obtained from various 2D and 3D experiments (with the third dimension representing the evolution as a function of strain or temperature) at different scales, showing that migration of boundaries or junctions are the essential processes governing the resulting nanostructure. Further, the interplay between these processes results in an interesting temperature effect, causing in some cases enormous grain elongations. In the second part, we focus on the question of what triggers boundary migration at comparably low deformation temperatures, an important question where controversial results can still be found. Recent experimental insights on this topic will be presented, which clearly indicate that plastic strain rather than the applied stress drives their motion.

2. Restoration during severe straining and the (unexpected) effect of temperature on grain shape

During heavy plastic deformation dislocation cell and grain sizes decrease continuously with strain, although the initially rapid size reduction tends to stagnate with strain [9]. Already early investigations on severely deformed single phase materials showed indeed that microstructural features (e.g. grain size or shape, and misorientation distribution) do not change on average any further once a sufficiently high strain is applied ($\varepsilon \sim 10$ for torsion at ambient temperature) [10–12]. As materials deformed to such strain levels can still be deformed further, structural recovery or restoration mechanism are required to keep structural features on average at a constant level. Despite knowing for a long time that such a steady state or saturation regime exists for single phase metals, the underlying necessary restoration or recovery processes remained unknown and controversially discussed. Accordingly, various different mechanisms have been proposed, see for instance Refs. [13–16].

Several years ago, results from 2D (3D) experiments on submicron sized copper and aluminum, respectively, allowed for the first time to unambiguously identify the actual mechanisms at the grain scale [17,18]. The structural evolution of samples already cold worked to severe strains was monitored quasi in-situ during additional strain increments. The experiments unambiguously revealed that migration of grain boundaries (GBs) and/or triple junctions (TJs) provokes a local collapse of grains, allowing adjacent grains to grow and maintain their average size during deformation. Therefore, the notion of saturation or steady state might be misleading, as the structure is not stable at all, but rather a dynamic equilibrium between (local) coarsening and refinement is achieved. It should be noted that migration of boundaries is often not observed over the entire boundary length. Rather, local bulges form that split apart neighboring grains, before being in some cases totally consumed (see figure 1). In any case, both processes, grain fragmentation or migration of junctions, can effectively reduce the grain length and ensure that average microstructural features are maintained. Thus the processes may be considered to be similar to that of conventional dynamic recrystallization, but occurring at much lower deformation temperatures (below 0.3 T_m). Due to the absence of significant thermal activation, the migration processes are thus often referred to as mechanically induced. Nevertheless, a clear thermal assistance can be determined down to extremely low deformation temperatures of 0.05 – 0.10 T_m [19].

As migration of GBs or TJs is certainly a thermally facilitated process, both mechanisms will occur at higher rates if the (deformation) temperature is increased. This is also reflected in an increase of the minimum grain dimensions for SPD at elevated deformation temperatures [19,20]. Additionally, more globular grain structures should evolve for an increase of the deformation temperature, a trend also observed in classic hot working experiments [9,21]. Similar behavior was also found for SPD deformed single phase materials, although mostly low melting point metals, such as aluminum alloys, were investigated in these studies, e.g. Refs. [22,23]. Interestingly, recent investigations on single phase materials deformed by high pressure torsion (HPT) up to strains of $\varepsilon \sim 100$ at different temperatures (between 0.05 and 0.35 T_m), suggest a different but general behavior [24,25]. For both tantalum (99.95 % purity) and nickel (99.99 %), the grains became more elongated with an increase of the deformation temperature. This trend continued up to deformation temperatures of about 0.2 – 0.25 T_m, before the generally expected trend of decreasing aspect ratios with deformation temperature could be observed. Interestingly, while the trends for nickel and tantalum were qualitatively similar, the absolute values of the resulting aspect ratios differed significantly.

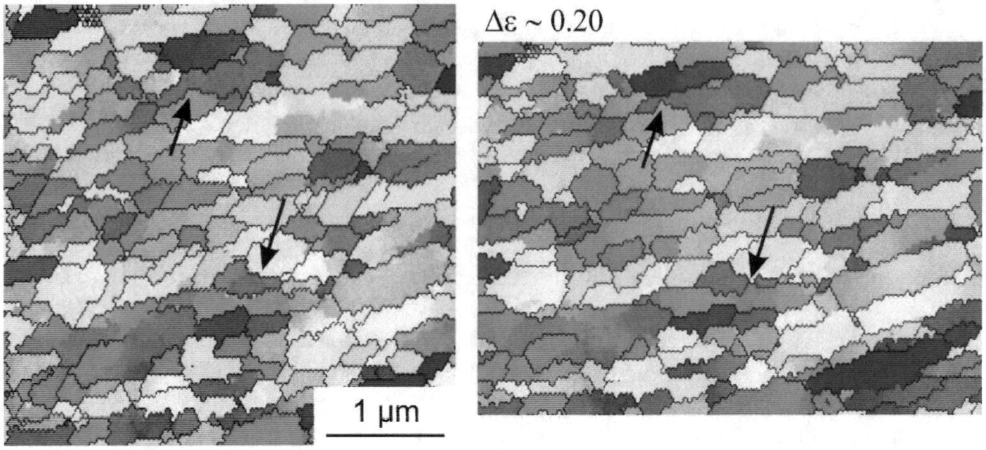

Figure 1. Microstructural evolution of ultra fine grained (UFG) nickel during an additional strain increment of $\varepsilon \sim 0.2$ by rolling at ambient temperature. Grain boundary migration is one of the main restoration mechanisms at severe strains. Boundaries do not necessarily move homogenously over the entire boundary length. In many cases bulges (an example is highlighted with arrows) form locally and effectively reduce the length of adjacent grains.

While for nickel maximum aspect ratios of about 4.5 were measured, the tantalum grains became exceptionally elongated at 673 K and reached values of almost 10, see figure 2a. Investigations on several other materials suggest that the trend of the aspect ratio as a function of the deformation temperature observed for nickel and tantalum is indeed a general one (figure 2b). Presumably this has been overlooked so far, as research has ben focused mainly on low melting point model materials for which room temperature deformation lies already well above 0.25 T_m, i.e. the temperature range where for most other pure metals maximum grain elongation was observed. It should be noted, that although the results were obtained on HPT deformed samples, similar trends are observable for other deformation routes as well, although the values of the aspect ratio may change, being higher in case of rolling compared to torsion, see figure 2b. In addition, alloying shifts the deformation temperature causing maximum grain elongation towards larger values, figure 2b.

Certainly understanding of this unexpected behavior would yield valuable insights into the underlying mechanisms. Based on evaluation of the microtextures a shift of the deformation mechanism can be excluded, and dislocation-based plasticity prevails for all investigated deformation temperatures [24]. A closer look at the various microstructures confirmed that the increase of grain elongation at intermediate deformation temperatures results from a significantly faster increase of the long grain dimensions compared to the shortest ones (i.e. in the case of the HPT disk along the axial direction) [24,25]. It is also evident that at moderately elevated deformation temperatures most of the grain boundaries appear straight, and the frequently observed fragmentation events from bulging boundary segments (figure 1) become widely absent with an increase of the deformation temperature, see figure 2a. This suggests that the formation of the bulges is an important mechanism for grain subdivision that prevents significant grain elongation at low deformation temperatures.

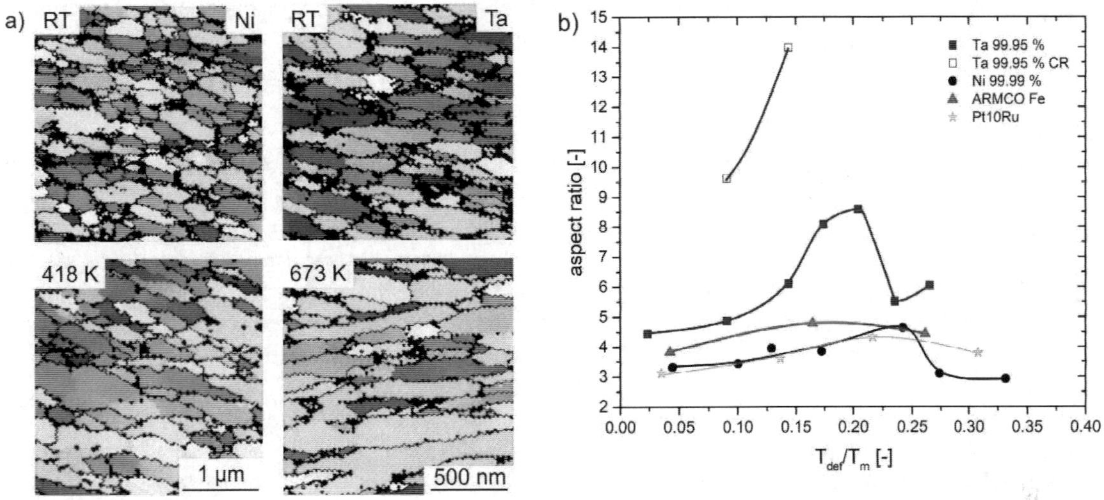

Figure 2. a) Representative microstructures of pure nickel and tantalum deformed at room temperature and the temperature (418 K and 673 K for nickel and tantalum, respectively) where maximum grain elongation was determined. b) Results obtained on a variety of materials under two deformation modes (HPT and rolling) confirm that the observed trends are indeed general ones. Image b) is taken from Ref. [26].

Formation of these bulges could be explained by connecting the presented results with in-situ TEM observations of mechanically induced grain boundary migration [27,28]. These experiments revealed that mechanically induced migration proceeds by the movement of individual GB steps, so called disconnections. These steps need not be pure steps, but can be associated with a dislocation, thus their movement will also realize a shear strain, referred to as coupled motion of a GB [27,28]. Of course their Burgers vector does not necessarily have to lie in the GB plane but could contain a climb component, necessitating elevated temperatures for their migration [28]. Such steps, with comparably low mobility, were found to hinder more mobile ones, resulting in the formation of huge macro-steps [27], having a similar appearance to the bulges observed during deformation of UFG structures especially at low deformation temperatures (figure 1). The observed increase of the aspect ratio with deformation temperature can thus be rationalized by increasing mobility of the steps within the boundaries. At elevated temperatures, the individual step mobility is increased, resulting not only in increased minimum grain dimensions (vertical axis in figure 2a), but as well in a reduced probability for macro-step formation, preventing frequent grain fragmentation, in line with the observed increase of the aspect ratio.

Interestingly, the deformation temperatures causing maximum grain aspect ratios, were also found to be in excellent agreement with those where the severely deformed structures started to coarsen during static annealing treatments [24,25]. The onset of structural coarsening in severely deformed metals was found to originate from migration of triple junctions [29]. For this reason, the reduction of the grain elongation above certain deformation temperatures (> 0.25 T_m for pure metals expected, compare figure 2) seems to be a consequence of the onset of triple junction migration, which effectively reduces the grain length.

While the description above provides a qualitative but general view on the important mechanisms that govern microstructural evolution at severe strains, the differences between the individual materials are astonishing, as seen in figure 2b. While measurable for iron or nickel, the effect is significantly more pronounced for the tantalum samples. It can thus not be explained by, or related to, the crystal structure. However, the absolute values of the measured aspect ratios are found to scale with the grain boundary energies that can be expected for the various materials [30], with larger grain aspects for metals having higher boundary energies. For pure metals the interfacial energies are also proportional to the grain boundary free volume [31,32], which in turn may explain why, for certain materials, intergranular mobility is enhanced. Nevertheless, despite being technically sound, this argument requires further research to be validated.

3. What triggers mechanically induced boundary migration?

In the previous section, we have summarized experimental findings, showing that the interplay between migration of GBs or the mobility of individual GB steps seem to determine the resulting microstructures after severe strains at moderate deformation temperatures (< 0.25 T_m). As diffusion processes are rather limited in this temperature regime, the processes are believed to be mainly mechanically triggered. While the description in the previous chapter provided a rather qualitative description of microstructural evolution and the responsible processes, quantitative predictions require a more fundamental understanding of what triggers the motion of grain boundaries. As described above, mechanically induced boundary migration proceeds by the movement of disconnections, which can couple with an applied stress field. Based on experimental evidence for shear coupling in bicrystal experiments and in-situ straining on UFG aluminium (e.g. Refs. [27,33]), most studies conclude that boundary migration is stress induced, becoming much easier in nanometals due to their enhanced yield stresses. Although technically sound, several studies also provide hints that migration is facilitated in highly strained regions, see for instance Refs. [34,35].

To provide further insights regarding this question, ultrafine-grained samples were subjected to cyclic loading at comparably large plastic strain amplitudes, as low cycle fatigue conditions are well known to provoke significant grain coarsening in nanostructured metals [36]. As severely deformed metals can be considered to deform in an almost ideally plastic manner (i.e. negligible work hardening), microstructural investigations on samples tested at different strain amplitudes, should allow for a conclusive statement regarding the effect of stress or strain on mechanically induced growth. However, in conventional fatigue experiments, tracking of the structural evolution is difficult as the position where grain growth and fatigue damage happens is hard to distinguish a priori. Due to these experimental constraints, a different testing routine was used. The nanostructured samples processed by HPT were directly subjected to cyclic torsion in the same setup [35]. After a fixed twist angle, the lower anvil is forced to change its rotation direction, mimicking LCF conditions, see the schematic in figure 3. The radial dependency of strain in torsion allows for the investigation of various strain amplitudes within one sample, while the applied hydrostatic pressure prevents sample failure and allows tracking of the microstructures up to an arbitrarily large number of cycles.

Experiments were conducted on ultra-fine grained nickel (99.99 %) at ambient temperature to avoid large thermal contributions and to probe the mechanically dominated regime. The nickel samples were subjected to a twist angle of five degrees for various numbers of cycles, and hardness and structural evolution were recorded continuously. As expected, the grains coarsened continuously with increasing number of cycles. However, within a couple of applied cycles, grain growth caused the formation of a shear band that had a wedge like shape spanning over the entire HPT disk, see figure 3. Interestingly, grain growth within the shear band was amplified compared to the regions outside the band, where the applied stresses are still high but where grain growth was only negligible. This unambiguously shows, that mechanically induced grain growth is promoted by the plastic strain. However, this finding may not be contradictive to the findings that GB motion proceeds by movement of disconnections. While nucleation of a disconnection at a GB would require rather high stress levels [37], they may be easily generated by interaction and absorption of lattice dislocations with/at GBs [28,38]. Thus it may not be surprising, that in highly strained regions, where such interactions also occur more frequently, GB migration rates are enhanced. Interestingly, the heavily coarsened grains inside the shear band region exhibited also preferred crystallographic orientations, different from the as HPT deformed (monotonic) state or outside the shear band [35], see figure 4. These orientations can accommodate cyclic slip on a single or two coplanar slip systems, while outside the shear band two non-coplanar systems are required

[39], with the latter one presumably retarding dislocation-GB interactions and so eventually the generation of disconnections.

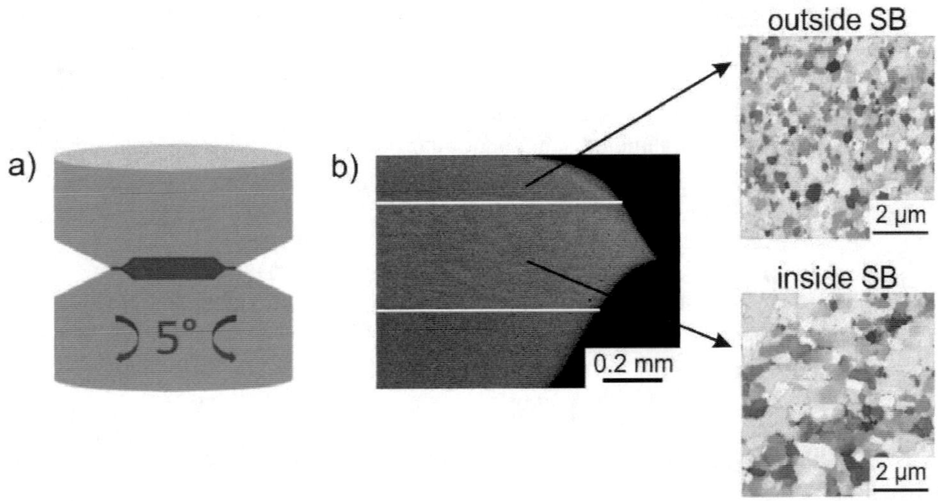

Figure 3. a) Schematics showing the experimental setup for cyclic HPT. b) Structure of ultra-fine grained nickel subjected to 50 cycles of cyclic HPT at a twist angle of five degrees at ambient temperature. A shear band developed throughout the HPT disk in which grain growth was amplified compared to the regions outside.

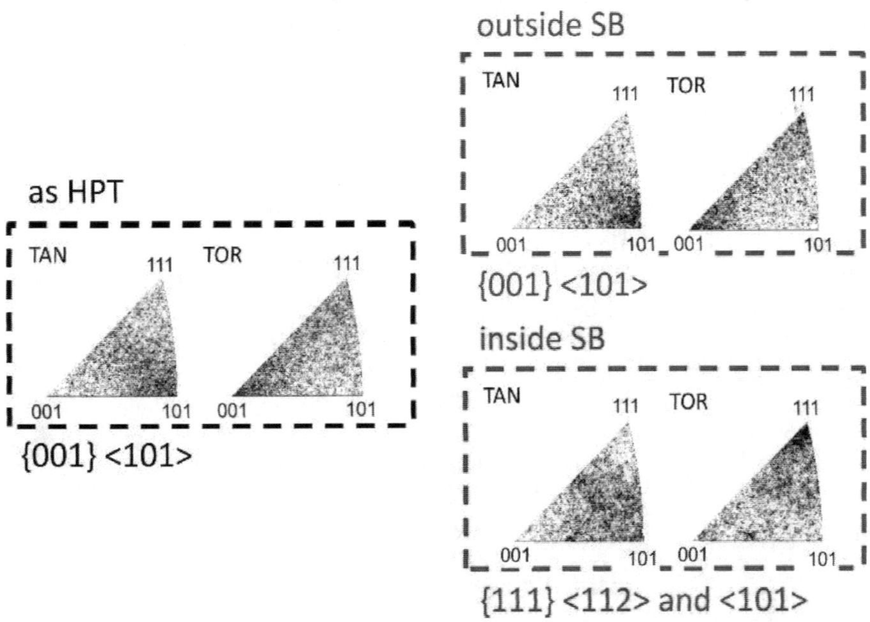

Figure 4. Inverse pole figure maps of the starting microstructure (monotonic HPT) and the grain structure outside and inside of the shear band (heavily coarsened region) plotted along the tangential (shear direction) and the axial direction (shear plane) of the HPT disk. It can be noticed that during cyclic HPT preferential crystallographic orientations develop, which can accommodate the cyclic strain on only one or two coplanar slip systems. Evaluated data is taken from Ref. [35].

4. Summary and conclusions

As a whole the experiments suggest that the main mechanisms controlling microstructural evolution at severe strains change with temperature. At low temperatures formation of macro-steps was found to be mainly responsible for grain fragmentation. Consequently, the enhanced step mobility at elevated deformation temperatures causes a significant increase of the resulting grain aspect ratios until frequent motion of triple junctions subdues this effect. Although at low deformation temperatures coupling with the applied stress field can force elementary boundary steps to migrate, experiments clearly show that boundary migration is amplified in highly strained regions, highlighting the importance of dislocation absorption at boundaries to initiate these processes.

Acknowledgements
Financial support by the European Research Council under ERC Grant Agreement No. 340185 USMS and by the Austrian Science Foundation (FWF) under project P25325-N20 is gratefully acknowledged.

References

[1] Divinski S V., Reglitz G, Golovin IS, Peterlechner M, Lapovok R, Estrin Y and Wilde G 2015 *Acta Mater.* **82** 11
[2] Huang X, Hansen N and Tsuji N 2006 *Science* **312** 249
[3] Rupert T J, Trelewicz J R and Schuh C A 2012 *J. Mater. Res.* **27** 1285
[4] Renk O, Hohenwarter A, Eder K, Kormout K S, Cairney J M and Pippan R 2015 *Scr. Mater.* **95** 27
[5] Pippan R and Hohenwarter A 2016 *Mater. Res. Lett.* **4** 127
[6] Hohenwarter A, Völker B, Kapp M W, Li Y, Goto S, Raabe D and Pippan R 2016 *Sci. Rep.* **6** 33228.
[7] Nikolić V, Wurster S, Firneis D and Pippan R 2018 *Int. J. Refract. Met. Hard. Mater.* **76** 214
[8] Estrin Y and Vinogradov A 2013 *Acta Mater.* **61** 782
[9] Hansen N 2001 *Metall. Mater. Trans. A.* **32** 2917
[10] Valiev R Z, Ivanisenko Y V., Rauch E F and Baudelet B 1996 *Acta Mater.* **44** 4705
[11] Wang Z C and Prangnell P B 2002 *Mater. Sci. Eng. A* **328** 87
[12] Hebesberger T, Stüwe H P, Vorhauer A, Wetscher F and Pippan R 2005 *Acta Mater.* **53** 393
[13] Mohamed F A 2003 *Acta Mater.* **51** 4107
[14] Pippan R, Scheriau S, Taylor A, Hafok M, Hohenwarter A and Bachmaier A 2010 *Annu. Rev. Mater. Res.* **40** 319
[15] Zhao Y H, Liao X Z, Zhu Y T, Horita Z and Langdon T G 2005 *Mater. Sci. Eng. A* **410–411** 188
[16] Edalati K, Akama D, Nishio A, Lee S, Yonenaga Y, Cubero-Sesin J M and Horita Z 2014 *Acta Mater.* **69** 68
[17] Renk O, Hohenwarter A, Wurster S and Pippan R 2014 *Acta Mater.* **77** 401
[18] Yu T, Hansen N, Huang X and Godfrey A 2014 *Mater. Res. Lett.* **2** 160
[19] Renk O and Pippan R 2018 *Scr. Mater.* **154** 212
[20] Vorhauer A and Pippan R 2008 *Metall. Mater. Trans. A* **39** 417
[21] Rollett A, Humphreys F, Rohrer G S and Hatherly M 2004 *Recrystallization and Related Annealing Phenomena*: Second Ed. (Amsterdam: Elsevier)
[22] Chen Y C, Huang Y Y, Chang C P and Kao P W 2003 *Acta Mater.* **51** 2005
[23] Wang Y Y, Sun P L, Kao P W and Chang C P 2004 *Scr. Mater.* **50** 613
[24] Renk O, Ghosh P and Pippan R 2017 *Scr. Mater.* **137** 60
[25] Renk O, Ghosh P and Pippan R. 2017 *IOP Conf. Ser.: Mater. Sci. Eng.* **219** 012037
[26] Renk O, Pippan R. 2019 accepted for publication in *Mater. Trans.* doi.org/10.2320/matertrans.MF201918
[27] Rajabzadeh A, Legros M, Combe N, Mompiou F and Molodov D A 2013 *Philos. Mag.* **93** 1299

[28] Rajabzadeh A, Mompiou F, Lartigue-Korinek S, Combe N, Legros M and Molodov D A 2014 *Acta Mater.* **77** 223
[29] Yu T, Hansen N and Huang X 2013 *Acta Mater.* **61** 6577
[30] Vitos L, Ruban A V, Skriver H L and Kollár J 1998 *Surf. Sci.* **411** 186
[31] Olmsted D L, Foiles S M and Holm E A 2009 *Acta Mater.* **57** 3694
[32] Scheiber D, Pippan R, Puschnig P and Romaner L 2016 *Model. Simul. Mater. Sci. Eng.* **24** 035013
[33] Winning M, Gottstein G and Shvindlerman L S 2001 *Acta Mater.* **49** 211
[34] Fang T H, Li W L, Tao N R and Lu K 2011 *Science* **331** 1587
[35] Kapp M W, Renk O, Leitner T, Ghosh P, Yang B and Pippan R 2017 *J. Mater. Res.* **32** 4317
[36] Mughrabi H and Höppel H W 2010 *Int. J. Fatigue* **32** 1413
[37] Rajabzadeh A, Mompiou F, Legros M and Combe N 2013 *Phys. Rev. Lett.* **110** 265507
[38] Priester L. 2013 *Grain Boundaries From Theory to Engineering* 1st ed. (Dordrecht: Springer Netherlands)
[39] Toth L S, Gilormini P and Jonas J J 1988 *Acta Metall.* **36** 3077

Analysis of in-grain orientation distributions in deformed aluminium using X-ray diffraction and finite elements

Loïc Renversade and Romain Quey
Mines Saint-Etienne, Univ Lyon, CNRS, UMR 5307 LGF, Centre SMS, F – 42023 Saint-Etienne, France

E-mail: romain.quey@mines-stetienne.fr

Abstract. The lattice orientation distributions developing inside individual grains of a polycrystalline aluminium sample subjected to plastic deformation are analysed. In the experiment, diffraction contrast tomography (DCT) and far-field 3D X-ray diffraction microscopy (3DXRD) are used to track the development of the orientation distributions of 466 grains of an aluminium polycrystal deformed in tension to successive strains of 1.0, 1.5, 2.0, 2.5 and 4.5%. A new method is presented and employed to determine the in-grain orientation distributions from broadening of the diffraction spots. The polycrystal deformation is then simulated using the crystal-plasticity finite element method, starting from the initial, DCT-determined microstructure. The experimental and numerical in-grain lattice orientation distributions are analysed and compared in terms of trends over all grains and grain by grain.

1. Introduction

In-grain lattice orientation distributions develop inside grains of polycrystals subjected to plastic deformation and are important for subsequent annealing. Historically, these orientation distributions have been analysed destructively, by 2D EBSD [1, 2]. They are now accessible non-destructively, in 3D, using synchrotron X-ray diffraction [3, 4, 5]. Numerically, they can be simulated by crystal plasticity finite element simulations [2, 6]. This work focuses on the orientation distributions developing in individual grains of an Al–0.3wt% Mn polycrystal deformed to a few percent in tension, using a synchrotron X-ray diffraction experiment and finite element simulations. The development of the orientation distributions is analysed in terms of (angular) extent and anisotropy, i.e. development of a preferential disorientation direction. In Section 2, we describe the experiment and a new post-processing method to determine orientation distributions from far-field 3DXRD diffraction (simply referred to as "3DXRD" in the following) data. In Section 3, we describe the finite element simulation. In Section 4, we compare experiment and simulation in terms of trends over all grains and grain by grain, before closing the article with conclusions in Section 5.

2. Experiment

2.1. Data acquisition

A high-purity Al–0.3wt% Mn alloy with an average grain size of about 130 μm and a relatively weak initial crystallographic texture was used. A tensile specimen with a square section of

1×1 mm was cut for the tensile test. A combination of near-field and far-field X-ray diffraction techniques were used (figure 1a). For the initial state, the polycrystal structure was mapped using DCT (figure 1b), and, in the deformed state, the grain orientations were followed by 3DXRD [3] (figures 1c,d). DCT was applied on a region 1.4 mm high, while 3DXRD was applied on a central region 0.55 mm high. In both cases, diffraction images were acquired with an integration step of 0.1° (in ω) while the sample was continuously rotated by 360°. Having a DCT region significantly larger than the 3DXRD region later made it possible to avoid any bias caused by the boundary conditions of the finite element simulation (applied to the top and bottom surfaces of the DCT polycrystal) on the rotations of the grains of the central region. 3DXRD was carried out at successive strains of 1%, 1.5%, 2%, 2.5% and 4.5%. As will be seen in the following, not only the overall shifts of the diffraction spots (which are related to the average grain rotations) but also their broadening (which is related to the development of orientation distributions) were used. The DCT polycrystal contained 1848 grains, and the region analysed by 3DXRD contained 824 grains, 466 of which were followed during the entire deformation and will be used in the following.

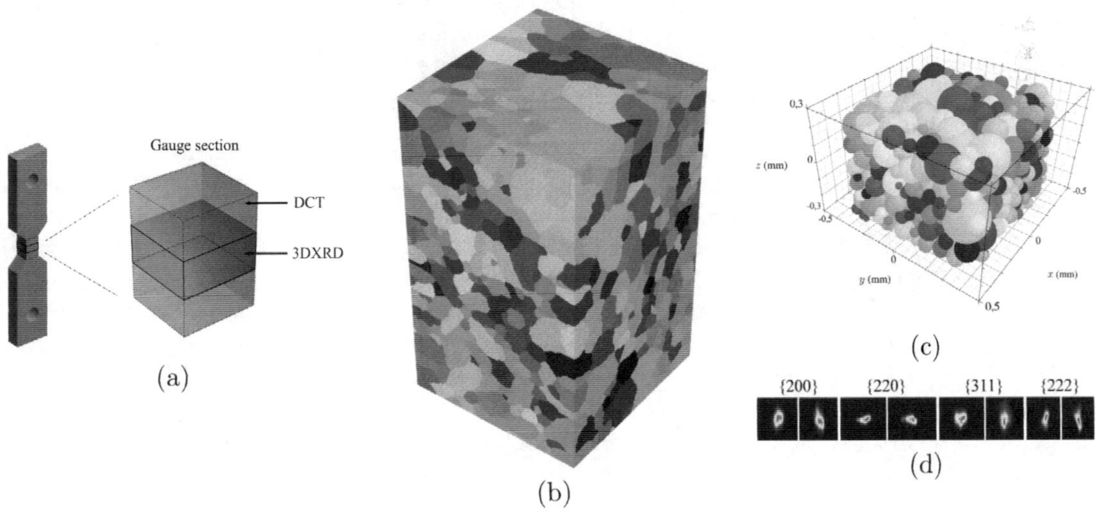

Figure 1. Principle and first results of the X-ray diffraction experiment. (a) Tensile specimen and observation regions, (b) initial microstructure obtained by DCT, (c) grains represented as spheres of equivalent volumes, obtained by 3DXRD, and (d) a series of diffraction spots of a particular grain (azimuthal projection, $\varepsilon = 4.5\%$).

2.2. Orientation distribution determination

The diffraction spots of a grain contain information on its lattice orientations (and elastic strains). As plastic deformation accumulates, diffraction spots broaden. Broadening occurs differently for different diffraction spots, but all diffraction spots result from the same orientation distribution. It should therefore be possible to determine an orientation distribution from the diffraction spots. Hansen et al. [7] proposed a direct, algebraic method to do so, but which, depending on the amount of data and unknowns, can become computationally very large and thus require the use of complex, iterative regularization methods. In this paper, we are mostly focused on the main properties of the orientation distributions, such as their angular extent and anisotropy (preferential disorientation directions). It should therefore be possible to consider a model orientation distribution, which takes these attributes into account using only a few parameters, and to determine the values of these parameters by optimization, by maximizing the similarity between the experimental diffraction spots and the diffraction spots generated from the model orientation distribution.

For a given grain, all available experimental diffraction spots (identified at the grain indexation stage) are used to determine the grain orientation distribution, which corresponds to 40–100 spots. Diffraction spots are intrinsically 3D, as they are 2D on the detector (of directions $\boldsymbol{u}$ and $\boldsymbol{v}$) but also span several diffraction images while rotating the sample (in ω). However, the diffraction spots can be projected in their corresponding azimuthal planes (η, ω) to retain only the lattice orientation information.

Simulated spots are generated from the model orientation distribution. In general, at low-to-moderate strains, most grains develop a "unimodal" orientation distribution [8, 2], which can be represented by a three-variate normal distribution,

$$P(\boldsymbol{w}) = \prod_{i=1}^{3} \frac{1}{\sqrt{2\,\pi\,\sigma_i{}^2}} \exp\left(-\frac{w_i'^2}{2\,\sigma_i{}^2}\right), \tag{1}$$

where w_i' ($i = 1, 2, 3$) are the disorientations along the principal directions of the distribution and form vector $\boldsymbol{w'}$. σ_i ($i = 1, 2, 3$) are the standard deviations along the three principal directions. The three principal directions of the distribution are related to the laboratory directions by

$$\boldsymbol{w'} = \boldsymbol{K}\,\boldsymbol{w}, \tag{2}$$

where $\boldsymbol{K}$ is a rotation matrix, and $\boldsymbol{w}$ is the disorientation vector expressed in the laboratory directions. It follows that P depends on six parameters, which are the three standard deviations ($\sigma_1 \geq \sigma_2 \geq \sigma_3$) and three parameters describing $\boldsymbol{K}$ (in practice, the three Rodrigues vector components are used). For specified values of these parameters, the orientation distribution is discretized on a grid of size $[-3\,\sigma_i, 3\sigma_i]^3$, typically using 80 bins in each direction (which corresponds to a total of 512,000 bins), and the corresponding diffraction spots are generated (in azimuthal projection) for all experimental diffraction spots.

For each grain, experimental and simulated diffraction spots are compared one-to-one in terms of the Pearson product-moment correlation coefficient, r_k. The similarity between all experimental and simulated diffraction spots is then simply measured by

$$O = \frac{1}{n}\sum_{k=1}^{n} r_k, \tag{3}$$

where n is the number of diffraction spots. To determine the parameters of the model orientation distribution ($\sigma_{1\text{–}3}$ and $\boldsymbol{K}$), O is then maximized by gradient-free optimization, using NLopt's subplex algorithm [9, 10]. In average over all grains, optimization resulted in a value of O of 0.84, which confirms that orientation distributions can be well-approximated by three-variate normal distributions and validates the whole methodology.

In the following, orientation distributions will be described and analysed in terms of two metrics. The first metric is the angular extent, θ_d, which corresponds to the average disorientation angle with respect to the average orientation. θ_d is not among the orientation distribution parameters used previously, but we consider an expression of the form

$$\theta_d = \frac{2}{3}\sqrt{\frac{8}{\pi}}\,(\sigma_1 + \sigma_2 + \sigma_3), \tag{4}$$

which is equal to the expression proposed in Ref. [8] for an isotropic distribution ($\sigma_1 = \sigma_2 = \sigma_3$). The second metric is the preferential disorientation direction, $\boldsymbol{v_1}$, which is provided by the first row of $\boldsymbol{K}$ (the vector associated to σ_1).

IOP Conf. Series: Materials Science and Engineering **580** (2019) 012022 doi:10.1088/1757-899X/580/1/012022

3. Simulation

The initial, DCT microstructure was meshed following the methodology detailed in Refs. [11, 12] and implemented in the free (open-source) Neper software package, see figure 2a. Grains were finely discretized into tetrahedral elements, with an average mesh density of 600 elements per grain. The crystal plasticity finite element simulation of the polycrystal deformation was then carried out using the FEpX software package [13, 14]. Plastic deformation is accommodated by slip on the 12 $\{1\,1\,1\}\langle 1\,1\,0\rangle$ slip systems. On each system (α), the slip rate, $\dot{\gamma}^\alpha$, is related to the resolved shear stress, τ^α, by

$$\dot{\gamma}^\alpha = \dot{\gamma}_0 \left| \frac{\tau^\alpha}{g^\alpha} \right|^{\frac{1}{m}} \mathrm{sgn}\,(\tau^\alpha), \qquad \dot{g}^\alpha = h_0 \left(\frac{g_s - g^\alpha}{g_s - g_0} \right)^{n'} \sum_\alpha |\dot{\gamma}^\alpha|. \tag{5}$$

The material parameters were identified from the experimental stress-strain behaviour, and their values are $\dot{\gamma}_0 = 1\ \mathrm{s}^{-1}$ (a convention), $m = 0.03$, $h_0 = 47$ MPa, $g_0 = 6$ MPa, $g_s = 455$ MPa and $n' = 2.6$. Figures 2b,c provide the rotation angles at $\varepsilon = 4.5\%$ (with respect to the initial orientation).

As in the experiment, the orientation distributions are analysed in terms of average extent (θ_d) and preferential disorientation direction ($\boldsymbol{v_1}$). For each grain, θ_d is readily computed from the grain elemental orientations, while the preferential disorientation direction ($\boldsymbol{v_1}$) is characterized by an eigendecomposition of the set of disorientations (represented as Rodrigues vectors) into eigenvectors ($\boldsymbol{v_i}$, $i = 1, 2, 3$) and eigenvalues ($\lambda_i = {\sigma_i}^2$, $\lambda_1 \geq \lambda_2 \geq \lambda_3$) [8]. Figure 2d shows the disorientation distribution corresponding to the grain of figure 2c, represented as Rodrigues vectors.

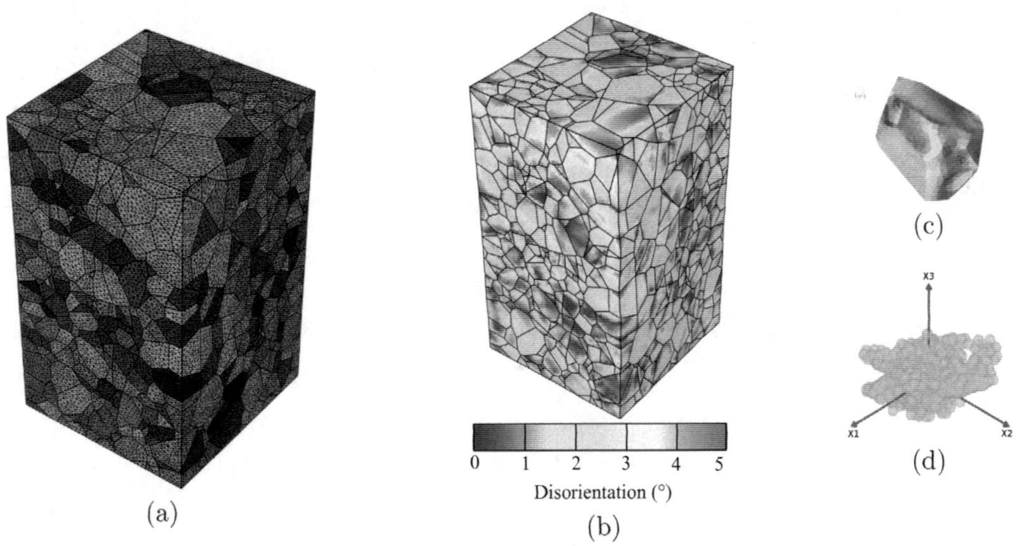

Figure 2. Principle and first results of the finite element simulation. (a) mesh, (b) lattice rotation angle at 4.5% (with respect to the initial orientation), (c) lattice rotation angle at 4.5% for a particular grain, and (d) corresponding disorientation distribution in Rodrigues space. Note the distribution anisotropy.

4. Results and discussion

The distributions over all grains of the average disorientation angles are represented in figure 3. For both experiment and simulation, average disorientation angles increase gradually during deformation, but also exhibit a high degree of variability, ranging, at $\varepsilon = 4.5\%$, in the experiment from 0.2° to 3.2° and in the simulation from 0.4° to 3.1°. Average values are, in the experiment, 0.18°, 0.24°, 0.31°, 0.34° and 0.61°, and, in the simulation, 0.17°, 0.27°, 0.37°, 0.46° and 0.77°. Figure 3c provides the correlation between the experimental and simulated average disorientation angles (θ_d), which show only moderate agreement.

The distributions over all grains of the preferential disorientation directions at successive strains are provided in figure 4 as equal-area projection onto the sample X–Y plane (Z = tensile direction), for both experiment and simulation. In the experiment, the preferential disorientation directions are first aligned with the tensile direction and then become perpendicular to it, mostly about direction X (figure 4a). In the simulation, the preferential disorientation directions are aligned with a direction close to X at all strains. (On figures 4a,b, the lack of uniaxial symmetry can simply be explained by the initial material texture.) A distribution of the preferential disorientation directions perpendicular to the tensile directions is consistent with an earlier study, made in hot plane strain compression, in which it was concluded that preferential disorientation directions align with the spin vectors of the most active slip systems [2]. In tension, most favorable slip systems have a spin vector oriented 90° away from Z, and so in the X–Y plane. Figure 4c provides the correlation between the experimental and simulated preferential disorientation directions at $\varepsilon = 4.5\%$, as the scalar product between the two. With an average value of 0.75, experimental and simulated preferential disorientation directions show

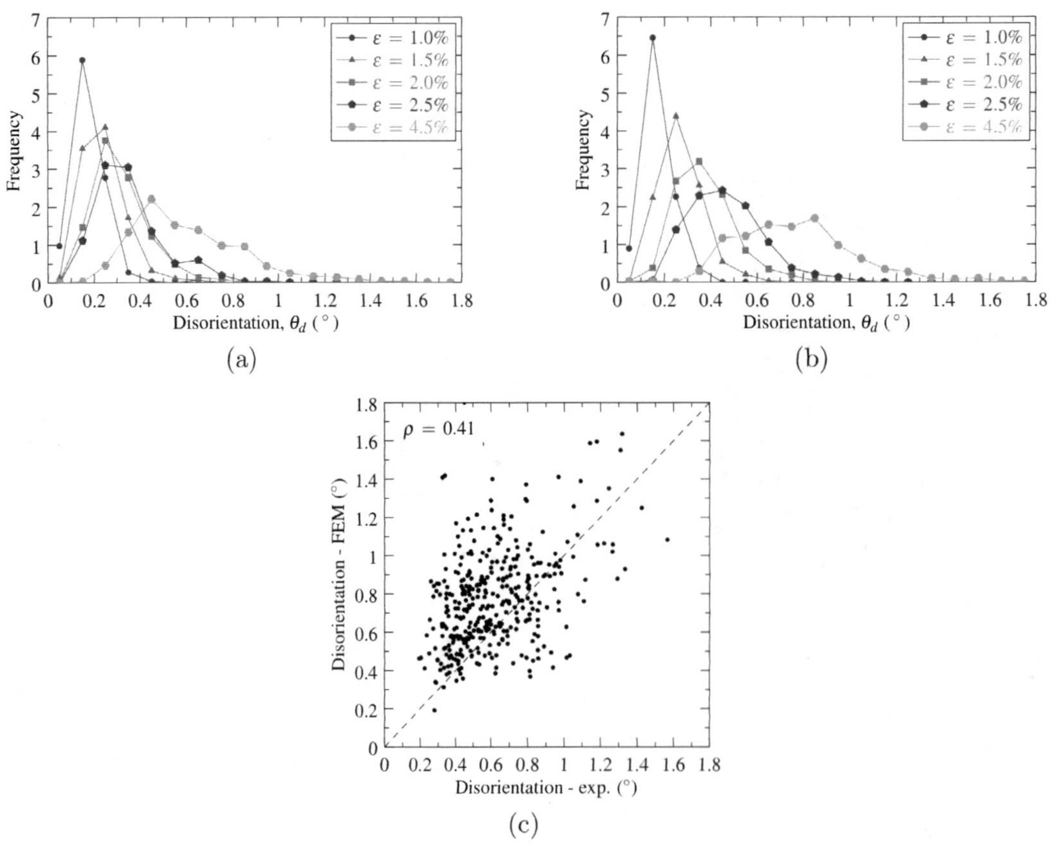

Figure 3. Distributions over all grains of the average disorientations, θ_d. (a) experiment, (b) simulation and (c) grain-by-grain comparison between experiment and simulation at $\varepsilon = 4.5\%$. ρ is the linear correlation coefficient.

an appreciable agreement.

A few more comments can be made about the comparison between experiment and simulation. In particular, it is seen that the grain-by-grain agreement on the average disorientation angles is relatively low (figure 3c) and that the distributions of the preferential disorientation directions at $\varepsilon = 1\%$ are different. This cannot be explained by experimental biases, such as a typical sample realignment toward the tensile direction at the beginning of deformation. Indeed, the average disorientation angles, as scalars, would not be affected, nor would be the in-grain preferential disorientation directions, as instantaneous, intragranular properties. (In contrast, sample realignment would affect the rotation directions with respect to the initial orientations.) Differences may actually be attributed to uncertainties on the crystal behaviour used in the simulation, whose parameters have been identified from the macroscopic behaviour. As for the distributions of the preferential disorientation directions at $\varepsilon = 1\%$, the differences may also be explained by the tendency of crystal-plasticity simulations to evolve microstructures too rapidly, as was also observed in [2].

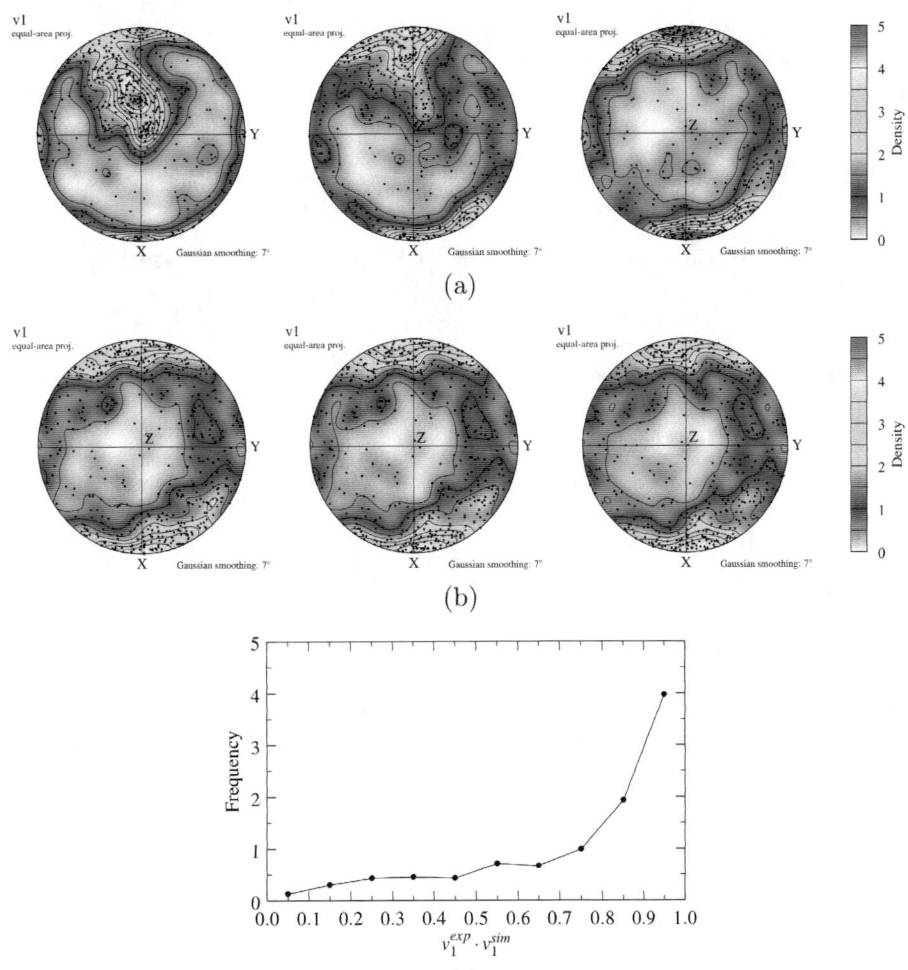

Figure 4. Distributions over all grains of the preferential disorientation directions, $\boldsymbol{v_1}$. (a) experiment and (b) simulation at strains of 1%, 2% and 4.5% (left to right). Z represents the tensile direction. (c) Distribution of the correlations between the experimental and simulated preferential disorientation directions ($\boldsymbol{v_1}^{\text{exp}}$ and $\boldsymbol{v_1}^{\text{sim}}$, respectively) at 4.5%.

5. Conclusions

The orientation distributions developed within individual grains of an aluminium polycrystal were experimentally characterized using a new far-field 3DXRD based method and numerically simulated using the finite element method. Results were obtained on a statistically-representative set of 466 grains. The new 3DXRD based method was particularly efficient in determining the orientation distribution from several tens of diffraction spots represented in azimuthal projection. The results, analysed in terms of angular extent (average disorientation angle, θ_d) and preferential disorientation direction ($\boldsymbol{v_1}$), were analysed. A high degree of variability among grains was found, which was well reproduced by the simulation. Trends over all grains were in agreement with previous observations made in hot plane strain compression [2]. In particular, preferential disorientation directions are found to align with the spin vectors of the typically most active slip systems. On a grain-by-grain basis, experiment and simulation were found in appreciable agreement on the preferential disorientation directions ($\boldsymbol{v_1}$).

6. Acknowledgments

W. Ludwig, J. Wright and A. Borbély are acknowledged for help during the experiment.

References

[1] Humphreys F and Bate P 2007 *Acta Mater.* **55** 5630–45
[2] Quey R, Driver J and Dawson P 2015 *J. Mech. Phys. Solids* **84** 506–27
[3] Poulsen H, Margulies L, Schmidt S and Winther G 2003 *Acta Mater.* **51** 3821–30
[4] Pokharel R, Lind J, Kanjarla A, Lebensohn R, Fai Li S, Kenesei P, Suter RM and Rollett A 2014 *Annu. Rev. Condens. Matter Phys* **5** 317–46
[5] Viganò N, Tanguy A, Hallais S, Dimanov A, Bornert M, Batenburg K and Ludwig W 2016 *Sci. Rep.* **6**
[6] Quey R, Dawson P and Driver J 2012 *J. Mech. Phys. Solids* **60** 509–24
[7] Hansen C, Sørensen H, Sükösd Z and Poulsen H 2009 *SIAM J. Imaging Sci.* **2** 593–613
[8] Glez J C and Driver J 2003 *Acta Mater.* **51** 2989–3003
[9] Johnson S *http://ab-initio.mit.edu/wiki/index.php/NLopt*
[10] Rowan T 1990 *Functional Stability Analysis of Numerical Algorithms* Ph.D. thesis
[11] Quey R and Renversade L 2018 *Comp. Methods Appl. Mech. Eng.* **330** 308–33
[12] Quey R, Dawson P and Barbe F 2011 *Comput. Methods Appl. Mech. Eng.* **200** 1729–45
[13] Marin E and Dawson P 1998 *Comput. Methods Appl. Mech. Eng.* **165** 1–21
[14] Dawson P and Boyce D 2015 *ArXiv* arXiv:1504.03296 [cond–mat.mtrl–sci]

Comparison of the Elastic-Plastic Transition in Nanocrystalline Nickel Pillars Fabricated by Electron Beam Lithography and Focused Ion Beam Methods

Matthew Sedore[1], Bradley J. Diak[1]*, Andrew Murray[2], Bruno Riel[2]

1. Mechanical and Materials Engineering, Queen's University, Kingston, Ontario, Canada, K7L3N6
2. Fibics Incorporated, 1431 Merivale Road, Ottawa, Ontario, K2E 0B9, Canada

*E-mail: diak@queensu.ca

Abstract. The elastic-plastic transition or yield behaviour was observed during compression of nanocrystalline nickel pillars fabricated by electroplating into electron beam lithography (EBL) forms or by focused ion beam (FIB) milling of the same electroplated structure. The experimental methodology allowed structures with different surface defect source/obstacles and similar bulk defects to be compared. Pillars with 1 to 2 um diameter cross-sections were made with height to diameter aspect ratios from 2:1 to 6:1, grain size of 91 ± 23 nm, and <110>, <111>, <100> textures in the growth direction. The pillars were compressed under a flat punch at constant loading rates from 1.5 to 30 μN/s using an instrumented indenter. EBL pillars generally showed a gradual elastic-plastic transition similar to yielding in bulk nickel followed by strain bursts. In contrast, FIB-milled pillars consistently showed stochastic yielding. The Young's modulus was 55 to 350 GPa for all pillars, agreeing with values in the literature. The yield stress at 0.2% offset strain for EBL pillars ranged from 480 to 1800 MPa, and 320 to 700 MPa for FIB-milled pillars that was narrowed to 550 to 650 MPa after subsequent annealing, with initial deformation localized near the tops of FIB-milled pillars due to geometrical tapering. EBL pillar shear yield strengths exceed the literature reported for Ni and appear to lose their size dependence.

1. Introduction

The size-effects and deformation response of micro-/nano-whiskers have been studied since the 1950-60s [1,2], but over the past 15 years the cross-over of techniques from the semi-conductor industry, i.e. electron beam lithography and focused ion beam (FIB), have led to a renewed focus on fundamental understanding of strengthening in micro- and nano-specimens [3–6]. Multiple studies have investigated size effects on yield strength [6,7] and ductility [4,8] as examples, with many of these studies illustrating a stochastic stress-strain response. The leading theory to explain the increased strengthening, "dislocation exhaustion and starvation", was presented by Greer and Nix [5] with much of the framework established by Uchic et al. [3,9]. Plastic deformation is facilitated by nucleating truncated sources within the core of larger specimens (> 1 μm) or surface nucleation events for smaller specimens (< 200 nm) [10]. The molecular simulation by Volkert and Lilleodden [11] supports the hypothesis of dislocation starvation, but the authors also remark that the fabrication methodology is responsible for differences in mechanical behaviour. Volkert and Lilleodden conclude that deformation is controlled by dislocation source activation where sources available decrease in micro/nano-specimens compared to bulk material. Potential microstructural differences between pillars such as impurities, surface damage and porosity act as sites for dislocation nucleation, affecting the yield and deformation behavior which should skew comparisons of the same material.

Both FIB-based and FIB-less techniques have been developed to fabricate microscopic specimens. FIB-milling methods were developed by Uchic et al. [3,9] and have been used to isolate regions of interest within a bulk material, but questions remain about the effect of damage. Kiener et al. [12] describe FIB damage as an artifact that is not easily removed, while others [13] claim that post-mill annealing of pillars can help restore the microstructure to a less-damaged state through recovery of direct ion-induced damage and implantation. Hofmann et al. [14] found that any FIB-imaging or milling causes large lattice distortions, even in very low ion doses, and that the microstructures is dominated with vacancies at low doses and extended dislocation networks that impose stresses and lattice strains far into the bulk material at larger ion doses. FIB-milled pillars also contain some degree of tapering from the annular milling process, resulting in an unequal distribution of stress concentrated at the smallest cross section at the top of pillars during loading. This specific geometrical artifact makes it difficult to separate end effects from non-perpendicular tip-pillar alignments. With all of the challenges using FIB, Burek and Greer [15] devised a FIB-less technique that uses electron beam lithography and electroplating to produce pillars without the disadvantage of FIB damage. Jennings et al. [16] observed no mechanical behaviour differences between grown electroplated copper pillars and pillars FIB-milled from bulk copper. The authors did not observe any size effects in mechanical properties, even though copper is well known for its ion milling anisotropy and therefore suffers inconsistent damage from varying milling speeds [12]. They suggest that the microstructural differences, i.e. zero to some dislocation density, would manifest the size effect, instead of intrinsic differences between bulk and electroplated copper. For reviews of pillar size effects see Greer and De Hosson [17] and the insightful last chapter of Kubin's book [18].

Recently, Papanikolaou et al. [19] presented a discrete dislocation plasticity model based upon source and obstacle strengths that starts to capture the range of yielding behaviours observed in the pillar literature. Particularly, the role of surface obstacles on surface sources seems to drive the differences in the size effects of yielding, where differences can be seen at ~1 μm. Therefore, it seems likely that pillars fabricated by electroplating within patterned substrates would be strengthened differently than the most carefully FIB-milled ones, yet the literature remains tentative in making this conclusion.

The present work compares the elastic-plastic transition behaviour of pristine, electrolytically-grown polycrystalline nickel pillars within a lithography-prepared template to pillars fabricated by FIB-milling, from the same electroplated material in order to evaluate the tendency for stochastic yielding. Nickel has a well understood slip geometry and dislocation behaviour, and well established electroplating recipes for producing fully dense, high purity material [20]. Furthermore, nickel has a high melting point of 1728 K such that at ambient temperatures, self-diffusion is limited during mechanical testing; is relatively strain rate insensitive [21]; and the surface is resistant to oxidation. The experiment here aims to fix the bulk defect level and modify the surface defects to develop a qualitative comparison to the predictions of Papanikolaou et al. [19].

2. Experimental Procedure

2.1 Pillar preparation

Three types of pillars were prepared: (1) EBL, (2) EBL + FIB-milled, and (3) FIB-milled. The targeted characteristic diameter or width was ~1 μm with height to diameter aspect ratio of 2:1 to 6:1, to avoid end effects and buckling, respectively [15]. All nickel pillars were derived from nickel grown on annealed, mechanically polished copper substrates. The substrates were rolled, annealed, mechanically ground, and then chemically polished using Mitchell's solution [22] to a smooth, flat substrate with a weak crystallographic texture and average grain size of 15-20 μm. Copper was selected to promote electrical conductivity during the electroplating process, but required a sputtered gold seed layer to promote nucleation. All pillar source material was plated together to ensure the same starting chemistry and similar microstructures. The procedural recipe for fabricating the pillars was developed by Sedore

[23]. A base nickel film of approximately 30 μm was first electroplated using a Watts bath electrolytic solution at 55°C, 525 mA/cm^2, and a flow rate recommended by International Micro Industries for their electroplater. The starting nickel film was grown to remove the effect of the softer copper substrate on the mechanical response. EBL pillars or a continuous film were grown on the nickel film after depositing another gold seed layer. A plasma clean was required on the EBL template to remove residue polymer before growing the EBL pillar.

EBL + FIB-milled pillars were obtained by shaping down larger circular cross-section EBL pillars to square ones with height to width ratios of 5:1 to 2:1 using a Micrion 2500 FIB with a 50 keV gallium source operating at ~100 pA. Additionally, 32 round pillars were milled directly from the substrate using a Zeiss Crossbeam 540 FIB-SEM operating at 30 keV with a gallium source, with one half milled to the gold seed layer at the pillar base and the other half over-milled past the substrate so that the gold layer was further up the height of the finished pillar creating bi-polycrystals. Rough milling was done at 7 nA, and shaping and polishing completed at 100 pA. All pillars were milled to approximately 1.1 to 1.4 μm in diameter and 3.95 to 5.5 μm tall with a taper of less than 5°. Approximately half of the EBL+FIB pillars with the gold near the base were mechanically compressed. The remainder were annealed for 1 h at 450°C in an argon filled tube, prior to mechanical testing to qualitatively evaluate its effect on FIB damage. The annealing kinetics were not investigated in this work.

2.2 Pillar characterization and mechanical testing
Pillars were sorted visually by SEM into three classes: good, acceptable, and unacceptable. All pillars start as virtual ideal objects with straight sidewalls, flat tops, and no shoulder at the base, which become less ideal after moving through the fabrication phases: tapers less than 5°, spherical caps smaller than 150% of the diameter, and target aspect ratios are deemed "good"; "acceptable" have tapers between 5 and 15°, and cap diameters from 150 to 300%. "Acceptable" pillars were used to calibrate the focal length and the cross hair alignment for the mechanical testing probe, as well as to test the stability of the system before testing good pillars.

Figure 1. From left to right, SEM secondary electron images of example EBL, EBL+FIB, FIB pillars. 10 kV, 45° tilt.

Mechanical testing was performed with a MicroMaterials Test Vantage nano-indenter equipped with a boron nitride 10 μm flat punch. The load controlled system has an estimated noise band of 0.05 mN. Thermal drift was kept below 0.5 nm/s as much as possible, but was not measured during compression tests in order to terminate the tests without unwanted deformation. Mechanical testing was performed at ambient temperatures of 25±0.1°C during summer months and 27±0.1°C during winter months to minimize temperature changes from cooling to heating of the room containing the apparatus to minimize thermal drift. All pillars were compressed at constant loading rates of 1.5 μN/s to 30 μN/s, and unloaded after different amounts of plastic reduction to observe pillar features from yielding events. Yield stress was defined at 0.2% offset strain.

3. Results and Discussion

3.1 Grain size, hardness and elasticity
The grain size was estimated to be 91 ± 23 nm from FIB secondary electron images of EBL cross-sections, and contained some finer growth twins. The method cannot discern similar orientations, so it is expected that the grain size is actually less than observed. Standard micro-hardness and nano-

indentation tests of the film using Vickers and Berkovich tips, respectively, measure an average hardness of 1.9 to 2.5 GPa. The grain sizes and hardness values agree well with the findings for electrodeposited nickel from Hughes et al. [24]. EBL pillars compressed in this study ranged in height from 2.1 to 5.5 µm with diameters of 0.7 to 1.8 µm. Similarly, FIB-milled pillars were 1.9 to 5.5 µm tall with diameters of 1 to 1.4 µm. The Young's modulius for the nickel substrate and pillars were determined from mechanical unloading at the end of indentation and compressions tests to be 55 to 350 GPa and 180 to 210 GPa, respectively, agreeing with the literature [7,8]. Using the elastic constants from Simmons and Wang [25], the Young's modulus for <100> to <111> nickel range from 129 to 350 GPa, respectively.

3.2 Elastic-Plastic Transition

Mechanical testing of EBL pillars resulted in a variety of different yield and deformation behaviours as shown in Figure 2. In cataloging these behaviours, it was important to identify reliable versus unreliable tests. Three types of unreliable tests include: (1) negative or oscillating displacement during loading caused by thermal drift; (2) rapid elastic loading above 10 mN in which the indenter tip completely pancakes the pillar and the tip contacts the substrate before any data is acquired; and (3) any test that shears the pillar off at the substrate, or buckles the pillar prematurely due to tip-pillar misalignment. All data from unreliable tests were omitted in the final comparison.

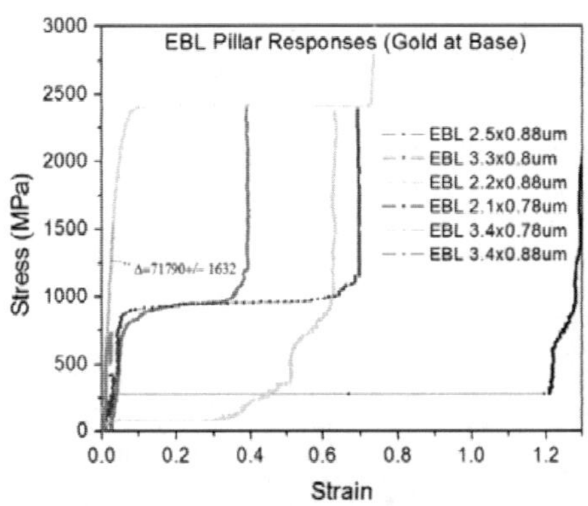

Figure 2. Engineering stress-strain curves of EBL grown pillars with gold at pillar-substrate interface, compressed at 15 µN/s. An elastic loading slope is shown.

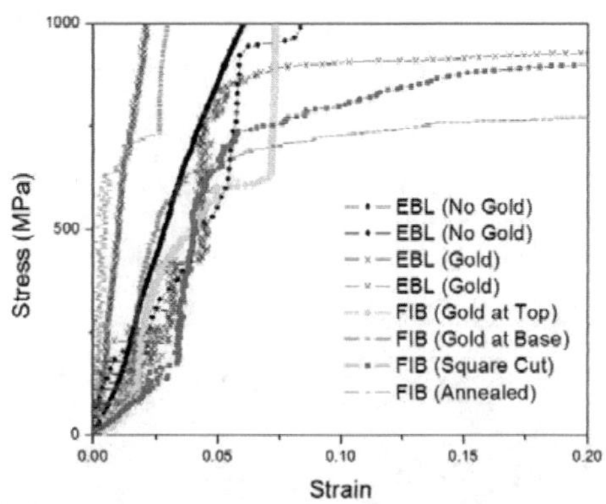

Figure 3. Engineering stress-strain curves highlighting the difference in yield point between EBL pillars and different FIB pillars at 15 µN/s.

The elastic-plastic response of EBL pillars gives a characteristically stochastic stress-strain curve during the initial stages of loading, followed by a linear elastic region gradually transitioning to plastic yielding, flow and ultimately failure. The initial stochastic behaviour is attributed to the flattening of a cap or smaller asperities on the top of the pillar that disappeared once the top became truly flat relative to the compression platen. The stress-strain behaviour after flattening was more typical of FCC bulk materials during load controlled compression tests in that there is a linear elastic region transitioning to plastic deformation through a region of continuous yielding in which barrelling of the specimens is commonly observed [26]. Disregarding the earliest yield events from suspected flattening of the pillar tops, the observed yield points of EBL pillars ranged widely from 780 to 1870 MPa using a 0.2% strain offset from the linear elastic region. Loading at different rates from 1.5 µN/s to 30 µN/s did not affect the observed yield behaviour as observed in bulk nickel [21]. It is presumed that EBL grown pillars contain few source dislocations; only one or two dislocations within the pillar should be expected, even when considering materials with high dislocation densities of $10^{14}/m^3$ in 2.5 μm^3 of pillar. Gold-less EBL pillars also exhibited the same deformation behaviour, but since the pillars had flat tops, no stochastic behaviour was present at low loads. This was because the cross-sectional area was almost identical throughout the entire pillar.

In contrast EBL + FIB pillars were observed to have lower yield strengths and deformed differently to the as-grown EBL pillars as shown by the stress-strain curves in Figure 3. Yield strengths of square pillars appeared lower at 320 to 625 MPa, but due to tapering and stress intensification at the square edges. The stress-strain curves show elastic loading followed by a distinct yield event in the form of a strain burst, followed by a series of elastic loading and inelastic strain burst events until catastrophic failure or the pillar is fully flattened on the substrate. Lee et al. [27] explain that strain bursts in FIB-milled objects are due to the nucleation of new or activation of grown-in dislocation sources that can occur at lower stresses in the presence of surface roughness or damaged volume from the FIB. A resulting back-stress halts further plastic deformation until the applied stress increases sufficiently for the dislocations to move, as detected by a sudden large displacement that stops when the dislocation source is exhausted. SEM images of plastically deformed pillars show that deformation initiates at the tops of pillars where the platen contacts the smallest cross sectional area and progresses downward as the pillar expands outward from Poisson ratio effects. This plastic-elastic interface moves Lüders-like down the pillar to accommodate increase in applied stress. In a few cases catastrophic failure occurred at the pillar base near the gold seed layer, but it was unclear if the failure was within the gold seed layer or at the nickel-gold interface.

Circular cross-section pillars FIB-milled from the electroplated nickel film behaved similarly to the square pillars. Initial elastic-plastic transitions varied from 630 to 700 MPa and prefaced stochastic bursts of plastic deformation. The circular cross-sectional area reduces macro stress concentrations at edges but the inevitable taper from annular milling still directs the elastic-plastic transition to initiate at the smallest cross-sectional area at the top of the pillar. Unloading such tests after the first few detectable yield events revealed pillars with bulges near the top of the pillar with detectable sample-scale shear bands on the outer surface at 45° to the loading axis (Figure 4), with each band likely corresponding to a strain burst in the respective stress-strain curve. It is unlikely that shearing occurred due to non-perpendicular tip-platen alignment. In addition to shearing, some pillars grown on the gold seed layer catastrophically failed at the pillar substrate interface at normal stresses above 1000 MPa. This effect was not observed in pillars grown without the gold seed layer, and will be discussed later.

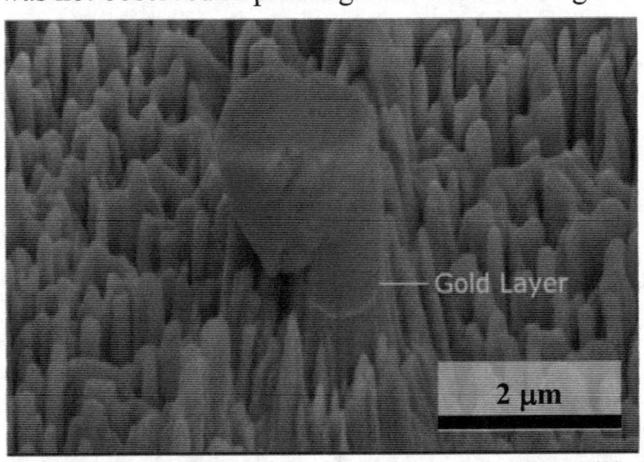

Figure 4. Secondary electron image of a FIB-milled nickel pillar with gold near base post deformation, illustrating shearing of the pillar from the top downward. 10 kV, 45° tilt.

Some FIB-milled pillars were over-etched below the gold layer, such that the second nickel growth step was near the top third of the pillar, and yielded between 350 and 500 MPa. Post-deformation imaging revealed localized yielding above the gold layer, similar to that of the standard FIB-milled pillars, but no discernible plastic deformation below the gold layer as seen in Figures 5A and B, until strains above 0.1 are reached. Discontinuity in the nickel grain structure and texture across the gold layer due to the stopping and restarting of the plating in-between gold deposition could affect the transition in plastic behaviour above and below the gold layer. According to Maurer et al. [28], {111} oriented nickel crystals are more likely to nucleate on flat surfaces while {110} crystals are more likely to nucleate when deposition sites exceed 1:1 height to diameter aspect ratios. Additionally, {100} grains were observed in high aspect ratio deposition sites. Moreover, growth of {110} and {100} oriented grains dominate over {111} growth which leads to columnar grains oriented along {110} and {100} as observed in the present nickel film. It would therefore seem likely that the nickel directly below the gold layer and the very top of the pillar would contain {110} and {100} oriented grains from growth dominance whereas the nickel directly above the gold layer would consist of more {111} oriented grains. The Schmid factor for an ideally oriented {111) grain is 0.272 versus 0.408 for {110} or {100} oriented grains. Ignoring grain shape and size effects, it is expected that the {100} and {110} grains near the pillar tip yield first as they have the higher Schmid factors, while any {111} grains just above the gold require a higher applied stress to yield. This behaviour was indirectly identified by immediately unloading after the yield event and observing a bulge at the pillar top ending just above the gold layer. The nickel above the gold uniformly spread outward instead of shearing, even at large displacements. This constraining effect indicates the nickel with an aspect ratio near 1.5:1 (height to diameter) above the gold is deforming non-uniformly from the bottom portion. The frictional forces that arise with such small aspect ratios can alter the deformation response of the pillar, restricting any horizontal sliding and in turn reduce the potential for shearing. The gold-nickel interface may also nucleate dislocations more easily to accommodate plastic strain, delaying yield in the lower part of the pillar. Finally, the gold seed layer has a higher atomic number than nickel and may enhance the back-scattered ion damage in the nickel above, and shield the ion penetration and damage to the underlying nickel. The complexities of the types of defects present from such interactions warrant further investigation.

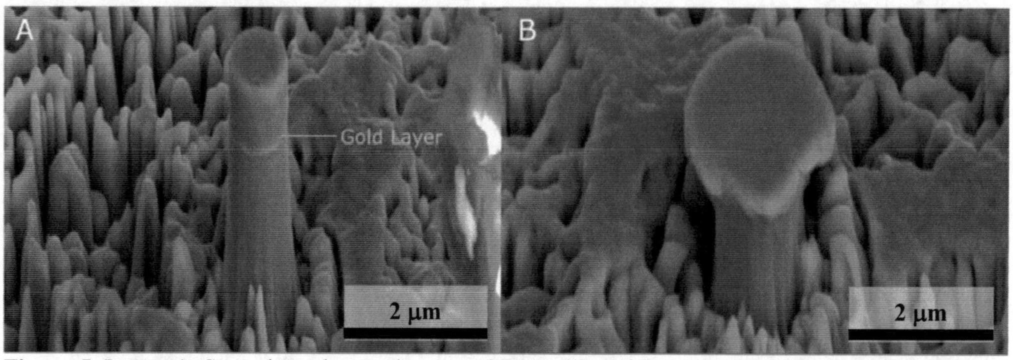

Figure 5 (bottom). Secondary electron images of FIB-milled nickel pillars with gold near the top. A: Compression stopped at yielding onset. B: Compressed to 40% strain. 10 kV, 45° tilt for both images.

Finally, FIB-milled pillars with gold at the column base were annealed at 450°C for 1 h in a sealed tube back-filled with argon. Mechanical testing of annealed pillars resulted in similar yield strengths as FIB-milled pillars. Yield strengths were calculated to be 550 to 650 MPa when considering the additional 20 to 40 nm oxide layer which formed, a decrease of 50 MPa from as-milled pillars of the same group. The yield strength was also calculated using pre-annealed measurements of cross sections and results in 580 to 700 MPa, which is similar to as-milled pillars. Young's modulus for the oxidized pillars was be 55 to 145 GPa which is similar to as-milled samples of the same group. Annealed pillars exhibited anisotropic bulging of the top during the initial stages of plastic deformation as seen in Figure 6A, but it remains unclear if the bulge located at the top is caused by end-effects, tapering or FIB damage. The oxide layer also does not appear to influence the plastic deformation behavior of the pillar. Remnants of a spalled oxide layer can be seen in Figure 6B. The oxide layer may, however, contribute to end effects and reduce pillar expansion due to Poisson effect. More importantly, the stochastic stress-strain behaviour seen in FIB pillars completely vanishes after annealing. Removing weak obstacles near the surface tends to allow more continuous source operation with more gradual yielding.

The quantifiable test for the size effect is in the yield stress. Figure 7 compares the shear stress at yield measured in this work versus pillar diameter for [111] and [269] nickel single crystals from Frick et al. [7] and Dimiduk et al. [8], respectively. The shear yield strengths of all pillars in the present work were determined from the axial yield stress after correction by the Taylor factor of 2.8 derived from average x-ray texture data. EBL pillars consistently had higher yield strengths compared to FIB-milled pillars, but these values are still less than the theoretical strength estimated to be 6.4 GPa to 11.6 GPa. Considering the Hall-Petch constants derived from Feaugas and Haddou [29], 1000 nm diameter to ~100

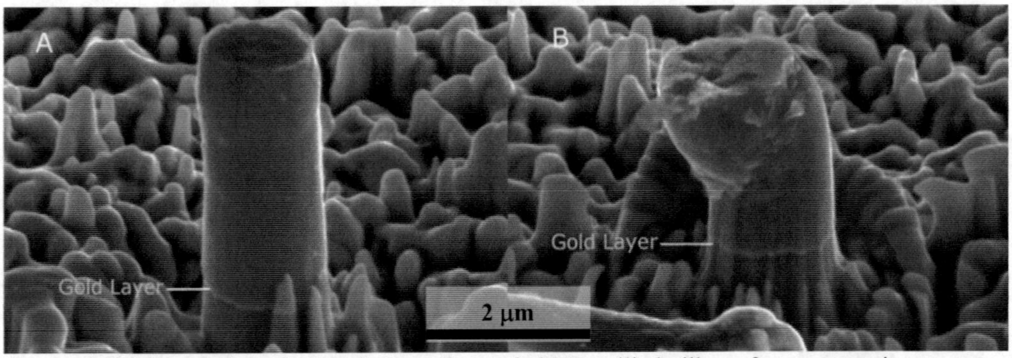

Figure 6. SEM secondary electron image of annealed FIB-milled pillars after compression tests at a loading rate of 15μN/s. Pillar A: Compressed until first elastic-plastic transition. Pillar B: Compressed until 35% strain. All images tilted 60°.

nm grain size gives a strengthening of 2.89 times, which can explain the average shift in the data upwards, but not the scatter. The range in the pillar data and the correlated elastic-plastic transition behaviour is more nuanced as discussed by Greer and De Hosson [17], and strictly correlated to the known specimen preparation procedures. For the small range of specimens diameters tested in the present study the highest strengths observed were for EBL pillars, which gradually yield followed by strain bursts, and show no size effect. Papanikolaou et al. [19] would explain this response by few weak bulk sources and no surface sources or obstacles. EBL+FIB introduce surface defect source/obstacles into the original plated nickel structures which lower the yield stress followed by stochastic plastic processes that drive the yield data down to the power-law line in Fig. 7. Annealing of FIB specimens did not change the yield strength significantly, but did remove the stochastic yield behaviour. The FIB-based fabrication therefore has a direct effect on the type of elastic-plastic transition behaviour.

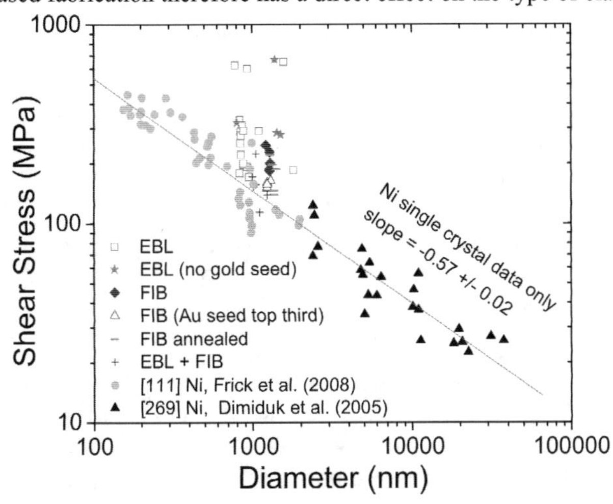

Figure 7. Logarithmic plot of the critical resolved shear stress (0.2% strain) versus pillar diameter of present nickel pillar compression results this study compared to single crystalline nickel [111] at 3% strain [7], and [269] [8] at 0.2% strain. The solid line is the best fit slope to the single crystal data only, and is in the range of the Papanikolaou et al. [19] prediction for bulk driven sources. The new polycrystalline data is mostly above this line.

4. Conclusions

Nickel pillars were fabricated by EBL and FIB methods of the same electroplated nickel to control initial chemistry, grain size, orientation and dislocation density. In load-controlled compression, EBL pillars yielded over a broad range of stresses from 480 to 1900 MPa with barrelling after further plastic deformation. In contrast, FIB pillars yielded at lower stresses (300 - 700 MPa) with square pillars yielding at lower stresses than round pillars due to geometric stress concentrations. The lower yield points for FIB compared to EBL are attributed to surface damage/defects from ion implantation that can act as nucleation sites and obstacles for dislocations. Plastic deformation initiated at the top of all pillars due to tapering from fabrication. FIB pillars exhibited strain bursts after yield followed by additional elastic loading in a stochastic way. Annealed FIB pillars yielded between 575 to 700 MPa, similar to the yield of FIB pillars, but without the stochastic response. It can be concluded that FIB-based fabrication determines the type and magnitude of the elastic-plastic transition, where there is a minimum strength power-law base-line for sample size effects in nickel above which knowing the details of the defect structure should make the elastic-plastic transition predictable.

Acknowledgments
Funding was supported by a NSERC Discovery Grant. All lithographic work was performed at Nanofabrication Kingston with guidance from R. Knobel and G. Gibson. The authors would also like to thank C. Cooney, C. MacDonald, J. Hiscocks, J. Wang and P. Changizian for technical support with the characterization methodologies used and the nano-indentation instrument.

References
[1] Brenner SS 1957 *J. Appl. Phys.* **28** 1023–26.
[2] Saimoto S 1960 *The Deformation of Copper Whiskers*, MASc Thesis (UBC, Canada).
[3] Uchic MD, Dimiduk DM, Florando JN and Nix WD 2004 *Science* **305** 986–89.
[4] Richter G, Hillerich K, Gianola DS, Mönig R, Kraft O and Volkert CA 2009 *Nano Lett.* **9** 3048–52.
[5] Greer JR, Nix WD 2005 *Appl. Phys. A Mater. Sci. Process.* **80** 1625 9.
[6] Abad OT, Wheeler J M, Michler J, Schneider AS and Arzt E 2016 *Acta Mater.* **103** 483–94.
[7] Frick CP, Clark BG, Orso S, Schneider AS and Arzt E 2008 *Mater. Sci. Eng. A* **489** 319–29.
[8] Dimiduk DM, Uchic MD and Parthasarathy TA 2005 *Acta Mater.* **53** 4065–77.
[9] Uchic MD, Wheeler IV R, Seekely MJ and Dimiduk D 2004 *Microsc. Microanal.* **10** 1126–7.
[10] Ryu I, Cai W, Nix WD and Gao H 2015 *Acta Mater.* **95** 176–83.
[11] Volkert CA and Lilleodden ET 2006 *Phil. Mag.* **86** 5567–79.
[12] Kiener D, Motz C, Rester M, Jenko M and Dehm G 2007 *Mater. Sci. Eng. A* **459** 262–72.
[13] Schilling A, Adams T, Bowman RM and Gregg JM 2007 *Nanotechnology* **18** 035301 1-5.
[14] Hofmann F, Tarleton E, Harder RJ, Phillips NW, Ma P-W, Clark JN, Robinson IK, Abbey B, Liu W and Beck CE 2017 *Sci. Rep.* **7** 45993.
[15] Burek MJ and Greer JR 2010 *Nano Lett.* **10** 69–76.
[16] Jennings AT, Burek MJ and Greer JR 2010 *Phys. Rev. Lett.* **104** 135503 1-4.
[17] Greer JR and De Hosson JT 2011 *Prog. Mater. Sci.* **56** 654–724.
[18] Kubin LP 2013 *Dislocations, Mesoscale Simulations and Plastic Flow* (Oxford Press, UK).
[19] Papanikolaou S, Song H and Van der Giessen E 2017 *J. Mech. Phys. Solids* **102** 17–29.
[20] Di Bari GA 2010 *Electrodeposition of Nickel* (Modern Electroplating, 5e) ed M Schlesinger and M Paunovic M (John Wiley & Sons, New York) 79–114.
[21] Sastry DH and Tangri K 1975 *Phil. Mag.* **32** 513–25.
[22] Hirsch PB and Mitchell TE 1967 *Can. J. Phys.* **45** 663–706.
[23] Sedore M 2016 *Yielding Behaviour of Electron Beam Lithography and Focused Ion Beam Made Nanocrystalline Micro/Nanopillars*, MASc Thesis (Queen's University, Canada).
[24] Hughes GD, Smith SD, Pande C S, Johnson HR and Armstrong RW 1986 *Scr. Metall.* **20** 93–7.
[25] Simmons G and Wang H *Single Crystal Elastic Constants and Calculated Aggregate Properties. A Handbook* 2e 1971 (MIT Press, Boston, MA).
[26] Thomason PF 1969 *Int. J. Mech. Sci.* **11**, 187–98.
[27] Lee S, Mordehai D, Rabkin E and Nix WD 2011 *J. Matls. Res.* **26** 1653-61.
[28] Maurer F, Brötz J, Karim S, Molares MET, Trautmann C and Fuess H 2007 *Nanotechn.* **18** 135709.
[29] Feaugas X and Haddou H 2003 *Metall. Trans. A* **34A** 2329-40.

IOP Conf. Series: Materials Science and Engineering **580** (2019) 012024 doi:10.1088/1757-899X/580/1/012024

Load transfer and strain gradients in pearlite investigated by means of in-situ neutron diffraction

Jeroen Tacq[1], **Bert Verlinden, Marc Seefeldt**

KU Leuven, Department of Materials Engineering, Kasteelpark Arenberg 44, 3001 Leuven, Belgium

1) now at: SIRRIS Gent, Technologiepark 48, 9052 Zwijnaarde, Belgium

Marc.Seefeldt@kuleuven.be

Abstract. In-situ neutron diffraction experiments have been performed during tensile testing of undeformed and cold rolled pearlite. The microstructure of the samples varied from spheroidised cementite to randomly oriented lamellae to lamellae aligned with the loading direction. Both the ferrite lattice strain and diffraction peak width have been analysed. Conclusions were drawn regarding the influence of cementite morphology on load transfer from ferrite to cementite. The ferrite lattice strain was found to decrease after yielding in a sample with aligned lamellae, whereas an increasing ferrite lattice strain would be expected. This might be explained based on the non-ideal nature of the lamellar microstructure. Upon unloading of the material, the diffraction peak width decreases, which may be attributed to strain gradients in the deformed pearlite.

1. Introduction

Fully pearlitic steels with lamellar microstructure are used for a wide range of applications including sawing wire, steel cords, and rails. All of these applications require the pearlite to be subjected to large plastic strains in the production process. During deformation, the large difference in mechanical properties between ferrite and cementite results in the transfer of load from ferrite to cementite. After unloading, residual stresses and strains will remain in the material, which will have a strong influence on the macroscopic material properties, such as strength and ductility.

The morphology of the harder cementite phase is expected to have a strong influence on the development of the internal strains. The current paper addresses this point by investigating the internal strain development in samples with various microstructures, using in-situ neutron diffraction. In addition to the internal strains, the behaviour of the diffraction peak width will also be studied to gain more insight into how the ferrite and cementite phases co-deform.

The development of internal strains during tensile deformation has already been studied for a number of pearlitic steels, having varying microstructures and process histories. The work by Oliver et al. [1] and Young et al. [2] on spheroidised steels, showed that the ferrite lattice strain in the loading direction suddenly decreases upon macroscopic yielding. This is attributed to a strong instantaneous load transfer from the plastifying ferrite matrix to the elastically deforming cementite particles. At larger tensile strains, the ferrite lattice strain again increases and continues to do so until fracture. Daymond and Priesmeyer [3] investigated a hypereutectoid steel, consisting of lamellar pearlite colonies with single phase ferrite at the colony boundaries. A drop in lattice strain at the yield point was not observed, and the ferrite lattice strain did not increase as much with increasing tensile deformation. Tomota et al. [4,5]

investigated an as-patented steel with a fully lamellar microstructure and observed a constant ferrite lattice strain in the axial direction, after yielding. These observations indicate that load transfer from cementite to ferrite is more pronounced for a lamellar microstructure. However, the disappearance of the lattice strain shift upon yielding also indicates that the load transfer is more gradual in a lamellar material. In deformed, swaged [5,6] or cold drawn [4], and subsequently stress relieved lamellar pearlite, the average axial ferrite internal strain after yielding decreases, instead of increasing or saturating, as was observed for undeformed samples.

The current study will investigate cold rolled pearlite rather than the cold drawn or swaged pearlite used previously. This has the advantage that the microstructure consists of ferrite and cementite lamellae aligned parallel to the rolling plane. This microstructure can be idealized as a single stack of flat, alternating ferrite and cementite sheets.

2. Experimental

2.1. Material

A plain carbon steel (AISI 1075/C75) of eutectoid composition, as specified in table 1, was used for this study. Four kinds of samples were produced and subjected to uniaxial tensile testing:

Table 1. Chemical composition of the AISI1075/C75 steel used in this investigation.

C	Mn	Si	P	S
0.76 wt%	0.65 wt%	0.21 wt%	≤ 0.04 wt%	0.016 wt%

(1) *As-patented.* Austenitized and isothermally transformed at 500°C. The microstructure consists of randomly oriented pearlite colonies.

(2) *As-rolled.* Isothermally transformed at 500°C and cold rolled to a strain of 1.1. The microstructure consists of pearlite colonies with their lamellae aligned with the rolling plane.

(3) *RS-relieved.* Cold rolled to a strain of 1.1, followed by a heat treatment for 15 min at 470°C in order to remove the residual stresses present after rolling. This was confirmed by X-ray diffraction. The heat treatment resulted in a reduction of the surface stress from 500 MPa to about 40 MPa. The deformation substructure may have relaxed to some extent, but the aligned, lamellar microstructure remains unchanged.

(4) *Spheroidised.* Cold rolled to a strain of 1.1, subsequently heat treated for 8.5 hours at 670°C and followed by overnight furnace cooling to room temperature. The microstructure consists of a fine dispersion of cementite particles.

2.2. Method

In-situ neutron diffraction measurements were carried out on the TKSN-400 (HK9) strain diffractometer at the Nuclear Physics Institute (NPI) in Řež near Prague. The neutrons fully penetrated the 1mm thick steel samples. Any macroscopic strain profiles are therefore averaged out and the internal micro-phase strains, resulting from the interaction between the two phases, could be determined. The lattice strains were measured in the loading direction.

All the samples were of dog bone shape. The gauge section had a width of 8 mm and a thickness of 1 mm. The tensile direction was parallel to the rolling direction.

In the elastic region, the tensile tests were performed in stress control. After macroscopic yielding, the tests were continued in strain control. Unloading and re-loading cycles were performed in stress control. The tensile test was stopped at each measurement point and the stress or strain held at a constant value throughout the entire measurement period (1.5 h). In strain control, this led to some stress relaxation (see figure 1). Within the first 10 min of measurement time, about 2/3 of the total drop in stress occurs. After that, the stress continues to decrease slowly until the end of the measurement.

Four ferrite diffraction peaks were measured (200, 211, 220 and 310). A 2-D detector with an angular range of about 10° was used. For each peak, the detector was positioned such that the peak was in the centre of the detector. The measurement time was 22 min per peak. The incoming slit was 14.6 mm high and 7.4 mm wide; the outgoing slit was 8 mm high and 3 mm wide. Care was taken to position the samples so that they were completely enclosed in the diffraction gauge volume, thereby preventing any artificial peak shifts.

3. Results

3.1. Macroscopic stress-strain response

The in-situ engineering stress-strain curves of all four samples are shown in figure 1. The as-patented sample has a sharp yield point, after which a significant amount of strain hardening takes place up to an engineering strain of about 5 %.

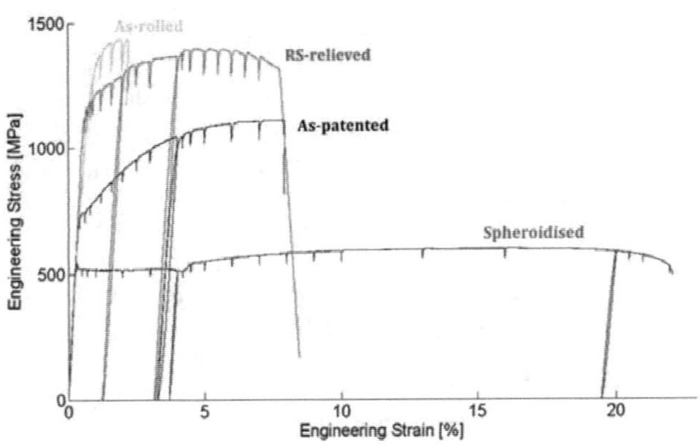

Figure 1. In-situ recorded stress-strain curves for all four samples studied. Some stress relaxation occurred during each neutron diffraction measurement.

The as-rolled sample reaches a much higher flow stress than the as-patented sample, but exhibits an extended micro-yield domain instead of a sharp yield point. This can be explained by the significant microstructural inhomogeneity observed in the rolled material. Another possible explanation is the development of intergranular residual stresses during prior cold rolling. The differences in residual stress among the differently oriented grains will result in different kinematic hardening contributions and therefore in a broad band of yield limits.

The RS-relieved sample yields at a smaller stress, but has a sharper yield transition as compared to the as-rolled one. On the other hand, it is not as sharp as for the as-patented sample. This indicates that the annealing treatment decreased the inhomogeneity to some extent, as well as the residual stresses. However, the inhomogeneity of the cementite morphology still results in a micro-plastic regime. A significant amount of strain hardening is also observed.

Finally, a completely different behaviour can be seen for the spheroidised sample. This sample exhibits a sharp yield point, followed by some yield softening. After that, a small amount of strain hardening occurs up to a strain of about 12 %. In the last stage of deformation, no strain hardening can be observed and the flow stress remains constant up to final ductile failure.

3.2. Ferrite lattice strains

The ferrite lattice strains in the tensile direction, which is also the rolling direction of the deformed samples, were measured for the 200, 211, 220 and 310 grain families. In figure 2 the evolution of ferrite lattice strain for the as-patented sample is plotted as a function of the tensile strain. The lattice plane spacing before the start of the tensile test was taken as a reference to calculate the strains. This means that any residual strains initially present in the material are neglected.

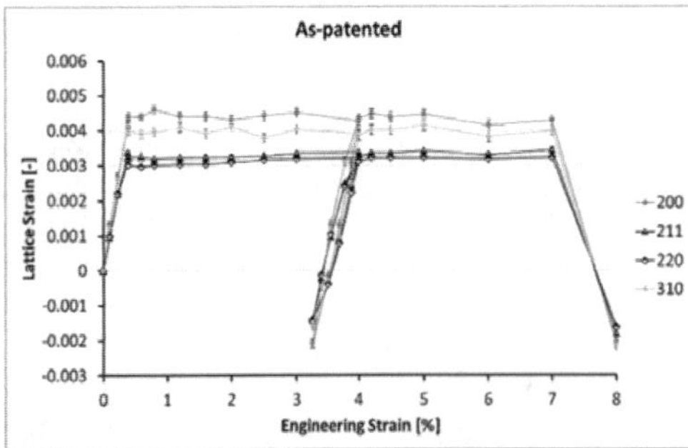

Figure 2. Ferrite lattice strain in the loading direction as a function of macroscopic tensile strain, for the as-patented sample and all probed reflections.

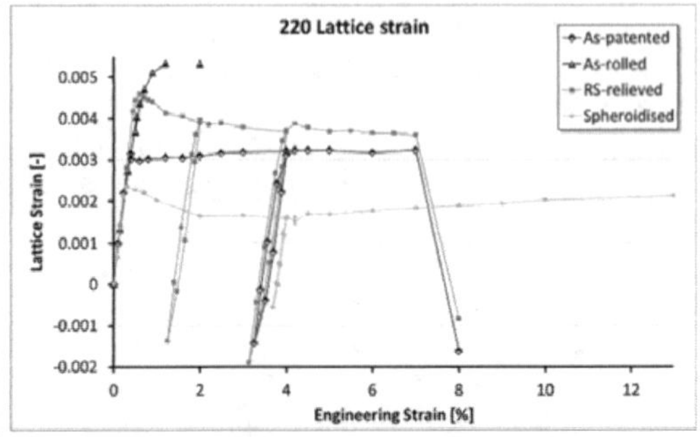

Figure 3. Ferrite 220 lattice strain in the loading direction, as a function of macroscopic tensile strain for all measured samples.

From the linear part of the applied stress vs. lattice strain plots, the diffraction elastic constants (DECs) of the various grain families can be determined. It can be seen that the different DECs result in the development of intergranular strains in the elastic region. After yielding, all grain families behave qualitatively the same and the ratios of the lattice strains do not change much anymore. Further intergranular strain development in the plastic region is limited. This is true for all samples with a lamellar microstructure.

The spheroidised sample behaves somewhat differently. In the elastic region the difference in DECs also leads to the development of intergranular strains, except that now the 310-DEC is smaller than the 200-DEC. This results in the 310 grain family exhibiting the largest lattice strain as opposed to the 200 grain family. After yielding, the ratios of the lattice strains also undergo a large change, suggesting that intergranular strain development is more significant as compared to the other samples.

The overall lattice strain in all four samples is well described by the strong 220-ferrite reflection. Comparing the different samples can therefore more easily be done by only focussing on the 220 grain family, as illustrated in figure 3. The ferrite lattice strain in the spheroidised sample drops significantly after yielding, up to a strain of about 2 %. After that point, the lattice strain continuously increases again up to the point where the ultimate tensile strength is reached. At that point, the lattice strain has again attained the same level as at the yield point.

The as-patented material behaves quite differently. A very small drop in the lattice strain can be observed upon yielding, after which the ferrite lattice strain remains virtually constant. The internal strain in the cold rolled, RS-relieved material decreases continuously after yielding. This agrees well with the results on cold drawn wire and swaged pearlite previously presented in literature [4-6].

A number of unloading-reloading cycles were also performed. For the spheroidised sample, both the unloading and reloading are almost linear. This is not true for the as-patented and RS-relieved samples. Their lattice strains exhibit a significant hysteresis which is of the same order of magnitude in both samples. The difference between the latter two samples can be found in the point where non-linearity starts. In the as-patented sample, it starts at the point where the ferrite lattice strain becomes zero. In the RS-relieved sample, it starts virtually at the very beginning of the unloading cycle.

3.3. Ferrite diffraction line widths

Together with the peak position, the peak width has also been determined. The line widths of the 220 ferrite diffraction peak, for all but the as-rolled sample, are shown in figure 4. The line widths are given as a function of measurement number in order to emphasize their evolution. Arrows indicate the end of the elastic regime. The start and end of each unloading-reloading cycle are highlighted using circles. The points where the macroscopic stress becomes zero is highlighted using a square.

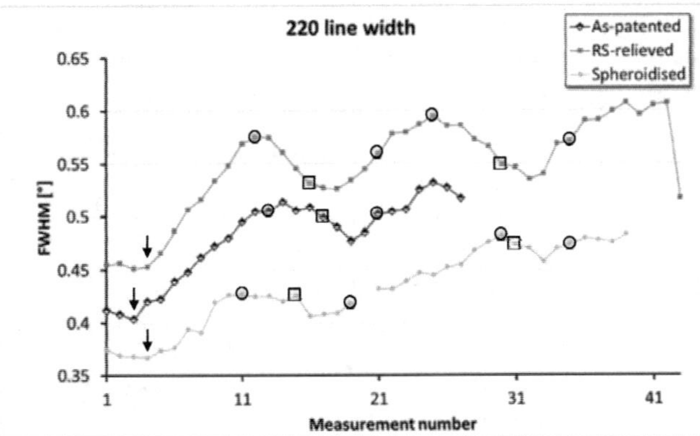

Figure 4. 220 ferrite diffraction line width as a function of measurement number. Arrows indicate the end of the elastic region. The start and end of the unloading-reloading cycles are indicated using circles. The points where the macroscopic stress becomes zero is indicated using a square.

In the elastic region the peak width does not change. After the onset of yielding, there is a significant amount of line broadening for all samples. With unloading, the strain broadening decreases significantly in the RS-relieved sample. This trend continues somewhat after the zero applied stress level is reached, and into the reloading section of the cycle. However, by the time the same stress level is reached as before unloading, the peak width has almost fully 'recovered'. A similar evolution is seen for the as-patented sample, although there the decreasing peak width is strangely concentrated in the reloading section of the cycle. No significant loss of peak width can be observed for the spheroidised sample.

4. Discussion

4.1. Influence of cementite morphology

The internal strain evolution in the spheroidised sample compares very well to the results presented, by for example, Oliver et al. [1], except for the fact that the lattice strain in their study drops much more quickly after yielding, probably because the microstructure was coarser and more homogeneous than in the present samples. The region in which the ferrite lattice strain is again increasing, coincides with a region of negligible macroscopic strain hardening. This suggests that while the ferrite phase is strain hardening, some of the load borne by the cementite phase is again transferred back to the ferrite.

The as-patented material strain hardens considerably while the internal ferrite lattice strain remains nearly constant. Therefore, the increase in strength of the composite material has to originate from the harder cementite phase carrying an increasing portion of the load.

The RS-relieved sample has been cold rolled and has its lamellae approximately aligned with the loading direction. After yielding, the ferrite lattice strain continuously decreases, while the sample does exhibit significant macroscopic strain hardening. Aligning the lamellae with the loading direction thus further increases load transfer to the stronger phase, as has also previously been suggested [4-6]. Lamellae alignment provides a significant contribution to the large strain work hardening.

In addition to load transfer from one phase to another, intergranular strains can also develop as a result of load transfer between differently oriented ferrite grains. The samples with a lamellar microstructure show very little intergranular strain development during plastic deformation. On the other hand, intergranular strains continue to increase in the spheroidised sample. The lamellar microstructure thus suppresses intergranular strain development to some extent because ferrite grain hardening is then governed by the morphological constraints rather than by the crystallographic orientation.

4.2. Decreasing lattice strain

In section 4.1 it has been established that the load transfer from ferrite to cementite becomes more efficient as the microstructure becomes more lamellar and has its lamellae aligned with the loading direction. However, one would still expect to see a continuous increase of the ferrite lattice strain, unless ferrite undergoes significant strain softening.

The RS-relieved sample has been deformed to a strain of 1.1 and has its lamella aligned with the rolling plane. It can therefore be idealized as a stack of alternating, flat ferrite and cementite sheets, parallel to the rolling plane. In order to maintain geometric compatibility between both phases, the total strain in both phases has to be the same. Such an assumption has often been made to describe the co-deformation of ferrite and cementite, as well as other materials with similar or whisker like microstructures, and will be called the iso-strain assumption [7-10].

In this simple case, and neglecting all constraints in any direction other than the loading direction, the stress in each phase at a certain imposed strain can be read from their individual in-situ flow curves. The stress response of the composite material is the volume averaged stress over the two phases. A simple model like this, results in a continuously increasing internal strain. Such a result has been observed by Sinclair et al. [10] for a Cu-Cr alloy with Cr whiskers embedded in a Cu matrix.

The fact that the internal strain in the ferrite phase of the RS-relieved sample continuously decreases past the yield point, indicates that the proposed idealisation of the microstructure and the iso-strain assumption, are oversimplifications. Another explanation would be that the ferrite undergoes strain softening during the tensile test, e.g., through shear banding after lamellae alignment.

Tomota et al. [4] suggest that the decrease in lattice strain results from a strong stress transfer from the ferrite phase to the cementite phase. Such a stress transfer also takes place in the proposed sandwich model, but the ferrite lattice strain continues to increase. To obtain a decreasing lattice strain, the cementite phase has to impose an additional compressive strain on ferrite. Such a mechanism has been described by Eshelby [11,12] for a hard and less deformable particle embedded in a deforming matrix and could apply in a microstructure of hard discontinuous particles rather than continuous lamellae.

The experimentally recorded stress-strain curve (figure 1) shows only one deflection point, suggesting that no cementite plasticity occurs up to fracture. Under the assumption of a perfect sandwich for the range of uniform elongation, the cementite phase would then have to deform elastically to a strain of 5 %, resulting in an internal cementite stress on the order of 11 GPa. Such a stress is unreasonably high, even for cementite, and has never been reported in literature, providing additional evidence that the sandwich representation and/or the iso-strain assumption do not hold in real pearlite. However, the lack of a second deflection point can also be explained by gradual plastification of cementite over the full strain interval of the tensile test (up to 8%). In light of the observed inhomogeneity of the microstructure, such gradual plastification of cementite might indeed be possible.

4.3. Analysis of reversible peak broadening

The ferrite peak width was seen to start increasing with the onset of plastic deformation (figure 4). However, a significant reduction in peak width is observed upon unloading. This observation is more pronounced in a material with aligned lamellae. In the following, possible contributions to peak broadening will be discussed in an effort to explain the experimental observations.

Inhomogeneous deformation on sample scale may lead to macroscopic strain gradients. Because the neutron diffraction signal comes from grains throughout the entire sample cross-section, the presence of a macroscopic strain profile would contribute to peak broadening. However, during a uniaxial tensile test the macroscopic deformation should be fairly homogeneous within the gauge volume.

Another possible contribution to peak broadening is size-broadening, i.e. reduction of the coherently diffracting domain size. However, the reduction in lamellae width during a tensile test will be very limited and break-up of the ferrite lamellae in smaller crystallographic domains only becomes significant at strains of 20-30 %. Additionally, the break-up of ferrite into smaller domains of homogeneous crystal orientation would not be expected to be reversible upon unloading.

That leaves only microstrains to explain the reversible peak broadening. Strain-broadening is due to inhomogeneities in the elastic strain distribution within the diffracting volume. Such inhomogeneities can result from the presence of dislocations, but can also be caused by elastic strain profiles within grains or by differences in strain among similarly oriented grains. Strain-broadening can at least partially be reversed and will now be discussed further.

Unloading can lead to some decrease in the dislocation content and therefore a reduction of the peak width. Dislocations of opposite sign can annihilate with each other, while dislocations of the same sign can arrange in walls to reduce their associated elastic strain fields. In a lamellar microstructure, dislocations are also known to be able to bend out from the interfaces upon deformation and can run back into the interfaces with unloading.

The peak width has been shown to be proportional to the square root of the dislocation density, $\sqrt{\rho}$, under a variety of different assumptions. The FWHM of the 110 peak for the RS-relieved sample increases from 0.45° before loading to 0.575° before the first unloading cycle. During the unloading cycle a minimum value of 0.525° is reached for the FWHM. About 40 % of the peak width gained during forward loading is thus lost. The same is true for the second unloading cycle. For the example described above this results in about 43 % of the dislocation density gained during plastic tensile straining being lost again during the unloading cycle. Such a strong decrease of the dislocation content seems unlikely: pinning, tangling and interlocking of dislocations would make it difficult for them to move away in to opposite direction once the external load has been removed.

The discussion in section 4.2 already suggested that the lamellar microstructure of a real pearlitic steel cannot be considered as a sandwich and that the two phases do not co-deform according to an iso-strain model. However, at the interfaces between ferrite and cementite, compatibility has to be maintained. The break-down of the iso-strain assumption then inevitably leads to the existence of total strain gradients in pearlite. Upon relaxation of the externally applied load, these strain gradients can also decrease in magnitude, resulting in a decreasing peak width. The presence of significant intergranular or interlamellar elastic strain gradients could thus explain the current experimental observations.

4.4. Strain gradients in pearlite

The above discussion leads to the conclusion that elastic strain gradients may be present in deformed pearlite. In general, strain gradients can exist between regions with different local flow stresses. The as-patented material will have colonies with lamellae differently oriented with respect to the applied deformation. This will lead to different deformation modes, ranging from near-uniaxial deformation in colonies with their lamellae aligned to the loading direction, to large shear deformations in a deck-of-cards type shearing of colonies with lamellae inclined to the loading direction [13-15]. As a result of these different deformation modes, strain gradients may develop among the colonies [16]. After lamellae alignment, the microstructure contains regions with more or less fragmented lamellae, again leading to local variations in load transfer, the pearlite flow stress and the introduction of strain gradients.

These strain gradients will have both a plastic and an elastic component. The plastic component of the gradients results in the introduction of additional "geometrically necessary" dislocations. This will provide an additional amount of strain hardening and increase the strength of pearlite. The differences in flow stress will also lead to differences in local elastic strain. This is the component of the strain-broadening which is probably more easily reversed.

Another kind of strain gradient can exist within the confines of a single lamella. Because of the break-down of the iso-strain assumption, both phases can deform at a different rate, leading to a strain mismatch between the two phases. Assuming that interface sliding is difficult due to the large area of the interface between two lamellae, a strain profile across the lamella is required in order to maintain compatibility at the interfaces. The gradient is then perpendicular to the plane of the lamellae, but is a gradient in the stress or strain component parallel to the plane of the lamellae and to the direction of mismatch between the two phases. As the average internal strain in the lamella decreases, so will the extent of the strain gradient. The gradient in elastic strain is also correlated to the thickness of the lamella. A small lamella thickness and a large average elastic strain, result in a stronger elastic strain gradient.

New experimental work suggests that a dislocation rich interface-region might develop along the ferrite-cementite phase boundaries at sufficiently large rolling strains. The resulting gradient in dislocation density causes a gradient in flow stress across the thickness of the ferrite lamellae. Such a gradient in the material properties will also cause a gradient in the elastic strain across the lamellae.

At small strains, the plastic strain gradients in ferrite might not yet be developed sufficiently to result in observable elastic strain gradients. This can be one of the reasons why the reversibility of the peak broadening is much less pronounced in the as-patented sample. The decrease of the peak width in the RS-relieved sample may also be more pronounced because of the alignment of the lamellae. The strain profile in lamellae aligned with the loading direction will be larger in this direction, which is also the direction probed by the neutrons. Relaxation of the external load and strain profile will thus cause a stronger decrease of the peak width in these lamellae.

5. Conclusions

Load transfer from the ferrite to the cementite phase becomes more efficient in a lamellar microstructure and with alignment of the lamellae along the loading direction. Alignment of the lamellae therefore results in a significant contribution to the large strain work hardening of pearlite.

The observed decreasing ferrite lattice strain in a sample with aligned lamellae is difficult to explain based on an idealised sandwich microstructure in which both phases co-deform under an iso-strain assumption. Pearlite should rather be considered as an inhomogeneous material in which both phases can deform by different amounts, even within the confines of a lamellar microstructure. As a result, both elastic and plastic strain gradients will be introduced in the material. These can explain the observed reversibility of the peak width upon unloading.

Acknowledgements

The authors would like to thank the beam line scientist J. Pilch of NPI Řež for his support. The work was supported by JT's fellowship of The Research Foundation - Flanders (FWO); the Belgian Science Policy BELSPO through the IAP project 7/21 "Multiscale Mechanics of Interface Dominated Materials" and the European Commission under the 7th Framework Programme through the 'Research Infrastructures' action of the 'Capacities' Programme, NMI3-II Grant number 283883. The research has been conducted at the CANAM (Center of Accelerators and Nuclear Analytical Methods LM2011019) infrastructure with a funding from the Ministry of Education, Youth and Sports of the Czech Republic.

References

[1] Oliver EC, Daymond MR and Withers PJ 2004 *Acta Mater.* **52** 1937
[2] Young ML, Almer JD, Daymond MR, Haeffner DR and Dunand DC 2007 *Acta Mater.* **55** 1999
[3] Daymond MR and Priesmeyer HG 2002 *Acta Mater.* **50** 1613
[4] Tomota Y, Lukáš P, Neov D, Harjo S and Abe YR 2003 *Acta Mater.* **51** 805
[5] Morooka S, Tomota Y, Suzuki T and Kamiyama T 2007 *Adv. Mater. Res.* **15-17** 912
[6] Kanie A, Tomota Y, Torii S and Kamiyama T 2004 *ISIJ Int.* **44** 1952
[7] Gil Sevillano J 1991 *J. Phys. III* **1** 967
[8] Watté P, Van Houtte P, Aernoudt E, Gil Sevillano J, Van Raemdonck W and Lefever I 1994 *Mat. Sci. Forum* **157-162** 1689
[9] Hu X, Van Houtte P, Liebeherr M, Walentek A, Seefeldt M and Vandekinderen H 2006 *Acta Mat.* **54** 1029
[10] Sinclair CW, Saada G and Embury JD 2006 *Phil. Mag.* **86** 4081
[11] Eshelby JD 1957 *Proc. R. Soc. London: Ser. A – Math. Phys. Sci.* **241** 376
[12] Eshelby JD 1959 *Proc. R. Soc. London: Ser. A – Math. Phys. Sci.* **252** 561
[13] Langford G 1977 *Metall. Trans. A* **8** 861
[14] Porter DA, Easterling KE and Smith GDW 1978 *Acta Metall.* **26** 1405
[15] Gil Sevillano J 2011 *Wire J. Int.* **44** 58
[16] Wang Y, Ohnuki T, Tomota Y, Harjo S and Ohmura T 2017 *Scripta Mater.* **140** 45

IOP Conf. Series: Materials Science and Engineering **580** (2019) 012025 doi:10.1088/1757-899X/580/1/012025

3D microstructural analysis in the steel industry

Akira Taniyama

Advanced Technology Research Laboratories, Nippon Steel Corporation, 1-8 Fuso-cho, Amagasaki, Hyogo, 660-0891, Japan

E-mail: taniyama.76c.akira@jp.nipponsteel.com

Abstract. Since steel products have complex microstructures and chemical composition to achieve high strength and high functionality, it is important to characterize the internal structure with analytic techniques and to be able to predict their properties using simulation techniques. It is now possible to characterize and analyze the structural arrangements in raw steel materials and steel products, including the shapes of defects such as voids and cracks, as well as the crystallographic grain structure, by use of powerful new 3D characterization and analysis techniques, which can provide information on the internal structure on the micron or even submicron scale. It is expected, therefore, that application of new methods and results should lead to improvements in the mechanical properties of steel products. In this article, 3D analysis of the internal structure in a duplex stainless steel, and of strain propagation in pure nickel are presented as examples of the industrial application of 3D microstructural analysis.

1. Introduction

Steel products have long been used as base products to support the infrastructure of human society. In the development of steel products, microstructure analysis provides important information in the quest to improve both mechanical properties and functionality. Since steel products have complex microstructure and chemical composition in order to achieve their high strength and superior functionality, it is important in research and development of steel products to be able to predict their properties using both analytic and simulation techniques. Such simulations require detailed knowledge of material parameters influencing internal structure formation and mechanical properties, such as the dislocation structure, the boundary structure, precipitates, and grain shapes, sizes and orientations, as well as crack and void shapes. In order to obtain these important properties, 2D techniques such as optical microscopy (OM), scanning electron microscopy (SEM), transmission electron microscopy (TEM) and atom probe field ion microscopy (AP-FIM) have been used for many years, and these techniques are complementarily used to obtain multi-scale information. Recently, the importance has been recognized of developing these 2D techniques into 3D techniques, such as X-ray computed tomography (CT), serial sectioning OM and SEM, TEM/STEM-CT and atom probe tomography (APT) for materials analysis, as shown in figure 1.

In this article, a 3D analysis of the internal structure in a duplex stainless steel, which has a complex phase domain shape, and of strain propagation in pure nickel are presented as examples of the industrial application of 3D microstructural analysis.

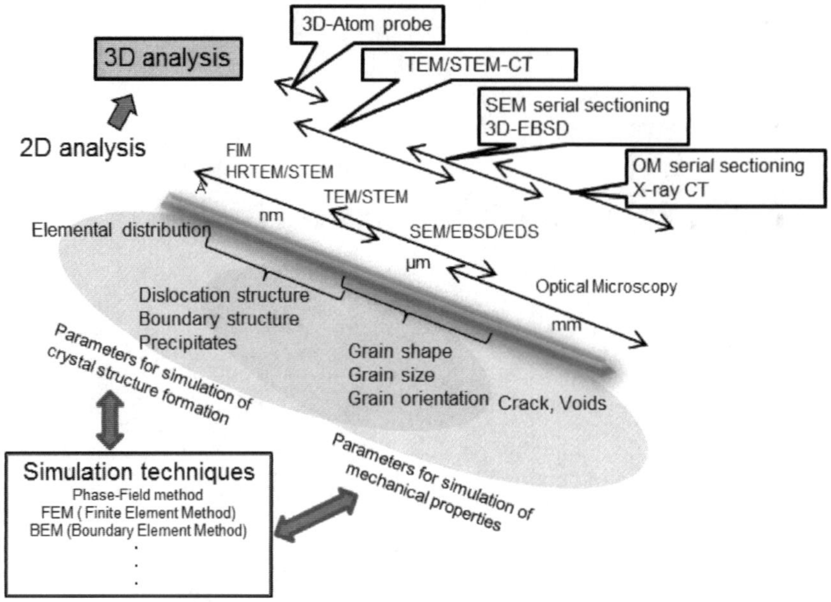

Figure 1 Recent trends in materials analysis.

2. 3D analysis techniques in the characterization of steels

Several 3D analysis techniques, such as serial sectioning and tomography techniques, have been developed to obtain the 3D surface morphology and internal structure. The simple way to take 3D structure images of objects is to use the serial sectioning technique. In this technique, a sample block is sliced by mechanical polishing and/or ion beam machining, and 2D images of cross-sections of the object are acquired by microscopic techniques, such as OM and SEM. Slicing of the sample block and acquiring of images are alternatively carried out, such that a 3D image can be reconstructed from a series of 2D images. Therefore, an important requirement in the technique is to cut each slice parallel to the sample block face, with each slice the same thickness, as precisely as possible. Furthermore, drift correction of 2D images is required in case the sample block moves during acquisition. Serial sectioning is a destructive analysis technique, and as such the sample block is lost disappeared after data acquisition. In contrast, tomography techniques using X-rays or TEM/STEM can be used to obtain 3D images of a sample block non-destructively, because the technique does not destroy the sample block during measurement. In these techniques, transmitted images are acquired by rotating a sample block around an axis perpendicular to the incident X-ray or electron beam. A 3D image of the sample is reconstructed with a series of the transmitted images by reconstruction software.

Figure 2 3D image of rolling contact fatigue cracks (light blue) around inclusions (yellow and orange) reconstructed with serial sectioning SEM images.

Since steel materials typically have a very fine internal structure, down to the micro- or sub micro-meter size, application of such 3D analysis techniques to observation of the fine structures in steels requires high spatial resolution. Serial

IOP Conf. Series: Materials Science and Engineering **580** (2019) 012025 doi:10.1088/1757-899X/580/1/012025

sectioning using a focused ion beam (FIB)/SEM system is useful to allow information to be obtained at the micro- or sub-micro meter scale, due to the precise sectioning that can be achieved using a FIB system and the high resolution available by SEM. Furthermore, EDS and EBSD measurement are also available in such a system [1]. Figure 2 shows an example of 3D image obtained by FIB/SEM. Thin fatigue cracks can be observed around inclusions in the steel sample. Tomography techniques using TEM/STEM can provide still higher resolution 3D images [2], though it is necessary to use a sample thin enough to be electron beam transparent. For a steel, this limits the sample thickness to a maximum of 300nm if information on the internal structure is required. As such it is difficult to obtain crystalline grain shapes in steels using this approach. Since X-rays can transmit through steel of almost 1 mm in thickness, it is possible to obtain non-destructively information on the crystalline grain structure, as well as the phase domain shape and non-metallic inclusions in bulk steel material using an X-ray tomography technique.

3. 3D internal structure analysis of duplex stainless steel

Microscopy, such as SEM and TEM, has been widely used to observe the internal micro-structure of steel materials. Although the obtained data show very useful and important 2D information about the internal structure, 3D information is also needed to provide a complete picture of the true structure in the material, such as the crystalline grain structure, the phase domain structure as well as non-metallic inclusions. In the following, an internal structure analysis of a duplex stainless steel with a complex structure of ferrite phase (bcc) and austenite phase (fcc) is described as an application of the 3D characterization of steel. As a destructive analysis, 3D-EBSD is used to obtain sub-micron scale 3D information. X-ray micro-CT is additionally used to obtain 3D information of a volume larger than that available using 3D-EBSD.

3.1. Experimental procedure

A hot-drawn duplex stainless was used as a sample in this experiment. For the 3D-EBSD measurement, cross-sectional FIB machining was applied to a 100μm×100μm area. The cross-section was sliced to 200nm in thickness. FIB machining and EBSD measurements were carried out alternately. Each EBSD measurement was carried out over a 60μm×60μm area. For the X-ray micro CT measurement, a needle shaped sample of 500μm in diameter and 10mm in length were prepared. The surface of the sample was polished by electrolytic polishing. A polychromatic X-ray beam was used to take transmitted X-ray images. The images were taken by rotating the sample through a 360 degrees rotation in a total of 3201 steps.

3.2. Results and discussion

Figure 3 shows 3D images the hot drawn duplex stainless steel obtained from the 3D-EBSD data. The

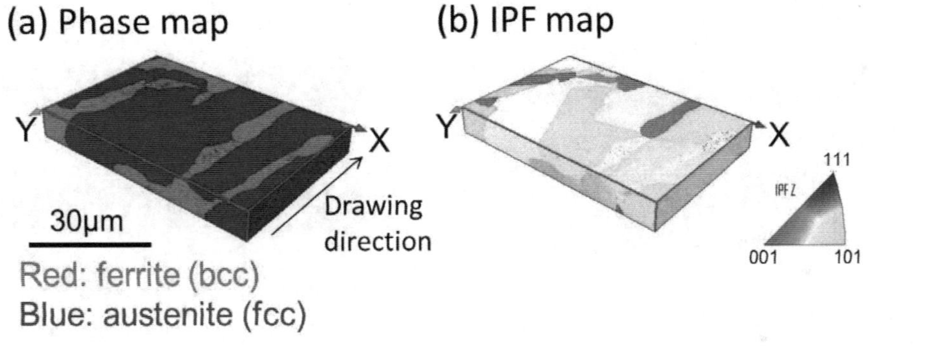

Figure 3 (a) 3D phase map and (b) IPF map of a duplex stainless steel taken using 3D-EBSD [3].

reconstructed volume is 56.8μm × 37.6μm × 6.6μm. Using information of the crystalline phase and orientation in obtained EBSD data, the 3D shape of the phase domains and of crystalline grains can be reconstructed as shown in figure 3(a) and 3(b), respectively. In figure 3(b), the orientations of the crystalline grains are indicated by inverse pole figure (IPF) coloring. Although the thickness of the obtained volume is too small to provide the complete shape of the crystalline grains of the ferrite and austenite phases, it is nevertheless possible to visualize the complicated 3D structure of the phase domain boundaries and grain boundaries by 3D-EBSD.

Figure 4(a) shows the crystalline structure for each grain. Twin boundaries are highlighted by red lines in the figure. Figure 4(b) shows a cross-sectional view of the section indicated by a yellow line in figure 4(a). The cross-section is perpendicular to the twin boundary. The twin boundary plane, indicated with a red arrow, lies approximately perpendicular to the surface of the observed volume. The twin boundary is not straight but has a complicated curvature. This is in fact an artifact of the serial sectioning technique. In order to reduce the influence of such artifacts, it is necessary to obtain a completely uniform thickness for each slice, with more precise drift correction between each slice. The angle between the plane orientation of twin boundary {111} and the normal orientation of the grains 1 and 2 is calculated to determine the plane orientation of the twin boundary. Table 1 shows calculated angular relationships. The angle between the twin boundary (1 1 -1) and the grains is approximately 90°. Therefore, the plane orientation of the twin boundary can be uniquely decided to be (1 1 -1).

As the austenite phase domain has a volume much larger than the volume that can be observed by 3D-EBSD, it is difficult to understand the whole of the domain. Therefore, absorption contrast tomography (ACT) using X-ray micro CT has also been used to obtain the whole shape of the austenite phase domain. Figure 5 shows cross-sectional views of the ACT image of the duplex stainless steel. The bright and dark contrast regions are identified as ferrite and austenite phases, respectively. Non-metallic inclusions are also observed as black particles. Figure 5(d) shows the 3D shape of the austenite phase domain extracted from figure 5(a). Non-metallic inclusions are also indicated in red in figure 5(d). It is revealed that austenite phase domain is elongated along the drawing direction, and each domain shows a cylindrical shape of 30 to 100μm in diameter and 100 to 500μm in length. Thus, it is certainly possible to obtain morphological information of phase domain in

Table 1 Angular relationship between plane orientation of twin boundary and surface normal direction of crystalline grain.

Twin boundary	Grain 1 [315]	Grain 2 [256]
(1 1 -1)	95.6°	85.9°
(1 -1 1)	46.9°	77.6°
(-1 1 1)	73.0°	49.9°
(1 1 1)	28.6°	21.1°

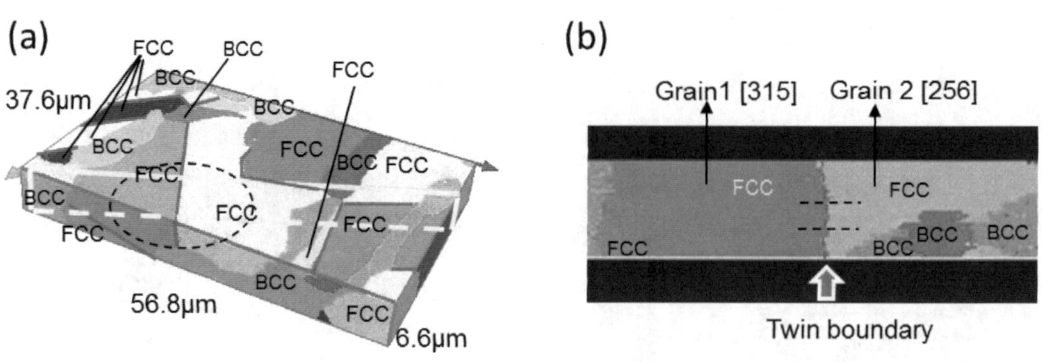

Figure 4 (a) Crystal structure for each crystalline grain and (b) cross-sectional view of a twin boundary.

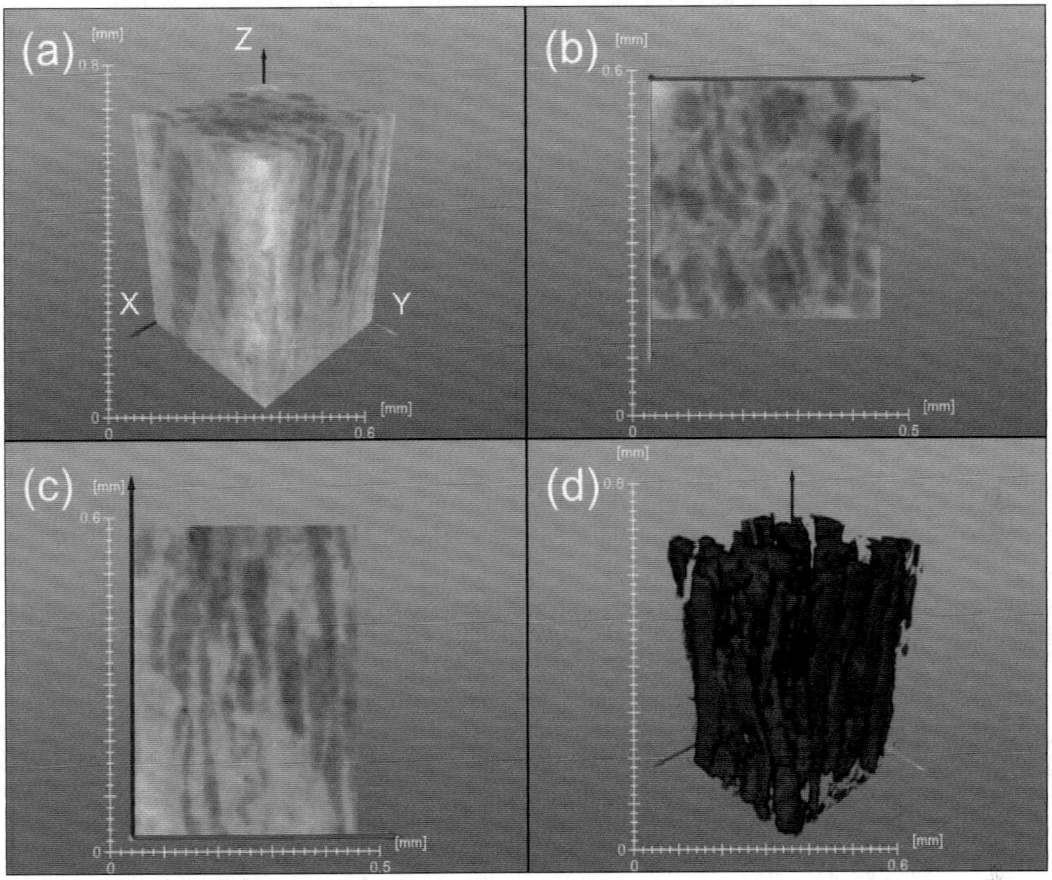

Figure 5 3D reconstructed images of a duplex stainless steel taken by X-ray CT: (a) volume rendering image, (b) X-Y cross-section, (c) X-Z cross-section, (d) extracted austenite phase domain.

metallic materials using ACT, but it is not possible with this technique to obtain crystallographic information, such as the grain shape, structure and orientation as can be achieved using 3D-EBSD. Recently, a new technique, diffraction contrast tomography (DCT), has been developed that allows the 3D shape and orientation of crystalline grains to be obtained [4-6]. Complementary use of ACT and DCT will be a powerful technique to observe the internal structure in metallic materials non-destructively.

4. Propagation of plastic deformation in pure Nickel taken by 3D-EBSD

The distribution of plastic and elastic deformation in crystals effects their recrystallization behavior. Therefore, deformation behavior has been studied by several simulation techniques, such as finite element modeling and crystal plasticity simulations. Actually, deformation in crystals propagates three dimensionally. As such 3D-visualization techniques are therefore required to obtain the deformation behaviour of crystals experimentally. In order to discuss deformation and fracture behavior of metallic materials, it is important to understand the propagation of plastic strain and deformation in a crystal grain or in the vicinity of a grain boundary, as materials typically consist of many crystal grains. The 3D-EBSD technique has been applied to analyze the propagation of plastic deformation in metallic materials [7].

4.1. Experimental procedure

A well-polished pure nickel plate (10mm × 10mm × 1mm) was annealed in a vacuum furnace for 3 hours at 973K, and then a small impression was made on a surface of the plate using a Vickers indenter, which has a diamond tip in the form of a square based pyramid. A weight of 10 g was loaded on the indenter. The depth of the impression was smaller than 2 μm under the loading condition. 3D-EBSD measurements were carried out with an integrated instrument combining FE-SEM, FIB and EBSD. In the instrument, cross-sectioning and orientation mapping are automatically acquired under control by dedicated control software for 3D imaging. Figure 6 shows snap-shot taken images during the EBSD measurement and FIB machining. EBSD maps of 30 μm × 30 μm area were obtained in the experiment. The step size of each EBSD map was 200 nm, and maps were taken using a slice thickness of 200 nm. In order to extract information of plastic deformation from EBSD maps, misorientation angles between the deformed and non-deformed areas were calculated using rotation matrix calculations [8].

4.2. Results and discussion

Figure 7 shows the distribution of plastic deformation in the nickel plate, where a small indent has been made in the vicinity of twin boundary. Figure 7(a) shows a 3D map reconstructed from a series of contour maps of misorientation angle obtained from the 3D-EBSD measurement results. In the 3D map it is seen that the nickel crystal is slightly deformed and still has a misorientation angle of several degrees at a depth of 10 μm from the surface. This indicates that the plastic deformation made with the indenter extends to a region five times as deep as the depth of the indent. Figure 7(b) shows a two-dimensional contour map of misorientation angle obtained at a depth of 1.8μm from the surface of the nickel plate. The deformed region around the impression has a misorientation angle larger than 5°. The region near the center of the impression, however, has a misorientation angle smaller than 1°. This indicates that severe plastic deformation occurs at in material touching the face of the pyramid shaped indenter, and that little deformation occurs at the apex of the indenter. Figures 7(c) and 7(d) show cross-sectional views across and along the twin boundary, respectively. Plastic deformation in the crystal shows anisotropic distribution, and propagation of the deformation is obstructed at the twin boundary and parallel to {111} planes. It is considered that relaxation of the plastic deformation caused by the slip planes, {111}, and the twining planes having the same index as the slip plane is responsible for the anisotropic propagation. Thus, 3D-EBSD is a powerful technique to visualize the

Figure 6 Snap-shot images taken during (a) EBSD measurement and (b) FIB machining [7].

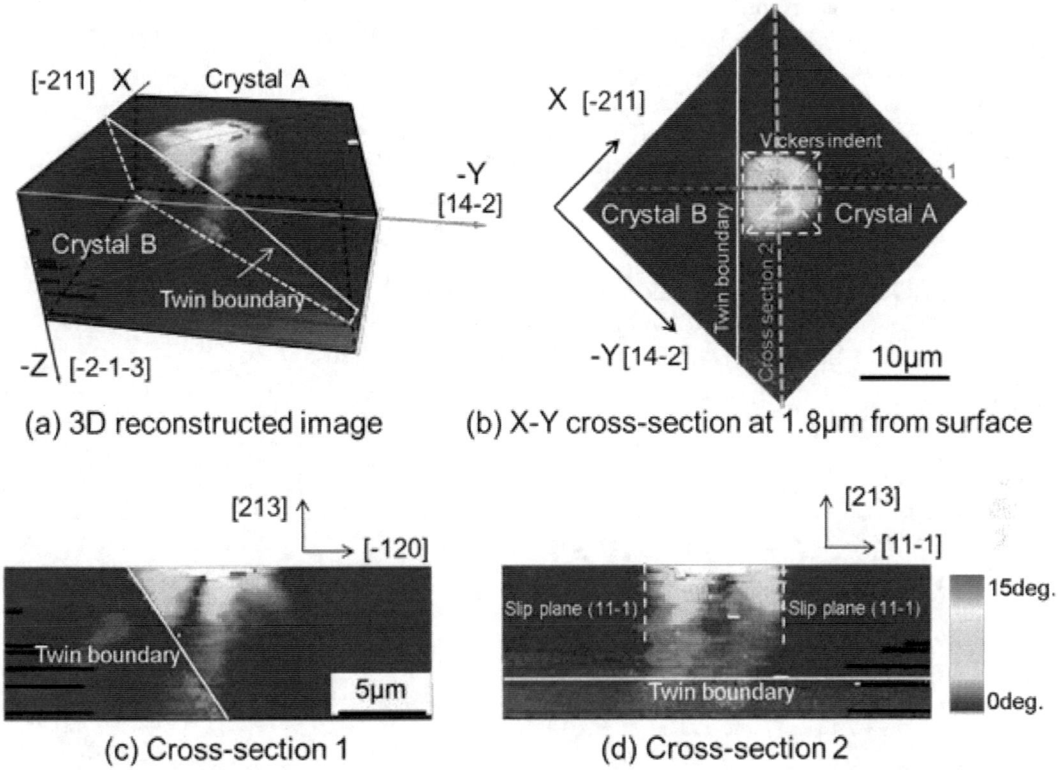

Figure 7 (a) 3D reconstructed image and (b) X-Y cross-sectional view of plastic strain distribution in the pure Ni plate indented by Vickers indentation. Cross-sectional views of (c) and (d) show the cross-sections along the red and yellow dotted lines indicated in (b), respectively [7].

3D propagation of plastic deformation and strain, and the obtained results should contribute to crystal plasticity studies.

5. Conclusion

3D analysis can provide a full picture of the morphology of the internal structure in steels, including the crystalline grain arrangement, the phase domains and the presence of non-metallic inclusions, as well as providing information on the propagation behavior of plastic deformation. In order to observe the evolution of such features and morphological changes during production processes, time-resolved 3D observations should be realized. Serial sectioning techniques can yield information at the micro- or sub micro-meter scale. It is difficult, however, to apply such techniques to in-situ observations, because serial sectioning is a destructive technique. Although tomography techniques can be used to observe a sample non-destructively, there are still issues regarding the ability for time-resolved observations, due to long acquisition times. Therefore, high intensity sources, such as synchrotron radiation, and high sensitivity detectors are required to improve the acquisition time. Collaborative analysis between in-situ observations and 3D analysis will be a key technique to observe the dynamical behavior of internal structure evolution and morphological changes, it is expected such combined data will support the development of advanced steel products in the near future.

References

[1] West G D and Thomson R C 2009 *Journal of Microscopy* **233** 442
[2] Kübel C, Voigt A, Schoenmakers R, Otten M, Su D, Lee T C, Carlsson A and Bradley J 2005 *Microscopy and Microanalysis* **11** 378
[3] Taniyama A 2010 *CAMP-ISIJ* **23** 471
[4] Ludwig W, Schmidt S, Lauridsen E M and Poulsen H F 2008 *J. Appl. Cryst.* **41** 302
[5] Johnson G, King A, Honnicke M, Marrow J and Ludwig W 2008 *J. Appl. Cryst.* **41** 310
[6] Ludwig W, King A, Reischig P, Herbig M, Lauridsen E M, Schmidt S, Proudhon H, Forest S, Cloetens P, Rolland du Roscoat S, Buffière J Y, Marrow T J and Poulsen H F 2009 *Mater. Sci. Eng.* **A 524** 69
[7] Taniyama A 2012 *Proc. of EMC2012 (Manchester)* PS2.2
[8] Randle V and Engler O 2000 *Introduction to Texture Analysis:Macrotexture, Microtexture and Orientation Mapping* (Overseas Publishers Association, Amsterdam)

IOP Conf. Series: Materials Science and Engineering **580** (2019) 012026 doi:10.1088/1757-899X/580/1/012026

The Coupling of Grain Growth and Twinning in FCC Metals

Spencer L Thomas[1], **Jian Han**[2,4], **David J Srolovitz**[2,3,4*]

[1] Department of Materials Science and Engineering, North Carolina State University, Raleigh, NC 27695 USA
[2] Department of Materials Science and Engineering, University of Pennsylvania, Philadelphia PA 19104 USA
[3] Department of Mechanical Engineering and Applied Mechanics, University of Pennsylvania, Philadelphia PA 19104 USA
[4] Department of Materials Science and Engineering, City University of Hong Kong, Hong Kong SAR, China

[*] srol@seas.upenn.edu; srol@cityu.edu.hk

Abstract. Coherent twin boundaries (CTBs) routinely form during the annealing of polycrystalline metals, in the absence of an applied stress. Molecular dynamics (MD) simulations of normal grain growth in nanocrystalline metals show such annealing twins as well the formation of twin junctions. MD simulations and theoretical analyses demonstrate how these junctions form and that their formation necessarily retards grain boundary (GB) migration. Both CTB and GB migration occurs via disconnection motion. We identify the types of disconnections important for CTB migration and show the disconnection pile-ups at TJs during GB migration are responsible for CTB formation in the vicinity of TJs. Analysis further shows that at least two types twinning partials are to be expected during TJ migration and that these give rise to the multiple twinning near migrating TJs observed in the MD simulations.

1. Introduction

Twinning is common in a wide range of materials and especially in face centered cubic (FCC) metals. In FCC crystals, coherent twin boundaries (CTBs) are $\Sigma3$ symmetric tilt grain boundaries (GBs) with a 60° rotation about a $\langle 110\rangle$ axis and are typically associated with very low energy. If we describe the FCC crystal structure as a stacked sequence of $\{111\}$ planes, ...ABCABCABC..., a CTB may be described as a ...ABCA**B**ACBA... (the CTB is at **B** here). CTBs commonly form during annealing or as a result of mechanical deformation and hence are often described as "annealing twins" or "deformation twins": the two types of CTBs are structurally identical. Here, we examine how CTBs form, move, extend and, especially, are coupled to microstructure evolution (GB migration).

Grain boundaries migrate through the motion of line defects, that are constrained to move in the GB plane. These line defects, known as disconnections, have both dislocation (Burgers vector **b**) and/or step (step height h) character. The set of possible disconnection modes $\{\mathbf{b}_m, h_m\}$ are fixed by bicrystallography; the important modes are determined by energetics and/or kinematics. Disconnection glide in the GB plane, translates the GB by a step height h and produces a shear displacement across the GB (i.e., displaces one crystal with respect to the other) by **b**. Since CTBs are GBs, they too move by disconnection motion. In most cases, the important disconnections for CTBs in FCC materials are $a/6\ \langle 112 \rangle$ Shockley partial dislocations, where a is the cubic lattice parameter.

We report on a series of molecular dynamics simulations of grain growth in an FCC polycrystal, that show disconnection formation and extension in the absence of an applied stress. These simulations also demonstrate that formation of CTB penta-twin junctions (five twin GBs meeting along a common $\langle 110 \rangle$ axis) and other CTB junctions are a necessary result of GB migration in polycrystals. Most such junctions retard GB migration. Then, we examine the roles played by disconnections in CTB nucleation and CTB migration. The main conclusions of this report are (i) that CTB nucleation during annealing is a natural consequence of disconnection-mediated GB migration in polycrystals, (ii) that twin junction formation occurs naturally during grain growth, and (iii) that such junctions limit GB migration. This report is partially based on our previously published results [1-4].

2. CTB Formation and Extension during Grain Growth

We performed large-scale MD simulations of nanocrystalline nickel to study the atomistic features of grain growth using an EAM potential [5] and the LAMMPS MD simulation software [6]. The melting point for this potential, T_m = 1590 K, was determined using the phase coexistence method [7]. The simulations were performed under periodic boundary conditions in cubic simulation cells of edge length ≈ 40 nm and initial mean grain size ≈ 4.5 nm. The starting microstructures were generated using a continuum grain growth simulation method [8]; the widely applied Voronoi tessellation approach to generating grain growth microstructures yields (i) grain size distributions unrepresentative of normal grain growth [9], (ii) perfectly flat GBs, and (iii) unrealistic triple junction (TJ) angles. Each grain was populated with atoms arranged in a perfect face centered cubic (FCC) crystal of randomly-chosen orientation. Atoms significantly closer to one another than the equilibrium 0 K nearest neighbor distance were removed.

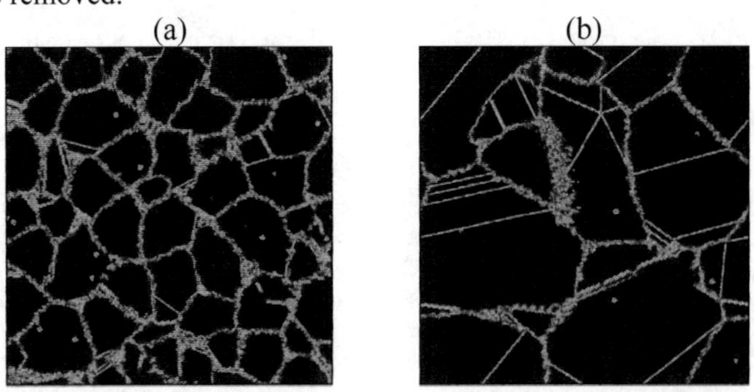

Figure 1. (a) Cross-section of the microstructure after initial relaxation and (b) after a 2.5 ns anneal. Atom colors are assigned based upon the centrosymmetry parameter. The thin, straight lines are CTB planes (with permission [1]).

Each initial atomic configuration was relaxed at 0 K using a conjugate gradient energy method and then MD annealed for 100 ps at 300 K to remove artifacts of the process that generated the initial configuration. A small number of twins were observed in the initial microstructure (see figure 1a). Annealing was performed in an NPT ensemble (Nosè-Hoover thermostat) for 2.5 ns with zero external stress at 1350 K (~$0.85T_m$). The simulation time and temperature were chosen to be sufficiently for significant grain growth without grains spanning the simulation cell and without GB melting. Visualization was performed using the OVITO atomic visualization software [10].

Many coherent twin boundaries (CTBs) form during grain growth (see figure 1b). Nearly all CTBs form at migrating TJs, often in tightly-packed arrays (see figure 2) or in junctions between multiple twins and/or GBs (see figure 3). Preferential formation of CTBs at TJs mirrors experimental observations [11], but why twins form at migrating TJs remains an outstanding question.

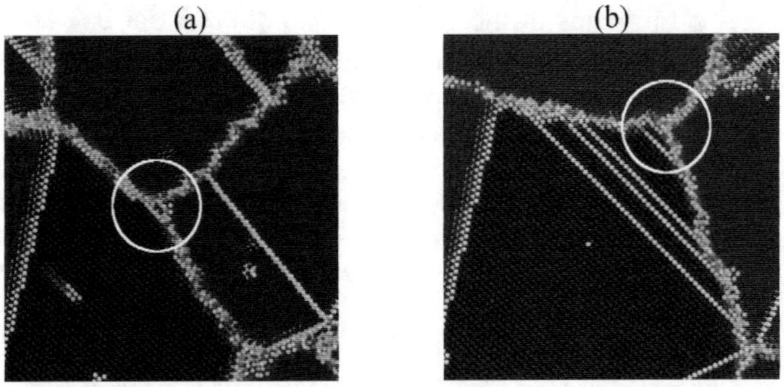

Figure 2. A series of twins emitted from a migrating TJ (circled in white).

Twin junctions have been observed to form in large networks during strain-engineered attention, with observations in nanowires [12,13], nanoparticles [14], and nanocrystalline (NC) metals (polycrystals with nanometer-scale grains) [15,16], but rarely in bulk coarse-grained materials [17]. The angle between CTB habit planes ({111} in FCC) is $\theta_A = \arccos(1/3) \approx 70.53^\circ$, just short of the 72° required to restore a perfect, unstrained crystal. This closure failure implies that penta-twins have disclination character. As disclination stress fields grow linearly with distance from the disclination line, the elastic energy is too large to accommodate except in very small crystals and/or in nano-scale polycrystals.

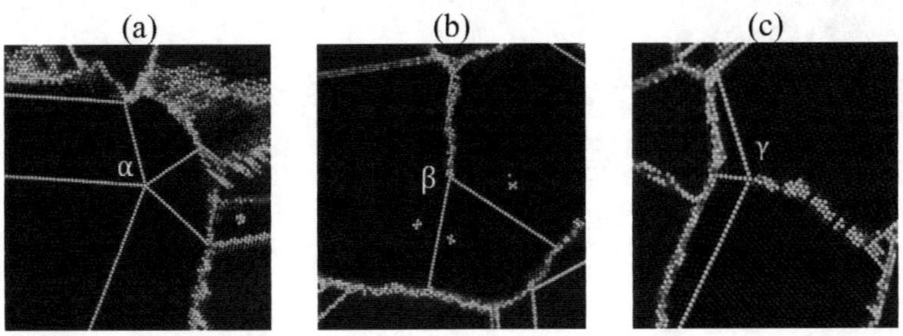

Figure 3. A penta-twin (α), a two CTB-GB junction (β) and three CTB-GB junction (γ) extracted from the nanocrystal grain growth simulation (with permission [1]).

CTB junctions arise naturally from GB migration in twinned grains. Consider a pair of twins on different planes intersecting a common GB. The CTBs extend as the GB migrates and eventually converge, leading to twin junction formation at the GB (see figure 4). As twins cannot terminate within a grain, the only possible CTB junctions consist of 5 twins (penta-twins) or two, three, or four CTBs and a GB; there are six distinct types of twin junctions - three are shown in figure 3 [1]. Twin junctions typically form along ⟨110⟩ axes in FCC metals.

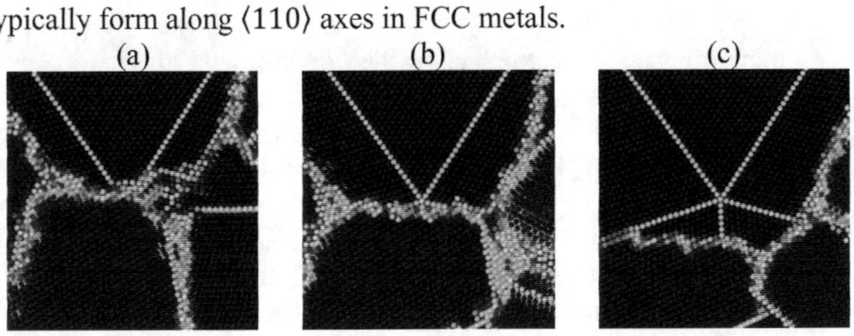

Figure 4. Penta-twin formed by collision of CTBs at a moving GB in a polycrystal (with permission [1]).

A migrating GB must form a twin junction to pull away from a 2-CTB collision (see Figs. 3 and 4). If the GB cannot form such a junction, it will be pinned, as in figure 5. Twin junction formation impedes GB migration as well, though this is less intuitive. Consider the formation of a new GB after a CTB collision (as in Figs. 3b and c). Further GB migration extends the new GB, increasing the energy of the system (proportional to the distance d from the junction to the original GB). As such, the junction imposes a new driving force in opposition to GB migration, that is proportional to the new GB energy. The effect on grain growth can be understood in terms of the von Neumann-Mullins isotropic grain growth law [18,19], which states (in 2D) that the rate of grain area change is $\dot{A} = (M\gamma\pi/3)(n-6)$, where M is the GB mobility, γ is the GB energy, and n is the number of GBs bounding the grain. For a shrinking grain, the effect of a (non-penta-) twin junction is to increase n by 1. This is significant; an initially 5-sided grain would cease shrinking entirely. Of course, this analysis of the effect on normal grain growth has been greatly simplified for clarity; including by the assumption of GB isotropy and the 2D nature of the analysis (this can be relaxed via the 3D extension of the von Neumann-Mullins expression [20]).

If a twin collision results in the formation of a penta-twin junction, the effect on the migration of the original GB is even more profound. This is because the penta-twin implies the existence of a disclination dipole (disclinations of strength $\omega = 2\pi - 5\arccos(1/3) \approx 7.356^\circ$ at the penta-twin core and the migrating GB). Further GB migration extends the dipole, resulting in a back force (per length parallel to the penta-twin) on the GB, $f(d) = -2K\omega^2 d\ [1 + \ln(R/d)]$, where $K = \mu/4\pi(1-\nu)$ (μ is shear modulus and ν is Poisson's ratio). If, however, the GB is perfectly slipping, the retarding force is $f(d) = -4\Lambda K\omega^2 d$ [1] (Λ = 0.593 - numerical integration). In this case, the force retarding the motion of the original GB is proportional to the distance to the penta-twin; much like a spring, pulling the GB back to the penta-twin junction.

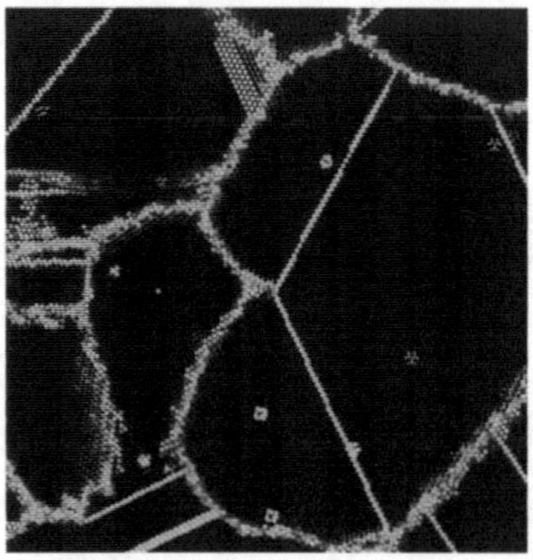

Figure 5. A GB pinned at a CTB collision point (center). The GB is severely distorted by the pinning junction (with permission [1]).

3. Disconnection description of CTB migration and CTB formation

Because CTBs are types of GBs, we can discuss GB and twin dynamics within the same mechanistic framework that is related to the crystal symmetry and the transformation that rotates one grain orientation into that of the grain across the GB. These symmetry considerations imply that there are line defects within the GB, known as disconnections [21,22,23], which have both dislocation and step character (i.e., a Burgers vector **b** and step height h). When a disconnection glides along a flat GB, one grain translates with respect to the other (translation vector **b**) and the GB migrates in the direction normal to the GB plane (by **h**). This corresponds to coupled GB sliding and migration.

Analysis of the translational symmetry associated with the bicrystallography of a CTB reveals the character of CTB disconnections. Figure 6a shows a CTB, where the lattice points in the upper/lower grains are shaded black/white. To analyse the bicrystallography, we interpenetrate the two grains to form a dichromatic pattern (figure 6b). If we translate the black/white grains with respect to one another by the green vector $\mathbf{b}_1$, the dichromatic pattern does not change, but the GB position shifts downwards by the blue vector $\mathbf{h}_{10}$ (Figs. 6b and c). The green vector $\mathbf{b}_1 = [1\bar{1}2]a/6$ and blue vector $\mathbf{h}_{10} = -[\bar{1}11]a/3$ are the Burgers vector and step height associated with CTB disconnection 1. This type of CTB disconnection is also known as a twining partial [24].

We note that, by the same Burgers vector $\mathbf{b}_1$ (see Figs. 6b and c), the GB can also shift upwards by the step $\mathbf{h}_{11}$. This implies that the bicrystallography of a CTB allows multiple disconnection modes; they are characterized by $\mathbf{b}_1$ and $\mathbf{h}_{1j} = \mathbf{h}_{10} + j\,\Delta\mathbf{h}$, where $\Delta\mathbf{h} = [\bar{1}11]a$ and j is an integer. We can also choose

$\mathbf{b}_2$ (see figure 6b) as the Burgers vector of a disconnection, which has the same set of step heights as the disconnection with Burgers vector $\mathbf{b}_1$. All possible CTB disconnection modes may be described as:

$$\mathbf{b}_1 = [1\bar{1}2]a/6, \mathbf{h}_{10} = -[\bar{1}11]a/3$$
$$\mathbf{b}_2 = -[211]a/6, \mathbf{h}_{20} = -[\bar{1}11]a/3,$$

where, in general, $\mathbf{b} = m\mathbf{b}_1 + n\mathbf{b}_2$, $\mathbf{h}_j = m\mathbf{h}_{10} + n\mathbf{h}_{20} + j\,\Delta\mathbf{h}$ and n, m and j are integers. The three twinning partials are ($\mathbf{b}_1$, $\mathbf{h}_{10}$), ($\mathbf{b}_2$, $\mathbf{h}_{20}$) and ($-\mathbf{b}_1 - \mathbf{b}_2$, $-\mathbf{h}_{10} - \mathbf{h}_{20} - \Delta\mathbf{h}$).

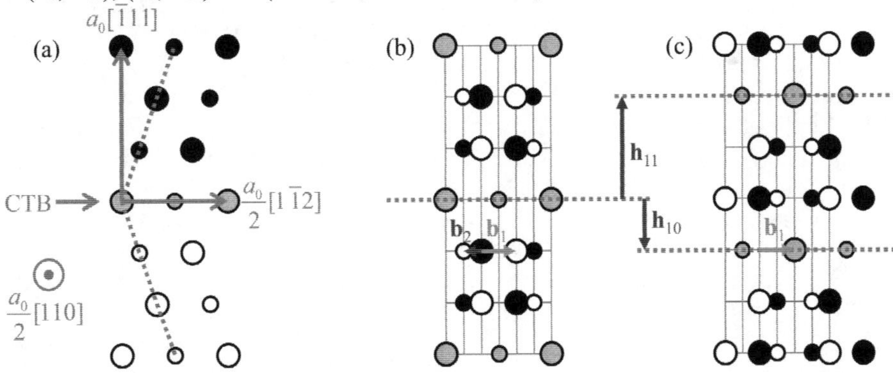

Figure 6. (a) CTB schematic. (b) Dichromatic pattern formed by interpenetration of upper (black) and lower (white) grains. Translation of the black with respect to the white grain by $\mathbf{b}_1$ produce pattern (c). The green vector denotes the Burgers vector of the CTB disconnection $\mathbf{b}_1$ and the purple vector denotes possible Burgers vector $\mathbf{b}_2$. The blue vectors denote two possible step heights $\mathbf{h}_{10}$ and $\mathbf{h}_{11}$ associated with $\mathbf{b}_1$.

Although many CTB disconnection modes are allowed by bicrystallography, they do not occur with equal probability. We expect that the most important are those associated with the lowest nucleation barriers. The nucleation barrier for disconnection mode ($\mathbf{b}$, $\mathbf{h}$) under shear stress (resolved to the CTB plane) $\boldsymbol{\tau}$ and chemical potential jump ψ (across the CTB) driving forces may be estimated as [3]

$$E^* = 2\Gamma_s|h| + 2Kb^2 \ln\left(\frac{\delta^*}{ea}\right), \tag{1}$$

where Γ_s is the step energy and $\delta^* = 2Kb^2/\boldsymbol{\tau}\cdot\mathbf{b} + \psi h$ is the critical disconnection separation.

The nucleation barriers for a set of disconnections under a shear stress $\boldsymbol{\tau}$ = 1 MPa $\mathbf{b}_1/|\mathbf{b}_1|$ are shown in figure 7a, where we set a = 3.615 Å, Γ_s = 0.59 J/m^2 [25], μ = 28.05 GPa, and ν = 0.35 based on an EAM copper potential [26]. Since $\boldsymbol{\tau}$ is parallel to $\mathbf{b}_1$, ($\mathbf{b}_1$, $\mathbf{h}_{10}$) will be the most easily activated mode. The second lowest barrier modes correspond to the other two (energetically degenerate) twinning partials ($\mathbf{b}_2$, $\mathbf{h}_{20}$) and ($\mathbf{b}_3$, $\mathbf{h}_{30}$) ≡ ($\mathbf{b}_1 + \mathbf{b}_2$, $\mathbf{h}_{10} + \mathbf{h}_{20} + \Delta\mathbf{h}$). When a chemical potential jump ψ = 1 MPa is applied across the CTB, the most easily activated disconnection is a pure step. The modes with the second lowest barrier correspond to the three twinning partials. Operation of either the pure-step mode or the three twinning partial mode leads to CTB migration with no shear parallel to the CTB plane. These results show that CTB migration can be controlled by different disconnections depending on the nature of the driving force. Plastic deformation of the polycrystal can drive lattice dislocations into the CTB which may be viewed as heterogeneous sources for disconnection-mediated CTB migration.

As described above, the dynamics of both CTBs and more general GBs by the migration of disconnections. Disconnections modes for more general GBs are discussed in Ref. [3]. In polycrystalline material, GBs are not of infinite extent or extend across an entire sample but rather are terminated by triple junctions. The three GBs meeting at a TJ must migrate and/or slide cooperatively. Since the mechanism of GB migration/sliding is the motion of disconnections, the motion of the triple junction involve reactions between disconnections of the three GBs. Disconnection reactions during TJ motion has been observed in both experiments [27,28] and MD simulations [4].

If the disconnection reactions are such that there is a Burgers vector accumulation at the TJ, the associated stress field will repel additional disconnections from the TJ and migration must stop. The condition that there is no net Burgers vector accumulation at the TJ can only be satisfied provided that the disconnection fluxes on the three GBs $J^{(i)}$ satisfy [3]

$$\begin{pmatrix} \mathbf{b}^{(1)} & \mathbf{b}^{(2)} & \mathbf{b}^{(3)} \\ h^{(1)}\sin\Theta^{(1)} & h^{(2)}\sin\Theta^{(2)} & h^{(2)}\sin\Theta^{(2)} \end{pmatrix} \begin{pmatrix} J^{(1)} \\ J^{(2)} \\ J^{(3)} \end{pmatrix} = 0, \quad (2)$$

where $\mathbf{b}^{(i)}$ and $h^{(i)}$ are the Burgers vector (column vector) and step height of disconnections on the ith GB, and $\Theta^{(i)}$ is the dihedral angle opposite the ith GB. The last row of this matrix corresponds to the condition that the three GBs meet at the TJ. Hence, effectively, we have 4 equations and 3 variables (the matrix is 4×3) and the only solution is that $J^{(i)} = 0$; i.e., there is no disconnection flux and no TJ migration. This implies that, in general, TJs cannot move and disconnections coming from the three GBs will accumulate/pile-up at the TJ and large TJ stresses will be generated at the TJ.

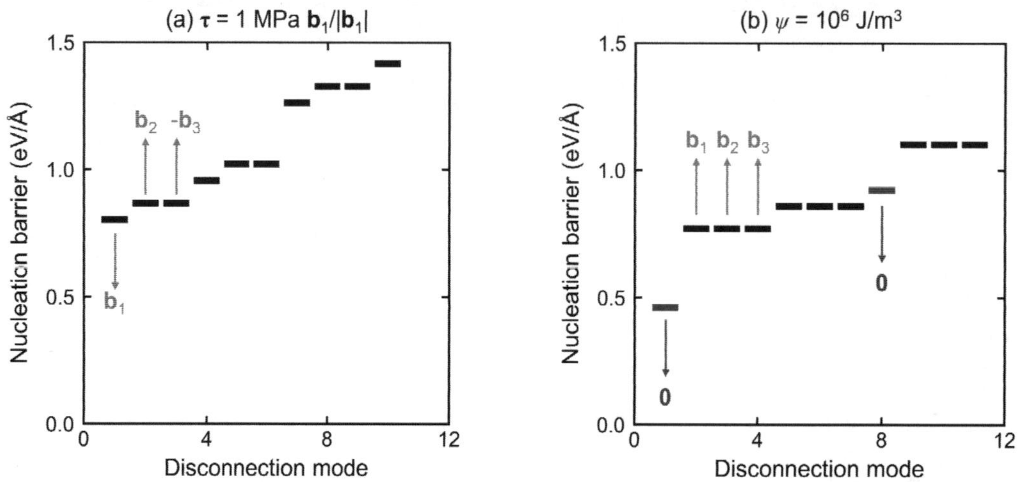

Figure 7. Nucleation barriers for a set of CTB disconnection modes: (CTB shear driven by a shear stress and (b) a CTB migration driven by a chemical potential jump.

Disconnection pile-ups at the TJs delimiting a GB may be described in the same manner as double-ended dislocation pile-ups. The stress-field at the end of such a pile-up at the TJ is proportional to the driving force for disconnection motion (if the GB were infinite) and the square root of the GB length (distance between TJs) L. If the driving force for disconnection migration is an applied stress τ then this gives the classical fracture mechanics result; i.e., the stress field at the TJ scales as $\sim \tau L^{1/2}$. If, on the other hand, the only driving force is capillarity (as in grain growth), then the stress field (associated with the disconnection pile-up) scales as $\sim \psi L^{1/2}$ (recall ψ is the jump in chemical potential across a GB). Since $\psi \sim \gamma\kappa$ (γ is the GB energy and κ is the mean GB curvature), $\kappa \sim 1/R$ (R is the grain size) and $L \sim R$, this implies that in grain growth there will be a stress generated at the TJ that $\sim \gamma/R^{1/2}$. In other words, grain growth generates large stresses at TJs and such stresses are larger in nanocrystalline materials than in large-grain polycrystals. Of course, nature abhors very large stresses, such those generated at the TJs in this manner. These stresses may be relaxed by formation of disconnections of other modes [3], dislocation plasticity within the grains or twinning (i.e., the formation of CTBs at the TJ).

We now consider twinning at the TJ as a means of relaxing the stress. Returning to the analysis of the TJ in terms of disconnections in Eq. (2), we now consider the possibility that the TJ may emit twinning partials dislocations. In particular, the fluxes of two twinning partials are associated with Burgers vectors $\mathbf{b}^{\text{T1}}$ and $\mathbf{b}^{\text{T2}}$ (i.e., J^{T1} and J^{T2}), such that

$$\begin{pmatrix} \mathbf{b}^{(1)} & \mathbf{b}^{(2)} & \mathbf{b}^{(3)} & \mathbf{b}^{\mathrm{T1}} & \mathbf{b}^{\mathrm{T2}} \\ h^{(1)}\sin\Theta^{(1)} & h^{(2)}\sin\Theta^{(2)} & h^{(2)}\sin\Theta^{(2)} & 0 & 0 \end{pmatrix} \begin{pmatrix} J^{(1)} \\ J^{(2)} \\ J^{(3)} \\ J^{\mathrm{T1}} \\ J^{\mathrm{T2}} \end{pmatrix} = 0. \quad (3)$$

The matrix on the left is 4×5, corresponding to 4 equations and 5 variables (the 5 fluxes). Now, there is an infinite set of fluxes that can satisfy this equation (rather than no solutions other than $J^{(i)} = 0$ as before). Note, that if we only consider one type of twin, say T1, the matrix is 4×4 and the only solution would, again be all $J = 0$.

This result implies that if there is only one disconnection mode operating in each GB (typical for low temperature) and there is no dislocation plasticity within the grains, then (at least) two types of twins are required for general TJ (and, hence, GB) migration. This is consistent with the grain growth simulations shown in figure 1 where a series of twins were frequently observed near migrating TJs [1].

4. Conclusions

The formation of coherent twin boundaries (CTBs) during the annealing of face centered cubic metals (in the absence of an applied stress) is widely observed in experiments and atomistic simulations. The molecular dynamics simulations of normal grain growth presented here showed: (1) CTBs readily form during grain growth in the absence of an applied stress and span between grain boundaries, (2) CTBs on non-parallel {111} that intersect a GB merge and form junctions as the GB migrates, (3) several types of junction form consisting of 2 or 3 CTBs and a GB or 5 CTBs in a penta-twin junction, and (4) sets of parallel CTBs often form at many migrating triple junctions (TJs). Theoretical analysis and MD simulations demonstrate that all CTB junctions retard (or stop) the migration of the GB from which the CTB junctions were formed and hence slow grain growth. Of the possible CTB junctions, penta-twin junctions are the most effective at retarding GB migration.

CTBs and more general GBs move via the migration of disconnections (line defects with both dislocation and step character) in their plane. The migration of CTBs can occur by disconnections motion that have twinning partial dislocation character or pure steps, depending on the nature of the driving force. The capillarity-driven migration of GBs during grain growth also occurs via the motion of disconnections. In a polycrystal, these disconnections pile-up at TJs creating large stresses in the vicinity of the TJs. Long-range TJ motion requires the relaxation of these stresses. This may occur via activation of higher energy disconnection modes, the emission/absorption of lattice dislocations, or twinning. In cases where the first two of these relaxation mechanisms do not operate (common in nanocrystalline metals), the large stresses associated with disconnection pile-ups lead to twinning at/near the migrating TJs. In cases where there is only one disconnection mode operating at the GBs meeting at the TJ, TJ migration requires at least two types of twinning modes (2 different twinning partials). This gives rise to the multiple twinning near migrating TJs observed in the MD simulations.

References

[1] Thomas SL, King AH and Srolovitz DJ 2016 *Acta Mater.* **113** 301
[2] Thomas SL, Chen K, Han J, Purohit PK and Srolovitz DJ 2017 *Nat. Comm.* **8** 1764
[3] Han J, Thomas SL and Srolovitz DJ 2018 *Prog. Mater. Sci.* **98** 386
[4] Thomas SL, Wei C, Han J, Xiang Y and Srolovitz DJ 2019 *Proc. Natl. Acad. Sci.* **116** 8756
[5] Mishin Y, Farkas D, Mehl MJ and Papaconstantopoulos DA 1999 *Phys. Rev. B* **59** 3393
[6] Plimpton S 1995 *J. Comp. Phys.* **117** 1
[7] Puna GP and Mishin Y 2009 *Phil. Mag.* **89** 3245
[8] Lazar EA, Mason JK, MacPherson RD and Srolovitz DJ 2011 *Acta Mater.* **59** 6837
[9] Mason JK, Lazar EA, MacPherson RD and Srolovitz DJ 2015 *Phys. Rev. E* **92** 063308
[10] Stukowski A 2010 *Model. Simul. Mater. Sci. Eng.* **18** 015012
[11] Lin B, Jin Y, Hefferan C, Li S, Lind J, Suter R, Bernacki M, Bozzolo N, Rollett A and Rohrer R

2015 *Acta Mater.* **99** 63
[12] Zhou GW, Zhang Z, Bai ZG, Feng SQ and Yu DP 1998 *Appl. Phys. Lett.* **73** 677
[13] Wu B, Heidelberg A, Boland JJ, Sader JE, Sun X and Li Y 2006 *Nano Lett.* **6** 468
[14] Hofmeister H 2004 *Fivefold twinned nanoparticles* in: *Encyclopedia of nanoscience and nanotechnology* (Stevenson Ranch: American Scientific Publishers) **3** 431
[15] Wada T, Negami T and Nishitani M 1994 *Appl. Phys. Lett.* **64** 333
[16] Narayan J, Srivatsa AR and Ravi KV 1989 *Appl. Phys. Lett.* **54** 1659
[17] Wang W, Lartigue-Korinek S, Brisset F, Helbert AL, Bourgon J and Baudin T 2015 *J. Mater. Sci.* **50** 2167
[18] von Neumann J 1952 *Metal interfaces*, ed C Herring (Cleveland: ASM) p 108
[19] Mullins WW 1956 *J. Appl. Phys.* **27** 900
[20] MacPherson RD and Srolovitz DJ 2007 *Nature* **446** 1053
[21] Bollmann W 1970 *Crystal defects and crystalline interfaces* (Berlin: Springer)
[22] Ashby MF 1972 *Surf. Sci.* **31** 498
[23] Hirth JP and Balluffi RW 1973 *Acta Metall.* **21** 929
[24] Anderson PM, Hirth JP and Lothe J 2017 *Theory of dislocations* (New York: Cambridge)
[25] Bulatov VV, Reed BW and Kumar M 2014 *Acta Mater.* **65** 161
[26] Mishin Y, Mehl MJ, Papaconstantopoulos DA, Voter AF and Kress JD 2001 *Phys. Rev. B* **63** 224106
[27] Miura S, Hashimoto S and Fujii TK1988 *Le J. Phys. Colloq.* **49** C5–599
[28] Zhu Q, Cao G, Wang J, Deng C,Li J, Zhang Z and Mao SX 2019 *Nat. Comm.* **10** 156

IOP Conf. Series: Materials Science and Engineering **580** (2019) 012027 doi:10.1088/1757-899X/580/1/012027

Effect of strain gradient on the microstructure and mechanical properties of pearlitic steel

Johan Ahlström[1] and Xiaodan Zhang[2]

[1] Division of Engineering Materials, Department of Industrial and Materials Science, Chalmers University of Technology, 412 96 Gothenburg, Sweden
[2] Section of Manufacturing Engineering, Department of Mechanical Engineering, Technical University of Denmark, 2800 Kgs, Lyngby, Denmark

E-mail: johan.ahlstrom@chalmers.se

Abstract. Pearlitic steels, with a combination of good strength and wear properties, are commonly used for railway rails. The passage of trains creates large shear strain gradients in the surface layer of rails. Knowledge of the microstructural evolution and material properties as related to the shear strain in this layer is therefore important for prediction both of crack evolution and fatigue life. A bi-axial torsion-compression machine was used to deform fully pearlitic R260 rail steel test bars to create a similar gradient structure. Uniaxial tension and compression tests were performed on these pre-deformed test bars to evaluate the mechanical properties of the material. The local microstructural parameters, such as thickness of the ferrite and cementite lamellae, the dislocation density in the ferrite lamellae, the interlamellar spacing and eth local hardness at different places across the diameter of the bars, as well as microstructural evolution across the radius, were characterized. An attempt to set up a correlation between the local microstructural parameters, hardness and the macro mechanical properties is made and discussed in the present study.

1. Introduction

Railway rails are commonly made from pearlitic steel, which provides a sufficiently wear resistant, stiff and strong structure at a reasonably low cost. In production, low grade rails are rolled to shape and left to air cool. Production routes for more advanced head-hardened pearlitic rails and bainitic rails are also emerging. Due to rolling contact during loading, railway rails in service accumulate large plastic strains in the top layer of the rail head. The cyclic plastic deformation also generates residual compressive stresses and leads to orientation of the microstructure. The accumulated strain gradient is large, and the surface layer often contains cracks (termed "head checks"), which make investigation of the elastic-plastic properties of material extracted from the field difficult.

Some established methods are available to reach large plastic strain levels similar to those found in the surface of railway rail heads during service. These include wire drawing [1-3] and high pressure torsion [4]. Previous investigations of rail steels deformed using these methods have concluded that anisotropy exists in both the plastic behavior (yield surface) [5] and fracture behavior [6]. However, a reliable evaluation of the local strength is still missing. The current study serves to investigate the influence of plastic strain gradients on the local microstructure and mechanical properties, especially from the small strain level. The results will provide the base for the extension of the present methodology to high strains.

2. Materials and experimental

The material studied is a R260 rail steel. The microstructure of this material is almost fully pearlitic (as shown in figure 1), with nominal chemical composition as shown in table 1.

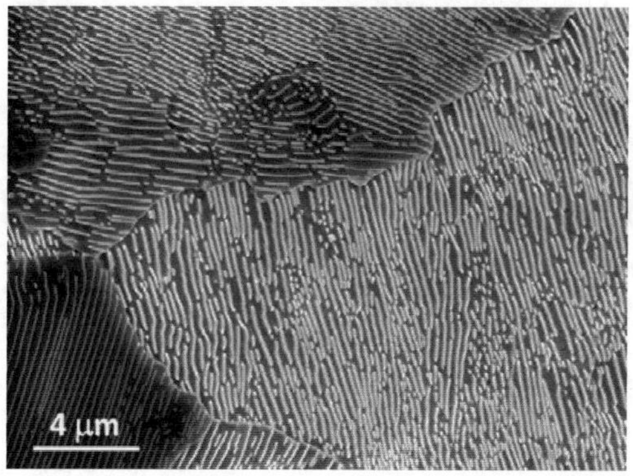

Figure 1. The pearlitic microstructure of the R260 rail steel

Table 1. Nominal chemical composition in wt-% according to ASTM E 572-13

C	Si	Mn	P	S	Cr	Al	V	N	Cu
0.72	0.31	1.04	0.006	0.010	0.02	<0.002	<0.005	0.006	0.018

2.1. Sample extraction and preparation

Test bars used to investigate the relationship between gradient structure and shear strain were extracted from rails about 20 mm below the surface of virgin rail heads. Test bars were deformed in a bi-axial machine using torsion-compression, described in detail in a previous study [7, 8]. Test bars with a gauge diameter of 10 mm were twisted 3 times (3 × 90°) using 500 MPa nominal compression load (i.e. constant force, not increasing as the diameter increases). The samples for the microstructure and hardness measurement on the longitudinal cross section were taken from the middle part of the gauge section of the bars. Scanning electron microscopy (SEM), and transmission electron microscopy (TEM) have been used to characterize the microstructure. Details of the sample preparation can be found elsewhere [9-13].

2.2. Strain calculation

It is necessary to handle the strain induced by the torsional loading from different perspectives for useful characterization of deformed material properties. Using a continuum approach, it can be assumed that each plane in the original gauge section remains planar, and that the total strain imposed is proportional to the radius of each material point. In a metal though, certain slip planes become more active, and there is a distribution of strain on each cross section. In a material with strong hardening capability these differences decrease and strains are distributed more evenly, while low hardening leads to strain localization and premature fracture.

A rough measure of shear strain can be acquired using outputs from the rotational and axial displacement sensors of the test system, compensating for elastic deflections in the load train and assuming an even strain distribution in the test bar gauge section. After straining, the mantle surface of the test bar can also be studied to determine the maximum shear strain based on flow lines [8].

The focus of this study will be on three places: the center, half radius and surface of the bars. The corresponding shear strain of these three positions are 0, 0.5 and 0.7, calculated using the formula $\gamma = \tan(90-\theta)\cdot r/R$ where θ is the measured angle at the outer surface of tested bars, r is the radius measured from the center line of the bars, and R is the final radius of the tested bars.

2.3. Microstructural characterization

Detailed microstructure characterization was carried out TEM (JEOL 2000FX) due to the fineness of the lamellar structure and sectioning problems related to the 3D lamellar pearlitic structure. Special care was taken to make sure the edge-on situation was used for determination of the parameters, such as the interlamellar spacing (ILS), the thickness of ferrite lamellae (F) and cementite lamellae (C) where ILS is the sum of F and C. Dislocation configurations were revealed by tilting the TEM foil and the densities were determined by the intersection method [1-3].

2.4. Micro-hardness measurements and mechanical testing

A Struers DuraScan 70 microhardness machine was used to conduct measurements on the test with shear strains of 0, 0.5 and 0.7 on the longitudinal cross section of the bars with a load of 500 g (HV0.5). An average value was calculated from ten indents. Tension and compress testing under strain-controlled mode was also carried out.

3. Results and discussion

3.1. Microstructure and hardness evolution

Figure 2 shows the microstructure evolution with increasing shear strain. It is clear that the number of shear bands increases with shear strain, evidenced by twisted lamellae in the microstructure. The detailed quantitative characterization of the microstructural parameters is shown in table 2. The key microstructural parameters such as ILS, F and C decrease by only several percent. However, the hardness shows a large increase with shear strain (see table 3), which indicates the change of another microstructural parameter, namely the dislocation density.

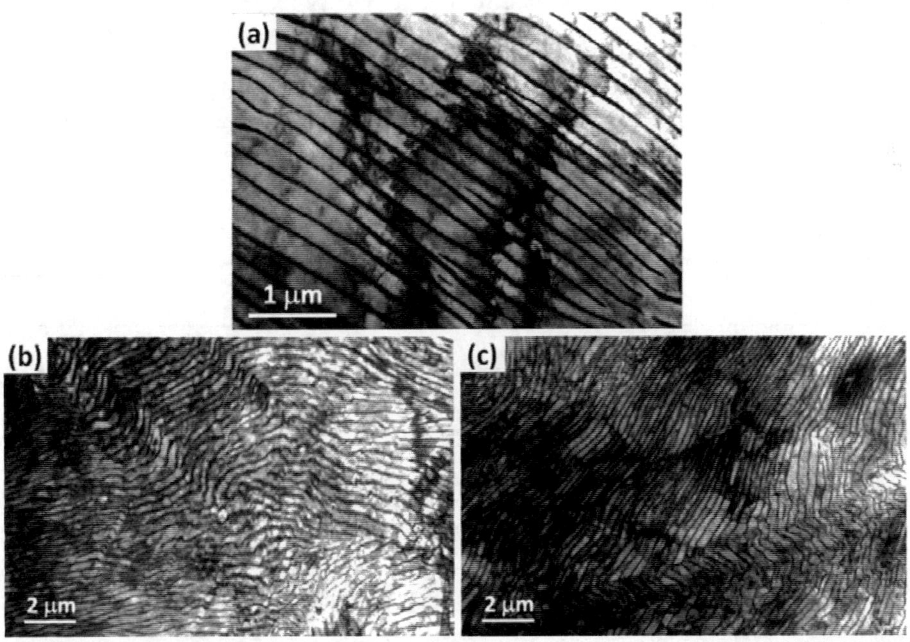

Figure 2. TEM micrographs at positions corresponding to strains of 0 (a), 0.48 (b) and 0.67 (c) of the 3-turns bar, show the alternating ferrite and cementite lamellae. The horizontal direction corresponds to the bar axis.

Table 2. ILS and thickness of ferrite (F) and cementite (C) versus shear strain [8]

Shear strain	ILS (nm)	F (nm)	C (nm)

0	233 ± 39	193 ± 30	40 ± 6.5
0.5	223 ± 40	184.1 ± 39	38.9 ± 8
0.7	224 ± 33	188.4 ± 35	35.6 ± 7.1

Table 3. Microhardness (HV0.5) versus shear strain [8]

Shear strain	Microhardness (HV0.5)
0	288 ± 9.8
0.5	337 ± 15.8
0.7	345 ± 14.8

3.2. Dislocation density

At the small strains investigated in the present study, shear-introduced twisting of the lamellar structure is observed (see figures 2b and 2c). This twisting-related reorientation of the lamellar structure to be vertical to the bar axis is just starting, and is not as severe as seen in pearlitic steel wires drawn to large strains [14, 15] which involves shear banding and curling [16, 17]. Such obvious heterogeneity in the heavily-drawn wires has not been observed in the present case at small strains. As such, the following analysis will be based on the assumption of a lamellar structure and the values of microstructural parameters including dislocation densities will be average values, which also allow the comparison of the present data with the reported data of deformed pearlite in the literature.

In a lamellar structure with alternating soft ferrite and hard cementite lamellae, the plastic deformation is largely controlled by the ferrite lamellae, where the slip initially starts in the ferrite and is then transferred into the cementite [18]. Table 4 shows the increase of dislocation density in the ferrite lamellae with increasing shear strain, while examples of the detailed dislocation configurations are shown in figure 3. The dislocations are of threading and tangled configuration, with bulging from the ferrite/cementite interfaces also seen. Threading dislocations were also observed in the original sample at a relative high dislocation density of 6.0×10^{13} m^{-2}. This high dislocation density may have its cause in the difference of the thermal expansion coefficient of the two phases in, in the semi-coherent interfaces and in elastic compatibility between the two phases [18-20].

Table 4. Dislocation density in the ferrite lamellae versus the shear strain [8]

Shear strain	Average dislocation density (m^{-2})
0	6.0×10^{13}
0.5	2.4×10^{14}
0.7	3.1×10^{14}

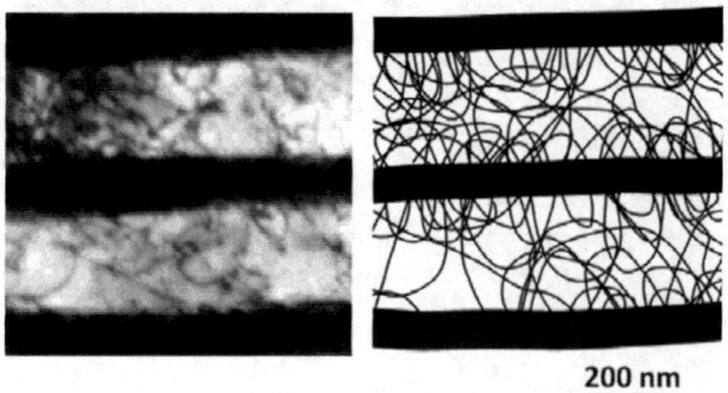

Figure 3. TEM micrograph and sketch showing dislocation configurations in the ferrite lamellae, where the thick black lines represent cementite lamellae and the thin black lines represent dislocations in the ferrite lamellae [8].

3.3. Structure-strength relationship

The quantified microstructural parameters including ILS, F and C (table 2) shows the changes are small for small shear strains in the present case. Accordingly, the hardness increase with shear strain has its major cause in the increase of dislocation density in the ferrite lamellae. The increment of the local strength can be evaluated through the calculated flow stresses from microstructural parameters deduced from the primary flow stress. The microstructural characterization suggests the following two strengthening mechanisms: boundary strengthening related to the ferrite thickness, F, where the cementite lamellae are taken as barriers for dislocation pile-ups, described by $\sigma(b) = k \cdot (2F)^{-0.5}$, where k is the Hall-Petch constant; and dislocation strengthening related to the dislocation density, described by $\sigma(\rho) = M\alpha Gb\sqrt{\rho}$, where M is the orientation factor, taken as $3, \alpha = 0.24$, $G = 78\ GPa$ is the shear modulus, and b is the burgers vector, taken as 0.248 nm.

On the assumption that these strength contributions can be additive, the flow stress can be expressed as:

$$\sigma = \sigma_0 + \sigma(b) + \sigma(\rho) \qquad (1)$$

where σ_0 is the friction stress, taken as 60 MPa [2, 3].

The above calculation of the local strength increment with the shear strain will be compared with the experimental evaluation from the hardness increment via the following formula:

$$\Delta\sigma_{HV} = \Delta HV_{0.5}/k_{HV} \qquad (2)$$

where $k_{HV} = 3.3 \pm 0.4$, consistent with hypo eutectoid steels [21], deformed pearlitic steel at low and medium strains [22], and deformed metals with lamellar dislocation structures at medium and large strains [23-26]. The comparison is shown in table 5. Taking into account the standard deviation, a good agreement is found between the measured strength increment by hardness and the calculated strength increase from quantified microstructural parameters.

Table 5. Local strength increase versus the shear strain [8]

Shear strain	$\Delta\sigma_{cal}$ (MPa)	$\Delta\sigma_{HV}$ (MPa)
0.5	117	146 ± 29
0.7	141	169 ± 44

4. Conclusion

To develop a microstructure-based methodology for evaluation of the local strength and mechanical response of rail surface layers after run-in, a bi-axial torsion-compression machine has been used to deform test bars of R260 pearlitic rail steel to low shear strains. The evolution of local microstructure and local strength with shear strain has been systematically characterized and quantified by TEM and microhardness. The following conclusions can be made:

- The interlamellar spacing, and the thickness of the ferrite and cementite lamellae, only show a small decrease up to a shear strain of 0.7.
- The observed hardness increase with the shear strain is due to the increase of dislocation density in the ferrite lamellae, where the main dislocation configurations are threading dislocations and dislocation tangles.
- Two strengthening mechanisms, namely boundary strengthening and dislocation strengthening, are proposed to account for the results, based on microstructural characterization. Based on the linear additivity of the contributions from these two strengthening mechanisms, good agreement is found between the measured strength increment seen in hardness measurements and the calculated strength increase determined using the quantified microstructural parameters.

Acknowledgements
J A gratefully thanks the National Centre of Excellence CHARMEC (Chalmers Railway Mechanics, www.charmec.chalmers.se). X Z acknowledges support from the European Research Council (ERC) under the European Union's Horizon 2020 research and innovation program (grant agreement No 788567-M4D).

References
[1] Zhang X, Hansen N, Godfrey A and Huang X 2018 *Mater Sci Technol* **34** 794
[2] Zhang X, Hansen N, Godfrey A and X Huang 2016 *Acta Mater.* **114** 176
[3] Zhang X, Godfrey A, Huang X, Hansen N and Liu Q 2011 *Acta Mater* **59** 3422
[4] Pippan R, Wetscher F, Hafok M, Vorhauer A and Sabirov I 2006 *Adv Eng Mater* **8** 1046
[5] Cvetkovski K and Ahlström J 2013 *Wear* **300** 200
[6] Kapp M, Hohenwarter A, Wurster S, Wang B and Pippan R 2016 Acta Mater 106 239
[7] Meyer K A, Nikas D and Ahlström J 2018 *Wear* **396–397** 12
[8] Nikas D, Zhang X and Ahlström J 2018 *Mater Sci Eng A* **737** 341
[9] Chen S, Hu J, Zhang X, Dong H and Cao W 2015 *J Iron Steel Res Int* **22** 1126
[10] Fan G H, Wang Q W, Du Y, Geng L, Hu W, Zhang X and Huang Y D 2014 *Mater Sci Eng A* **590** 318
[11] Zhang X D, Liu W, Godfrey A and Liu Q 2009 *Metall. Mater. Trans. A* **40** 2171
[12] Zhang X, Godfrey A, Huang X, Hansen N, Liu W and Liu Q 2009 *30th Risø international symposium on materials science: nanostructured metals – fundamentals to applications* (Risø, Denmark) p 409
[13] Zhang X, Godfrey A, Hansen N, Huang X, Liu W and Liu Q 2010 *Mater Charac* **61** 65
[14] Zhang X, Hansen N, Godfrey A and Huang X 2012 *33rd Risø international symposium on materials science: Nanometals – status and perspectives* (Risø, Denmark) p 407
[15] Zhang X, Hansen N, Godfrey A and Huang X 2014 *35th Risø International Symposium on Materials Science: New Frontiers of Nanometals* (Risø, Denmark) p 153
[16] Langford G 1977 *Metall Trans A* **8** 861
[17] Zhang X, Godfrey A, Hansen N, Huang X, Liu W and Liu Q 2010 *Mater. Charac* **61** 65
[18] Zhang X D, Godfrey A, Liu W and Liu Q 2011 *Mater. Sci. Techno* **27** 562
[19] Dollar M, Bernstein I M and Thompson A W 1988 *Acta Metall* **36** 311
[20] Zhang X, Godfrey A, Hansen N and Huang X 2013 *Acta Mater* **61**, 4898
[21] Pavlina E J and Van Tyne C J 2008 *J Mater Eng Perform* **17** 888
[22] Zhang X 2009 Quantitative Investigation of Microstructural Evolution During the Cold Wire-drawing of a Pearlitic Steel Wire and its Relationship with Mechanical Properties, Tsinghua University, Beijing
[23] Meyers M A and Chawla K K 2009 *Mechanical Behavior of Materials*, Cambridge University Press, Cambridge, p 223
[24] Zhang X, Hansen N, Gao Y and Huang X 2012 *Acta Mater* **60** 5933
[25] Zhang X, Hansen N and Nielsen C V 2017 *Local microstructure and flow stress in deformed metals* in: IOP Conference Ser: Mater Sci Eng, **219** 012053
[26] Zhang X, Nielsen C V, Hansen N, Silva C M A and Martins P A F 2019 *Int. J. Plast.* **115** 93

IOP Conf. Series: Materials Science and Engineering **580** (2019) 012028 doi:10.1088/1757-899X/580/1/012028

Comparing local deformation measurements to predictions from crystal plasticity during reverse loading of an aerospace alloy

M D Atkinson[a], R A Lebensohn[b] and J Quinta da Fonseca[a]

[a] School of Materials, The University of Manchester, Oxford Road, Manchester, M13 9PL, UK
[b] Los Alamos National Laboratory, MS B216, Los Alamos, NM 87545, USA

michael.atkinson@manchester.ac.uk

Abstract. Cyclic loading is of great importance in aerospace applications but the origins of the change in flow stress with load path are poorly understood and difficult to predict. Many crystal plasticity models are capable of modelling macroscopic flow behaviour during non-monotonic load paths, such as the Bauschinger effect often seen when reverse loading. Although these models represent the microstructure explicitly, it remains unclear whether they capture reversibility at the local microstructural scale, or simply fit the macroscopic hardening response better by using more fitting parameters. Here we present experimental surface deformation data acquired using high resolution digital image correlation (HR-DIC) during a reverse load cycle. With resolution greater than 200nm, these experiments reveal the discrete nature of deformation in the form of localised crystallographic slip bands. These results are then compared statistically, over many hundreds of grains, to crystal plasticity simulations of reverse loading. Simulation results are taken from both finite element method (CPFEM) and fast Fourier transform (CPFFT) encompassing the two popular numerical techniques for full-field crystal plasticity models.

1. Introduction

To further develop physical models including details of material microstructure, it is critical that their predictions are compared to experimental results at the scale of the microstructure. High resolution digital image correlation (HR-DIC) is one such experimental technique, able to provide full-field strain measurements with sub-micron resolution at the surface of a material. Over the last 10 years, the technique has developed to where it can now be routinely applied to many metallic materials [1–3], as well as polymers and ceramics [4]. Although typically monotonic deformation is studied, there are no restrictions to loading condition applied, allowing local strain measurements during complex load cases.

Non-monotonic load cases, such as strain-rate jumps or reversal of load direction, are key benchmarks for any physical based model. Complex interactions between microstructure features lead to, for example, the Bauschinger effect seen when reverse loading. For microstructure-based models to successfully simulate these load cases, the correct local interaction must occur to produce macroscopic response, rather than simply changing the macroscopic response by including more parameters. One such class of model is full-field crystal plasticity, which explicitly include microstructure details, such as grain structure and crystallographic slip. These full-field models can either be included with the finite element method (FEM) [5–8] to solve for stress equilibrium and strain compatibility, or methods based on the use of fast Fourier transforms (FFT) [9–11]. Each solution method has relative advantages, with

FEM able to be applied to complex domain shapes and with arbitrary boundary conditions. Whereas FFT methods have lower computational demand and creation of synthetic polycrystalline volumes is simpler, as the domain does not need to be meshed.

In this paper we will present local strain measurements, from HR-DIC, taken during a load reversal. The load path is then simulated using both a CPFEM and CPFFT model, where a volume element is created with properties representative of the material microstructure. This provides predictions of local strain, that can be compared to the experimental values. We aim to consider the mechanisms of accommodating reverse strain and if these are seen in the models.

2. Methods and material

The material studied here is a solid solution strengthened nickel superalloy Haynes 625, consisting of a face centred cubic (FCC) crystal structure and with nominal composition given in Table 1. Received as a hot-rolled and mill-annealed plate, to ensure no influence of prior deformation, the material was further annealed at 870°C for 1 hour and then air cooled. The material is made up of roughly equiaxed grains containing annealing twins and with no significant texture present. A flat dogbone shaped mechanical sample was produced by electrical discharge machining (EDM) and polished to EBSD quality on both flat faces. Patterning for HR-DIC was then performed on one face using the gold remodelling technique, with details of the process given by Di Gioacchino and Quinta da Fonseca [1]. The pattern produced was imaged using a Tescan Mira 3 scanning electron microscope (SEM) collecting the backscattered electron signal, providing high contrast between the gold pattern and nickel substrate. Images were acquired using 10 kV beam voltage and 20 μm field width over a grid with 20 % overlap, then subsequently stitched using ImageJ [12,13] to produce a single image.

A Deben 2 kN tension/compression stage was used to mechanically load the sample at a strain rate of $1x10^{-4}$ s^{-1}. The stage was fitted with an anti-buckling fixture to prevent sample buckling during compression. Prior to deformation, the HR-DIC pattern was imaged and then a single reverse load cycle was applied, interrupted at three points for imaging. The reverse cycle started with compression and was interrupted at macroscopic plastic strains of -0.014, 0.003 and 0.012, where imaging was performed in the unloaded state. These three steps represent forward compression loading, reversal to close to 0 net strain and reverse tension strain to a similar magnitude as in forward direction. They will be referred to as compression, reversal and tension step respectively throughout. A commercial software package, LaVision DaVis [14], was employed for the digital image correlation, where an algorithm based on FFT cross-correlation of successively smaller image sub regions is used to calculate surface displacement. Strain maps are then produced by differentiating the displacement using a finite difference approach.

Simulation of the reverse load path was performed using two crystal plasticity (CP) models, one using the FEM to solve for equilibrium and the second using an FFT-based algorithm. The CPFEM model used was developed by Dawson et. al. with detailed description of the implementation available [5–7]. The CPFFT model was developed by Lebensohn et. al. and again a detailed description of the implementation is available [9]. Both models are elasto-viscoplastic type, employing Hookean elasticity and a power law flow rule of the form proposed by Peirce, Asaro and Needleman [8]. Work hardening evolution laws did not match between the original models, so the CPFFT work hardening model was modified to match that in the CPFEM model and the same values used for the parameters in both simulations. 12 octahedral (111)<01-1> slip systems were included in the simulations and the hardening parameters determined by fitting to a macroscopic stress-strain in compression. Finally, anisotropic elastic compliance values were taken from the literature [15].

A polycrystalline volume element was produced using the Voronoi tessellation software Neper [16,17], where a two stage process was used to produce a volume with grain size and morphology representative of the material. In stage one, a weighted Voronoi tessellation is created with a size

Table 1. Nominal composition of Haynes 625 in weight percent.

Ni	Co	Fe	Cr	Mo	Nb+Ta	Mn	Si	Al	Ti	C
Balance	1	5	21	9	3.7	0.5	0.5	0.4	0.4	0.10

distribution that matches the test material, as measured by an EBSD scan including over 5000 grains. Here only the sizes of primary grains, not annealing twins, are considered by excluding twin boundaries in calculating the experimental size distribution. In stage two, a subset of the primary Voronoi cells are split along straight boundaries to produce the morphology of annealing twins. Each primary cell was assigned a random orientation and the straight boundaries were along a randomly selected (111) plane for the assigned orientation. Each twin was assigned an orientation that is a 60° around the same [111] direction of the boundary. This process reproduces both the morphology of an annealing twin, with straight boundaries, and the misorientation relation between twin and parent grains. The volume is then discretised by two methods. For the FEM simulation, the volume is a meshed with tetrahedral second-order elements that conform to grain boundaries in the tessellation. For the FFT simulation, the volume is rasterised to a regular grid. Both discretisation processes were performed so that each node in the mesh and Fourier point in the raster represent a similar volume.

3. Results and Discussion

3.1. Experimental

Local strain maps produced using HR-DIC are included as Figure 1, all in-plane components of strain are available and here strain in the loading direction (axial) is displayed, where the loading direction is

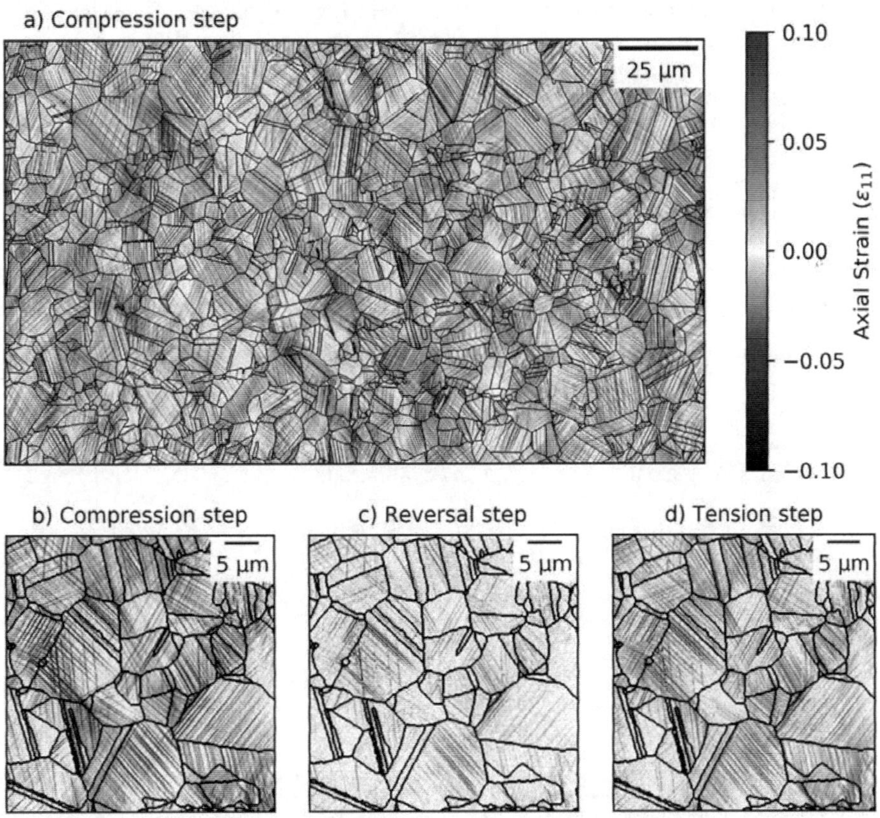

Figure 1. Axial strain maps produced using HR-DIC for 3 points during strain reversal where loading direction is horizontal. a) Entire mapped region following forward compressive deformation, b) a sub region of a, c) a sub region of the reversal step strain map where macroscopic strain has been returned approximately 0 and d) a sub region of the tension step strain map where magnitude of macroscopic strain is approximately equal to the compressive step but now in tension.

IOP Conf. Series: Materials Science and Engineering **580** (2019) 012028 doi:10.1088/1757-899X/580/1/012028

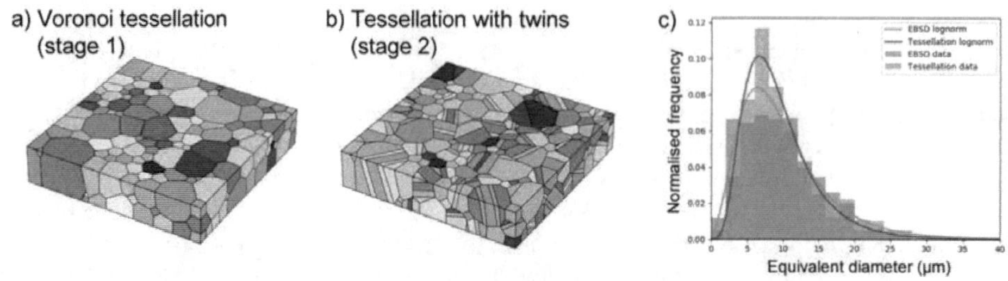

Figure 2. Volume element produced to represent the microstructure of alloy 625. a) Weighted Voronoi tessellation with grain size distribution fit to material with twin boundaries excluded, b) tessellation with annealing twins included by adding straight boundaries along (111) planes and c) comparison of grain size distributions (stage 1) for surface grains. Note that grain colours are assigned randomly.

horizontal. Figure 1a shows the entire mapped region for the compression step. The region covers an area of 220x120 μm^2, including approximately 1500 grains with equivalent diameter greater than 1 μm. Grain boundaries derived from the aligned EBSD map are shown as black lines. Figure 1b-d display a sub region of the mapped region for each deformation step, compression, reversal and tension respectively. These better display the 117 nm spatial resolution of the HR-DIC data, where deformation at this resolution is seen to localise into slip bands that are aligned with (111) slip planes in this FCC material. The HR-DIC maps cover a large number of grains, allowing statistical comparisons to simulation data but also at a resolution high enough to capture the crystallographic strain localisation.

For the forward compression step, only compressive local strains are seen (Figure 1a-b), however when load direction is reversed and material returned to 0 net macroscopic strain, both compressive and tensile local strains are now seen (Figure 1c). When further reverse tensile deformation is applied, Figure 1d, compressive deformation can still be seen in the local strain map, despite now the macroscopic strain being equal in magnitude to in the first compression step. In the regions of the tension map where compressive deformation is seen, there is a change to the slip band structures from those initially formed in the compression step. In many cases the density of slip bands has increased and others are intermittent, with breaks in the path seen during compression. Suggesting that, in the higher density regions, reverse deformation is accommodated by the formation of new slip bands, rather than reversing deformation on the bands formed during forward loading. Welsch, et. al. [18] presented a similar mechanism to explain the work hardening of a high Mn steel, where slip bands become saturated with dislocations as strain is accrued by backstresses from the dislocations preventing the operation of a dislocation source, and is further stabilised by slip across the band on other slip planes. A new band is then nucleated, by activation of a new dislocation source in adjacent material and continues to accommodate deformation. Here we

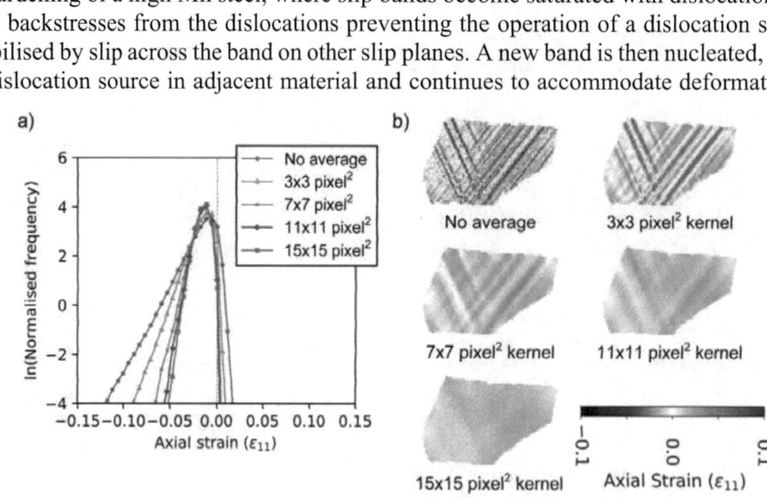

Figure 3. The effect of kernel averaging and kernel size on HR-DIC data where a) are strain distributions from the compression step and b) are single grains from the same step.

suggest that bands saturated and stabilised in the forward direction remain in the material and reverse deformation is accommodated by the nucleation of new slip bands in adjacent material.

3.2. Simulations and comparison to experimental data

The synthetic microstructure produced for crystal plasticity simulation is included as Figure 2, where Figure 2a shows the Voronoi tessellation of primary cells (stage 1) and Figure 2b then shows the final tessellation with a subset of the primary cells split along straight boundaries to produce annealing twins. Figure 2c then shows a comparison of grain size distribution as measured by EBSD and the grain sizes measured at the surface of the synthetic volume.

As both models used here are based on continuum mechanics, careful consideration must be taken when comparing HR-DIC experiments, where strain is seen to localise into slip bands which is not included as a deformation mechanism in continuum crystal plasticity simulations. Here we will present a first simple approach to smoothing HR-DIC strain data using kernel averaging to compare to simulation predictions. **Error! Reference source not found.** shows the influence of kernel averaging, with increasing kernel size, on the local strain from the compression step. **Error! Reference source not found.**a shows this for the strain distribution of the entire map, and **Error! Reference source not found.**b shows the effect on a single grain. As the kernel size increases, the distribution tail of compressive strain is seen to contract, with the high strain magnitude slip band features smoothed, as seen in the single grain maps. To choose the correct size of kernel for comparison, the length scale introduced to the simulation volume during the grain size optimisation is used. Using this length scale, the kernel containing 7x7 pixel^2 best approximates the area represented by a single node and Fourier point in the FEM and FFT simulation respectively on the surfaces of the volume.

Axial strain distributions from the experiment are now compared to the two from simulations in Figure 4, where a log frequency scale is used to highlight changes in strain localisation in the tail of the

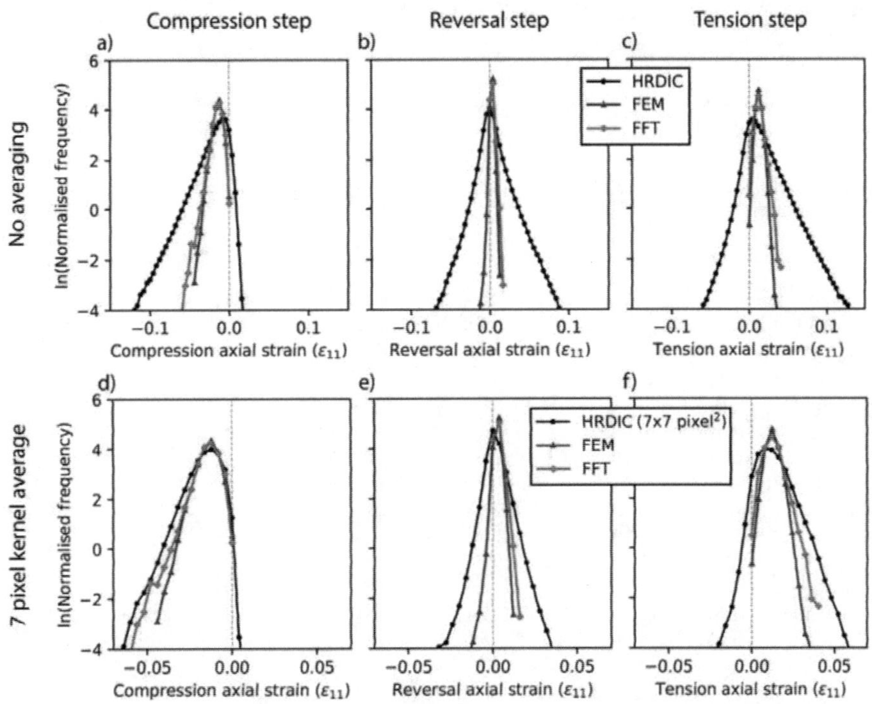

Figure 4 – Comparison of measured local strain distributions with those predicted by the FEM and FFT based models. The first row (a-c) compares the raw experimental data for 3 deformation steps and the second row (d-f) compares the experimental data with 7x7 pixel^2 kernel averaging applied.

distributions. In the figure, the first row (a-c) compares the raw experimental data for 3 deformation steps and the second row (d-f) compares the experimental data with 7x7 pixel^2 kernel averaging applied. When comparing simulation to the raw HR-DIC data, higher local strains are seen at all strain steps in the experiment. This is related to the formation of slip bands in the material that are not predicted by the model. With the kernel average applied, the distribution of strain is predicted quite well for the compression step (Figure 4d), however for the reversal and tension step the localised strains seen in the experiment are still higher than predicted by the model (Figure 4e-f). In all cases the FFT-based model predicts strains up to higher magnitude in the current loading direction than for FEM, relating to the different approach to solving for stress equilibrium. In FEM stress equilibrium is solved in a weak form, whereas in the FFT-based both equilibrium and compatibility are solved in strong form, meaning that peak stresses are higher in the FFT-based simulation, leading to concentration to larger strains. The larger spread in strains seen in the experiment after reversal suggests that the mechanism by which reverse strain is accommodated is not captured in the models. As seen in the strain maps (Figure 1), load reversal can include interaction and nucleation of slip bands and as slip bands do not develop in these models, their effects on the strain distribution on reversal is not predicted.

4. Conclusion

The experimental results show when loading direction is reversed the strain localisations produced during forward loading do not reverse in all cases, with refined slip band structures seen after reversal. The two crystal plasticity models predict strain localisation well for forward loading, once consideration is made of the fact that slip bands do not form in these models. However, local strain is not predicted well following a reversal. This suggests that the discreet nature of slip and arrangement into slip band structures are important when modelling local strains. At the microstructure scale, interactions between slip bands and formation of refined band structure on reversal influences the local strains and omission from crystal plasticity models limits the validity of predicted local strains following a load reversal.

Acknowledgements
The authors would like to thank the Engineering and Physical Sciences Research Council UK (EPSRC) for funding the study through the Centre for Doctoral Training in Advanced Metallic Systems (EP/G036950/1). The authors are grateful to Rolls-Royce Plc for their funding and support through the EPSRC strategic partnership (EP/M005607/1). The authors also thank P. Dawson for making his CPFEM model available and to R. Quey for the open source software Neper.

References
[1] Di Gioacchino F and Quinta da Fonseca J 2013 *Exp. Mech.* **53** 743–54
[2] Orozco-Caballero A, Lunt D, Robson J D and Quinta da Fonseca J 2017 *Acta Mater.* **133** 367–79
[3] Lunt D, Orozco-Caballero A, Thomas R, Honniball P, Frankel P, Preuss M and Quinta da Fonseca J 2018 *Mater. Charact.* **139** 355–63
[4] Montgomery C B, Koohbor B and Sottos N R 2019 *Exp. Mech.*
[5] Dawson P R and Boyce D E 2015 *arXiv [cond-mat.mtrl-sci]* 1–42
[6] Marin E B and Dawson P R 1998 *Comput. Methods Appl. Mech. Eng.* **165** 1–21
[7] Marin E B and Dawson P R 1998 *Comput. Methods Appl. Mech. Eng.* **165** 23–41
[8] Peirce D, Asaro R J and Needleman A 1983 *Acta Metall.* **31** 1951–76
[9] Lebensohn R A, Kanjarla A K and Eisenlohr P 2012 *Int. J. Plast.* **32–33** 59–69
[10] Maiti T and Eisenlohr P 2018 *Scr. Mater.* **145** 37–40
[11] Shanthraj P, Eisenlohr P, Diehl M and Roters F 2015 *Int. J. Plast.* **66** 31–45
[12] Schindelin J, Rueden C T, Hiner M C and Eliceiri K W 2015 *Mol. Reprod. Dev.* **82** 518–29
[13] Preibisch S, Saalfeld S and Tomancak P 2009 *Bioinformatics* **25** 1463–5
[14] LaVision GmbH
[15] Wang Z, Stoica A D, Ma D and Beese A M 2016 *Mater. Sci. Eng. A* **674** 406–12
[16] Quey R, Dawson P R and Barbe F 2011 *Comput. Methods Appl. Mech. Eng.* **200** 1729–45
[17] Quey R and Renversade L 2018 *Comput. Methods Appl. Mech. Eng.* **330** 308–33
[18] Welsch E, Ponge D, Hafez Haghighat S M, Sandlöbes S, Choi P, Herbig M, Zaefferer S and Raabe D 2016 *Acta Mater.* **116** 188-99

IOP Conf. Series: Materials Science and Engineering **580** (2019) 012029 doi:10.1088/1757-899X/580/1/012029

Effect of heterogeneous laser surface treatment on mechanical properties of interstitial free steel

Pengzhan Cai[1], Xiaojin Lian[1], Zhixiang Tang[1], Ling Zhang[1,2,*], Tingting Wang[3], Linjiang Chai[3,*], Guilin Wu[1]

[1] International Joint Laboratory for Light Alloys (MOE), College of Materials Science and Engineering, Chongqing University, Chongqing, 400045, China

[2] Electron Microscopy Center of Chongqing University, Chongqing University, Chongqing 400044, P.R. China

[3] College of Materials Science and Engineering, Chongqing University of Technology, Chongqing 400054, China

* Corresponding author: zhangling2014@cqu.edu.cn, chailinjiang@cqut.edu.cn

* To whom any correspondence should be addressed.

Abstract. Laser surface treatment was applied to a Ti-added interstitial-free steel by linear-scanning. The distance between each linear scan was 1.5 mm and the width of the laser-modified zone for each scan was 1 mm. These process parameters result in a heterogeneous surface, containing hard laser-modified lamella separated by non-treated material. It is found that after the treatment, higher strength and good ductility can be obtained. Electron channelling contrast imaging was used to investigate the structure of the sample, which was then compared with the untreated one to reveal the origin of the improved mechanical behaviour of the laser surface-treated steel.

1. Introduction

Recently, heterogeneous structured bulk metals have been developed with both high strength and good ductility [1-6]. For example, a material as strong as ultrafine-grained metal and as ductile as its conventional coarse-grained counterpart has been reported in a heterogeneous lamella-structured Ti with soft micro-grained lamella embedded in a hard ultrafine-grained lamella matrix [7]. Similar results have been found in single-phase metals such as interstitial-free (IF) steel [8], Cu [9], AZ31 Mg alloy [10] and some multi-layered materials, such as copper with bronze [11]. Suppression of instabilities during tensile testing is speculated to be an effective way to retain ductility without sacrificing the strength of ultrafine/nano grained structures [5, 12]. On the other hand, by generating severe plastic deformation in sub-surface layers through surface modification, such as surface mechanical attrition treatment (SMAT) [13], surface mechanical grinding treatment (SMGT) [14-16], shot peening [17], ultrasonic shot peening [18], and laser-shock peening (LSP) [19], the introduction of a gradient structure [20] into metals significantly improved the overall properties of materials. Mechanical incompatibility through a induced strain gradient and complex stress states has been suggested as the main reason for the improved properties [20].

It is of great interest to investigate the mechanical properties of only surface-treated heterogeneous lamella structured metals. Lu and coworkers [21] reported that by topologically controlling the planar structure of pure Cu sheets through partial SMAT processing, both high strength and high ductility could be achieved. The hard and soft zones were however, created randomly by this method. It is desirable therefore to find other methods to form surface heterogeneous lamella structures. Among the various surface treatment approaches, laser surface treatment (LST) is very promising as it is very flexible and can deal with complex shapes by controlling the laser process parameters. In the present study, an IF steel was processed by incorporating a simple layered pattern by LST. The laser surface treated of a coarse-grained sample showed excellent properties, which will be discussed in detail.

2. Experimental

1 mm thick Ti-added (0.0012 wt%C - 0.061 wt%Ti) commercial IF steel sheet was used in the present study. The fully recrystallized grain size was about 20 μm. Laser surface treatments were directly performed on both sides of dog-bone shaped tensile samples. The tensile samples had a gauge length of 10 mm and a width of 5 mm. The sample surfaces were polished with a final grinding paper of 4000# SiC and carefully cleaned before laser surface treatment. A pulsed 600W Nd:YAG laser apparatus was used. The designed pattern and samples are shown in figure 1. The distance between each laser track is 1.5 mm, with a width of the laser track of ~ 1 mm. The laser power was 100 W with corresponding power and energy densities of 127.4 W/mm^2 and 12.5 J/mm^2, respectively. Other parameters used are a pulse duration of 5 ms, a pulse frequency of 20 Hz, a beam diameter of 1 mm, a beam travel speed of 10 mm/s and a defocusing amount of 2 mm. During the laser treatment, a continuous flow of Ar gas was applied to avoid oxidation and the sample was held by a home-made clamping arrangement.

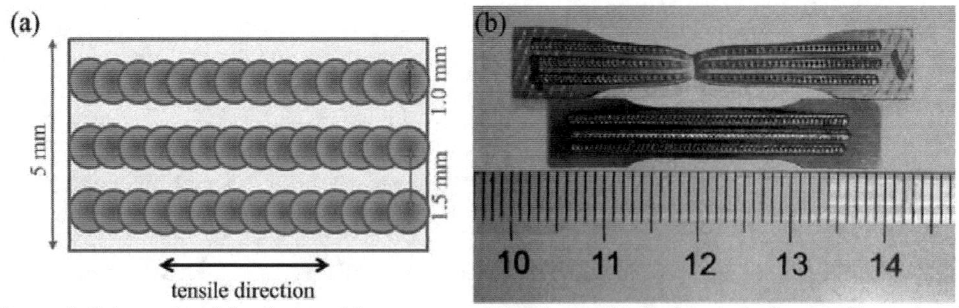

Figure 1. Schematic of the designed laser pattern (a) and samples before and after tensile testing (b).

Tensile tests were performed at a loading rate of 0.5 mm/min. The pulsed laser treated sample surfaces were investigated before and after tensile testing. The cross-sectional area perpendicular to the tensile direction was cut out after tensile test for microstructural characterization. The surface close to the facture was denoted as "sample after tensile test" and the other side, close to the clamping region, was denoted as "sample before tensile test". Electron channeling contrast imaging (ECC imaging) in a field emission gun scanning electron microscope (JEOL 7800) was used. The cross-section samples were polished by grinding paper and then electro-polished in a mixed solution of 90 ml methanol and 10 ml perchloric acid (at 30 °C and 15 V for 2min). Hardness across the laser-modified zones from the sample surface to center was measured using a Vickers indentation tester (HVS-1000) at a load of 250 mN for 10s.

3. Results and discussion

Figure 2(a) is an overview of the surface appearance after laser processing. Figure 2(b) is a low magnification image of a laser track. The sample surface is found to be relatively flat after the LST, except for small protrusions at the center (melting zone: MZ) and "water ripples" at the edge of each treated circle (heat affected zone: HAZ) [22].

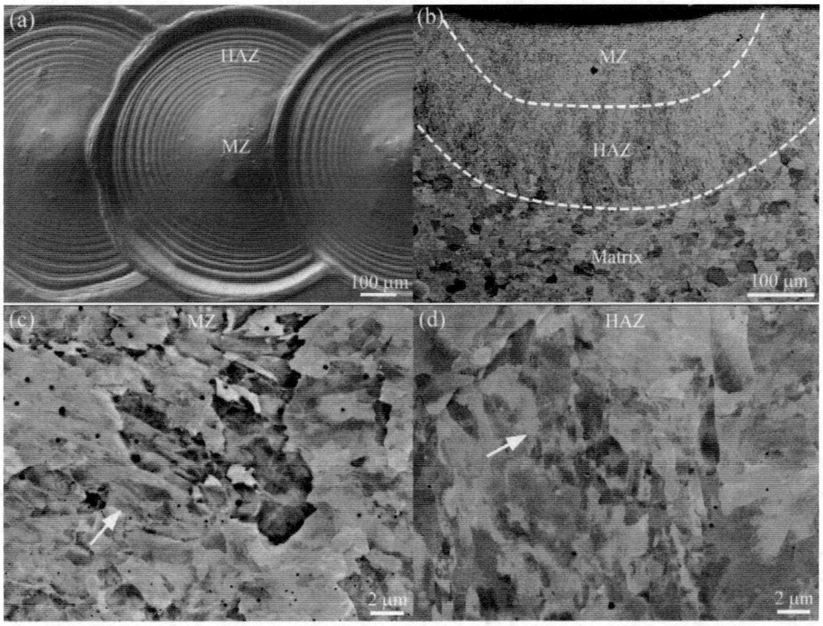

Figure 2. SEM images of the laser processed sample before tensile testings: (a) surface with one laser track; (b-d) cross-section. (b) low magnification overview, (c) high magnification of the MZ region and (d) high magnification of the HAZ region.

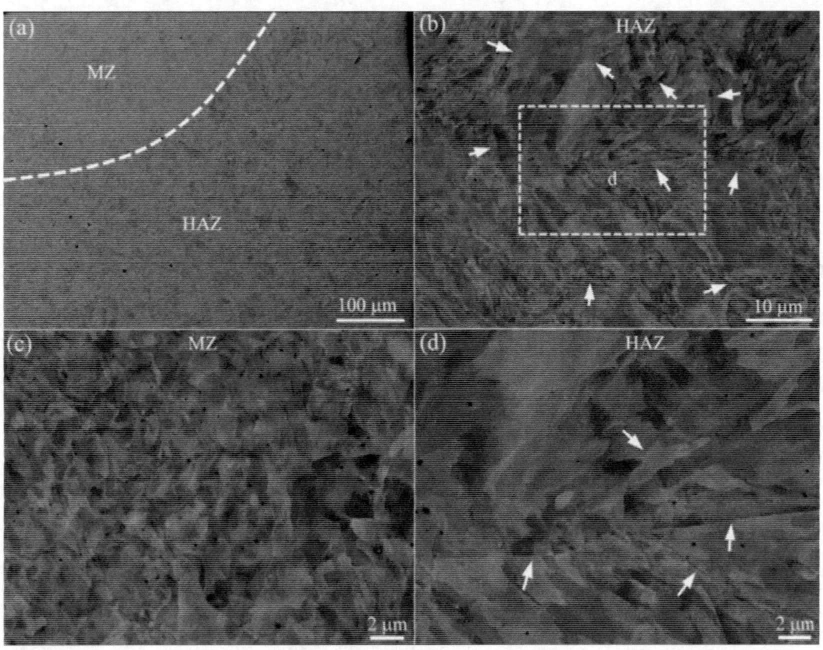

Figure 3. SEM images of the cross-section of laser processed sample after tensile testing: (a) overview; (b) and (d) the HAZ region; (c) the MZ region.

IOP Conf. Series: Materials Science and Engineering **580** (2019) 012029 doi:10.1088/1757-899X/580/1/012029

Cross-sectional microstructures of laser treated samples were examined in detail by ECC and representative images are shown in figure 2(b-d). Figure 2(b) is a low-magnification image where a laser-modified zone can be found. A hemispherical zone containing a molten zone (MZ) and a heat affected zone (HAZ) with a length of ~ 900 μm and a depth of ~ 300 μm can be seen. Representative microstructures in the MZ and HAZ regions are shown in (c) and (d), respectively. The equiaxed matrix was completely replaced by an ultra-fine columnar structure (about hundreds of nanometers in width and several micrometers in length) in the MZ region (figure 2(c)), and a fine columnar structure in the HAZ region, produced by the heat lost along the matrix during solidification. Substructures with alternating contrast in one grain as indicated by arrows were revealed by ECC. These substructures are on the micro-scale. Small pores (dark spots) can be seen in figure 2(c) and (d), which may be formed during solidification. After tensile deformation, it is found that the image contrast between the MZ and HAZ regions become more obvious, as seen in figure 3(a). The MZ region is more homogeneous than that before tensile (figure 3(c)), while "Van Gogh's Sky-like" microstructures [23, 24] can be seen in the HAZ (figure 3(b)). At higher magnification, columnar blocks inside each grain are clear. The possible initial grain boundaries are indicated in (b) and (d) by arrows. In general, after tensile testing, the microstructures of the treated zone become finer.

The hardness variation across a laser-affected zone is shown in figure 4, taking the middle of the MZ region as the zero-point. The hardness measurements were performed from the sample surface to the center across the whole laser-treated zone. It is found that the highest hardness region (MZ) has a size of about 150 μm in depth and 600 μm in width, in agreement with the ECC observations. At a depth of about 200 ~ 250 μm, the hardness drops from ~ 200 Hv to ~ 150 Hv and the matrix has a hardness of about 100 Hv.

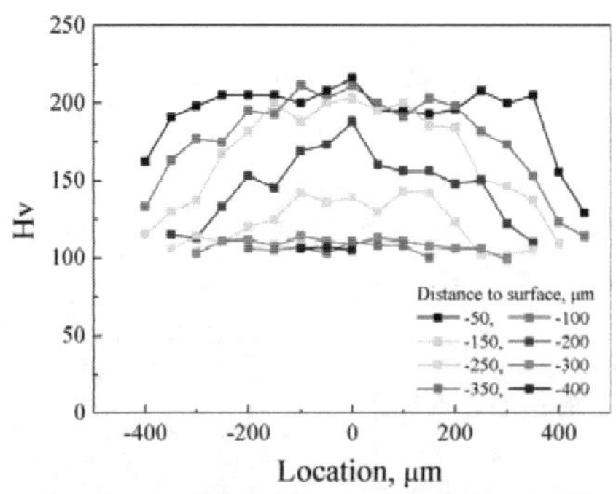

Figure 4. Hardness variation across a laser-affected zone, lateral coordinate indicates the location of the indents, "0" is the center and negative value are for the left part of the laser affected zone. The "distance to surface" takes the sample surface as the zero point.

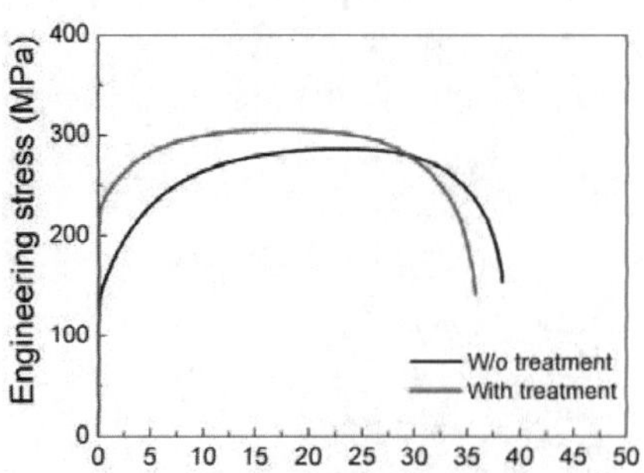

Figure 5. Stress-strain curves of IF steel with and without laser surface treatment

Figure 5 shows stress-strain curves of the two types of samples. It is clear that the strength of the laser treated sample is higher than that of the untreated one. At the same time, the elongation (especially the total elongation) does not decrease too much. A summary of the average properties (averaged over 3 samples) can be found in table 1. About 60% increase in yield strength (YS) and ~ 6% increase in ultimate tensile strength (UTS) is obtained by the laser surface treatment. The uniform elongation (UE) is ~ 5% lower than without laser treatment while the decrease in total elongation is ~ 2%. The effective hardness of the whole sample was calculated based on a rule of mixtures (ROM, Hv_{ROM}), taking the hardness of the MZ, HAZ and matrix as 200 Hv, 150 Hv and 100 Hv, respectively, and considering the corresponding volumes of each region (6 treated zones on both sizes). The volume fraction of the laser treated zones (V_{LST}) were simplified as cuboids, which are a little larger than the real volume fractions.

$$Hv_{ROM} = V_{matrix} \times Hv_{matrix} + V_{HAZ} \times Hv_{HAZ} + V_{MZ} \times Hv_{MZ}$$

$$= \frac{5\times1-0.8\times0.25\times6}{5\times1} \times 100 + \frac{0.8\times0.25\times6-0.6\times0.15\times6}{5\times1} \times 150 + \frac{0.6\times0.15\times6}{5\times1} \times 200 = 117.4$$

It is found that the relative hardness increase, △Hv, is much smaller than that of △YS, which means that some other strengthening mechanisms are expected to be active. This is similar to observations in other heterogeneous lamella structured materials [7]. The much finer microstructure in the treated zones and the Van Gogh's Sky-like microstructure in the HAZ might be the reason, and related to the accommodation of incompatibility between the hard laser treated zones and the soft matrix.

Table 1 Mechanical properties of the tested samples and calculated contribution of laser treatment

	YS, MPa		UTS, MPa		UE, %		TE, %		V_{LST}, %	Hv or Hv_{ROM}	△Hv, %
	Ave.	std.	Ave.	std.	Ave.	std.	Ave.	std.			
Without treatment	139.3	4.2	280.5	6.3	23.3	0.5	39	0.8	-	100	-
With treatment	224.2	5.6	299.4	6.2	18.7	1.6	37.5	2.5	**24**	117.4	**11.7**

Compared with the partially SMATed Cu [21], where the treated depth is only 40 μm, it is obvious that the laser treated sample has larger modification volumes. However, the yield strength of the partially SMATed Cu was more than twice of the annealed Cu, with little decrease in uniform elongation. There may be several reasons for the difference between the present sample and the partially SMATed Cu. First of all, in the SMATed Cu, not only dislocations but also nanotwins develop, as Cu is prone to form deformation twins. Secondly, the distance between the laser-treated tracks is on the mm scale, and such a large distance may limit the contribution of the heterogeneous structure in the present sample.

4. Summary

Laser surface treatment has been applied to an IF steel to develop a surface heterogeneous lamella structure and investigate its effect on mechanical properties. It is found that after the treatment, a zone of about 150 μm in depth and 400 μm in width is obtained due to sample melting (molten zone). The total laser treated zone is about 300 μm in depth and 900 μm in width, including the heat affected zone. After such a treatment, higher strength and similar ductility can be obtained compared to the non-treated sample.

Acknowledgements

The authors wish to thank the financial support of the Natural Science Foundation of China (NSFC, Grants No. 51671039), State Key Research and Development Program of MOST of China

(2016YFB0700403). L. Zhang would like to thank Prof. Niels Hansen and Prof. Dorte Juul Jensen for useful discussions.

References

[1] Wei Y, Li Y, Zhu L, Liu Y, Lei X, Wang G, Wu Y, Mi Z, Liu J, Wang H and Gao H. *Nat. Commun.* 2014 **5** 3580.
[2] Ma X, Huang C, Moering J, Ruppert M, Höppel H W, Göken M, Narayan J and Zhu Y. *Acta Mater.*2016 **116** 43-52.
[3] Wu X L, Jiang P, Chen L, Zhang J F, Yuan F P and Zhu Y T. *Mater. Res. Lett.* 2014 **2** 185-191.
[4] Yang M, Pan Y, Yuan F, Zhu Y and Wu X. *Mater. Res. Lett.* 2016 **4** 145-151.
[5] Ma E. *JOM* 2006 **58** 49-53.
[6] Zhao Y, Topping T, Bingert J F, Thornton J J, Dangelewicz A M, Li Y, Liu W, Zhu Y, Zhou Y and Lavernia E J. *Adv. Mater.* 2008 **20** 3028-3033.
[7] Wu X, Yang M, Yuan F, Wu G, Wei Y, Huang X and Zhu Y. *Proc Natl Acad Sci USA* 2015 **112** 14501-14505.
[8] Zhang L, Chen Z, Wang Y, Ma G, Huang T, Wu G and Jensen D J. *Scripta Mater.* 2017 **141** 111-114.
[9] Liu H S, Zhang B and Zhang G P. *Scripta Mater.* 2011 **65** 891-894.
[10] Luo X, Huang T, Wang Y, Xin Y and Wu G. *Sci. Rep.* 2019 **9** 5428.
[11] Ma X L, Huang C X, Xu W Z, Zhou H, Wu X L and Zhu Y T. *Scripta Mater.* 2015 **103** 57-60.
[12] Ma E. *Scripta Mater.* 2003 **49** 663-668.
[13] Tao N R, Wang Z B, Tong W P, Sui M L, Lu J and Lu K. *Acta Mater.* 2002 **50** 4603-4616.
[14] Fang T H, Li W L, Tao N R and Lu K. *Science* 2011 **331** 1587-1590.
[15] Liu X C, Zhang H W and Lu K. *Scripta Mater.* 2015 **95** 54-57.
[16] Wu X, Jiang P, Chen L, Yuan F and Zhu Y T. *Proc Natl Acad Sci USA* 2014 **111** 7197-7201.
[17] Zhang X, Hansen N, Gao Y and Huang X. *Acta Mater.* 2012 **60** 5933-5943.
[18] Liu G, Lu J and Lu K. *Mater. Sci. Eng. A* 2000 **286** 91-95.
[19] Ye C, Suslov S, Fei X and Cheng G J. *Acta Mater.* 2011 **59** 7219-7227.
[20] Lu K. *Science* 2014 **345** 1455-1456.
[21] Liu X, Wu K, Wu G, Gao Y, Zhu L, Lu Y and Lu J. *Scripta Mater.* 2016 **124** 103-107.
[22] Chai L, Wu H, Wang S, Chen K, Wang T and Xia J. *Mater. Chem. Phys.* 2017 **198** 303-309.
[23] Emura S, Tsuzaki K and Tsuchiya K. *Mater. Sci. Eng. A* 2010 **528** 355-362.
[24] Naka S, Marty M, Thomas M and Khan T. *Mater. Sci. Eng. A* 1995 **192/193** 69-76.

IOP Conf. Series: Materials Science and Engineering **580** (2019) 012030 doi:10.1088/1757-899X/580/1/012030

An efficient method to improve the spatial resolution of laboratory X-ray diffraction contrast tomography

H Fang[1], D Juul Jensen[1] and Y Zhang[1]

[1]Department of Mechanical Engineering, Technical University of Denmark, 2800 Kgs. Lyngby, Denmark.

E-mail: hfang@mek.dtu.dk

Abstract. Lab-based diffraction contrast tomography (LabDCT) has recently enabled non-destructive 3D characterization of the crystallographic orientations and grain morphologies in bulk materials. Despite the wide accessibility and availability of lab-based X-rays, the current spatial resolution of LabDCT is only about 20 μm and has to be improved to make this technique a more versatile tool. Conventional LabDCT takes advantage of the Laue focusing effect, *i.e.* the diffracted beam from a sample illuminated by an incoming polychromatic beam, is focused when the source and the detector are symmetrically positioned. In this study we have investigated the possibility of increasing the spatial resolution of LabDCT by placing the detector at a distance beyond that required for Laue focusing. We first developed a forward projection simulation tool to elucidate the effects of placing the detector at a larger distance. Next, we performed LabDCT measurements on a partially recrystallized pure aluminum sample at different sample-to-detector distances. The results show that the diffraction spots can be magnified by a factor of about three compared to the current LabDCT setup. The benefits and limitations resulting from increasing the sample-to-detector distance are discussed.

1. Introduction

The properties of most polycrystalline materials, such as metals and ceramics, depend to a large extent on the size, shape and crystallographic orientation of the crystalline grains that constitute such materials. At present, the most widely used technique to characterize grain shapes and orientations is electron backscatter diffraction (EBSD) [1, 2]. However, the electrons limit the characterization to only 2D and requires a destruction of the sample when combined with serial sectioning to achieve mapping in 3D. Recently, 3D mapping of grain orientations with a resolution down to 1 nm using transmission electron microscopy (TEM) has been demonstrated on an aluminum sample with a thickness of about 150 nm [3]. Despite the high resolution of this technique, the probed sample volume is restricted to the nanometer regime. To enable non-destructive characterization of grain structures in 3D for polycrystalline materials at the micron and submicron length scales, several techniques at synchrotron radiation sources have been developed over the past two decades. The main examples of these techniques are differential aperture X-ray microscopy (DAXM) [4], three-dimensional X-ray diffraction microscopy (3DXRD) [5] and diffraction contrast tomography (DCT) [6]. Although all these techniques are very useful for determination of the grain structure in 3D, they require an intense X-ray beam that is available only at synchrotron radiation sources. This restriction places a serious limit in accessibility for a wide range of experiments, *e.g.* when many repeated measurements, or very time consuming ones, are needed.

A transition to the use of a laboratory X-ray source for DCT has been recently made. It has been demonstrated that lab-based diffraction contrast tomography (LabDCT) is able to determine non-destructively the size, position, shape and orientation of grains in 3D [7, 8]. In the current LabDCT set-up, the sample and the detector are placed such that the sample-to-source distance L_{ss} equals the sample-to-detector distance L_{sd}, which focuses the diffracted pattern of the grains into lines on the detector. This so-called Laue focusing does not provide any geometrical magnification of the diffraction spots. The current spatial resolution that can be achieved with this set-up is rather limited (about 20 μm), which has to be improved to make LabDCT a more versatile tool for studying a wide range of materials science problems.

In the current work, we explore the potential for improving the resolution of LabDCT by increasing the ratio of L_{sd}/L_{ss}, thereby geometrically magnifying the diffraction spots. To achieve this aim, we first develop a computational tool to simulate forward projections of an already-known 3D grain structure and investigate the effect of L_{sd}/L_{ss} on features of the diffraction spots to verify our ideas. Then the diffraction projections of a partially recrystallized pure aluminum sample are measured using LabDCT at different ratios of L_{sd}/L_{ss}. The resulting diffraction spots for the same grains are compared for different L_{sd}/L_{ss} and validated by the forward simulations. The effect of increasing L_{sd}/L_{ss} on the geometrical magnification is also evaluated and demonstrated to be promising for increasing the resolution. The benefits and limitations by increasing the ratio of L_{sd}/L_{ss} are also discussed.

2. Simulations of forward projections

Methods for computation of forward projections for synchrotron 3DXRD and for the variant synchrotron DCT, where both use a parallel and monochromatic beam, have been described elsewhere [5, 9, 10]. However, the LabDCT is characterized by use of a cone-shaped polychromatic X-ray beam. This means that the formation of forward projections in LabDCT has some distinct features differing from those of synchrotron DCT. Here we briefly describe our simulations of the forward projections for LabDCT.

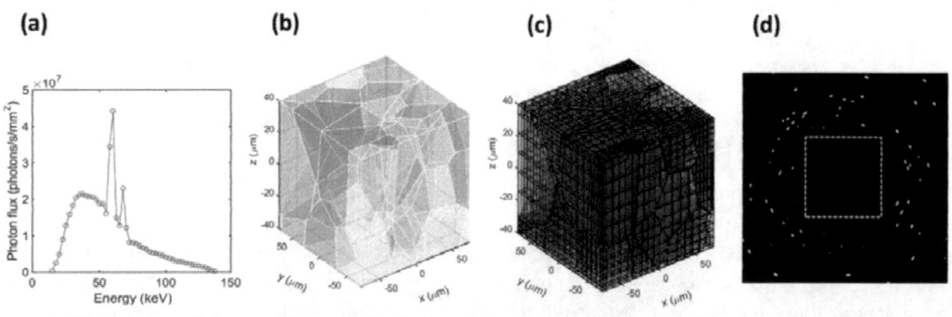

Figure 1. Illustration of the workflow of the forward simulations of projections for a sample with known grain morphologies and orientations in 3D. (a) Spectrum of the laboratory X-ray source using a tungsten target; (b) 3D visualization of a virtual alpha iron sample of a rectangular shape ($L_x \times L_y \times L_z = 150 \times 150 \times 80$ μm^3); (c) polyhedral mesh of the sample, and (d) simulated projection at a rotation angle $\omega = 0°$ and $L_{ss} = L_{sd} = 14.0$ mm with the edge of beam-stop marked by the dashed line.

The laboratory X-ray source is assumed to be a point as the X-ray beam is usually focused to a size of about 4 μm that is negligible compared to L_{ss} and L_{sd}. The profile of the X-ray spectrum emitted from a tungsten target is shown in figure 1a. Figure 1b shows a virtual polycrystalline alpha iron sample with randomly generated grain orientations that is used as here as the input. The simulations can also be carried out using 3D maps of grains from any available data source as an input. The virtual sample is then discretized into polyhedral cells in 3D, shown in figure 1c, to accelerate the computation. In this way we can determine the position, shape and intensity of diffraction spots from each polyhedral. By collecting the diffraction spots for all polyhedral cells, one can derive projection images at any rotation angle ω. An example is shown in figure 1d. The current simulations were performed using MATLAB® and the projections were exported in the form of 16-bit grey images.

3. Experimental

A partially recrystallized pure aluminium (99.996 wt.% Al) sample was selected to demonstrate the magnification of diffraction spots by increasing L_{sd}/L_{ss} in the LabDCT set-up. A rectangular sample ($6.0 \times 4.0 \times 1.3$ mm^3) was cut from a 12% cold-rolled pure Al plate and electro-polished. An Vickers hardness indent was made on the plane defined by the rolling direction (RD) and transverse direction (TD) to stimulate formation of nuclei upon annealing. More detailed descriptions of how the indentation was made can be found elsewhere [11]. The sample was then annealed at temperatures in the range from 275 °C to 340 °C for a total time of 50 min. After annealing the deformed structure is partially recrystallized and new grains with an average grain size of about 100 μm are found adjacent to the indentation site. This partially recrystallized sample was characterized by LabDCT.

Table 1. Values of L_{ss} and L_{sd} for LabDCT measurements of the partially recrystallized Al sample

Scan	L_{ss} (mm)	L_{sd} (mm)	L_{sd} / L_{ss}
I	14.0	14.0	1.00
II	11.0	18.0	1.64
III	11.0	24.0	2.18

The LabDCT measurements were performed on a Zeiss Xradia 520 Versa X-ray microscope. A detector (2032×2032) with an effective pixel size of 3.36 μm was used for recording the diffraction patterns. Three different ratios of L_{ss} and L_{sd} were used, as listed in table 1. The exposure time for each projection was 600 s. For reconstruction of the grains in 3D, the sample was rotated through 360° with a step of 2° during the scan. The GrainMapper3D™ software was used for indexing and mapping of grains from the LabDCT projections. ParaView was employed to visualize the grain structure.

4. Results and discussion

4.1. Simulations of forward projections

Figure 2 shows simulated forward projections at two different ratios of L_{sd}/L_{ss} and a rotation angle of 30° for the virtual alpha iron sample with the grain structure shown in figure 1b. By increasing L_{sd}/L_{ss} from 1 to 1.64, the diffraction spots formed for lattice planes of the same grain are geometrically magnified. For example, the 002 diffraction spot of grain 32 (grain diameter = 61.9 μm) is magnified by a factor of 1.91, while a magnification of 1.76 is found for the $11\bar{2}$ spot of grain 17 (grain diameter = 56.7 μm). This magnification is not symmetric. The diffractions spots are magnified more in the radial direction (a factor of about 1.6) than in the azimuthal direction (a factor of about 1.2) as L_{sd}/L_{ss} is increased from 1 to 1.64.

The field of view (FOV) captured in the projection is reduced by increasing L_{sd}/L_{ss}, as is clearly seen in figure 2. This is because the accessible maximum scattering angle $2\theta_{max}$ is decreased. However, some diffraction spots blocked by the beam-stop at $L_{sd}/L_{ss} = 1$ can be seen at $L_{sd}/L_{ss} = 1.64$. Examples of these diffraction spots are the $0\bar{1}\bar{1}$ spot of grain 3 (grain diameter = 29.1 μm^3) and the 200 spot of grain 30 (grain diameter = 29.6 μm) marked by triangles in figure 2b. This demonstrates that although increasing L_{sd}/L_{ss} results in a reduced FOV it also increases the likelihood to obtain diffraction signal from smaller 2θ angles on the detector.

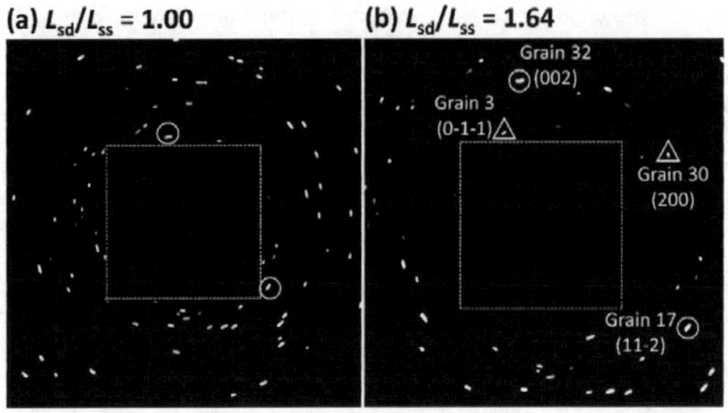

Figure 2. Simulated forward projections at $\omega = 30°$ of a virtual alpha iron sample consisting of 48 grains with an average diameter of 35 μm and randomly generated orientations for different L_{sd}/L_{ss} ratios. (a) L_{ss} and L_{sd} = 14.0 mm; (b) L_{ss} = 14.0 mm and L_{sd} = 22.9 mm. Circles mark the diffraction spots seen on both images while triangles mark the diffraction spots seen only in (b). The grain identification number and Miller indices of the marked spots are also shown. The edges of the beam-stop are marked by dashed lines.

4.2. LabDCT measurements at different values of L_{sd}/L_{ss}

Figure 3 shows the 3D grain structure reconstructed from LabDCT measurement at $L_{ss} = L_{sd}$ = 14.0 mm. All the grains except grain 2 are recrystallized. Grain 2 is only recovered and has an internal substructure with a spread of orientations. For convenience we use at present the average orientation Euler angles to represent this grain as an input for forward simulation.

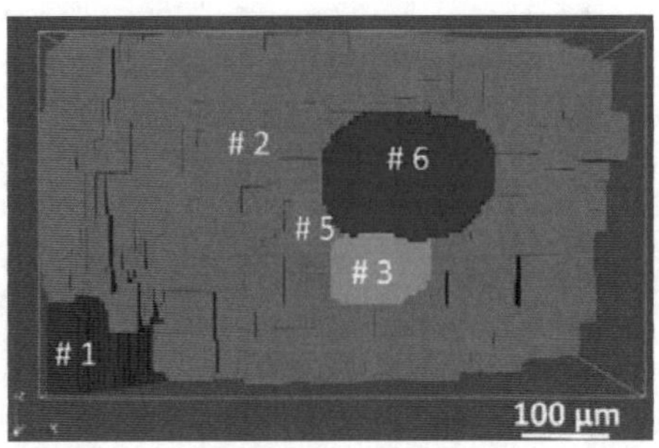

Figure 3. A front view of the 3D grain map reconstructed from the LabDCT measurement at L_{ss} = L_{sd} = 14.0 mm. The grain numbers are marked. Grain 4 is not seen because it is located behind the junctions of grains 3, 5 and 6. Random colors are used.

Figure 4 compares the projections of diffraction patterns at a rotation angle of ω = -146° from both the LabDCT measurements and forward simulations for L_{sd}/L_{ss} = 1.00, 1.64 and 2.18 using an input sample shown in figure 3. At this rotation angle diffraction spots for all grains in the sample can be found. Comparing figures 4a with 4b, it is seen that the common diffraction spots are significantly magnified as L_{sd}/L_{ss} is increased to 1.64, as also shown in the forward simulations in figures 4d and 4e. For L_{sd}/L_{ss} = 2.18 the diffraction spots shown in projection for L_{sd}/L_{ss} = 1.00 are mostly missing, except the four diffraction spots, $\bar{1}1\bar{3}$ of grain 6, $\bar{1}1\bar{3}$ and $1\bar{3}\bar{1}$ of grain 2 and $1\bar{1}1$ of grain 5, which are partially present and significantly magnified (see figures 4c and 4f). It can also be seen that new diffraction spots are seen on the detector, such as the $\bar{2}20$ spot of grain 1. The magnifications on average are about 2.4 at L_{sd}/L_{ss} = 1.64 and 3.5 at L_{sd}/L_{ss} = 2.18.

Figure 4 also shows that the positions and shapes of the simulated diffraction spots for the recrystallized grains are in very good agreement with the measurements. However, in figure 4a some relatively large diffraction spots, marked by circles, do not appear in figure 4d. Since the forward simulations uses inputs from the reconstructed grain structure, the simulated projection should ideally be identical to the measured one. These missing spots in the simulation suggest that one or more grains have not been indexed during reconstruction using the GrainMapper^{TM} software. The shapes of the diffraction spots for grain 2 are also different. In the LabDCT measurements the diffraction spots from grain 2 are much larger and more blurred than those shown in the simulated projections. This can be understood though as only the average orientation was used for grain 2 in the simulations, whereas this grain is in fact a recovered deformed grain and has significant variation in interior orientations. The comparison between the simulated and the measured projections demonstrates nevertheless that the present forward simulation is a very useful tool for evaluating the reconstruction results.

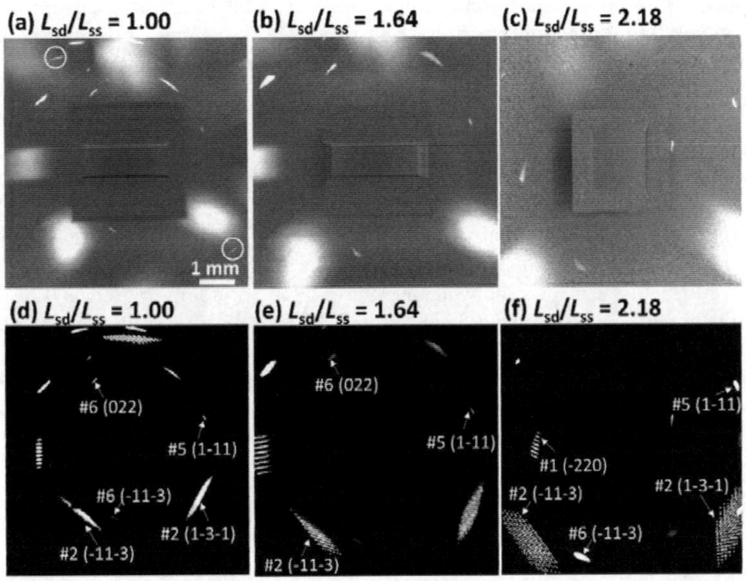

Figure 4. Projections of diffraction patterns rendered by LabDCT measurements (a, b and c) and from forward simulations (d, e and f) for L_{sd}/L_{ss} = 1.00, 1.64 and 2.18. The scale bar shown in (a) applies to all projections. Missing spots in the simulations are marked by circles in (a). Indices of the diffraction spots are shown in the simulated projections. The edges of the beam-stop are not shown in the simulated projections for clarity.

4.3. Discussion

4.3.1. Benefits of increasing L_{sd}/L_{ss}. A geometrical magnification is demonstrated by the projections from both the experimental LabDCT measurements and from the forward simulations. In total, we compared the size of 97 diffraction spots for different rotation angles at L_{sd}/L_{ss} = 1 and 1.64. The results show that the spot size is magnified by a factor between 1.2 and 9.0 depending on the diffraction planes and volumes of the grains. On average a value of 2.2 is found. Further increase of L_{sd}/L_{ss} enhances this magnification. However, such an increase also results in a lack of sufficient statistics for the diffraction spots, as most diffraction spots recorded at L_{sd}/L_{ss} = 1 cannot be captured when L_{sd}/L_{ss} > 2. A detector with a larger area would be needed for such analysis using larger values of L_{sd}/L_{ss}. Nonetheless, figure 4 indicates that a magnification factor of about 3.5 can be reached when L_{sd}/L_{ss} is close to 2.18. It is also necessary to evaluate how such a magnification affects the quality of the magnified diffraction spots, including sharpness and blurring, which should not be compromised to ensure a proper grain mapping. To evaluate the sharpness, we compared the full width half maximum (FWHM) of the derivative of the histogram profile across the edges of diffraction spots from LabDCT for the three scan conditions (I, II and III in table 1). We found that the FWHM values are almost equal for the three ratios of L_{sd}/L_{ss}. The variations of grey values for each diffraction spot are also calculated to evaluate the blurring. The values fluctuate around 0.05 ± 0.02 as a function of spot size and no significant difference is found as L_{sd}/L_{ss} is increased. Therefore, it is concluded that increasing L_{sd}/L_{ss} does not compromise the quality of the diffraction spots, demonstrating that increasing L_{sd}/L_{ss} can be an efficient method to improve the spatial resolution for LabDCT.

4.3.2. Limitations of increasing L_{sd}/L_{ss}. The maximum accessible scattering angle $2\theta_{max}$ decreases from about 19° to 9° as L_{sd}/L_{ss} is increased from 1 to 2.73, assuming a fixed detector size. This imposes an upper limit for increasing L_{sd}/L_{ss}. According to our simulations, $2\theta_{max}$ should be larger than 10° to ensure a sufficient number of pronounced diffraction spots that can be easily segmented from the background noise. Therefore, with the present detector L_{sd}/L_{ss} should be in the range 1.50 ~ 2.45 to achieve at the same time both a considerable magnification and a satisfactory number of diffraction spots. Of course this limitation can be overcome by using a detector with a larger recording area.

5. Conclusions

In this study we have investigated the possibility of increasing L_{sd}/L_{ss} as a method to improve the spatial resolution of LabDCT. A forward simulation tool has been developed and applied to simulate LabDCT projections of a virtual sample and a real sample. The forward simulations show that increasing L_{sd}/L_{ss} is very effective in achieving a geometrical magnification of diffraction spots. Experimental LabDCT measurements confirm this result and show that the image quality is compromised as L_{sd}/L_{ss} is increased. Magnification factors of between 2 and 3.5 are found for $1.64 < L_{sd}/L_{ss} < 2.18$ compared to that under the Laue focusing condition $L_{sd}/L_{ss} = 1$. The agreement and disagreement between the simulated diffraction patterns and experimental LabDCT measurements both provide further insights into 3D grain mapping with more fine details. The present study demonstrates that increasing L_{sd}/L_{ss} can be an efficient method for improving the spatial resolution for the LabDCT.

Acknowledgements

The authors thank the financial support by the European Research Council (ERC) under the European Union's Horizon 2020 research and innovation programme (M4D – grant agreement No. 788567).

References

[1] Venables J A and Harland C J 1973 *Philos. Mag.* **27** 1193
[2] Krieger Lassen N C, Juul Jensen D and Conradsen K 1994 *Acta Cryst.* **A50** 741
[3] Liu H H, Schmidt S, Poulsen H F, Godfrey A, Liu Z Q, Sharon J A and Huang X 2011 *Science* **332** 833
[4] Larson B C, Yang W, Ice G E, Budai J D and Tischler J Z 2002 *Nature* **415** 887
[5] Poulsen H F 2004 *Three-dimensional X-ray diffraction microscopy: mapping polycrystals and their dynamics*. Springer Tracts in Modern Physics vol. 205. (Berlin: Springer) p 25
[6] Ludwig W, Schmidt S, Lauridsen E M and Poulsen H F 2008 *J. Appl. Cryst.* **41** 302
[7] King A, Reischig P, Adrien J and Ludwig W 2013 *J. Appl. Cryst.* **46** 1734
[8] Sun J, Zhang Y, Lyckegaard A, Bachmann F, Lauridsen E M and Juul Jensen D 2019 *Scrip. Mater.* **163** 77
[9] Oddershede J, Schmidt S, Poulsen H F, Sorensen H O, Wright J and Reimers W 2010 *J. Appl. Cryst.* **43** 539
[10] Sørensen H O, Schmidt S, Wright J P, Vaughan G, Techert S, Garman E F, Oddershede J, Davvasambu J, Paithankar K S, Gundlach C and Poulsen H F 2012 *Zeitschrift für Kristallographie Cryst. Mater.* **227** 63
[11] Xu C, Zhang Y, Godfrey A W, Wu G, Liu W, Tischler J Z, Liu Q and Juul Jensen D 2017 *Sci. Rep.* **7** 42508

In-situ study of microstructural evolution and local strain distribution during tensile loading of near-micrometre grain size aluminium

W Q Gao[1], C L Zhang[1], S Q Zhang[1] and A Godfrey[1]

[1] Key Laboratory of Advanced Materials (MOE), School of Materials Science and Engineering, Tsinghua University, Beijing 100084, China

E-mail: awgodfrey@mail.tsinghua.edu.cn

Abstract. A combined digital image correlation (DIC) and electron backscatter diffraction (EBSD) study of local strain distribution and microstructural evolution during tensile loading has been carried on samples of near-micrometre grain size Al prepared by spark plasma sintering. To achieve the required spatial resolution and allow combined DIC and EBSD measurements a method of surface patterning using colloidal silica was developed. The DIC results reveal a wide variety in heterogeneity of plastic strain, both in terms of the average strain experienced by each grain and regarding the extent and nature of strain localization within individual grains. No obvious correlation is found between the heterogeneity in plastic strain and either Schmid factor, grain size or the pattern of local grain rotations. The experiments also allow investigation of the pattern of crystal rotation during tension in near-micrometer grain size samples and reveal differences in the mean grain rotation and the in-grain orientation spread.

1. Introduction

In order to understand and control deformation inhomogeneity of materials, it is essential to understand comprehensively the relationship between the deformation inhomogeneity of materials at the grain scale and the link to micro-scale deformation mechanisms. The strain heterogeneity and microstructural inhomogeneity during deformation are interrelated and in single phase materials may depend on the grain orientation, and well as on the size and characteristics (size, orientation, shape) of neighboring grains [1]. The near-micrometer grain size regime is of particular interest with regard to deformation heterogeneity as it has been reported that samples with grain sizes in this regime may exhibit unusual mechanical properties, such as hardening by annealing and softening by deformation, and an obvious yield-drop phenomena [2,3]. These properties are a reflection of a transition in the deformation mechanisms as the average grain size is reduced to a few micrometers. It is of interest therefore to investigate the strain heterogeneity in samples with near-micrometer grain size, as such samples will typically contain a population of grains with sizes that straddle this transition regime. It has been shown that for such experiments samples of Al prepared by spark plasma sintering (SPS) are well suited, as in such samples the grains are in a fully recrystallized condition with an approximately random texture and a low dislocation density [4].

Digital image correlation (DIC) is a powerful technique employed for investigating the surface deformation and provides information regarding full-field displacements and strains [5]. Recent developments in achieving fine-scale speckle patterns, by remodeling of a deposited layer of gold [6] have shown the potential for this method to achieve sub-micron resolution in deformed metals. Electron backscatter diffraction (EBSD) is now a standard laboratory technique for mapping crystal orientations on a sample surface, and as a scanning electron microscope microscope (SEM) based technique, allows the possibility for in-situ studies of a range of metallurgical phenomena, including phase transformations [7], recrystallization and grain growth [8], and plastic deformation [9,10]. Although the combination of both DIC and EBSD for in-situ studies of plastic deformation and strain heterogeneity is therefore of great interest, a challenge remains in obtaining fine-scale speckle patterns for DIC that do not interfere adversely with the EBSD signal.

2. Material and methods

Samples of Al with an average grain size of 4.4 μm were prepared by using the SPS technique. For the SPS samples, spherical Al spherical powder with an average particle size of 5μm was sintered to disks of diameter 20 mm and height 6 mm following the procedure described in [4]. Dog-bone shape tensile samples were cut from the as-sintered disks by electron-discharge machining. The width, gauge length and thickness of the tensile samples were 1.8, 9 and 0.6 mm, respectively.

Prior to tensile deformation, the specimens were manually ground to 4000-grit silicon carbide paper, and then electrochemically polished in a 10% perchloric acid and 90% alcohol solution to achieve a flat polished surface free of mechanical damage and residual deformation. To achieve a uniform distribution of suitable DIC markers a beaker of colloidal silica solution containing particles with a mean size of 50 nm was first given an ultrasonic treatment for 40 mins, after which a small amount was dropped onto the sample surface. Indents were added to the tensile sample to allow easy identification of the area of interest, and to allow re-alignment of the sample between the alternating EBSD and DIC measurements.

Tensile deformation was carried out using a home-built push-to-pull stage, equipped with a load cell to record the instantaneous force during tensile loading. The in-situ observations were conducted on a TESCAN MIRA 3 LMH equipped an CMOS-based EBSD detector from Oxford Instruments operating at 20KV. Secondary electron (SE) images were used for the DIC measurements, with the region of interest for DIC covering an area of 69.2 x 69.2μm^2 (microscope magnification of x3000) at a pixel size 4096 x 4096 pixels. To capture the same area in EBSD maps of size 80 x 80 μm^2 were collected using a step-size of 0.3μm. Images for DIC analysis were collected before tensile testing (0% strain) and at tensile strains of 0.45%, 1.6%, 3.3%, 5.8% and 8.9%. EBSD data were acquired at strains of 0%, 1.6%, 5.8% and 8.9%. For the DIC calculations the VIC 2D software package was used.

3. Results

A SEM micrograph of the area investigated in study is shown in figure 1(a), with figures 1(b,c) showing higher magnification images of selected areas to illustrate the DIC marker distribution over a range of length scale. A uniform distribution of fine-scale markers is achieved, with the marker diameter ranging from 20 – 200 nm. The coverage of the markers is such that they do not adversely affect EBSD data collected from the same area. Although some aggregation of the larger oxide particles is seen at grain boundaries, this does not adversely influence the EBSD data, as in general boundary regions are associated with lower quality indexing. For the DIC calculations a test-box (subset) size of 91 x 91$pixel^2$ (1.5 x 1.5μm^2) was used, as illustrated in figure 1(c), using a step-size of 3 pixels, providing sufficient resolution to study strain heterogeneity during tensile deformation on a fine scale within each grain.

Strain-stress curves obtained by different methods are shown in figure 2. The red markers show values obtained from the load-cell (stress) and movement of hardness indents (strain) during loading, while the blue markers show the tensile strain along the loading direction obtained from the DIC measurements (averaged over the entire mapped area). The black line reproduces the tensile stress-strain curve for a sample prepared with similar Al powder as measured using a standard universal testing frame at an initial strain rate of 1.6 x $10^{-3}s^{-1}$ [4]. A good agreement in seen between all three sets of data.

Maps showing the variation in the strain component along the tensile direction (e_{xx}) at strains of 0.45%, 1.6% and 8.9% are shown in figure 3(a-c). For convenience black lines are overlaid on figures

3(b,c) to show the positions of the grain boundaries (as determined from the EBSD measurements). Note that in these figures, and in the EBSD maps the tensile direction is along the horizontal direction. Figure 3(d) shows an electron channeling contrast (ECC) image of the result of the tracked area after the final deformation step (8.9%), in which slip bands can be clearly seen in a number of grains.

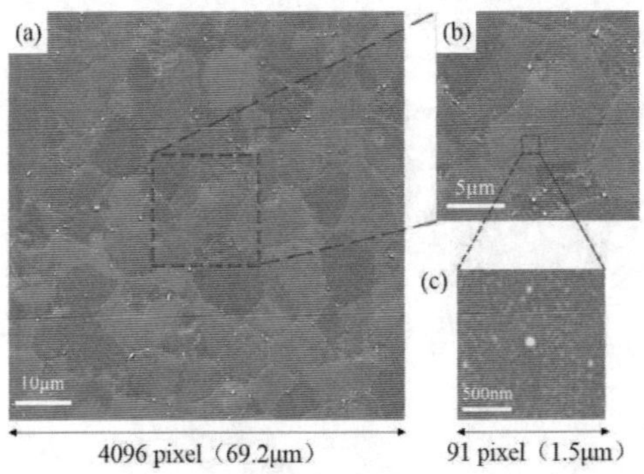

Figure 1. SEM images of the examined area in this study; (a) entire examined area in this study; (b) magnified image showing a range of SiO_2 particle sizes; (c) further enlarged image showing the fine markers used for DIC and their number and distribution within the test-box size of 91 x 91 pixel^2.

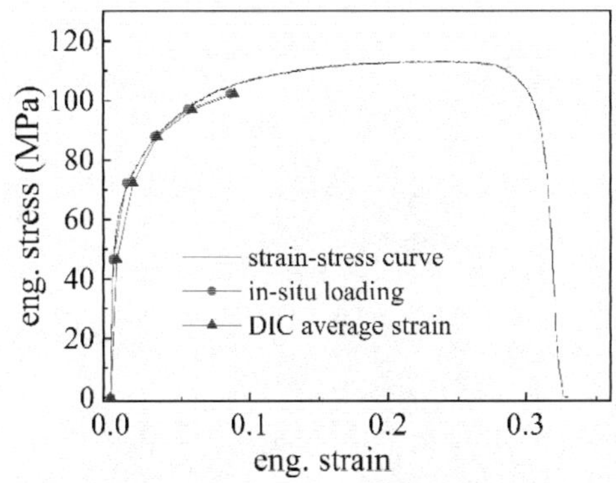

Figure 2. Stress-strain data during tensile deformation: red points represent the strain obtained during in-situ loading; blue points are plotted using the average tensile strain obtained from DIC calculations. The black line shows the strain-stress curve for a similar grain size sample using a standard test frame [4]

The DIC data reveal the clear presence of a variety of strain heterogeneity during tensile deformation. Except in a few cases, the highest local strain at each strain step is preferentially concentrated in larger size grains, while the strain associated with small grains is in general low. Not only is the strain uneven among different grains, but also a heterogeneous distribution of strain is seen inside most grains. In some grains significant strain accumulation is found at, or close to, grain boundaries and triple points, while in some areas bands of locally higher strain are observed to pass through grain boundaries and into grain interiors. In other grains, the slip concentration is seen within the grain but without transfer of strain across the grain boundaries. The pattern of strain inhomogeneity in the examined area of the sample becomes increasingly clear with increasing strain. Although at the lowest strain investigated (0.45%) the pattern of strain heterogeneity is rather weak, the areas with higher local higher strain are seen to match those after deformation to larger strains, supporting the validity of the DIC data at the lowest strain. In general the highest local strains at each strain step measured are approximately 2 – 3 times larger than the average strain.

EBSD maps illustrating the microstructure before and after tensile deformation to a strain of 8.9% are presented in figure 4. The inverse pole figure maps (figures 4a,d) reveal at least qualitatively an increase in orientation spread within each grain. The heterogeneity of grain subdivision, is revealed in more detail in figures 4(b,e), which plots for each grain the misorientation angle between each pixel and the average orientation of the grain. In some grains the tensile deformation results in a significant variation in orientation across each grain, while in other grains, including some of larger size, a much smaller orientation variation is seen. Figures 4(c,e) show the kernel average misorientation (KAM) at each map pixel, calculated using a 3 x 3 kernel with an upper cut-off angle of 4°. This map in effect displays a smoothed version of a boundary misorientation map, and picks up local orientation gradients within the angular resolution of the technique, and as such is commonly used to estimate the geometrically necessary dislocation density at each pixel. The maps reveal the formation inside some grains of low angle boundaries and also the development of orientation gradients close to grain boundaries and in some cases in grain interior regions.

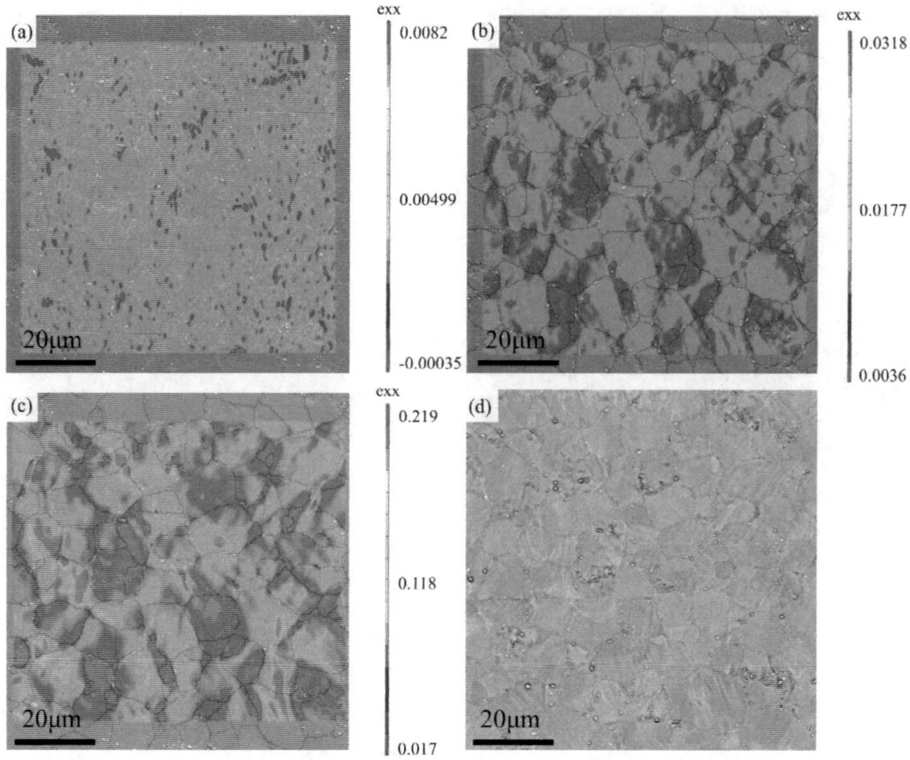

Figure 3. DIC results showing the variation of tensile (e_{xx}) strain of after loading to (a) 0.45%, (b) 1.6% and (c) 8.9%. Grain boundary positions (obtained from EBSD mapping) are overlaid in black in (b,c); (d) electron channeling contrast image of the examined area after 8.9% tensile deformation.

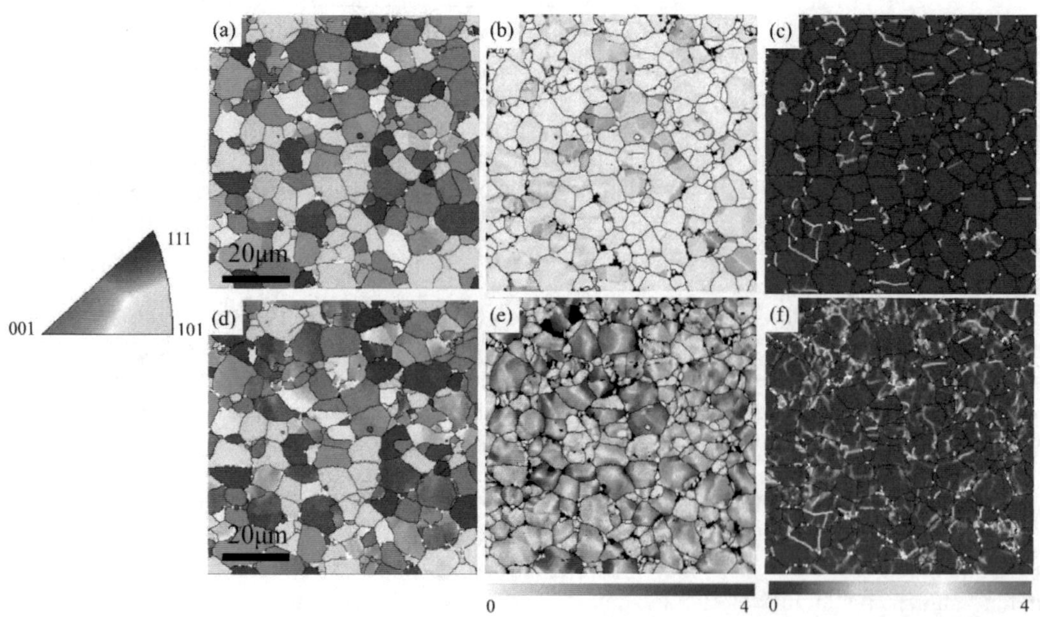

Figure 4. EBSD data for the examined area before deformation (a-c) and after deformation to 8.9% strain (d-f); inverse pole figure coloring showing tensile axis direction; (b,e) deviation angle to grain average orientation (c,f) 3 x 3 kernel average misorientation angle maps.

IOP Conf. Series: Materials Science and Engineering **580** (2019) 012031 doi:10.1088/1757-899X/580/1/012031

4. Discussion

An advantage of the present experiment, where colloidal silica markers are used for DIC, is that information both on the local deformation and the microstructural evolution and crystal rotations can be obtained on from the same region on a sample surface, providing a rich data set for analysis of deformation heterogeneity. Here we focus on two aspects, namely the Schmid factor and size of each grain. The Schmid factor provides a measure of the ease of primary slip system activation, and it has been reported that grains with high Schmid factor (SF) in FCC materials with conventional grain sizes are prone to slip and strain localization [11].

Some example data is shown in figure 5. Grains 96, 52, and 67 have a similar Schmidt factor (SF = 0.41 – 0.42), but clearly the very small grain (#96) sustains less local strain than the two larger grains. Grains 27, 28, 29 have a similar SF (0.45), with significantly less plastic strain taken by grain 28. This grain has a smaller size than the other two grain, but the strain pattern may also be affected by the cluster of smaller un-numbered grains lying between grains 27 and 28. Schmid factor and grain size do not appear to be the only determinant of local strain heterogeneity: grain 50 has a higher SF (0.47) than grains 27-29, but the average local strain for this grain is relatively low and uniform. Further analysis of the data, including both grain orientation, size and neighbors is underway, noting that in some cases transmission of localized strain between grains may also play role. Although not examined in detail in this work, it is worth highlighting that a comparison of the KAM map in figure 4f and the local strain map at the same strain (figure 3d) shows very few similarities.

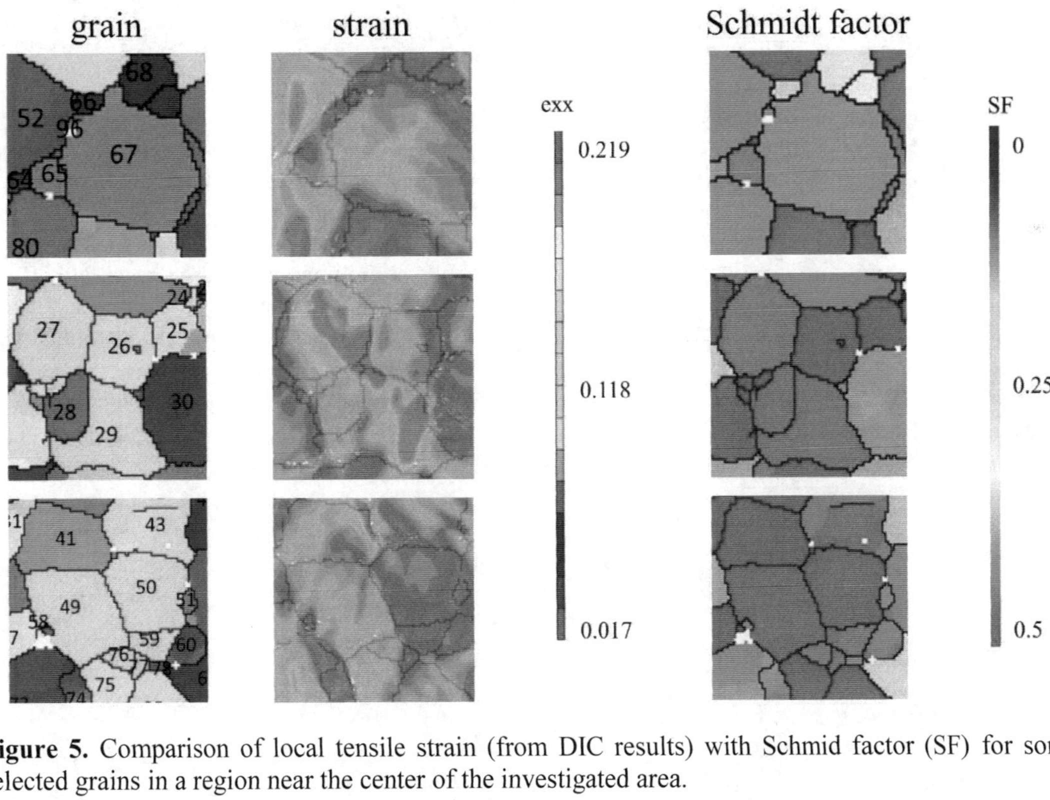

Figure 5. Comparison of local tensile strain (from DIC results) with Schmid factor (SF) for some selected grains in a region near the center of the investigated area.

The in-situ nature of the experiment also allows the crystal rotation of each grain to be determined. An example of the data available is shown in figure 6, which plots both the crystal direction parallel to the tensile axis for the average orientation of grain 41 (marked in figure 5) after each deformation step, and also shows evolution of the spread of crystal directions parallel to the tensile axis within the grain. For this grain an overall crystal rotation towards the <100> corner of the unit triangle is observed, consistent with X-ray synchrotron observations of grain rotation in the bulk of larger size grains with this initial orientation [12]. The results here also show that the orientation spread within for grain develops in a different manner, with the spread approximately perpendicular to the overall lattice rotation.

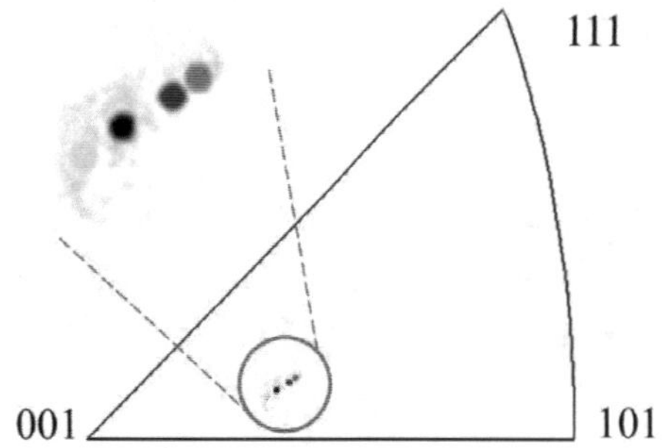

Figure 6. Inverse pole figure showing crystal rotation during tensile deformation for grain 41 (see figure 5). Red, blue, black and green points correspond to the grain average orientation after 0%, 1.1% , 5.5% and 8.9% tensile strain, respectively; smaller points in yellow show the evolution of the spread in tensile axis within this grain at each strain level.

5. Summary

By careful application colloidal silica (SiO_2) markers allow combined high resolution DIC and EBSD investigation, where the spatial resolution is sufficient to follow in-situ tensile deformation of samples with near-micrometer grain size. In the present study, where Al prepared by SPS has been investigated, no significant degradation is seen after tensile straining to 8.9% in either the surface markers or in the EBSD data, where the same area was mapped at tensile strains of 0%, 1.6%, 5.8% and 8.9%.

The DIC results show wide variety in heterogeneity of plastic strain, both in terms of the average strain experienced by each grain and regarding the extent and nature of strain localization within individual grains. No obvious correlation between the heterogeneity in plastic strain and either Schmid factor or grain size is found. Similarly, the observed pattern of local strain heterogeneity differs from the variation in local misorientation (as viewed in a KAM map), highlighting the fact that such maps are not in general a proxy for plastic strain. The present 2D results are limited in that the stress state differs to that in the bulk, and the true grain size is unknown. A parallel X-ray synchrotron investigation is underway where lattice rotations in bulk grains can be obtained, albeit at a lower spatial resolution and without accompanying DIC data for plastic strain heterogeneity.

Acknowledgments

Financial support from the National Natural Science Foundation of China (project number 51671113) is gratefully acknowledged.

References

[1] Wang L and Barabash R I 2011 *Metall. Mater. Trans.* A **42** 626–35
[2] Huang X, Hansen N and Tsuji N 2006 *Sci.* **312** 249–51
[3] Tsuji N, Ito Y, Saito Y and Minamino Y 2002 *Scr. Mater.* **47** 893–9
[4] Le G M, Godfrey A and Hansen N 2013 *Mater. Des.* **49** 360–7
[5] Pan B, Qian K and Xie H 2009 *Meas. Sci. Technol.* **20** 62001
[6] Gioacchino F D and Fonseca J Q D 2013 *Exp. Mech.* **53** 743–54
[7] Kirch D M, Ziemons A and Burlet T 2008 *Rev. Sci. Instrum.* **79** 4545
[8] Hurley P J and Humphreys F J 2010 *J. Microsc.* **213** 225–34
[9] Chen P, Mao S C and Liu Y 2013 *Mater. Sci. Eng.* A **580** 114–24
[10] Schwartz A J, Kumar M and Adams B L 2012 *Electron Backscatter Diffraction in Materials Science* (Boston, MA: Springer)
[11] Liu J H, Vanderesse N and Stinville J C 2019 *Acta Mater.* **169** 260–274
[12] Winther G, Margulies L, Schmidt S, Poulsen H F 2004 *Acta Mater.* **52** 2863–72

IOP Conf. Series: Materials Science and Engineering **580** (2019) 012032 doi:10.1088/1757-899X/580/1/012032

Imaging modalities at the Swedish Materials Science beamline at PETRA III

Z Hegedüs,[1] T Müller,[1] J Hektor,[1] E Larsson,[2] T Bäcker,[1] S Haas,[1] ALC Conceiçao,[1] S Gutschmidt[1] and U Lienert[1]

[1] Deutsches Elektronen-Synchrotron (DESY), Notkestr. 85, D-22607 Hamburg, Germany
[2] RISE Research Institutes of Sweden, Ideon Beta 5, SE-223 70 Lund, Sweden

E-mail: zoltan.hegedues@desy.de

Abstract. High-energy synchrotron radiation has been demonstrated to be a powerful tool for materials characterization. The development of novel methodologies is still ongoing, driven by major technological advances regarding the available source brilliance and efficient large area detectors. The Swedish Materials Science beamline at PETRA III is dedicated to materials characterization by high-energy X-rays and scheduled to enter into user operation starting August 2019. The beamline has been designed in particular for the combination of two complementary techniques: wide and small angle scattering and imaging. The beamline design is presented briefly and the different techniques are reviewed with regard to the contrast mechanisms and the ability to obtain spatially resolved information.

1. Introduction

High-energy synchrotron radiation fills the important spatial and temporal resolution gap between electron microscopy and neutron diffraction, enabling the in situ investigation of bulk materials during e.g. thermo-mechanical processing. By combining different contrast mechanisms (e.g. attenuation, small and wide angle scattering) it has been demonstrated that important microstructure attributes such as crystallographic phase identification, orientation distributions, stress/strain state, size distributions and dislocation densities can be obtained. Initially, only averages over the illuminated sample volume were evaluated, but with the improving ability to provide intense micro-focused beams, various imaging methodologies have been developed exploiting tomographic reconstruction principles and/or scanning narrow illumination volumes across the sample. In some cases, area detectors can be employed for the simultaneous acquisition of scattering from many volume elements and therefore dramatically accelerate three-dimensional mapping as compared to point-by-point scanning. However, the applicability of specific methodologies is dependent on suitable microstructures and sample geometries.

PETRA III is currently the most brilliant high-energy X-ray source worldwide and the Swedish Materials Science beamline (SMS) has been designed in particular to perform multi-modal experiments providing spatial resolution. A number of techniques have been selected for implementation and in the following the status, underlying principles, potentials, and limitations are described briefly.

2. Beamline description

The SMS beamline comprises of a side branch (P21.1) dedicated for the study of diffuse scattering from single crystals, amorphous and nanocrystalline materials and a longer inline branch (P21.2) devoted to the combination of wide (WAXS) and small-angle X-ray scattering (SAXS) and imaging techniques with

IOP Conf. Series: Materials Science and Engineering **580** (2019) 012032 doi:10.1088/1757-899X/580/1/012032

both broad and narrow bandwidth beams. In the following only techniques relevant to the characterization of polycrystalline bulk materials at P21.2 will be discussed.

P21.2 is served by an in-vacuum undulator, whose odd harmonics have a sharp central cone and are ideal for focusing and for small beam applications, while the donut shape of the slightly de-tuned even harmonics serve as a suitable source for large-beam experiments such as imaging.

A broad band double Laue monochromator, 100 meters downstream of the source, provides monochromatic ($\Delta E/E \approx 10^{-3}$) radiation ensuring high photon flux throughout the whole 40–150 keV energy range. Optionally, after collimation, a high resolution four bounce channel-cut monochromator can be used for experiments demanding very narrow $\Delta E/E \approx 10^{-4}$ energy distribution. The monochromatic beam may then be focused from several different locations with 1D or 2D compound refractive lenses (CRL) depending on the application. The largest unfocused beam size is approximately 2×6 mm (v×h, FWHM) and the smallest achieved focus size (by CRLs) is about 1×5 µm at the sample position.

3. Grain-resolved diffraction

Grain-resolved diffraction refers to situations where a sufficiently small number of grains (or subgrains) are illuminated by the incident beam, such that their diffraction spots do not extensively overlap on the detector [1]. The probability of spot overlap increases roughly with the square of the orientation spread (mosaicity) within the grains. Under favorable conditions, up to about one thousand grains can be illuminated simultaneously. Series of diffraction images are acquired while the sample is being rotated and the recorded reflections are assigned to individual grains by suitable software.

Three different methodologies can be delineated depending on the sample-to-detector distance (cf. figure 1a). If the sample-to-detector distance is comparable to the sample dimensions ('near-field'), then the location of diffraction peaks on the detector is mostly dependent on the position of the diffracting units within the sample (at the cost of reduced strain sensitivity). By forward- or back-projection of the detected intensity distributions, crystallographic orientation maps of the illuminated sample volume can be evaluated (comparable to electron back-scatter diffraction surface orientation maps) [2]. The spatial resolution is limited by the detector resolution and typically about 2 µm can be achieved. The same detector design can be used for tomography, but thin scintillators are required resulting in poor detection efficiency and longer exposure times (see also section 4).

In the 'far-field' regime the sample-to-detector distance (typically about 1 m) is large compared to the sample diameter but small enough to still capture several complete diffraction rings on a large area detector. In general, no space constraints are posed for sample environments. Averaged quantities such as the position, size, crystallographic orientation, and the complete elastic strain tensor can be extracted. A test experiment[1] was performed on a well-annealed NiBi sample of about 1.1 mm diameter with approximately 40 µm average grain size [3]. The full reconstructed volume with the 953 indexed nickel grains is shown in figure 1b. Using the full available beam intensity, a rotation speed of 25 °/s would have been required to prevent detector saturation. Since the grain indexing is not yet limited by counting statistics even faster rotation should be feasible such that, in favorable conditions, one indexing scan could be accomplished per second. While the grain specific data enable the investigation of grain-to-grain interactions, it has also been demonstrated that by local averaging over an appropriate number of grains three-dimensional strain/stress mappings can be obtained [4]. If applicable, the grain-resolved technique is much faster than 3D-scanning of a local gauge volume as produced by a conical slit cell (cf. section 5.2).

The far-field detector setting can also be combined with a focused incident beam [5]. Spot overlap is minimized and intra-granular orientation variations can be resolved. However, a large number of images must be recorded (cf. section 5).

[1] The experimental details were as follows: beam energy: 59.87 keV, sample to detector distance: 984 mm, angular step size: 0.1°, exposure time: 0.2 s, beam attenuation factor: 54, detector: Varex XRD 4343CT, pixel size 150 µm, number of pixels 2880×2880

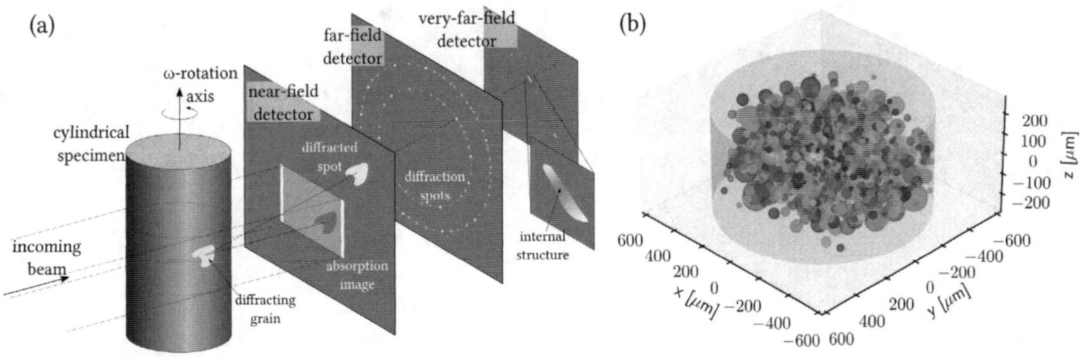

Figure 1. a) schematic illustration of the different measurement setups showing the near-field, the far-field and the very-far-field detectors and the detector images. The figure is not to scale; b) 953 indexed nickel grains in a NiBi sample as color coded spheres—the color and the radius indicate the orientation and the size of the grains, respectively and the transparent cylinder shows the dimensions of the sample.

By increasing the sample-to-detector distance even further (typically about 4 m, 'very-far-field') the reciprocal space resolution can be increased such that the intrinsic peak profiles from metallic materials with low dislocation density can be characterized (at the cost of covering only small sections of a Debye–Scherrer ring) [6]. It is noted that the possibility of beam focusing with a very large focal length of about 40 m maintains the high reciprocal space resolution. The energy bandwidth can be cut down accordingly by the channel cut monochromator (cf. section 2).

4. Synchrotron micro-tomography

In synchrotron micro-tomography (SRμCT) a 2D projection image of the sample is acquired at each angular (ω) rotation-step, usually over a range of 180 degrees. The acquired projection images constitute the integral of the linear X-ray attenuation coefficient (μ) along the path of the X-rays. By utilizing either analytical or regularized iterative reconstruction methods, a 3D map of μ can be reconstructed for the entire illuminated volume. Transmitted X-rays undergo both absorption and phase shift, therefore the reconstructed μ is a combination of absorption, phase shift and their mixture. At small sample-to-detector distances—typically in the order of centimeters—the reconstructed μ primarily consists of the absorption part, which is optimal for strongly absorbing samples (high-Z elements). At high energies, however, μ becomes small for most elements, therefore the absorption contrast gets weaker. By increasing the sample-to-detector distance, the phase contrast contribution becomes more pronounced, which is advantageous for the investigation of poorly absorbing or phase samples and interfaces. On the other hand, an untreated phase contribution smears out the resolution at sharp interfaces. By using phase retrieval algorithms (PhR) the absorption and phase contributions may be separated. One such algorithm is Paganin PhR [7], which suppresses the absorption contribution resulting in predominant phase contrast.

For materials science related research, the combination of SRμCT with other X-ray techniques e.g. WAXS, SAXS or grain-resolved diffraction is particularly promising for revealing crystallographic information about a sample. However, this versatility comes at a cost as the currently available diffractometer at P21.2 is primarily optimized for interface diffraction and has a slight run-out error upon rotation.

The run-out error was characterized[2] with a precisely manufactured steel ball immersed in glue in a Kapton tube. The projected images of the steel ball were fitted and the center positions were then used to perform post-alignment of the individual images, thus producing optimal slice reconstructions. Since the run-out error proved to be reproducible, a correction in the TomoPy [8] reconstruction workflow has been

[2] The scanning parameters were the following; energy: 46 keV, sample-to-detector distance: 300 mm, exposure time: 0.4 s, sample rotation: 180°, no. projections: 1800, effective pixel size: 0.72 μm.

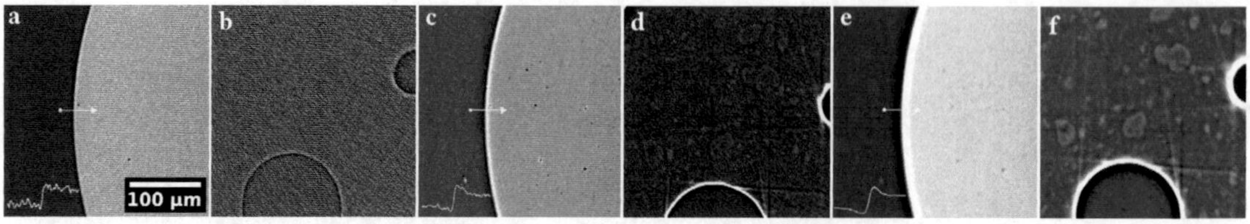

Figure 2. Tomographic reconstructions of an absorption sample (strongly absorbing steel ball—a, c and e) and a phase sample (poorly absorbing glue—b, d and f). Subfigures a and b were measured at a small (35 mm) sample-to-detector distance (i.e. no phase effects), while the other four subfigures were recorded at a sample-to-detector distance of 300 mm. The yellow insets indicate the profiles at the edge of the steel ball showing the effect of the phase as overshoot at the edge of the steel ball.

implemented and can be applied to correct the diffractometer imperfections for other measurements.

In SRµCT a common choice of detector is a scintillator combined with a 45° mirror, an objective and a CCD or CMOS camera. This setup has the advantage that the scintillator is more or less independent from the other parts of the detector, therefore can be changed upon demand. This is important, as the particular type and thickness of the scintillator determines the resolution and the required measurement time. The thicker the scintillator, the better the detection efficiency, but on the other hand the larger the smearing effect in the scintillator material. The small source size of PETRA III and the large distance from the source at P21.2 results in a particularly small virtual source size, therefore within reasonable sample-to-detector distances the resolution is determined solely by the detector—scintillator and camera.[3]

The effect of the sample-to-detector distance (determining the degree of phase contrast) and the comparison of the analytical GridRec [9] reconstruction with and without PhR is shown in figure 2. The GridRec reconstruction works well for the strongly absorbing steel ball (figure 2a), but due to the lack of absorption contrast it performs poorly for the glue (figure 2b). Phase effects—even reconstructed without PhR—reveal tiny details (inclusions) in the steel ball (figure 2c), but only slightly increase the reconstruction quality for the glue (figure 2d). Using PhR exposes details in the glue (figure 2f), but at the same time hides small details in the steel ball (compare figure 2c and figure 2e).

5. Pencil-beam techniques

SRµCT (section 4) provides information on the 3D structure of multiphase materials, but lacks crystallographic information. Grain-resolved diffraction (section 3) requires a limited number of illuminated grains (to avoid spot overlap on the detector), which is often exceeded for fine grained microstructures. In these cases, pencil-beam techniques (when the beam size is comparable to the characteristic length in the sample, e.g. grain size[4]) provide a unique tool to obtain local crystallographic information.

This technique employs scanning of a micrometer-sized beam across the sample cross-section with a diffraction or scattering pattern recorded at each scan step. This results in local information with a spatial resolution of the step and beam size perpendicular to the beam direction. Thus, structural gradients can be investigated by scanning along a line (1D profile) or across an area (2D maps) as long as the material is homogeneous in at least one direction, which is then oriented parallel to the incident beam. Good examples for this kind of profiling are materials after any kind of surface treatment (mechanical, chemical or irradiation), resulting in a gradient structure from the surface towards the bulk of the sample.

[3] With a 300 µm horizontal source size and 150 m source-to-sample distance, the detector may be placed as far as 500 mm from the sample before a virtual 1 µm resolution is worsened by source size smearing.

[4] For WAXS CT the beam size should be significantly larger than the grain size such that the reconstructed voxels still contain a statistically relevant number of grains.

5.1. Pencil-beam tomography

Many materials are heterogeneous in all three spatial directions and profiling results in a diffraction signal, averaged along the incident beam direction. In this case, the combination of pencil beam mapping at various ω angles (cf. figure 1a) and tomographic reconstruction provides diffraction information in all three (spatial) dimensions. This method was proposed already in the 1980s [10], but reasonable measurement times and spatial resolution have only been achieved using the brilliant beams of third generation synchrotrons and focusing optics.

Compared to absorption tomography, a diffraction / scattering computed tomography dataset does not consist of a stack of 2D projections recorded at different rotation angles, but rather of a stack of grids of 2D diffraction / scattering patterns. Thus, large number of patterns has to be recorded and evaluated requiring high photon flux and automated data processing to obtain reasonable measurement times and analysis efforts. Furthermore, additional steps are required in the reconstruction process compared to absorption tomography. In most studies, each 2D pattern is integrated over the azimuthal angle to obtain 1D profiles (intensity as a function of the scattering vector / scattering angle), which is, however, only possible for rotationally symmetric diffraction / scattering patterns [11]. From these profiles, the intensity for each scattering vector bin is reconstructed using well-known tomographic reconstruction techniques (cf. section 4). Compared to absorption tomography, the computing time is multiplied by the number of scattering vector bins for reconstruction of the complete diffraction / scattering patterns. For this reason, diffraction studies often focus on the reconstruction of integrated intensities of a few peaks, indicating for example the spatial distribution of certain phases [12, 13].

Pencil-beam tomography can be used both with WAXS and with SAXS. While the actual measurements (except the larger sample-to-detector distance for SAXS) and the reconstruction are very similar, structural information from different length scales is obtained. While the WAXS signal yields information on the atomic level such as crystallographic phases, lattice parameters and crystallographic defects (dislocations, stacking faults), SAXS probes particle sizes, shapes and orientations in the nano- to mesoscopic range (up to about 1 μm). By placing multiple detectors at different distances to the sample, simultaneous SAXS/WAXS measurements provide a characterization technique covering a broad length scale [14].

Diffraction computed tomography has been used to characterize a large number of different complex materials. For example, Rietveld analysis of reconstructed 1D diffraction patterns has been used to determine quantitative phase compositions [15], atomic bond lengths [16] or average crystallite sizes [17] for each voxel. This shows that many evaluation techniques, that are known from conventional diffraction analysis, can be utilized in combination with pencil-beam XRD-CT scans to obtain a comprehensive understanding of micro- and particularly nanostructures in 3D. Modern instruments provide full XRD-CT scans in significantly less than one hour per slice, enabling in-situ experiments [15, 17]. The beam size in such experiments is always a compromise between high spatial resolution (small beam size), manageable measurement time and sufficient number of crystallites in diffraction conditions (large beam), since smooth Debye–Scherrer rings are required for many analysis techniques (except single grain techniques, cf. section 3). Texture distorts the quantitative analysis of reconstructed diffraction patterns, but usually still permits the determination of qualitative phase distributions [18]. Spotty Deybe–Scherrer rings due to a limited number of diffracting crystallites result in artifacts in the reconstructed patterns, since individual grains coming to diffraction condition have significant impact on the integrated pattern. Thus, the beam size should be large enough to interact with a sufficient number of crystallites. Distinct strong spots from few large grains or a second coarse-grained phase can be removed via filtering to avoid reconstruction artefacts [18].

P21.2 is well suited for pencil-beam tomography, since the typical beam size of a few microns to a few tens of microns can be easily achieved and adjusted by corresponding positioning of the CRLs. The interface diffractometer provides all degrees of freedom required for such measurements.

5.2. Post-specimen apertures

Whereas XRD-CT is a powerful tool as long as azimuthal integration of the 2D diffraction patterns is possible, it becomes limited as soon as gradients along the azimuth are present. The most prominent examples are texture and residual stresses. Thus, direct restriction of the gauge volume via post-specimen apertures is often more effective compared to tomographic techniques for such measurements, in particular if only 1D or 2D slices are of interest.

The most widely used type of post-specimen apertures are conical slit cells. They consist of a strong absorber plate with conical openings in the size of a few tens of microns [19]. Only diffracted beams from a certain position in the sample along the beam can pass the conical slit, whereas those from other positions are absorbed by the plate. Thus, the gauge volume is limited in beam direction. The length of the gauge volume depends on the diffraction angle and usually counts several hundred microns [20, 21]. If needed the gauge length can be reduced via sample translation and reconstruction techniques and is then only limited by the width of the post-aperture slit [22]. Conical slits have been used to investigate 3D spatial distributions of residual stresses and textures in various materials [19–21].

One disadvantage of conical slits is that a conical slit is limited to the investigation of a pre-defined crystallographic symmetry group due to the fixed distance of its openings. Alternative aperture designs, such as spiderweb [23] slits have been proposed, but are not yet regularly applied.

Measurements with post-specimen apertures have not been conducted at P21.2 yet. However, the beamline is generally well suited for such experiments, which will be tested in the near future.

6. Conclusion and outlook

It has been discussed that P21.2 is capable of using several multi-dimensional imaging techniques. The combination of these techniques enables the investigation of structural characteristics from the Angstrom regime up to the millimeter scale, with spatial resolution down to the micrometer range. Future developments include the installation of a second diffractometer, optimized for rapid and precise rotation. With the advancement of the beamline control system and the incorporation of new fast detectors a considerable increase of the measurement speed is expected.

References

[1] Poulsen H F 2004 *Three-Dimensional X-Ray Diffraction Microscopy* (Berlin Heidelberg: Springer)
[2] Renversade L, Quey R, Ludwig W, Menasche D, Maddali S et al. 2016 *IUCrJ* **3** 32
[3] Hefferan C M, Li S F, Lind J, Lienert U and Rollett A D 2009 In *C. Sci. Press*, vol. 14, 209
[4] Oddershede J, Camin B, Schmidt S, Mikkelsen L P, Sørensen H O et al. 2012 *Acta Mater.* **60** 3570
[5] Hektor J, Hall S, Henningsson N, Engqvist J, Ristinmaa M et al. 2019 *Materials (Basel).* **12** 446
[6] Jakobsen B, Poulsen H F, Lienert U, Almer J, Shastri S D et al. 2006 *Science (80-.).* **312** 889
[7] Paganin D, Mayo S C, Gureyev T E, Miller P R and Wilkins S W 2002 *J. Microsc.* **206** 33
[8] Gürsoy D, De Carlo F, Xiao X and Jacobsen C 2014 *J. Synchrotron Radiat.* **21** 1188
[9] Dowd B A, Campbell G H, Marr R B, Nagarkar V V, Tipnis S V et al. 1999 In Bonse U, ed., *Dev. X-Ray Tomogr. II*, 224
[10] Harding G and Kosanetzky J 1985 *Phys. Med. Biol.* **30** 183
[11] Schroer C G, Kuhlmann M, Roth S V, Gehrke R, Stribeck N et al. 2006 *Appl. Phys. Lett.* **88** 1
[12] Stock S, De Carlo F and Almer J 2008 *J. Struct. Biol.* **161** 144
[13] Stock S R and Almer J D 2012 *J. Appl. Crystallogr.* **45** 1077
[14] Daniels J E, Pontoni D, Hoo R P and Honkimäki V 2010 *J. Synchrotron Radiat.* **17** 473
[15] Li T, Heenan T M M, Rabuni M F, Wang B, Farandos N M et al. 2019 *Nat. Commun.* **10** 1497
[16] Bleuet P, Welcomme E, Dooryhée E, Susini J, Hodeau J L and Walter P 2008 *Nat. Mater.* **7** 468
[17] Vamvakeros A, Jacques S D M, Di Michiel M, Matras D, Middelkoop V et al. 2018 *Nat. Commun.* **9** 4751
[18] Álvarez-Murga M, Bleuet P and Hodeau J L 2012 *J. Appl. Crystallogr.* **45** 1109
[19] Nielsen S F, Wolf A, Poulsen H F, Ohler M, Lienert U and Owen R A 2000 *J. Synchrotron Radiat.* **7** 103
[20] Poulsen H F, Garbe S, Lorentzen T, Juul Jensen D, Poulsen F W et al. 1997 *J. Synchrotron Radiat.* **4** 147
[21] Staron P, Fischer T, Keckes J, Schratter S, Hatzenbichler T et al. 2013 *Mater. Sci. Forum* **768-769** 72
[22] Lienert U, Martins R, Grigull S, Pinkerton M, Poulsen H and Kvick Å 1999 *MRS Proc.* **590** 241
[23] Sinsheimer J, Bouet N, Ghose S, Dooryhee E and Conley R 2016 *J. Synchrotron Radiat.* **23** 1296

IOP Conf. Series: Materials Science and Engineering **580** (2019) 012033 doi:10.1088/1757-899X/580/1/012033

Visualization of low-misorientation dislocation structures from orientation data using customized All-Euler maps

Chuanshi Hong

Department of Mechanical Engineering, Technical University of Denmark, DK-2800, Kgs. Lyngby, Denmark

chho@mek.dtu.dk

Abstract. A method for visualization of low-misorientation dislocation structures from orientation data obtained by electron backscatter diffraction (EBSD) in scanning electron microscopy is presented. The method is termed "customized All-Euler maps". The microstructure of high purity (99.996 wt.%) aluminum lightly rolled to a thickness reduction of 12% is presented as a case study. Dislocation structures with misorientations across dislocation boundaries approaching the orientation precision of standard EBSD (~0.5°) are revealed using the customized All-Euler maps. Cautions and limitations in using such maps are discussed.

1. Introduction

Low-misorientation dislocation structures, in which the misorientation angles across dislocation boundaries are often as low as 0.5° or even lower, are of both engineering and scientific significance, as they are the prominent microstructural features of lightly deformed metals and provide important insights into the early stage of the evolution of deformation-induced dislocation structures [1–5]. Although transmission electron microscopy (TEM) can reveal details of dislocation boundaries in lightly deformed metals [6–8], the technique has several main limitations — e.g. the field of view is relatively small, and the operation is often labor intensive. Electron backscatter diffraction (EBSD) in the scanning electron microscope (SEM), which typically has a precision in orientation measurement of around 0.5°, is widely utilized to characterize the microstructure of deformed metals. Since the orientation precision of EBSD is challenged in studies of low-misorientation dislocation structures, the capability of standard EBSD in resolving such structures needs to be examined. Furthermore, proper visualization of the orientation data becomes critical.

In the present work low-misorientation dislocation structures in a lightly rolled Al of high purity is characterized by EBSD, and optimal visualization of the orientations is explored by employing an in-house developed color scheme termed customized All-Euler maps.

2. Material and methods

Aluminum with a high purity of 99.996% was used in this study. The sample, in the form of a sheet of 94 × 46 × 3 mm^3, was first annealed at 600 °C for 7 days, resulting in grain sizes ranging from a few to tens of millimeters. The sample was then rolled at room temperature to a thickness reduction of 12%. Details of the material and rolling process is available in [9,10]. Metallographic samples were prepared from the section containing the normal and the rolling directions, i.e. the ND-RD section. SEM-EBSD characterization was carried out on a Zeiss Supra 35 microscope, using a step size of 0.5 μm.

Figure 1 demonstrates the principle of customized All-Euler maps in comparison with standard All-Euler maps produced by the commercial EBSD software Channel 5. In standard All-Euler maps (figure 1a$_1$–a$_2$), the red, green and blue color components (r, g and b) cover the full ranges of the three Euler angles for aluminum, with ϕ_1 being 0–360° and Φ and ϕ_2 both being 0–90°. Here while Φ and ϕ_2 have reduced ranges by considering the symmetry of a cube crystal, ϕ_1 does not has a reduced range because by default the sample symmetry is not considered during EBSD data collection. For a specific EBSD map with low-misorientation dislocation structures (figure 1a$_2$), the actual distribution of Euler angles can be much narrower than the full range (figure 1b$_1$), which means the standard All-Euler maps only utilize a small portion of the color space. As a result, the microstructure is only weakly revealed due to the low color contrast (figure 1a$_2$). The idea of customized All-Euler maps is to make full use of the color space by making the r, g and b color components cover only the actual distribution of Euler angles. For instance, in figure 1b$_1$, the r, g and b color components represent the ranges of ϕ_1 = 305.2±7.6°, Φ = 25.9±7.6° and ϕ_2 = 62.5±7.6°. The center of the ranges corresponds to the mean orientation of the map, which has (ϕ_1, Φ, ϕ_2) = (305.2°, 25.9°, 62.5°), and the size of the ranges equals to 3 times the largest standard deviation of the three Euler angles of the map. The three Euler angles are set to have the same size of ranges to ensure their variations have a similar impact on the resulting contrast. This results in much-improved contrast in the customized All-Euler map in figure 1b$_2$, in which a set of planar dislocation structures running from lower-left to upper-right (marked by the solid line) is most apparent whilst another set running from lower-right to upper-left (marked by the dashed line) is also identified.

The customized All-Euler maps in this work were generated using in-house developed Matlab codes. A previous example of this coloring technique can be found in a different application [11], where the local lattice rotations around voids in a shock-loaded Al single crystal were analyzed.

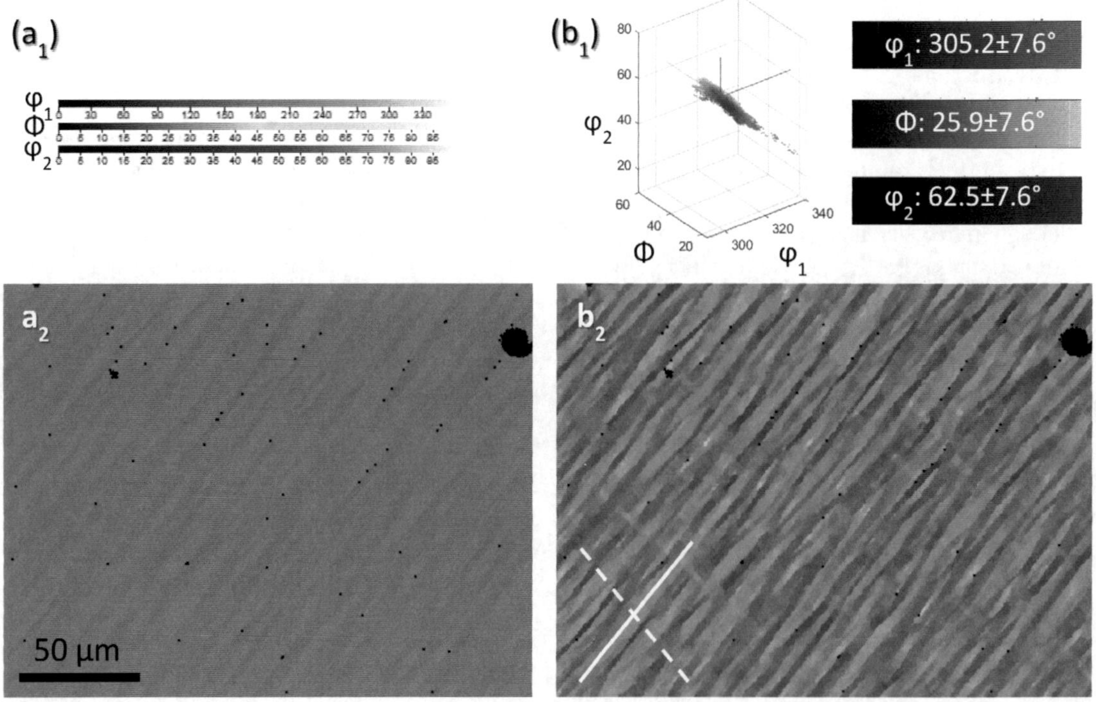

Figure 1. (a$_1$ & a$_2$) Color scheme and a standard All-Euler map. (b$_1$ & b$_2$) Color scheme and a customized All-Euler map of the same region. Refer to the text for a detailed description of the color schemes. The solid line and the dashed one in b$_2$ mark the direction of a profound set of planar dislocation structures and that of a less well-developed set, respectively.

3. Results and discussion

By using the customized All-Euler map method described above, it is found to be possible to visualize dislocation structures with low misorientations approaching the orientation precision (~0.5°) of standard EBSD. Figure 2 shows a region of such structures visualized by various techniques for comparison.

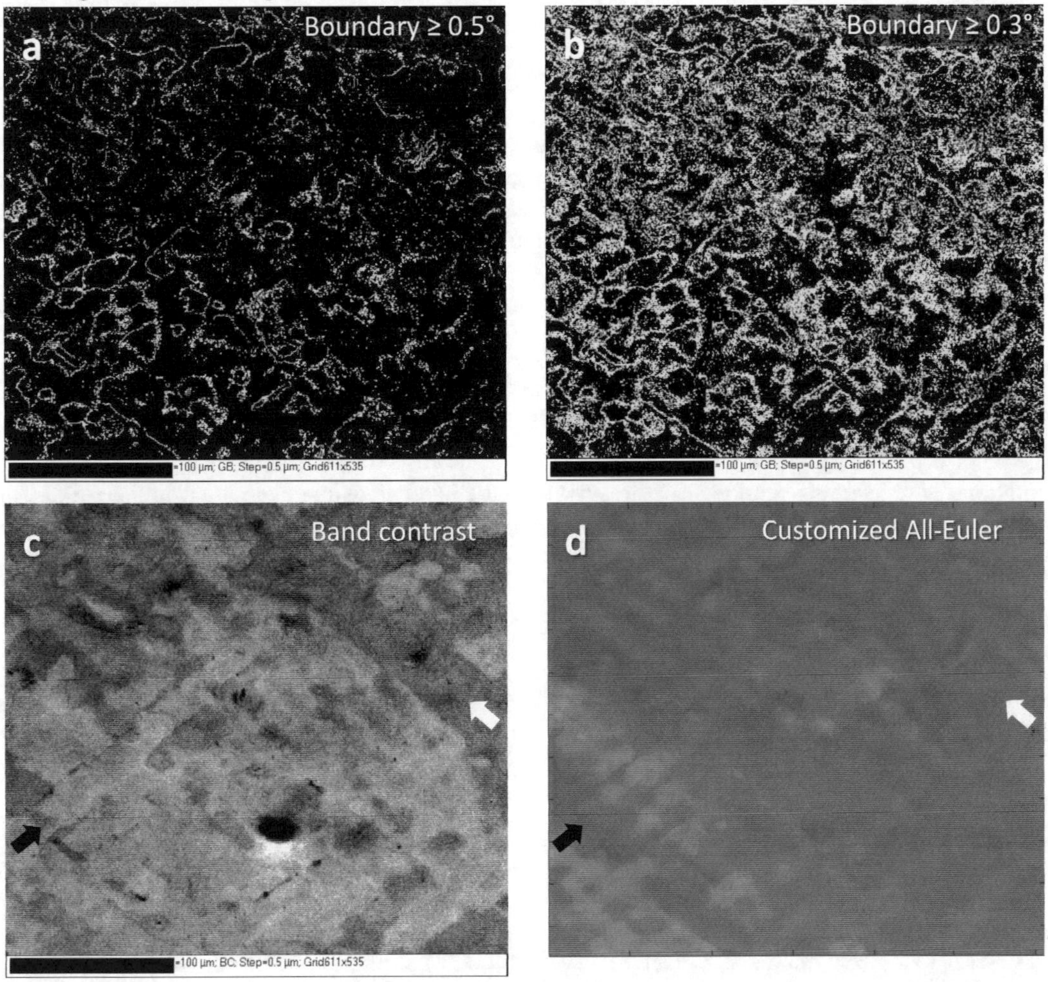

Figure 2. A comparison between different visualization techniques of the same orientation data containing low-misorientation dislocation structures. (a) Boundary detection with a threshold of 0.5°. (b) Boundary detection with a threshold of 0.3°. (c) Band contrast. Figures a–c are produced using the Channel 5 software. (d) Customized All-Euler map. All figures are of the same magnification, and the scale bars correspond to 100 µm.

Figure 2a and b show dislocation boundaries detected by the software Channel 5, using a threshold of 0.5° and 0.3°, respectively. As shown in figure 2a, boundary detection with a threshold of 0.5° reveals incompletely the microstructure in part of the region, which seems to be cell-like. Random noise is already noticeable over the entire map, suggesting the limit of the reliability of boundary detection is being approached. A further decrease in the threshold angle for boundary detection simply increases the noise level with negligible gain in the microstructural feature detection. Figure 2b shows boundary detection with 0.3°, which contains high level of noise but has basically the same microstructural information as figure 2a. The band contrast map from Channel 5 (figure 2c) shows little of the dislocation structures; only two microbands with different alignments are noticeable, as indicated by the black and white arrows. By contrast, the customized All-Euler map (figure 2d) shows clearly two intersecting sets of planar dislocation structures. Note that the dislocation boundaries in figure 2a mostly correspond to the orange-colored cells that have relatively well-defined boundaries, and the two microbands in figure 2c are found in figure 2d, both of which validate the potential of the customized All Euler Map method for detection of the microstructural features observed.

Figure 2 shows a clear advantage of the customized All-Euler map (figure 2d) over boundary detection (figure 2a) in terms of the visualization of low-misorientation dislocation structures. The results show that when the orientation precision of EBSD is being approached, boundary detection is very sensitive to the uncertainties in the measurement of the individual orientations, whilst the customized All-Euler maps are much less sensitive. This is attributed to the different principles for revealing microstructures between these two techniques. While boundary detection measures the differences between each pair of neighboring pixels and thus highlights measurement errors, customized All-Euler maps show the property of a group of pixels allowing human visual perception and are therefore more tolerant of individual uncertainties.

In figure 1 the orientation is relatively homogeneous over the entire region investigated, and the spread of orientation is mainly associated with misorientations associated with dislocation structures, which corresponds to a narrow distribution in Euler space. In this case the result is good contrast over the whole area in the customized All-Euler maps.

Under certain conditions, however, continuous orientation variations over a large length scale exist in the deformed microstructure [12,13], which can 'overwhelm' the relatively small local variations in orientations associated dislocation structures, making the latter invisible even in customized All-Euler maps. An example of such is shown in figure 3.

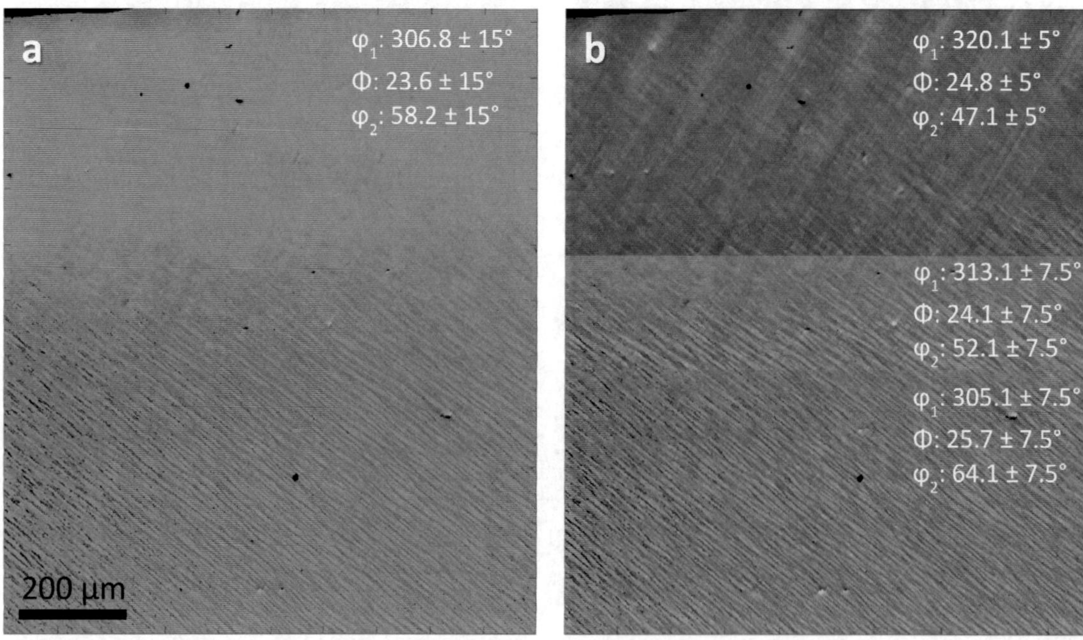

Figure 3. Customized All-Euler maps of a region with a continuous variation in orientation over a large length scale. (a) The map uses the distribution of Euler angles of the entire region. (b) The map is divided into three sub-regions, each of which uses the distribution of Euler angles in the sub-region.

In figure 3a, the orientation gradually changes by 5–9° between the top and the bottom parts of the region investigated. The upper part covers a height of 500–600 µm while the size of bottom part is even larger. The customized All-Euler map in figure 3a only shows the difference between the top and the bottom parts, and does not reveal much of the dislocation structure. In this case, the image contrast can be largely improved by dividing the map into smaller sub-maps of relatively homogeneous orientations.

By dividing figure 3a into three regions, as shown in figure 3b, the image contrast can be optimized for each region using the local distribution of Euler angles, resulting in much improved visualization of the underlying microstructures.

Any degeneracy in Euler space should be avoided in customized All-Euler maps. One such degeneracy is due to the selection of different equivalent variants of the same orientation during data acquisition of EBSD, which is removed during post-processing of the data by forcing all pixels of each given grain to choose the same variant. Another type of degeneracy is the discontinuity related to the periodic boundaries of the Euler angles at 0° or 360°. Both can be fixed by choosing values that are closest to the Euler angles assigned to the mean orientation of each grain. For instance, if the Euler angle ϕ_1 of the mean orientation of a given grain is $\phi_{1m} = 1°$, then in this grain a pixel that has $\phi_1 = 359°$ will be assigned with a new value, $\phi_1 = 359°-360° = -1°$, to ensure that ϕ_1 has a continuous distribution that centers around ϕ_{1m}.

The Euler space is known to have a singularity for $\Phi = 0°$, corresponding to the situation where the z direction of sample reference frame, z_s, coincides with the z direction of the crystal reference frame, z_c. This singularity can be avoided by choosing a different but equivalent assignment of the x, y and z directions to the crystal reference frame, which will result in a new equivalent set of Euler angles, whenever the mean orientation of a given grain has an Euler angle Φ that is very close to 0°.

Another factor to be considered is the non-linear nature of the Euler space. By avoiding the singularity problem, the most distorted region in the Euler space is already avoided. Moreover, in the study of low-misorientation dislocation structures, since the orientation of each map is only distributed over a small volume of Euler space, the effect of the non-linearity is expected to be further reduced in such small volumes.

The morphology of dislocation structures is closely related to the crystal orientation and reflects the slip systems that are active during deformation [2–4]. The morphology of dislocation structures has also been shown to have an impact on boundary migration during recrystallization of deformed metals [14–17]. Improved visualization of dislocation structures using customized All-Euler map is therefore of value for such studies. While the present work does not include quantification of the visualized dislocation structures, it has been shown that the subgrain structures can be detected by further data processing, e.g. using an orientation averaging technique [18,19].

4. Conclusions

As shown for high purity Al (99.996 wt.%) rolled by a thickness reduction of 12%, customized All-Euler maps are capable of visualizing dislocation structures with boundary misorientations approaching the orientation precision of standard EBSD (~0.5°). In cases where cumulative orientation variations exist over a large length scale, it is shown that the image contrast is largely improved by dividing the map into smaller regions of relatively homogeneous orientations and making customized All-Euler maps for each region. Artificial discontinuities in Euler space can be removed in customized All-Euler maps. Also, the singularity in Euler space can be avoided by choosing a different but equivalent set of Euler angles.

Acknowledgement

This work is supported by funding from the European Research Council (ERC) under the European Union's Horizon 2020 research and innovation program (grant agreement No 788567). The author thanks Dr. Chaoling Xu for providing the sample for this study and Prof. Dorte Juul Jensen for helpful discussions.

References

[1] Hansen N 2001 New discoveries in deformed metals *Metall. Mater. Trans. A* **32** 2917–35
[2] Huang X and Winther G 2007 Dislocation structures. Part I. Grain orientation dependence *Philos. Mag.* **87** 5189–214
[3] Winther G and Huang X 2007 Dislocation structures. Part II. Slip system dependence *Philos. Mag.*

87 5215–35
[4] Liu Q, Juul Jensen D and Hansen N 1998 Effect of grain orientation on deformation structure in cold-rolled polycrystalline aluminium *Acta Mater.* **46** 5819–38
[5] Zhang C L, Zhang Y B, Wu G L, Liu W, Juul Jensen D and Godfrey A 2017 Quantification of deformation microstructure at ultra-low tensile strain in pure Al prepared by spark plasma sintering *IOP Conf. Ser. Mater. Sci. Eng. Pap.* **219** 012050
[6] Wei Y L, Godfrey A, Liu W, Liu Q, Huang X, Hansen N and Winther G 2011 Dislocations, boundaries and slip systems in cube grains of rolled aluminium *Scr. Mater.* **65** 355–8
[7] Hong C, Huang X and Winther G 2013 Dislocation content of geometrically necessary boundaries aligned with slip planes in rolled aluminium *Philos. Mag.* **93** 3118–41
[8] McCabe R J, Misra A and Mitchell T E 2004 Experimentally determined content of a geometrically necessary dislocation boundary in copper *Acta Mater.* **52** 705–14
[9] Xu C, Zhang Y, Lin F, Wu G, Liu Q and Juul Jensen D 2016 Crystallographic analysis of nucleation at hardness indentations in high-purity aluminum *Metall. Mater. Trans. A* **47A** 1–8
[10] Xu C, Zhang Y, Godfrey A, Wu G, Liu W, Tischler J Z, Liu Q and Juul Jensen D 2017 Direct observation of nucleation in the bulk of an opaque sample *Sci. Rep.* **7** 42508
[11] Hong C, Fæster S, Hansen N, Huang X and Barabash R I 2017 Non-spherical voids and lattice reorientation patterning in a shock-loaded Al single crystal *Acta Mater.* **134** 16–30
[12] Liu Q, Wert J and Hansen N 2000 Location-dependent lattice rotation and shear strain in rolled aluminium single crystals of cube and Goss orientations *Acta Mater.* **48** 4267–79
[13] Liu Q and Hansen N 1998 Macroscopic and microscopic subdivison of a cold-rolled aluminium single crystal of cubic orientation *Proc. R. Soc. A Math. Phys. Eng. Sci.* **454** 2555–92
[14] Zhang Y B, Budai J D, Tischler J Z, Liu W, Xu R, Homer E R, Godfrey A and Juul Jensen D 2017 Boundary migration in a 3D deformed microstructure inside an opaque sample *Sci. Rep.* **7** 4423
[15] Zhang Y, Godfrey A and Juul Jensen D 2014 In-Situ investigation of local boundary migration during recrystallization *Metall. Mater. Trans. A* **45** 2899–905
[16] Juul Jensen D, Lin F X, Zhang Y B and Zhang Y H 2013 Importance of local structural variations on recrystallization *Mater. Sci. Forum* **753** 37–41
[17] Schmidt S, Nielsen S F, Gundlach C, Margulies L, Huang X and Juul Jensen D 2004 Watching the growth of bulk grains during recrystallization of deformed metals *Science* **305** 229–32
[18] Humphreys F J, Bate P S and Hurley P J 2001 Orientation averaging of electron backscattered diffraction data *J. Microsc.* **201** 50–8
[19] Hurley P J and Humphreys F J 2002 Characterizing the deformed state in Al-0.1 Mg alloy using high-resolution electron backscattered diffraction *J. Microsc.* **205** 218–25

IOP Conf. Series: Materials Science and Engineering **580** (2019) 012034 doi:10.1088/1757-899X/580/1/012034

Study of potential recrystallization nuclei in the cold-rolled microstructure of an electrical steel by electron backscatter diffraction

WC Hsu, L Chang* and PW Kao

Department of Materials and Optoelectronic Science, National Sun Yat-sen University, Kaohsiung, Taiwan, R. O. C.

lwchang@mail.nsysu.edu.tw (L. Chang)

Abstract. Strain induced boundary migration (SIBM) is widely recognized to be the main mechanism corresponding to recrystallization in metals of low or medium amount of deformation. The nucleation process in recrystallization of heavily deformed metals is, however, less understood for key issues such as how the nuclei are formed and where they are located. The evolution of texture from deformation to recrystallization cannot thus be explained well. In this study, an attempt was conducted to identify the potential recrystallization nuclei (PRN) in the cold-rolled microstructure of an electrical steel by electron backscatter diffraction (EBSD). This work demonstrates that the SIBM mechanism can be the main mechanism corresponding to recrystallization nucleation in a heavily deformed steel. The same area in both the cold-rolled and partially recrystallized states was analysed using EBSD to obtain the correlation between the recrystallized grains and the cold-rolled substructures. The deformed grains having high stored energy were first selected, followed by identifying substructures surrounded entirely or partly by high angle boundaries in these heavily deformed grains. The deformation induced grains having low intra-granular misorientation of 2° or less exhibit a very similar orientation distribution to that of the recrystallized grains and are proposed to be the PRN in subsequent recrystallization.

1. Introduction

The process of recrystallization of plastically deformed metals and alloys is extremely important for controlling the texture, and therefore optimizing the corresponding physical and mechanical properties. In recrystallization, the key issue concerning the formation of new grains (or nuclei) has not been explained well. It is well recognized nowadays that the new grains indeed grow from small regions, such as subgrains and cells that are likely surrounded by high angle grain boundaries [1]. According to this idea, the orientations of the recrystallization nuclei are inherent in the deformed microstructure. The small regions that are capable of growing into new grains are therefore called the potential recrystallization nuclei (PRN). Once the PRN can be identified in the deformed microstructure, the texture of a deformed metal after recrystallization can be predicted accordingly.

Strain induced boundary migration (SIBM) has been widely accepted as the main mechanism corresponding to recrystallization in metals subjected to a low or medium amount of deformation [2, 3]. According to this mechanism, a high angle boundary (HAB) present prior to deformation, can bulge out from the low stored energy side toward the high stored energy side. However, the nucleation process in heavily deformed metals was suggested to rely on different mechanisms, such as subgrain coalescence and subgrain growth [1, 4]. Based on the fact that more and more HABs can be generated in the grain interiors with increasing the deformation, the role of SIBM might have to be re-examined for recrystallization in heavily deformed metals [5, 6]. In our previous work [7], electron backscatter diffraction (EBSD) observations confirmed that recrystallization was preferentially evolved from areas having high local misorientation. The recrystallization nuclei in a partially recrystallized specimen were observed to be directly related to the areas having high kernel average misorientation (KAM) values in the deformed grains. However, the KAM method is a pixel-based technique and cannot be related directly to the PRN. In this study, we report a subsequent effort to identify the PRN from the cold-rolled microstructure based on a modified SIBM model.

2. Experimental methods

An electrical steel after cold rolling was used for EBSD analyses. The composition and hot-rolling conditions of the steel are reported elsewhere [7]. After cold-rolling from 10 mm to 3 mm, the microstructure of the rolled plate was analyzed by EBSD in at least six areas of 230 × 172 μm^2 each, with a step size of 0.25 μm. The cold-rolled sheets were also annealed at 903-983 K for one minute to reach a partly or fully recrystallized state. The microstructure and texture prior to, and after annealing, were analyzed by EBSD for the same area. Some EBSD orientation image maps (OIMs) were acquired at a smaller step size of 0.05-0.1 μm to reveal the detailed microstructure in the cold-rolled state. The EBSD analyses were carried out using an Oxford Nordlys detector mounted on a Zeiss Supra 55 scanning electron microscope. EBSD data were analyzed using the Channel 5 software.

3. The modified SIBM model

Due to the heavy deformation, one or several subgrains in a cold rolled plate can be surrounded partly or entirely by HABs, as shown schematically in Figure. 1(a), and are called the deformation induced grains (DIGs). The pressure driving a HAB to migrate outward is expressed as [5, 6]:

$$\Delta P = P_E - P_I - P_\gamma \tag{1}$$

where $P_E=2\gamma_E/\lambda_E$ is the outward drag force due to the sub-boundaries in front of the HAB, $P_I=2\gamma_I/\lambda_I$ is the inward drag force due to the sub-boundaries behind the HAB, and P_γ is the inward drag force due to the curvature of the HAB. Also γ_E (γ_I) is the surface tension of the sub-boundaries in front of (behind) the HAB and λ_E (λ_I) is the average size of the subgrains in front of (behind) the HAB. Figure. 1(a) shows that R_0 is the plane radius of a HAB cap which is free to bulge and x is the height of the cap. The existence of the DIGs, which are surrounded by deformation induced HABs was examined by EBSD for the steel studied in this paper. Figure. 1(b) shows the EBSD OIM (ND inverse pole figure coloring) acquired with viewing direction along the transverse direction (TD) of the cold rolled plate in a fine step size of 0.05 μm. The corresponding boundary map is shown in Figure. 1(c), indicating that HABs (red lines) are distributed in the interior of a deformed, <111>//ND grain. Some subgrains, surrounded mostly by HABs and free from sub-boundaries internally, can be identified such as A1 and A2 labeled in Figure. 1(c) for examples. Subgrain groups, having larger sizes, surrounded by HABs are also observed such as B1-B3 in Figure. 1(c). Low angle boundaries are distributed in the interiors of B1-B3. The EBSD result verifies the existence of DIGs in the deformed microstructure. Accordingly, there are three factors which determine whether any given DIGs can be a PRN. First, the outward drag force P_E or the stored energy in the surrounding area is large. Second, the internal drag force P_I is small, and finally, the HAB that is free to bulge exhibits a relatively large radius of curvature.

4. EBSD Results and discussion

Figure. 2(a) shows a ND OIM acquired from TD. HABs (>15°) are highlighted in dark lines and distributed heterogeneously in the deformed structure. Some deformed grains are free of intragranular

HABs but others contain HABs with a moderate to high density. The deformation heterogeneity can be demonstrated clearly by constructing a map of high misorientation regions (HMRs, see Figure. 2(b)) according to the suggestion given by Mishin et al. [8]. It is noted that all the intragranular HABs are located in the HMRs and the crystal orientations of HMRs were found to be mainly in the γ-fiber orientations with a high intensity near {111}<112>. Since results in Figure. 1 demonstrate the existence of DIGs, the DIGs in the analyzed areas are identified using a 15°-5° criterion and illustrated in Figure. 2(c). These DIGs are mainly distributed along the original grain boundaries and on shear bands in the grain interiors. It is noted that the DIGs possess orientations which are different from those of the cold-rolled matrix. Figure. 2(d) shows the OIM of the recrystallized grains in the same area after the steel was annealed at 923 K for 60 s. The recrystallization fraction of this area is 25%. The figure indicates that the regions having high density of DIGs are mostly replaced by recrystallized grains with similar orientations after annealing. On the other hand, those regions containing no DIGs remain unrecrystallized.

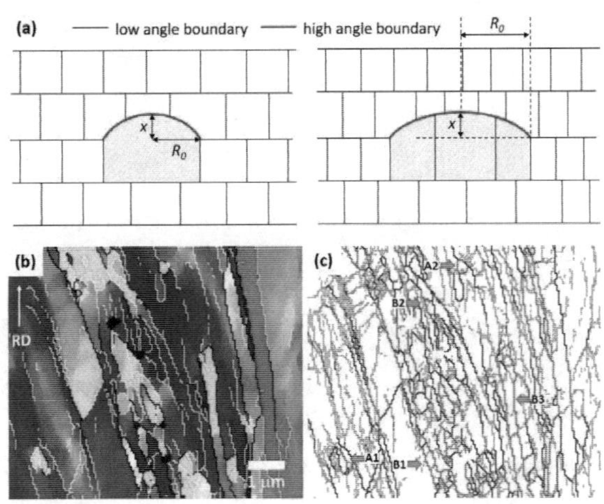

Figure. 1 (a) A schematic drawing for the deformation induced grains (DIGs) which are grayed out, and EBSD (b) ND OIM + boundary map (white: 5-15°, dark: >15°) and (c) boundary map (gray: 1-5°, dark: 5-15°, red: >15°) for the 70% cold-rolled electrical steel showing the observed DIGs. The scan step size was 0.05 μm.

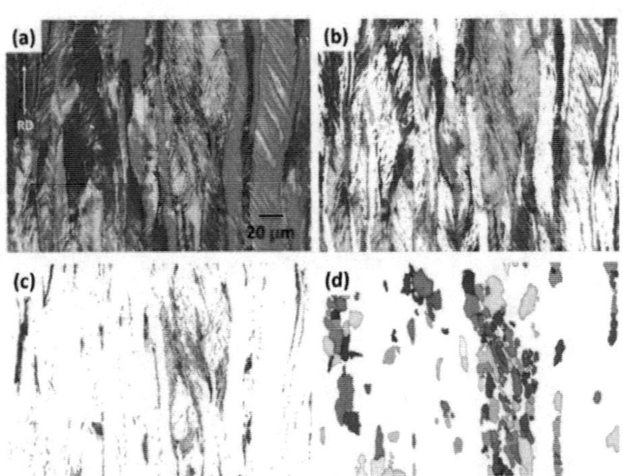

Figure. 2 EBSD (a) ND OIM + boundary map (dark: >15°), (b) HMR map and (c) DIG map for the 70% cold-rolled electrical steel, and (d) OIM showing the recrystallized grains for the steel annealed at 903 K for one minute. The scan step size was 0.25 μm.

Figure. 3 shows the φ_2=45° ODF sections derived from EBSD OIMs for the cold rolled and annealed (923 K for 60 s) plate. The ODF section for the cold-rolled plate (Figure. 3(a)) shows a strong α-fiber, mainly from {001} < 110 > through {112}<110> to {111}<110>, and a strong γ-fiber ({111}//ND) both of which are typical for low carbon steels. The ODF section for the HMRs in the cold-rolled structure is shown in Figure. 3(b), confirming that the HMRs are mainly in the γ-fiber. The orientations of the DIGs (see Figure. 3(c)) distribute rather randomly over the γ-fiber, the θ-fiber (<100>//ND), the partial <110>//TD fiber and the partial <110>//RD fiber. Finally, the ODF section for the recrystallized grains in the annealed sample is shown in Figure. 3(d). For this annealed state (923 K, 60 s), the average recrystallization fraction is 40 % and the recrystallization nucleation is nearly saturated. It has been reported that the recrystallization proceeds by continuous growth of the recrystallized grains into the remaining deformed grains with low stored energies as the samples were annealed at higher temperatures [9]. It is worth pointing out that the orientation distribution of the DIGs in the cold-rolled structure (Figure. 3(c)) is qualitatively similar to that of the recrystallized grains (Figure. 3(d)), but the recrystallized grains show much higher intensities in the cube ({100}<001>), Goss ({110}<001>) and {111}<11-2> orientations and lower intensities in the rotate cube ({100}<011>) and {110}<1-10> orientations.

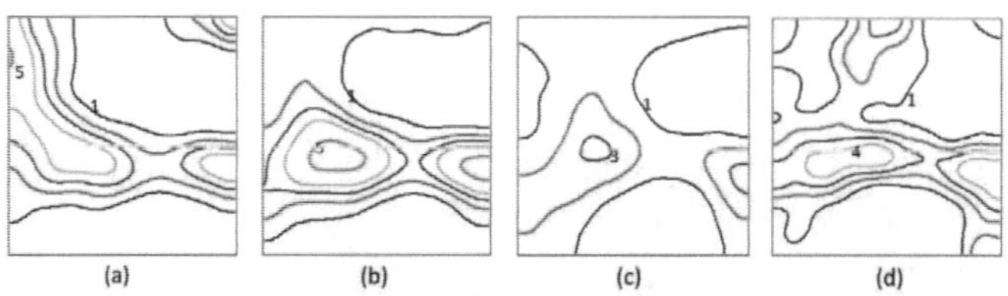

Figure. 3 ODF φ_2=45°sections derived from 6 EBSD OIMs for each of the following conditions: (a) the cold rolled plate, (b) the HMRs in the cold rolled plate, (c) the DIGs in the cold rolled plate, and (d) the recrystallized grains in the annealed (923 K for 60 s) plate.

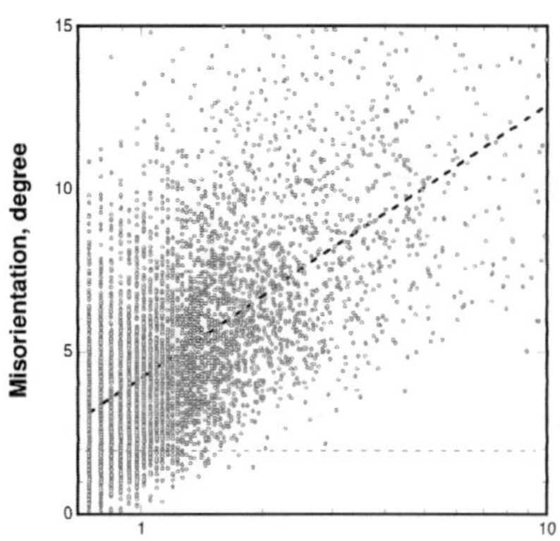

Figure. 4 Distribution of misorientation as a function of the equivalent diameter of DIGs. Data are collected from three 230 × 172 μm² OIMs.

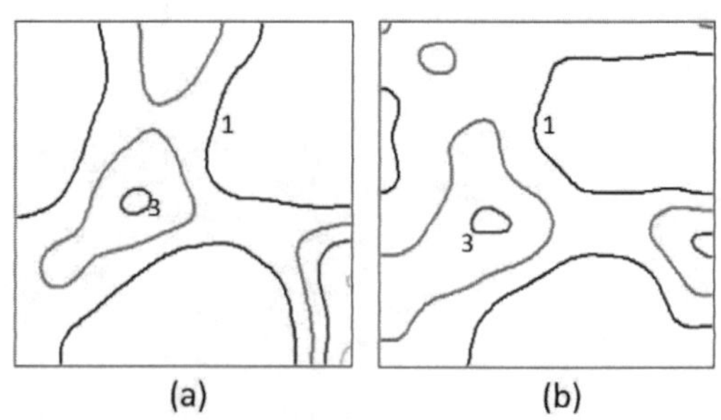

Figure. 5 ODF φ_2=45°sections derived from 6 EBSD OIMs for the DIGs (a) having low misorientation values (<2°) and (b) having large sizes (>3 μm) in cold rolled plate.

According to the discussion above, the DIGs are all distributed in the HMRs having high stored energies. The high stored energies of the matrix generate high outward drag forces on the HABs surrounding the DIGs. Accordingly, as long as the inward drag forces remain low, the DIGs can grow into nuclei during annealing. It is, however, tedious to examine the drag force, ΔP, of each DIG to distinguish whether it is a PRN or not. Alternatively, a proper criterion may be chosen to identify the PRN in a statistical manner. Figure 4 shows the intra-granular misorientation as a function of the equivalent diameter each DIG. The intra-granular misorientation is calculated by the Channel 5 software based on the grain orientation spread (GOS) principle [10], and is used to evaluate the inward drag force of the DIGs. Ideally, large DIGs having low intra-granular misorientation values are ideally suited as PRN for recrystallization. However, Figure 4 shows that the misorientation increases with increasing DIG size. A higher misorientation means a higher energy of the substructure and therefore also of the inward drag force. On the other hand, only a portion of the DIGs has intra-granular misorientation values less than 2 degrees. All of these are indeed less than 2 μm in diameter. The traditional theory based on subgrain growth, however, suggests that substructures having large sizes have an advantage doing growth competition. To verify these two ideas, the ODF sections of DIGs having low intra-granular misorientation (<2°) and of those having large size (> 3 μm) are shown in Figure 5. It is evident that the ODF of DIGs having low intra-granular misorientation shows stronger cube, Goss and {111}<112> orientations compared to that of all DIGs, and is therefore closer to that of the recrystallized grains. On the other hand, the ODF of the large-size DIGs exhibits a high density of rotated cube orientations and low densities of cube and Goss orientations. As a result, the DIGs having low interior misorientation should be the PRN.

In order to confirm the inference given above, the drag force as a function of the misorientation angle (θ_E) of the sub-boundaries surrounding DIGs is calculated for small DIGs which are free from interior sub-boundaries and large DIGs having a high intra-granular misorientation (θ_I) of 5°. The results are shown in Figure. 6. For the small DIGs, the drag force is positive as the surrounding substructures have boundaries whose misorientation angles are larger than 3° for 4 μm-diameter DIGs and 6° for 2 μm-diameter DIGs. These values are reasonable for steels rolled to 70% [12]. Moreover, the drag force due to the curvature effect decreases with increasing size of the DIGs. On the other hand, the large DIGs with average diameters of 6-10 μm do not have a positive drag force unless the outside substructures exhibit a large misorientation of 7° or above. The small DIGs with a low intra-granular misorientation can therefore grow continuously on annealing, whereas the growth of large-size DIGs is easily be hindered by surrounding substructures having relatively low stored energies.

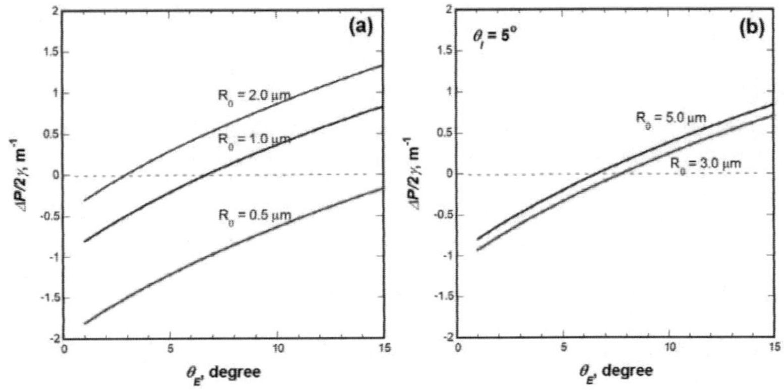

Figure. 6 Outward drag force as a function of the misorientation angle (θ_E) of the substructures outside the DIGs for (a) small DIGs which are free from interior sub-boundaries and (b) large DIGs having an intra-granular misorientation (θ_I) of 5°.

Table 1 lists the total number and average number density of the DIGs, PRN and recrystallized grains (RXG) after annealing at 903 K and 923 K obtained from the EBSD analyses. The results indicate that one in four of the DIGs is a PRN. It also seems that one out of two PRNs will grow successfully into a recrystallization grain. However, it should be noted that the present result is obtained from a two dimensional EBSD analysis, whereas the recrystallized grains shown in the annealed samples can result from PRN hidden below the surface that could not be detected. As a result, it is reasonable to believe that the number of PRN analyzed using the method suggested in this study is approximately one order of magnitude higher than the recrystallization nuclei actually formed. The overestimation of the PRN can be attributed to several reasons. First, the outward drag force, which determines the mobility of the HAB, is not considered in our analysis, but can indeed be different for each PRN. Second, this drag force varies with the movement of the HAB. Third, nearby PRN of similar orientations might impinge with each other and then grow more slowly due to the orientation pinning effect [12]. All these factors might cause a portion of PRNs to be merged on growth competition. Further effort is needed on this issue. Moreover, it worth emphasizing that the orientation distribution of the PRN will never be exactly the same as that of the recrystallized grains even if the PRN can be identified more accurately, since they will grow to recrystallized grains with very different sizes.

Table. 1 Total number and average number density of the DIGs, PRN and recrystallized grains (RXG) after annealing at 903 K and 923 K from the EBSD analyses.

	DIG	PRN (<2°)	RXG (22%)	RXG (40%)
Number of OIMs	14	14	9	6
Total number	24294	6603	1120	1314
Number density, 1000 μm^{-2}	42	12	3	6

5. Conclusion

This work demonstrates that strain induced boundary migration can be the main mechanism for recrystallization nucleation in a heavily deformed steel. The same area in both the cold-rolled and partially recrystallized states was analyzed using EBSD to obtain the correlation between the recrystallized grains and the cold-rolled substructures. The deformed grains having high stored energies were first selected, followed by identifying the substructures surrounded entirely or partly by high angle boundaries in these heavily deformed grains. The deformation induced grains, having low misorientation of 2° or less, exhibit very similar orientation distribution to that of the recrystallized grains and are proposed to be the potential recrystallization nuclei in subsequent recrystallization.

Acknowledgement

This work is supported by the Ministry of Science and Technology, R. O. C. (MOST-106-2622-8-006-001 and MOST 107-2622-8-006-015). We also acknowledges China Steel Corporation for the assistance of sample preparation.

References

[1] Doherty RD et al 1997 *Mater. Sci. Eng.* A **238** 219
[2] Beck PA and Sperry PR 1950 *J. Appl. Phys.* **21** 150
[3] Bailey JE and Hirsch PB 1962 *Proc. R. Soc.* A **267** 11
[4] Jones AR, Ralph B and Hansen N 1979 *Proc. R. Soc. Lond.* A **368** 345
[5] Bate P and Hutchinson B 1997 *Scripta Mater.* **36** 195
[6] Higginson R and Bate P 1999 *Acta Mater.* **47** 1079
[7] Hsu WC, Chang L, Sun PL, Ho N, Kao PW and Hsiao IC 2015 *ISIJ Int.* **55** 2212
[8] Mishin OV, Godfrey A and Jensen DJ 2009 *Electron Backscatter Diffraction in Materials Science* ed AJ Schwartz et al (New York:Springer) pp 263-275
[9] Hsu WC, Chang L, Kao PW and Hsiao IC 2018 *ISIJ Int.* **58** 958
[10] Allain-Bonasso N, Wagner F, Berbenni S and Field DP 2012 *Mater. Sci. Eng.* A **548** 56
[11] Li BL, Godfrey A, Meng QC, Liu Q and Hansen N 2004 *Acta Mater.* **52** 1069
[12] Magnusson H, Jensen DJ and Hutchinson B 2001 *Scripta Mater.* **44** 435

Microstructure and strength of weldment in Pt20Rh alloys dispersion-strengthened by ZrO_2 particles

Faping Hu[1], Ziyang Wang[1], Tianbo Yu[2], Zonglun Yang[3], Guobing Wei[1], Xuan Luo[1], Niels Hansen[1,4], Xiaoxu Huang[1,2], Weidong Xie[1*]

[1] International Joint Laboratory for Light Alloys (MOE), College of Materials Science and Engineering, Chongqing University, Chongqing 400044, China;
[2] Department of Mechanical Engineering, Technical University of Denmark, DK-2800 Kgs. Lyngby, Denmark
[3] Chongqing Polycomp International Corporation, Chongqing 400044, China
[4] Technical University of Denmark, Risø Campus, DK-4000 Roskilde, Denmark

E-mail: wdxie@cqu.edu.cn

Abstract. A series of Pt20Rh alloys dispersion-strengthened by different volume fractions of ZrO_2 particles have been prepared by internal oxidation of Pt20Rh powder followed by consolidation and hot forging. The alloys were cold rolled to sheets, followed by annealing. The effects of welding on the microstructure and mechanical properties were also investigated. The microstructure of the rolled sheets showed a lamellar structure and the yield stress was significantly increased with increasing volume fraction of dispersion particles. After welding, the yield stress at room temperature was almost unchanged, although a small decrease was observed at 1000 °C. However, the work hardening capacity and ultimate tensile stress were significantly reduced in the welded sample due to grain coarsening and to increased pore density in the fusion zone.

1. Introduction

Platinum is characterized by a high melting point combined with excellent chemical inertia [1-3]. These properties are particularly useful in the glass industry where platinum is the main metal for manufacturing bushing, i.e. a platinum container used in the manufacture of continuous glass fibers. The bushing is exposed to an aggressive service environment (high temperature range from 1100 °C to 1400 °C, alternating loads with a stress between 1 MPa to 40 MPa, contact with molten glass and air, etc.) for typically more than one year [3-5]. However, pure platinum is relatively soft at room temperature, and even softer after high temperature exposure due to grain growth.

An innovative preparation route that combines powder metallurgy and internal oxidation has been proven to be very effective for strengthening of platinum-based alloys [4, 6], in which the alloys are produced by consolidation of milled powder that has been internally oxidized. To widen the industrial application of the material its weldability is important. For example, in the fiber glass industry, the bushing is a rectangular open topped box having a base with typically 1000 – 4000 precisely shaped orifices with bore diameters from 1.1 mm up to 2.5 mm and lengths 2.25 times their bore diameters [5]. To construct this bushing the base and the orifices are fabricated separately, followed by a manual welding during processing. As a result, the weldment will be composed of different zones with different structure and mechanical properties. These zones can be characterized as: (i) the fusion zone (FZ); (ii) the heat affected zone (HAZ); and (iii) unaffected base metal (BM). In combination, the presence of these zones may weaken the material by reducing fracture toughness and high temperature strength.

In the present work dispersion-strengthened platinum alloys with various volume fractions of ZrO_2 have been produced by internal oxidation-powder metallurgy processes and arc welded to explore the effects of welding on the microstructure and strength properties.

2. Experimental procedures

Table 1 shows the nominal composition of investigated alloys. Sheets of Pt20Rh, 1 mm in thickness, containing different amounts of ZrO_2, were prepared by internal oxidation-powder metallurgy processes as described in a previous paper [6] and were welded along the transverse direction (TD). The welding parameters (manual arc welding with no filler metal) are given in table 2.

Table 1. Composition of the investigated alloys (in weight %).

Alloys	Zr	Y	Others	Rh	Pt
A	–	0.015	≤0.01	20	Bal.
B	0.05	0.015	≤0.01	20	Bal.
C	0.10	0.015	≤0.01	20	Bal.
D	0.30	0.015	≤0.01	20	Bal.

Table 2. Welding parameters.

Weld parameters	
Argon flow	3-4 L/min
Current	30-50 A
Weld speed	12-17 cm/min

The weldments and their unwelded BM were machined into dog-bone shape tensile samples (gauge section of 20 mm, 10 mm and 1 mm in length, width and thickness, respectively) taken along the rolling direction. Tensile tests were conducted at an initial strain rate of $8.3 \cdot 10^{-3}s^{-1}$ at RT and of $8.3 \cdot 10^{-5}s^{-1}$at 1000 °C. The samples were cut so that about 3 mm of weld line was located in the mid-gauge position of the tensile specimen for each weldment.

To characterize the microstructure of the samples, scanning electron microscopy (FEI Nova 400 FEG, and Zeiss Supra 35 equipped with electron backscatter diffraction detector) and transmission electron microscopy (JEOL JEM-2100) were carried out. The samples for SEM observations were prepared by mechanical grinding and polishing. The sizes and area of the pores were measured by the Image-Pro Plus6 (IPP) software. Mechanical grinding and ion thinning were used in the preparation of TEM samples. For simplicity, the alloys are referred to hereafter as A (0 wt%Zr), B (0.05 wt%Zr), C (0.1 wt%Zr) and D (0.3 wt%Zr).

3. Results

3.1 Mechanical properties

All weldments fractured at the welding line, both during tensile testing at room temperature and at 1000 °C. Figure 1 shows the engineering stress-engineering strain curves at room temperature, which indicates that the yield stress (0.2% proof, $\sigma_{0.2}$) and the ultimate tensile stress (UTS) of the BM are significantly increased with increasing ZrO_2 content, accompanied by a slight decrease in total elongation. After welding, the initial part of the tensile curve of each weldment was basically the same as its BM, but failure occurred at a much lower strain in the weldment than in the BM. As a result, the $\sigma_{0.2}$ yield stress of each weldment is almost the same as its BM, but the total elongation and UTS of each weldment are clearly reduced compared with the BM, although the elongation of weldment is still acceptable. The measured values of $\sigma_{0.2}$, UTS and total elongation of the investigated samples are shown in table 3.

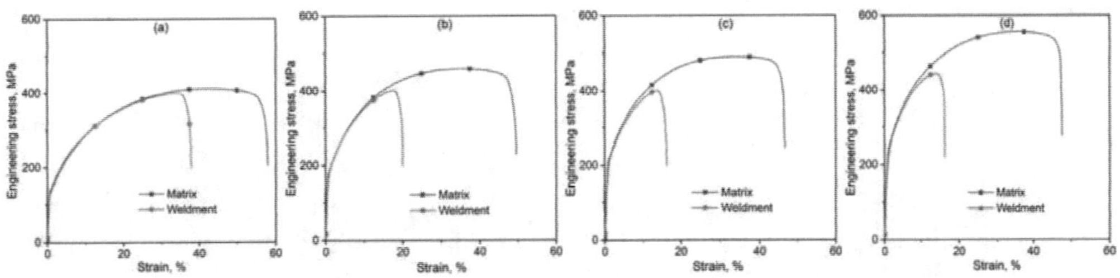

Figure 1. Stress-strain curves of samples A (a), B (b), C (c) and D (d) and their welded counterparts at room temperature.

Figure 2 shows the engineering stress-strain curves at 1000°C, indicating that the strength and elongation of both weldments and their BMs are significantly decreased, compared to their room-temperature counterparts. After welding, the elongation to failure for the weldments is dramatically reduced compared to the corresponding BM, ranging from 7% to 2% with increasing ZrO_2 particle concentration. Furthermore, the $\sigma_{0.2}$ of the BM and of the weldments are significantly raised with increasing ZrO_2 content, and the $\sigma_{0.2}$ of each weldment is close to that of its BM although a slight decrease was observed in samples C and D, which are similar with room-temperature tensile results.

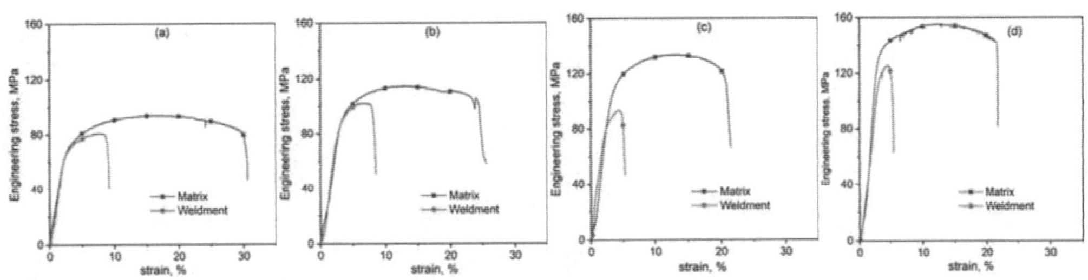

Figure 2. Engineering stress-strain curves of samples A (a), B (b), C (c) and D (d) and their weldments during tensile testing at 1000 °C.

Table 3. Tensile properties (BM/weldment) of the investigated samples at RT and at 1000 °C.

Alloy	$\sigma_{0.2}$, MPa		UTS, MPa		Total elongation, %	
	RT	1000 °C	RT	1000 °C	RT	1000 °C
A	143/146	67/66	411/400	91/81	57/37	28/7
B	200/196	89/87	459/402	114/102	49/19	22/6
C	228/218	99/82	491/401	137/94	46/15	17/3
D	256/236	124/113	555/445	152/124	46/15	18/2

3.2 Microstructure

ZrO_2-strengthened Pt20Rh alloys possess an excellent thermal stability and basically no recrystallization occurs after annealing at 1150 ° C for 30 min, as shown in figure 3a, where the morphology of the BM was characterized by high angle boundaries (HABs) approximately parallel to TD and low angle boundaries (LABs) distributed between the HABs, indicating a recovered, lamellar structure. After welding, two new zones were developed, namely the HAZ and FZ, as shown in figures 3b and 3c. The grains in the HAZ are fully recrystallized with a slightly elongated shape along TD, while the grains in FZ are as-cast dendrites. Furthermore, a large number of LABs are observed in both the HAZ and FZ, but their distribution is non-uniform.

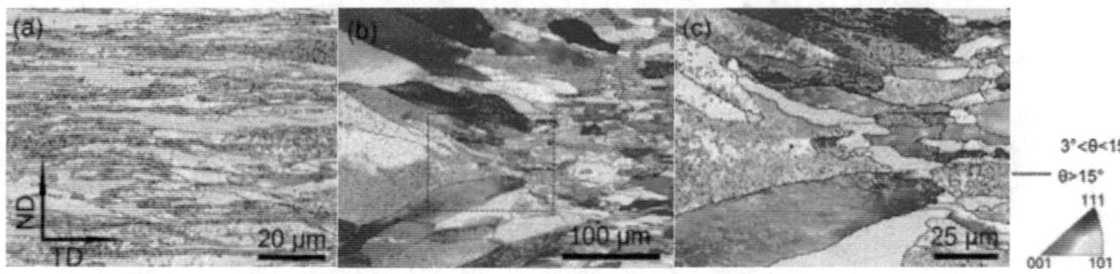

Figure 3. EBSD inverse pole figure (IPF) with respect to the RD of sample C: (a) the BM; (b) the transition region between HAZ (right) and FZ (left) ; (c) magnified area of the rectangle marked by dotted lines in (b). Black lines and white lines indicate HABs and LABs, respectively.

Due to the powder metallurgy process, the formation of pores within the alloys seems inevitable. Figure 4 shows the morphology of the pores in the BM (a, c and e) and FZ (b, d and f) of the investigated alloys, and figure 5 shows distributions of pore diameters. The pores are approximately spherical and basically uniformly distributed. The porosity in the BM (FZ) for samples A, C and D are 0.18% (0.41%), 0.26% (0.54%), and 0.68% (1.29%), respectively. It is obvious that both the porosity and the pore size increase with increasing ZrO_2 particle content in the BM regions. After welding, the porosity in the FZ was approximately doubled for each sample. The frequency of smaller pores (0 - 0.2 μm) significantly decreased while that of larger pores (>0.2 μm) increased.

Figure 6 shows the representative particle morphology in samples A (a) and C (b) in the FZ. Spherical particles were found in these two alloys and no obvious aggregation was detected. In our recent studies, the particles in sample A have been identified as yttria, while the particles in the other alloys (samples B, C and D) were identified as ZrO_2 with either a monoclinic structure or a tetragonal structure.

4. Discussion

Based on the results of the tensile tests, the Pt20Rh strengthened by various ZrO_2 particle concentrations basically maintained their yield stress after fusion welding both at room temperature and at 1000 °C, while the work hardening capability was dramatically decreased, especially at 1000 °C. The microstructural analysis showed that three aspects may be responsible for these phenomena.

The BM regions of the investigated alloys have a recovered structure with lamellar HABs that are divided by LABs. During arc welding, the metal in the FZ underwent a melting process followed by solidification and developed into a cast structure characterized by coarse grains that weaken the alloy according to Hall-Petch relationship [7]. However, based on EBSD results, there are also numerous LABs formed within the resolidified grains, which may be ascribed to the high welding temperature [8] and good electrical conductivity and therefore rapid solidification. These LABs can impede dislocation movement and strengthen the alloys [9]. It should be noted that the distribution of LABs is non-uniform and some coarse grains containing a few LABs are weaker. These weaker grains may lead to premature fracture during deformation and cause failure of the whole sample.

Pores within a metal are usually detrimental to its strength and ductility because they reduce the stressed area and cause stress concentration, both of which effects are enhanced with increasing pore size and porosity. Furthermore, pore shape also affects the mechanical properties of a metal with usually a smaller dihedral angle resulting in easier cracking, due to increased stress concentration. Fortunately, the pores in the investigated alloys are relatively fine, spherical and homogeneously distributed, though they are increased in number and enlarged with increasing particle content or after welding. Thus, at the initial stage of deformation, these fine pores can act as obstacles for dislocations motion. With increasing deformation, however, the pores may gradually become crack sources due to stress concentrations caused by accumulative dislocations around the pores.

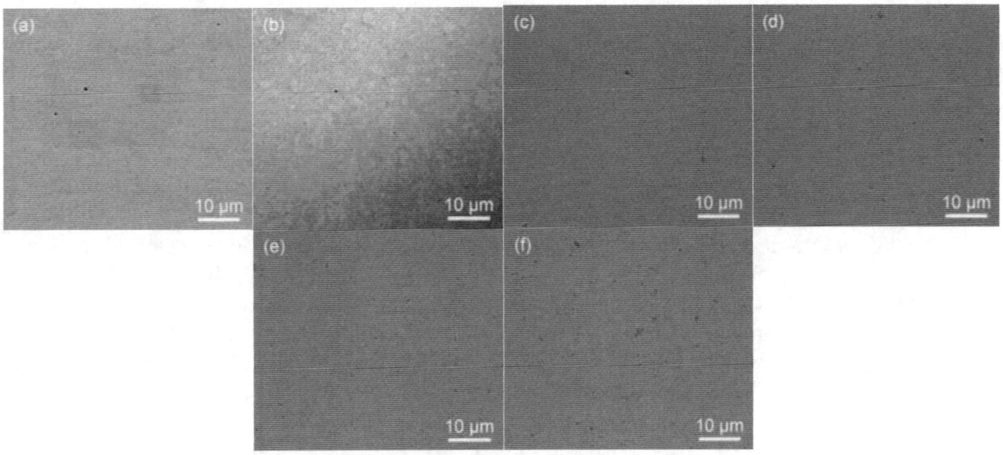

Figure 4. SEM images showing the microstructure in the HAZ and FZ of sample A (a, b), sample C (c, d) and sample D (e, f).

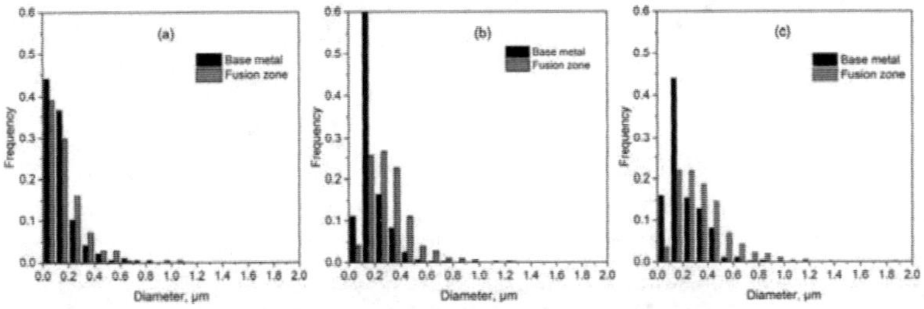

Figure 5. Pore size distributions in the BM and FZ of samples A (a), C (c) and D (c).

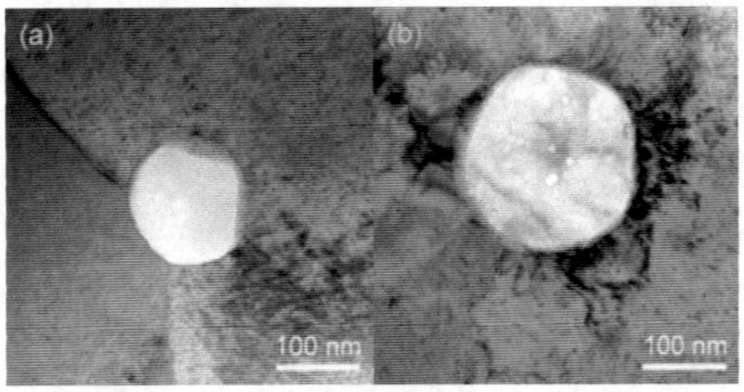

Figure 6. TEM bright field images showing particles in sample A (a) and C (b).

Impenetrable particles in a metal can effectively strengthen the metal both at room temperature and at high temperature by hindering dislocation motion. To pass through the particles, a dislocation must bow out between the particles, therefore increasing the yield stress of the metal, which is well described by the Orowan equation, where the increment of yield stress is positively correlated to the particle spacing [10]. Thus, as shown in figures 1 and 2, the yield stress of the investigated alloys was significantly increased with increasing impenetrable ZrO_2 particle concentration. After welding, the particles basically maintained their spherical shape and no obvious segregation was detected. This may be an important reason to explain why the yield stress of the alloys is basically unchanged after welding.

5. Conclusions

Pt20Rh alloys strengthened by ZrO_2 particles with various volume fractions have been prepared and welded. The effects of welding on microstructure and tensile properties at room temperature and at 1000 °C have been investigated. The conclusions are as follows.

1. The yield stress of Pt20Rh significantly increased with increasing ZrO_2 particle content. After welding, each weldment basically maintained its yield stress compared to the BM both at room temperature and at 1000 °C, while its work hardening capability was significantly decreased.
2. After welding, the initial recovered structure developed into an as-cast coarse grain structure in the FZ which was divided by non-uniform LABs caused by rapid cooling. Fine, spherical and homogeneously distributed pores were formed during the powder metallurgy process, whose area and size were increased with increasing particle content and after welding. Particles were observed to be spherical in the investigated alloys without obvious segregation.
3. LABs and uniformly distributed particles in the FZ should be responsible for the basically unchanged yield stress before and after welding, while the significant decline in total elongation and UTS may be the result of coarsened grains and increased pore density in the FZ.

Acknowledgements

Financial support from the Chongqing Science and Technology Bureau of China through Grant cstc2017zdcy-zdyfX0070 and cstc2018jszx-cyzdX0138 as well as the European Research Council (ERC) under the European Union's Horizon 2020 research and innovation program through Grant agreement No 788567 is gratefully acknowledged.

Reference

[1] Hu B X, Ning Y and Chen L 2012 *Platin. Met. Rev.* **56** 40.
[2] Chen H, Xie W and Liu W T 2018 *Vacuum* **160** 445.
[3] Selman G L, Day J G and Bourne A A 2018 *Platin. Met. Rev.* **18** 46.
[4] Maruyama K, Yamasaki H and Hamada T 2009 *Sci. Eng, A* **510** 312.
[5] Stokes J 1987 *Platin. Met. Rev.* **31** 54.
[6] Dai Y, Ma Q and Liu W T 2018 *Mater. Sci. Tech.* **34** 654.
[7] Hansen N 2013 *Scripta Mater.* **51** 801.
[8] Okamoto H 2013 *J. Phase Equilib. Diff.* **34** 176.
[9] Huang T, Shuai L and Wakeel A 2018 *Acta Mater.* **156** 369.
[10] Hansen N 1997 *Acta Metal.* **25** 863.

IOP Conf. Series: Materials Science and Engineering **580** (2019) 012036 doi:10.1088/1757-899X/580/1/012036

In-situ neutron diffraction study on the deformation of a TRIP-assisted multi-phase steel composed of ferrite, austenite and martensite

A Lavakumar[1], M H Park[2], S Gao[1], A Shibata[1,2], Y Okitsu[3], W Gong[1,4], S Harjo[4] and N Tsuji[1,2]

[1] Department of Materials Science and Engineering, Kyoto University, Yoshida-honmachi, Sakyo-ku, Kyoto 606-8501, Japan.
[2] Elements Strategy Initiative for Structural Materials (ESISM), Kyoto university, Yoshida-honmachi, Sakyo-ku, Kyoto 606-8501, Japan.
[3] Automobile R&D center, Honda R&D Co., Ltd. 4930 Shimotakanezawa, Haga-machi, Tochigi 321-3393, Japan.
[4] J-PARC Center, Japan Atomic Energy Agency, 2-4 Shirakata, Tokai-mura, Naka-gun, Ibaraki, 319-1195, Japan

Corresponding author: avala.lavakumar.63n@st.kyoto-u.ac.jp

Abstract. Multi-phase steels showing transformation induced plasticity (TRIP), can exhibit an excellent combination of high strength and good ductility by the aid of martensitic transformation during deformation. Even though TRIP-assisted multi-phase steels have been widely used in industry, the role of each phase in the enhancement of mechanical properties is still unclear given their complicated microstructures. In order to understand better the nature of the TRIP effect, the mechanical interaction between different phases at the micro-scale should be clarified. In the present study, the mechanical behavior of a transformation induced plasticity (TRIP) assisted multi-phase steel, has been characterized by *in-situ* neutron diffraction during tensile testing. The result of strain partitioning between the different phases obtained from the *in-situ* neutron analysis revealed that the martensite phase took much more elastic strain than the ferrite and retained austenite phases, which suggests that the work hardening behavior in the present steel is affected by the higher load borne by deformation-induced martensite.

1. Introduction

Transformation induced plasticity (TRIP) assisted steels are used in the automotive sector on account of their excellent strength to elongation ratio [1] and generally possess a complex microstructure composed of ferrite, martensite, bainite and retained austenite [2]. In particular, retained austenite transforms into martensite during deformation, which is commonly known as the TRIP phenomenon [3]. The good combination of mechanical properties of TRIP assisted steels is due to enhanced strain hardening by the transformation [4]. Many experimental studies [2, 5–9] have focused on the development of TRIP assisted steels, as well as on understanding their deformation mechanisms. Evaluation of deformation mechanisms in TRIP assisted steels is often challenging, due to their complicated microstructure and the dynamic transformations that occur during deformation. Therefore, the present work particularly aims to study the deformation mechanisms that take place during tensile loading of TRIP assisted steels.

Recently the development of *in-situ* diffraction techniques [4,10] enables a better understanding of various processes that take place during deformation of multi-phase microstructures, such as stress partitioning between the different phases. In the present study, we try to understand the effects of stress partitioning in different phases and the evolution of martensite on the strain hardening behaviour of a TRIP assisted multi-phase steel by the use of an *in-situ* neutron diffraction technique.

2. Experimental procedure

The nominal chemical composition of the TRIP assisted multi-phase steel studied is Fe-1.6Mn-1.4Si-2.0Ni-0.4C-0.004P-0.017Nb (in weight percent). The steel was prepared by vacuum induction melting and casting followed by hot rolling, warm rolling and inter-critical annealing. The thermo-mechanical processing route adopted in the present study is given in figure 1. Microstructural observations were carried out in a field-emission type scanning electron microscope (FE-SEM, JEOL 7100) equipped with an electron back-scattering diffraction (EBSD) detector. *In-situ* neutron diffraction measurements were collected during tensile testing at room temperature at the BL-19 TAKUMI beamline in J-PARC. The tensile loading direction was parallel to the rolling direction of the steel sheet. For the *in-situ* neutron diffraction experiments, using a white neutron beam, a time-of-flight (TOF) method was used, with a continuous range of velocities and variable wavelengths. By estimating the flight times of the detected neutrons, the wavelengths can be determined and thus neutron diffraction profiles acquired. The tensile specimen was mounted horizontally in a tensile loading machine at a 45° angle to the incident neutron beam. Two neutron detectors were fixed an angles ±90° to the incident neutron beam to detect the diffracted patterns from crystallographic planes perpendicular or parallel to the tensile direction, respectively. A more detailed description of the neutron diffraction technique at the TAKUMI beamline can be found in former literature [4, 8]. The tensile specimens, with a gauge length of 10 mm, a width of 3 mm and thickness of 1 mm, were prepared using electrical discharge machining. Tensile tests of the specimens were carried out at an initial strain rate of $8.3 \times 10^{-4}\,s^{-1}$ at room temperature (RT) up to fracture, in a step-load controlled manner, with additional holding for 300 seconds at selected stress levels. The volume fraction of martensite was determined from the integrated intensities of the diffraction profiles.

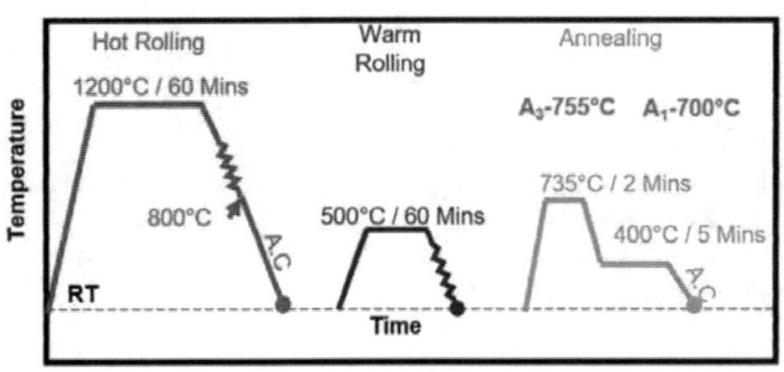

Figure 1. Schematic illustration showing the thermo-mechanical processing route used in the present study.

3. Results and discussion

3.1 Microstructural characteristics and tensile properties

Figure 2(a) shows a SEM-EBSD microstructure image of the present steel, containing both ferrite and retained austenite (highlighted in green in the figure) before the tensile test. From the micrograph, it is evident that two phases are homogeneously distributed in the microstructure. The average grain size of each phase was estimated to be ~ 0.75 µm. The microstructure contains a higher fraction of ferrite, (nearly 60%) than of retained austenite. The volume fraction and lattice parameters of the retained austenite and ferrite were estimated from neutron diffraction patterns obtained in the *in-situ* tests, using the Rietveld refinement method.

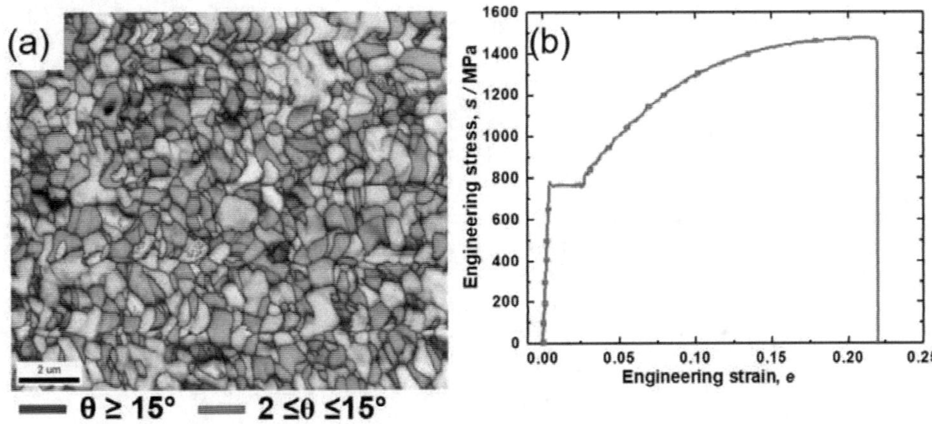

Figure 2. (a) SEM-EBSD superposed image quality (IQ) and phase maps, showing ferrite and retained austenite microstructure in the present steel before tensile testing. Retained austenite is colored green. Blue and red lines in the micrograph represents high angle grain boundaries with θ > 15° and low angle grain boundaries with 2 ≤θ ≤15°. (b) Engineering stress-strain curve of the present steel. Red marks on the stress-strain curve represents the points, at which *in-situ* neutron diffraction profiles were taken.

Figure 2(b) shows the engineering stress-strain curve of the present steel. Plastic deformation started at a yield strength of 770 MPa. The material was then strain hardened significantly to reach an ultimate tensile strength of 1475 MPa. The material also showed good tensile ductility, where the uniform elongation and total elongation were 21% and 22%, respectively. In figure 2 (b), the red marks on the stress-strain curve indicate the stresses at which *in-situ* neutron diffraction profiles were taken.

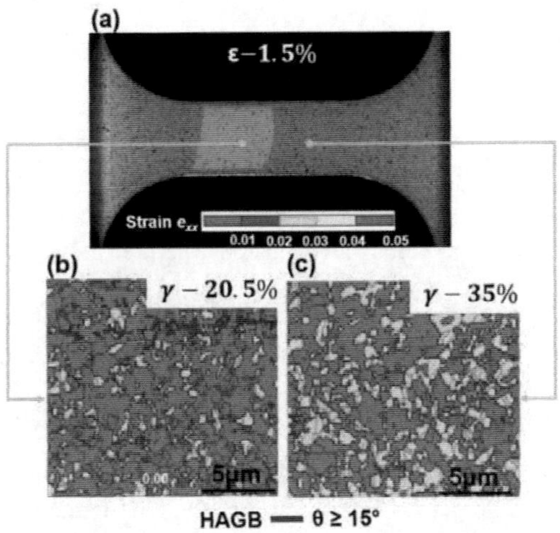

Figure 3. (a) A tensile specimen showing Lüders deformation. Observation taken at a tensile strain of 1.5%. (b, c) SEM-EBSD phase maps taken in areas (b) within the Lüders band and (c) outside of the Lüders band.

From figure 2(b) and figure 3 it can be seen that Lüders deformation occurs after a slight yield-drop at beginning of plastic deformation in the steel examined in this study. This is a common phenomenon when dislocation motion is strongly suppressed by solute pinning of dislocations in steels. It should be noted that ultrafine-grained materials generally show limited uniform elongation due to early plastic instability [11]. However, in the present material a significant amount of uniform elongation was observed after Lüders deformation. This is attributed to the deformation induced martensitic transformation, which postpones necking and increases the uniform elongation of the steel. It should also be noted that during Lüders deformation some of retained austenite transforms to martensite (shown in figure 3).

3.2 Stress partitioning between phases and its contribution to work-hardening

Figure 4 shows *in-situ* neutron diffraction patterns, corresponding to crystallographic planes perpendicular to the tensile direction, before and during deformation. It can be observed from figure 4 that the intensities of the retained austenite peaks decreased after plastic deformation, which implied that metastable austenite transformed to martensite during tensile deformation. Shifts of the diffraction peaks are also visible.

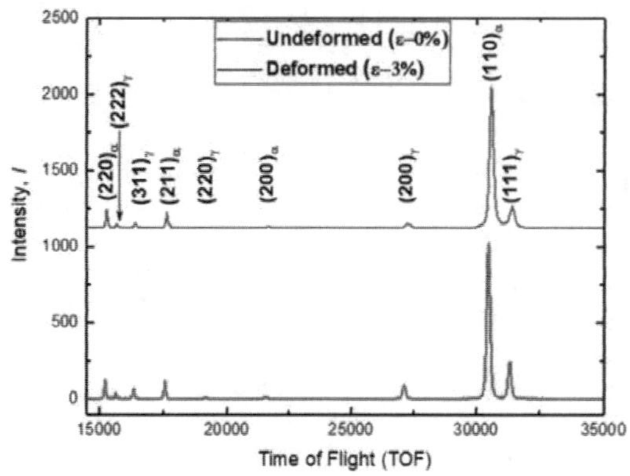

Figure 4. Typical diffraction patterns of the present steel measured before deformation (red) and during deformation (blue) at a strain of 3.0%

Figure 5 shows the evolution of lattice strain in each of the different phases with increasing applied stress. The lattice strains for ferrite, retained austenite and martensite were calculated by the use of the following equation,

$$\varepsilon_{hkl} = \frac{d_{hkl} - d_o}{d_o}$$

where ε_{hkl} is the lattice strain of a specific (hkl) plane, and d_o and d_{hkl} are the interplanar spacing of the specific lattice plane (hkl) in undeformed condition and deformed condition, respectively. It should be noted that initially martensite was not present in the microstructure. Therefore, the lattice strain of martensite in the undeformed condition was estimated according to the method described in the literature [4].

The macroscopic deformation causes different stress and strain responses in different phases. Accordingly, deformation gradients develop in the microstructure. From figure 5, it can be observed that initially ferrite and retained austenite were elastically deformed, with plastic deformation starting later. The lattice strains in both phases continuously increased in elastic regions with the applied stress. As mentioned above, at the beginning of deformation Lüders elongation was observed in the tensile specimen. Because of the resulting macroscopically inhomogeneous deformation (Lüders deformation), lattice strains in the plastic region were measured after the Lüders deformation. The lattice strain of ferrite continuously increased with applied stress, which indicated that ferrite was plastically deformed. A sudden drop in the lattice strain of the austenite phase was observed after the Lüders deformation. This might result from the evolution of compressive stresses in retained austenite due to the deformation-induced martensitic transformation. The lattice strain of martensite was estimated after the Lüders deformation. The lattice strains of martensite continuously increased with applied stress, and showed higher values than those of ferrite and austenite. The results suggest that martensite takes a higher phase stress than ferrite and austenite in this steel.

Figure 6 shows changes in the volume fraction of martensite and the incremental strain hardening parameter with respect to true strain. In general, most materials have a constant value of the incremental strain hardening parameter "*n*", which determined as, $n = \partial(\ln \sigma)/\partial(\ln \varepsilon)$. This was, however, not the case in the present TRIP-assisted steel, due to progressive evolution of the martensitic phase. Therefore, the "*n*" values can be taken as a parameter to understand the strain hardening ability of the present material (TRIP-assisted steel). At an initial stage, the strain hardening parameter steeply decreased due to the Lüders deformation. Although some retained austenite transformed into martensite even in that range, it was mostly compensated by the inhomogeneous deformation. After the Lüders deformation, austenite increasingly transformed into martensite, which resulted in increase of the strain hardening parameter to the maximum value at around a true strain of 0.07. With further deformation the strain hardening parameter decreased, corresponding to the fact that the most of austenite in the microstructure had fully transformed to martensite. The continuous evolution of the martensite phase developed inhomogeneous stress fields in the microstructure, which might impede dislocation motion thereby resulting in a higher work hardening rate. The stress partitioning from the softer ferrite to the stronger martensite at higher tensile strains also improves the strain hardening ability of the material. This can be considered as the reason for the enhanced strain hardening in the present TRIP steel.

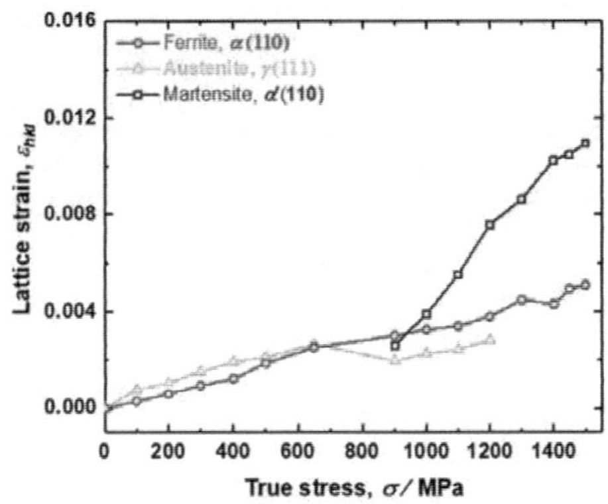

Figure 5. Change of lattice strains for the different phases during tensile deformation, obtained by the *in-situ* neutron diffraction. The lattice strains for ferrite (α), retained austenite (γ), and martensite (α′) are colored in blue, red and green, respectively.

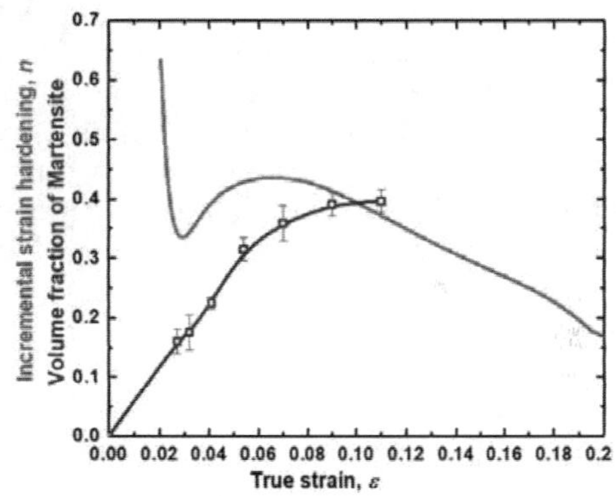

Figure 6. Change of incremental strain hardening parameter, *n* and evolution of martensite phase fraction with applied strain in during tensile deformation.

4. Conclusions

In this study, *in-situ* neutron diffraction analysis during tensile deformation of a TRIP assisted multi-phase steel has been carried out. The steel contained two phase microstructure composed of ferrite and retained austenite before tensile testing. During tensile deformation, metastable austenite transformed to martensite and resulting in TRIP phenomena. At the initial stages of deformation (ε <0.11) significant load partitioning was observed between the ferrite, austenite and martensite phases. At later stages, the stronger martensite phase takes a higher load than the softer ferrite phase. It can be concluded that the enhanced work-hardening in the present steel can be attributed to the evolution of deformation-induced martensite and the resultant load sharing mechanism among the complex micro-constituents.

IOP Conf. Series: Materials Science and Engineering **580** (2019) 012036 doi:10.1088/1757-899X/580/1/012036

References

[1] Militzer M 2002 A Synchrotron Look at Steel *Science (80-.).* **298** 975–6
[2] Jacques P J 2004 Transformation-induced plasticity for high strength formable steels *Curr. Opin. Solid State Mater. Sci.* **8** 259–65
[3] Zackay V, Parker E R, Fahr D and Busch R 1967 The enhancement of ductility in high-strength steels *Trans. ASM*
[4] Harjo S, Tsuchida N, Abe J and Gong W 2017 Martensite phase stress and the strengthening mechanism in TRIP steel by neutron diffraction *Sci. Rep.* **7** 15149
[5] Zhang J, Hao S, Jiang D, Huan Y, Cui L, Liu Y, Yang H and Ren Y 2017 In situ synchrotron high-energy X-ray diffraction study of microscopic deformation behavior of a hard-soft dual phase composite containing phase transforming matrix *Acta Mater.* **130** 297–309
[6] Song C, Yu H, Lu J, Zhou T and Yang S 2018 Stress partitioning among ferrite, martensite and retained austenite of a TRIP-assisted multiphase steel: An in-situ high-energy X-ray diffraction study *Mater. Sci. Eng. A* **726** 1–9
[7] Li N, Wang Y D, Liu W J, An Z N, Liu J P, Su R, Li J and Liaw P K 2014 In situ X-ray microdiffraction study of deformation-induced phase transformation in 304 austenitic stainless steel *Acta Mater.* **64** 12–23
[8] Asoo K, Tomota Y, Harjo S and Okitsu Y 2011 Tensile Behavior of a TRIP-aided Ultra-fine Grained Steel Studied by Neutron Diffraction *ISIJ Int.* **51** 145–50
[9] Lee P D, Ennis B L, Hanlon D N, Azeem M A, Rowley D, Jimenez-Melero E, Daisenberger D, Krugla M and Atzema E H 2016 Metastable austenite driven work-hardening behaviour in a TRIP-assisted dual phase steel *Int. J. Plast.* **88** 126–39
[10] Jacques P, Furnémont Q, Mertens A and Delannay F 2001 On the sources of work hardening in multiphase steels assisted by transformation-induced plasticity *Philos. Mag. A* **81** 1789–812
[11] Tsuji N, Ito Y, Saito Y and Minamino Y 2002 Strength and ductility of ultrafine grained aluminum and iron produced by ARB and annealing *Scr. Mater.* **47** 893–9

Quantification of recrystallization kinetics in industrially processed 5xxx aluminum alloy sheet

X C Lei[1], R E Sanders[1], X F Yang[1] and D Juul Jensen[1, 2]

[1]International Joint Laboratory for Light Alloys (Ministry of Education), College of Materials Science and Engineering, Chongqing University, Chongqing 400044, China
[2] Department of Mechanical Engineering, Technical University of Denmark, DK-2800 Lyngby, Denmark

E-mail: 510744188@qq.com

Abstract. The annealing kinetics of two aluminum alloys AA5xxx is investigated. The volume fraction of recrystallized material, V_V, is determined for a series of partly recrystallized samples both by hardness measurements and by electron back scattering diffraction (EBSD). Because of extended recovery overlapping with recrystallization, in particular in AA5182, it is not straightforward to estimate V_V from the hardness measurements. Two methods are suggested: one relying solely on the hardness data, and one in which EBSD data are used to estimate the hardness at the onset of recrystallization. Effects of the choice of method and of parameters on the resulting annealing kinetics are analysed, and the potential and limitations of the two methods are discussed.

1. Introduction

The 5xxx alloys were initially developed in the 1930's in response to demands for sheet materials with higher strengths, good formability, and high levels of corrosion resistance and weldability. In this study two 5xxx aluminum alloys (AA5182 and AA5657) with markedly different microstructures are investigated. In the automobile industry, AA5182-O sheets are used for deep drawn body-parts, including inner door or hood parts [1]. On the other hand, AA5657-H24, a higher base purity alloy, is widely used in bright and anodized sheet products[2].

The chemical compositions of the two alloys are given in table 1. Due to the very low solubility of Si in high Mg alloys, Mg_2Si is often present in the microstructure of AA5182 as a major constituent. Large (1-10 μm) constituent phases containing Fe, Mn and Si such as $Al_{12}(Fe, Mn)_3Si$, $Al_6(Fe, Mn)$ and Al_3Fe, which are found in most commercial alloys with >0.05%Fe [2]. During homogenization at 500-530°C, fine Mn-rich dispersoids [3] of about 0.1~0.4 μm are formed in the AA5182 alloy. These particles increase the strength of the alloy and serve to retard recrystallization and grain growth during annealing. While the two alloys selected for the present study represent considerable differences in deformed microstructure, annealing is of key importance for industrial thermomechanical processing for both alloys.

The overall aim of this study is to quantify the *recrystallization* kinetics. The classic model describing transformation kinetics was proposed by Johnson and Mehl [4], Avrami [5], and Kolmogorov [6] for phase transformations, but is generally also used for recrystallization. In this so-called JMAK model, the increase of the recrystallized volume fraction (Vv) during isothermal annealing is expressed as:

$$Vv = 1 - \exp(-kt^n) \quad (1)$$

$$n = \frac{\log(-\ln(1 - Vv)) - \log k}{\log t} \quad (2)$$

where t is the annealing time, and k and n are two parameters describing the kinetics, where n is often termed the Avrami exponent. If (-ln(1-Vv)) is plotted versus t on double logarithm scale a straight line should thus be obtained with n as the slope (see equation (2)). In many cases the equations (1-2) are found to fit experimented data points nicely [7-9], while in others no fit, or a fit to more than one straight line can be observed [10, 11]. In the present work we shall use equations (1-2) to analyse the recrystallization kinetics based on Vv values experimentally determined by electron backscatter diffraction (EBSD) and hardness measurements. Even simple analysis of the EBSD results gives a precise measure of Vv [12, 13]. The hardness data reflect all the mechanisms occurring during the annealing, i.e., recovery, recrystallization, grain growth and also precipitation if that takes place during the annealing. The specific aim of this work is to establish methods to deduce Vv from these more complex hardness data.

2. Experimental

2.1. Materials description

Two commercial aluminum alloys, AA5182 and AA5657, in the form of sheets are used for the present study. The two alloys were supplied by Novelis Inc. The AA5182 sheet was first hot rolled to 4 mm, and subsequently cold rolled to 1 mm (75% reduction) on industrial rolling mills. The AA5657 sheet was first hot rolled to 6 mm, then cold rolled to 3.5 mm followed by inter-annealing at 320-345°C for 100 minutes. Finally the sheet was cold rolled to 0.76 mm (78% reduction) followed by 235-240°C batch annealing (recovery treatment) for 3 hours. In this study, specimens were cut from the two alloys sheets and annealed isothermally at 245°C and 300°C respectively for various durations, ranging from 5 minutes to 1440 minutes in an air furnace. Thereby a series of partially recrystallized specimens was obtained.

Table 1. Chemical compositions of the AA5182 and AA5657 alloys.

Alloy	Mg	Mn	Fe	Si	Cu	Cr	Al
AA5182	4.50	0.26	0.16	0.09	0.05	0.04	Bal.
AA5657	0.6-1.0	0.03	0.10	0.08	0.10	0.00	Bal.

2.2. Hardness measurements

The Vickers microhardness was determined using a HMV-G microhardness instrument. The specimens were prepared by mechanical grinding to 4000 grit SiC abrasive paper. The hardness in the rolling plane containing the rolling direction (RD) and the transverse direction (TD) was measured applying a load of 50 g for 10 seconds.

2.3. EBSD measurements

A TESCAN MIRA3 scanning electron microscope equipped with an Oxford Instruments Channel 5.0 EBSD system was used to characterize the microstructures. The microscope was operated at 20 kV. The longitudinal plane containing the rolling direction (RD) and the normal direction (ND) was inspected. EBSD specimens were prepared by mechanical grinding and subsequent polishing with a standard colloidal alumina suspension solution with mean particle size of 30 nm. The EBSD specimens were finally electropolished in a 10% perchloric acid and ethanol solution at 12°C for 10-15 seconds with an electric potential difference of 20 V. The step sizes for EBSD scanning were chosen to be 0.5 μm and 1 μm for the AA5182 and AA5657 alloys, respectively. Based on EBSD maps, the recrystallized volume fraction (Vv) was calculated by the equation:

$$V\text{v} = \frac{\sum \lambda_i}{L} \tag{3}$$

where λ_i is the recrystallized grain intercept length, i.e., the chord length of grain i intersected by a test line where L is the total test line length. Fifteen lines along ND with total line lengths of 3600 μm and 10000 μm were used for each EBSD map for the AA5182 and AA5657 alloys, respectively.

3. Results and discussion

In the deformed state AA5182 has a Vickers hardness of ~117 HV which, during annealing, decreases to ~79 HV for the almost fully recrystallized state. The results are shown in figure 1(a). Initially the hardness decreases at a slow rate from 5 minutes to 60 minutes, then at a higher rate. The hardness evolution for AA5657 is shown in figure 1(b). Compared to AA5182, it is clear that the hardness drop before initiation of recrystallization is less significant (this is partly due to the already recovered structure in the initial state), and the total reduction in hardness is only ~20 HV, whereas in AA5182 it is ~40 HV.

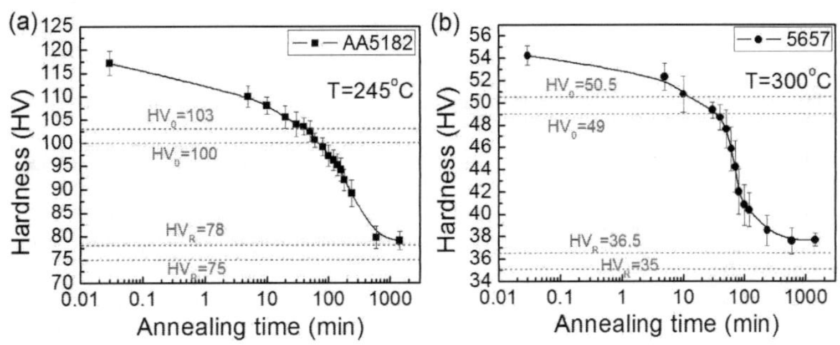

Figure 1. Vickers microhardness vs. time during isothermal annealing of (a) AA5182 annealed at 245°C and (b) AA5657 annealed at 300°C.

Figure 2. IPF maps from EBSD characterization of partially recrystallized AA5182 annealed at 245°C for (a) 60 minutes (Vv = 0.05); (b) 160 minutes (Vv = 0.26); (c) 1440 minutes (Vv = 0.96).

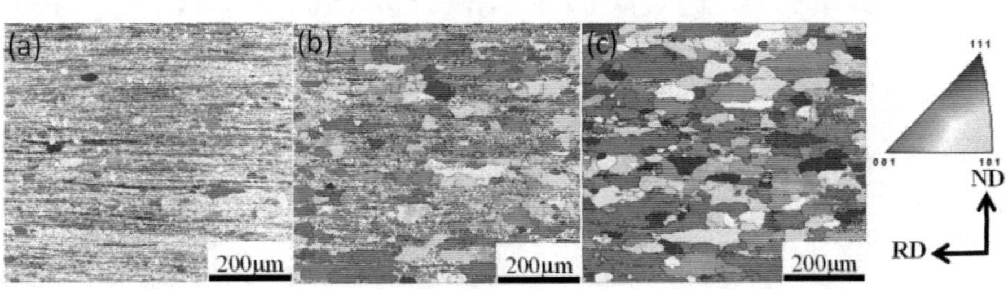

Figure 3. IPF maps from EBSD characterization of partially recrystallized AA5657 annealed at 300°C for (a) 30 minutes (Vv = 0.05); (b) 80 minutes (Vv = 0.52); (c) 240 minutes (Vv = 0.95).

Images of the microstructures of the two alloys as observed by EBSD are shown in figures 2 and 3. The recrystallized grains are almost equiaxed with an average size of ~10 μm. In the AA5657 alloy the recrystallized grains are elongated along RD with an average size of ~60 μm (see figure 3). After 240 minutes annealing, some grain growth has occurred even though the microstructure is not fully recrystallized.

The volume fraction of recrystallized material, *V*v, is readily calculated from the EBSD data based on equation (3). The most common way to quantify recrystallization is to fit the *V*v data by the JMAK equation (1). The results, shown in figure 4, show 2 distinct stages, i.e., one *n* value cannot describe the present data well for either of the alloys. A similar result is also seen for recrystallization of other materials, e.g., copper, iron and steel [11, 14, 15]. In the following, we shall analyse the dominant stage, i.e., the part with the highest *n* value.

From the hardness data, *V*v can be estimated by the following equation:

$$V\text{v} = \frac{HV_0 - HV}{HV_0 - HV_R} \quad (4)$$

where HV is the hardness at a certain annealing time, HV_0 is the hardness after recovery and HV_R is the hardness when the sample is fully recrystallized. A particular challenge here is how to determine HV_0, in particular for the AA5182 alloy in which the transition from recovery to recrystallization is very subtle. Actually, for this alloy it is also not straightforward to determine exactly the value of HV_R.

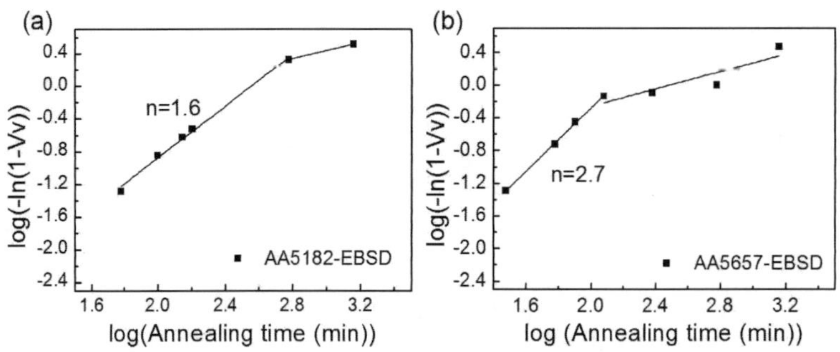

Figure 4. Avrami plot of the evolution of the recrystallized volume fraction (*V*v) based on EBSD data for both alloys.

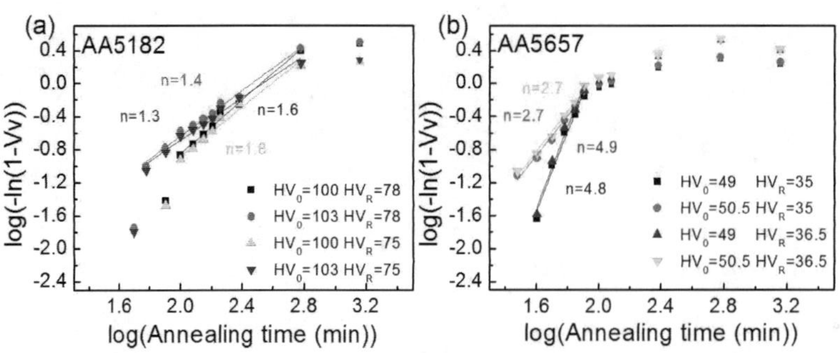

Figure 5. Avrami plot of the evolution of the recrystallized volume fraction (*V*v) determined by a series of HV_0 and HV_R values based on hardness data for both alloys.

To illustrate how large an effect the choice of the HV_0 and HV_R values has on the analysis of the recrystallization kinetics, the results for a series of HV_0 and HV_R values are plotted in figure 5. The HV_0 and HV_R values are shown by dashed lines in figure 1. Even though these values are not very different, they have a huge effect on the *n* values, in particular for the AA5657 alloy in which the *n* value varies from 4.9 to 2.7. This clearly illustrates how critical it is to properly determine the two hardness values of HV_0 and HV_R is, if hardness data are to be used to quantify recrystallization kinetics. Two methods to avoid such problem are suggested in the following.

Method I. In this method it is assumed that recovery is largely complete when recrystallization starts, and the *V*v values determined by EBSD are used to fit HV_0 and HV_R by linear fitting. The fitted values together with the calculated *V*v*(hardness)* by equation (4) are given in figure 6. It is obvious that this method gives excellent agreement between *V*v*(EBSD)* and *V*v*(hardness)*. However, the approach is impractical because of the many EBSD measurements needed. We therefore also tried to fit HV_0 and HV_R only using 3 EBSD data points, corresponding to *V*v of 0.05, 0.26, 0.96 and 0.05, 0.52, 0.95 for AA5182 and AA5657 respectively. Quite good agreements with the much more complete fitting approach are found. This is illustrated by Avrami plots in figure 6. The good agreement suggests that appropriate values for HV_0 and HV_R can be obtained based on only 3 fitting points as long as these are well distributed over the recrystallization interval.

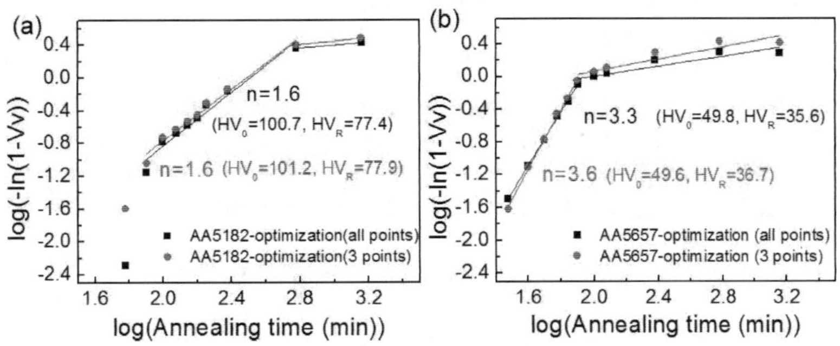

Figure 6. Avrami plot of the evolution of the recrystallized volume fraction (*V*v) based on hardness data by 2 series of HV_0 and HV_R values fitted from *all* EBSD points and *3* EBSD points respectively.

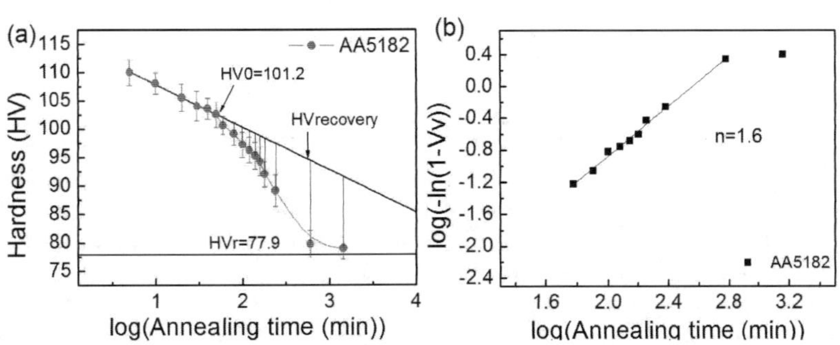

Figure 7. (a) An estimate of a 'continued' *HVrecovery* curve and (b) Avrami plot of the evolution of recrystallized volume fraction (*V*v) of the AA5182 alloy based on 'continued' recovery.

Method II. For an alloy such as AA5182 it is not clear if recovery continues during recrystallization in the not yet recrystallized regions. Concurrent recovery and recrystallization has, for example, been observed in iron [16] and in copper [17, 18]. For the present AA5182 alloy, we find that the initial drop in hardness is related only to recovery. When the hardness values are plotted as a function of the log annealing time, a straight line is observed to fit these initial points well (see figure 7(a)). Assuming that recovery continues to follow this line while recrystallization takes place, *V*v may be determined by the equation (5), where *HVrecovery* is determined from the straight line extrapolated to the relevant annealing times. The Avrami plot is shown in figure 7(b). The agreement with method I is excellent.

$$Vv = \frac{HVrecovery - HV}{HVrecovery - HV_R} \tag{5}$$

4. Conclusions and outlook

The annealing kinetics of two alloys of the 5xxx series have been analysed by EBSD and hardness measurements. Two methods have been proposed by which the hardness data can be converted to give an adequate description of the recrystallization kinetics. It is found that:

- The recrystallization kinetics cannot be described by a single Avrami exponent for either of the two alloys.
- When analysed based on hardness data, the resulting Avrami exponents depend critically on the choice of transition points from recovery to recrystallization.
- Both methods suggested for conversion of the hardness values to Vv values are adequate. Method I requires normalization based on EBSD maps for at least 3 recrystallization conditions, while method II relies solely on the hardness measurements but requires many data points to be measured in the annealing time interval when only recovery (and not recrystallization) takes place.

It is of interest to notice that the resulting Avrami exponents for the alloys are significantly different: 1.6 in the AA5182 alloy and 2.7 in the AA5657 alloy. This is likely to be because of different nucleation and growth mechanisms, which are related to the difference in deformation microstructures and particle contents in the two alloys. A further investigation of this is planned for a next publication.

Acknowledgments

This research is funded by the National Natural Science Foundation of China (Grant No.51571046, No.51421001), and the Project No. 2018CDJDCL0019 supported by the Fundamental Research Funds for the Central Universities. Furthermore we appreciate the support from the 111 Project (B16007) by the Ministry of Education and the State Administration of Foreign Experts Affairs of China, and D J J acknowledge support from the European Research Council (ERC grant agreement No.788567) for part of the work.

References

[1] Engler O, Kuhnke K and Hasenclever J 2017 *J. Alloys Compd.* **728** 669-81
[2] Sanders R E, Baumann S F and Stumpf H C 1989 *Treatise on Materials Science and Technology.* (Academic Press: New York) Vol 31 p 65.
[3] Ratchev P, Verlinden B and Houtte P V 1995 *Acta Matal. Mater.* **43** 621-29
[4] Johnson W A and Mehl R F 1939 *Trans. AIME.* **135** 416-43
[5] Avrami M 1939 *J. Chem. Phys.* **7** 1103–12
[6] Kolmogorov A N 1937 *Ivz. Akad. Nauk. SSSR. Ser. Mater.* **3** 355-59
[7] Price C W 1985 *Scripta Metal. Mater.* **19** 669-73
[8] Lü Y P, Molodov D A and Gottstein G 2011 *Acta Mater.* **59** 3229-43
[9] Chao H Y, Sun H F, Chen W Z and Wang E D 2011 *Mater. Charact.* **62** 312-20
[10] Lin F X, Zhang Y B, Pantleon W and Juul Jensen D 2018 *Metal Mater. Trans A.* **49** 5246-58
[11] Jägle E A and Mittemeijer E J 2012 *Metal Mater. Trans A.* **43** 1117-31
[12] Juul Jensen D 1992 *Scripta Metal. Mater.* **27** 1551-56
[13] Wu G L and Juul Jensen D 2008 *Mater. Charact.* **59** 794-800
[14] Lin F X, Zhang Y B, Tao N, Pantleon W and Juul Jensen D 2014 *Acta Mater.* **72** 252-61
[15] Samajdar I, Verlinden B, Van Houtte P and Vanderschueren D 1997 *Scripta Mater.* **37** 869-74
[16] Vandermeer R A and Rath B B 1989 *Mater. Trans A.* **20** 391-401
[17] Vandermeer R A, Juul Jensen D and Woldt E 1997 *Metal Mater. Trans A.* **28** 749-54
[18] Woldt E and Juul Jensen D 1995 *Metal Mater. Trans A.* **26** 1717-24

2D/3D local strain analysis of layered metal composites with a strength-ductility synergy

D Y Li[1], M Huang[1], Y P Xia[1], C Xu[1], W M Gan[2], L Geng[1] and G H Fan[1,3]

[1]School of Materials Science and Engineering, Harbin Institute of Technology, Harbin, 150001, P.R. China

[2]Institute of Materials Research, Helmholtz-Zentrum Geesthacht, Garching, D-85748 Germany

[3]Research Center for Light & High-performance Materials, Nanjing Tech University, Nanjing, 211816, P.R. China

E-mail: ghfan@hit.edu.cn (G.H. Fan)

Abstract. A long-standing question is why layered structures can overcome the strength-ductility trade-off and achieve synergy in this relationship. In this paper, layered Ti/Al, Ti/Ti metal materials are taken as examples to investigate the influence of a heterogeneous layered structure on their deformation behavior from a viewpoint of the local strain. In-situ neutron diffraction, DIC and synchrotron radiation tomography were performed to obtain lattice strain evolution (elastic stage), local strain evolution (plastic stage), and crack initiation and propagation (fracture stage), respectively. It is found that strain delocalization is the key to achieving a strength-ductility synergy in layered materials, and that layered structures can influence the local strains from start to failure, which improves strain compatibility between the component layers. In-situ tracking of local strains based on 2D/3D characterization methods during entire deformation process deepens our understanding of the deformation behavior of layered materials.

1. Introduction

For metal structural materials, the strength and ductility are mutually exclusive. In recent years, researchers have found that heterogeneous structures can help to overcome the strength-ductility trade-off and achieve synergy in this relationship. To reveal the processes of strengthening and toughening, various deformation mechanisms have been proposed, such as stress/strain partitioning[1],delayed necking [2] and back stress strengthening[3]. In fact, these deformation mechanisms are all attributable to redistribution of stress and strain[4]. It is well known that the local failure of materials mostly derives from local strain overload, and strain localization has always been linked with plastic instability, which results in necking and fracture of materials [5]. Therefore, in-situ tracking of the local strain during deformation is critical for investigating the deformation behavior of materials.

The previous methods for strain analysis mainly depend on characterization of the dislocation structure, shear bands and surface morphology of materials before and after deformation[6]. Generally, transmission electron microscopy (TEM) and scanning electron microscopy (SEM) have been used to characterize dislocation structures and slip bands, respectively. Additionally, the electron backscatter diffraction (EBSD)technique can also provide strain information based on Kikuchi bands and local misorientation [7]. However, these analysis methods focus on analysis of plastic strain, ignoring analysis of elastic strain. Moreover, most of these methods are limited to the microscopic scale, and it is difficult to perform quantitative measurement of both elastic and plastic strains. Recently, 2D/3D advanced characterization techniques provide a possibility for carrying out elastic and plastic strain analysis during the entire deformation process. For example, (i) elastic strains from the distortion of crystal lattice will result in a change in lattice parameter, which can be obtained by neutron diffraction[8]; (ii) plastic strains can be tracked in-situ by the digital image correlation (DIC) technique[9]; and (iii) the morphology and evolution of cracks can be obtained from synchrotron radiation tomography[10].

In this work, layered Ti/Al and Ti/Ti metal composites (Ti/Al and Ti/Ti LMCs) are taken as examples to investigate the essential mechanisms of the strength-ductility synergy in terms of strain distribution.

2. Experimental methods

The Ti and Al sheets were alternatively stacked and hot-pressed at 500 °C for 1 h in vacuum to obtain good interface bonding. Then, the hot-pressed Ti/Al LMC was hot-rolled by multiple rolling passes at 500 °C with a thickness reduction of 40%. Finally, the hot-rolled Ti/Al LMC was annealed at 500 °C for 1 h. The Ti/Ti LMC was fabricated using a hot-pressing process. Ti (TA1) sheets with different thicknesses of 100 μm and 20 μm were pretreated using 10 vol. % HF, and alternatively stacked, and then hot pressed at 700°C for 2h.

The microstructures of the Ti/Al and Ti/Ti LMCs were analyzed by SEM (Zeiss, Merlin Compact) equipped with an Oxford Instruments HKLEBSD system. The specimens for EBSD were polished by electro-polishing. A step size of 2.0μmwas used to acquire the EBSD data.

In-situ neutron diffraction experiments on the materials science diffractometer STRESS-SPEC at the neutron source FRM II, Germany was carried out to track the strain evolution of Ti/Al LMC during the elastic deformation region. A tensile specimen with length, width and thickness of 80mm, 3.4mm and 3mm was prepared by electrical discharge machining. The wavelength of the neutron beam about 1.68 Å and a measured beam size of $5\times5\times5mm^3$ were used for this tensile test. A tensile strain rate of $\sim10^{-3}s^{-1}$was selected. Each diffraction pattern was captured at an interval of 1~5min.

The DIC technique was applied to characterize the local strain evolution of the LMCs during plastic deformation. An in-situ tensile test was performed on FEI HELLOS Naonolab600i scanning electron microscope equipped with a micro-tensile stage (SEM tester 1000 EBSD).A tensile strain rate of 5×10^{-4} s^{-1} was selected. The specimens for tensile tests were cut to a size of $10\times1.5\times2$ mm^3, corresponding to length, width and thickness, respectively.

A quasi in-situ SR-CT experiment was performed using the BL13W1 beamline station of the Shanghai Synchrotron Radiation Facility (SSRF) to analyze the crack initiation and propagation in LMCs during the fracture process. The tensile specimen for SR-CT was prepared with a diameter of 1.5 mm and a height of 20 mm. The SR-CT test was carried out using amicro-tensile stage (Kammrath-Weiss), to different macrostrains of 0%, 3.0%, 5.0%, 10.0%, 20.0%, respectively. The tensile strain rate was about $5\times10^{-4}s^{-1}$.

3. Results

Figure 1a shows an inverse pole figure (IPF) map displaying the ND of the Ti/Al LMC which reveals a multilayered structure with a Ti or Al thickness of 100μm. The pole figures (PFs) show that the Ti component of the Ti/Al LMC exhibits a strong basal texture with <0001>//ND, as shown in figure 1b, and the Al layer show a random orientation corresponding to the IPF view. The tensile stress-strain curve of the Ti/Al LMC was shown in the previous work [11]. Our previous results show that the Ti/Al LMC can combine strength of Ti and ductility of Al, and exhibits an outstanding elongation which exceeds that of either the Ti or Al component [1].Lattice strain analysis obtained by neutron diffraction is useful to understand the elastic deformation of the Ti/Al LMC. Figure 1c shows diffraction patterns before and after deformation. Taking $\{100\}_{Ti}$as an example, the shift of diffraction peak can be determined, as shown in figure 1d. The applied stress-lattice strain curve is plotted in figure 1e. In general, lattice strain deviates from a linear trend, indicating the initiation of yielding[1]. The lattice strain evolution of the Ti/Al LMC shows that the deformation process of the LMC is divided into three stage: an elastic stage, an elastic-plastic stage and a plastic stage. In the elastic stage, the Ti and Al layers maintain an equal strain condition, whereas in the elastic-plastic stage, the Al layer starts to yield while the Ti layer still deforms elastically. Once the Ti layer yields, the Ti and Al layers both enter the plastic deformation stage.

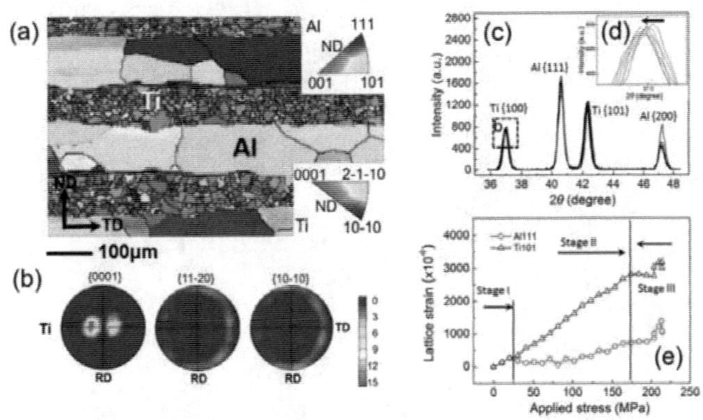

Figure. 1 (a) IPF map (ND coloring) of a Ti/Al LMC; (b) the $\{0001\}$, $\{11\bar{2}0\}$, $\{10\bar{1}0\}$ pole figures of Ti layers in the Ti/Al LMC; (c) Diffraction patterns of the Ti/Al LMC at different applied stress levels; (d) Magnified region in (c); (e) Lattice strain *vs* applied stress curves [11]. The lattice strain evolution of $\{111\}_{Al}$ and $\{101\}_{Ti}$ in the Ti/Al LMC are shown.

Local strain evolution obtained from DIC is analyzed to reveal the plastic deformation process of the LMC samples. Figure 2 shows the local strain distribution of the Ti/Al LMC at various macro strains, where the loading direction is parallel to the TD. It is apparent that the strain localization preferentially occurs in the hard Ti layer, while the adjacent Al layer shows a uniform distribution of local strain, as shown in figure 2a~d. The non-synchronous deformation behaviors of Ti and Al layers result in a stress/strain partitioning. As the macrostrain increases, strain localization of the Ti layer is intensified and gradually accumulates at the interface (figure 2e). However, these localized strains can be transferred by the adjacent Al layers, presenting a strain delocalization trend (figure 2f). Careful examination of the strain evolution of the Ti/Al LMC indicates that its outstanding elongation derives from the strain delocalization.

The same strain transfer behavior is also found in the Ti/TiLMC system. Figure 3a shows the microstructure of the Ti/TiLMC, consisting of alternating fine-grain layers and coarse-grain layers with thicknesses of 20 μm and 100 μm, respectively. The average grain sizes in these layers are about12μm and 55μm, respectively, as shown in figure 3b. The observation of local strain evolution (figure 3c) suggests that strain localization preferentially occurs in the fine-grain layer, but that localized strain can be transferred to the adjacent coarse-grain layer through the Ti/Ti interface, which effectively relieves the strain localization of the fine-grain layer.

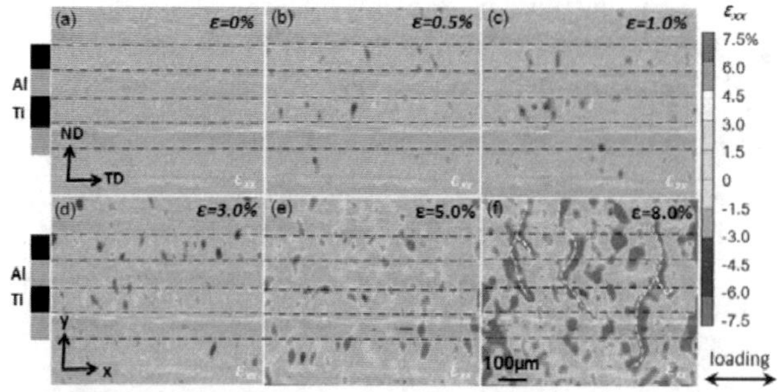

Figure. 2 Local ε_{xx} strain distributions of the Ti/Al LMC obtained by DIC at different macrostrains: (a) 0%; (b) 0.5%; (c) 1.0%; (d) 3.0%; (e) 5.0%; (f) 8.0%. The tensile direction is along the TD.

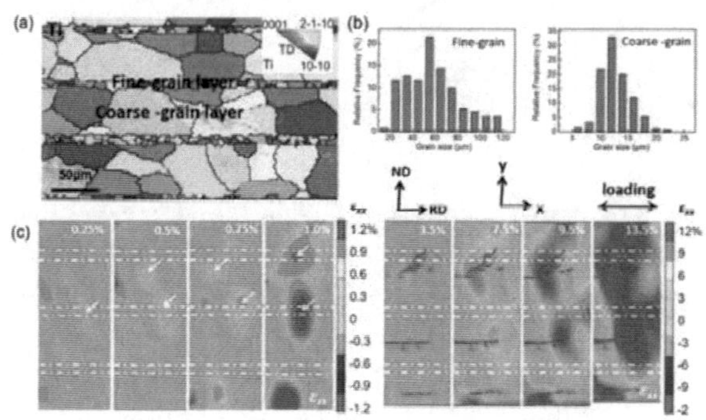

Figure. 3 (a) The IPF map of the Ti/Ti LMC with alternant fine-grain and coarse-grain layers; (b) the distribution of grain size; (c) local ε_{xx} strain evolution of Ti/Ti LMC at different macro strains.

The crack initiation and propagation behavior of the Ti/Al LMC during the fracture process was investigated by a combination of in-situ loading and SR-CT. Figure 4 shows the morphology and distributions of micro-cracks at various macro strains. Some detailed results found include: (i) cracks easily nucleate at the Ti/Al interface at a low macro strain (figure 4a~b); (ii)in addition, a large number of diffuse micro-cracks are found in the Al layer, and the quantity of these micro-cracks increases as with increasing macro strain(figure 4c~e);(iii)although local micro-cracks appear early, the propagation of these micro-cracks can be suppressed by the layered structure, so that they are restricted from evolving into a main crack causing fracture of the LMC. In order to further reveal the fracture mechanism in LMCs, the crack morphology in the Al layer and near the interface obtained at a macro strain of ~10.0% are further analyzed, as shown in figure 4f~j.The interfacial cracks are connected together into a sheet shape, whereas the cracks in the Al layer are separated with different morphologies.

4. Discussion

LMCs can combine the mechanical properties of each component layer to achieve a synergy between strength and plasticity. Taking the Ti/Al LMC as an example, its deformation process is distinctly divided into three stages: an elastic stage, an elastic-plastic stage and a plastic stage [11]. The strain analysis during the whole deformation process is beneficial to the understanding of the deformation behavior of the LMC.

In the elastic stage, both Ti and Al layers deform elastically. The lattice strain evolution (figure1e) shows that the Ti layer and Al layer maintain equal lattice strains in the layered structure, and the difference in Young modulus between the layers determines that they will undertake different applied stresses, leading to stress partitioning[12].

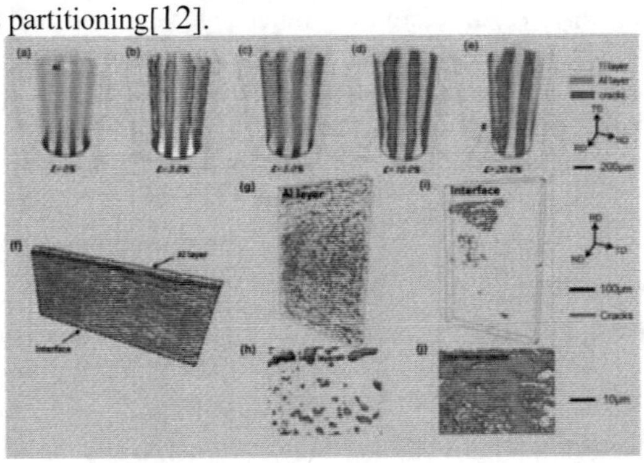

Figure. 4 Morphology and distributions of cracks in the Ti/Al LMC at different macrostrains: (a) 0%; (b) 3.0%; (c) 5.0%; (d) 10.0%; (e) 20.0% [11]. Sub-figure (f) shows an analysis of the crack morphology in the Al layer (g) and near the interface (i). Magnified regions of (g) and (i) are shown in (h) and (j).

In elastic-plastic stage, the Al layer preferentially yields due to the low yield strength, while the Ti layer still maintains elastic deformation. At this point, the continuous lattice strain increment of the Ti layer (figure1e) indicates that the applied stress is effectively transferred to the Ti layer, further enhancing the stress partitioning [1]. The partitioning of applied stress facilitates the activation of plastic deformation in the Ti layer, which improves deformation compatibility between the Ti and Al layers.

In the plastic stage, both the Ti and Al layers deform plastically. The 2D strain analysis obtained by DIC (figure2) provides a visible way to understand the plastic deformation process of the LMC. Strain localization occurs in the Ti layer during the early plastic deformation (figure2b), while the Al layer shows strain uniformity. This may be related to the following factors: Firstly, prismatic slip is dominant in Ti layer, which causes strain localization; Secondly, the layered structure redistributes the applied stress during the deformation, leading to an accumulation of internal stress between Ti and Al[13], which affects their deformation behavior. Specifically, this internal stress converts the uniaxial applied stress into a triaxial stress state[14], and preferentially constrains the deformation of the Al layer. Moreover, at a high macrostrain level, the strain localization of the Ti layer is gradually transferred to the adjacent Al layer through the interface, presenting a strain delocalization. In other words, the layered structure enhances the deformation compatibility between the Ti and Al layers[15].

In terms of the fracture process, due to the difference between mechanical properties of Ti and Al, their deformation incompatibility causes an internal stress to accumulate at the interface, so that ata low macro strain level, cracks preferentially nucleate at the interface. However, the layered structure can constrain the propagation of such interfacial cracks, which is mainly attributed to two aspects: (i) the crack tips can be blunted by plastic deformation of the Ti and Al component layers, which reduces the driving force for crack propagation[16]; (ii) a constraint effect due to the layered structure changes the stress states in layered composites, which increases the resistance to crack propagation[17]. In addition, it is found that a large number of micro-cracks occur in the Al layer at a macro strain of 5%. On the one hand, these micro-cracks provide a certain plastic strain of Al layer, while on the other hand, they alleviate stress concentration in the Al layer. Therefore, the initiation and propagation of cracks indirectly affect the internal stress state and the redistribution of strain localization in layered Ti-Al metal composites, thereby affecting its overall plastic deformation behavior and resulting in ultrahigh ductility.

5. Conclusions

In the study the deformation of Ti/Al and Ti/Ti LMC have been taken as examples, and various2D/3D strain characterization techniques used to reveal the mechanisms leading to strength-ductility synergy in these systems. It was found that strain delocalization is the key to achieving strength-ductility synergy of layered materials, and that layered structure can influence local strains and change the stress state in LMCs over the entire deformation process, which improves strain compatibility between the component layers. The layered structure maintains an iso-strain condition, which results in stress partitioning in both components during elastic deformation and ensures good deformation compatibility during plastic deformation through the redistribution of local strain, by mechanisms such as strain-transfer between the component layers and crack blunting by plastic deformation of the component layers.

Acknowledgements
This work was financially supported by the National Natural Science Foundation of China (Grant Nos. 51571070 and 51571071), and the Major Project Development Fund (Grant No. ZDXMPY20180103)

References
[1] Ojima M, Inoue J, Nambu S, Xu P, Akita K, Suzuki H and Koseki T 2012 *Scr. Mater.* **66** 139-42.
[2] Liu H S, Zhang B and Zhang G P 2011 *Scr. Mater.* **64** 13-16.
[3] Wu X L, Yang M X, Yuan F P, Wu G L, Wei Y J, Huang X X and Zhu Y T 2015 *Proc. Natl. Acad. Sci. U.S.A.* **112** 14501-5.
[4] Su R, Li L, Nie Z H, Zhang Q H, Wang Y D, Zhou X T, Wu Y D, Hui X D, Li X J and Wang M G 2014 *Appl. Phys. Lett.* **105** 221907.
[5] Ashby M F 1970 *Philos. Mag.* **21** 399-424.
[6] Antolovich S D and Armstrong R W 2014 *Prog. Mater. Sci.* **59** 1-160.
[7] Wright S I, Nowell M M and Field D P 2011 *Microsc. Microanal.* **17** 316-29.
[8] Tomota Y, Lukas P, Harjo S, Park J H, Tsuchida N and Neov D 2003 *Acta. Mater.* **51** 819-30.
[9] Hsu W N, Polatidis E, Šmíd M, Casati N, Van Petegem S and Van Swygenhoven H 2018 *Acta. Mater.* **144** 874-83.
[10] Wu S C, Xiao T Q and Withers P J 2017 *Eng. Fract. Mech.* **182** 127-56.
[11] Huang M, Xu C, Fan G H, Maawad E, Gan W M, Geng L, Lin F X, Tang G Z, Wu H, Du Y, Li D Y, Miao K S, Zhang T T, Yang X S, Xia Y P, Cao G J, Kang H J, Wang T M, Xiao T Q and Xie H L 2018 *Acta. Mater.* **153** 235-49.
[12] Seok M Y, Lee J-A L, D-H Lee, Ramamurty U, Nambu S, Koseki T and Jang J-I 2016 *Acta. Mater.* **121** 164-72.
[13] Ma L L, Wang L, Nie Z H, Wang F C, Xue Y F, Zhou J L, Cao T Q, Wang Y D and Ren Y 2017 *Acta. Mater.* **128** 12-21.
[14] Fan G H, Geng L, Wu H, Miao K S, Cui X P, Kang H J, Wang T M, Xie H L and Xiao T Q 2017 *Scr. Mater.* **135** 63-67.
[15] Kim J G, Baek S M, Lee H H, Chin K-G, Lee S and Kim H S 2018 *Acta. Mater.* **147** 304-12.
[16] Pineau A, Benzerga A A and Pardoen T 2016 *Acta. Mater.* **107** 424-83.
[17] Lafarie-Frenot M C, Henaff-Gardin C and Gamby D 2001 *Compos. Sci. Technol.* **61** 2327-36.

IOP Conf. Series: Materials Science and Engineering **580** (2019) 012039 doi:10.1088/1757-899X/580/1/012039

Quantification of microstructure in a eutectic high entropy alloy $AlCoCrFeNi_{2.1}$

Adrianna Lozinko [1], Oleg V. Mishin [2], Tianbo Yu [2], Uta Klement [1], Sheng Guo [1], Yubin Zhang [2*]

[1]Department of Industrial and Materials Science, Chalmers University of Technology, SE-412 96 Gothenburg, Sweden

[2] Department of Mechanical Engineering, Technical University of Denmark, DK 2800 Kgs. Lyngby, Denmark

Email: yubz@mek.dtu.dk

Abstract. Eutectic high entropy alloys (EHEAs) are a new class of metallic alloys with good mechanical properties at various temperatures. In the present investigation, microstructural parameters such as the volume fraction of two phases (FCC ($L1_2$) and BCC (B2)) forming the eutectic, the orientation relationship between these phases and interphase boundary spacings in as-cast EHEA $AlCoCrFeNi_{2.1}$ were quantified using electron microscopy. It is found that the two phases have a Kurdjumov-Sachs orientation relationship, i.e., $\{111\}_{FCC} \parallel \{110\}_{BCC}$ and $\langle 110\rangle_{FCC} \parallel \langle 111\rangle_{BCC}$. It is also found that both regular semi-coherent lamellar and irregular curved interphase boundaries are present within individual eutectic colonies. The habit planes for the semi-coherent lamellar interfaces are {224} and {123} for the FCC and BCC phases, respectively. Quantitative microstructural analysis shows that the ratio of volume fractions of the FCC and BCC phases within the regular lamellar regions differs to that within irregular regions, which suggests a local chemistry difference between the two regions. Finally, the solidification process of EHEAs is discussed, and possible ways to optimize the mechanical properties by microstructural design are suggested.

1. Introduction

High entropy alloys (HEAs), also known as multi-principal-element alloys, became well-known firstly in 2004 [1,2], and have now become new research frontier in the field of metallic alloys. HEAs contain at least four principal elements, which are often mixed in equiatomic or close to equiatomic ratios. HEAs open up a massive and unexplored compositional space, beyond that previously defined by traditional alloy design strategies, leading to the development of novel metallic materials with unprecedented structural and functional properties. For example, exceptional fracture toughness was revealed in a CoCrFeMnNi alloy at cryogenic temperatures [3], and some metastable high-entropy dual-phase alloys were observed to overcome the strength-ductility trade-off [4]. Additionally, novel multicomponent nanoparticles for catalysis [5], new superconducting materials [6], and even superhard ceramic materials [7] have been obtained based on the HEA concept.

Recently, a new class of metallic alloys, eutectic HEAs (EHEAs) has been developed, containing a mixture of the BCC and FCC phases. These alloys demonstrate high strength and ductility up to 700 °C [8]. Among the EHEAs that have been developed so far, the first proposed EHEA, $AlCoCrFeNi_{2.1}$, has attracted the most attention [9–11]. However, a detailed description of the microstructure has still not been provided, and the solidification process for this alloy is still poorly understood.

In the present study, the microstructure of this $AlCoCrFeNi_{2.1}$ EHEA is characterized using scanning electron microscopy (SEM), electron backscatter diffraction (EBSD), and transmission electron microscopy (TEM). Microstructural parameters, including the spatial distribution and volume fractions of the phases, and the spacing and habit plane of the semi-coherent interphase boundaries are quantified. The quantitative description of the microstructure provides a better understanding of the solidification process as well as a link to account for the mechanical properties. The results obtained are essential for developing advanced computer models for improved design of HEAs.

2. Experimental

The $AlCoCrFeNi_{2.1}$ (elements in atomic ratios) alloy was prepared through an arc melting route, using high purity (> 99.9 %) metals, in a Ti-gettered high-purity argon atmosphere. Melting was repeated at least five times on flipped samples to improve the homogeneity of the ingot. Finally, the ingot was suction-cast into a copper mold with $10 \times 10 \times 60$ mm^3 dimensions.

The crystal structure and phases were characterized by X-ray diffraction, using a Bruker AXS D8 Advance diffractometer with Cr-Kα radiation, operated at 35 kV with a beam current of 50 mA. The scanning range was 35 – 158 °, with a 0.025 ° step size and a scanning speed of 0.15 °/min.

Samples were mechanically polished using standard polishing suspensions, with particle sizes down to 1 μm, followed by chemical polishing using an OP-S solution. SEM and EBSD analysis were performed with a Zeiss Leo Gemini 1550 field emission gun scanning electron microscope. For EBSD, an acceleration voltage of 20 kV and a step size of 75 nm were used. For TEM observations, a thin foil was prepared by twin-jet electrochemical polishing in a 1:3 $HClO_4$:C_2H_5OH solution at 0 °C and 12 V. The foil was investigated in a JEOL 2100 microscope.

3. Results

3.1. Microstructural characterization

EBSD maps and TEM images showing the microstructure of the EHEA are given in figure 1. A very complex structure is seen in the EBSD map. The microstructure consists of ordered FCC ($L1_2$) and ordered BCC (B2) phases, shown in figure 1a in red and blue, respectively. The volume fractions of the FCC and BCC phases are about 69% and 31%, respectively. The lattice parameters determined from X-ray diffraction are 0.3590 nm for the FCC phase and 0.2873 nm for the BCC phase.

Two distinct features are seen within individual eutectic colonies: regular lamellae with straight interphase boundaries; and fragmented regions with irregular curved boundaries (see figure 1a-c). The lamellae are in most cases connected to the colony boundaries at both sides (see the white lines in figure 1a), while the irregular curved boundaries tend to be aligned perpendicular to the lamellar boundaries (see e.g. the regions marked by the yellow arrows in figure 1a and 1b). Each lamellar bundle consists of typically four to six thin lamellae of each phase, and some of the lamellae can be 60 μm long. Serrations or jogs also appear in the lamellar regions. In some cases, the long lamellae are interrupted, especially for the BCC phase (see e.g. the one marked by the white arrow in figure 1a).

Twin boundaries are seen within the FCC phase, highlighted in green in figure 1a. Most colonies contain only one set of straight interphase boundaries, while a few colonies containing twin variants can have multiple sets of straight interphase boundaries (see e.g. the colony A in figure 1a). Misfit dislocations are present along the straight interphase boundaries, i.e., these interfaces are semi-coherent (see figure 1c).

Microstructural quantification based on the EBSD maps indicates that the volume fractions of the lamellar and irregular regions are about 42% and 58%, respectively (see table 1). The relative volume fraction of the FCC phase is slightly higher in the irregular regions ($42/58 \approx 72\%$) than in the regular lamellar regions ($29/42 \approx 69\%$). The average width of the well-defined lamellae is finer than the average width within the irregular regions for both phases (see table 1). There is also a difference in the interphase boundary spacing between the FCC and BCC phases: the former is roughly twice the latter.

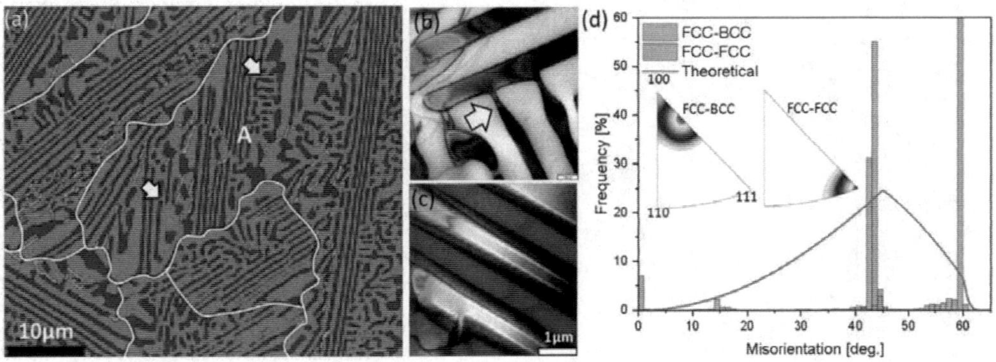

Figure 1. Microstructure of the as-cast $AlCoCrFeNi_{2.1}$ EHEA. (a) EBSD phase map, where the FCC and BCC phases are shown in red and blue, respectively. FCC twin boundaries are shown in green, while the white lines mark the eutectic colony boundaries. (b) and (c) TEM images. Semi-coherent straight interphase boundaries with misfit dislocations are seen in (c). (d) misorientation (angle-axis pair) distribution for interphase boundaries for boundaries within the FCC phase. The white and yellow arrows in (a) and (b) mark an interrupted BCC lamella and dendrite-like features, respectively.

Table 1. Microstructural parameters for the FCC and BCC phases in two regions with different morphologies in the EBSD map in figure 1a.

Phase	Lamellar region		Irregular region	
	Average width (μm)	Phase fraction (%)	Average width (μm)	Phase fraction (%)
FCC	0.70	29	1.16	42
BCC	0.35	13	0.59	16
All	-	42	-	58

In spite of the fragmented structure within the colonies, all the fragments have a nearly identical crystallographic orientation for the BCC phase, and similarly nearly identical for the FCC except for one twin variant (see e.g. figure 2a-c). The misorientation between the BCC and FCC phases within each colony is about 43°⟨114⟩ (see figure 1d). From the pole figures for the two phases in figure 2b and 2c, it can be seen that one {111} plane in the FCC phase for each twin is parallel to one {110} plane in the BCC phase, and similarly one {110} plane in the FCC phase is parallel to one {111} plane in the BCC phase. This corresponds to the well-known Kurdjumov-Sachs (K-S) orientation relationship, and this relationship is observed for both lamellar and irregular regions.

To determine the crystallographic plane for the straight interphase boundaries, Kikuchi diffraction patterns of the two phases in a colony were recorded in using TEM. As shown in figure 3, when the interface is edge-on, the {112} and {224} planes of the FCC phase and the {123} plane of the BCC phase are almost parallel to the interphase boundary traces. This result is further supported by the analysis based on the EBSD and SEM data (see figure 2d and 2e).

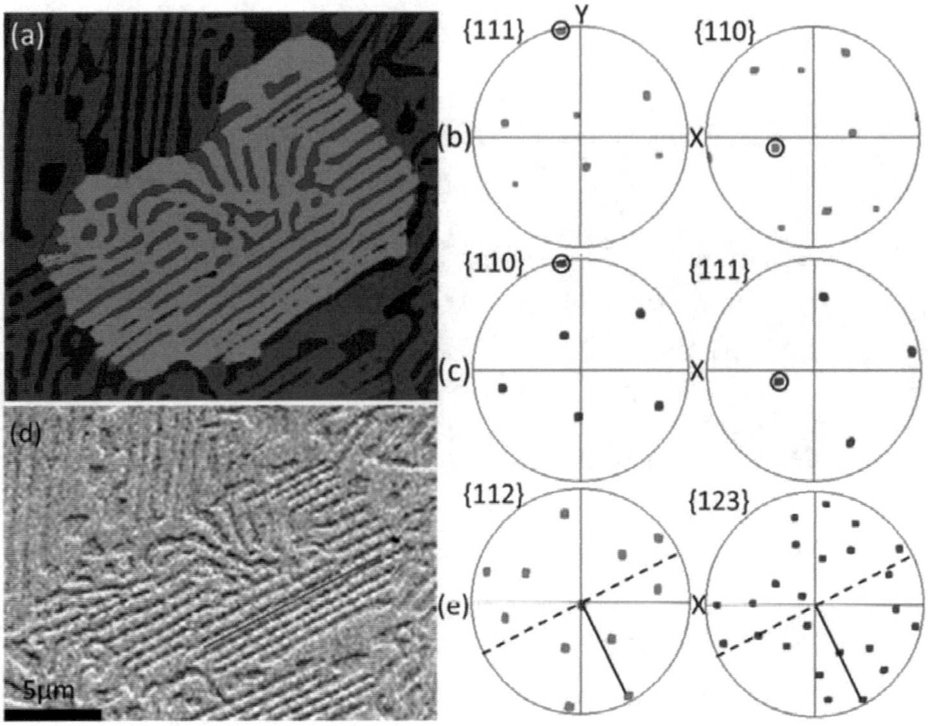

Figure 2. Analysis of orientation relationship between the FCC and BCC phases. (a) EBSD phase map for one highlighted colony, where the FCC and BCC phases are shown in red and blue, respectively. FCC twin boundaries are shown in green. (b) and (c) {111} and {110} pole figures showing the orientation relationship between the FCC and BCC phases. (d) SEM image taken in the same region as in (a). (e) {112} and {123} pole figures for the FCC and BCC phases, respectively, showing the crystallographic plane that is parallel to the lamellae direction (marked by the red line in (d)). Pole figures with red and blue colours are from FCC and BCC phases, respectively.

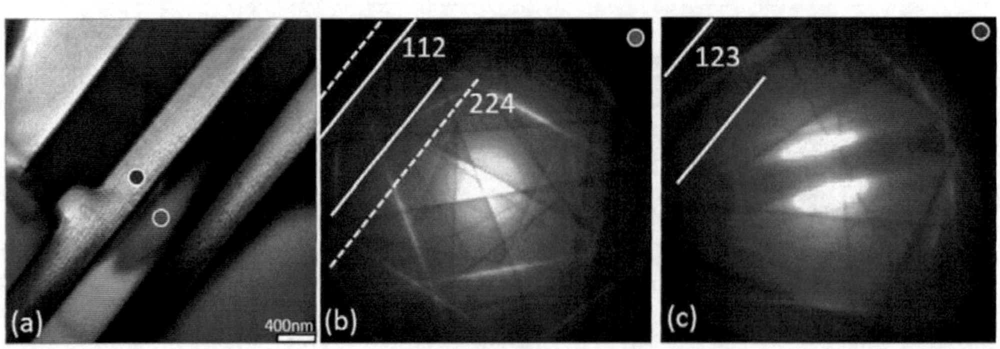

Figure 3. TEM analysis of interphase boundaries in a lamellar region: (a) TEM image, where red and blue dots mark FCC and BCC lamellae, respectively. (b) and (c) are Kikuchi patterns from the FCC and BCC lamellae, respectively. It is seen that the Kikuchi lines are parallel to the straight interface in (a).

4. Discussion

In the present EHEA alloy, the FCC and BCC phases form cooperatively during solidification. Based on different types of eutectic reactions, a distinction has been made between *normal* and *anomalous* eutectic alloys. *Normal* alloys are distinguished by well-defined lamellar structures and orientation relationships, while *anomalous* alloys show typically an irregular structure with no preferred orientation relationship [12]. The present results show that there is a clear orientation relationship between the two phases in $AlCoCrFeNi_{2.1}$. Therefore, it is concluded that this EHEA is a *normal* eutectic alloy. The two phases have a Kurdjumov-Sachs orientation relationship, i.e., $\{111\}_{FCC} \parallel \{110\}_{BCC}$ and $\langle 110\rangle_{FCC} \parallel \langle 111\rangle_{BCC}$, which agrees with the conclusions of Ref. [13].

The TEM analysis shows the presence of semi-coherent interphase boundaries, and that these boundaries are parallel to the FCC {112}/{224} planes and BCC {123} planes (see figure 3). The lattice plane spacings for the FCC {224} and BCC {123} planes are quite similar, 0.0733 nm and 0.0768 nm, respectively. The habit planes are therefore likely to be {224} and {123} for the FCC and BCC phases, respectively. The misfit between the two planes is about 5%, which leads to the formation of misfit dislocations to accommodate the misfit strain (see figure 1c).

The results also show that the irregular regions develop on both sides of the lamellar regions. It is suggested that the irregular structure is typically formed because the growth rates of the two phases are different: one phase solidifies faster than the other phase. For the present case, the FCC phase solidifies faster than the BCC phase, leading to the formation of local dendrite-like structures along the BCC lamellae (see figures 1a and 1b) and to the formation of the interrupted BCC lamellae. Additionally, a higher volume fraction of the FCC phase can contribute to the formation of the irregular structure [12]. For volume fractions of the FCC and BCC phases in the irregular regions of greater than 2:1, the balance for the cooperative growth between FCC and BCC with the expected volume fraction ratio of 2:1 can be broken.

Based on the observed results, the solidification process of the $AlCoCrFeNi_{2.1}$ EHEA can be briefly described as follows:

i. The two phases nucleate and grow cooperatively along the low energy semi-coherent interfaces (habit plane). The volume fraction ratio between the FCC and BCC phase is 2:1.
ii. During solidification, the FCC phase grows faster than the BCC phase, leading to the formation of interrupted lamellae and dendrite-like structures in the BCC phase.
iii. Stacking faults can occur in the FCC phase, which leads to the formation of twins predominantly perpendicular to the lamellar direction (see figure 1a).

Based on the present results, it cannot be concluded how the local chemical segregation affects the growth of the two phases. Assuming that the chemical content of the FCC and BCC phases in the regular and irregular regions is identical, the higher FCC fraction in the irregular regions implies local chemical segregation, i.e., more elements that compose the FCC phase segregate in the liquid and form a higher fraction of this phase in the irregular regions. Further experimental characterization of the local chemistry and simulations such as phase field modeling are required to understand this process.

From the materials design point of view, lamellar structures with coherent/semi-coherent boundaries can be beneficial for mechanical properties [14], as the lamellae typically deform cooperatively, which significantly increases the strength. On the other hand, the low energy structure of the coherent or semi-coherent boundaries can absorb more dislocations, which can improve material ductility. In addition, flat interphase boundaries are usually extraordinarily stable at elevated temperatures because of the low energy structure and lack of capillary driving force for migration. In contrast, the non-coherent interphase boundaries can be expected to behave like conventional grain boundaries with less stability at high temperatures. For the present alloy, the average interphase boundary spacing for the irregular regions is larger than that for the lamellar regions, implying that the strength in the irregular regions is lower than that of the lamellar regions. Apparently, to improve mechanical properties, the processing should be optimized to eliminate the formation of the irregular regions. Two possible ways are proposed here:

i. To change the chemical composition according to the chemistry in the lamellar regions, so that chemical segregation will not occur during the eutectic reaction and only the lamellar structure will be formed during the eutectic reaction.
ii. To increase the cooling rate in order to suppress the local segregation and to refine the lamellar boundary spacing, which can significantly improve the mechanical properties.

5. Conclusions

In the present paper, the microstructure of the as-cast eutectic high entropy alloy $AlCoCrFeNi_{2.1}$ has been quantified in detail using electron microscopy. The results show that this alloy consists of two phases, FCC and BCC, with volume fractions of 69% and 31%, respectively. Within individual eutectic colonies, the microstructure can be divided into two regions with distinct features: a regular lamellar structure in the center and an irregular structure on both sides of the lamellar region. The orientation relationship for these two types of structures is however identical, and is described by the Kurdjumov-Sachs orientation relationship, i.e., $\{111\}_{FCC} \parallel \{110\}_{BCC}$ and $\langle 110\rangle_{FCC} \parallel \langle 111\rangle_{BCC}$. The habit planes for the lamellar interfaces are determined to be {224} and {123} for the FCC and BCC phases, respectively. Microstructural analysis shows that the ratio of the volume fractions of the FCC and BCC phases is higher in the irregular regions than in the lamellar regions. This suggests that local chemical segregation is more likely to occur during the solidification process, which can be the main reason for the development of the irregular regions.

Acknowledgements
AL and SG thank the financial support from the Swedish Research Council (2015-04087). The authors acknowledge additional funding from InterReg. TY and YZ thank further the financial support by the European Research Council (ERC) under the European Union's Horizon 2020 research and innovation programme (M4D – grant agreement No. 788567).

References

[1] Yeh J W, Chen S K, Lin S J, Gan J Y, Chin T S, Shun T T, Tsau C H and Chang S Y 2004 *Adv. Eng. Mater.* **6** 299–303
[2] Cantor B, Chang I T H, Knight P and Vincent A J B 2004 *Mater. Sci. Eng. A* **375–377** 213–8
[3] Gludovatz B, Hohenwarter A, Catoor D, Chang E H, George E P and Ritchie R O 2014 *Science.* **345** 1153–8
[4] Li Z, Pradeep K G, Deng Y, Raabe D and Tasan C C 2016 *Nature* **534** 227-30
[5] Yao Y, Huang Z, Xie P, Lacey S D, Jacob R J, Xie H, Chen F, Nie A, Pu T, Rehwoldt M, Yu D, Zachariah M R, Wang C, Shahbazian-Yassar R, Li J and Hu L 2018 *Science.* **359** 1489–94
[6] Koželj P, Vrtnik S, Jelen A, Jazbec S, Jagličić Z, Maiti S, Feuerbacher M, Steurer W and Dolinšek J 2014 *Phys. Rev. Lett.* **113** 1–5
[7] Sarker P, Harrington T, Toher C, Oses C, Samiee M, Maria J P, Brenner D W, Vecchio K S and Curtarolo S 2018 *Nat. Commun.* **9** 1–10
[8] Lu Y, Dong Y, Guo S, Jiang L, Kang H, Wang T, Wen B, Wang Z, Jie J, Cao Z, Ruan H and Li T 2014 *Sci. Rep.* **4** 6200
[9] Lu Y, Jiang H, Guo S, Wang T, Cao Z and Li T 2017 *Intermetallics* **91** 124–8
[10] Wani I S, Bhattacharjee T, Sheikh S, Bhattacharjee P P, Guo S and Tsuji N 2016 *Mater. Sci. Eng. A* **675** 99–109
[11] Shi P, Ren W, Zheng T, Ren Z, Hou X, Peng J, Hu P, Gao Y, Zhong Y and Liaw P K 2019 *Nat. Commun.* **10** 1–8
[12] Glicksman M E 2011 *Principles of Solidification: An Introduction to Modern Casting and Crystal Growth Concepts* (New York, NY: Springer New York)
[13] Wang Q, Lu Y, Yu Q and Zhang Z 2018 *Sci. Rep.* **8** 14910
[14] Lu K, Lu K, Lu L and Suresh S 2009 *Science.* **349** 349–53

Grain size effect on deformation mechanisms in Mg-3Gd

X Luo [1], J Q Luo [1], Z Q Feng [1], F P Hu [1], T Yu [2], C Berggreen [2], G L Wu [1], N Hansen [1,3] and X Huang [1,2]

[1] Joint International Laboratory for Light Alloys (MOE), College of Materials Science and Engineering, Chongqing University, Chongqing 400044, China
[2] Department of Mechanical Engineering, Technical University of Denmark, DK-2800, Kgs, Lyngby, Denmark
[3] Technical University of Denmark, Risø Campus, DK-4000 Roskilde, Denmark

Email: wugl@cqu.edu.cn

Abstract. Mg-3Gd (wt. %) samples were prepared by accumulative roll-bonding (ARB) and subsequent annealing to produce samples with average grain sizes in a wide range from 3.3 μm to 45.1 μm. These samples were tensile tested at room temperature. Dislocation structures and deformation twins in the tested tensile samples were characterized by transmission electron microscopy. Dislocations with different types of Burgers vector were identified based on two-beam diffraction contrast experiments. The results revealed that deformation twins and <a> dislocations are dominant in coarse-grained samples while <c+a> and <a> dislocations characterize the fine grain deformation structure. The transition of deformation mechanisms and its implications for microstructural design and property optimization are discussed.

1. Introduction

Compared to cubic metals, the lower crystallographic symmetry of hexagonal close-packed (HCP) magnesium with c/a ratio of 1.624 results in high plastic anisotropy and unique mechanical characteristics [1, 2]. The high anisotropy in magnesium alloys usually results in a high critical resolved shear stress (CRSS) ratio between non-basal and basal slip, and basal slip and twinning act as dominant deformation modes [1-4]. However, some non-basal slip is expected, as it is necessary to accommodate the strain along c-axis in order to achieve the five independent deformation systems required for an arbitrary deformation [5, 6]. Therefore, the activation of non-basal slip systems is extremely important for highly ductile magnesium alloys.

The addition of rare earth elements, such as Gd and Y, into magnesium can change the CRSS and the core structure for <c+a> dislocations [7], thereby improving the ability to activate <c+a> slip [8]. In contrast to the effect of rare earth elements in magnesium alloys [9], current experimental and simulated results for fine-grained magnesium alloys with high ductility suggest that more controversy lies in understanding the grain size effect [10]. The deformation mechanisms in coarse-grained magnesium alloys at room temperature are dominated by basal <a> slip and twinning [1, 2]. However, due to the restraint of deformation twinning in fine-grained magnesium alloys, it is considered more slip systems are likely to be activated [11, 12]. Therefore, our hypothesis is that the control of grain size provides the opportunity to activate multiple slip systems. In this work, we prepared coarse-grained and fine-grained Mg-3Gd samples by accumulative roll-bonding (ARB) and subsequent annealing, and carried out

IOP Conf. Series: Materials Science and Engineering **580** (2019) 012040 doi:10.1088/1757-899X/580/1/012040

detailed electron microscopy observations to identify the grain size effect on the activation of <a> slip, <c+a> slip and twinning.

2. Experimental

An Mg-3%Gd ingot with a diameter of 100 mm was prepared by semi-continuous casting, followed by homogenization treatment at 450 °C for 24 h. The homogenized ingot was then subjected to hot extrusion to a 3 mm thick plate, hot rolling to 1 mm sheet in thickness at 450 °C, and finally annealing at 500 °C for 3 h. After annealing the initial sheet samples show a coarse grain structure and a weak basal texture. The annealed sheet was sparking-machined into specimens with a length of 70 mm and a width of 25 mm, which were then processed by the ARB technique [13] for two cycles. The ARB samples were finally annealed between 310 °C and 500 °C for 1 h to prepare samples with average grain sizes ranging from 3.3 μm to 45.1 μm.

Dog-boned tensile specimens with a gauge length of 13 mm and cross section of 5×1 mm^2 were cut from the annealed samples. Uniaxial tensile tests were performed in a Shimadzu AGX-50 testing frame at a strain rate of 8.3×10^{-4} s^{-1} at room temperature. The tensile direction was parallel to the rolling direction (RD) of the ARB-processed sheets.

The microstructures of the annealed samples were characterized by electron backscatter diffraction (EBSD) in a Zeiss Auriga scanning electron microscope equipped with an Oxford Aztec EBSD detector. The samples for EBSD analyses were first mechanically polished and then electrochemically polished in the Struers AC2 solution. Tensile tested samples were also characterized in a JEM-2100 transmission electron microscope (TEM). TEM foil samples were cut in longitudinal section (RD-ND section; ND being the normal direction) from the uniformly elongated area of each tested tensile samples. The foils were prepared by ion milling after mechanical grinding to 100 μm thickness. Dislocation contents of individual grains were analyzed in the TEM under two-beam diffraction conditions using different diffraction vectors ***g***.

3. Results and discussion

Figure 1 shows EBSD boundary maps of the annealed and initial samples. After annealing for 1 h at 310 °C and 360 °C, the material is fully recrystallized samples with average grain sizes of 3.3 μm and 6.3 μm, as shown in figures 1a and b. Figure 1c shows the microstructure of the initial sample with an average grain size of 45.1 μm. Note that low concentration of low-angle boundaries (marked by green lines) are observed within recrystallized grains.

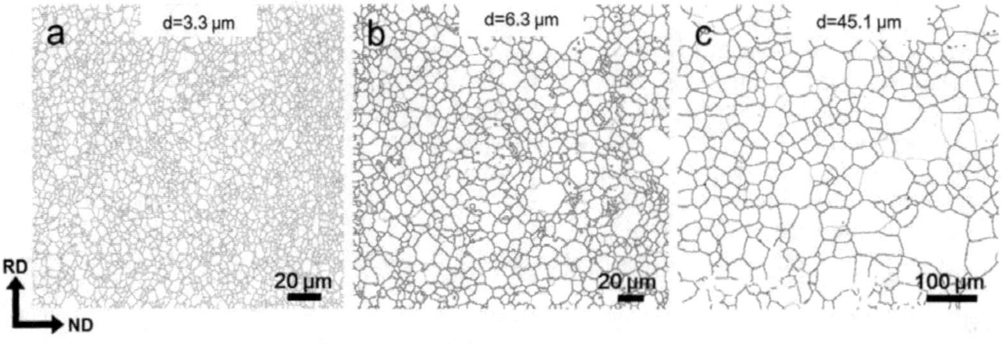

Figure 1. EBSD boundary maps showing the grain structures of (a, b) two fine-grained samples produced by ARB processing followed by annealing for 1 h at (a) 310 °C and (b) 360 °C, and (c) the initial sheet sample, where high-angle boundaries above 15° are drawn in black, and low-angle boundaries between 2° and 15° are drawn in green.

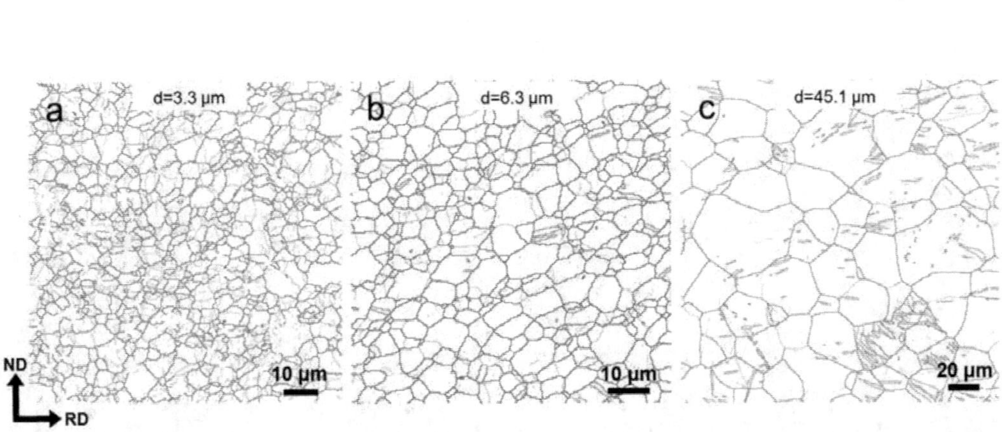

Figure 2. EBSD boundary maps showing microstructures of 5% tensile deformed samples (a) 310 °C 1h, (b) 360 °C 1h and (c) initial sample, where high-angle grain boundaries above 15° are drawn in black, low-angle boundaries between 2° and 15° are drawn in green, and twin boundaries are drawn in red.

EBSD results of fine-grained and coarse-grained samples after deformation to a tensile strain of 5% are shown in figure 2, where the boundaries with misorientation angles above 15° (high angle boundaries; HABs), between 2° to 15° (low angle boundaries; LABs) and twin boundaries 86° (TBs) are indicated by black, green and red lines, respectively. As shown in figures 2a and b, a high concentration of LABs (marked by green lines) is found in fine grains. In contrast within coarse grains in figure 2c, many twins (marked by red lines) are formed, with only a low concentration of LABs.

Figure 3 illustrates TEM images of a fine-grained sample tensile-deformed to failure. Many dislocations were activated during tensile deformation. In figure 3a, the two-beam dark field (TBDF) TEM image with g=[0002] near the [01$\bar{1}$0] zone axis shows a mass of visible dislocations inside the grain, and all these dislocations show a c-component Burgers vector, e.g. <c+a> and <c>. The TBDF TEM image using g=[2$\bar{1}\bar{1}$0] near the [01$\bar{1}$0] zone axis also shows lots of dislocations (see figure 3b), and these dislocations are identified as <c+a> (marked by red dashed arrows) and <a> type (marked by white dashed arrows). Moreover, all dislocations are visible in figure 3c. Note the basal plane is indicated by white solid line in figure 3c. In order to distinguish different dislocation types clearly, a sketch stored all dislocations remained in this deformed grain is given in figure 3d. Red and white lines in figure 3d represent the <c+a> and <a> dislocations respectively. After careful observation, it is revealed that dislocation-based deformation mechanisms including <c+a> and <a> slip dominate in fine-grained samples during tensile deformation.

Figure 4 illustrates TEM images of coarse-grained sample tensile deformed to failure showing typical twin and dislocation structures. Based on selected area electron diffraction (SAED) analysis (see inset in figure 4a), it was found that deformation twinning and parallel stacking faults (SFs marked by red solid arrows) were activated. Note the basal plane is indicated by white solid line. Parallel <a> dislocations are also found and are indicated by white dashed arrows in figure 4a. A sketch showing dislocations and SFs corresponding to the rectangular region in figure 4a is given in figure 4b. White and red solid lines in figure 4b represent <a> dislocations and SFs respectively. From the TEM observations described above, it is seen that deformation mechanisms including deformation twinning and <a> slip dominate in coarse-grained samples during tensile deformation.

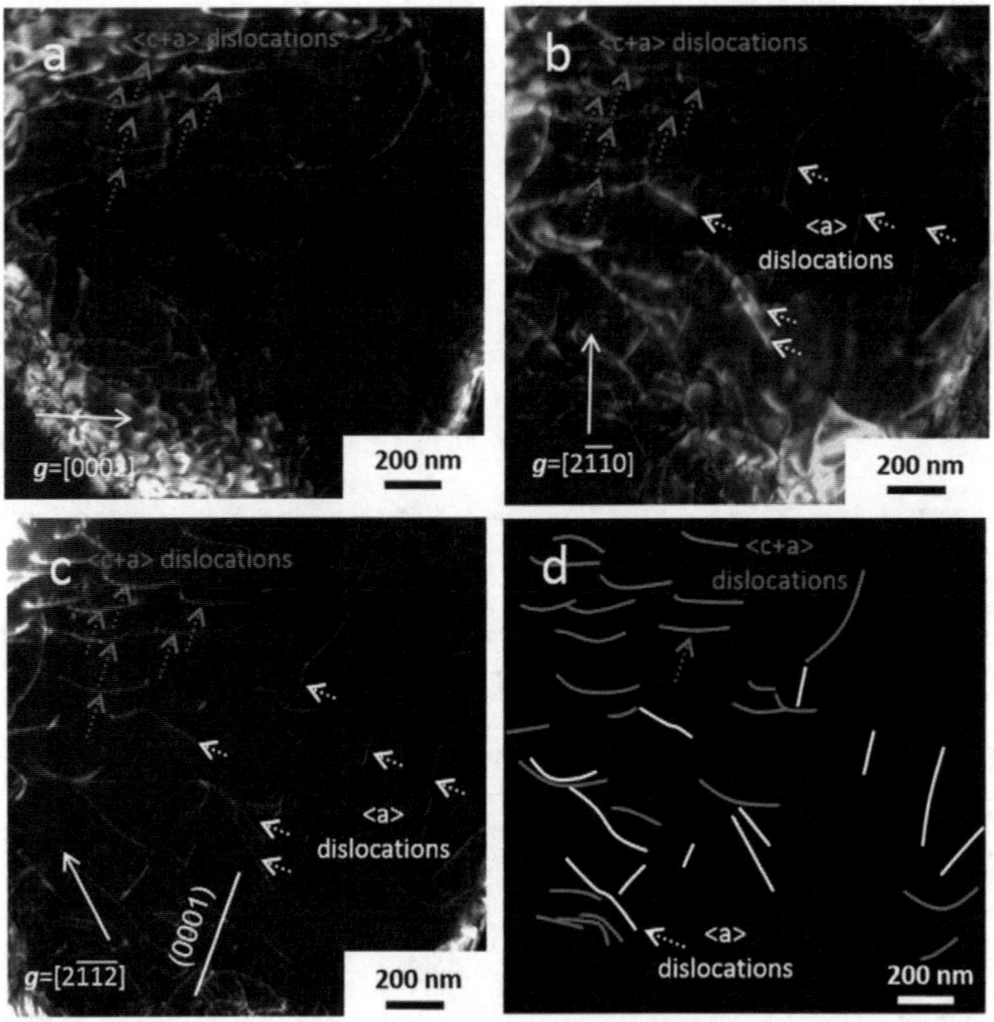

Figure 3. TEM images of fine-grained sample tensile deformed to failure; (a, b, and c) are TBDF images with g=[0002], g=[2$\bar{1}\bar{1}$0] and g=[2$\bar{1}\bar{1}\bar{2}$]], respectively, near the [01$\bar{1}$0] zone axis; (d) a sketch showing all dislocations identified in this grain.

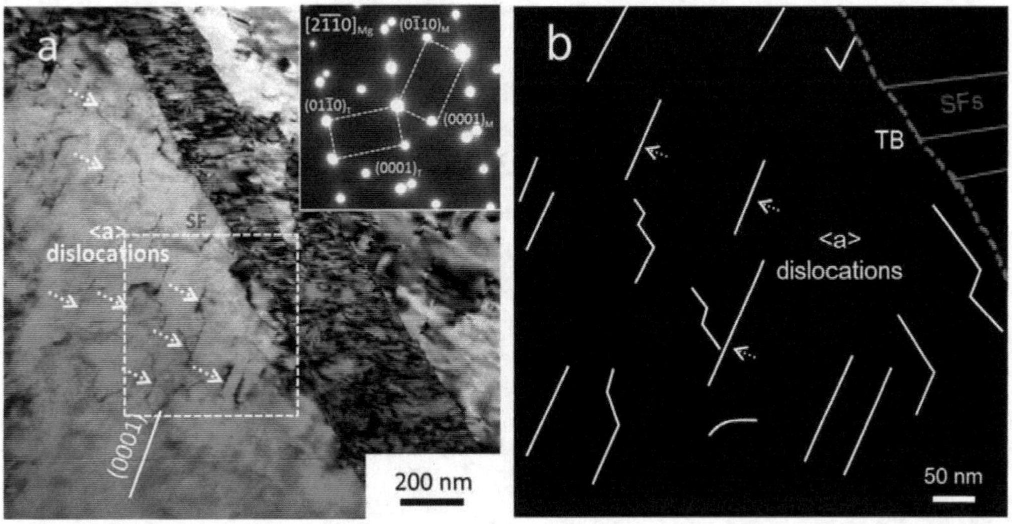

Figure 4. TEM image of coarse-grained sample tensile deformed to failure showing the activation of deformation twinning and <a> dislocations.

In summary, a transition in deformation mechanisms from deformation twinning and <a> slip in coarse grain structures to <a> and <c+a> slip in fine grain structure was confirmed by both EBSD and TEM observations. In contrast with normal <a> slip in conventional coarse-grained magnesium alloys, <c+a> slip shows stronger multiplication, interaction and cross-slip ability, which not only lead to a high work hardening rate but also improve slip distance and dislocation density. Furthermore, the activation of multiple slip systems in fine-grained samples may transform the typically highly anisotropic deformation of magnesium alloys to a more homogeneous deformation mode, similar to fcc and bcc metals, such as Al and IF steel [14, 15]. The abnormally active <c+a> slip in fine grains help, therefore, to ensure homogeneous deformation and satisfy the von-Mises condition during plastic forming of magnesium alloys.

4. Conclusions

The grain size dependence of the transition in deformation mechanisms, from twinning and <a> slip to <c+a> and <a> slip based plasticity, has been investigated by tensile deformation of Mg-3Gd with a grain size in the range 3.3 to 45.1 um. A transition marking change of dislocation types from <a> in a twinned structure at large grain size, to <a> and <c+a> in small grains was confirmed. The transition of deformation mechanisms and its implications for microstructural and property optimization highlight the scope for advanced design of magnesium alloys through structural refinement to a fine grain size.

Acknowledgements

This work was sponsored by the State Key Research and Development Program of Ministry of Science and Technology (MOST) of China (2016YFB0700401, 2016YFB0700403) and National Natural Science Foundation of China (No. 51471039, 51501022, 51327805). Part of this research was conducted using mechanical testing equipment from DTU Center for Advanced Structural and Material Testing (CASMAT), Grant No. VKR023193 from Villum Fonden. TY thanks the European Research Council (ERC) under the European Union's Horizon 2020 research and innovation programme (grant agreement No 788567). NH thanks the support of the "111" Project (B16007) by the Ministry of Education and the State Administration of Foreign Experts Affairs of China.

References

[1] Christian J W and Mahajan S 1995 *Prog. Mater. Sci.* **39** 1-157
[2] Agnew S R and Duygulu Ö 2005 *Int. J. Plasticity* **21** 1161-93
[3] Chapuis A and Driver J H 2011 *Acta Mater.* **59** 1986-94
[4] Staroselsky A and Anand L 2003 *Int. J. Plasticity* **19** 1843-64
[5] Koike J, Kobayashi T, Mukai T, Watanabe H, Suzuki M, Maruyama K and Higashi K 2003 *Acta Mater.* **51** 2055-65
[6] Zhang D, Jiang L, Schoenung J M, Mahajan S and Lavernia E J 2015 *Philos. Mag.* **95** 3823-44
[7] Ding Z, Liu W, Sun H, Li S, Zhang D and Zhu Y 2018 *Acta Mater.* **146** 265-72
[8] Sandlobes S, Zaefferer S, Schestakow I, Yi S and Gonzalez-Martinez R 2011 *Acta Mater.* **59** 429-39
[9] Basu I, Pradeep K G, Mießen C, Barrales-Mora L A and Al-Samman T 2016 *Acta Mater.* **116** 77-94
[10] Wu Z, Ahmad R, Yin B, Sandlöbes S and Curtin W A 2018 *Science* **359** 447-52
[11] Meyers M A, Vöhringer O and Lubarda V A 2001 *Acta mater.* **49** 4025-39
[12] Fan H, Aubry S, Arsenlis A and El-Awady J A 2016 *Scripta Mater.* **112** 50-53
[13] Saito Y, Utsunomiya H and Tsuji N 1999 *Acta Mater.* **47** 579-83
[14] Kamikawa N, Huang X, Tsuji N and Hansen N 2009 *Acta Mater.* **57** 4198-208
[15] Huang X, Kamikawa N and Hansen N 2010 *J. Mater. Sci.* **45** 4761-69

IOP Conf. Series: Materials Science and Engineering **580** (2019) 012041 doi:10.1088/1757-899X/580/1/012041

Microstructural evolution of Ta-4%W during cold rolling

Guoqiang Ma[1*], Andrew Godfrey[2], Qiang Chen[3], Guilin Wu[1], Darcy A. Hughes[4], Niels Hansen[1,5]
[1]International Joint Laboratory for Light Alloys (MOE), College of Materials Science and Engineering, Chongqing University, Chongqing, 400044, China
[2]Key Laboratory of Advanced Materials (MOE), School of Materials Science and Engineering, Tsinghua University, Beijing 100084, China
[3]Southwest Technology and Engineering Research Institute, Chongqing 400039, China
[4]Fremont, CA, USA
[5]Risø Campus, Technical University of Denmark, DK-4000 Roskilde, Denmark

maerkkdp@cqu.edu.cn

Abstract. Tantalum alloys are regarded as materials with great potential for challenging environments due to their exceptional mechanical properties, especially at high temperatures. To analyse deformation mechanisms and the relationship between microstructure and strength, characterization of the deformation microstructure of Ta-4%W during cold rolling from 10% to 50% has been carried out by transmission electron microscopy combined with electron backscatter diffraction investigations. The microstructural evolution of Ta-4%W was found to follow the same general framework as FCC metals (Al, Cu, Ni) and IF steel, with added complexity due to the additional constraints on dislocation motion.

1. Introduction

The characterization of deformation microstructures and the quantification of microstructure parameters form the basis for an analysis of deformation mechanisms and the relationship between microstructure and strength. Extensive studies based on a structural analysis by transmission electron microscopy have covered FCC metals such as Al, Cu and Ni [1-3], whereas investigations of BCC metals have been more limited, with the exception of interstitial free (IF) steels [4-7]. To supplement these studies we have chosen to investigate the evolution of the deformation microstructure during cold rolling of the refractory metal alloy Ta-4%W, with a high melting point (3290 K for pure Ta).

To allow a comparison with previous work we use the same nomenclature as previously introduced based on grain subdivision [8-10] and the hypothesis that dislocation configurations follow the low energy dislocation structure (LEDS) principle [11], i.e., the energy per unit line length of stored dislocations is the minimum accessible for the material conditions.

2. Experiment

The original material used in present work is a Ta-4%W alloy, which was prepared by electron beam melting followed by hot forging and annealing to obtain a recrystallized rod of 50 mm in diameter and with an average grain size of 26 μm. Slabs 2 mm in thickness were cut perpendicular to the axis of the rod and rolled to reductions of 10%, 30% and 50% respectively at room temperature. The ratio of l/d (where l is the contact length between the plate and the roll, and d is the average thickness of the sample) was kept approximately at 2.5 during the rolling process.

The microstructures were characterized by electron backscatter diffraction (EBSD) and in the transmission electron microscope (TEM). Samples for EBSD observation were polished in a 1:9 (volume fraction) $HF:H_2SO_4$ solution at room temperature at current density of 0.1 A/cm^2 for 9 minutes, and studied using a JEOL 7800F scanning electron microscope (SEM) operated at 20 kV, equipped with an Oxford Instruments AZtec EBSD system [12]. Foils for TEM observation were prepared by twin jet technique in a 1:5:94 (volume fraction) $HF:H_2SO_4:CH_3OH$ solution at -30°C and studied using a JEOL JEM2100 electron microscope operated at 200 kV [13]. All the EBSD and TEM observations were conducted in the longitudinal section (the section containing the rolling direction, RD and the normal direction, ND).

3. Results and discussion

3.1. EBSD observations

Figure 1 shows example EBSD orientation maps of the Ta-4%W alloy at different strain levels. Equiaxed grains with a large distribution in grain size were observed in the initial sample (see figure 1a). After 10% cold rolling (see figure 1b), no clear structures can be detected by EBSD, but it can be seen that within each grain the orientations have changed in different regions. After 30% cold rolling (see figure 1c), a similar deformation microstructure to the one observed at 10% rolling is seen. Parallel groups of low angle dislocation boundaries are observed more frequently in some grains with those boundaries inclined to the RD at certain angles. After 50% cold rolling (see figure 1d), more well-defined dislocation boundaries have formed and the grains are clearly elongated along the RD. Some grains with just a few boundaries visible using EBSD are present. Note that boundaries with misorientation angles lower than 2° are not marked due to the resolution limit of the EBSD technique.

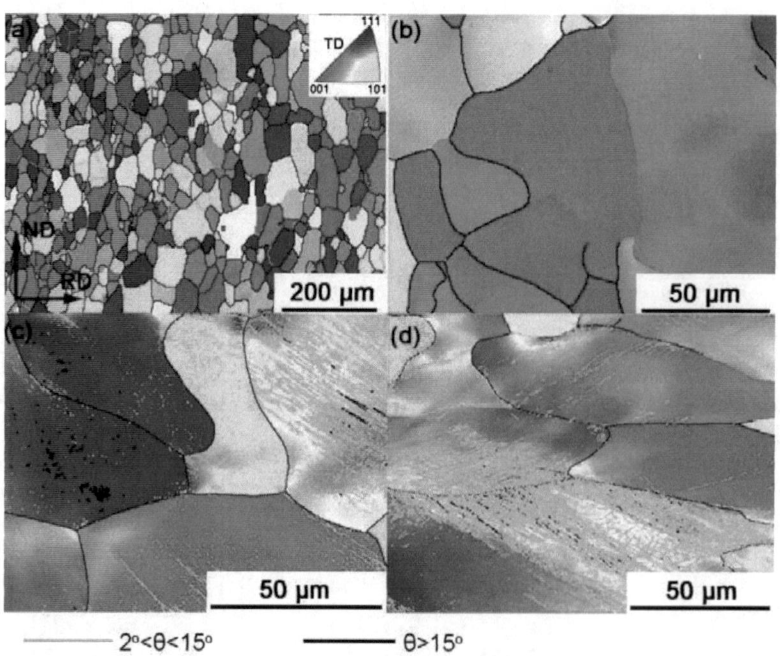

Figure 1. EBSD orientation maps of the Ta-4%W alloy: (a) initial sample, (b) 10% cold rolling, (c) 30% cold rolling and (d) 50% cold rolling. Note the parallel lines representing dislocation boundaries forming within grains (c and d).

3.2. *TEM observations*

Figure 2 shows example TEM images of the initial Ta-4%W alloy. There are a few dislocations in the grains and the grain boundaries are sharp, showing that the samples are fully recrystallized.

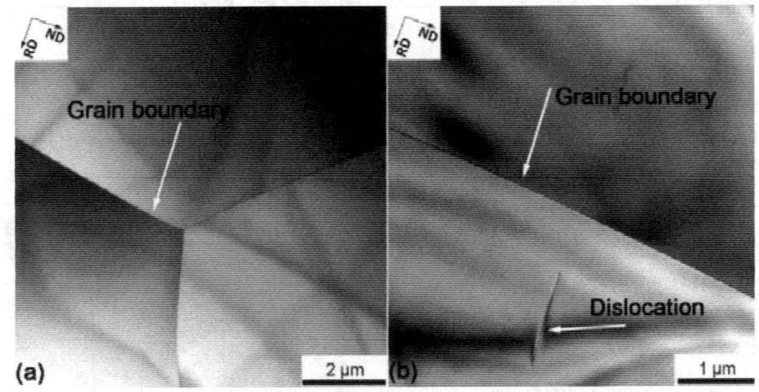

Figure 2. TEM images of the initial Ta-4%W alloy: (a) region 1, (b) region 2.

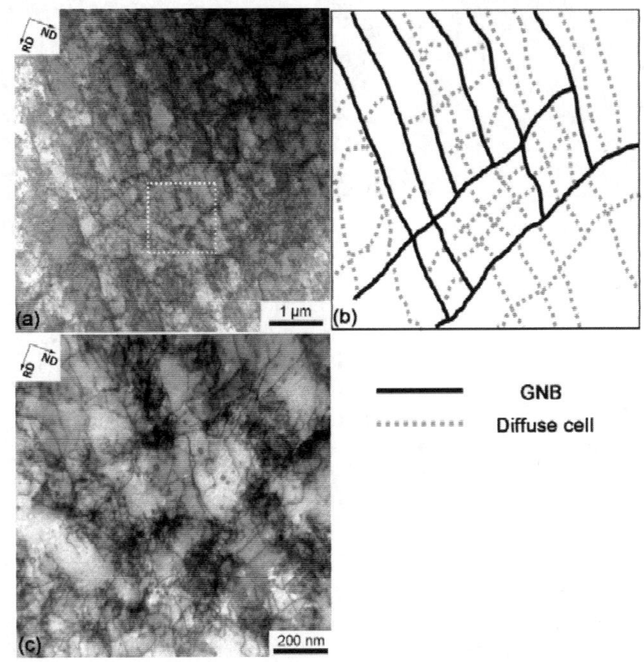

Figure 3. Microstructures of the Ta-4%W alloy after 10% cold rolling: (a) TEM image, (b) sketch, and (c) detailed observation of dashed box in (a).

Figure 3 shows the microstructures of the Ta-4%W alloy after 10% cold rolling as seen in the TEM. The grains are subdivided into misoriented regions with diffuse dislocation boundaries formed between these regions. Since some boundaries are so diffuse that the boundary dislocations cannot be distinguished from their surrounding, a line of contrast change was used as an indication of boundary position. Detailed observations show that a well developed cell structure has not yet developed between

IOP Conf. Series: Materials Science and Engineering **580** (2019) 012041 doi:10.1088/1757-899X/580/1/012041

the longer boundaries (see figure 3c). Individual dislocations cluster together with dislocation free regions between the clusters, in contrast to the fairly uniform distribution of dislocations in a Taylor lattice found in non-cell forming metals at low strain [4, 10]. The dark and light contrast in the TEM images, figure 3c, suggests that this structure is a nascent cell structure likely formed as an initial Taylor lattice makes a transition to a well developed cell structure with increasing strain. The sketch (figure 3b) clearly shows the long boundaries (solid black lines) and nascent cell boundaries (dashed grey lines). To better confirm this idea, TEM observations using different **g**-vectors were conducted, as shown in figure 4. Tilting the sample to different diffraction **g**-vectors showed that this substructure contains mixed dislocations and multiple Burgers vectors. It should be noted that the crystallographic alignment of these structures is easily found when tilting the sample to a **g**-vector of [12-1] from a **g**-vector of [101]. This alignment matches the type of alignment that would be found in a prior Taylor lattice [4, 10].

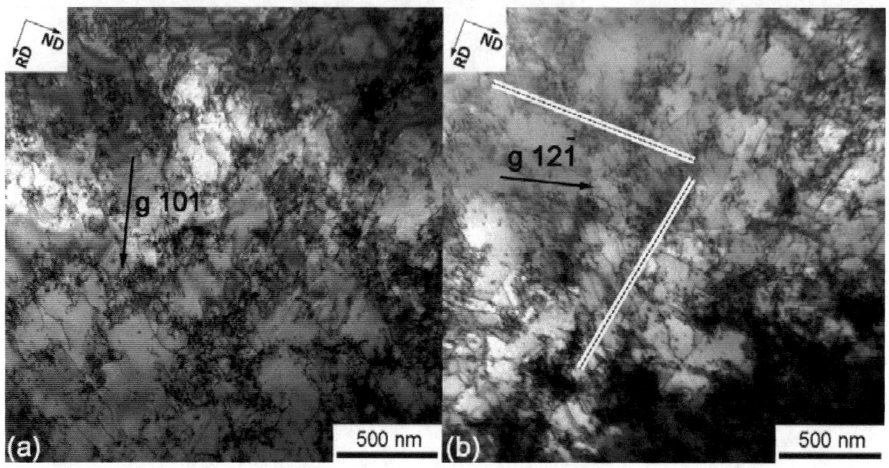

Figure 4. TEM observations of Ta-4%W alloy after 10% cold rolling with different **g** vectors: (a) **g**=[101] and (b) **g** =[12-1], in each case with a [-111] beam direction (BD).

Previous studies [1, 2, 6] of cell forming metals, such as Al, Ni and IF steel, have illustrated the microstructural evolution during plastic deformation, which includes cell block formation at low strain. The sharp cell boundaries (inside the cell blocks) are termed incidental dislocation boundaries (IDBs) and are formed by random trapping of dislocations. With increasing strain, geometrically necessary dislocation boundaries (GNBs) develop and form cell blocks, which accommodate the strain using fewer slip systems than required by the Taylor criterion. For non-cell forming metals, the framework for continuous grain subdivision is analogous to that observed in cell forming metals, however, the long dense dislocation walls (DDWs) and MBs are also referred to as domain boundaries (DBs), since domains of Taylor lattices exist instead of blocks of cells. The differences between DDWs and DBs are initially that DBs may be more diffuse with only slight orientation changes across them, as well as, maintaining a crystallographic orientation, while DDWs may form either near crystallographic planes or in a macroscopic direction. The major similarity between cell forming metals and non-cell forming metals is that both the DDWs and DBs are GNBs that accommodate differences in lattice rotation in adjoining volumes [14], and both IDB and Taylor lattices are incidental dislocation structures formed by random trapping of dislocations.

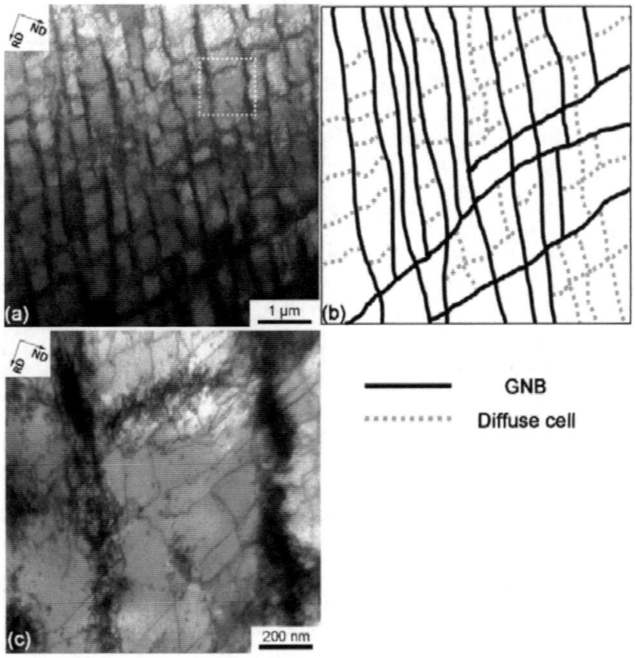

Figure 5. Microstructures of the Ta-4%W alloy after 30% cold rolling: (a) TEM image, (b) sketch, and (c) detailed observation of dashed box in (a).

Figure 5 shows the microstructures of the Ta-4%W alloy after 30% cold rolling. Similar deformation microstructures to those seen in the 10% cold-rolled sample are seen. However, many GNBs are sharper and more pronounced. In some regions the diffuse cell structures are also much more developed (see figure 5c).

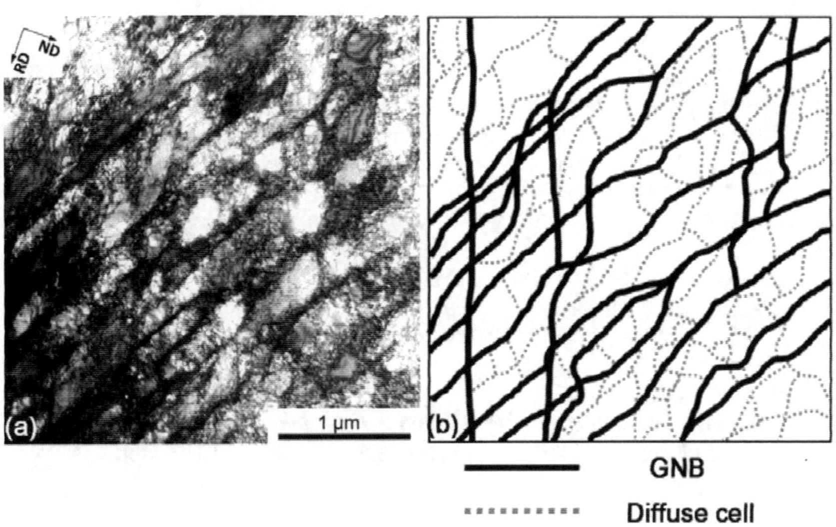

Figure 6. Microstructures of the Ta-4%W alloy after 50% cold rolling: (a) TEM image and (b) sketch.

Figure 6 shows the microstructures of the Ta-4%W alloy after 50% cold rolling. At this medium-level strain, both the GNBs and cell boundaries are well-defined and easily identified. The morphology of the microstructure has thus undergone a transition and resembles the structure of a cell forming metal.

The formation of Taylor lattices and diffuse cells is mainly caused by a large constraint on the three-dimensional dislocation mobility. In this study, the rolling was carried out at room temperature, i.e. a homologous temperature of $0.1T_m$ (melting point in K) for the Ta-4% W alloy compared to 0.32 T_m for Al and 0.17 T_m for Ni. The low homologous temperature slows down thermally activated dislocation processes that enable three dimensional motion. The effect of W in solid solution is more complex as clusters of W may increase the friction stress while single atoms may facilitate the movement of screw dislocations [4, 5, 15]. The Ta-4% W alloy is thus likely to evolve a deformation structure similar to other non-cell forming metals deformed at low strain (e.g., 10% cold rolling). With increasing strain the density of dislocations increases so that their self-stresses provide a driving force for the structure to rearrange into a cell block structure. All of these observed structures can be characterized as LEDS. The formation of clear well defined GNBs and IDBs, however, indicates a continuous evolution to ever lower energies per unit dislocation line length with increasing strain.

4. Conclusions

The analysis of the deformation microstructure evolution during cold-rolling of Ta-4%W follows previous studies of FCC metals (Al, Cu, Ni) and IF steel. However the high melting point, and the presence of W as a solute, reduces the three dimensional dislocation mobility. A structure consisting of of Taylor lattices and diffuse cells delineated by domain boundaries is formed instead of a typical cell block structure at small strains. These structural changes will significantly affect the thermo-mechanical properties of the Ta-4%W alloy and are under investigation.

Acknowledgements

This work was supported by the National Natural Science Foundation of China (51421001) and State Key Research and Development Program of China (2016YFB0700403)

References

[1] Liu Q and Hansen N 1995 *Scripta Metall. Mater.* **32** 1289
[2] Hughes D A and Hansen N 1993 *Metall. Trans. A* **24** 2022
[3] Huang X 1998 *Scripta Mater.* **38** 1697
[4] Spitzig W A and Mitchell T E 1966 *Acta metal.* **14** 1311
[5] Hughes D A 1994 *Evolution Refract. Metals Alloys* ed Dalder E N C, Grobstein T and Olsen C S Warrendale The Minerals, Metals & Materials Society 219
[6] Li B L, Godfrey A, Meng Q C, Liu Q and Hansen N 2004 *Acta Mater.* **52** 1069
[7] Haldar A, Huang X, Leffers T, Hansen N and Ray R K 2004 *Acta Mater.* **52** 5405
[8] Hansen N 2001 *Metall. Mater. Trans. A* **32** 2917
[9] Hansen N and Barlow C Y 2014 *Phys. Metall. (Fifth Edition)* ed Hono D E L Oxford Elsevier 1681
[10] Hughes D A 1993 *Acta Metall. Mater.* **41** 1421
[11] Kuhlmann-Wilsdorf D 1989 *Mater. Sci. Eng. A* **113** 1
[12] Zhang J, Ma G Q, Godfrey A, Shu D Y, Chen Q and Wu G L 2017 *IOP. Conf. Ser. Mater. Sci. Eng.* **219** 012051
[13] Ma G Q, He Q Y, Luo X, Wu G L and Chen Q 2019 *Materials* **12** 117
[14] Hughes D A, Hansen N and Bammann D J 2003 *Scripta Mater.* **48** 147
[15] Kioussis N G and Ghoniem N M 2010 *J. Comput. Theor. Nanos.* **7** 1317

IOP Conf. Series: Materials Science and Engineering **580** (2019) 012042 doi:10.1088/1757-899X/580/1/012042

Comparison of local stress and strain in a heterogeneously-compressed AA 1050 ring by electron microscopy, microhardness and finite element modelling

C V Nielsen[1], N Hansen[2], X Zhang[1]

[1] Section of Manufacturing Engineering, Department of Mechanical Engineering, Technical University of Denmark, Produktionstorvet 425, 2800 Kgs. Lyngby, Denmark
[2] Risø Campus, Technical University of Denmark, DK-4000 Roskilde, Denmark

Email: xzha@mek.dtu.dk

Abstract. An investigation comparing the local stress and strain in a centimeter-scale heterogeneously-compressed AA 1050 ring under nearly full sticking conditions has been carried out. Three complementary techniques were used, covering multiple length scales from nanometers to tens of millimeters, namely electron microscopy (electron backscatter diffraction), microhardness, and finite element modelling. To underpin the present findings on the centimeter-scale heterogeneous structure, the same techniques were applied in an analysis of the homogeneous microstructure resulting from compression of cylinders of the same original hot-extruded AA 1050 material. The analyses support the idea that the local-region microstructure in a heterogeneous structure evolves in accordance with universal mechanisms and principles established for the deformation microstructure of polycrystalline metals.

1. Introduction

Heterogeneity is ubiquitous in the industrial production of metallic components, due to friction, wear and the heterogeneous distribution of stress and strain. A heterogeneous structure may sometimes also be induced deliberately, by gross plastic deformation to strain-harden certain regions, or by surface deformation to increase the fatigue life or wear resistance of a component. However, metals and alloys undergoing thermo-mechanical plastic deformation are typically investigated on the assumption that the process and the resulting microstructure are homogeneous, leading to formulation of the universal principles and mechanisms for the deformation microstructure evolution of polycrystalline metals in the last several decades [1].

For investigation of the mechanical behavior of heterogeneity in deformed metals, a methodology is required to account for variations of structural parameters and strength on multiple length scales from the nanometer to the macroscopic scale. In the present study, three complementary characterization techniques have been used to investigate local stress and strain in a heterogeneously-compressed AA 1050 ring under nearly full sticking conditions: (i) finite element modelling (FEM), to evaluate the local stress and strain during forming from the macroscale; (ii) electron microscopy, with spatial resolution of nanometer scale, to characterize the local structure and structural parameters; and (iii) microhardness, with the indent spatial resolution at the micrometer scale to measure the local strength. The relationship between the structural parameters (average spacing between low angle and high angle boundaries) and the stress and strain for the local areas in the ring with a heterogeneous structure are formulated based on homogeneously compressed aluminum cylinders to obtain corresponding values of boundary spacing, stress and strain, on the hypothesis that the characteristics of local regions in a heterogeneous structure may evolve in accordance with universal principles and mechanisms for the deformation microstructure evolution of polycrystalline metals.

2. Experimental

An Ø25 mm AA 1050 hot extruded bar was used for homogeneously compressed cylinders and heterogeneously compressed rings. The center cross-section of the rings and cylinders has been examined in the present study. The sample microstructure was characterized by scanning electron microscopy (SEM), electron backscatter diffraction (EBSD) and transmission electron microscopy (TEM). The details for sample preparation can be found elsewhere [2-9]. The foils were made via a double-jet electro polishing technique applied in recent studies [10-12] and TEM observation was carried out using a JEOL 2000FX TEM at 200 kV. According to the deformation structure fineness, 1000 nm, 200 nm and 120 nm were chosen as step sizes for the EBSD scans [13, 14]. The Vickers microhardness, HV0.05, was measured using a Struers Durascan microindentor with a load of 50 g.

The AA 1050 bar has a <100> texture typical of hot-extruded pure Al, an average grain size of 47 μm with a mixture of large and small grains, a dislocation density of ~ 1×10^{13} m^{-2} in the grains and intermetallic AlFeSi and AlFeSiMn particles with a size ranging from 200 nm to 2 μm.

3. Homogeneous deformation of solid cylinders

Homogeneous compression of solid cylinders to six different strains, 0.31, 0.70, 0.96, 1.00, 1.44 and 1.90, has been achieved. The homogeneity of deformation was confirmed with a good approximation by microhardness testing (HV0.05) on the cross-section, with the difference between the highest and lowest hardness of ΔHV = 3. The flow curve of the original AA 1050 cylinder was determined by compression testing and fitted by a Hollomon expression as $\sigma = 126\varepsilon^{0.23}$. To lay the foundation to evaluate the local flow stress from the microhardness and effectivestrain, the relationship between the microhardness and flow stress was fitted by the relationships:

$$HV = 7.4 - 16.31 \times \sigma + 15.56 \times \sigma^2 \qquad \varepsilon: 0 - 0.31 \tag{1a}$$

$$HV = 4.08 - 0.87 \times \sigma + 0.28 \times \sigma^2 \qquad \varepsilon: 0.31 - 0.96 \tag{1b}$$

$$HV = 3.5 \times \sigma \qquad \varepsilon > 0.96 \tag{1c}$$

where HV is the microhardness, ε is the compression strain and σ is the flow stress.

Figure 1 shows the microstructure evolution to a finer and finer scale with increasing strain. The grains are subdivided by low angle dislocation boundaries at small strains, supplemented with high angle boundaries at large strains. The dislocation cell boundaries rotate towards the compression plane and form a lamellar structure with increasing strain, as shown in figure 1d.

When focusing on the strength-structure relationship, two structural parameters can be identified as strengthening parameters: the spacing (D) between, and the misorientation angle (θ) across, the deformation-induced boundaries. The classification of boundary misorientation angles into low (θ < 15°) and high (θ ≥ 15°) angles defines the two main strengthening mechanisms: Taylor/dislocation strengthening [15], and Hall-Petch/boundary strengthening [16, 17]. The flow stress can then be estimated as the sum of these two contributions with the addition of the friction stress and a contribution from dispersed particles in AA 1050. With a modified Hall-Petch slope [18, 19], the structure-strength relationship can be written:

$$\sigma - \sigma_{00} - \sigma_{particle} = \left[k_{HP}\sqrt{f^{HAB}} + M\alpha G\sqrt{3b\theta^{LAB}(1 - f^{HAB})}\right] D_{av}{}^{-0.5} \tag{2}$$

$$\sigma - \sigma_{00} - \sigma_{particle} = k_2 D_{av}{}^{-0.5} \tag{3}$$

where f^{HAB} is the fraction of high angle boundaries, θ^{LAB} is the average misorientation of low angle boundaries, D_{av} is the average spacing between all boundaries measured along random intersection lines, σ_{00} is the friction stress for pure aluminum, and $\sigma_{particle}$ is the strength contribution from particles. The value of $\sigma_{00} + \sigma_{particle}$ is estimated to be ~ 32 MPa [14].

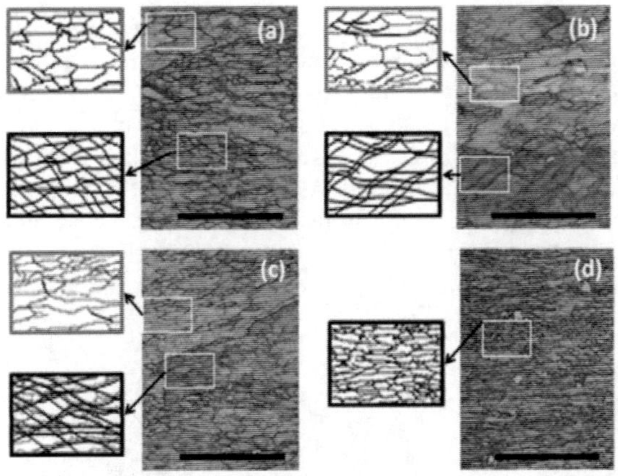

Figure 1. EBSD micrographs showing dislocation structures of the homogeneously compressed AA 1050 cylinders at strains of: (a) 0.7, (b) 0.96, (c) 1.44 and (d) 1.9. [14]

The misorientation across the dislocation boundaries can be obtained boundary by boundary in the EBSD data. The boundary misorientation angle distributions are shown in figure 2. Based on the microstructural parameters from figure 1 and combined with equation (3), the value of k_2 for homogeneously compressed AA 1050 is determined as $k_2 = 98 \pm 7.8$ MPa $\mu m^{0.5}$.

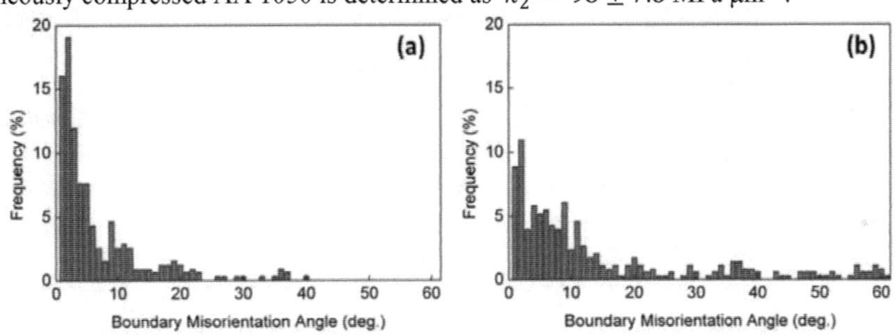

Figure 2. Frequency of boundary misorientation angles for homogeneously compressed cylinders at strains of 1.44 (a) and 1.90 (b)

4. Ring compression with heterogeneous deformation

In order to introduce heterogeneous deformation, rough tools with surface roughness $R_a = 0.48\ \mu m$ and poor lubrication with kerosene are chosen, with the friction factor, $f = 1$, close to the full sticking condition. The initial rings are made with the typical ratio among the outer diameter $D_0 = 18\ mm$, the inner diameter d_0, and the height h_0, $6:3:2$. Each ring was compressed to a final height reduction of 54.7% in 6 steps. Afterwards, the ring test experiments are analyzed by axisymmetric finite element modeling (FEM), with the underlying calculations built on the irreducible flow formulation by energy minimization. The tool is assumed to be perfectly rigid and the material is assumed to be rigid-plastic. For more details on the numerical implementation, see [20]. The friction model employed between the workpiece material and the tool surface is [21],

$$\tau = f\alpha k \tag{4}$$

where f is the friction factor to be calibrated, α is the ratio of real to apparent contact area, and k is the instantaneous shear flow stress obtained by the finite element simulation, taking strain hardening into account. The real contact area is determined by an analytical expression suitable for finite element implementation [22].

The effective strain distribution from FEM modelling is shown in figure 3, where seven representative areas with significant variations are highlighted. These heterogeneous microstructures illustrate pronounced characteristics: (i) equiaxed subgrains surrounded by low angle dislocation boundaries (Area 1); (ii) equiaxed dislocation cells in the original grains [23, 24] (Area 2); (iii) slightly deformed areas where the original grain shape is preserved in the sticky zone (Area 3); (iv) parallel lamellar dislocation structures with shear bands, representing the typical deformation microstructure at high strains (Area 4); (v) inclined dislocation cell structures with one set of dense dislocation walls (DDWs) mixed with equiaxed dislocation cells (Area 5); (vi) inclined dislocation cell structures with two sets of DDWs dividing the original grains (Area 6); and (vii) compressed original grains with elongated dislocation boundaries in the grains (Area 7). The average spacing between all boundaries, D_{av}, in each of these areas is shown in table 1. Note that Area 3 is almost undeformed, containing only the original grain boundaries, so the method based on deformation sub-grain structures is not applied here.

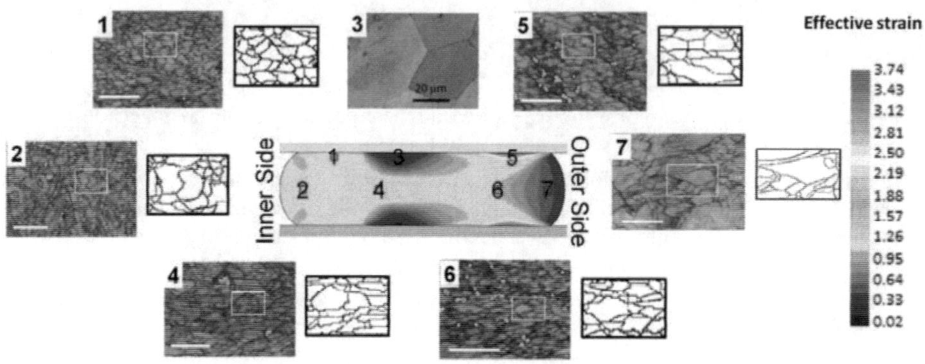

Figure 3. Effective strain distribution from FEM modelling of the ring with numbers 1-7 representing seven areas for the microstructural and hardness analyses. EBSD micrographs and sketches show heterogeneous dislocation structures in the seven areas [14, 25].

Table 1. Average spacing (D_{av}) in each representative area in the compressed AA 1050 ring [25]

Position	Area 1	Area 2	Area 4	Area 5	Area 6	Area 7
D_{av} (μm)	0.95	1.05	0.74	0.84	0.85	1.21

For the seven areas, the calculated flow stresses (σ_{cal}) using equation (2) in section 3, simulated strains, the calculated flow stresses ($\sigma_{hardness}$) from equation (1) and simulated flow stresses ($\sigma = 126\varepsilon^{0.23}$) are shown in table 2. The numbers are in good agreement taking into account an estimated average standard deviation of around 6 %.

Table 2. Simulated strain, calculated flow stresses (σ_{cal} and $\sigma_{hardness}$) and simulated flow stress (σ_{simul}) of the seven representative areas in the ring. [14, 25]

Area	**1**	**2**	**3**	**4**	**5**	**6**	**7**
ε_{simul}	1.8	1.06	0.03	1.84	2.55	1.13	0.41
σ_{cal} **(MPa)**	132.7± 8.0	127.1 ± 7.6	-	145.5 ± 9.1	138.7 ± 8.5	137.9± 8.5	120.8 ± 7.1
$\sigma_{hardness}$ **(MPa)**	145.9± 8.0	130.2 ± 8.4	46.3 ± 4.2	132.4± 8.4	139.7 ± 8.4	132.4± 8.4	98.4 ± 7.8
σ_{simul} **(MPa)**	144.2	127.7	54.9	145	156.3	129.6	102.6

5. Discussion

The evolution of microstructure and mechanical properties during plastic deformation in the homogeneous cylinders and heterogeneous ring has been investigated extensively in aluminum over several length scales from the sample scale (~ 10 mm) to the microscopic dimensions (~ several hundred nanometers), by a combination of numerical simulation, hardness testing, and electron microscopy in the present study. The microstructural evolution during plastic deformation can be characterized as a structural subdivision to a finer and finer scale with an increase of strain, where subdividing boundaries are low, medium and high angle boundaries forming different structural morphologies (see section 3 and 4). In this process the low energy dislocation structure (LEDS) principle is applied [26, 27], whereby due to a line energy reduction per unit length of dislocation line, dislocation boundary/cell formation is favored over a random distribution of dislocations. In parallel, long range stress fields are reduced, recently observed in copper by 3D X-ray diffraction [28].

The flow stress is expressed as a sum of dislocation strengthening and Hall-Petch strengthening, taking D_{av} as the sole strengthening parameter, and not taking into account that dislocation strengthening is isotropic and Hall-Petch strengthening may be anisotropic. Besides being empirical, this parameter can be derived based on physical mechanisms such as dislocation and boundary strengthening (see section 3). D_{av} is therefore considered an acceptable parameter for an analysis by EBSD of the local flow stresses in the compressed ring. This follows our hypothesis suggesting that small local regions in a heterogeneous microstructure can be analyzed in parallel with the analysis of a macroscopic homogeneous deformation structure. This experimental basis is therefore expanded by including structural data obtained by TEM of aluminum AA 1200 (99%) [29] homogeneously deformed by cold rolling to high strains. The microstructural analysis allows the local stress and strain to be estimated by equation (3) and figure 4a. These local stresses and strains are plotted in figure 4b, also including 7 points from Ring 2 pre-strained to an equivalent strain of 1 before compression, compared with Ring 1 in the present study. Figure 4b illustrates a high work hardening rate in stage III at low and medium strains followed by a low-hardening-rate in stage IV. A comparison of figure 4b and the Hollomon fitting curve for homogeneous compression testing of the solid cylinders shows the same work hardening exponent and a similar strength coefficient. This good agreement suggests that the local stress and strain in small regions of a heterogeneous structure can be analyzed as if the region had been homogeneously deformed. This hypothesis may also rationalize the assumption of isotropic hardening in the numerical simulation and is underpinned by the good agreement between calculated and simulated stresses.

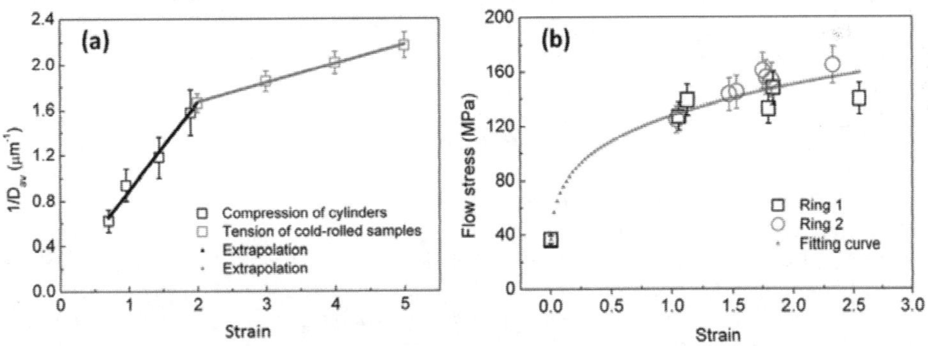

Figure 4. (a) $1/D_{av}$ versus strain including data from cold-rolled samples from [27] and (b) σ_{cal} versus ε_{cal} with fitting curve of $= 128\varepsilon^{0.23}$, including the same analysis of local stress and strain in Ring 2 pre-strained to the equivalent strain of 1 compared with the present Ring 1 [14].

6. Conclusion

An investigation comparing the local stress and strain in a heterogeneously-compressed AA 1050 ring under nearly full sticking conditions was performed by applying three complementary techniques covering multiple length scales from centimeters to nanometers, namely finite element modelling, electron backscatter diffraction, and microhardness. The same techniques were applied in an analysis of microstructure in homogeneously compressed AA 1050 cylinders to underpin the findings in the centimeter-scale heterogeneous ring. The good agreement between the fitted stress-strain curve from heterogeneous local areas in the ring and the Hollomon fitting curve obtained from homogeneous compression of solid cylinders suggests that the local stress and strain in small regions of a heterogeneous structure can be analyzed by considering that the region has been homogeneously deformed, and supports the hypothesis that the microstructure in local regions of a heterogeneous structure evolves in accordance with universal principles established for homogeneously deformed polycrystalline metals.

Acknowledgements
X Zhang acknowledge support from the European Research Council (ERC) under the European Union's Horizon 2020 research and innovation program (grant agreement No 788567-M4D).

References

[1] Hughes D A and Hansen N 2004. *Plastic Deformation Structures*. ASM Handbook, ASM International, Materials Park, Ohio, USA, pp 192
[2] Zhang X, Godfrey A, Huang X, Hansen N, Liu W and Liu Q 2009 *30th Risø International Symposium on Materials Science: Nanostructured metals: Fundamentals to applications (Risø, Denmark)* pp 409
[3] Fan G H, Wang Q W, Du Y, Geng L, Hu W, Zhang X and Huang Y D 2014 *Mater. Sci. Eng. A* **590** 318
[4] Zhang X, Hansen N, Godfrey A and Huang X 2018 *Mater. Sci. Techno.* **34** 794.
[5] Zhang X, Godfrey A, Huang X, Hansen N, Liu W and Liu Q 2014 *35th Risø International Symposium on Materials Science: nanostructured metals – new frontiers of nanometals* (Roskilde: Technical University of Denmark) p 153
[6] Zhang X, Godfrey A, Winther G, Hansen N and Huang X 2012 *Mater. Charac.* **70** 21
[7] Zhang X D, Liu W, Godfrey A and Liu Q 2009 *Metall. Mater. Trans. A* **40** 2171
[8] Yang R, Zhang X, Mallipeddi D, Angelou N, Toftegaard H L, Li Y, Ahlström J, Lorentzen L, Wu G and X Huang 2017 *IOP Conf. Ser.: Mater. Sci. Eng.* 219 012046
[9] Zhang X, Hansen N, Godfrey A and Huang X 2012 *33rd Risø international symposium on materials science: Nanometals – status and perspectives* (Risø, Denmark) p 407
[10] Chen S, Hu J, Zhang X, Dong H and Cao W 2015 *Journal of Iron and Steel Research International* **22** 1126
[11] Wang B, Liu B, Zhang X and Gu J 2018 *Surf. Coat. Technol.* **356** 89
[12] Nikas D, Zhang X and Ahlström J 2018 *Mater. Sci. Eng. A* **737** 341
[13] Zhang X D, Godfrey A, Liu W and Liu Q 2011 *Mater. Sci. Techno.* **27** 562
[14] Zhang X, Nielsen C V, Hansen N, Silva C M A and Martins P A F 2019 *Int. J. Plast.* **115** 93
[15] Taylor G L 1934 *Proc. R. Soc. Lond.*. **145** 362
[16] Hall E O 1951 *Proc. R. Soc. Lond. B.* **64** 747
[17] Petch N J 1953 *J. Inst Steel Inst.* **174** 25
[18] Hansen N 2004 *Scripta Mater.* **51** 801
[19] Zhang X, Hansen N, Gao Y and Huang X 2012 *Acta Mater.* **60** 5933
[20] Nielsen CV, Zhang W, Alves L M, Bay N and Martins P A F 2013 *Modeling of thermoelectro-mechanical manufacturing processes with applications in metal forming and resistance welding*(London: Springer-Verlag)
[21] Wanheim T and Bay N 1978 *Annals of the CIRP* **27** 189
[22] Bay N 1987 *Journal of Mechanical Working Technology* **14** 203
[23] Huang X and Hansen N 1997 *Scripta Mater.* **37** 1
[24] Le G M, Godfrey A, Hong C S, Huang X and Winther G 2012 *Scripta Mater.* **66** 359
[25] Zhang X, Hansen N and Nielsen C V 2017 *IOP Conf. Ser.: Mater. Sci. Eng.* **219** 012053
[26] Kuhlmann-Wilsdorf D 1989 *Mater. Sci. Eng.* **113** 1
[27] Kuhlmann-Wilsdorf D 1989 *ICSMA 8* **1** 221
[28] Levine L E, Larsen B C, Yang W, Kassner M E, Tischler J Z, Delos-reyes M A, Fields R J and Liu W 2006 *Nat. Mater.* **5** 619
[29] Liu Q, Huang X, Lloyd D J and Hansen N 2002 *Acta Mater.* **50** 3789

Decoding stress-strain diagrams to identify the evolution of crystal defects and their obstacle strengths using constitutive relations analyses

S Saimoto[1*], M R Langille[2]

1. Mechanical and Materials Engineering, Queen's University, Kingston, Canada
2. Univ. Grenoble Alpes, CNRS, Grenoble INP, SIMaP, F-38000 Grenoble, France

* E-mail: shigeo.saimoto@queensu.ca

Abstract. Recent advances in microstructural characterization from the mesoscopic and microscopic scales to the nanometric and atomic scale have revealed precise details regarding the formation of nanoparticles responsible for increased strength of alloys. The size and distribution of such features have been modelled ad hoc with limited success to replicate the work-hardening behaviour. The missing issue is that of including also the evolution of the crystal defects that arise as a result of plastic flow, such as the volume fraction of nano-voids and rotated lattice structures as observed by small angle X-ray scattering (SAXS). The constitutive relation analyses (CRA) approach assumes that all shape change is due to dislocation motion, and through use of the Taylor slip model, a functional relation has been formulated to replicate the measured curve in terms of shear stress and shear strain. The CRA parameters can be correlated to the evolving deformation products by quantitatively deriving a relation for the obstacle strength factor, α, using the strain dependence of the mean slip distance, λ. Hence, the evolution of different crystal defects as straining proceeds can be assessed.

1. Introduction

The strength and ductility of an alloy can be defined by the yield stress, which corresponds to a dislocation-line force sufficient to cause inelastic motion of dislocations in the crystal matrix comprised of solutes and dispersed particles, and also by the subsequent strain to attain the ultimate tensile strength (UTS), which is determined by the nature of work-hardening. The main focus of study over many decades has been to characterize the size, shape and distribution of dispersed particles and their capacity to resist shearing. This depends on whether the particles are coherent or incoherent with the matrix crystal, which in-turn is dependent on their size and shape. Modern advances in electron imaging, together with the ability to characterize solute elements by use of atomic probe tomography have enabled the characterization of particles from the microscopic scale down to the atomistic scale, as demonstrated in many reviews [1, 2, 3]. The current challenge is to correlate the statistical response of the defined microstructure to the mechanical properties. The approaches used to perform this task generally rely on application of the additive rule, whereby the various strengthening mechanisms are summed and superposed with a work-hardening relation for strain evolution [4, 5]. The issue with this approach is that the defect species that are generated during plastic flow may change, such that the evolution of weaker defects results in a lower work-hardening coefficient and hence a reduction in the strain to attain the Considère relation, i.e., a lower ductility. A different approach is to reverse this process, whereby a functional constitutive relation is developed to replicate the measured stress-strain (σ - ε) curve, and from the parameters characterizing the work-hardening process to identify the evolving obstacles responsible for hindering dislocation motion over the applicable strain-range. The constitutive relation analyses (CRA) approach, developed over the last decade, has been applied to decode the σ - ε diagram, and the present work extends the analyses to determine the obstacle strength factor, α, of the evolving defect structures as function of strain. Thus the analyses elucidate the defect types and their generation process.

2. Theory: Functional constitutive relation, mean slip distance and the derivation of α_D

The basis of the Saimoto-Van Houtte constitutive relation derivation is calculation of the micro-slip areas required to comply with the imposed strain using Taylor slip analyses [6]. Each slip patch that is bounded by dislocation length can be calculated using a mean slip distance λ together with its bounded area. Thus, the total energy to create the total dislocation lengths can be equated with the retained or stored dislocation lengths calculated using the Livingston relation [7]. For this the flow shear stress $\tau = \alpha \mu b \rho^{1/2}$, wherein α is the obstacle strength factor, μ is the shear modulus, b is the Burgers vector, and $\rho = 1/\ell^2$, where ℓ is the forest dislocation spacing. Previous statistical analysis has shown that the intersection density ρ (number per unit area) is related to the total line length, ρ_t, as $\rho_t = 2\rho$ (length per unit volume) [8]. Since the stored energy is typically less than 10% of the expended energy required for the specific shape change, an annihilation factor A was introduced to balance the equation. The resulting model relation correlating the shear stress τ to shear strain γ was in the power-law form earlier empirically determined by Hollomon [9]. However for this model at least two curves are required to replicate the measured data after conversion of tensile ($\sigma - \varepsilon$) data to that in terms of shear strain ($\tau - \gamma$). For this reason the intersection of the first curve, characterized by coefficient β_1, with that of β_2 is disjointed, with the transition strain interval at the intersection point of τ_3 and γ_3. An iterative fitting process can be used to find the two fitting curves with least error such that the transition zone is minimized. The following analyses will show this disruption is due to modelling limitations. The fact that this model captures the essential aspects of plastic flow was demonstrated and correlated to the kinetics of thermally activated plastic flow [10]. The final result was that normalization of the ($\tau - \gamma$) curve using τ_3 and γ_3 for identical specimens tested at the same strain rate at different temperatures resulted in the data collapsing onto a master curve independent of temperature.

The mean slip distance λ (equation (1)), which was earlier defined for slip-based work-hardening models by Nabarro et al. [8] can be precisely determined for each curve by using a calibration factor ϕ [11] derived from $\theta_{cal} = \partial\tau/\partial\gamma$ at $\varepsilon_{0.2\%}$, given as $1/\phi = (\mu/\theta_{cal})/ 2\alpha$. Many trials have shown that use of the engineering standard proof strain of 0.2 % is convenient. However if a yield point phenomenon exists, the fitted data points are determined from near the ultimate tensile stress (UTS) back to the yield point elongation (YPE) position, and this truncated data set is used for the fitting. Following this procedure, the β_1 curve is back-extrapolated to 0.2 % to determine the calibration term. Hence the newly defined yield stress σ_0^{final} correlates to the friction stress for plastic flow beyond the YPE position. Since the fitting is performed simultaneously for both curves, the defined value of σ_0^{final} is common to both fit curves.

$$\lambda = \frac{\phi b \mu^2}{2\, \tau \partial\tau/\partial\gamma|_{T,\dot{\gamma}}} \qquad (1)$$

The derived constitutive relation given as equation (2) with fit parameters, σ_0^{final}, P/A, β and C_1 from $\lambda = C_1\tau^\beta$. The A-factor accounts for the athermal dynamic annihilation of dislocations during plastic flow to result in reduced stored work. After converting the (σ – ε) polynomial fits to $\tau = (\sigma - \sigma_0^{final})/M$ and γ = M ε, in which M is the Taylor factor, two loci, β_1 and β_2, are created in combination with the back-extrapolated value of $\sigma_0^{final.}$. From the value of the term in brackets in equation (2), given in units of $MPa^{(2+\beta)}$, P/A is determined, wherein C_1 is derived from a plot of log λ versus log τ. Note that β values are derived from the fitting process and not from the log λ – log τ plot. In this relation, P is a ratio (y/x) derived by assuming the new segment lengths of expanding dislocations is given as yλ and the encircling swept area as $x\lambda^2$

$$\tau = \left[\left(\frac{P}{4A}\right)(2+\beta)(\alpha\mu b)^2\frac{1}{C_1 b}\right]^{\frac{1}{2+\beta}}\gamma^{\frac{1}{2+\beta}} \tag{2}$$

In order to derive a microstructure-based relation, equation (2) is modified by substituting $\lambda = C_1\,\tau^\beta$:

$$\tau^2 = \left[\left(\frac{P}{4A}\right)(2+\beta)(\alpha\mu b)^2\frac{1}{\lambda b}\right]\gamma \tag{3}$$

This λ-based relation as for equation (2) encompasses modelling parameters P/A and α^2. However the model only requires that the product (P/A) α^2 be kept constant. In cases where α is a well-established value it saves computing time to assume a value, which can be checked after completion of the constitutive relations analyses (CRA) described above. However the operative strength factor denoted as α_D correlated to the model relation can be derived, point by point, as follows. Differentiation of equation (3) results in equation (4), into which equation (1) is substituted. Simplifying results in equation (5).

$$2\tau\frac{\partial\tau}{\partial\gamma} = \left[\left(\frac{P}{4A}\right)(2+\beta)(\alpha\mu b)^2\frac{1}{b}\right]\frac{\partial\left(\frac{\gamma}{\lambda}\right)}{\partial\gamma} \tag{4}$$

$$\alpha_D = \left(\frac{\mu}{2\theta_{cal}}\right)\left[\left(\frac{P}{4A}\alpha^2\right)(2+\beta)\right]\left(1-\frac{\partial\ln\lambda}{\partial\ln\gamma}\right)_{\text{any }\gamma} \tag{5}$$

The limitation to the CRA modelling method becomes apparent as follows. The varying term in equation (5) is given as $\left(1-\frac{\partial\ln\lambda}{\partial\ln\gamma}\right)$, which can be derived by differentiation of the natural logarithm of equation (3) with respect to ∂ln γ, whereby $\left(1-\frac{\partial\ln\lambda}{\partial\ln\gamma}\right) = 2\left(\frac{\partial\ln\tau}{\partial\ln\gamma}\right) = \left(\frac{2}{2+\beta}\right)$. The factor $\left(\frac{2}{2+\beta}\right)$ results from differentiating equation (2). Hence CRA modelling results in α_D being constant with strain, which is consistent with the modelling procedure where α was kept constant at 0.4 for aluminum (Al). The cause of this discrepancy is that the relation $\lambda = C_1\,\tau^\beta$ does not exactly replicate log λ versus log τ plot, which shows that the fitted curve β only approximates the measured values and smooths out undulations. For mesoscale modelling this practice is acceptable but for detailed examination of the microstructure, $\left(\frac{\partial\ln\lambda}{\partial\ln\gamma}\right)$ is required at any given γ and is derived from ln λ versus ln γ plot, resulting in α_D versus γ. Moreover a new criterion for initiation of incipient diffuse necking is predicted, namely if $\left(1-\frac{\partial\ln\lambda}{\partial\ln\gamma}\right)$ becomes zero, the effective α becomes zero, implying that laminar slip initiates strain localization. The following data analyses show that this destabilizing process is incurred as the Considere condition is approached.

The first term in equation 5 ($\mu/2\,\theta_{cal}$) is derived from CRA as $\alpha/\phi = (\mu/\theta_{cal})/2$, where α is the prior established value. Whether this choice is consistent with the model can be checked after completion of CRA process as described above. This is a short-cut process since permitting α to become a free parameter during the fitting procedure entails much longer computing time but the resulting {(P/A) α^2} remains the same. On the other hand, λ is determined point by point at any given γ, but ∂λ/λ is independent of α due to cancellation. Thus, α_D is dependent on the change in (τ ∂τ/∂γ) factor from the τ – γ curve and the calibration slope, which may lead to ambiguity if a yield phenomenon exists. However the slope at back-extrapolated $\varepsilon_{0.2\%}$ is based on the slip model and not empirical, and hence should encompass the high strain deformation mode. Hence CRA determines the value of σ_0^{final} which separates the yield phenomenon from the friction stress applicable beyond the YPE.

IOP Conf. Series: Materials Science and Engineering **580** (2019) 012043 doi:10.1088/1757-899X/580/1/012043

3. Experimental

The prime data acquisition method used for the CRA is standard ASTM tensile testing using digitalized recording of load and extension performed on any universal testing system. In the current case, testing to failure was performed at 5 x 10^{-4} s^{-1} using a screw-driven Instron 5969 machine equipped with a 50 kN load cell. The strain measurements were recorded using a clip-on 50.8 mm 2630-115 extensometer as previously described [12]. All mechanical tests were performed at 22 °C. The specimen was prepared as previously reported from extrusion channels of AA6063 in the shape of 3-cell trapezoid with overall dimensions of 78 mm x 143 mm x 66 mm with gauge size of 50 mm x 12 mm x 2mm. The machined surfaces were finished using 600-grit SiC paper and heat treated at 185 °C according to industrial practice for 5 h (T6) to attain optimum strength after water quenching (WQ) from 540 °C. Table 1 lists the composition of the extrusions as specified by the supplier. A detailed description of the data acquisition for small angle scattering (SAXS), performed at the Cornell High Energy Synchrotron Source G1 beamline, and the transmission electron microscopy (TEM) studies at McMaster University, are reported elsewhere [12, 13].

Table 1. As-received AA6063 composition in weight %

Al	Si	Mg	Cu	Fe	Mn	Ti
Bal.	0.4	0.49	0.01	0.16	0.029	0.01

4. Result and discussion

4.1 Decoding of optimum aged AA6063 extrusion specimens (5 h at 185°C)

The CRA procedure separates the yield phenomenon from the work-hardening beyond YPE such that σ_0^{final} defines the friction stress primarily due to the dispersed aged products that underwent shearing during yielding and subsequent straining as shown in figure 1 [12].

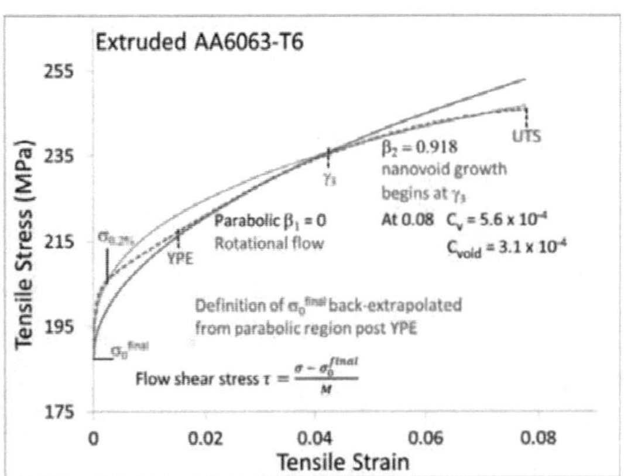

Figure 1. Stress-strain diagram of extruded and optimum aged AA6063-T6 determined at 296 K and strain rate of 5 x 10^{-4} s^{-1} from [12]. The original data (dashed curve) is optimum fitted using CRA to form the β_1 and β_2 loci. The back-extrapolated yield stress common to both modelled loci is σ_0^{final}.

As modelled by Foreman and Makin [14], shearing of needle-shaped nano-precipitates which align with the cube axis of the matrix, follows a percolation process whereby subsequent hardening is due to the increase in forest dislocations due to multiple slip such that more or less homogeneous microstructure results at YPE. However if other larger unshearable micro-meter (μm) sized particles exist, at a specific shear stress due to stress concentration in the adjacent region, new dislocation arrays arise to maintain strain compatibility by forming rotated structures, as modelled by Brown [15]. This model predicts a parabolic behaviour that can be observed using CRA, and from the fit parameters in table 2 the size of the particles can be assessed [12]. The formation of a rotated lattice structure changes after some strain due to changes in local Schmid factor. Thus a counter rotation can occur resulting in a region of duplex slip which generates various deformation debris, such as dislocation loops and vacancies, resulting in the change in work-hardening coefficient that is represented by the third deformation stage in figure 1. The formation of rotated structures has been shown to occur using SAXS and the nucleation of nano-voids by vacancies has also been observed. The measured volume fraction of nano-voids C_{void} is comparable to that of the calculated volume fraction of vacancies C_V using the CRA parameters [13]. To extend this decoding process, one way would be to identify the obstacle strength factor in each of the three deformation stages as presented in the next section.

Table 2: Input and fit parameters from CRA of AA6063 aged at 185 °C for 5 h

Treatment	α*	μ (MPa)	P/A	σ_o^{final} (MPa)	$\sigma_{0.2\%}$ (MPa)	β_1	$C_{I\ \beta 1}$ (μm)	β_2	$C_{I\ \beta 2}$ (nm)	τ_3 (MPa)	γ_3
AQ	0.4	32012	0.069	30.4	41.7	-0.27	0.95	1.10	21.3	24.8	0.175
T6 As-Rec	0.4	32012	0.084	187.4	205.1	0.00	0.97	0.91	110.1	16.0	0.126
T6 Control	0.4	32012	0.072	195.6	211.1	0.00	1.23	0.49	443.2	12.3	0.109

* α for forest dislocations was used for initial modelling.

4.2 Identification of microstructure components using the derived alpha factor, α_D

In order to determine α_D according to equation (5), a plot of ln λ versus ln γ is shown in figure 2, from which α_D versus γ in figure 3 can be derived. The α_D-profile most similar to that of pure Al at 78 K [16] is that for WQ specimen in figure 3, which shows that α_D decreases from near 0.4 to below 0.2 near γ_3, where an abrupt change occurs due to the switch in the fit-locus as described earlier. Thus this transition strain interval should be ignored. Beyond γ_3, the profile dips to near zero and increases again to near 0.4. It must be kept in mind that for a two obstacle model $\tau = \tau_{For} + \tau_{Deb}$, wherein the debris contribution is only about 1/9 of the forest one. However since α_D is primarily a measure of the most rapidly increasing defect density, the low value for vacancies and nano-defects are reasonable. It can be shown that this strain corresponds to the formation of a cell structure, at which point λ becomes constant as indicated in figure 2. This observation indicates that the λ parameter as function of stress or strain is more sensitive to microstructure evolution than the model prediction of a constant α_D. As the cell structure takes form, dislocation density in the walls start to increase and α_D attains a value of 0.4 again prior to the start of diffuse necking, and instability due to slip band formation starts to occur.

For the optimum aged specimen for which $\beta_1 = 0$, α should be 0.4 as the model demands. However during initial yielding, CRA, as indicated in figure 1, separates the yield phenomenon from the higher strain plastic flow. Consideration of the back-extrapolated value of α_D to $\gamma = 0.006$, the proof strain, suggests that the nano-precipitates are stronger and requires larger α as predicted [15]. The higher efficiency of the quenching procedure in the laboratory compared to that in industry is represented by α_{yield} for the T6 control being larger that of the T6 As-Rec sample. At γ_3 the plastic response changes to that of deformation around micro-scale particles. At this point the aforementioned duplex slip occurs with generation of nano-defects. Thus the effective alpha decreases. Due to space restriction, extensive discussion is not possible but beyond γ_3, the derived α should apply to the secondary much weaker defects. Since its density is responsible for only about 1/10 of the flow stress, the main load carrying structure is the forest density, which is approaching saturation and does not detectably affect the operative strength factor due to the high formation rate of nano-defects. On the other hand, the model encompasses the total flow stress and hence the predicted α_D is constant.

5. Conclusion

The present analyses of optimum aged AA6063, analogous to that of pure Al at 78 K, has demonstrated that nano- and micro-particles embedded in the matrix gives rise to more complex α_D profiles. Defect identification is difficult since the portion due to forest dislocation density needs to be accounted for as approximately 90 % of the flow stress. Applying the derived value of α_D to the remaining 10 % of flow stress should allow prediction of the undulating densities of nano-defects as function of strain, which implies their dynamic generation and annihilation.

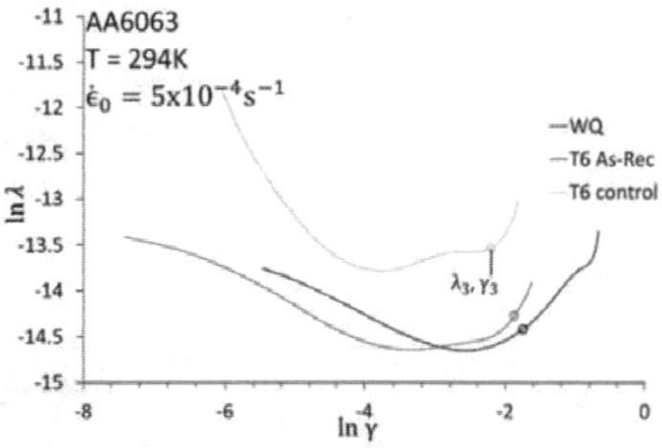

Figure 2. Natural logarithm of λ versus ln γ for three heat treatments of AA6063.

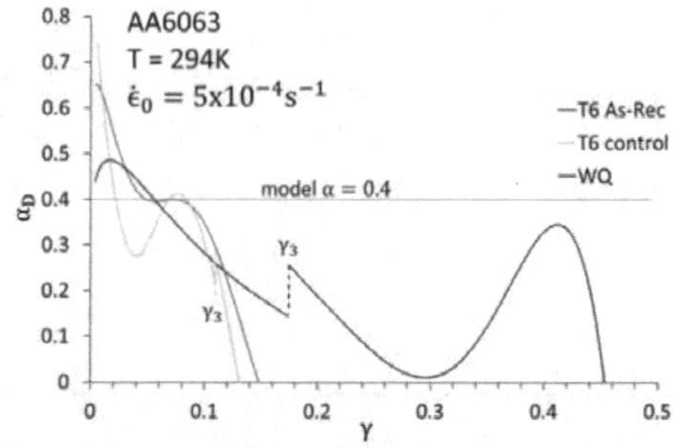

Figure 3. Derived obstacle strength, α_D, versus γ using slopes in figure 2 and equation (5) showing the deviation from the model value of 0.4. Note that the α_D magnitude at γ = 0.006 (proof strain) for the T6 control specimen is about 0.75 compared to 0.63 for the T6 As-Rec specimen.

Acknowledgements
The authors thank the Natural Sciences and Engineering Research Council of Canada for support and to Queen's University for the use of facilities. We are indebted to Professor B.J. Diak for consultation.

References

[1] Edwards G A, Stiller K and Dunlop G L and Cooper M J 1998 *Acta Mater.* **46** 3893-04.
[2] De Geuser F, Dorin T, Lefebre W, Gault B and Deschamps A 2014 *Mater. Sci. Forum* **794-796** 926-32.
[3] Buchanan K, Colas K, Ribis J, Lopez A and Garnier J 2017 *Acta Mater.* **132** 209-27.
[4] Fisk M, Ion J C and Lindgren L-E 2014 *Comp. Mater. Sci.* **82** 531-9.
[5] Myhr O R, Grong O and Schafer C 2015 *Metall. Mater. Trans. A* **46A** 6018-39.
[6] Saimoto S and Van Houtte P 2011 *Acta Mater.* **59** 602-12.
[7] Livingston J D 1962 *Acta Metall.* **10** 229-239.
[8] Nabarro F R N, Basinski Z S and Holt D B 1964 *Adv. Phys.* **13** 193-323.
[9] Hollomon J H 1945 *Trans. AIME* **162** 268-290.
[10] Saimoto S 2019 *Acta Mater.* **174** 43-52.
[11] Saimoto S 2006 *Philos. Mag.* **86** 4213-4233.
[12] Saimoto S, Singh M A, Langille M R, Levesque J, Inal K, Niewczas M and Woll A R 2018 *Mater. Sci. Eng. A* **709** 9-16.
[13] Singh M A, Saimoto S, Langille M R, Levesque J, Inal K and Woll A R 2017 *Phil. Mag.* **97** 2496-513.
[14] Brown L M, 1997 *Phil Trans. R. Soc. Lond. A* **355** 1979-90.
[15] Foreman A J E and Makin M J 1967 *Can. J. Phys.* **45** 511-7; 1966 *Phil. Mag.* **14** 911-24.
[16] Saimoto S, Langille M R and Niewczas M 2018 *Mater. Sci. Forum* **941** 2270-7.

IOP Conf. Series: Materials Science and Engineering **580** (2019) 012044 doi:10.1088/1757-899X/580/1/012044

Recovery kinetics in Al-0.3%Cu after heavy rolling

L F Shuai[1,4], T L Huang[1,2*], G L Wu[1,2], N Hansen[1,3], X Huang[1,4], T Yu[4]
[1] Joint International Laboratory for Light Alloys (MOE), College of Materials Science and Engineering, Chongqing University, Chongqing, 400045, China
[2] Electron Microscopy Center of Chongqing University, Chongqing University, Chongqing 400045, China
[3] Risø Campus, Technical University of Denmark, DK-4000 Roskilde, Denmark
[4] Department of Mechanical Engineering, Technical University of Denmark, DK-2800 Kgs. Lyngby, Denmark

E-mail: huangtl@cqu.edu.cn

Abstract. The recovery kinetics of a nanoscale lamellar structure Al-0.3%Cu alloy, which was produced by 98% cold rolling, during annealing at temperatures from 100 to 200 °C have been analyzed in this study. Microstructural characterization using electron channelling contrast in the scanning electron microscope shows a uniform coarsening of the lamellar structure and gradual precipitation of second phase particles (Al_2Cu) during annealing at these temperatures. Fitting of the microhardness values for samples annealed for between 4 and 4096 minutes at different temperatures with a recovery model reveals a change, in the apparent activation energy from 79 kJ/mol at the beginning to 105 kJ/mol at the end of recovery annealing, which leads to a discussion of the recovery mechanisms.

1. Introduction

It is well-known that the nanostructured metals produced by high strain deformation have high strength and low ductility. There is a possibility to improve the balance between strength and ductility by recovery annealing [1]. However, such nanostructured metals have high stored energies, which serve as the driving force for recovery and recrystallization and reduce their thermal stability [2,3]. Moreover, in order to avoid losing much strength, the annealing conditions have to be restricted to thermal recovery to avoid occurrence of recrystallization. Therefore, studies of recovery kinetics in nanostructured metals are of great importance.

Recently, the recovery kinetics of commercial purity aluminum deformed to true strain of 4 and 5.5 have been analyzed [4,5]. Based on hardness measurements supplemented by microstructural characterization, a unified model was developed for recovery annealing. The analysis shows that the apparent activation energy increases as recovery proceeds towards recrystallization. The presence of a small amount of impurities and precipitates may have a strong effect on the recovery kinetics. In this study, a nanostructured Al-0.3%Cu alloy with a high purity matrix Al was designed to investigate the effect of addition of a small amount of Cu on the recovery kinetics.

2. Experimental

The Al-0.3%Cu used in this study was produced by adding 0.3%wt oxygen-free high conductivity copper into a high purity (99.999%) Al. An as-cast ingot with 1.5–2 mm large grains was forged at 200

°C to obtain a 50 mm thick plate and then cold rolled by multiple passes to a final thickness of 1 mm, corresponding to a true strain of 4.5 (98% reduction in thickness).

A number of smaller samples cut from the 1 mm thick plate were isothermal annealed at temperatures of either 100, 125, 150 or 175 °C over time intervals ranging from 4 min to 4096 min in a tube furnace. The heat treatments allowed microstructural coarsening to be studied from its initiation to the beginning of recrystallization. To avoid the effect of inhomogeneity that may be present through the sample thickness on the recovery kinetics study, all examinations were conducted at layer corresponding to the center of the cold rolled sheet.

Vickers microhardness tests were carried out on the polished surface of all samples with a load of 200 g and a holding time of 10 seconds. The hardness value at each condition was taken as the average of at least ten measurements.

The microstructures of the deformed and annealed samples were characterized by electron channeling contrast (ECC) in a JEOL 7800F scanning electron microscope (SEM) operated at 20 kV. Specimens from the longitudinal section containing the rolling direction (RD) and the normal direction (ND) were prepared using mechanical polishing followed by electropolishing.

3. Results and Discussion

3.1. Structural coarsening and precipitation

Lamellar boundaries with an average boundary spacing of 200 nm constitute the deformed microstructure, the vast majority of which are free of particles, although a few small particles were observed lying at the lamellar boundaries (figure 1), as reported in previous studies [6,7].

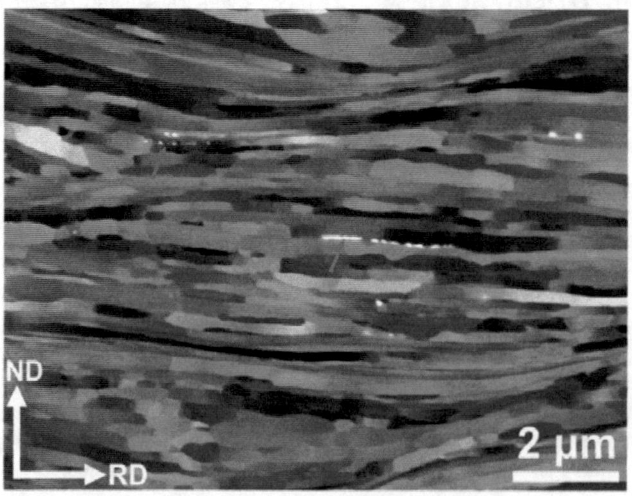

Figure 1. SEM-BSE image showing the lamellar structure of 98% cold rolled Al-0.3%Cu where two examples of particles are pointed by red arrows.

The lamellar structure coarsened uniformly without a significant change in the lamellar morphology after annealing under the conditions used in this study. Such a coarsening pattern is consistent with the triple junction motion mechanism, which was suggested to be the key mechanism for uniform coarsening of lamellar structures [4,8]. It was found that, the boundary spacing along the ND is almost unchanged and the lamellar boundaries are as extended as those in the deformed structure after annealing at 100 °C for 1 hour (figure 2a). Their thermal stability is attributed to Cu segregation to structural defects as dislocations, dislocation boundaries and high angle boundaries [6,9]. Figure 2c shows the microstructure after annealing at 150 °C for 1 hour where the lamellar boundary spacing increased

IOP Conf. Series: Materials Science and Engineering **580** (2019) 012044 doi:10.1088/1757-899X/580/1/012044

obviously and locally some equiaxed structures can be seen. As the annealing temperature is increased to 175 °C, the recovered structure is more equiaxed as a whole, and a few recrystallized grains also appear in the recovered lamellar structure, as shown in figure 2d. From larger EBSD maps of the sample annealed at 175 °C for 1 hour, it was found that the distribution of the recrystallized grains in the structure is heterogeneous and their volume fraction is less than 15%.

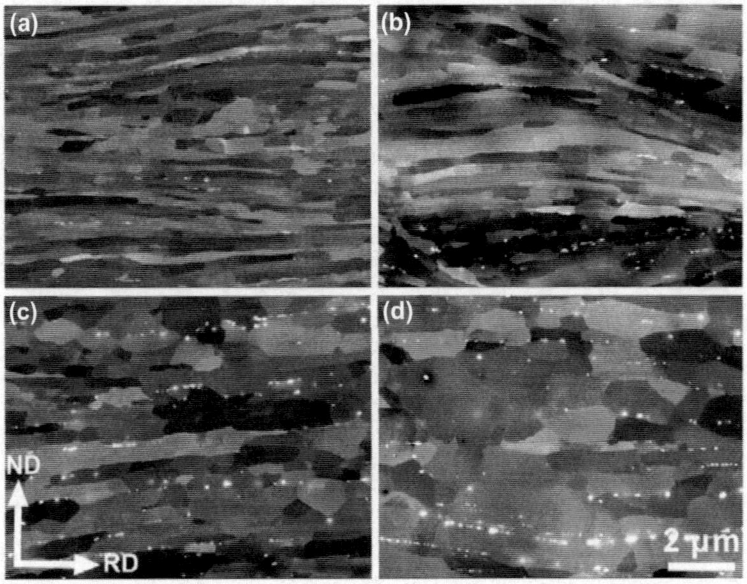

Figure 2. SEM-BSE image showing lamellar structures in samples after annealing at (a) 100 °C, (b) 125 °C, (c) 150 °C, (d) 175 °C for 1 h. One recrystallized grain is pointed by a red arrow.

Precipitation of Al_2Cu precipitates also occurred during the annealing treatments. As shown in figure 2, the number and size of precipitates in the lamellar structure both increase with increasing annealing temperature. Most of them are arranged along the lamellar boundaries, while others lie inside the recovered grains.

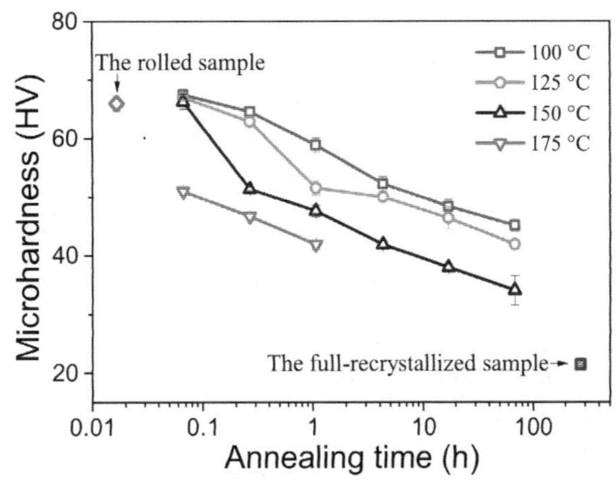

Figure 3. Microhardness variation of 98% cold rolled Al-0.3%Cu after isothermal annealing. Error bars show the standard deviations, and some are smaller than the symbols used.

IOP Conf. Series: Materials Science and Engineering **580** (2019) 012044 doi:10.1088/1757-899X/580/1/012044

3.2. Microhardness

The plot in figure 3 presents the microhardness variation with annealing time. The red and blue data points represent the hardness values of the deformed sample and the fully recrystallized sample, respectively. Compared to the deformed sample, there is a slight increase in the microhardness during the early stage of annealing at 100 °C and 125 °C. The hardness of the sample annealed at 175 °C for 4 min is clearly below the value of the deformed sample because of pronounced coarsening of the lamellar structure. These curves maintain a slow descending trend with increasing annealing time at different temperatures. This is due to the fact that the softening resulting from coarsening of the lamellar structures overshadows the effects of precipitation hardening.

3.3. Activation energy and recovery mechanisms

In order to analyze the recovery kinetics of the Al-0.3%Cu after heavy rolling, the following model developed by Vandermeer and Hansen was applied [10]:

$$\frac{df^2}{dt} = -f^2 K_0 \exp\left(-\frac{Q_0 - \beta P_0 f^2}{RT}\right) \tag{1}$$

where f^2 is assumed to be proportional to the stored energy of the recovered sample, which can be expressed as:

$$f^2 = \left(\frac{H - H_r}{H_d - H_r}\right)^2 \tag{2}$$

where H_d , H_r and H correspond to the hardness of the deformed sample, of the completely recrystallized sample and of the recovered sample, respectively. P_0 is the deformation stored energy before annealing, t is the annealing time, R is the universal gas constant, T is the absolute temperature, Q_0 is the activation energy at the end of recovery, and K_0, Q_0 and β are parameters associated with the operative recovery mechanisms.

The integration of the differential equation, equation (1), can be solved with the aid of exponential integrals, resulting in the following relationship:

$$E_1\left\{\frac{\beta P_0 f^2}{RT}\right\} = E_1\left\{\frac{\beta P_0}{RT}\right\} + t t_0^{-1} \tag{3}$$

where the $E_1\{\}$ are exponential integrals of the quantities inside the brackets; t_0^{-1} is related to annealing temperature and can be expressed as

$$t_0^{-1} = K_0 \exp\left(-\frac{Q_0}{RT}\right) \tag{4}$$

where Q_0 and the apparent activation energy Q_{app} are related according to:

$$Q_{app} = Q_0 - \beta P_0 f^2 \tag{5}$$

In this study, the microstructural characterization indicates that all the samples are in recovered state, except that for a few recrystallized grains in the recovered structure after annealing at 175 °C for 1 hour. Moreover, precipitation appears to have a small effect on the hardness. Therefore, all the hardness data are used in the recovery analysis.

In order to determine the apparent activation energy, it is reasonable to assume that the same recovery mechanisms operate at different annealing temperatures at a specified recovery stage (i.e. the same hardness). In other words, the fitting parameters K_0, Q_0 and β (and thus $\beta P_0/R$) are temperature-independent. For each annealing temperature, a curve of $t_0^{-1} vs \beta P_0/R$ is calculated for each (t, f^2) pair after the values of (t, f^2) pairs from experimental are inserted into equation (3). Then a maximum convergence point ($\beta P_0/R, t_0^{-1}$) is determined manually by superimposing curves of all annealing times.

In this way different values of $\beta P_0/R$ are calculated for different temperatures in the model. Therefore, it is necessary to implement a second fitting by taking the average of $\beta P_0/R$ from the first fitting for all temperatures [5]. Such a procedure enables accurate estimation of the apparent activation energy and significantly improves the model applicability.

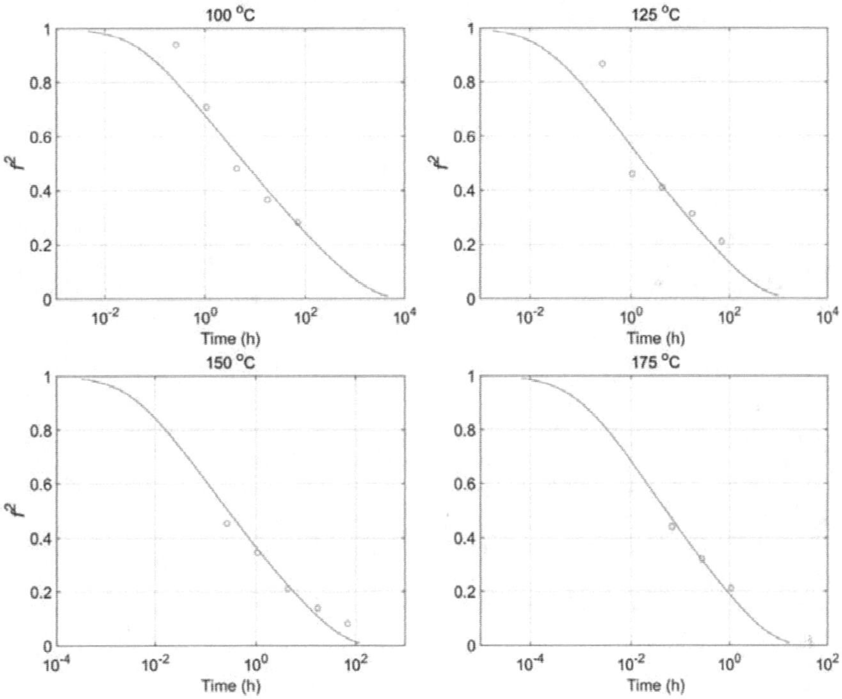

Figure 4. Isothermal kinetics of recovery in 98% cold rolled Al-0.3%Cu.

The recovery kinetics curves fitted collectively with the average value of $\beta P_0/R$ for different temperatures are drawn in figure 4. The average value of $\beta P_0/R$ is 2840 K. It is found that there are good fits between the experimental data and the recovery model. According to equation (4), there is an Arrhenius-type relationship between the model parameter t_0^{-1} and the annealing temperature. This relationship is plotted in figure 5 by using the values of t_0^{-1} obtained from the second fittings. From the slope of linear fitting in figure 5, the apparent activation energy at the end of recovery (i.e. Q_0) can be estimated as 105 kJ/mol, and the activation energy at the beginning of recovery (Q_{app} when $f=1$) can be also calculated as 79 kJ/mol by taking the average value of $\beta P_0/R$. The recovery rate constant K_0 can be estimated as 2×10^{11} h^{-1} from the intercept of the linear fitting.

It is found that the apparent activation energy increases from 79 kJ/mol to 105 kJ/mol as the recovery process proceeds. The lower activation energy at the beginning of recovery may be related to the high stored energy from the contribution of the high density of structural defects introduced by heavy deformation. The annihilation of those defects such as vacancies and dislocations leads to the increase in the apparent activation energy as recovery proceeds. Structural coarsening depends strongly on the solute element added in the material. However, the apparent activation energy at the end of recovery (105 kJ/mol) is lower than the activation energy (130 kJ/mol) for diffusion of solute atom Cu in Al lattice. The decrease may be attributed to Al_2Cu precipitation along boundaries accompanying structural coarsening by, e.g. triple junction motion, which reduces Cu segregation at boundaries and reduces solute drag for boundary migration.

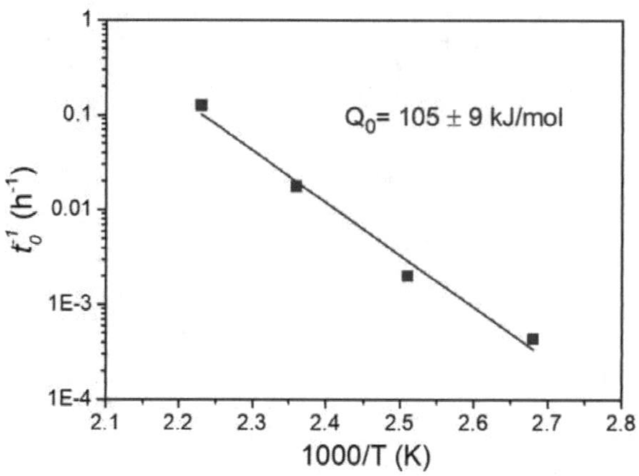

Figure 5. Temperature dependence of recovery kinetics in 98% cold rolled Al-0.3%Cu.

4. Conclusions

Recovery kinetics in Al-0.3%Cu after 98% cold rolling have been analyzed using a recovery model based on microhardness measurements [5,10] and SEM-ECC characterization. It is found that:

1. Deformed lamellar structures undergo a uniform coarsening during recovery annealing and this coarsening pattern is consistent with operation of triple junction motion.
2. The hardness variation is dominated by the boundary structural coarsening while Al_2Cu precipitation only has a minor effect, and therefore the recovery model based on hardness measurement is applicable to the current material.
3. The apparent activation energy can be estimated from the model and is found to increase from 79 kJ/mol at the beginning of recovery to 105 kJ/mol at the end of recovery.

Acknowledgments

The authors wish to thank the financial support of the State Key Research and Development Program of MOST of China (2016YFB0700400), Natural Science Foundation of China (NSFC, Grants No. 51421001). NH thanks the support of the 111 Project (B16007) by the Ministry of Education and the State Administration of Foreign Experts Affairs, China. TY thanks the support of the European Research Council (ERC) under the European Union's Horizon 2020 research and innovation programme (ERC Advanced grant – M4D/grant agreement number 788567).

References

[1] Huang X 2009 *Scripta Mater.* **60** 1078-82
[2] Hansen N, Huang X, Møller M G and Godfrey A 2008 *J Mater. Sci.* **43** 6254-9
[3] Mishin O V and Godfrey A 2008 *Metall. Mater. Trans. A* **39** 2923
[4] Yu T B, Hansen N and Huang X 2013 *Acta Mater.* **61** 6577-86
[5] Yu T B and Hansen N 2016 *Metall. Mater. Trans. A* **47** 4189-96
[6] Huang T L, Shuai L F, Wakeel A, Wu G L, Hansen N and Huang X 2018 *Acta Mater.* **156** 369
[7] Shuai L F, Huang T L, Wu G L, Huang X and Mishin O V 2018 *J. Alloys Compd.* **749** 399-405
[8] Yu T B, Hughes D A, Hansen N and Huang X 2015 *Acta Mater.* **86** 269-78
[9] Shuai L F, Huang T L, Wu G L, Hansen N and Huang X 2017 *Risoe International Symposium on Materials Science* **219**
[10] Vandermeer R A and Hansen N 2008 *Acta Mater.* **56** 5719-27

IOP Conf. Series: Materials Science and Engineering **580** (2019) 012045 doi:10.1088/1757-899X/580/1/012045

Heterogeneous microstructure of low-carbon lath martensite with continuous yielding behavior in Fe-C-Mn alloys

M Sugiyama[1], G Sawa[1,2], K Hata[2], N Maruyama[2]

[1] Department of Materials Science, Osaka University, Nippon Steel Research Alliance Labs., 2-1 Yamadaoka, Suita 565-0871, JAPAN
[2] Advanced Technology Research laboratories, Nippon Steel Corporation, 20-1 Shintomi, Futtsu 293-8511, JAPAN
e-mail: sugiyama@mat.eng.osaka-u.ac.jp

Abstract. High strength steels have been rapidly developed for several kinds of application fields such as light weight automobile bodies, multistory buildings, and oil well tubulars. With an increase of strength, the use of the low-carbon lath martensite phase puts this component as the main microstructure, or a part of the dual phase one. However, the martensite is a non-equilibrium phase, and the microstructure is also a heterogeneous composite-like one including of fine carbides and retained austenite. In order to understand the hierarchical structure, multi-scale microscopic techniques have been utilized. Internal twins in a lath and some large lath domains with fine carbides are revealed by scanning electron microscope (SEM) and transmission electron microscope (TEM) observations, although a conventional low-carbon martensite is known to include mainly dislocations in the lath structure. The heterogeneous microstructure is discussed based on a model of the local difference of transformation temperature of each lath block segment during cooling. On the plasticity of low-carbon lath martensite, the continuous yielding behavior is one characteristic property, and the elastic limit is obtained by the cyclic tensile test. The heterogeneous deformation behavior is detected as a periodic strain partitioning among lath blocks in a grain by digital imaging correlation analysis using *in situ* SEM tensile testing.

1. Introduction

The majority of steels for automobile bodies and undercarriage components have been replaced by high strength steels with more than twice the strength as previous ones [1]. With further increasing demand of high strength steels in the market, the practical use of martensite phase is expected in material design of high strength microstructure. Although the martensite phase is already utilized in a dual phase and transformation-induced plasticity steels, and others, most products including the martensite phase are used after a tempering treatment, and it is still difficult to control the mechanical properties of fresh martensite quenched in low carbon steels.

From the viewpoint of industrial application, the low carbon lath martensite is most attractive microstructure with high strength and good weldability, however, the ductility is not sufficient. Fundamental knowledge of low carbon lath martensite is required to improve mechanical properties such as the yielding characteristics and elongation. It is well known that the morphology and internal structure of martensite are changed with carbon content in Fe-C alloys [2-3], and the crystal orientation relationship between the matrix and lath are made clear under several transformation conditions [4-6]. When the structure of the low carbon steels is considered, the microstructure may be categorized by three different strategies. The first factor is the mixing of the different phases, composed of the martensite, retained austenite, proeutectoid ferrite, and pearlite, with dependence of cooling velocity from austenite phase, because the Ms temperature of lath martensite is relatively high, e.g. around 673K. The second factor is the composite structure of the lath microstructure and carbides including carbon clusters, because the Mf temperature is higher than room temperature, and carbon atoms can move even at the room temperature. The third factor is the hierarchical microstructure of lath martensite itself, such as packets, blocks, sub-blocks, and the 'laths'.

Recently, the morphology and crystallography of low carbon lath martensite have been studied using 2D and 3D EBSD analysis, and it is made clear using variant analysis that some of characteristic variants are formed in a same block and packet, based on the K-S orientation relationship [7-10]. It is further revealed that the sub-block boundaries are formed in a block, in addition to the conventional block regions. On the strengthening mechanism, both block and sub-block boundaries act as a barrier for dislocation motion as seen in uniaxial micro-tensile tests [11]. It is also reported in Fe-0.13C-5.1Ni(wt%) that the lath martensite microstructures are composed of coarse laths of various sizes embedded in a matrix of conventional thin laths. Judging from the morphology, dislocation density and the growth area connected to the prior austenite grain boundary, it has been proposed that the coarse lath domain is firstly transformed at around the Ms temperature from the soft austenite matrix, which is reasonable regarding the intrinsic martensitic transformation sequence on the heterogeneity [12].

The main purpose of the present study is to make clear the real lath microstructure in Fe-C-Mn alloys, especially in the low carbon content region, based on the microstructure reported using lath martensite having a low Ms temperature below room temperature in Fe-Ni-C alloys. The defect structure of the lath martensite has been investigated by TEM observations. The heterogeneous microstructure of lath martensite is further discussed with region showing the gradual yielding behavior.

2. Experimental procedure

An ingot of 50kg in weight was melted in a laboratory vacuum furnace. After hot forming, a homogenizing heat treatment was carried out at 1373K for 1hrs, and the slab roughly hot-rolled from 30 mm to 3.4 mm in thickness. After removing the surface decarbonized layer, the plate was cold rolled to 1.2 mm in thickness. Two kinds of model plates with compositions Fe-0.07C-1.0Mn(wt%) and Fe-1.0C-2.0Mn(wt%) alloys were prepared by almost similar procedures. The small amount of additional Mn has an effect to control the ferrite nose in the TTT diagram. Each sample plate are given different heat treatments and cooling conditions to obtain the lath martensite microstructure.

Two kinds of different sized samples for tensile tests were cut by electric discharge machine, one for the tensile tests in an Instron-type universal testing machine and the other for the *in situ* tensile testing in a SEM. A suitable sample surface was prepared by electron polishing using 95% acetic acid and 5% percholoric acid for SEM observations. To obtain the strain mapping, the sample surface was etched in 1% nital, and *in situ* SEM observation carried out. The strain mapping with a change of displacement position of each surface point (digital imaging correlation method) has been carried out in SEM,. The SEM used is a Hitachi S-5000 equipped with an EBSD detector for crystal orientation analysis. The TEM observation has been done using thin foil samples after electron polishing by conventional FE-TEM with an accelerating voltage of 200kV.

3. Results

The tensile strength of the samples used is about 1100MPa, and the elongation is 5~6%. There is no clear yield point on the stress-strain curves obtained from all samples, resulting in a continuous hardening behavior on the curves under tensile testing at room temperature. With a cyclic tensile test by 20MPa each, it is confirmed in Fe-0.07C-1Mn(wt%) alloy that the initial deformation is elastic and the elastic limit appears at around 180~200MPa, as seen in figure 1(a)(b). Each stress-strain curve is superimposed in the same graph of figure 1(a), and the stress and strain profiles are plotted independently as a function of test time in figure.1(b). Although the negative value of the stress is artefact in figure 1(b), the stress value goes back to zero until stress condition of 160MPa when it is unloaded. Then, the plastic strain is residual more than the 180MPa stress condition, as described in figure 1(b).

IOP Conf. Series: Materials Science and Engineering **580** (2019) 012045 doi:10.1088/1757-899X/580/1/012045

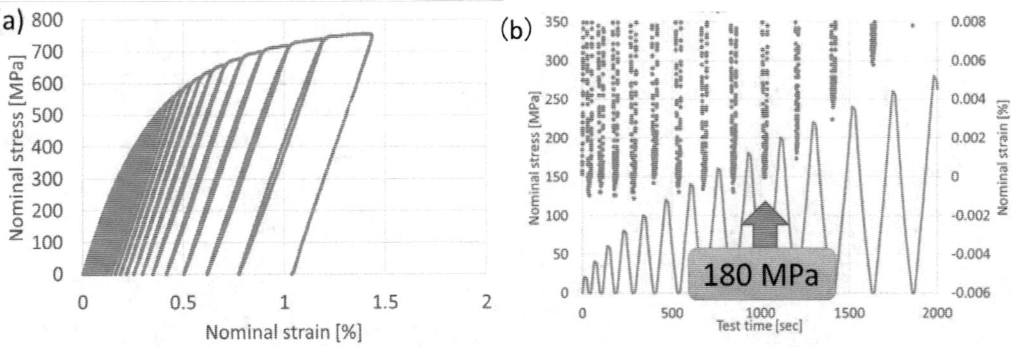

Fig.1 Stress-strain curves obtained by cyclic tensile test with increasing of 20MPa step; Superimposed S-S curves in (a), and replotted stress and strain values with test time in (b).

The microstructure of lath martensite has been investigated by SEM and TEM observations. There is a slight amount of ferrite grains at the prior austenite boundary in Fe-0.07C-1Mn(wt%), but the main part of the microstructure consists of the hierarchical structure such as packets, blocks and laths in the prior austenite grain. Figure 2 is a typical back scattering electron (BSE) image taken by SEM; showing packets and blocks in a prior austenite grain. The prior austenite grain size is about 80 μm, the block and its inner lath microstructure are observed in region A and B, and large lath domains with more 5μm in width are observed in region C.

Figure 3 is an enlarged BSE image of area A in figure 2. As indicated by the scale bar of 0.5 μm, a characteristic lath structure of 0.2 μm in width is shown in blocks indicated by P and Q. It is noted that the different variant of R is observed in the area adjacent to Q, although Q and R is considered as a same block region. The width of regions S and T is estimated at about 2～4 μm, and there is no lath structure, where small precipitates such as carbides are observed. In the narrow region of U, there is a fine stripe contrast showing an internal twin, which is clearly identified by the following observations.

The internal defects of lath martensite in low carbon steels have been studied by TEM observations. The typical microstructure of the lath martensite contains a high density of dislocations and tempered fine carbides in both of the Fe-0.07C-1Mn and the Fe-0.1-2Mn alloys. In both samples, however, internal twins are also observed together with dislocations, as seen in figure 4(a)(b). The thickness of internal twins is several tens of nm, and some of them are penetrated from end to end in a lath domain,

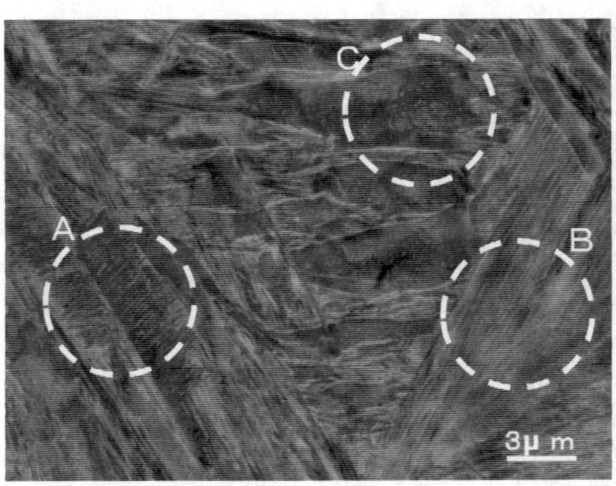

Fig.2 BSE image showing several types of block microstructure in Fe-0.07C-1Mn steel.

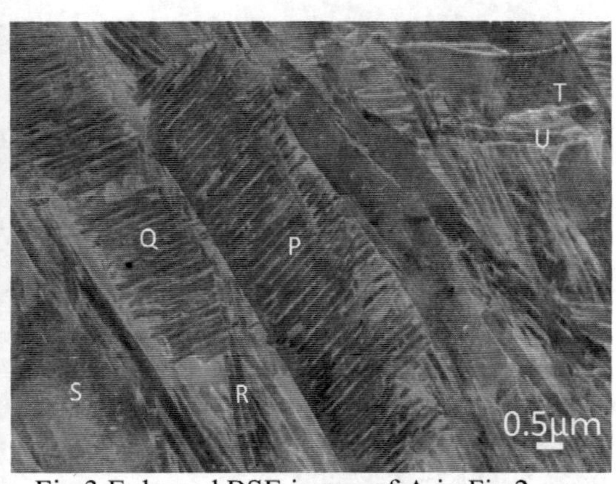

Fig.3 Enlarged BSE image of A in Fig.2. Fine lath in block is observed in P and Q.

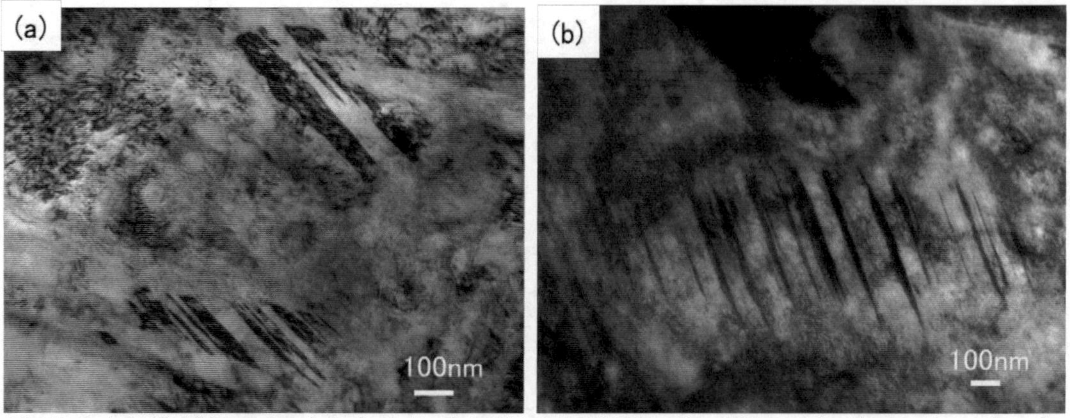

Fig.4 Bright field images taken from Fe-0.1wt%-2.0wt%Mn alloy. Internal twins are introduced by penetration into a lath domain (a) and by a part of distribution in a lath domain (b).

as shown in bottom area of figure 4(a), and others are introduced inside a lath structure, as seen in figure 4(b). Although it is difficult to estimate the amount of internal twins quantitatively, the internal twins are sometimes observed even in the Fe-0.07%C-1%Mn alloy. In comparison with the amount of internal twinning in samples quenched with 500℃/sec and 1500℃/sec cooling rate, the amount of internal twins tends to increase in the sample quenched by faster cooling speed. With slight composition difference between Fe-0.07%C-1%Mn and Fe-0.1%C-2%Mn alloys, the internal twins in lath martensite are observed more in Fe-0.1%C-2%Mn. With increasing Mn content, the ferrite nose leads to the delayed direction on the phase transformation, which gives a similar effect to increasing the cooling speed.

The nature of the internal twins is investigated by conventional diffraction analysis under TEM observations, as seen in figure 5. Figure 5(a) is a dislocation structure of a lath martensite, in which the dashed line indicates the lath boundary. When the sample is tilted in the same area, the internal twins appear in the lath microstructure, as shown in figure 5(b). The internal twins are always coexistent with the dislocations in the low lath martensite. The tilting experiment is important to confirm the internal twins under TEM observation. By the diffraction analysis shown in figure 5(c)(d), the twin plane is

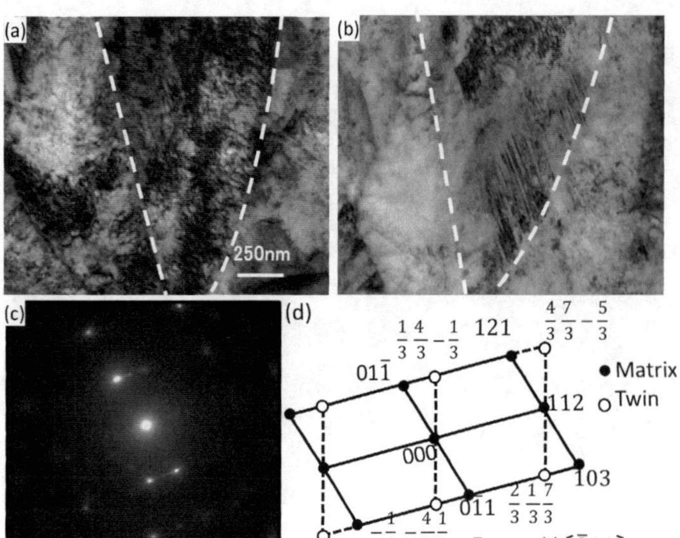

Fig.5 A series of bright field images and diffraction patterns.
(a) Each lath variant shows lots of dislocations.
(b) After tilting of same area, a fine internal twin structure appears inside a lath.
(c) Selected area diffraction pattern taken from an internal twined region. Spots from the twin indicate strong streaks perpendicular to twin plane.
(d) Schematic diagram of diffraction pattern, in which the incident beam direction is parallel to [-311] .

IOP Conf. Series: Materials Science and Engineering **580** (2019) 012045 doi:10.1088/1757-899X/580/1/012045

identified to be (112), in which the twin plane is satisfied with the edge-on condition with an incident beam direction of [-311]. The thickness of the twins is very thin such as several tenth of nm, resulting in the strong streaks on the diffraction pattern of figure 5(c).

As already seen in figure 1, the clear yielding point on the stress-strain curve is not observed for low carbon lath martensite. That indicates a heterogeneous deformation process in the early stage of deformation. The initial deformation behavior is considered to be very sensitive to the microstructure. The digital imaging correlation is a powerful technique to identify the morphological change in lath microstructure. The markers for measurement of small displacements uses directly the surface roughness of lath martensite after a nital etching treatment. An initial microstructure of lath martensite has been observed by SEM in Fe-0.07wt%C-1.0wt%Mn alloy, and the sample is deformed using a tensile test stage in the SEM until 0.4% strain, then the secondary electron image is taken from the same area, as indicated in figure 6(a). Based on the two micrographs, the strain mapping is calculated as seen in figure 6(b). It is made clear by the mapping data that the color change showing initial strain is distributed by the unit dimension of block noted by yellow arrows in lath microstructure.

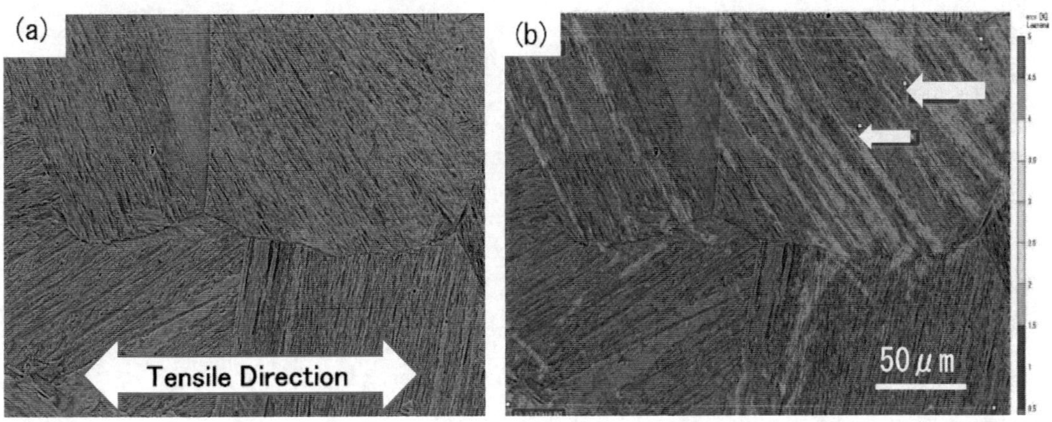

Fig.6 *In situ* SEM image taken from areas in Fe-0.07%C-1%Mn under tensile condition of 0.4%. (b) is the strain mapping of same area of (a) estimated by digital imaging correlation analysis.

4. Discussion

It is characteristic that the internal twins are observed even in a low carbon lath martensite with less than 0.1wt%C in Fe-C-Mn alloys. In Fe-C alloy systems, the martensite microstructure is of lath type in the carbon range of less than 0.6wt%, and the lens and thin plate martensite having internal twin structure appear in Fe-C(>0.6wt%), whose Ms temperature is lower than 623K. The density of twins changes from a partial to a whole region of lens martensite on decreasing of the Ms temperature, the internal twins are introduced instead of the dislocation, as a result of the lattice invariant shear of (101)[10-1]. Thus, the internal twins of lath martensite in the low carbon martensite are considered to be introduced as the lattice invariant shear in the area transformed at low temperature below 623K. Since the Ms temperatures of Fe-0.07wt%C-1wt%Mn and Fe-0.1%C-2%Mn(wt%) are estimated to be 739K and 689K by the following equation; Ms=546exp(-1.362C)-30.4Mn-11Si [13], the Mf temperatures are expected to be lower than 623K. If the lath martensite with internal twins is transformed at low temperature in comparison with surrounding lath martensite area, the austenite matrix will be plastically deformed by the dilation component of the surrounding transformation. This is the reason why the internal twins in low carbon martensite are always coexistent with dislocations. On the other hand, there are lots of carbide in the large lath martensite, and it may be considered that the large lath area is transformed at high temperature, as proposed by Morsodrof [12]. This will be further studied in detail.

On the issue of mechanical properties, it is important whether the internal twins play a role of obstacles for dislocation motion, resulting in a negative factor to improve the ductility. The amount of internal twins in low carbon lath martensite is however not so high as affect the mechanical properties. The existence is nevertheless valuable for consideration of the transformation mechanism of lath martensite.

The elastic limit showing round shaped yielding behavior is estimated by the cyclic tensile test in the present study. There are some softened areas in the microstructure where dislocations will start to move at low stress. It is calculated by Allen that local softening phenomenon leads to round shaped stress-strain curve [14]. The local softened area in a lath microstructure and different transformation temperature of each lath domain cause the different strength of each block domain. In the present study, the initial deformation of low carbon martensite occurs at the unit of a block dimension. Because of the resolution limit of the present DIC method, it is difficult to make clear the exact locally location of the softened region, but the heterogeneous deformation behavior takes place at each block microstructure. Since the block is a group of almost same lath orientation, it is considered that the local softening regions originated from carbon segregation around dislocations, or precipitation of carbides and/or clusters are similarly distributed among blocks on cooling and auto tempering.

5. Conclusions

The microstructure of lath martensite in Fe-lowC-Mn alloys have been investigated by several kinds of microscopy in a multi-scale dimension study, and the following conclusions are obtained.

(1) The multi-scale microscopic observations reveal that the microstructure of lath martensite in low carbon steels consists of heterogeneous lattice defects such as dislocations, internal twins, and different types of carbides, although the internal defect content of lath martensite is considered to be mainly dislocations in Fe-C-Mn alloys.

(2) A large size of lath martensite is often observed under the SEM observations, which is of similar dimension to the conventional block region defined as the hierarchical structure of low carbon lath martensite.

(3) The characteristic heterogeneous microstructure is related to the continuous yielding behavior on the stress-strain curve. It is confirmed by the DIC method that the initial deformation under tensile test in SEM is occurred by the different magnitude among each block regions.

Acknowledgement

A part of this work was supported by "Advanced Characterization Nanotechnology Platform, Nanotechnology Platform Program of the Ministry of Education, Culture, Sports, Science and Technology (MEXT), Japan" at the Research Center for Ultra-High Voltage Electron Microscopy (Nanotechnology Open Facilities) in Osaka University.

References

[1] Senuma T 2001 *ISIJ Int.* **41** 520-532
[2] Krauss G 1999 *Mat. Sci. and Eng.* 40-57
[3] Hutchinson B, Hagstrom J, Karlsson O, Lindell D, Tornberg M, Lindberg F and Thuvander M 2011 *Acta Mater.* **59** 5845-5858
[4] Wakasa K and Wayman C M 1981 *Acta Metall.* **29** 973-990
[5] Sandvik B P J and Wayman C M 1983 *Metall Trans.* **A14** 809-822
[6] Kelly P M, Jostsons A and Blake R G 1990 *Acta Metall. Mater.* **38** 1075-1081
[7] Morito S, Tanaka H, Konishi R, Furuhara T and Maki T 2003 *Acta Mater.* **51** 1789-1799
[8] Kitahara H, Ueji R, Tsuji N and Minamino Y 2006 *Acta Metall.* **54** 1279-1288
[9] Morito S, Huang X, Furuhara T, Maki T and Hansen N 2006 *Acta Metall.* **54** 5323-5331
[10] Kinney C C, Pytlewski K R, Khachaturyan A G and Morris Jr. J W 2014 *Acta Mater.* **69** 372-385
[11] Mine Y, Takashima H, Matsuda M and Takashima K 2013 *Mater. Sci. Eng.* A **560** 535-544
[12] Morsdorf L, Tasan C C, Ponge D and Raabe D 2015 *Acta Mater.* **95** 366-377
[13] Perlade A, Bouaziz O and Furnémont Q 2003 *Mater. Sci. and Eng.* **A356** 145-152
[14] Allain S, Bouaziz O and Takahashi M 2012 *ISIJ Int.* **52** 717-722

IOP Conf. Series: Materials Science and Engineering **580** (2019) 012046 doi:10.1088/1757-899X/580/1/012046

Statistics and Reproducibility of Grain Morphologies and Crystallographic Orientations Mapped by Laboratory Diffraction Contrast Tomography

Jun Sun[1*], **Jette Oddershede**[1], **Florian Bachmann**[1], **Hrishikesh Bale**[2], **Erik Lauridsen**[1]

[1]Xnovo Technology ApS, Theilgaards Alle 9, 1.th., 4600 Køge, Denmark

[2]Carl Zeiss X-ray Microscopy Inc., 4385 Hopyard Rd., Pleasanton 94588, CA, USA

[*]jsun@xnovotech.com

Abstract. Laboratory diffraction contrast tomography (LabDCT) enables a user to reconstruct 3D grain maps of polycrystalline materials non-destructively. For each grain, the morphology and crystallographic orientation, as well as derived properties such as grain boundary properties can be determined. Through two application examples this paper demonstrates the capabilities and potential of the current LabDCT implementation. Firstly, for well-annealed grain structures the reproducibility of LabDCT for more than 95% of the grains was found to be 5 μm on grain center-of-mass positions and 0.02° on orientations, while 90% of the grain boundary locations are determined with an accuracy better than 4 μm. The second example highlights the available statistics on thousands of grains, as well as the complementarity between LabDCT and absorption contrast tomography, readily available due to the integration of LabDCT on a commercial X-ray microscope

1. Introduction

The majority of engineering materials are polycrystalline. Their properties and performance strongly depend on the 3D microstructure in a sense that the individual grains, their crystallographic orientations, sizes and shapes as well as their spatial distribution ultimately determine the behavior of the material during processing and use. Characterizing such microstructures through X-ray diffraction contrast tomography was for a long time only possible at large-scale synchrotron facilities [1–5].

Laboratory diffraction contrast tomography (LabDCT) makes use of high-resolution diffraction images acquired on a ZEISS Xradia Versa X-ray microscope [6,7]. The diffraction signals are based on polychromatic X-rays and acquired in a special Laue-focusing geometry for improved signal to noise ratio and spot separation, see Figure 1 for a schematic of the experimental setup. The unique nondestructive 3D crystallographic imaging capabilities of LabDCT complement the structural data obtained by traditional absorption-based tomography, and together they provide an unprecedented insight into the structure of materials.

Recent developments of the GrainMapper3D reconstruction algorithm behind the LabDCT technique have extended capabilities to include full reconstruction of the 3D grain structure, including both grain morphology and crystallographic orientation, thereby making the LabDCT more comparable to conventional 3D-EBSD data – while still supporting 4D time-dependent studies [8]. From data on both the grain morphology and orientation, it is possible to extract the full macro parameters describing the grain boundary character, opening up new possibilities for statistical studies of grain boundary properties [9,10]. The wide accessibility of LabDCT allows for non-destructive, time evolution studies of dynamic microstructural evolution and materials characterization [11–13].

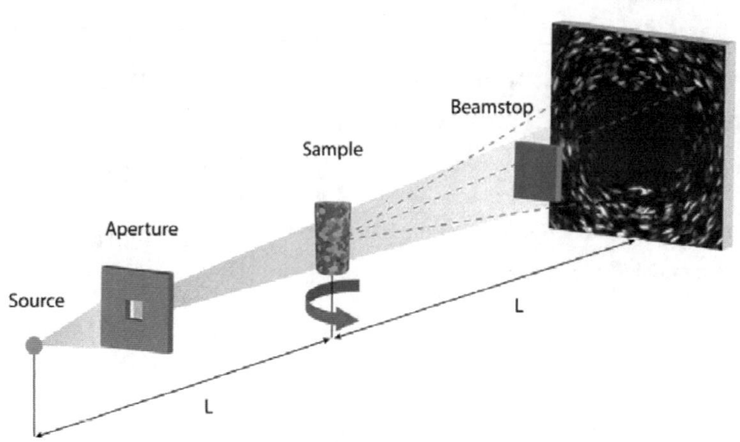

Figure 1. Schematic of the LabDCT setup.

In the present paper the current capabilities of LabDCT will be illustrated through two selected application examples. The sample materials have been chosen to serve two main purposes: 1) to validate the quality of the grain maps, particularly in terms of reproducibility of the technique, and (2) to demonstrate the plethora of statistically significant information available from a standard LabDCT experiment. The presentation of the results will be accompanied by a discussion of related materials science applications and an outlook towards the future of LabDCT.

2. Results and Discussion

2.1. Beta-Ti alloy

In order to quantify the reproducibility of LabDCT, data for the same β-Ti sample was collected on two different ZEISS X-ray microscopes. The reconstructed grain maps were registered and statistics on both matching grain pairs and boundary locations were extracted.

2.1.1. Experimental. Details on composition and processing of the metastable beta titanium alloy Tiβ21S sample material has been reported elsewhere [6]. A cylindrical sample with a diameter of 300 μm was used for the LabDCT data collections carried out on two different X-ray microscopes, each with a data collection time of 15 h.

2.1.2. Reproducibility. The reconstructed grain maps (2.5 μm voxels) based on the data collected for the β-Ti sample using two different ZEISS Xradia Versa X-ray microscopes are shown in Figure 2, inverse pole figure (IPF) colour-coded according to the cylinder axis direction ((a) and (b)) and voxel-by-voxel completeness ((c) and (d)). The completeness is defined as the ratio between observed and forward projected signals [8]. From the grain maps it is obvious that the illuminated volume in (a) and (c) is slightly larger than in (b) and (d), hence the volume in (a) and (c), the source, was registered against the volume in (b) and (d), the target.

IOP Conf. Series: Materials Science and Engineering **580** (2019) 012046 doi:10.1088/1757-899X/580/1/012046

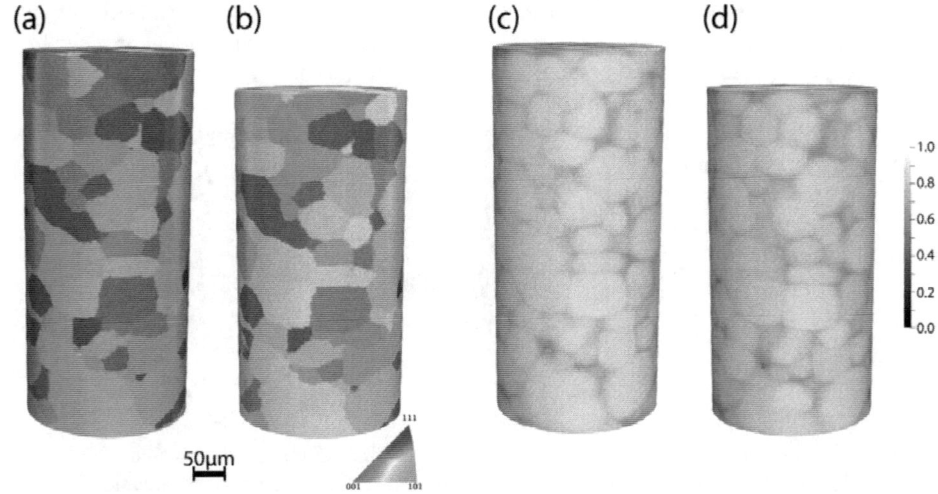

Figure 2. β-Ti grain maps measured on the same sample using two different instruments, referred to as the source (a,c) and target (b,d) volumes. The maps are registered and rendered using an IPF colouring based on the cylinder axis direction (a,b) and a greyscale colouring based on completeness (c,d).

Of the 509 grains in the target volume, 499 can be matched in the source volume within a distance tolerance of 25 μm and a misorientation criteria of 0.05°. Half of the ten grains in the target volume that cannot be matched in the source volume have mean completeness values less than 0.60; eight of the grains have equivalent sphere diameters less than 20 μm, and the remaining two less than 30 μm. Six of the grains are located at the ill-determined top surface of the target volume, and an additional two are at the very surface of the sample. In short, the majority of the unmatched grains in the target volume are considered as low-fidelity grains by several different measures.

The distributions of completeness-weighted center-of-mass distances (mean 2.3 μm) and misorientations (mean 0.010°) between the matching grain pairs can be seen in Figure 3. A total of 98% of the matched grain pairs fall within a misorientation criteria of 0.02° (twice the mean), while 96% of the pairs have center-of-mass distances less than 5 μm (2 voxels). The five matched grain-pairs with completeness-weighted center-of-mass distances larger than 15 μm are all located at the top surface of the target volume. At this surface the risk of comparing partly illuminated target grains with fully illuminated source grains is high, hence the observed center-of-mass discrepancies are not surprising.

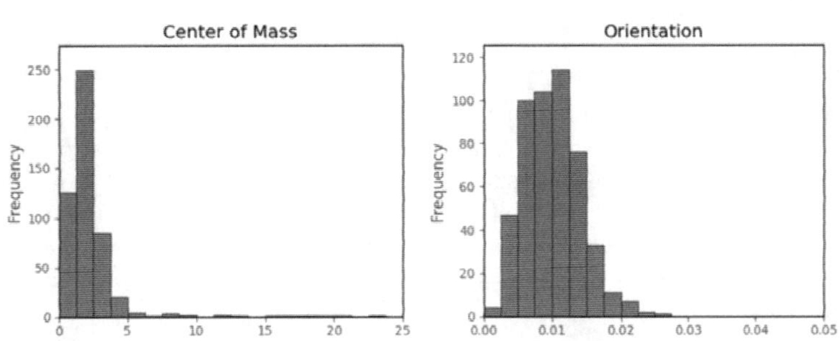

Figure 3. β-Ti histograms for matching grain pairs between the reconstructions of the two different data sets.

The reproducibility in grain boundary location between the two maps was quantified using the method proposed by Ludwig et al. [4]. Based on five slices distributed 100 μm apart along the cylinder axis, the central one of which is shown in Figure 4, it was found that the mean distance in grain boundary location is 2.1 μm, while 90% of the grain boundaries fall within a distance of 4 μm.

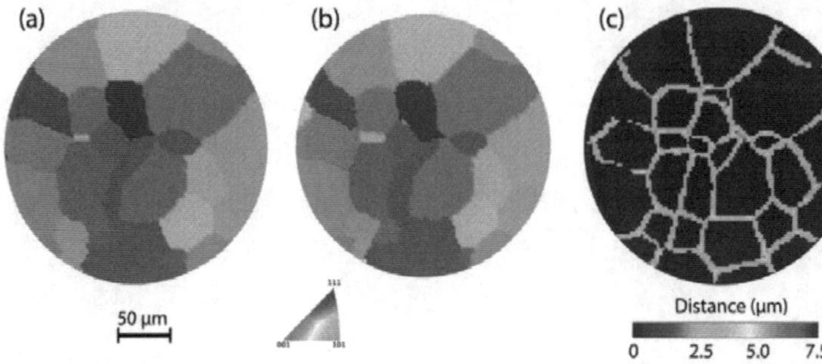

Figure 4. β-Ti slices through the registered source (a) and target (b) volumes and the corresponding difference map (c) showing reproducibility in grain boundary location.

2.2. AlCu alloy

The second material chosen was an aluminium sample with grain boundaries decorated with copper. The large number of grains and grain boundaries mapped in this sample serves as an excellent example of how large 3D volumes can be covered with LabDCT, and what this means in terms of statistics. Furthermore, the complementarity between LabDCT and absorption contrast tomography (ACT) is nicely illustrated by the comparison of the copper network along the aluminium grain boundaries measured by ACT and the grain boundaries reconstructed from the LabDCT data.

2.2.1. Experimental. The AlCu sample material was processed as previously reported [14] to reach an average grain size of 170 μm and into cylindrical specimen with a diameter of 1.4 mm. A volume of interest covering the full cross section and 2.4 mm along the length of the sample was mapped by ACT and LabDCT, and was then reconstructed with a voxel size of 5 μm using GrainMapper3D as described in [8]. The total data collection time was 18 h.

2.2.2. Large grain statistics. Mapping representative sample volumes are of the utmost importance for investigations of materials microstructure and properties. For example, drawing conclusions based on a selection of grains with a different texture compared to the overall sample texture may lead to flawed conclusions, and if one is looking for rare events such as abnormal grain growth, sample fracture or the behaviour of specific types of grain boundaries, then the mapped sample volume must be large enough to capture sufficiently such events.

Figure 5 (a) shows the grain map of the AlCu sample comprising 1930 grains IPF colour-coded according to the cylinder axis direction of the sample, while Figure 5 (b) shows the more than 10,000 corresponding grain boundaries colour-coded according to misorientation angle. Figure 5 (c) and (d) show additional statistics for the 1930 grains in the AlCu sample, namely the grain size distribution and the distribution of faces per grain, respectively. The grain size distribution has a mean of 137 μm and shows fewer small grains than expected. The average number of faces per grain in the AlCu sample was determined as 11.6 based on the LabDCT grain map, as compared to the predicted value of 13.4. The current implementation of LabDCT supports grain sizes down to 10-40 μm, mainly limited by the low brilliance of laboratory X-ray sources, which is somewhat reflected in Figure 5(c). Increasing the sensitivity of LabDCT towards smaller grain sizes is, however, a key focus area in the continued development of the technique to broaden its application space.

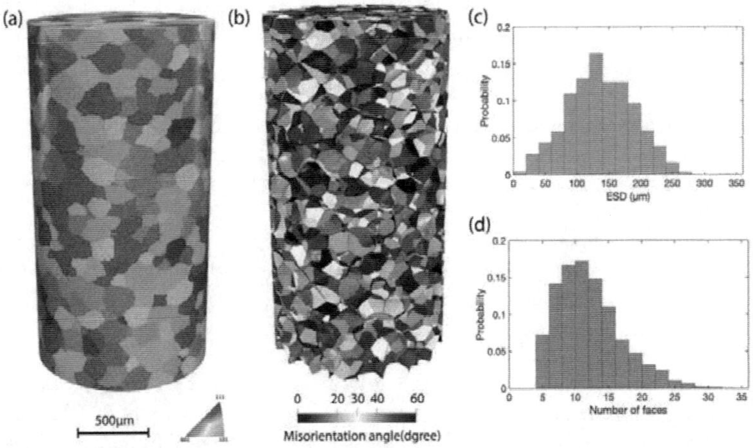

Figure 5. AlCu grain map IPF colour-coded according to along the cylinder axis direction (a), and corresponding grain boundaries colour-coded by misorientation (b); derived distributions of equivalent sphere diameters (c) and number of faces per grain (d).

2.2.3. Correlating LabDCT and ACT. The complementarity of LabDCT and ACT can be used to provide unprecedented insight into the structure of materials. In the case of the AlCu sample the two different imaging modalities can be exploited to give two independent measures of the grain boundary locations. Figure 6 (a) shows a 2D slice of the LabDCT grain map colour coded by IPF along the cylinder axis overlaid by the ACT reconstruction, highlighting the Cu-rich grain boundaries in bright colours. In general, the plot shows good agreement along grain boundaries, some of which are more decorated than others. Figure 6 (b) shows a selection of grains (highlighted by the red box in (a)) in 3D extended out of the ACT-determined volume (colours inverted as compared to (a)). The greenish colours indicate that all of the grains have a crystallographic <101> axis closely aligned with the cylinder axis of the sample. However, the grain boundaries between the selected grains, visualised in 3D and colour coded by grain boundary misorientations in Figure 6 (c), clearly show that the selected grains are rotated differently about the aligned crystallographic <101> axes due to the presence of high-angle grain boundaries.

Figure 6. AlCu 2D slice comparing LabDCT and ACT (a), a selection of grains in the slice shown in 3D against the ACT data (b), and grain boundaries between these coloured by misorientation (c). LabDCT colour code in (a) and (b) is IPF. For (a) the Cu-rich grain boundaries in the ACT reconstruction are shown in bright colours, while the ACT colour scale is inverted for (b) and (c).

2.2.4. Grain boundary character. The intricate 3D nature of the grain boundary network is clearly illustrated by both Figure 5 (b) and Figure 6 (c). In addition to the misorientations, grain boundary normal information can be extracted from the 3D grain maps, allowing the user to experimentally determine all five parameters needed for a full grain boundary description and subsequent analysis [15]. The grain boundary characteristics derived from LabDCT grain maps have already been used for a number of materials science applications. Keinan et al. [9] found a clear correlation between the grain boundary character measured by LabDCT and the distribution of impurities imaged by ACT in polycrystalline silicon for photovoltaic applications. This may eventually lead to better solar cells where the grain boundary types showing most of the performance-degrading impurities are suppressed in the manufacturing process. Sun et al. [10] showed that the penetration of liquid gallium into a matrix of aluminium grains depends strongly on the grain boundary energies, by visualising the grain boundary wetting by ACT and deriving grain boundary energies from the LabDCT grain maps. This result may help prevent liquid metal embrittlement, which causes serious problems for processes such as welding and galvanizing, as well as in nuclear reactors with a liquid metal spallation target.

3. Conclusions

In the present paper we have demonstrated the current capabilities of LabDCT in terms of grain morphology and orientations, as well as grain boundary properties derived from these. New encouraging results on the reproducibility of LabDCT based on comparing grain maps of the same sample measured on two different machines have been presented. The grain statistics on thousands of grains have been highlighted, along with the complementarity between LabDCT and ACT readily available due to the integration of LabDCT on a commercial X-ray microscope.

References

[1] Lauridsen E M, Schmidt S, Suter R M and Poulsen H F 2001 *J. Appl. Crystallogr.* **34** 744–50
[2] Poulsen H F, Nielsen S F, Lauridsen E M, Schmidt S and Suter R M 2001 *J. Appl. Crystallogr.* **34** 751–6
[3] Ludwig W, Schmidt S, Lauridsen E M and Poulsen H F 2008 *J. Appl. Crystallogr.* **41** 302–9
[4] Ludwig W, Reischig P, King A, Herbig M, Lauridsen E M, Johnson G, Marrow T J and Buffière J-Y 2009 *Rev. Sci. Instrum.* **80** 033905
[5] Li S F and Suter R M 2013 *J. Appl. Crystallogr.* **46** 512–24
[6] McDonald S A, Reischig P, Holzner C, Lauridsen E M, Withers P J, Merkle A P and Feser M 2015 *Sci. Rep.* **5** 14665
[7] Holzner C, Lavery L, Bale H, Merkle A, McDonald S, Withers P, Zhang Y, Jensen D J, Kimura M, Lyckegaard A, Reischig P and Lauridsen E M 2016 *Micros. Today* **24** 34–43
[8] Bachmann F, Bale H, Gueninchault N, Holzner C and Lauridsen E M 2019 *J. Appl. Crystallogr.* in press DOI: 10.1107/S1600576719005442
[9] Keinan R, Bale H, Gueninchault N, Lauridsen E M and Shahani A J 2018 *Acta Mater.* **148** 225–34
[10] Sun J, Zhang Y, Lyckegaard A, Bachmann F, Lauridsen E M and Juul Jensen D 2019 *Scr. Mater.* **163** 77–81
[11] McDonald S A, Holzner C, Lauridsen E M, Reischig P, Merkle A P and Withers P J 2017 *Sci. Rep.* **7** 5251
[12] Sun J, Lyckegaard A, Zhang Y B, Sahi C A, Patterson B R, Bachmann F, Gueninchault N, Bale H, Holzner C, Lauridsen E and Juul Jensen D 2017 *IOP Conf. Ser. Mater. Sci. Eng.* **219** 012039
[13] Patterson B R, Dehoff R T, Sahi C A, Sun J, Oddershede J, Bachmann F and Lauridsen E 2019 *ibid*
[14] Dake J M, Oddershede J, Henning O S, Werz T, Shatto J C and Uesugi K 2016 *PNAS PLUS* **113** 5998–6007
[15] Rohrer G S, Saylor D M, Dasher B El, Adams B L, Rollett A D and Wynblatt P 2004 *Zeitschrift für Met.* **95** 197–214

IOP Conf. Series: Materials Science and Engineering **580** (2019) 012047 doi:10.1088/1757-899X/580/1/012047

Aging of 3D-printed maraging steel

E G Thorsteinsdóttir[1], D S Primdahl[2], Y B Zhang[1], D Juul Jensen[1] and T Yu[1]

[1] Department of Mechanical Engineering, Technical University of Denmark, DK-2800 Kgs. Lyngby, Denmark
[2] Grundfos A/S, DK-8850 Bjerringbro, Denmark

tiyu@mek.dtu.dk

Abstract. The effects of aging on the microstructural development and mechanical properties of a 3D-printed maraging steel have been investigated. The microstructures within printed, solution treated and aged samples are characterized by X-ray tomography and optical microscopy, as well as by scanning and transmission electron microscopy. The mechanical properties are estimated based on hardness measurements. The distribution of voids is found to be inhomogeneous, and even though the samples are 99.9% dense in the sample interior, the remaining voids affect significantly the hardness in the aged condition. In addition to precipitation, the aging treatment leads to formation of about 12% of reverted austenite. Together this leads to a strengthening of the material. Solution treatment results in homogenization of the microstructures such that after aging the solution treated samples are found to have higher strength than the samples that are just aged after printing. These results are discussed and compared to the aging behaviour of conventionally manufactured samples.

1. Introduction

Additive manufacturing in the form of 3D printing is a new popular manufacturing process, currently undergoing rapid growth [1]. Huge investments worldwide have been made, related to the development of new machines and processes, as well as to the production of suitable powders and education of users. The development of this technique has been fast, and today 3D printing is not only used for prototypes but also for manufacture of components. It is expected that 3D printing will be used even more extensively in the future, both to re-design current products and for new designs based on the possibilities in 3D printing.

One of the materials often used in 3D printing is maraging steel. This is a popular material because of its good weldability and superior mechanical properties, which make it ideal for tool manufacturing. Maraging steel is strengthened through aging, leading to precipitation of intermetallic particles [2]. Important parameters are the aging temperature and time, which are known to have a large influence on the mechanical properties [3].

In conventional manufacturing, maraging steels are typically solution treated [3], which is generally not the case for 3D-printed material. It is, therefore, not surprising that similar aging of 3D-printed and conventionally manufactured samples does not result in the same mechanical properties. An additional complication is that 3D-printed samples may contain fractions of retained austenite and reverted austenite after aging. It has been suggested that this austenite may contribute to the less good mechanical properties of 3D-printed samples compared to equivalent ones manufactured conventionally, where austenite reversion typically only occurs at higher temperatures [4,5].

The aim of the present work is to investigate the effects of aging on the microstructural development and the mechanical properties of 3D-printed maraging steel samples and to compare these effects with those for conventionally manufactured samples. Finally post-printing solution treatment is also considered, and effects of such treatments are discussed.

2. Experimental Details

The samples were printed by selective laser melting using a dual laser system (two 400 W lasers) according to standard industrial practice. Spherical shaped magaing steel powder with size range 10 - 45 μm was used for printing. In total 38 samples with size 3 × 6 × 12 mm^3 and 10 samples with size 22 × 200 × 5.3 mm^3 were printed. For comparison, a rod of maraging steel with a similar nominal chemical composition (table 1) was used.

Table 1. Chemical composition of the maraging steel rod (in wt.%).

Ni	Mo	Co	Ti	Cr	S	P	Mn	Si	C	Fe
17.0-19.0	4.5-5.2	8.5-10.0	0.8-1.2	≤0.25	≤0.01	≤0.01	≤0.15	≤0.1	≤0.03	Bal.

Aging was carried out at 480, 500, 520 and 540 °C for 1 to 10 h. For additional 3D-printed samples, a solution treatment at 850 °C for 1 h was also carried out before aging at 500 °C. All samples were air cooled.

The mechanical properties were quantified by Vickers hardness measurements using a load of 500 g for 10 s in the longitudinal section (containing the building direction). The reported hardness values are averaged over at least 6 measurements.

The microstructures in the longitudinal section were characterized by optical microscopy, scanning electron microscopy (SEM) and transmission electron microscopy (TEM). Additionally absorption contrast X-ray tomography was used to detect voids non-destructively in 3D. For optical microscopy and SEM observations, samples were mechanically polished followed by etching in a 2% Nital solution. For electron backscatter diffraction (EBSD) studies, samples were mechanically polished using OP-U colloidal silica suspension (40 nm particle size) for the final step. For the TEM observations, thin foils were prepared using a modified window technique.

The X-ray tomography characterization was performed using a Zeiss Xradia 520 Versa X-ray microscope. A detector (2032 × 2032 pixels) with pixel size of 3.36 μm was used for recording absorption patterns. For determination of the distribution of voids in 3D, a 3 × 3 × 6 mm^3 sample aged at 500 °C for 6 h was used. During the high-resolution X-ray measurements, only one quarter of the cross section was analysed, in the top part of the sample (ca. 1.5 × 1.5 × 1.5 mm^3), and the sample was rotated through 360° in 4051 steps. The resulting spatial resolution is estimated to be 2 μm (voxel size 1.556 μm). Avizo was employed to visualize the voids.

3. Results and discussion

The distribution of voids is shown in figure 1. The voids are primarily distributed in the sub-surface layers (50-150 μm from surface). In the outer zone the voids are many and relatively large (see zone 1 in figure 1b). Here the sample density is only 97.9% while in the 2 inner zones the density is 99.9%. That the outer surface is more porous is in agreement with general observations for 3D-printed samples and is related to the lack of subsequent re-melting for these outer layers. For the 2 inner zones it is worth noticing that the sizes (equivalent spherical radii) of the voids in zone 2 (see figure 1b) are large, up to 20 μm, while in zone 3 the sizes do not exceed 15 μm and most of the voids are below 5 μm.

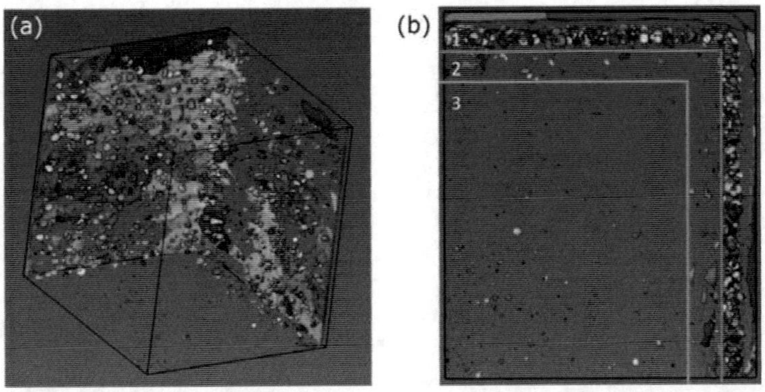

Figure 1. The distribution of voids inside a 3D-printed sample – only a quarter of the top part is shown (ca. 1.5 × 1.5 × 1.5 mm^3). The voids are shown in random colours, and some part of the surface is also coloured. a) Full 3D view, b) 2D projection, seen from the top.

The microstructures observed by optical microscopy are shown in figure 2. The meltpools are clearly visible in the as-printed and in the aged conditions, see figure 2a and b. In the samples that were solution treated before aging (see figure 2c) the microstructure is homogenized and the meltpools are no longer visible. However, in this condition a lath martensitic structure is visible, which is typical in maraging steels after conventional processing (including solution treatment and aging). This conventional martensitic structure is not observed if the sample is just aged without solution treatment after printing (see figure 2b).

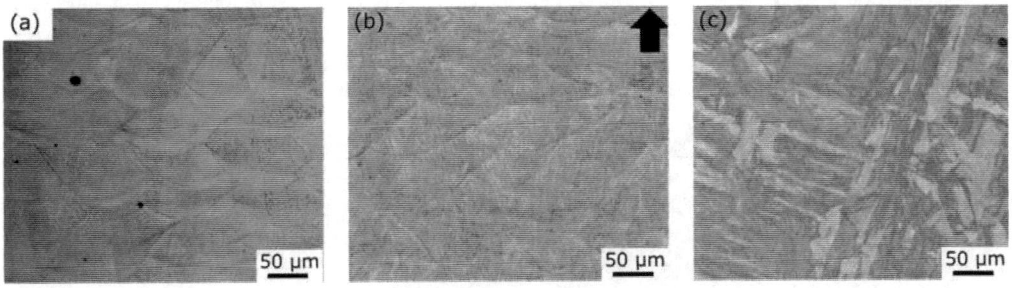

Figure 2. Optical micrographs of the printed and heat treated samples. a) in the as printed condition, b) aged at 480 °C for 1 h and c) solution treated at 850°C and aged at 480 °C for 6 h. The arrow indicates the common building direction.

Much more detail about the microstructure is revealed by SEM and TEM. High magnification investigations by SEM electron channelling contrast (ECC) imaging reveal the cell structure of the printed sample (figure 3a). Both equiaxed cells and elongated cell structures are observed. This is in agreement with literature results [4,6-8] and is consistent with EBSD characterization of the present sample, suggesting these cells are bounded by low angle dislocation boundaries formed during cooling. After aging, the structure is dominated by extended features which by EBSD are revealed to be austenite (see figures 3c and 4). After solution treatment, the printing-induced cells are no longer observed, and instead the typical martensitic structure is clearly visible (figure 3b). Aging of this sample results in precipitation (see figure 3d), as in conventionally manufactured maraging steel samples. No austenite is apparent in the solution treated and aged samples.

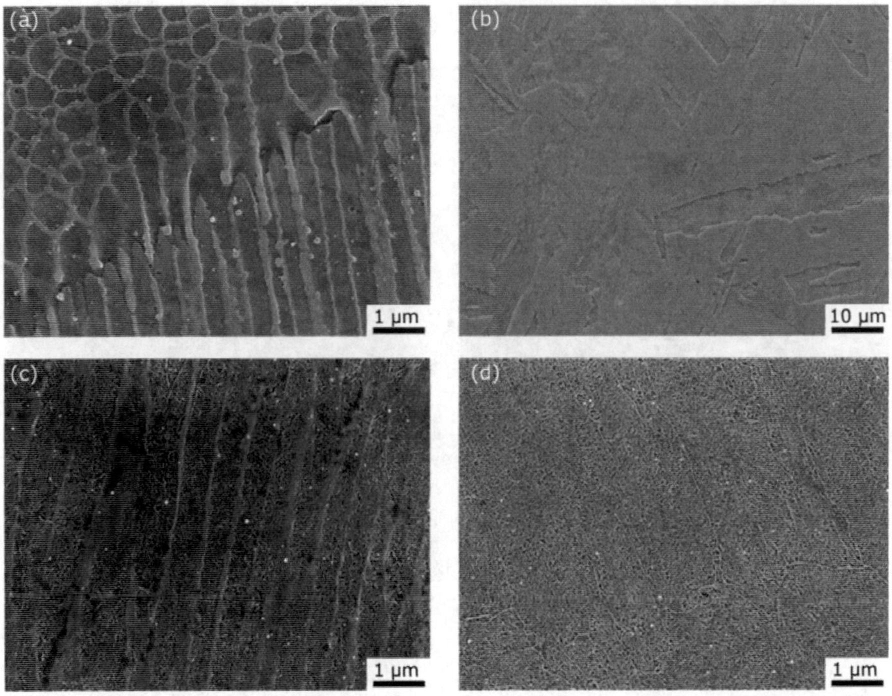

Figure 3. SEM micrographs of printed and heat treated samples. a) the as-printed condition, b) solution treated at 850 °C for 1 h, c) aged at 500 °C for 6 h and d) solution treated at 850 °C for 1 h and aged at 500 °C for 6 h.

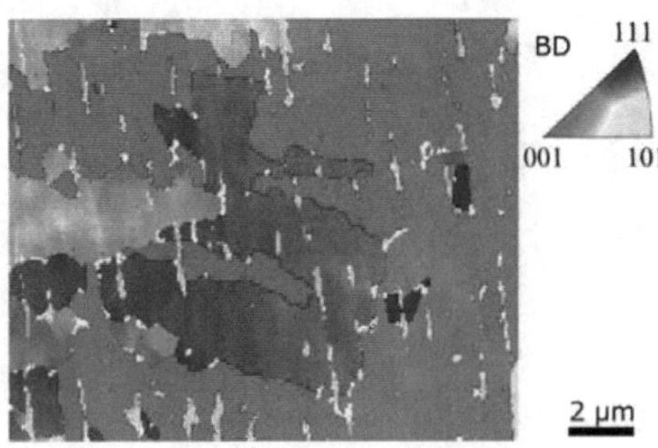

Figure 4. Orientation map obtained by EBSD for a 3D-printed sample aged at 500 °C for 6 h. The colour is based on the crystallographic orientation of the building direction (BD). The elongated features shown in green are austenite.

The microstructures of the as-printed and aged samples were investigated in further detail by TEM. As can be seen from figure 5, the as-printed sample has a dislocation structure with no visible evidence of precipitates. A few titanium oxides are observed, which are expected to originate from the printing process. After aging, precipitation is obvious (see figure 5b and c), and the dislocation structure observed in the printed condition is removed by the annealing. The larger elongated regions are found to be austenite, in agreement with the SEM EBSD results.

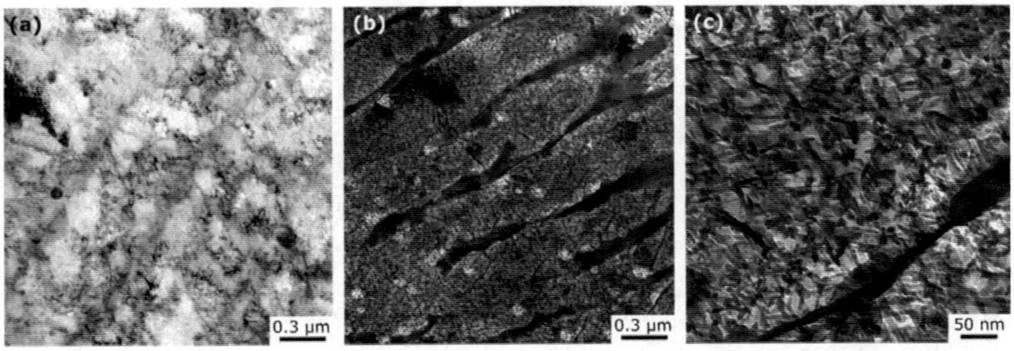

Figure 5. TEM micrographs of printed and aged samples. a) as-printed condition, b) aged at 500 °C for 6 h, c) higher magnification micrograph of the sample shown in b).

The microstructural observations reported above have significant effects on the mechanical properties. The Vickers hardness for all the printed, aged and solution treated samples as well as the conventional ones are reported in table 2. After solution treatment, the 3D-printed sample and the conventionally manufactured sample have a similar hardness. After aging, however, the hardness of the 3D-printed samples is lower than that of the conventional samples. It should be noted that the hardness measurements were carried out in zone 3 of the printed samples (see figure 1b), i.e. where the void density is very low. However, as demonstrated here, the voids still have a significant effect on the mechanical properties after aging although their effect is small in the solution treated condition.

Table 2. Averaged Vickers hardness values (the standard deviation is typically ~10).

	3D-printed	3D-printed solution treated	Conventional solution treated
Initial	377	324	329
480 °C 1/3/6 h	569/608/630	-/-/-	628/680/698
500 °C 1/3/6/10 h	598/615/633/600	621/637/642/-	649/679/688/-
520 °C 1/3/6/10 h	604/617/616/618	-/-/-	665/674/678/-

The aging treatment results in strengthening for all the samples. The highest hardness values of approximately 700 Vickers are obtained for the conventional sample. The aged 3D-printed samples have values in the range 600 – 640 Vickers, and the values after 10 h annealing at 540 °C and 520 °C indicate over-aging. The maximum hardness in this series of samples is observed after 6 h at 500 °C. Generally, it is reported in the literature that the additional solution treatment before aging of maraging steels leads to a reduction in yield strength and an improvement in elongation [2,6,9]. For the present 3D-printed samples, we do not observe any significant drop in hardness resulting from the additional solution treatment before aging. In fact, the solution treated sample aged at 500 °C for 6 h has the highest hardness among all the 3D-printed samples. This may be related to the fact that the printed and aged samples contain a significant fraction of soft austenite.

When relating the hardness results to the microstructural observations, it is clear that the precipitation that takes place during the aging treatment strengthens significantly the printed samples, as it does also for conventional processing. However, the peak aging condition appears to be at a slightly higher temperature (same annealing time) for the printed samples than for the conventional samples.

4. Summary

In the present work, aging of 3D-printed maraging steel samples has been compared to that of conventionally manufactured samples. It has been demonstrated that the 3D printing has a large effect on the microstructural evolution and on the hardness. Even the inner parts of the printed samples contain voids of sizes up to 15 μm, and although the fraction of voids is very low, it affects the hardness of the printed samples, especially in the aged condition. As in the conventionally manufactured samples, aging of the printed samples leads to precipitation and thus strengthening. However in printed and aged samples, a significant fraction of austenite is observed to be present, which may be avoided if the printed samples are solution treated before the aging. Depending on the application, the heat treatment of 3D-printed maraging steel components can be optimized, but the optimization parameters may be quite different from those for conventionally manufactured maraging steel.

Acknowledgement

DJJ, YBZ and TY thank the European Research Council (ERC) under the European Union's Horizon 2020 research and innovation programme (M4D – grant agreement No. 788567) for supporting part of this work.

References

[1] Becker T and Dimitrov D 2015 *Rapid Prototyping Journal*, **22** 487-494
[2] Kempen K, Yasa E, Thijs L, Kruth J P and Van Humbeeck J 2011 *Physics Procedia* **12** 255-263
[3] Mouritz A 2012 *Introduction to aerospace materials* (Cambridge: Woodhead Publishing Limited) p 244-246
[4] Guo W F, Guo C and Zhu Q 2018 *Materials Science Forum* **941** 2160-2166
[5] Lima Filho V X, Barros I F and Abreu H F 2017 *Materials Research* **20** 10-14
[6] Casati R, Lemke J N, Vedani M 2017 *La Metallurgia Italiana* **1** 11-20
[7] Bai Y, Yang Y, Wang D and Zhang M 2017 *Materials Science & Engineering A* **703** 116-123
[8] Mutua J, Nakata S, Onda T and Chen Z C 2018 *Materials & Design* **139** 486-497
[9] Tan C, Zhou K, Ma W, Zhang P, Liu M and Kuang T 2017 *Materials & Design* **134** 23-34

IOP Conf. Series: Materials Science and Engineering **580** (2019) 012048 doi:10.1088/1757-899X/580/1/012048

Extraction of grain boundary curvature from voxel-based representations of polycrystalline microstructures

Mingyan Wang, Shuang Wu and Carl E. Krill III

Institute of Functional Nanosystems, Ulm University, Albert-Einstein-Allee 47, 89081 Ulm, Germany

E-mail: mingyan.wang@uni-ulm.de

Abstract. Extracting the curvature of grain boundaries from 3D voxel-based representations of microstructure is challenging. Previous investigations have concluded that the accuracy of curvature estimation depends strongly on the degree of smoothing after a microstructure has been meshed. We find that best results are obtained when the degree of smoothing is determined on a boundary-by-boundary basis. Evaluating a voxelized representation of a Reuleaux tetrahedron of known boundary curvature, we establish that, for a given boundary, the optimal smoothing power minimizes the coefficient of variation (CV) of local curvature. We apply the minimum-CV criterion to the extraction of grain boundary curvatures from a measurement of polycrystalline Al-5 wt% Cu performed by 3D x-ray diffraction (3DXRD) microscopy. The results show that grain boundaries in this sample are surprisingly flat. Moreover, boundaries of larger area manifest lower curvature, while higher curvature values are observed solely for boundaries of smaller area.

1. Introduction

Grain boundary (GB) curvature is one of the most important morphological properties of any given microstructure. The product of GB energy and curvature serves as the driving force for grain growth [1], but it is the curvature alone that determines the direction in which a GB migrates. Now that polycrystalline microstructures can readily be characterized in 3D using techniques like focused ion beam nanotomography or 3D x-ray diffraction (3DXRD) microscopy [2–5], increased attention is being devoted to the extraction of accurate boundary curvature values from experimental data. The output of such characterization techniques is a voxel-based representation of microstructure, which presents difficulties for accurate curvature estimation. On the one hand, the boundaries themselves are specified in an implicit manner—as interfaces between 3D grain volumes—and, on the other hand, these interfaces exhibit roughness at the length scale of the underlying grid; the latter propagates, in turn, as errors in curvature estimation. Both issues must be addressed before any attempt can be made to relate extracted boundary curvature values to the properties of a given specimen.

In this regard, the software package DREAM.3D [6] is a promising tool, offering built-in functions for meshing grain interfaces and smoothing the mesh. It is necessary, however, for the user to tune various parameters in these functions, such as the smoothing power. The optimization of smoothing parameters was investigated by Zhong *et al.* [7], who applied

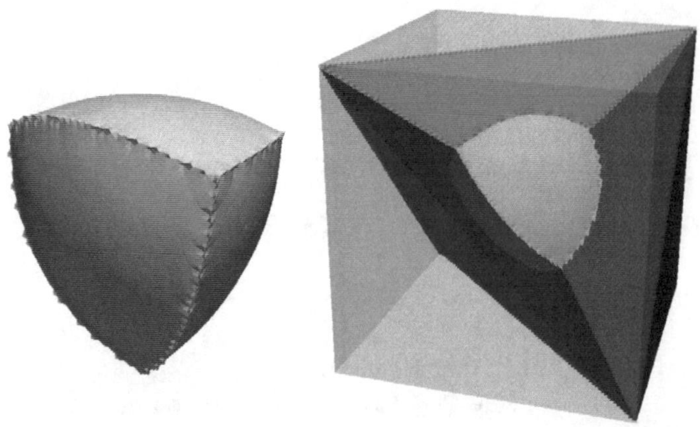

Figure 1. *(left)* A Reuleaux tetrahedron has four spherical surfaces, each with a curvature of $-1/R$, where R denotes the distance between vertices. *(right)* A Reuleaux tetrahedron embedded in a cubic box, with planar boundaries forming triple lines along edges of the tetrahedron.

DREAM.3D to simulated spherical grains having known boundary curvatures. These authors concluded that smaller test grains were better processed using lower smoothing powers, whereas larger spheres required higher smoothing powers. However, in real microstructures, the optimal smoothing power for a given boundary was not found to be correlated to the size of the grains meeting at that boundary [7]. Apparently, the optimal smoothing power is a property of individual boundaries rather than entire grains. Nevertheless, the authors of Ref. [7] obtained satisfactory results using a set of smoothing parameters that worked well for the majority of GBs in an experimental dataset.

In this work, we propose a method for determining the optimal smoothing power on a boundary-by-boundary basis, based on findings derived from a voxel-based test structure similar in shape to a typical grain [8]. Then we apply the method to a polycrystalline Al-5 wt% Cu specimen that was characterized by 3DXRD microscopy to determine the curvature of the sample's boundaries and evaluate its dependence on boundary area.

2. Methods

To analyze voxel-based representations of microstructure, we employ DREAM.3D's built-in function "Quick Surface Mesh" to generate a surface mesh of the imported data, after which all GBs are represented as surfaces composed of triangles. Vestiges of the voxelized nature of the input data can be seen in the "stepped" morphology of the GBs. Therefore, some form of smoothing must be carried out to recover the intrinsic shape of each boundary. Smoothing is implemented in DREAM.3D following the method described in Ref. [9]. The principal curvatures κ_1 and κ_2 are the eigenvalues of the Weingarten matrix [10]. The mean curvature H is defined as $(\kappa_1 + \kappa_2)/2$. In the DREAM.3D smoothing algorithm, the smoothing power is determined by the number of smoothing iterations, N, as well as by the smoothing factor λ, which governs how far a vertex is permitted to move from its original location during a single smoothing iteration. The final smoothing power is quantified by the product $N\lambda$.

3. Results

3.1. Determination of boundary curvatures on the test structure

Instead of the spherical grain shape used in Ref. [7], we validate our curvature analysis strategy against a test grain having the shape of a Reuleaux tetrahedron. The latter is defined by the intersection of four spheres of radius R that are centered on the vertices of a regular tetrahedron of side length R (figure 1). By construction, the curvature of each surface of the Reuleaux tetrahedron is $-1/R$. We embedded the tetrahedral grain in a cubic box, dividing the surrounding space into four regions of equal volume by placing planar boundaries along diagonals of the cube faces and extending them inwards until they intersect the Reuleaux tetrahedron along

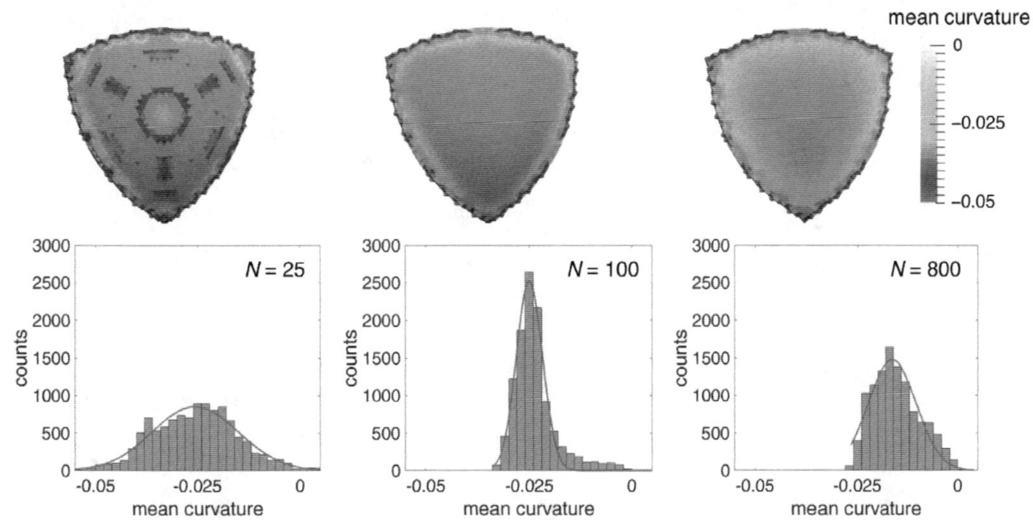

Figure 2. Influence of smoothing power on curvature estimation for the test structure with $R = 40$. The smoothing parameter N takes values of 25, 100 and 800 from left to right. Images in the top row show values of local mean curvature H for one of the surfaces of the Reuleaux tetrahedron. Corresponding distributions of H are plotted below, after excluding locations adjacent to triple lines; red curves are fits of Guassian functions to the histograms.

its edges. All dihedral angles between boundaries meeting at triple lines are 120°, resembling the dihedral angles observed in real microstructures.

This construction was then mapped onto a cubic grid and imported into DREAM.3D. The voxel side length was kept at 1 for all test structures examined in this study. After the meshing step, the triangular mesh was subjected to DREAM.3D's smoothing algorithm, whereby λ was set to 0.2 for each vertex of a surface triangle but reduced to 0.1 for vertices located at a triple line or quadruple point (to suppress shrinkage of the central grain during smoothing).

We first test the influence of smoothing power on the mesh by varying the number of iterations N (holding λ at the values specified above). In order to cover a wide range of smoothing power, N is set to 25, 50, 100, 200, 400 and 800. The influence of N is shown in figure 2 for one of the curved surfaces of the test structure with $R = 40$. Here, the surface is colored such that patches with extracted curvatures close to the true value ($-1/40 = -0.025$) are green, while white and red patches denote under- or over-estimations of the magnitude of curvature, respectively. Curvature estimation is highly inaccurate close to triple lines, as evidenced by the preponderance of red patches at the edges of each surface. This finding is not surprising, given the abrupt change in surface orientation that occurs at a triple line. After excluding the latter regions, we generate histograms of the local curvature value H for the remaining triangles of the GB mesh. The mean and standard deviation of these histograms are found to vary strongly with N. When $N = 25$, the overall degree of smoothing is inadequate, as indicated by the presence of reddish patches in the boundary interior. Raising N to 100 improves the uniformity of the extracted curvature values, but progressing to still higher values of N results in oversmoothing, with an increasing fraction of the boundary taking on a curvature close to zero.

To quantify these observations, we fit Gaussian functions to the histograms of figure 2, yielding average boundary curvatures $\langle H \rangle$ of -0.0256, -0.0250 and -0.0163 for $N = 25$, 100 and 800, respectively; the corresponding relative errors are 2.4%, 0% and -34.8%. From this we conclude that the optimal smoothing parameter N_{opt} is approximately 100 when $R = 40$.

Intuitively, one might expect N_{opt} to be related to the resolution with which the surface in question is voxelized [7]—*e.g.*, to the ratio between the size R and the voxel size. Higher

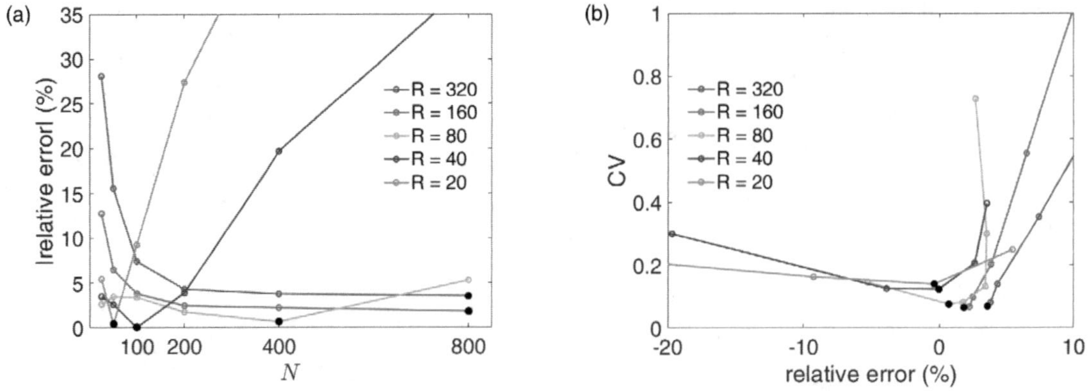

Figure 3. (a) Absolute value of the relative error (RE) in estimated boundary curvature, plotted against N for different sizes R of the test structure. Points of minimum RE are marked by filled black circles. (b) Coefficient of variation (CV) plotted against RE for various R values. The black circles in (b) correspond to the points of minimum RE marked in (a).

resolution ought to yield a more accurate representation of grain boundaries, thereby better preserving their intrinsic curvature. We test this notion by allowing R to take values of 20, 40, 80, 160 and 320, varying N in each case from 25 to 800. In figure 3(a), we plot the absolute value of the relative error (RE) in curvature against N for each R. As in the previous example (figure 2), the RE is found to be large when N is either too small or too large. For each R, the optimal choice for N is marked in figure 3(a) by a filled black circle. Clearly, N_{opt} varies with the size of the tetrahedron, tending toward larger values with increasing R. This is consistent with the conclusion drawn by Zhong *et al.* [7] from spherical test structures of various diameters.

In a real microstructure, however, the effect of smoothing on a specific boundary is not directly correlated with the grain size [7]: after all, it is possible for any surface of a chosen grain to be highly curved or nearly flat. Since N_{opt} cannot be chosen based on grain size, we must formulate a different strategy for optimizing the smoothing power—preferably one that can be applied to grain boundaries on an individual basis.

In all curvature histograms that we extracted from surfaces of the test structure, N_{opt} was found to generate not only the smallest RE but also the narrowest distribution. To quantify the latter, we use the coefficient of variation (CV), which is defined as the ratio of the standard deviation σ to the absolute value of the mean $|\mu|$, both extracted from a Gaussian fit. In figure 3(b) we plot CV against RE for different sizes of the Reuleaux tetrahedron. For each R, the point of minimum CV is marked by a filled black circle. These points coincide with the points of lowest RE in figure 3(a). We conclude that the minimum-CV criterion identifies smoothing conditions that lead to curvature estimation with a relative error below 5%.

3.2. Boundary curvatures of polycrystalline Al-5 wt% Cu

The curvature analysis procedure established in the previous section is now applied to an experimental dataset. A cylindrical sample of diameter 1.4 mm was cut by spark erosion from an Al-5 wt% Cu plate that had been cold rolled to 50% reduction in thickness, followed by heat treatment at 640°C for 20 min. The microstructure of this specimen was mapped in 3D by 3DXRD microscopy performed at the synchrotron radiation facility SPring-8. The morphology of individual grains was reconstructed from near-field diffraction patterns using the reconstruction routine described in Refs. [5,11,12]. The voxel size of the dataset is 5 μm.

Figure 4 shows a 3D mapping of the sample. The internal microstructure of the sample is visible in a cross section, in which GBs are marked in black: most of the latter are curved, but

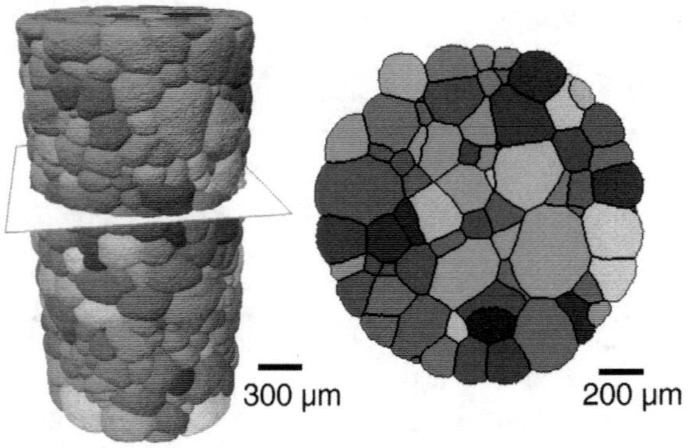

Figure 4. 3D mapping of polycrystalline Al-5 wt% Cu, performed by 3DXRD microscopy. Grain colors are indexed to the local crystallographic orientation via the components of the Rodrigues vector. The cross section shows the sample's internal microstructure, with grain boundaries traced in black.

some are nearly flat. We mapped a total of 942 grains and 5082 grain boundaries. Adopting the previous sections's method of curvature extraction, we plot the resulting curvature values as the blue histogram in figure 5. Most grain boundaries in the specimen are found to have rather small average curvatures, as the median of the curvature distribution is just $0.0016\,\mu\text{m}^{-1}$.

4. Discussion

In Ref. [7], Zhong *et al.* held N and λ fixed while extracting the curvature of GBs from voxel-based measurements of austenitic and ferrite steel microstructures. Their choice of optimal smoothing power is equivalent to $N = 50$ in our study. In figure 5 we compare our variable-N analysis of boundary curvature in polycrystalline Al-5 wt% Cu to curvatures determined using $N = 50$. The median value is nearly identical in the two cases ($0.0017\,\mu\text{m}^{-1}$ for $N = 50$ vs. $0.0016\,\mu\text{m}^{-1}$ for N_{opt}), but there is a difference for boundaries with $\langle H \rangle > 0.002\,\mu\text{m}^{-1}$, which the fixed-$N$ approach systematically undersmooths, leading to an increased number of boundaries having larger curvature values compared to the distribution produced by our procedure.

The median value for $\langle H \rangle$ of $0.0016\,\mu\text{m}^{-1}$ corresponds to a median radius of curvature of $\sim 630\,\mu\text{m}$, which is 7.5 times larger than the specimen's average (equivalent-sphere) grain radius of $84\,\mu\text{m}$. We conclude that GBs in this sample are much flatter on average than would be expected simply by equating the average boundary curvature to the inverse average grain radius.

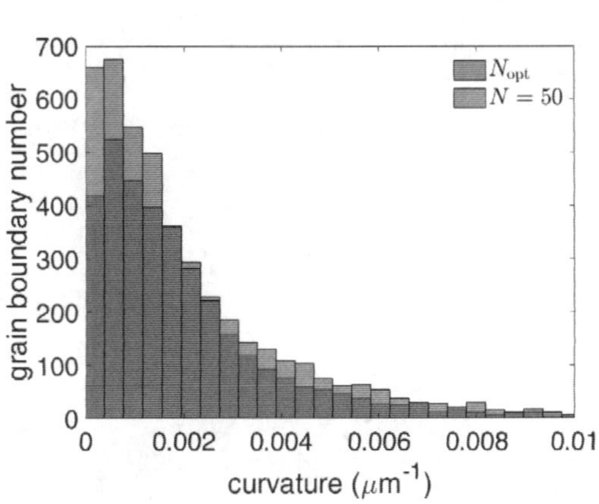

Figure 5. Distribution of grain boundary curvatures $\langle H \rangle$ in Al-5 wt% Cu, extracted using $N = 50$ for all boundaries (orange bars) versus each boundary's N_{opt} value (blue bars).

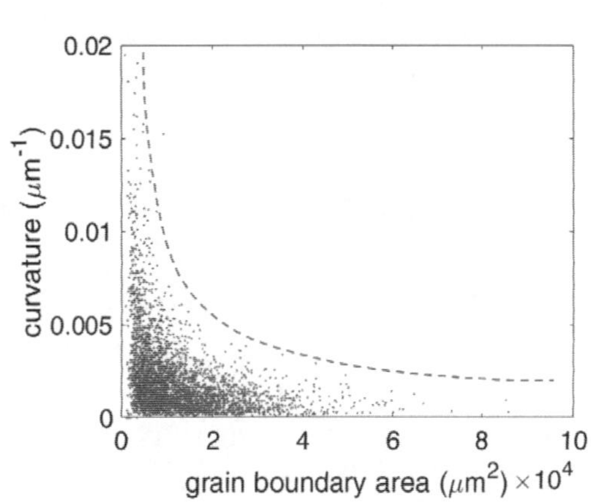

Figure 6. Boundary curvature $\langle H \rangle$ plotted as a function of boundary area in Al-5 wt% Cu. The red dashed line marks the approximate upper bound of the data.

This finding agrees with the qualitative impression given by the cross section in figure 4, in which several nearly flat boundaries are visible.

Finally, in figure 6 we examine the relationship between GB curvature and boundary area. Although most curvature values are small, regardless of the boundary area, there is a clear trend (highlighted by the red dashed line) for boundaries of larger area to have smaller curvature. Likewise, whenever a boundary's curvature is large, the corresponding area is small.

5. Conclusions

The software package DREAM.3D provides built-in functions for extracting grain boundary curvatures from voxel-based representations of 3D microstructure, but accurate values are obtained only as long as appropriate parameters are determined for the smoothing step that follows meshing of the input data. Using test structures based on a Reuleaux tetrahedron of various sizes, we validate a new method for choosing the best-performing smoothing conditions, finding that the most accurate curvature values are obtained when the coefficient of variation of a boundary's curvature distribution is minimized. The method is applied to a voxel-based 3D mapping of polycrystalline Al-5 wt% Cu that was acquired using x-ray microscopy. In this sample, we find most grain boundaries to be relatively flat, with a median value of only $0.0016\,\mu\text{m}^{-1}$ for the overall distribution of boundary curvatures. We also discover that grain boundaries of larger area tend to manifest smaller curvature values, and highly curved boundaries are always small in area.

Acknowledgments

We thank Dimitri Molodov (RWTH Aachen) for graciously providing the Al-5 wt% Cu sample. We are likewise grateful to the Japan Synchrotron Radiation Research Institute for the allotment of beam time on beamline BL20XU of SPring-8 (Proposal 2015A1580).

References

[1] Burke J and Turnbull D 1952 *Progress in Metal Physics* **3** 220–92

[2] Bhandari Y, Sarkar S, Groeber M, Uchic M D, Dimiduk D M and Ghosh S 2007 *Computational Materials Science* **41** 222–35

[3] Rowenhorst D J, Gupta A, Feng C R and Spanos G 2006 *Scripta Materialia* **55** 11–6

[4] Poulsen H F 2004 *Three-dimensional X-ray Diffraction Microscopy: Mapping Polycrystals and their Dynamics* vol 205 (Berlin: Springer-Verlag)

[5] Schmidt S, Olsen U, Poulsen H F, Soerensen H O, Lauridsen E M, Margulies L, Maurice C and Jensen D J 2008 *Scripta Materialia* **59** 491–94

[6] Groeber M A and Jackson M A 2014 *Integrating Materials and Manufacturing Innovation* **3** 5

[7] Zhong X, Rowenhorst D J, Beladi H and Rohrer G S 2017 *Acta Materialia* **123** 136–45

[8] Wu S 2017 *Analysis of Boundary Curvature in Voxel-Based Representations of Polycrystalline Microstructures* Master's thesis Ulm University

[9] Field D A 1988 *Communications in Applied Numerical Methods* **4** 709–12

[10] Goldfeather J and Interrante V 2004 *ACM Transactions on Graphics* **23** 45–63

[11] Schmidt S 2014 *Journal of Applied Crystallography* **47** 276–84

[12] Dake J M, Oddershede J, Srensen H O, Werz T, Shatto J C, Uesugi K, Schmidt S and Krill III C E 2016 *Proceedings of the National Academy of Sciences* **113** E5998–6006

Local residual stress in partially recrystallized iron characterized using high resolution electron backscatter diffraction

T Wang [1], D Juul Jensen [1], A Godfrey [2], Y Zhang [1*]

[1] Department of Mechanical Engineering, Technical University of Denmark, Kgs. Lyngby 2800, Denmark

[2] Key Laboratory of Advanced Materials (MOE), School of Materials Science and Engineering, Tsinghua University, Beijing 100084, China.

*Email address: yubz@mek.dtu.dk

Abstract. Recent synchrotron measurements have shown the presence of local residual stress within nearly defect-free grains in partially recrystallized samples. A detailed quantification of such local stress is very important for understanding local boundary migration during recrystallization and for tailoring the microstructure of materials. In this study, we utilize high resolution electron backscatter diffraction (HR-EBSD, Wilkinson's method) to investigate the local residual stress in recrystallized grains in a sample of partially recrystallized iron. The effects of reference selection on the strain distribution are also discussed.

1. Introduction

During recrystallization of deformed metals, new nuclei/grains form and grow by boundary migration 'consuming' the surrounding deformation matrix. It is generally assumed that recrystallized grains are largely defect-free and stress-free. However, very recent studies using advanced synchrotron techniques have shown the presence of ultra-low angle boundaries inside recrystallized grains of partially recrystallized Al [1], as well as the presence of local residual stress within recrystallized grains in partially recrystallized Ni [2] and iron [3]. How the internal structure and the local residual stress develop during recrystallization and how they may affect the recrystallization process have not yet been studied.

One reason for this is the lack of a characterization technique that can characterize both crystallographic orientations and elastic strain/stress with high precision. Although advanced synchrotron techniques are powerful, enabling non-destructive characterization of both crystallographic orientation and elastic strain in bulk samples, the limited access to these techniques prevents systematic studies, which are essential for understanding the effects of different thermomechanical treatments on the formation of residual stress, from being carried out. Alternatively, electron microscopy, specifically high resolution electron backscatter diffraction (HR-EBSD), has been demonstrated to be capable of characterization of both the microstructure and local elastic strain in detail [4-7]. For example, with HR-EBSD, elastic strains have been determined in Si_{1-x} Ge_x with micrometer spatial resolution [7]. The cross-correlation principle of HR-EBSD allows detection of small lattice rotations with in the best case high angular resolution down to 0.01° [8]. Additionally, with HR-EBSD the full strain tensor can be readily determined, based on which the stress tensor can be derived, while it is still rather difficult to determine full strain tensor with synchrotron techniques. These features, together with the ease of access to EBSD systems make HR-EBSD a powerful technique for studying the development of local strain (both plastic and elastic) during thermomechanical processing. Nevertheless, it has been pointed out that absolute stress values may not be accurately obtained by this method in some cases and additionally the effect of reference selection needs further evaluation [9, 10].

The aim of the present paper is to explore the possibilities of using HR-EBSD for characterization of the local microstructure and elastic strain and stress within individual recrystallized grains in partially recrystallized pure iron. The effects of choosing different reference patterns on the strain calculation are also studied, and the mechanisms leading to formation of the local residual stress are briefly discussed.

2. Experimental

A pure iron sheet was cold rolled to a reduction of 50% in thickness followed by annealing at 550 °C for 10 minutes to obtain a partially recrystallized structure. After annealing, the sample was water quenched. The sample was about 13% recrystallized with an average grain size for the recrystallized grains of about 16 μm after annealing. The sample longitudinal section (defined by the sample normal direction (ND) and rolling direction (RD)) was mechanically polished, followed by electropolishing using an A2 solution.

The sample was characterized using a Zeiss Supra 35 scanning electron microscope equipped with an EBSD acquisition system. For data collection and preliminary data analysis the software package, Channel 5, was used. During EBSD mapping, high quality Kikuchi patterns within a recrystallized grain were obtained without binning, and were saved on the local hard disk. For the HR-EBSD characterization the raw patterns were analyzed using the commercial software package, CrossCourt3, from which elastic strain and lattice rotation angles for each pixel in the recrystallized grain were calculated.

3. Results and discussion

An image of the grain selected for analysis using HR-EBSD is shown in figure 1. The figure shows an inverse pole figure (IPF) map, a kernel average misorientation (KAM) map and an image quality (IQ) map. A low angle boundary with misorientation of 1.0 ± 0.3° is seen in the center-left part of the recrystallized grains. To avoid effects of this low angle boundary on the determination of residual stress, only the major part of the grain with a grain size of ~10 μm is included in the analysis. For accurate determination of stress and strain values, the selection of a proper reference Kikuchi pattern is critical, and this should be from a strain-free volume, as displacements between all the Kikuchi patterns and the selected reference pattern are used for calculation of the strain distribution. To be able to evaluate the importance of the choice of reference pattern, in the present work we select two patterns: one from a pixel with low KAM value (marked with a black diamond in figure 1b) and one with high IQ (marked with a white diamond in figure 1c). These two choices are both considered appropriate for references and are defined as references 1 and reference 2 in the following.

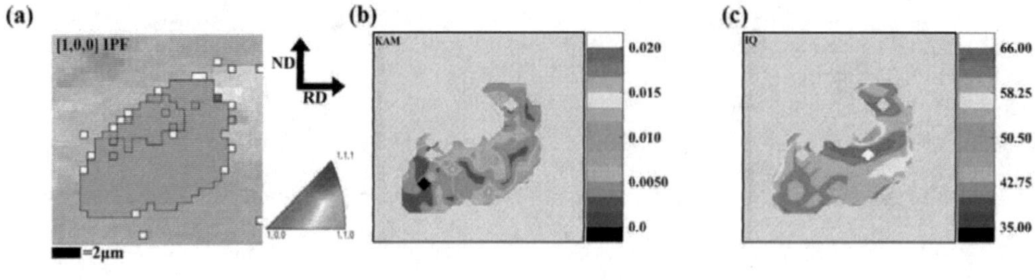

Figure 1. HR-EBSD parameters for the selected recrystallized grain: (a) IPF map showing the whole grain including the central left region which is separated from the major part of the grain by a low angle boundary; (b) KAM map and (c) IQ map of the major part of the grain. The grey areas, including the two small regions inside the analyzed grain, represent not analyzed pixels.

IOP Conf. Series: Materials Science and Engineering **580** (2019) 012049 doi:10.1088/1757-899X/580/1/012049

Figure 2a and 2b show the spatial distribution of the ε_{33} strain component (along the sample transverse direction (TD); normal to the characterization surface) within the selected grain determined based on the two reference Kikuchi patterns. The presence of local residual strain within the recrystallized grain is evident using both two reference Kikuchi patterns; ε_{33} strain is distributed through the whole grain, showing positive values located within the grain interior and along the grain boundary, while extreme values are only observed at some places close to the grain boundary, where the data are very likely to be outliers. The extreme values are at the same positions for both references, and the majority of the two spatial distributions are similar, though within some regions the strain values are smaller using reference pattern 1 compared to reference pattern 2 (an example region is marked with a red block in figures 2a and 2b), while in others areas the opposite is true (i.e., larger values using reference pattern 1 compared to reference pattern 2; an example is marked with a black block in figures 2a and 2b).

The relative frequency of ε_{33} strain values is shown in figures 2c and 2d, determined using the two reference Kikuchi patterns. The overwhelming majority of the strain values are positive, with no significant difference found between the two distributions. Excluding the outliers, the variation of the ε_{33} component within this grain is about 1×10^{-3} for both cases, which agrees well with the result obtained using synchrotron micro-diffraction [3].

Generally, the elastic strain values determined based on the two references are roughly comparable. This could be due to the fact that both reference Kikuchi patterns are from recrystallized grains and have high quality (see figure 1c). This differs from the situation reported in Ref. [9, 10] where Kikuchi patterns from volumes with elastic/plastic strain are used as references. In the following, only the results obtained based on reference 1 are presented.

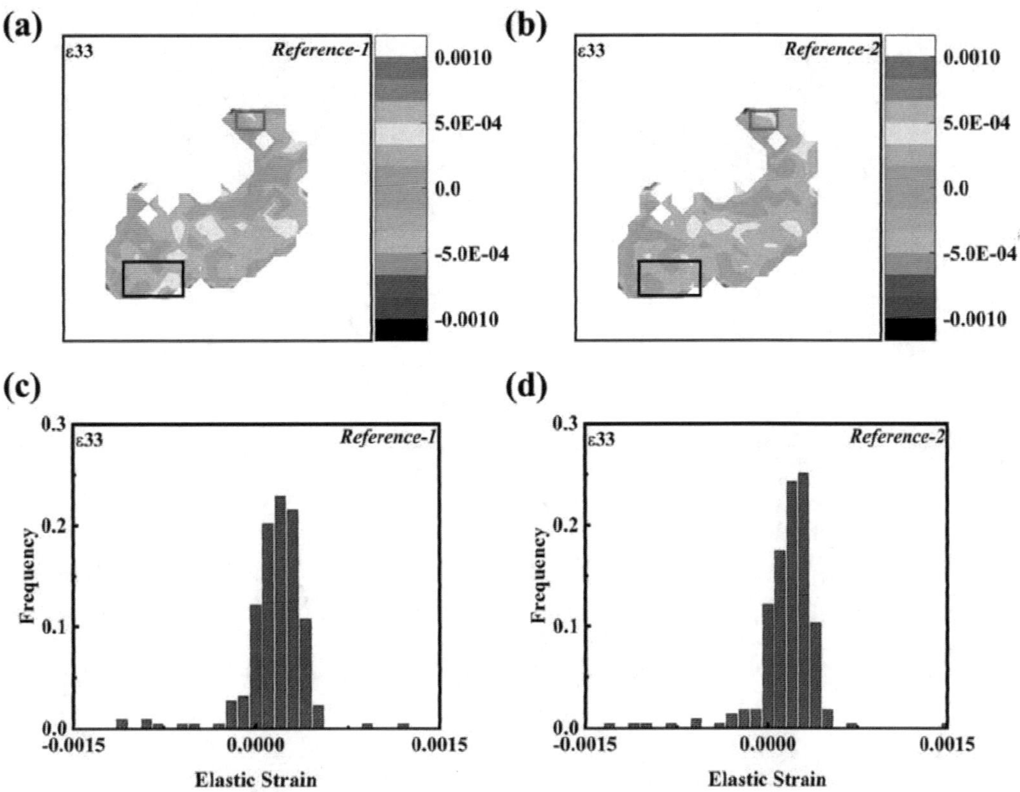

Figure 2. Maps showing the spatial distribution of the ε_{33} strain component determined using (a) reference pattern 1 and (b) reference pattern 2. Black and red boxes mark areas where the strain values are smaller and larger, respectively, determined using reference pattern 1 compared to reference pattern 2. Relative frequency of residual strain determined using (c) reference 1 and (d) reference 2.

Figure 3 presents maps showing the distribution of orientation deviation, shear strain, and normal strain based on the use of reference 1. A gradual rotation of orientation, with angle up to 0.29°, is observed in the regions adjacent to the boundary between the selected grain and the center-left grain, see figure 3a. The low angle boundary between the two grains has also some impact on the local distribution of the strain components (see figures 3b and 3c). Except for some extreme values at the grain boundary, the majority of the shear strain components are small (close to zero), whereas the magnitude of the normal strain components is relatively high, with maximum close to 1×10^{-3}. In contrast to the positive strain values of ε_{33}, the values of ε_{11} strain (along the sample RD) are mainly negative, while values of the ε_{22} strain (along the sample ND) span both positive and negative values in the range $-1 - 1 \times 10^{-3}$. No spatial correlation between the different strain components is found.

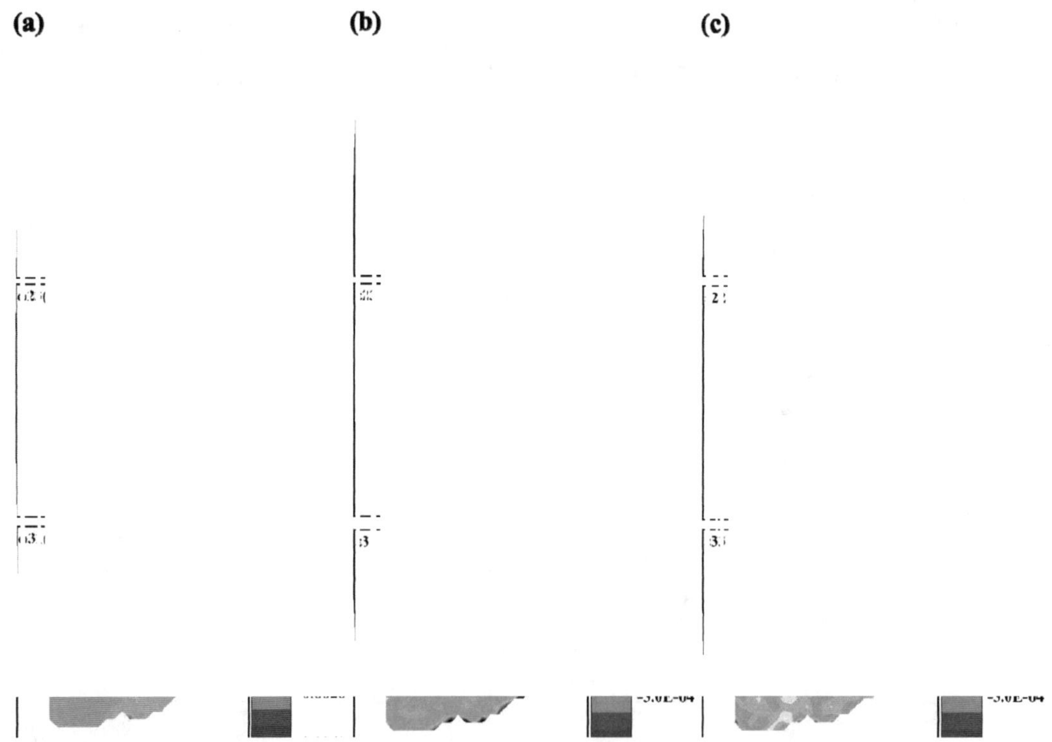

Figure 3. Maps showing the spatial distribution of (a) orientation deviation, (b) shear strain, and (c) normal strain components, in each case calculated using reference 1.

Figure 4 presents maps showing the distribution von Mises Stress, shear stress, and normal stress, determined using reference 1. As shown in figure 4a, the von Mises Stress, calculated based on the stress of all components, reveals the presence of local residual stress variations, with a mean of 227 MPa within the recrystallized grain. This stress level is comparable to the yield stress of the recrystallized grains, 235 MPa, as estimated using a Hall-Petch equation [11,12]:

$$\sigma_y = \sigma_0 + k_y/\sqrt{d} \tag{1}$$

where σ_y is the yield stress, σ_0 is a friction stress of 50 MPa [13], k_y is the Hall-Petch slope of 740 MPa*$\mu m^{0.5}$ [14], and d taken as a grain diameter of 16 μm.

IOP Conf. Series: Materials Science and Engineering **580** (2019) 012049 doi:10.1088/1757-899X/580/1/012049

In accordance with the strain distributions, the local residual stress within the selected grain matches the tendency of the distributed gradients in elastic strains (see figures 4b and 4c), despite the fact that the σ_{33} components are zero everywhere, resulting from the assumption in the software that the out-of-plane stress is precisely zero at the free sample surface.

The present results evidently show the presence of local residual stress/strain and orientation variations within recrystallized grains, in agreement with findings from synchrotron experiments [1-3]. The strain values of the ε_{33} component obtained from HR-EBSD is also similar to that from synchrotron micro-diffraction experiments on the same sample [3]. However, the HR-EBSD measurements are conducted at the free sample surface, where the stress state can be different from that within the bulk. It is important therefore to study the stress state inside the bulk grains using synchrotron techniques, e.g. micro-diffraction, in the future.

Nevertheless, the results show evidently that the sign of the strain/stress components along different sample directions are distinctly different, suggesting that the formation of local residual stress may be related to the sample macroscopic strain. The local strain variation within the recrystallized grain also suggests that the local strain variation in the surrounding deformation matrix may have an influence. Statistically sound data collected on many more recrystallized grains using HR-EBSD and synchrotron techniques are required in the future to clarify the mechanisms leading to the formation of local residual stress and their effects on the recrystallization process. The orientation variation data obtained with HR-EBSD (see figure 3a) can also be used to determine the local dislocation tensor and thus local geometrically necessary dislocation density [15, 16], which will also be conducted in the near future.

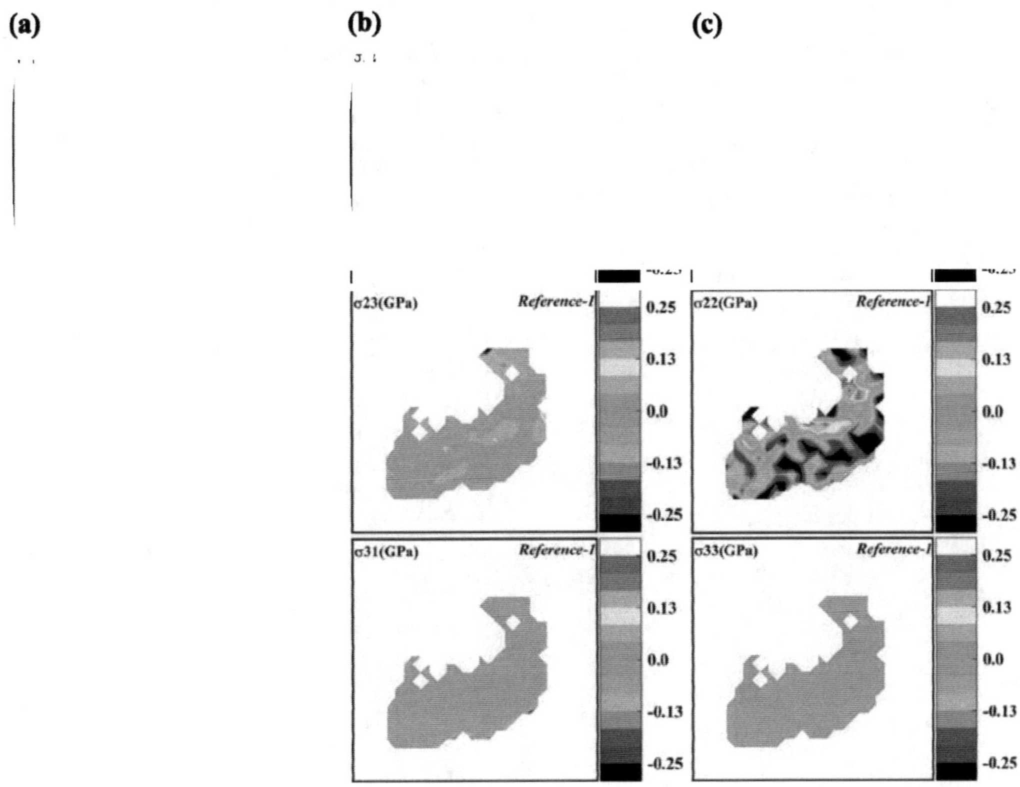

Figure 4. Maps showing the distributions of (a) von Mises Stress, (b) shear stress, and (c) normal stress measured using reference 1.

4. Summary

In the present paper, HR-EBSD has been used to measure the local residual elastic strain and orientation variations in a recrystallized grain within a partially recrystallized iron sample. The measurements evidently show the presence of local residual elastic strains inside the recrystallized grain. The results further show that the normal strain values are higher than those of the shear strain, and that the sign of the three normal strain components is distinctly different. The strain values vary and the sign of the stress values can change from positive to negative locally within the recrystallized grain. The results suggest that both the macroscopic sample strain and the local plastic strain of the deformed grains around the recrystallized grains may have an impact on the formation of local residual stress.

Acknowledgements
The authors acknowledge the financial support by the Danish Council for Independent Research Fund (8022-00340B).

Reference

[1] Ahl S R, Simons H, Zhang Y, Detlefs C, Stohr F, Jakobsen A C, Juul Jensen D and Poulsen H F 2017 *Scr. Mater.* **139** 87
[2] Zhang Y and Barabash R 2019 *Quantum Beam Sci.* **3(1)** 6
[3] Zhang Y, Yu T, Godfrey A, Xu R and Juul Jensen D 2019 *in preparation*
[4] Arsenault RJ and Taya M 1987 *Acta Mater.* **35** 651
[5] Zhang Y, Andriollo T, Faester S, Barabash R, Xu R, Tiedje N, Thorborg J, Hattel J, Juul Jensen D and Hansen N 2019 *Acta Mater.* **167** 221
[6] Troost K, van der Sluis P and Gravesteijn D 1993 *Appl. Phys. Lett.* **62** 1110
[7] Wilkinson A 1996 *Ultramicroscopy* **62(4)** 237
[8] Britton T and Wilkinson A 2011 *Ultramicroscopy* **111** 1395
[9] Mikami Y, Oda K, Kamaya M and Mochizuki M 2015 *Mater. Sci. Eng., A* **647** 256
[10] Koga N, Nakada N, Tsuchiyama T, Takaki S, Ojima M and Adachi Y 2012 *Scr. Mater.* **67** 400–403
[11] Hall E O 1951 *Proc. Phys. Soc. London Sect. B.* **64** 747–753
[12] Petch N J 1953 *J. Iron Steel Inst.* **174** 25–28
[13] Jago R A and Hansen N 1986 *Acta metall.* **34** 1711–1720
[14] Ruan Q, Yang M, Liu W and Godfrey A 2019 *Mater. Sci. Eng. A* **744** 764–772
[15] Ruggles T J and Fullwood D T 2013 *Ultramicroscopy* **133** 8-15
[16] Konijnenberg P J, Zaefferer S and Raabe D 2015 *Acta Mater.* **99** 402-414

IOP Conf. Series: Materials Science and Engineering **580** (2019) 012050 doi:10.1088/1757-899X/580/1/012050

Simultaneous enhancement of strength and ductility in a fine-grained Fe-30Mn-0.11C steel at 77 K

Y H Wang[1,2*], J M Kang[1], X M Chen[1], T Yu[2], N Hansen[3,4] and X Huang[2,4]
[1]National Engineering Research Center for Equipment and Technology of Cold Strip Rolling, Yanshan University, Qinhuangdao 066004, P. R. China
[2]Department of Mechanical Engineering, Technical University of Denmark, DK-2800 Kgs. Lyngby, Denmark
[3]Risø Campus, Technical University of Denmark, DK-4000 Roskilde, Denmark
[4]College of Materials Science and Engineering, Chongqing University, 400044 Chongqing, P. R. China

Email: yhwang@ysu.edu.cn

Abstract. High Mn austenitic steels are a class of materials for cryogenic applications. Previous studies on Fe-34.5Mn-0.04C austenitic steels have shown that refining grain size can suppress the formation of deformation-induced ε and α' martensite and improve the tensile strength and ductility at temperatures down to 123 K. Here we report a new observation of simultaneous increase of strength and ductility at liquid nitrogen temperature (77 K) in a fine grained Fe-30Mn-0.11C (wt. %) steel where the addition of 0.11%C stabilizes the austenite. This improvement is attributed to complete suppression of martensitic transformation and enhanced occurrence of deformation twinning.

1. Introduction

High manganese austenitic steels have been widely studied for their outstanding tensile strength and ductility [1-3]. Some high Mn austenitic steels have also been investigated for cryogenic applications [4, 5]. However, ε-martensitic transformation may occur in this class of high Mn steels during plastic deformation, which often causes premature fracture through the interactions of ε-martensite/ε-martensite, ε-martensite/annealing twin boundaries, and ε-martensite/grain boundaries. In our previous studies on Fe-34Mn-0.04C austenitic steels it has been shown that refining the grain size can suppress martensitic transformation at grain boundaries and in the grain interior regions, and as a result the tensile strength and ductility is significantly improved at temperatures even down to 123 K. However, at 93 K, the Fe-34.5Mn-0.04C steel showed a decrease in elongation and tensile strength due to premature fracture caused by the formation of deformation-induced α′-martensite [6] and ε-martensite [7].

There are two effective methods to suppress the ε-martensitic transformation. One method is to refine the austenite grain size [8]. A previous study has shown that the embrittlement was eliminated by grain refinement to 460 nm in Fe-17Mn-0.6C steel [9]. The other method is to add alloying elements such as carbon [10], aluminium [11] and manganese [12, 13] that are austenite stabilizers. It is possible to obtain stable austenite even at cryogenic temperatures when enough carbon and manganese are added. An early investigation has reported that for the binary Fe-Mn system, when the Mn content increases to 30~40%, the dominant plastic deformation mechanism is dislocation slip and martensitic transformation is suppressed. When the Mn content was increased to 36%, no ε-martensite was observed in samples tensile tested at room temperature, and only 4.2% ε-martensite was formed after tensile deformation to a strain of more than 60% at liquid nitrogen temperature. To enhance the stability of austenite, we added 0.11%C to an Fe-30Mn alloy to produce an Fe-30Mn-0.11C austenitic steel. A fully recrystallized fine-grained sample with a mean grain size of 5.6 μm was prepared by conventional cold rolling and annealing. The tensile properties of this fine-grained steel were tested at 293 K and 77K. The strength and ductility at 77K are increased simultaneously and significantly as compared with the results at room temperature. This enhancement is attributed to the complete elimination of grain boundary ε-martensite and the enhanced formation of deformation twins within the grains.

2. Experimental

The steel studied was melted using a vacuum induction technique and had a main chemical composition of Fe-30Mn-0.11C (wt.%). The ingot was heated to 1000 °C and held for 1 h, and then forged at 800~1000 °C to produce a 20 mm thick plate. The plate was cold rolled by 50% in thickness reduction. Then, annealing was carried out at 700 °C for 2h to obtain a fully recrystallized grain structure with a mean grain size of 5.6 μm.

Tensile specimens with a gauge length of 10 mm, a width of 1.8 mm and a thickness of 1.8 mm were prepared from the cold rolled and annealed sample. Uniaxial tensile tests with the tensile axis aligned along the rolling direction (RD) were carried on an MTS tensile testing machine equipped with a cryogenic chamber. Tensile tests were carried out at an initial strain rate of 10^{-3} s^{-1} at RT and 77 K.

A field emission scanning electron microscope (FE-SEM) equipped with a backscattered electron (BSE) detector and an electron backscatter diffraction (EBSD) system was employed for characterization of the microstructure and fracture morphology.

3. Results and discussion

3.1. Microstructure

Figure 1a shows the microstructure after 50% cold rolling. The original austenite grains are elongated along the RD and deformation twins are formed within many of the grains. X-ray diffraction (XRD) analysis shows diffraction peaks corresponding to austenite (Figure 1b). However it is seen that the γ (200) diffraction peak is very weak, whereas the γ (111) diffraction peak is very strong, suggesting the evolution of a texture in the material.

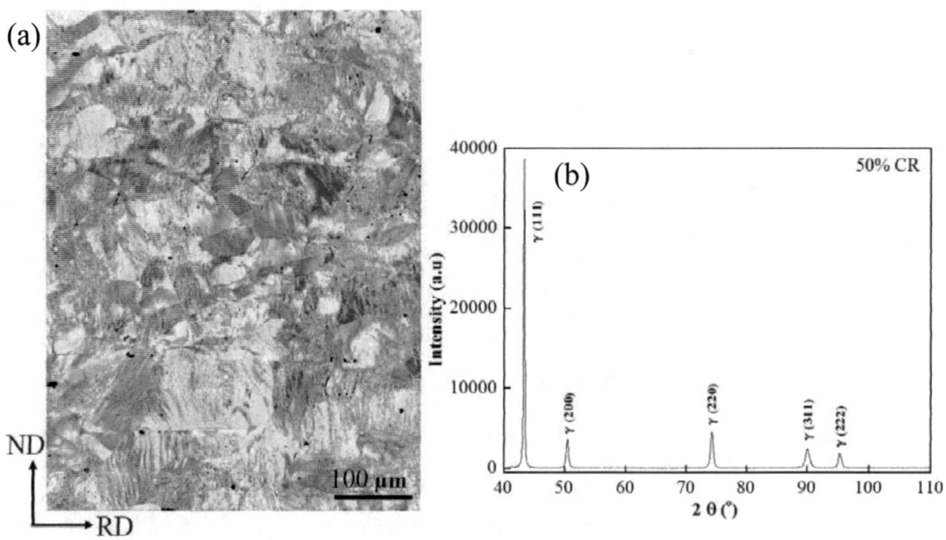

Figure 1. (a) EBSD map of the sample cold rolled to 50% showing the morphology of deformed grains and the formation of deformation twins in the grains. (b) XRD pattern showing austenite single phase.

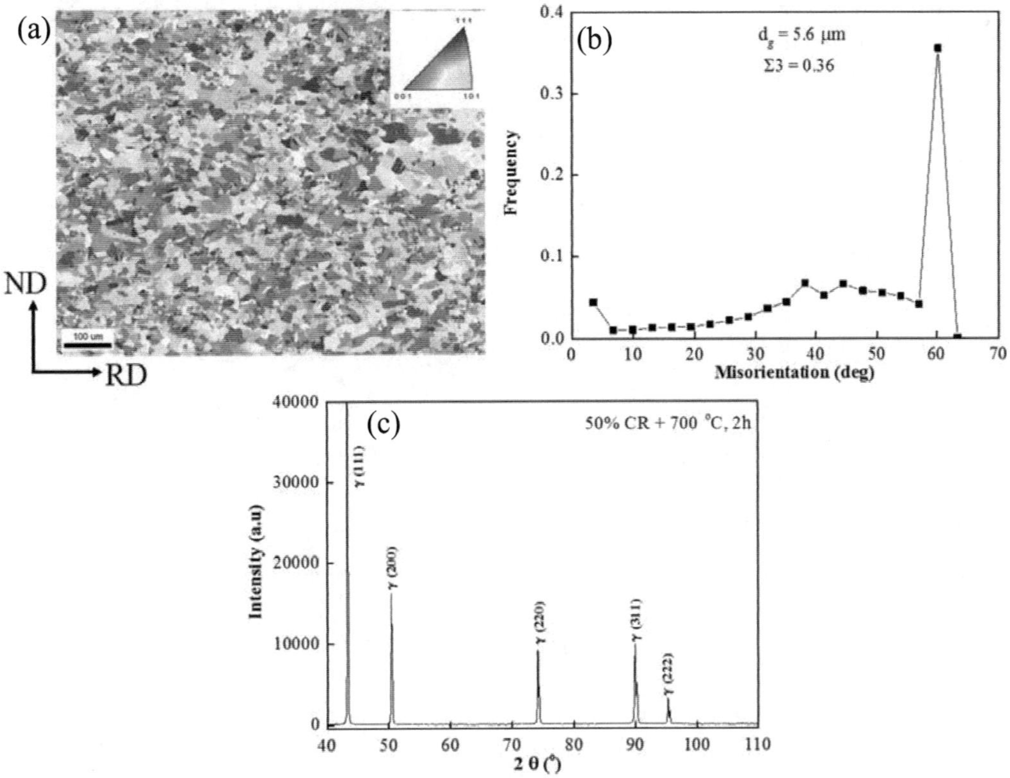

Figure 2. Microstructural analysis of the sample cold rolled to 50% thickness followed by annealing at 700 °C for 2 h. (a) EBSD orientation map showing the recrystallized grain structure containing annealing twins, (b) misorientation distribution and (c) XRD pattern.

Cold rolling to 50% thickness followed by annealing for 2h at 700 °C led to full recrystallization, resulting in the formation of a fine grain structure containing many annealing twins (figure 2a). The area fraction of Σ3 boundaries was measured to be 0.36 (figure 2b). By taking the twin boundaries into account, the mean grain size was measured to be 5.6 μm. XRD measurement shows a single austenite phase (figure 2c), with no detection of either ε-martensite nor α′-martensite. It is concluded that a fully recrystallized fine grain structure is obtained in Fe-30Mn-0.11C steel through conventional cold-rolling and annealing treatment.

3.2. Tensile properties and fracture morphologies

Figure 3a shows the engineering stress-strain curves of Fe-30Mn-0.11C steel tested at room temperature (293 K) and 77 K. At 293 K, the Fe-30Mn-0.11C steel shows yield and tensile strengths of 230 MPa and 531 MPa, respectively, along with a total elongation of 65%. At 77 K, higher yield and tensile strengths (360 and 860 MPa) and higher total elongation (84%) are observed.

The work-hardening rates of the two tested Fe-30Mn-0.11C samples were calculated from the corresponding true stress-true stain curves and are shown in figure 3b. Both samples exhibit good work hardening capacity, leading to the occurrence of tensile instability following the Considère's criterion:

$$\frac{d\sigma}{d\varepsilon} = \sigma \tag{1}$$

where σ and ε are the true stress and true strain, respectively. Note that a higher work hardening rate over the entire tensile strain range is observed at 77 K than 293 K, resulting in a larger uniform elongation at 77 K.

Figure 3a also shows stress-strain curves of the Fe-34.5Mn-0.04C steel tested at 293 K, 193 K, 123 K and 93 K from literature [6]. At 293 K, the yield strength, ultimate tensile strength and uniform elongation are 274 MPa, 564 MPa and 45%, respectively. With decreasing temperature, the yield strength gradually increases, but the tensile strength and the uniform elongation do not show a continuous increase with deceasing temperature. At 193K, the uniform elongation reaches a maximum value of about 54%; at lower temperatures the uniform elongation decreases and premature fracture takes place at 93 K with a total elongation less of than 20% [6]. Clearly the present alloy shows an enhanced combination of strength and ductility compared with the Fe-34.5Mn-0.04C alloy at cryogenic temperatures.

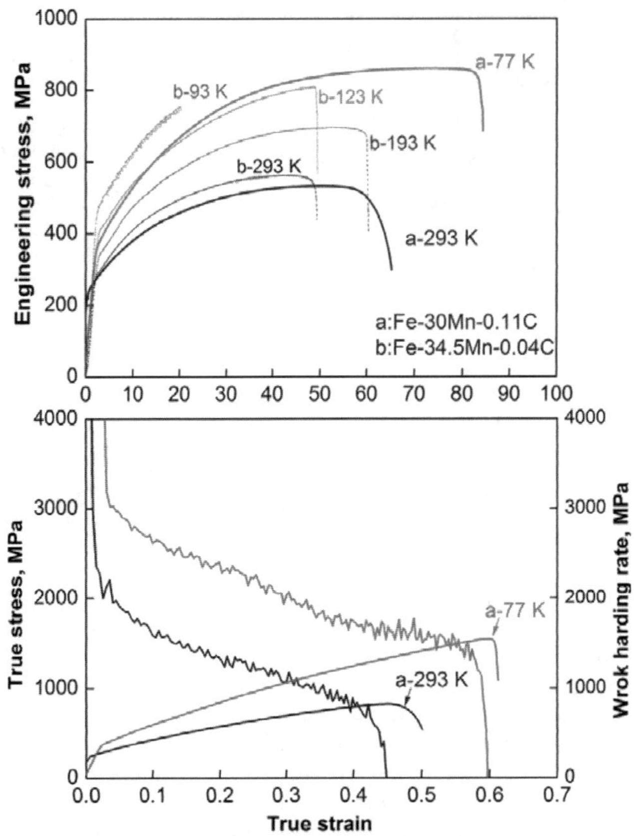

Figure 3. (a) Engineering stress-strain curve of the sample with a mean grain size of 5.6 μm of Fe-30Mn-0.11C steel and 3.8 μm of Fe-34.5Mn-0.04C steel, and (b) the corresponding work hardening rate of Fe-30Mn-0.11C.

Typical SEM micrographs obtained from the fracture surfaces of samples tensile tested at 293 K and 77 K are presented in figure 4. The fracture surface at both temperatures is mostly composed of dimples, showing ductile fracture.

3.3 Microstructural evolution after tensile-tested at 77 K

Figure 5 a-c shows inverse pole figure (IPF), image quality (IQ), and phase maps constructed from the EBSD data, as observed in longitudinal-section in the area beneath tensile-fractured surface of Fe-30Mn-0.11C steel tested at 77 K. Deformations twins are observed in many grains (see figure 5a and b) with no α′-martensite or ε-martensite detected. The suppression of martensite formation and the formation of deformation twins are considered to contribute to the enhancement of both strength and ductility at 77 K.

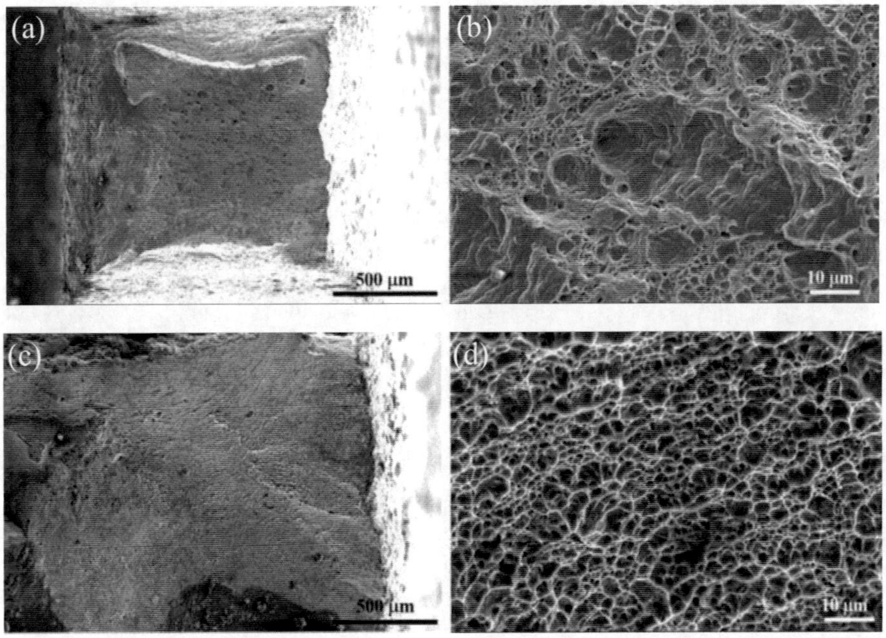

Figure 4. SEM fractographs of the tensile specimen fractured at (a) and (b) 293 K and (c) and (d) 77 K.

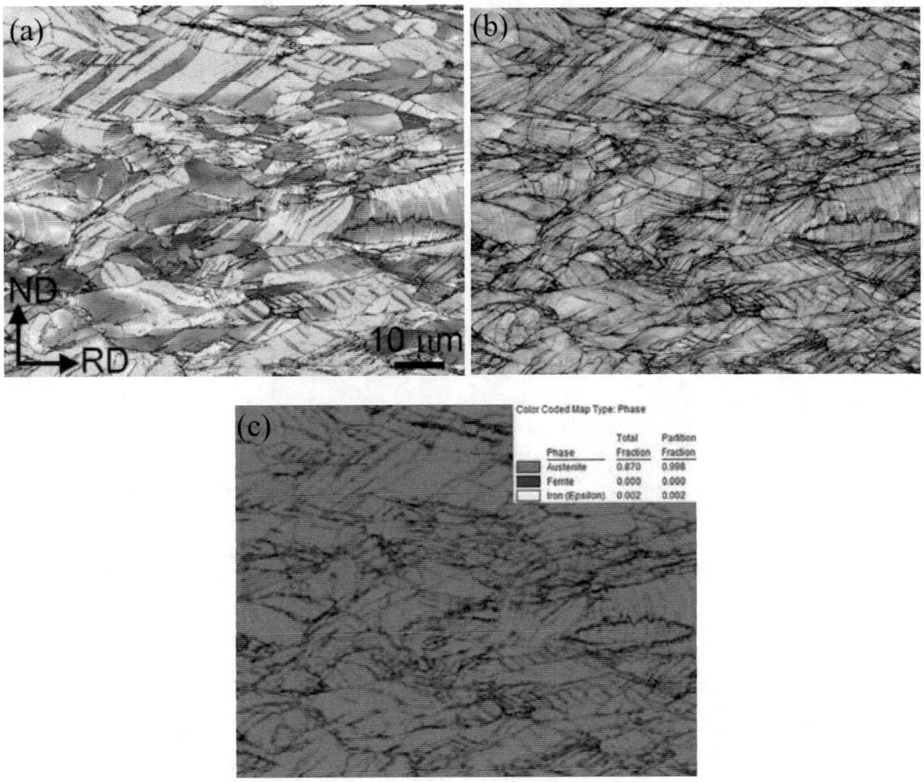

Figure 5 Inverse pole figure (IPF) map (a), image quality (IQ) map (b) and phase map (c) from an EBSD scan of the Fe-30Mn-0.11C steel after tensile-tested at 77 K.

4. Summary

A fine grained Fe-30Mn-0.11C steel with a mean grain size of 5.6 μm was successfully produced through traditional cold rolling and annealing. The strength and ductility was found to increased simultaneously with decreasing temperature in the range from 293 K to 77 K. This enhancement is attributed to full suppression of martensitic transformation and the activation of deformation-induced twins within the grains at cryogenic temperatures.

Acknowledgments

The authors gratefully acknowledge support from the National Nature Science Foundation of China (Grant No. 51871194) and the National Natural Foundation of Hebei Province, China (Grant No. E2018203312). XH thanks the support of State Key Research and Development Program of MOST of China (2016YFB0700403, 2016YFB0700401). NH thanks the support of the 111 Project (B16007) by the Ministry of Education and the State Administration of Foreign Experts Affairs of China.

References

[1] Bouaziz O, Allain S, Scott C P, Cugy P and Barbier D 2011 *Current Opinion in Solid State and Materials Science* 15(4) 141-168.
[2] Wang Y H, Kang J M, Peng Y, Wang T S, Hansen N and Huang X 2018 *Scr. Mater.* 155 41-5.
[3] Wang Y H, Kang J M, Peng Y, Wang T S, Hanse N n and Huang X 2018 *Journal of Materials Science & Technology* 34(10) 1939-1943.
[4] Grässel O, Kruge L r, Frommeyer G and Meyer L W 2000 *International Journal of Plasticity* 16 1391-1409.
[5] Wang Y H, Shi B D, He Y M, Zhang H W, Peng Y and Wang T S 2018 *Materials* 11(2).
[6] Wang Y H, Peng Y, He Y M, Wan g T S and Huang X 2014 *Proceedings of the 35th Risø Internaional Symposium on Materials Science*. 471-474.
[7] Wang Y H, Kang J M, Chen X M, Yu T B, Hansen N and Huang X X 2019 unpublished.
[8] Koyama M, Lee T, Lee C S and Tsuzaki K 2013 *Materials & Design* 49 234-241.
[9] Koyama M, Sawaguchi T, Lee T, Lee C S and Tsuzaki K 2011 *Materials Science and Engineering: A* 528(24) 7310-7316.
[10] Seol J B, Jung J E, Jang Y W and Park C G 2013 *Acta Materialia* 61(2) 558-578.
[11] Sohn S S, Hong S, Lee J, Suh B C, Kim S K, Lee B J, Kim N J and Lee S 2015 *Acta Mater*100 39-52.
[12] Tomota Y, Strum M and Jr. J W Morris 1988 *Metallurgical Transactions A* (19A) 1563-1568.
[13] Tomota Y, Strum M and Jr. J W Morris 1986 *Metallurgical Transactions A* (17A) 537-547.

Ultrafine-grained Tungsten by High-Pressure Torsion – Bulk precursor versus powder processing route

M Wurmshuber[1,*], S Doppermann[1], S Wurster[2] and D Kiener[1]

[1] Department of Materials Science, Montanuniversität Leoben, Jahnstraße 12, 8700 Leoben, Austria
[2] Erich-Schmid Institute for Materials Science, Austrian Academy of Sciences, Jahnstraße 12, 8700 Leoben, Austria
[*] corresponding author (michael.wurmshuber@unileoben.ac.at; +43 3842 804 409)

Abstract. The continuous enhancements and developments in the field of power engineering, as well as the uprising of nuclear fusion technology, demand novel high performance materials featuring exceptional strength and damage tolerance as well as durability in harsh environments. Ultra-fine grained bulk materials fabricated by high-pressure torsion, exhibiting a grain size less than 500 nm are promising candidates for these applications. Tungsten, the material of choice for plasma-facing materials in fusion reactors, is expected to exhibit even more enhanced properties by precise doping with impurity atoms, strengthening grain boundary cohesion. In order to allow this meticulous control of chemical composition, in-house mixing of the raw material powders is preferable to use of commercially available alloys. Several challenges arise in powder processing of tungsten via high-pressure torsion, originating in the intrinsic strength and high melting point of the material, and in the affinity of the powder to oxygen. Strategies to overcome these problems will be addressed in this work. Furthermore, we compare ultra-fine grained tungsten produced from a bulk precursor to that from the developed powder approach regarding microstructural features, hardness and rate-sensitive properties. The powder route showed promising and widely comparable results to the material processed from bulk tungsten, rendering it an effective way for fabricating ultra-strong tungsten, while keeping the additional ability to accurately control chemical composition and tailor grain boundary segregation states.

1. Motivation

When it comes to implementing new ground-breaking technologies or optimizing and improving existing concepts in the sectors of power engineering, nuclear engineering, armored protection or other safety relevant applications, materials fulfilling the required standards are often the limiting factor. Ultrastrong and damage tolerant materials are of demand in these fields of application: two characteristics that are deemed mutually exclusive in many conventional materials. Nanostructured and ultra-fine grained (ufg) metals are a promising material class to combine high strength with enhanced damage tolerance [1,2], while also featuring additional beneficial properties such as radiation tolerant behavior [3,4]. While tungsten is frequently considered as candidate material for the mentioned high performance applications due to its physical properties [5–7], a rather low fracture toughness and high ductile-brittle transition temperature often impedes the utilization of tungsten-based materials. The grain boundaries are found to be one of the weak links. However, ab-initio simulations revealed that by controlling grain boundary chemistry, the grain boundary cohesion can be increased [8–10]. The consequences are repression of intercrystalline fracture, the preferred fracture mode in nanostructured materials [1,2], based on which an enhanced fracture toughness can be expected [10]. Severe plastic deformation techniques, such as high-pressure torsion (HPT), have gained a lot of interest in recent years, as they introduce a high amount of deformation and microstructural refinement to materials, creating ultra-fine grained or nanocrystalline bulk materials. During HPT, a millimeter-sized, disk-shaped specimen is shear-deformed between two anvils under high confining pressure (several GPa). The applied von Mises-strain during HPT can be calculated using the number of rotations, the distance from the center of the disk and the thickness of the sample. Further details of the HPT process are reported elsewhere [11–13]. In order to control chemical composition precisely, a powder approach is necessary. Yet, before the chemical composition of tungsten can be altered, it must first be proven that it is possible to fabricate pure ufg tungsten with satisfactory properties from a powder starting material. Since several challenges arise with powder processing of tungsten at low temperatures using HPT, this work will address these problems and ways to mediate them. Finally, ufg tungsten samples produced from powder and a bulk precursor are compared regarding their mechanical properties.

2. Challenges with powder processing of ufg W via HPT

In order to realize systematic doping of tungsten grain boundaries, the local chemical composition of the material has to be tuned precisely. To investigate the effect of impurity atoms on the material properties thoroughly, a large number of samples with varying doping concentrations must be fabricated and characterized. Therefore, an in-house fabrication method is essential to provide a fast and easy route for processing such doped material samples, as well as to ensure access to crucial processing parameters. A powder metallurgical approach seems to be the best choice to realize this. When it comes to high-performance materials, their extraordinary properties that are desired for many of their applications often provide the biggest challenges and limitations in terms of fabrication. Tungsten is yet another example of this phenomenon, as its exceptional intrinsic strength and high melting point result in several problems that have to be tackled when processing ufg tungsten from powders on a laboratory scale. The most severe challenges that need to be handled will be addressed in this section:

2.1 Oxidation and contamination of powders

Oxidation and contamination is not only a problem for tungsten powder in particular, but for most metal powders in general. The small particle size of the powders (< 100 μm) leads to a high contact area with the surrounding atmosphere. In air and under humid conditions this can result in severe oxidation or contamination of powder particle surfaces. After processing from oxidized powder, the final components exhibit not only an oxidation layer on the surface, which could be removed rather easily, but also within the material, leading to considerable and most likely unwanted changes in material properties. Furthermore, the high hardness of metal oxides can lead to problems with compacting the powder particles or during later sintering, where oxides usually act as diffusion barriers.

In the case of W, oxidation by air already occurs at room temperature (RT). The oxidation rate accelerates significantly with increasing temperature and humidity. The oxide formed is always WO_3, which becomes volatile at temperatures above 750°C, leading to consumption of the material [6,14].

As powder oxidation has been a well known problem for decades, several solutions exist today. The most convenient one is storing and handling all powders within a so-called glovebox, a sealed container with a desired inert atmosphere (e.g. argon) inside. It is crucial that all powders are opened and manipulated inside the box using the designated gloves and that the oxygen and humidity level is kept to a minimum. For transporting the powder mixture outside of the box to the compacting facility (in this

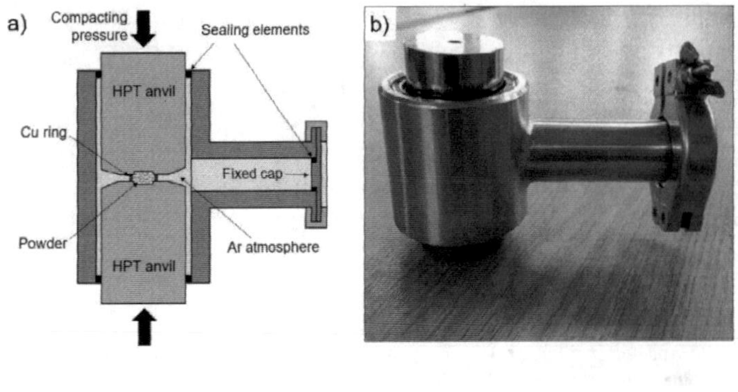

Figure 1. (a) Sketch and (b) photograph of sealed chamber to allow powder to be transported in local argon atmosphere to the HPT device.

IOP Conf. Series: Materials Science and Engineering **580** (2019) 012051 doi:10.1088/1757-899X/580/1/012051

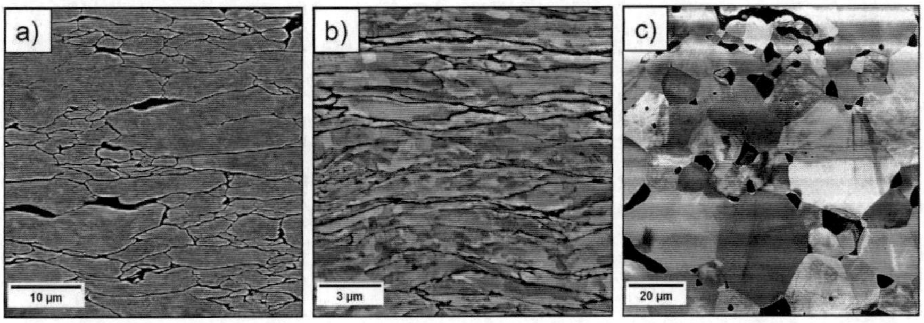

Figure 2. SEM micrographs (back-scattered mode) of (a) compacted W powder, (b) compacted and deformed W and (c) compacted and sintered W.

case the HPT tool) a small sealed container was utilized that has to be assembled together with the HPT anvils and the powder inside the glovebox (figure 1). This way the powder mixture can be transported from to the HPT tool in a local argon atmosphere and compacted without unwanted oxidation.

2.2 Compacting W powder

Due to the easy availability in the lab and the high achievable uniaxial pressures, the HPT tool can also serve as a compacting press for material powders [15,16]. The pressures reached plus a short additional shear straining is sufficient to compact the powders and close pores for most metals. Materials showing high hardness and limited ductility, such as tungsten, typically do not compact well, even under high nominal pressure. Figure 2a shows a RT compact of W powder by HPT (nominal pressure of 12 GPa). The initial powder particles changed shape due to the high pressure and applied strain, but no connection formed between the particles. The compaction is enough for the samples to be handled, but subsequent deformation by HPT will only lead to the majority of the particles shearing apart without any proper plastic deformation. The result of this deformation is a material containing lots of microcracks and pores along the initial powder particle interfaces (figure 2b). A significant improvement in interparticular bonding is expected from annealing the compacted sample, which is discussed in the next subsection.

2.3 Sintering W green compacts

Solid state sintering is a popular method to reduce porosity and increase density of powder compacted specimens of high-melting materials, such as refractory metals and ceramics. The usual temperature range for sintering is 0.6 – 0.8 * T_{melt} [K] [5,6]. At these elevated temperatures diffusion is enhanced, leading to a redistribution of material and closure of pores. The driving force for this process is the reduction of surface area within the material [5,6]. Conventional sintering temperatures of W are rather high (above 2000°C) due to the high melting point [5,6]. Furthermore, sintering has to be performed either in a reducing gas atmosphere or in vacuum, as oxidation is a serious problem at these temperatures.

In this work, a vacuum furnace with a maximum operating temperature of 1600°C was used for sintering the compacted tungsten samples. The microstructure after sintering is displayed in figure 2c. Naturally, the sintering process leads to substantial amount of grain growth, but this should be of no concern, as the subsequent HPT step will re-introduce grain refinement. There is still a large amount of porosity present due to the comparably low sintering temperature and the absence of any pressure applied to the material. However, it is evident that the particles have merged together and formed large grains. Therefore, severe plastic deformation applied during HPT should be able to close the residual pores and deform the material sufficiently to reduce the grain size to the ufg regime. In general, it is beneficial to work with relatively small powders (< 10 µm), as they show enhanced surface diffusion, improving the densification by sintering considerably.

2.4 Severe plastic deformation of W

While all the limitations from a powder starting material can be handled by adapting the approaches described in the previous subsections, HPT processing of pure W, even from a bulk precursor, is not a straightforward task. High strength, low ductility and the high melting point are again an issue when it comes to severe plastic deformation of tungsten. The presence of dislocation plasticity is advantageous in order to allow the necessary deformation and grain refinement [13]. For tungsten, a body-centered cubic (BCC) metal, dislocation plasticity is fully thermally activated at ~ 460°C (0.2 * T_{melt} [K]) [17,18].

On the contrary, in order to deform a material using HPT, the hardness of the anvils (~850 HV; tool steel Böhler S390) has to be somewhat higher than the sample [12]. Heating to a temperature above 400°C leads to considerable softening of the anvils. Therefore, a trade-off in deformation temperature is necessary to find the ideal conditions for severe plastic deformation of W. Increasing the applied nominal pressure by decreasing the sample radius also results in better deformability of tungsten.

Even by considering all these aspects, the maximum number of rotations possible by HPT in this work did not exceed 1.5 turns, the equivalent of ~1800 % strain, as the material becomes too hard from grain refinement to be further processed by the available anvils.

3. Fabrication route for ufg W from powders

The raw tungsten powder (particle size 2 μm, purity 99.97%) provided by Plansee SE was opened, stored and handled only in argon atmosphere within a glovebox (M. Braun LABstar Glove Box Workstation). A copper ring is glued around the cavity of the HPT anvil (diameter 6 mm, depth 0.15 mm) to allow more powder to be filled in the cavity without spilling. Subsequently, the sealed mini-chamber is assembled around the powder filled anvil and its counterpart within the glovebox (figure 1). The chamber is then inserted in a HPT tool [12] and the powder sample is compacted under a pressure of 12 GPa and deformed for 30-40 seconds at a speed of 0.2 rpm. Afterwards, the mini-chamber is opened and the residual copper around the sample removed. Subsequently, the compacted sample is annealed in a vacuum furnace (Leybold Heraeus PD 1000) at 1600°C for 7 hours at a pressure lower than 10^{-4} mbar. In addition, HPT samples have been manufactured from ultra-high purity tungsten bulk material (99.9999% purity, Plansee SE). The material disks were deformed using the HPT tool with a pressure of 12 GPa at 300°C and 400°C for the maximum amount of turns possible (1 – 1.5 turns).

4. Material characterization

After processing, the microstructure of all samples was investigated in the tangential direction using a field-emission scanning electron microscope (SEM; LEO type 1525). Vickers microhardness was measured along the radius with a load of 500 g (HV0.5) using a Buehler Micromet 5104 machine. Nanoindentation tests were performed with a KLA G200 Nanoindenter using a Berkovich tip and the continuous stiffness measurement (CSM) method [19,20]. Strain rate jump tests [21,22] were conducted with the same nanoindentation setup to investigate strain rate sensitivity and activation volume.

5. Results and discussion

5.1 Microstructure

Micrographs of all materials at a radius of 3 mm (outer edge of the HPT deformed disk, corresponding to strains of ~ 10-17) were recorded using the SEM in backscattered electron mode and are displayed in figure 3. It is apparent from the micrographs that both bulk and powder processed materials exhibit a similar microstructure and grain sizes of ~110-160 nm for each deformation temperature, as detailed in table 1. No pores or oxide layers are visible in the samples fabricated from powders.

5.2 Microhardness

Microhardness testing is a fast method to assess the gradient in mechanical properties in HPT deformed materials to check whether the deformation was high enough to reach the saturation state, i.e. the point where grain size cannot be refined any further due to a dynamic equilibrium of dislocation generation

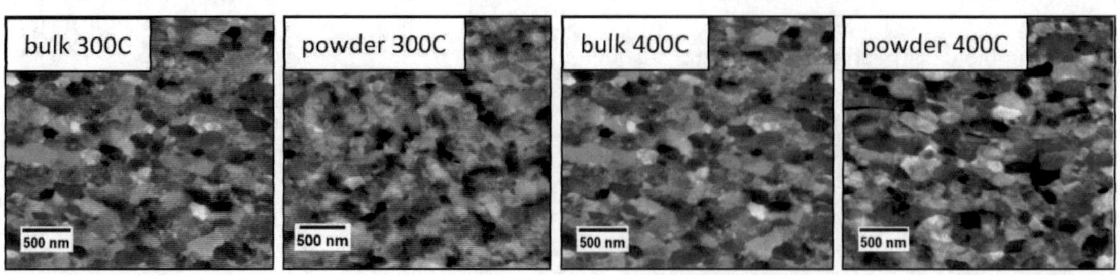

Figure 3. SEM images (backscattered electron mode, same magnification for all images) of resulting microstructures of bulk and powder processed samples after HPT deformation at different temperatures.

Table 1. Results from microstructural evaluation, microhardness and nanoindentation tests. If not stated otherwise, the values were measured at a radius of 3 mm on the sample disk.

Deformation temperature	300°C		400°C	
Precursor	bulk	powder	bulk	powder
Grain size @ 3 mm [nm]	112 ± 19	127 ± 25	143 ± 26	158 ± 36
Grain size @ 1 mm [nm]	173 ± 39	163 ± 46	219 ± 66	256 ± 78
Microhardness [HV0.5]	884	940	871	850
Nanohardness [GPa]	11.17 ± 0.28	11.02 ± 0.16	10.27 ± 0.08	9.96 ± 0.56
Elastic modulus [GPa]	394 ± 8	410 ± 4	402 ± 4	408 ± 16
Strain rate sensitivity [-]	0.016 ± 0.002	0.017 ± 0.002	0.016 ± 0.002	0.015 ± 0.002
Activation volume [b³]	6.78 ± 0.89	6.21 ± 0.69	7.22 ± 0.99	7.59 ± 0.82

and annihilation [13]. The trend of measured microhardness with applied von Mises-strain for all material samples is displayed in figure 4a. All samples could be deformed for between 1 and 1.5 turns, corresponding to strains of ~10-17. One can observe that the hardness is still increasing at the highest achieved strains, indicating that the saturation regime was not reached for any material and deformation temperature. The hardness gradient over applied strain for both samples deformed at 400°C and for the powder sample deformed at 300°C are in good agreement. The bulk sample deformed at 300°C shows a higher hardness for low strains, which could be due to possible differences in the initial microstructure. At high strains, this sample shows a similar hardness trend as all other samples.

To allow a better correlation with the mechanical properties gained from microhardness tests, the grain sizes were evaluated at a radius of 1 mm and 3 mm using the grain intercept method (see table 1). When plotting the hardness versus the inverse square root of the grain size (figure 4b), one can see that all materials follow a clear linear trend in accordance with the Hall-Petch relationship [23,24]. The slope of the Hall-Petch line k_y = 8.7 kg/mm$^{-3/2}$ is found to be in good agreement with literature (10 kg/mm$^{-3/2}$) [25]. Serious contamination or residual pores in the powder processed material should have led to a pronounced deviation from the Hall-Petch behavior of the bulk samples. This is an important finding, as it proves that by adapting the enhancements to the powder processing route presented in Section 2, the successful fabrication of ufg W with nanostructured grains and a comparable hardness-microstructure relationship than processed from bulk precursor material is possible.

5.3 Nanoindentation tests

Nanoindentation provides a locally resolved evaluation of both plastic (hardness) and elastic (modulus) properties of a material. By superimposing a sinusoidal signal on the conventional load signal, a continuous evaluation of hardness and modulus with indentation depth is possible (CSM method) [20]. The results from CSM nanoindentation tests are displayed in table 1. All samples were indented at least 6 times at a radius of 3 mm, where the finest grain size and highest hardness values are present. The hardness values from nanoindentation were found to be ~10 – 11 GPa. The measured elastic moduli lie in the range for the modulus of tungsten reported in literature of 390 – 410 GPa [6,26]. Body-centered

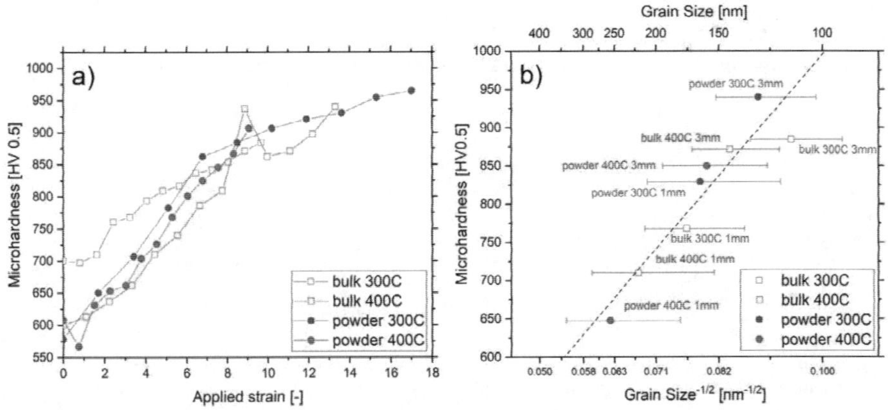

Figure 4. (a) Microhardness progression with applied HPT strain. (b) Hall-Petch plot of bulk and powder processed samples.

cubic and ultra-fine grained metals exhibit different mechanical properties dependent on the strain rate they are loaded with, a phenomenon known as strain-rate sensitivity (SRS) [27]. The origin of this material behavior is believed to lie in the intrinsic deformation behavior of the BCC lattice and the increased amount of dislocation-grain boundary interactions [27]. Dislocation pinning by impurites can also have a significant influence on the SRS. The SRS exponent and the activation volume for plastic deformation can be obtained from nanoindentation strain rate jump tests [21,22].

Calculated values of the SRS and activation volume obtained from strain-rate jump tests for each sample are given in table 1. All samples show similar dependence on strain rate, indicating the same deformation mechanisms in all tested materials. Moreover, the similar values of SRS suggest that the powder processed samples do not exhibit a significantly increased amount of impurities compared to the samples processed from bulk W. The magnitude of the measured activation volume (around 7 b^3) lies in the regime of grain boundary-dislocation interactions [1,21,27]. However, as previous experiments on single crystalline and ufg W have reveiled, this value for activation volume corresponds also to the activation energy for the kink-pair mechanism, which is believed to be the dominant deformation mechanism for this material and testing temperature [18].

6. Summary and outlook

In conclusion, we established a powder processing route for ufg W resulting in similar microstructural evolution, mechanical properties, deformation behavior and microstrucure-property relationships as ufg W fabricated from a bulk starting material. Although the powder processing route is more time-consuming, the ability to control chemical composition in terms of precise doping with impurity or alloying elements clearly underlines its benefits. The developed fabrication route can now be used to tune grain boundary segregation states and explore the effect of doping of grain boundaries in ufg W on the investigated properties and on fracture behavior and performance in harsh environments, with the goal of fabricating novel high-performance materials for application in future engineering technologies.

Acknowledgements

The authors acknowledge funding by the European Research Council under Grant number: 771146 (TOUGHIT). Financial support under the scope of the COMET program (A2.12) within the K2 Center “IC-MPPE” (Project No 859480) is gratefully acknowledged.

References

[1] Meyers M A, Mishra A and Benson D J 2006 *Prog. Mater. Sci.* **51** 427–556
[2] Hohenwarter A and Pippan R 2015 *Philos. Trans. R. Soc. A*
[3] Wurster S and Pippan R 2009 *Scr. Mater.* **60** 1083–7
[4] Wurmshuber M, Frazer D, Bachmaier A, Wang Y, Hosemann P and Kiener D 2018 *Mater. Des.* **160** 1148–57
[5] Pink E and Bartha L 1989 *The metallurgy of doped/non-sag tungsten* (Essex: Elsevier Science Publishers Ltd)
[6] Lassner E and Schubert W-D 1999 *Tungsten. Properties, Chemistry, Technology of the Element, Alloys, and Chemical Compounds* (New York: Kluwer Academic / Plenum Publishers)
[7] Neu R, Dux R, Kallenbach A, Pütterich T, Balden M, Fuchs J C, Herrmann A, Maggi C F et al. 2005 *Nucl. Fusion* **45** 209–18
[8] Scheiber D, Pippan R, Puschnig P and Romaner L 2016 *Model. Simul. Mater. Sci. Eng.* **24** 85009
[9] Scheiber D, Pippan R, Puschnig P, Ruban A and Romaner L 2016 *Int. J. Refract. Met. Hard Mater.* **60** 75–81
[10] Leitner K, Scheiber D, Jakob S, Primig S, Clemens H, Povoden-Karadeniz E and Romaner L 2018 *Mater. Des.* **142** 36–43
[11] Valiev R Z, Islamgaliev R K and Alexandrov I V. 2000 *Prog. Mater. Sci.* **45** 103–89
[12] Pippan R, Scheriau S, Hohenwarter A and Hafok M 2008 *Mater. Sci. Forum* **584–586** 16–21
[13] Pippan R, Scheriau S, Taylor A, Hafok M, Hohenwarter A and Bachmaier A 2010 *Annu. Rev. Mater. Res.* **40** 319–43
[14] Warren A, Nylund A and Olefjord I 1996 *Int. J. Refract. Met. Hard Mater.* **14** 345–53
[15] Zhilyaev A P and Langdon T G 2008 *Prog. Mater. Sci.* **53** 893–979
[16] Bachmaier A and Pippan R 2013 *Int. Mater. Rev.* **58** 41–62
[17] Kraft O and Arzt E 2009 *Phys. Rev. Lett.* **103** 105501
[18] Kiener D, Fritz R, Alfreider M, Leitner A, Pippan R and Maier-Kiener V 2019 *Acta Mater.* **166** 687–701
[19] Pharr G M and Oliver W C 1992 *J. Mater. Res.* **7** 1564–83
[20] Hay J, Agee P and Herbert E 2010 *Exp. Tech.* **34** 86–94
[21] Maier-Kiener V and Durst K 2017 *Jom* **69** 2246–55
[22] Maier V, Durst K, Mueller J, Backes B, Höppel H W and Göken M 2011 *J. Mater. Res.* **26** 1421–30
[23] Hall E O 1951 *Proc. Phys. Soc.* **64** 747–53
[24] Petch N J 1953 *J. Iron Steel Inst.* **174** 25–8
[25] Vashi U K, Armstrong R W and Zima G E 1970 Powder *Met. Trans.* **1** 1769–71
[26] Grünwald E, Nuster R, Treml R, Kiener D, Paltauf G and Brunner R 2015 *Mater. Today Proc.* **2** 4289–94
[27] Caillard D and Martin J L 2003 *Thermally activated mechanisms in crystal plasticity* (Kidlington: Elsevier Ltd.)

IOP Conf. Series: Materials Science and Engineering **580** (2019) 012052 doi:10.1088/1757-899X/580/1/012052

Recent progress in dilatometry for quantitative analysis of precipitation kinetics

Roland Würschum, Robert Enzinger, Elisabeth Hengge, Wolfgang Sprengel

Institute of Materials Physics, Graz University of Technology, Petersgasse 16, 8010 Graz, Austria

E-mail: wuerschum@tugraz.at

Abstract. A high-stability non-contact dilatometer based on two-beam Michelson interferometry was developed recently, enabling isothermal measurements of relative length changes in the sub-10^{-5} regime with minimized drift over times as long as 10^6 s. This advanced technique opens up novel potentials for quantitative studies of precipitation phenomena, which are associated with tiny relative length changes that have not been accessible before. Using a dilute Al-Mg-Si alloy as a case study, the amount of the metastable coherent β''- and of the semi-coherent β'-phase, as well as the amount of the stable β-phase could be determined quantitatively from isothermal length change measurements. This is accomplished by the distinct length change features that arise from the volume excess due to the precipitates, on the one hand, and the contraction of the matrix upon precipitation of solute atoms, on the other hand. Moreover, reaction-rate analyses of isothermal length change variations yield the kinetic parameters for the formation and transformation of the precipitates.

1. Introduction: Dilatometry as tool in materials science

The principle of dilatometry, i.e., of measuring volume changes caused by physical or chemical processes, dates back to the nineteenth century [1]. Meanwhile, dilatometry has developed to a powerful toolbox with a number of measurement routines applicable to various phenomena in materials science.

Concerning lattice defects, dilatometric measurements of the macroscopic length change $\Delta L/L$ in combination with X-ray diffraction for determining the change of lattice constant $\Delta a/a$, yields access to the concentration of thermal lattice vacancies in pure metals [2, 3]. For high vacancy concentrations and low vacancy diffusivities as in various intermetallic compounds, both the vacancy formation and migration enthalpies can be determined by isothermally monitoring the length change which is associated with vacancy equilibration after a temperature jump [4, 5].

Furthermore, deformation-induced lattice defects are detectable by means of their excess volume since this gives rise to a length contraction upon annealing with respect to a defect-free reference sample [6]. In favourable cases when it is possible to assign the length change in a temperature window to the annealing out of a certain defect type, rather specific information for this defect can be obtained. For instance, for ultrafine-grained Ni and Cu prepared by extreme plastic deformation, a structural key quantity of grain boundaries, namely the grain boundary excess volume, could be determined in this way [7, 8]. In the same manner the concentration of deformation excess vacancies became measurable by dilatometry for ultrafine-grained Ni [9, 10]. Moreover, as dilatometry by its nature can be applied as function of orientation, the anisotropic behaviour of vacancy annealing, which arises from structural anisotropy in this material, allowed conclusions to be made regarding the vacancy formation and relaxation volume [9].

For studying transformation processes in alloys dilatometry has also turned out to be a powerful tool [11, 12, 13, 14, 15, 16]. For example, dilatometry is capable of studying the precipitation kinetics in shape memory alloys [12] or determining the induced dislocation density in bainitic steel [13]. Furthermore, precipitation in Al-alloys with a high concentration of alloying elements can also be monitored by dilatometry [14, 15].

However, the amount of precipitation has been quantitatively studied by dilatometry so far only for cases where the associated relative length change exceeded 10^{-4} and, moreover, when this took place on short time scales of the order of minutes. Therefore, most relevant alloy systems, such as Al alloys of the 6xxx series with low amounts of Mg and Si, until recently have not been amenable to study by dilatometry so far due to the low amount of precipitation and due to the long time scale necessary to reproduce practical relevant processing conditions. In order to get access to this regime, a highly stable non-contact dilatometer was developed that enables isothermal measurements of relative length change measurements in the sub-10^{-5} regime with minimized drift over times as long as 10^6 s [17]. Using commercial grade Al-Mg-Si alloy (EN AW-6060) as model system, the potential of this technique has been demonstrated recently [18, 19]. A summary will be given in the following, after a brief description of the chosen alloy system.

2. Short literature survey on precipitation processes in Al-Mg-Si

The Al-Mg-Si alloy EN AW-6060 was chosen as model system because of the complex precipitation sequence, involving metastable phases, and because of the comprehensive research done before on the precipitation processes in this alloy (for reviews see, e.g., Refs. [20, 21]). Summarizing briefly, precipitation proceeds via three metastable phases (clustering, β'', β') before the final equilibrium β-precipitates are formed. Initial clustering is a highly complex multistage process [22], presumably starting with single element clusters which transform into co-clusters [23]. The formation of the β''-phase takes place via one or more transition stages, designated, e.g., GP-II, pre-β'', initial-β'' [24, 25]. Coherency between β''-precipitates and the Al-matrix causes hardening. The associated strain field strongly influences the crystal structure of β'' which most likely is monoclinic $Mg_4Al_3Si_4$ [26, 27]. Subsequently, the semicoherent metastable β'-phase (hexagonal Mg_9Si_5 [28]) is formed associated with a loss in hardness. Additional metastable phases are reported (named U1, U2, B or A, B, C), the number densities of which usually are much lower [29]. The finally formed equilibrium β- phase is Mg_2Si with the Ca_2F-structure [30].

3. Experimental procedure

The length change measurements reported here were performed by means of a non-contact dilatometer based on two-beam Michelson interferometry. The dilatometer, with a self-developed housing and furnace, optimized for high stability, is described in detail elsewhere [17]. The cylindrically shaped samples (length: 20 mm, diameter: 5.7 mm) were solution annealed in the dilatometer at 540 °C (30 min), subsequently quenched in a He gas stream, then held at 32 °C for 4 min, and finally rapidly heated at a rate of 100 °C/min to the temperature of isothermal precipitation treatment between 170 °C and 260 °C.

4. Results and discussion

4.1. Resolving β''- and β'-precipitation by dilatometry

Figure 1 shows the dependence of the relative length changes $\Delta L/L$ on time t for various isothermal aging temperatures between 180 °C and 240 °C. As outlined in detail elsewhere (see [18]), the clear time correlation of length change and hardness during aging indicates that the length increase arises from the formation of the metastable coherent β''-phase and the length decrease from the formation of the metastable semicoherent β'-phase. It is a highly remarkable fact that the two phases can be

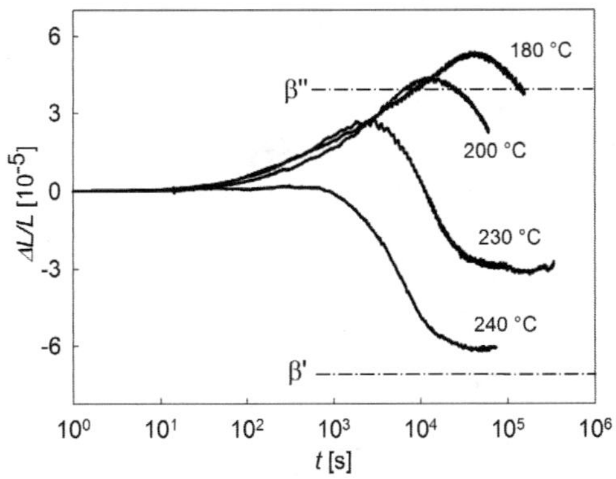

Figure 1. Dependence of relative length change $\Delta L/L$ on time t measured at the quoted annealing temperatures. The horizontal lines mark the quantitatively estimated length changes arising from β''- and β'-formation (see Table 1).

Precipitate		β''	β'
Composition		$Mg_4Al_3Si_4$	Mg_9Si_5
$(\Delta L/L)_{prec}$	$[10^{-4}]$	3.47	5.23
$(\Delta L/L)_{matrix}$	$[10^{-4}]$	−3.08	−5.93
$(\Delta L/L)_{calc}$	$[10^{-4}]$	0.39	−0.71
$(\Delta L/L)_{exp}$	$[10^{-4}]$	0.53	−0.61

Table 1. Relative length change associated with precipitation of β'' and β'. $(\Delta L/L)_{matrix}$: matrix contraction; $(\Delta L/L)_{prec}$: expansion upon precipitation; $(\Delta L/L)_{calc}$: sum; $(\Delta L/L)_{exp}$: measured value deduced from the extrema of the $\Delta L/L$ curves. For details see ref. [18].

distinguished by the sign of the length change. A quantitative estimation shows that this arises from the sensitive balance between the matrix contraction $(\Delta L/L)_{matrix}$ and the expansion $(\Delta L/L)_{prec}$ associated with the volume difference between the forming precipitates and the supersaturated solid solution (see Table 1).

Since the matrix expansion due to dissolved Mg is remarkably stronger than the matrix contraction due to dissolved Si, the extraction of Mg and Si from the matrix results for both β'' ($Mg_4Al_3Si_4$) and β' (Mg_9Si_5) in a matrix contraction $(\Delta L/L)_{matrix}$. However, the contraction for β' is much stronger due to the higher relative amount of Mg extracted from the matrix (see Table 1). Although, the expansion $(\Delta L/L)_{prec}$ due to precipitates for β' is larger than that for β'', the calculated sum $(\Delta L/L)_{calc} = (\Delta L/L)_{matrix} + (\Delta L/L)_{prec}$ results in a contraction for β' due to the strong matrix contraction, in comparsion to β'' where the expansion $(\Delta L/L)_{prec}$ exceeds the net value of the matrix contraction. The values $(\Delta L/L)_{calc}$ quantitatively estimated from the fraction of precipitates and the crystallographic data (see [18]) fairly well reproduce the experimentally observed values $(\Delta L/L)_{exp}$ (see Table 1). This does not only hold for the metastable precipitates β'' and β', but also for the stable β-phase formed at higher temperatures (see [18]). We therefore can conclude that dilatometry enables a *quantitative* volumetric identification of precipitate phases in dilute alloys.

4.2. Precipitation kinetics from analysis of isothermal length change

Knowing the absolute amount of precipitates as function of time and temperature (see above), also the underlying kinetics of the precipitation processes can be determined. The formation of β'' is described by JMAK-type kinetics (Avrami exponent of $n = 1$) combined with a second term describing the $\beta'' \to \beta'$ transformation. For higher temperatures in addition to the $\beta'' \to \beta'$ transformation, also a direct formation of β'-precipitates has to be taken into account. This yields the following set of three

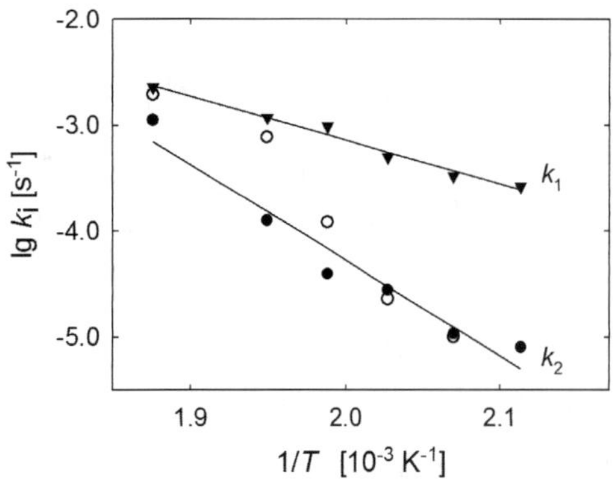

Figure 2. Arrhenius plot representation of reaction rates k_1 (▼), k_2 (•), k_3 (∘) determined from fitting the $\Delta L/L$-curves according to the solution of rate model (equations (4–6)).

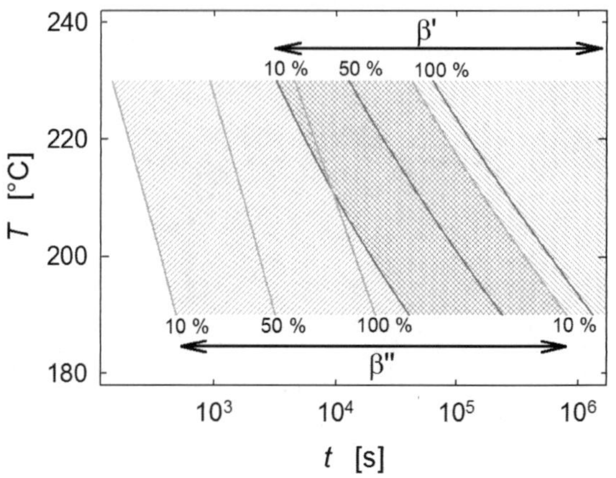

Figure 3. Time-temperature-precipitation (TTP) characteristic for the formation and dissolution of β'' and β'. The iso-concentration lines are calculated from the solution of the rate model (equations (4–6)).

coupled rate equations:

$$\dot{c}_{\beta''}(t) = k_1\Big\{c_0 - \Big[c_{\beta''}(t) + c_{\beta'}(t) + c_{\beta'_{dir}}(t)\Big]\Big\} - k_2 c_{\beta''}(t)\,, \quad (1)$$

$$\dot{c}_{\beta'}(t) = k_2 c_{\beta''}(t)\,, \quad (2)$$

$$\dot{c}_{\beta'_{dir}}(t) = k_3\Big\{c_0 - \Big[c_{\beta''}(t) + c_{\beta'}(t) + c_{\beta'_{dir}}(t)\Big]\Big\} \quad (3)$$

for the atomic fractions of the precipitate phases with c_0 as the maximum amount finally attained. Here, k_1 denotes the rate for formation of the precipitate phase β'', k_2 that for formation of β' from β'', and k_3 that for direct formation of β'.

With the initial conditions of a vanishing amount of precipitates for $t = 0$, the solution of the set of rate equations yields the following temporal evolutions:

$$c_{\beta''}(t) = c_0 \frac{k_1}{k_2 - (k_1 + k_3)} \Big\{\exp(-(k_1 + k_3)\,t) - \exp(-k_2\,t)\Big\}\,, \quad (4)$$

$$c_{\beta'}(t) = c_0 \frac{1}{k_2 - (k_1 + k_3)} \Big\{-\frac{k_2 k_1}{k_1 + k_3} \exp\Big(-(k_1 + k_3)\,t\Big) + k_1 \exp(-k_2 t)\Big\} + c_0 \frac{k_1}{k_1 + k_3}\,, \quad (5)$$

$$c_{\beta'_{dir}}(t) = c_0 \frac{k_3}{k_1 + k_3} \Big\{1 - \exp(-(k_1 + k_3)\,t)\Big\}\,. \quad (6)$$

By means of this set of solutions, the experimental isothermal $\Delta L/L(t)$-curves can be fitted (see [19]). For each temperature the three rate constants k_i ($i = 1, 2, 3$) as obtained from the fit are shown in the Arrhenius plot (figure 2). Linear fits of the temperature variations $k_1(T)$ and $k_2(T)$ according to Arrhenius-type behaviour yields the corresponding activation energies $Q_1 = (0.90 \pm 0.30)$ eV and $Q_2 = (1.42 \pm 0.28)$ eV. A comparison of the temperature variations of k_2 and k_3 shows that for temperatures above 200 °C the rate of direct β'-formation exceeds that of $\beta'' \rightarrow \beta'$ transformation. The deduced activation energies are similar to those found earlier for Al-Si-Mg alloys [15]; they also can be considered as characteristic for metastable precipitates in Al-alloys [31, 32].

Analyzing isothermal measurements as in the present case compared to non-isothermal ones has the distinct advantage that the resulting kinetic parameters, such as the activation energies, only vary insignificantly with the reaction model [33]. JMAK-theory with an Avrami exponent of $n = 1$, applied here, was found to be suitable giving analytical solutions even in the present case where formation of the precipitate phases is interdependent and occurring in parallel. Comparative kinetic analyses with various reaction models yielded similar results.

4.3. Time-temperature-precipitation (TTP) diagrams

Dilatometry data have been used previously to construct transformation diagrams both for ferrous metals [34] and for non-ferrous metals [35], including Al alloys. However, measurements of this kind have so far been restricted to relatively highly alloyed systems where the volume changes upon precipitation are much larger compared to those studied here. By means of the kinetic parameters obtained from the analysis of the $\Delta L/L(t)$ curves, the concentrations of β'' and β' for each time and temperature can be calculated from the solution of the rate model (equations (4–6)). This allows to construct an important part of the technologically most relevant TTP-diagram of this alloy as shown in figure 3.

The hatched area marks the parts where the concentration of one or both of the phases exceeds 10 % of the respective maximum concentration. As expected, with increasing temperature the transformations are shifted to shorter times. Within the temperature–time regime where β' has not yet started to form at a noticeable amount, the iso-concentration lines of β'' show an almost linear behaviour related to the logarithmic time scale.

The crosshatched area marks the overlapping regime where the concentrations of both phases exceed 10 %. As can be discerned from the crossing of the iso-concentration lines 100 % β'' and 10 % β', for low temperatures, and correspondingly long times, a noticeable amount of β' is formed only after the maximum amount of β'' has been reached. For high temperatures and correspondingly fast kinetics, the formation of β' in appreciable amounts sets in prior to reaching the maximum amount of β''.

5. Outlook

As demonstrated by the present case study, the recent progress in dilatometry regarding sensitivity and stability opens up a novel pathway for quantitative analysis of complex precipitation kinetics in alloys. The method not only appears to be particularly relevant for elucidating precipitation processes in the technologically important class of light-weight alloys, but also for other materials classes, such as, e.g., maraging steels. The dilatometer system described was recently mounted in an improved temperature-stabilized enclosure ($\Delta T \leq 0.2$ °C). With this additional stability improvement long-time processes associated with even smaller length changes, occurring as for instance upon natural aging of Al alloys, are within reach.

References

[1] Fizeau M H 1864 *Ann. Chim. Phys.* **(4) 2** 143
[2] Simmons R and Balluffi R 1960 *Phys. Rev.* **117** 52–61 cited By 342 URL `https://journals.aps.org/pr/abstract/10.1103/PhysRev.117.52`
[3] Hehenkamp T, Berger W, Kluin J E, Lüdecke C and Wolff J 1992 *Phys. Rev. B* **45**(5) 1998–2003 URL `https://link.aps.org/doi/10.1103/PhysRevB.45.1998`
[4] Van Ommen A and Miranda J 1981 *Phil. Mag. A* **43** 387–407 URL `https://doi.org/10.1080/01418618108239417`
[5] Schaefer H E, Frenner K and Würschum R 1999 *Phys. Rev. Lett.* **82**(5) 948–951 URL `https://link.aps.org/doi/10.1103/PhysRevLett.82.948`
[6] Oberdorfer B, Lorenzoni B, Unger K, Sprengel W, Zehetbauer M, Pippan R and Würschum R 2010 *Scripta Mater.* **63** 452 – 455 URL `https://doi.org/10.1016/j.scriptamat.2010.05.007`
[7] Steyskal E M, Oberdorfer B, Sprengel W, Zehetbauer M, Pippan R and Würschum R 2012 *Phys. Rev. Lett.* **108** 055504 URL `https://link.aps.org/doi/10.1103/PhysRevLett.108.055504`
[8] Oberdorfer B, Setman D, Steyskal E M, Hohenwarter A, Sprengel W, Zehetbauer M, Pippan R and Würschum R 2014 *Acta Mater.* **68** 189 – 195 URL `https://doi.org/10.1016/j.actamat.2013.12.036`

[9] Kotzurek J A, Steyskal E M, Oberdorfer B, Hohenwarter A, Pippan R, Sprengel W and Würschum R 2016 *Appl. Phys. Lett.* **109** 021906 URL https://doi.org/10.1063/1.4958895
[10] Kotzurek J A, Sprengel W, Krystian M, Simic S, Pölt P, Hohenwarter A, Pippan R and Würschum R 2017 *Internat. J. Mater. Res.* **108** 81–88 URL https://www.hanser-elibrary.com/doi/10.3139/146.111463
[11] Edmonds D 1973 *Metall. Trans.* **4** 2527–2533 URL https://link.springer.com/article/10.1007/BF02644254
[12] Recarte V, Pérez-Sáez R, No M and Juan S 1997 *J. Phys. IV France* **07** C5–329 – C5–334 URL https://doi.org/10.1051/jp4:1997552
[13] Garcia-Mateo C, Caballero F, Capdevila C and de Andres C G 2009 *Scripta Mater.* **61** 855 – 858 URL https://doi.org/10.1016/j.scriptamat.2009.07.013
[14] Lasagni F, Dumont M, Salamida C, Acuna J and Degischer H P 2009 *Int. J. Mater. Res.* **100** 1005–1013 URL https://doi.org/10.3139/146.110145
[15] Daoudi M I, Triki A, Redjaimia A and Yamina C 2014 *Thermochim. Acta* **577** 5–10 URL https://doi.org/10.1016/j.tca.2013.12.007
[16] Milkereit B, Reich M and Kessler O 2017 *Mater. Sci. Forum* **877** 147–152 URL https://doi.org/10.4028/www.scientific.net/MSF.877.147
[17] Luckabauer M, Sprengel W and Würschum R 2016 *Rev. Sci. Instr.* **87** 075116 URL http://aip.scitation.org/doi/abs/10.1063/1.4959200
[18] Hengge E, Enzinger R, Luckabauer M, Sprengel W and Würschum R 2018 *Phil. Mag. Letters* **98** 301–309 URL https://doi.org/10.1080/09500839.2018.1542170
[19] Enzinger R, Hengge E, Sprengel W and Würschum R 2019 *J. Mater. Sci.* **54** 5083–5091 URL https://doi.org/10.1007/s10853-018-03210-z
[20] Ravi C and Wolverton C 2004 *Acta Mater.* **52** 4213–4227 URL https://doi.org/10.1016/j.actamat.2004.05.037
[21] van Huis M, Chen J, Zandbergen H and Sluiter M 2006 *Acta Mater.* **54** 2945–2955 ISSN 1359-6454 URL https://doi.org/10.1016/j.actamat.2006.02.034
[22] Torsæter M, Hasting H S, Lefebvre W, Marioara C D, Walmsley J C, Andersen S J and Holmestad R 2010 *J. Appl. Phys.* **108** 073527 URL https://aip.scitation.org/doi/10.1063/1.3481090
[23] Murayama M and Hono K 1999 *Acta Mater.* **47** 1537–1548 URL https://doi.org/10.1016/S1359-6454(99)00033-6
[24] Matsuda K, Ikeno S, Gamada H, Fujii K, Uetani Y, Sato T and Kamio A 1998 *Metall. Mater. Trans. A* **29** 1161–1167 URL https://doi.org/10.1007/s11661-998-0242-7
[25] Marioara C, Andersen S, Jansen J and Zandbergen H 2001 *Acta Mater.* **49** 321–328 URL https://doi.org/10.1016/S1359-6454(00)00302-5
[26] Ninive P H, Strandlie A, Gulbrandsen-Dahl S, Lefebvre W, Marioara C D, Andersen S J, Friis J, Holmestad R and Løvvik O M 2014 *Acta Mater.* **69** 126–134 URL https://doi.org/10.1016/j.actamat.2014.01.052
[27] Ehlers F, Dumoulin S and Holmestad R 2014 *Comput. Mater. Sci.* **91** 200–210 URL https://doi.org/10.1016/j.commatsci.2014.04.060
[28] Vissers R, van Huis M, Jansen J, Zandbergen H, Marioara C and Andersen S 2007 *Acta Mater.* **55** 3815–3823 URL https://doi.org/10.1016/j.actamat.2007.02.032
[29] Matsuda K, Sakaguchi Y, Miyata Y, Uetani Y, Sato T, Kamio A and Ikeno S 2000 *J. Mater. Sci.* **35** 179–189 URL https://doi.org/10.1023/A:1004769305736
[30] Geisler A H and Hill J K 1948 *Acta Crystal.* **1** 238–252 URL https://doi.org/10.1107/S0365110X48000661
[31] Dorward R C 1973 *Metall. Trans.* **4** 507–512 ISSN 1543-1916 URL https://doi.org/10.1007/BF02648703
[32] Ovono D O, Guillot I and Massinon D 2007 *J. Alloys & Comp.* **432** 241–246 URL https://doi.org/10.1016/j.jallcom.2006.05.132
[33] Vyazovkin S and Wight C A 1998 *Internat. Rev. Phys. Chem.* **17** 407–433 URL https://doi.org/10.1080/014423598230108
[34] Vander Voort G F (ed) 1991 *Atlas of Time-Temperature Diagrams for Irons and Steels* (ASM International)
[35] Vander Voort G F (ed) 1991 *Atlas of Time-Temperature Diagrams for Nonferrous Alloys* (ASM International)

IOP Conf. Series: Materials Science and Engineering **580** (2019) 012053 doi:10.1088/1757-899X/580/1/012053

Deformation microstructures and strength of face-centered cubic high/medium entropy alloys

Shuhei Yoshida[1*], Takuto Ikeuchi[1], Yu Bai[1,2], Akinobu Shibata[1,2], Niels Hansen[3], Xiaoxu Huang[4], Nobuhiro Tsuji[1,2]

1. Department of Materials Science and Engineering, Kyoto University, 606-8501 Kyoto, Japan
2. Elements Strategy Initiative for Structural Materials, Kyoto University, 606-8501 Kyoto, Japan
3. Technical University of Denmark, Risø campus, DK-4000 Roskilde, Denmark
4. Department of Mechanical Engineering, Technical University of Denmark, DK-2800 Kgs. Lyngby, Denmark

*Corresponding author's e-mail address: yoshida.shuhei.26x@st.kyoto-u.ac.jp

Abstract. In this study we report the deformation microstructures and strength of medium entropy alloys (MEAs) and high entropy alloys (HEAs), which are defined as alloys composed of four or less, and five or more principal elements, respectively, with (near-) equi-atomic concentrations. The friction stress (fundamental resistance to dislocation motion in the crystal lattice) and Hall-Petch relationship of various MEAs (CoCrFeNi, CoCrNi, etc.), taken as subsystems of the equi-atomic CoCrFeMnNi HEA, were precisely measured at room temperature. Experimental values of the friction stresses were found to fit with a theoretical model proposed by Toda-Caraballo *et al.* very well, which indicates that the strength of the alloys is closely related to a heterogeneously distorted crystal lattice. At the same time, values of the average lattice distortion in the alloys were found to be comparable to those in some dilute alloys, contradicting the belief that "severe" lattice distortion is a reason for the higher strength than in dilute systems. Finally, a strengthening mechanism due to element-element interactions was proposed as an additional mechanism in FCC HEAs and MEAs.

1. Introduction

Conventionally, metallic materials have been designed simply based on one kind of solvent element, such as Fe, Al, etc., where a small amount of alloying elements, such as C, Cr, Cu, etc., are added to achieve superior properties. In contrast, in 2004, Yeh *et al.* [1] and Cantor *et al.* [2] independently proposed the concept of high entropy alloys (HEAs), which are defined as alloys composed of five or more alloying elements with (near-) equimolar composition, focusing on the center part of a phase diagrams. Since then, discoveries of various kinds of new HEAs and of their exceptional physical properties have been reported extensively [3]. Furthermore, as derivatives of HEAs, a concept of medium entropy alloys (MEAs), composed of four or fewer alloying elements with (near-) equimolar concentration was also proposed. Some of these MEAs have been found to possess similar, or even superior, physical properties compared with HEAs. For example, Gludovatz *et al.* investigated mechanical properties and fracture toughness of equiatomic CoCrFeMnNi HEA [4] and CoCrNi MEA [5] with FCC single-phase and found that these materials showed excellent resistance to fracture.

Although many experimental evidences have shown their outstanding mechanical properties, the underlying physics of strengthening effect is still unclear.

In this study, we have investigated the effect of alloying elements on the strength of HEAs and MEAs, focusing particularly on their dislocation microstructures. The friction stress, i.e., the fundamental resistance to dislocation motion, of various kinds of HEAs and MEAs were obtained from intercepts of Hall-Petch relationships of their yield strength in tensile tests. Two kinds of theoretical model were adapted to understand the strengthening effect in the HEAs and MEAs. In addition, their deformation microstructures upon yielding were observed by scanning transmission electron microscopy (STEM) under diffraction contrast conditions. Finally, we discuss the strengthening mechanism that dominates the mechanical behavior of HEAs and MEAs.

2. Experimental methods

Ingots of Ni-60Co (= $Co_{60}Ni_{40}$), CoCrFeNi, CoFeMnNi, CoFeNi, CoMnNi, FeMnNi, $(CoNi)_{80}Cr_{20}$, and $Co_{20}(CrNi)_{80}$ were produced by vacuum arc-melting of constituent pure-metal elements with purities of more than 99.9 at.% and casting in a water-cooled copper mold under a protective Ar atmosphere. To remove macro-segregation of constituent elements, the ingots were subsequently cold-rolled by about 90% reduction in thickness and homogenized at 1200 °C for 12 h under a vacuum environment using a tube furnace. Afterwards, disc-shaped specimens with a diameter of 10 mm and a thickness of 0.8 mm were cut from the homogenized materials for high-pressure torsion (HPT).

The HPT was performed at room temperature under a pressure of 6 GPa at a speed of 0.2 rpm (rotation per minute). The total rotation angle applied was 1800° (5 rotations). After the heavy deformation by HPT, the materials were annealed at various temperatures ranging from 600 °C to 1000 °C for 20 s to 7.2 ks to obtain specimens with a wide range of mean grain sizes. Tensile tests using small-sized specimens cut from the HPT processed specimens (gauge length: 2 mm, gauge cross section: 1 mm × 0.5 mm) were performed at room temperature at a quasi-static strain rate of 8.3×10^{-4} s^{-1} to characterize mechanical properties of the materials. Tensile deformation of Ni-60Co and $Co_{20}(CrNi)_{80}$ with a mean grain size of 2 μm was interrupted at a nominal strain of $e \sim 2\%$ to investigate deformation microstructures after yielding. The displacement of the gauge section was precisely measured by a CCD video camera extensometer, and the strain was calculated by a digital image correlation (DIC) technique.

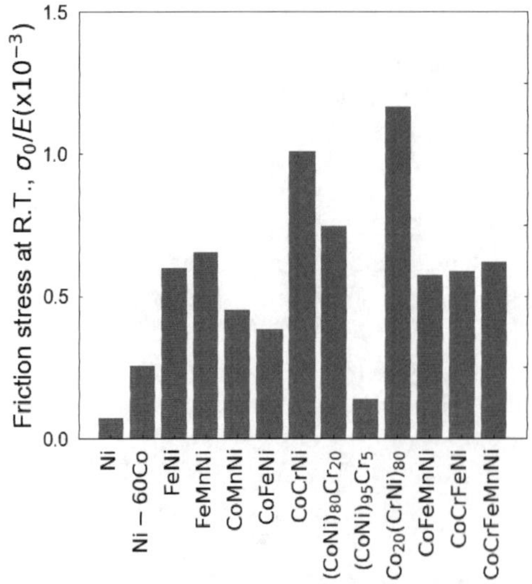

Figure 1: Friction stresses (σ_0) of various alloys at room temperature (R.T.). The values of friction stress were extracted from the Hall-Petch relationship and normalized by their Young's modulus (E) to eliminate the effect of elastic modulus.

Grain sizes of the tensile specimens were measured by an intercept method based on microstructural images obtained by a JEOL, JSM-7800F field emission (FE) scanning electron microscope (SEM), equipped with a backscattered electron (BSE) detector. Deformation microstructures were observed by using STEM (JEOL, JEM-2100) equipped with bright field (BF) and annular dark field (ADF) detectors. The STEM observations were conducted under diffraction contrast conditions (so-called diffraction contrast imaging (DCI)-STEM). Experimental details of the technique are explained in Ref. [6]. In each material, more than 20 grains from two thin-foils were observed to confirm the reproducibility of the results.

3. Results and discussion

Figure 1 shows experimentally obtained friction stresses, σ_0, of the alloys at room temperature, together with the data for pure Ni [7], CoCrNi MEA [8], and CoCrFeMnNi HEA [9], all normalized by Young's modulus, E, to eliminate the effect of elasticity. The σ_0 values were obtained by fitting the yield strength (0.2% proof stress), σ_Y, and mean grain sizes, d, of the materials according to the Hall-Petch relationship [10,11]:

$$\sigma_Y = \sigma_0 + kd^{-\frac{1}{2}}, \tag{1}$$

where k is a constant (the Hall-Petch slope). Note that more than 10 data points with a wide range of grain sizes were used for the fitting of each material. It was found that a simple increase in the number of alloying elements did not always result in an increase in the friction stress.

Toda-Caraballo *et al.* [12,13] modified a well-known solid solution strengthening model originally proposed by Labusch [14]. In their model, the material strength is expressed as

$$\frac{\tau(0\ \mathrm{K})}{G} \propto \varepsilon_{\mathrm{ave}}^{\frac{4}{3}}. \tag{2}$$

where $\tau(0\ \mathrm{K})$ is the critical resolved shear stress (CRSS) for dislocation slip at 0 K, and G is the shear modulus. In the model $\varepsilon_{\mathrm{ave}}$ is an average atomic size misfit parameter evaluated by simple equations, of

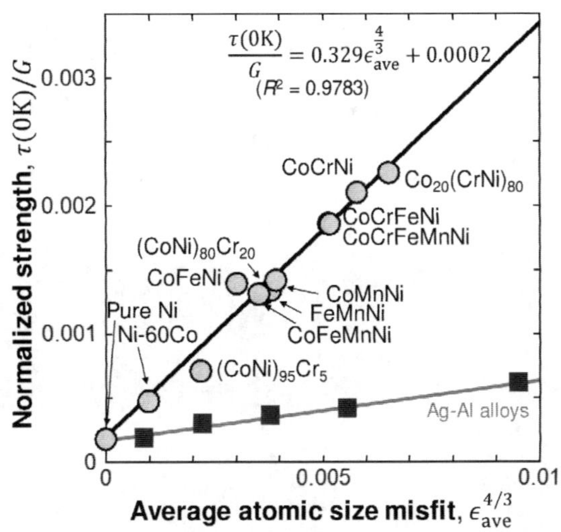

Figure 2: Relationship between the average atomic size misfit parameter (the average lattice distortion) calculated by the Toda-Caraballo model and the strength (the friction stress at 0 K divided by the Taylor factor of random texture, $M = 3.06$) normalized by shear modulus (G) of the alloys studied in the present study. The solid black line is a fitting line determined based on the model proposed by Toda-Caraballo (equation (2)). Similar plots for the data of Ag-Al binary alloys extracted from previous literature is also shown in the figure.

which details can be found in Ref. [12,13]. Wu *et al.* [15] already reported the shear modulus, G, and temperature dependence of the yield strength in various HEAs and MEAs.

In this study, the experimental values of σ_0 were extrapolated to 0 K by using the results reported by Wu *et al.* [15], assuming that the Hall-Petch slope was independent of temperature in low-temperature region. Next, τ(0 K) was determined by dividing σ_0 at 0 K by the Taylor factor for FCC materials having random texture ($M = 3.06$). Figure 2 shows the relationship between the average atomic size misfit and τ(0 K) normalized by the shear modulus of the HEAs and MEAs, together with that of Ag-Al binary alloys for comparison [16]. It was found that equation (2) fitted very well with the experimental data points of various HEAs and MEAs. This indicates that crystal lattices of equi-atomic multicomponent alloys are heterogeneously distorted due to the difference in atomic sizes of the constituent alloying elements. Accordingly, dislocations in such crystal lattices are considered to be always affected by the elastic fields originating from the lattice distortions. It is also very interesting that values of the average atomic size misfits for the dilute alloys are distributed over a wide range up to 0.01, which is comparable to the distribution of the misfit values for the MEAs and HEA, despite the fact that it has been believed that the reason for the high strength of HEAs is due to the "severe" lattice distortion effect caused by many kinds of elements. It should be noted, on the other hand, that the strength of the MEAs and HEAs were found to be much higher than that of the dilute alloys even at an equivalent average lattice distortion (average atomic size misfit). This suggests that the high strength of HEAs and MEAs cannot be simply explained only by the effect of lattice distortion.

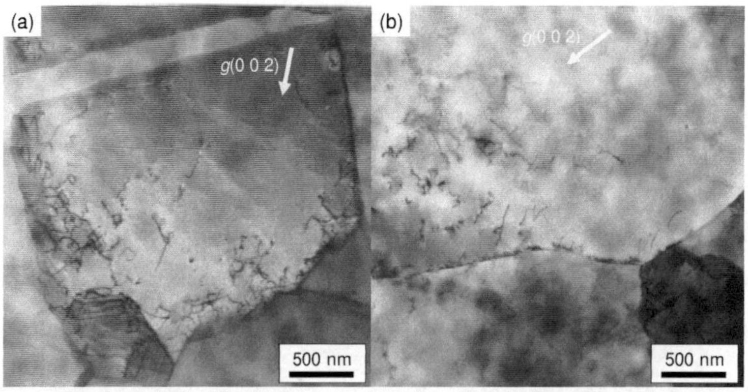

Figure 3: (a) Representative DCI-STEM image showing the dislocation microstructure in Ni-60Co alloy tensile-deformed to $e \sim 2\%$. (b) DCI-STEM micrographs taken at different position in the same sample.

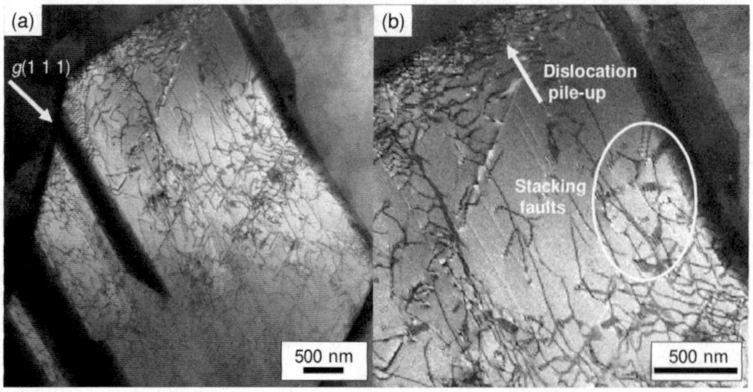

Figure 4: (a) Representative DCI-STEM image showing the dislocation microstructure in the $Co_{20}(CrNi)_{80}$ alloy tensile-deformed to $e \sim 2\%$. (b) Enlarged image of (a).

To investigate the yielding mechanism of the alloys having different friction stresses, we observed dislocation microstructures of Ni-60Co (Low σ_0) and $Co_{20}(CrNi)_{80}$ (High σ_0), alloys showing the lowest and highest friction stress among the HEAs and MEAs studied, by the DCI-STEM technique. Figure 3 (a) and (b) are representative microstructures of Ni-60Co tensile-deformed to $e \sim 2\%$. The images show dislocations randomly-tangled near grain boundaries. On the other hand, figure 4 (a) shows representative microstructures of $Co_{20}(CrNi)_{80}$ deformed to $e \sim 2\%$. It can be seen that the dislocation density in $Co_{20}(CrNi)_{80}$ is higher than that in Ni-60Co, and $Co_{20}(CrNi)_{80}$ showed some dislocations forming planar structures. In addition, the enlarged image (figure 4 (b)) shows some stacking faults and dislocations piled up at grain boundaries. Investigation of the effect of grain orientation is our on-going work and will be presented in a future publication. Although the yielding mechanism of both alloys seemed to be dislocation slip, the difference in dislocation microstructures was confirmed to depend on the friction stress. The planarity of dislocation microstructures in FCC metals is often attributed to (1) low stacking fault energy or (2) short-range ordering (SRO) and short-range clustering (SRC) [17].

Based on the above experimental facts, we propose a new strengthening mechanism by element-element interaction originating from weak elemental inhomogeneity (deviation from "random" distribution) in equi-atomic or near-equi-atomic alloys, i.e., SRO and SRC, for understanding the additional strengthening mechanism. In general, the SRO and SRC states are responsible for the strength of solid solutions, which is not simply because of the increased value of ε_{ave}. When a dislocation passes through an ordered or clustered region, the stable arrangement of elements (SRO or SRC state) is destroyed by shifting one atomic plane along the Burger's vector, which causes extra resistance to the dislocation motion. In dilute alloy systems, the change in the atomic arrangement around solute atoms before and after a dislocation passes through rarely happens because the majority of atoms are the solvent element, and such an effect on the strength is limited. In this case, it is enough to consider only the effect of lattice distortion, as Labusch did [14]. On the other hand, in high-alloy systems, even weak element-element interactions* (SRO or SRC) can have a significant impact on the strength since the concentration of alloying elements is high [18]. (*Note that we name the interaction among alloying elements as element-element interaction, different from "solute-solute interaction", because it is difficult to distinguish between solutes and solvent in HEAs and MEAs.) The destruction of the stable arrangement of elements can happen everywhere and will lead to larger fluctuations of local potential energy for dislocation segments, resulting in an additional strengthening effect in high-alloy systems. The importance of element-element interactions was also demonstrated by Zhang *et al.* [19] very recently, based on their theoretical modeling of SRO in HEAs. This effect of weak elemental inhomogeneity has been omitted in conventional dislocation theory, but the present results suggest that the strengthening effect of element-element interactions cannot be neglected in high-alloy systems such as HEAs and MEAs because of their particularly high concentration of alloying elements.

4. Conclusion

In the present study, we precisely obtained the friction stress and Hall-Petch relationship of various MEAs and HEAs at room temperature. Experimental values of the friction stresses were found to fit with a theoretical model proposed by Toda-Caraballo *et al.* very well. This indicates that the strength of the alloys is closely related to the crystal lattice distortion. At the same time, values of the average lattice distortion in the alloys were found to be comparable to those in some dilute alloys, although "severe" lattice distortion had been believed as a reason for the higher strength than dilute systems. The results of DCI-STEM observations showed a planar morphology of the dislocation microstructure in $Co_{20}(CrNi)_{80}$ with higher σ_0 while randomly tangled dislocations were observed in the Ni-60Co alloy with lower σ_0. From those results, a strengthening mechanism based on element-element interactions was proposed as an additional mechanism in FCC HEAs and MEAs.

Acknowledgments
This work was financially supported by the Elements Strategy Initiative for Structural Materials (ESISM), the Grant-in-Aid for Scientific Research on Innovative Area "High Entropy Alloys" (No.

18H05455), the Grant-in-Aid for Scientific Research (S) (No. 15H05767) and the Grant-in-Aid for JSPS Research Fellow (No. 18J20766), all through the Ministry of Education, Culture, Sports, Science and Technology (MEXT), Japan. The supports are gratefully appreciated. The authors also would like to thank Dr. Tianbo Yu and Ms. Gitte Christiansen of the Technical University of Denmark for technical assistance related to electron microscopy.

References

[1] Yeh J-W, Chen S-K, Lin S-J, Gan J-Y, Chin T-S, Shun T-T, Tsau C-H and Chang S-Y 2004 Nanostructured high-entropy alloys with multiple principal elements: novel alloy design concepts and outcomes *Adv. Eng. Mater.* **6** 299–303

[2] Cantor B, Chang I T H, Knight P and Vincent A J B 2004 Microstructural development in equiatomic multicomponent alloys *Mater. Sci. Eng. A* **375–377** 213–8

[3] Miracle D B and Senkov O N 2017 A critical review of high entropy alloys and related concepts *Acta Mater.* **122** 448–511

[4] Zhang Z, Mao M M, Wang J, Gludovatz B, Zhang Z, Mao S X, George E P, Yu Q and Ritchie R O 2015 Nanoscale origins of the damage tolerance of the high-entropy alloy CrMnFeCoNi. *Nat. Commun.* **6** 10143

[5] Gludovatz B, Hohenwarter A, Thurston K V S, Bei H, Wu Z, George E P and Ritchie R O 2016 Exceptional damage-tolerance of a medium-entropy alloy CrCoNi at cryogenic temperatures *Nat. Commun.* **7** 10602

[6] Zhu Y, Ophus C, Toloczko M B and Edwards D J 2018 Towards bend-contour-free dislocation imaging via diffraction contrast STEM *Ultramicroscopy* **193** 12–23

[7] Keller C and Hug E 2008 Hall–Petch behaviour of Ni polycrystals with a few grains per thickness *Mater. Lett.* **62** 1718–20

[8] Yoshida S, Bhattacharjee T, Bai Y and Tsuji N 2017 Friction stress and Hall-Petch relationship in CoCrNi equi-atomic medium entropy alloy processed by severe plastic deformation and subsequent annealing *Scr. Mater.* **134** 33–6

[9] Otto F, Dlouh A, Somsen C, Bei H, Eggeler G and George E P 2013 The influences of temperature and microstructure on the tensile properties of a CoCrFeMnNi high-entropy alloy *Acta Mater.* **61** 5743–55

[10] Hall E O 1951 The deformation and ageing of mild steel III Discussion of results *Proc. Phys. Soc. Sect. B* **64** 747

[11] Petch N J 1953 The Cleavage Strength of Polycrystals *J. Iron Steel Inst.* **174** 25–8

[12] Toda-Caraballo I, Wrobel J S, Dudarev S L, Nguyen-Manh D and Rivera-Diaz-Del-Castillo P E J 2015 Interatomic spacing distribution in multicomponent alloys *Acta Mater.* **97** 156–69

[13] Toda-Caraballo I 2017 A general formulation for solid solution hardening effect in multicomponent alloys *Scr. Mater.* **127** 113–7

[14] Labusch. R 1970 A statistical theory of solid solution hardening *Phys. status solidi* **41** 659–69

[15] Wu Z, Bei H, Pharr G M and George E P 2014 Temperature dependence of the mechanical properties of equiatomic solid solution alloys with face-centered cubic crystal structures *Acta Mater.* **81** 428–41

[16] Butt M Z and Feltham P 1993 Solid-solution hardening *J. Mater. Sci.* **28** 2557–76

[17] Gerold V and Karnthaler H P 1989 On the origin of planar slip in f.c.c. alloys *Acta Metall.* **37** 2177–83

[18] Yoshida S, Ikeuchi T, Bhattacharjee T, Bai Y, Shibata A and Tsuji N 2019 Effect of elemental combination on friction stress and Hall-Petch relationship in face-centered cubic high / medium entropy alloys *Acta Mater.* **171** 201–15

[19] Zhang L, Xiang Y, Han J and Srolovitz D J 2019 The effect of randomness on the strength of high-entropy alloys *Acta Mater.* **166** 424–34

A unified model for microstructural coarsening of deformed metals

T Yu[1]

[1] Department of Mechanical Engineering, Technical University of Denmark, DK-2800, Kgs. Lyngby, Denmark

Email: tiyu@mek.dtu.dk

Abstract. In this work the kinetics of continuous coarsening of deformation microstructures are analysed. The coarsening examples span a large range in structural scale and driving force, including subgrain growth in a deformed subgrain structure, uniform coarsening in a deformed lamellar structure, and grain growth in a deformed nanocrystalline structure. The unified coarsening model for deformation microstructures considers the change of the apparent activation energy during coarsening and successfully describes the kinetics in each of the given coarsening examples. It is also found that the increase in the apparent activation energy dominates the coarsening kinetics and leads to a significant decrease in the coarsening rate as coarsening proceeds.

1. Introduction

In the last three decades, great effort has been put into producing nanocrystalline (grain size below 100 nm) and nanostructured (grain size typically 100-1000 nm) materials by plastic deformation to large strains [1-4]. Strong nanocrystalline and nanostructured materials can now be produced by a number of routes, but their thermal stability and formability are critical for industrial application. After deformation to a large strain, the structure typically has a high stored energy, which serves as the driving force for dislocation recovery and structural coarsening, leading to a reduction in strength and enhancement of formability. Thus, there is a possibility to optimize the properties of nanocrystalline and nanostructured materials by annealing. However, the annealing conditions must be defined accurately to avoid excessive loss of strength, for example due to recrystallization [5].

In the present work, a recently proposed unified coarsening model is introduced and applied to three examples of thermally activated continuous coarsening. This new model is then compared with a standard model for grain growth.

2. Unified coarsening model

A unified coarsening model for deformation microstructures was recently proposed that considers the variation of the apparent activation energy during coarsening [6]. Following a classical recovery rate equation and taking the stored boundary energy as the source of driving force, the rate equation for structural coarsening can be expressed as

$$\frac{dD}{dt} = k_1 D \exp(\frac{k_2}{DT}) . \quad (1)$$

By the aid of exponential integrals, the integrated form can be expressed as

$$Ei(-\frac{k_2}{DT}) - Ei(-\frac{k_2}{D_0 T}) = -k_1 t\,, \tag{2}$$

where $E_i(\,)$ is the exponential integral of the quantity inside the bracket, D_0 is the initial grain size (or boundary spacing), D is the grain size after annealing, t is the annealing time, T is the absolute temperature, k_2 is a constant to be fitted, and k_1 is temperature dependent, which can be written as

$$k_1 = K_0 \exp(-\frac{Q_0}{RT})\,, \tag{3}$$

where K_0 is a constant to be fitted, R is the gas constant, and Q_0 is the extrapolated activation energy at the end of coarsening. The three fitting parameters K_0, Q_0 and k_2 are associated with the operative coarsening mechanisms. The apparent activation energy at any stage of coarsening (i.e. at grain size D) can be subsequently obtained as

$$Q_{app} = Q_0 - \frac{k_2 R}{D}\,. \tag{4}$$

Equation (2) describes the continuous coarsening of a deformation structure during isothermal annealing. In order to estimate the model parameters k_1 and k_2, coarsening data, i.e. (t, D) pairs at one annealing temperature, are inserted into equation (2) and a curve of k_1 vs k_2 is calculated for each (t, D) pair. A maximum convergence point (k_2, k_1) is then determined manually by superimposing k_1 vs k_2 curves of all annealing times at that temperature (see figure 1). When such a procedure is carried out for other temperatures, different convergence points (k_2, k_1) can be obtained, and the estimated temperature independent constant k_2 may vary. Therefore an important subsequent procedure is to use an average k_2 for all temperatures to re-fit the coarsening data [6]. Such a collective fitting can reduce the fitting error significantly, especially for estimating the activation energy Q_0 based on equation (3).

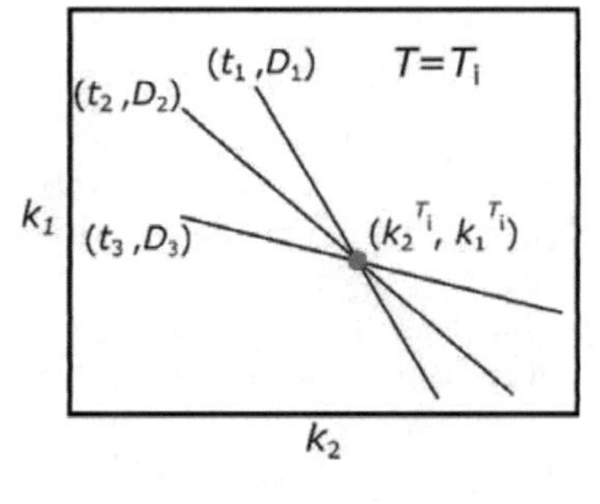

Figure 1. Sketch showing the process of determining the convergence point (k_2, k_1) at temperature T_i using coarsening data (t_1, D_1), (t_2, D_2) and (t_3, D_3). The curves are almost straight in a semi-log plot (k_1 on a logarithmic scale).

3. Examples of microstructural coarsening

In order to test the model over a wide range of structural scales, three representative well-characterized deformation microstructures from the micrometre scale to the nanometre scale were chosen: (i) a subgrain structure, in an Al-0.05 wt.% Si single crystal channel die compressed by 70% at room temperature [7]; (ii) a lamellar structure, in commercial purity Al (AA1050) deformed by cold rolling to a true strain of 5.5 [8]; and (iii) a nanocrystalline structure, in high purity W powder, ball-milled at room temperature in an Ar atmosphere [9]. The coarsening data of these examples are shown in figure 2, as well as the fitted curves based on the model, i.e. equation (2).

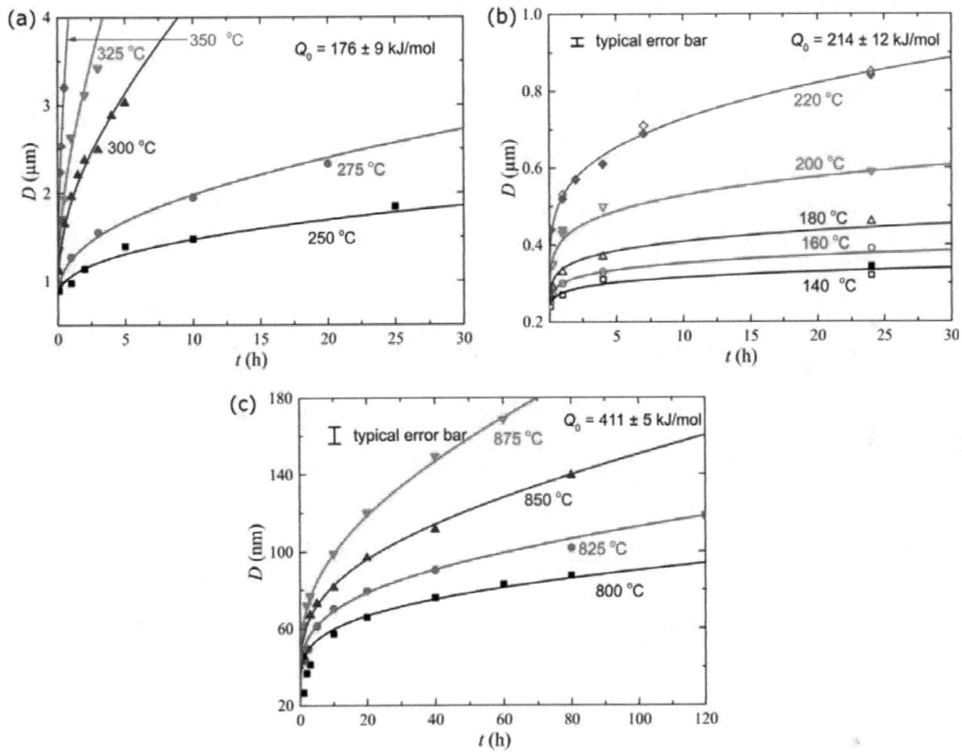

Figure 2. Coarsening data of deformation microstructures fitted by the unified coarsening model. (a) Subgrain growth of Al-0.05 wt.% Si [7]; (b) coarsening of a AA1050 lamellar structure [8]; (c) grain growth of high purity nanocrystalline W [9].

4. Apparent activation energy

The apparent activation energy is an important parameter in the coarsening kinetics. It provides an indication of the underlying rate-controlling mechanism and also reveals the extent of the temperature dependence of coarsening. In the following, the apparent activation energy will be directly calculated based on the experimental data and compared with the modelling results.

To calculate the apparent activation energy, one has to assume the same coarsening mechanisms at different annealing temperatures, i.e. the same apparent activation energy at a given grain size independent of temperature. Moreover, as an approximation, the apparent activation energy is taken to be constant in a small coarsening window (i.e. a small grain size range, e.g. from 0.6 μm to 0.7 μm). These assumptions, together with the Arrhenius equation, enable a stepwise analysis of the apparent activation energy as demonstrated recently [6]. As a result, one may draw horizontal lines in figure 2, and then calculate the apparent activation energy for each coarsening window by comparing the annealing times (from interpolation of experimental data) at different temperatures based on the following equation

$$\ln t = c + \frac{Q_{app}^{step}}{RT}, \quad (5)$$

where both c and the stepwise apparent activation energy Q_{app}^{step} are grain-size dependent fitting parameters. Such a procedure is analogous to that for analysing recrystallization kinetics (see e.g. [10]), where the whole process is typically considered as one window.

Based on figure 2 and equation (5), the directly calculated stepwise apparent activation energies for the three examples are shown in figure 3 as discrete data points, all of which agree with the smooth curves obtained from the unified coarsening model. The apparent activation energies increase during coarsening in all three cases. The estimated values at the final stages of coarsening are in agreement with some well-defined physical mechanisms, whereas in the early stages they are reduced by the presence of short-circuit diffusion paths in the deformation microstructures. For coarsening of the Al-0.05% Si subgrain structure, the estimated apparent activation energy is consistent with lattice diffusion of Si in Al, for which the activation energy is in the range of 129-154 kJ/mol [11], suggesting that solute drag may be the rate controlling mechanism. For coarsening of the AA1010 deformed lamellar structure, the apparent activation energy is consistent with diffusion of Fe in the bulk of Al, where the activation energy is reported to be in the range of 183 - 259 kJ/mol [12,13], also pointing to an important effect of solute drag. For grain growth of the W nanocrystalline structure, the apparent activation energy is close to that for grain boundary diffusion of W, 381 kJ/mol [14], suggesting that diffusion of W atoms along the boundaries and across the boundaries may require similar activation energies, and also that solute drag may not be important due to the negligible concentration of substitutional solutes.

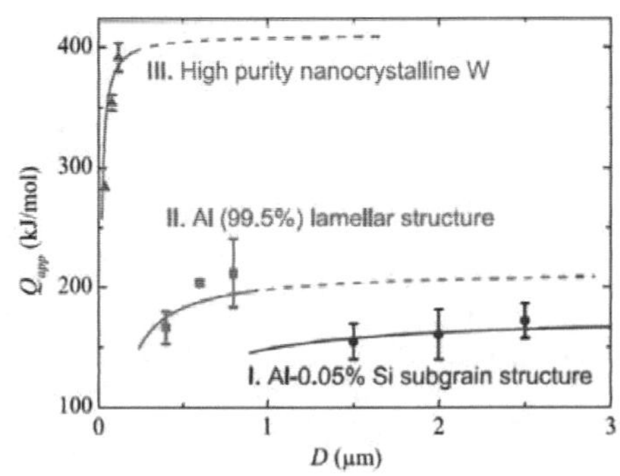

Figure 3. Increase of the apparent activation energy during coarsening of three deformation microstructures [6]. The smooth curves obtained from the unified coarsening model agree with the results (discrete points) directly calculated based on equation (5).

5. Model comparison

In the literature, the continuous coarsening of deformation microstructures has been commonly analysed in analogy to grain growth in fully recrystallized coarse-grained materials [5]. The corresponding kinetics equation is generally expressed as

$$D^n - D_0^n = k_0 t \exp(-\frac{Q}{RT}), \tag{6}$$

where Q is the activation energy, n is the growth exponent ($n = 2$ in the ideal case of grain growth) and k_0 is a kinetics constant. In this equation, the activation energy Q is assumed to be constant during coarsening and may be estimated from isochronal annealing data in a $\ln(D^n\text{-}D_0^n)$ vs T^{-1} plot.

When equation (6) is used to analyse coarsening of deformation microstructures, the growth component n varies significantly with annealing temperature [7]. For the example of the deformed lamellar structure (figure 2b), the growth component increases from $n = 7.4$ at 220 °C to $n = 14.8$ at 140 °C. In other words, equation (6) with a fixed growth component cannot fit the coarsening data at different annealing temperatures.

If one ignores the large variation in the growth exponent n in equation (6) and assigns the ideal growth exponent $n = 2$ for all temperatures, then a good linear fitting may be still achieved in a $\ln(D^2-D_0^2)$ vs T^{-1} plot based on isochronal annealing data (since the quality of fitting for isochronal annealing is independent of n when D_0 is much smaller than D). This is demonstrated in figure 4 for the example of the deformed lamellar structure. For both 1 h and 24 h isochronal annealing, good linear fittings are obtained, although the estimated activation energies (dependent on the chosen growth exponent n) are much lower than those obtained with the unified coarsening model. It should be noted that such low activation energies for coarsening of deformation microstructures are frequently reported in the literature based on equation (6), leading to confusion in the understanding of coarsening kinetics and boundary mobility.

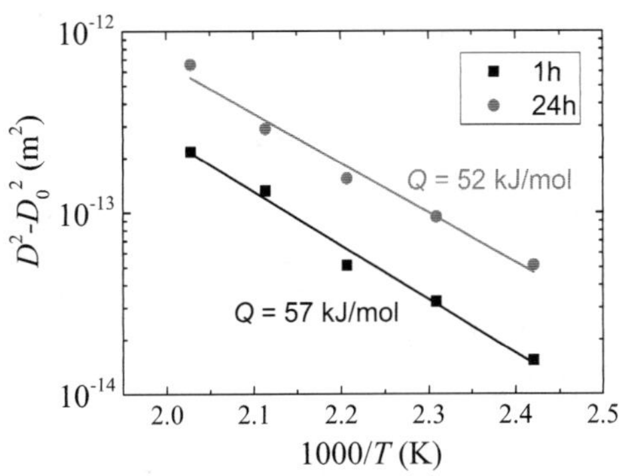

Figure 4. Estimated activation energy based on equation (6) for coarsening of the AA1050 deformed lamellar microstructure.

Based on the unified coarsening model, the variation of the apparent activation energy during coarsening at different temperatures can be estimated. Combining equations (2) and (4), one gets the following relationship between the annealing time and the apparent activation energy

$$t = \frac{1}{k_1}\left[Ei(-\frac{k_2}{D_0 T}) - Ei(-\frac{Q_0 - Q_{app}}{RT}) \right]. \tag{7}$$

The corresponding variation of the apparent activation energy during annealing of the deformed Al lamellar microstructure is shown in figure 5. For a higher temperature after a given annealing time (e.g. 1 h, see figure 5), coarsening has reached a later stage, which is much slower and associated with a higher apparent activation energy. Consequently, the coarsening rate may appear only weakly dependent on the temperature during isochronal annealing. This explains the artificially low activation energy estimated from equation (6) (see figure 4).

Equation (6) with a fixed growth component cannot be used to describe the coarsening at different annealing temperatures, and the error is increased when a larger temperature range has to be covered. For estimation of the apparent activation energy based on equation (6), the error is increased when the chosen growth component n has a larger deviation from the fitted growth component, which varies with temperature.

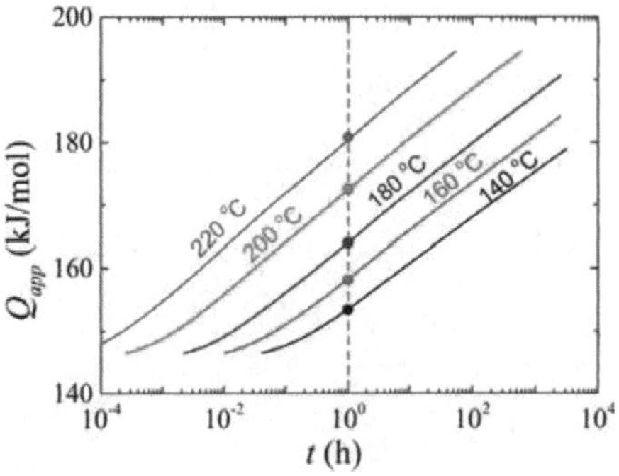

Figure 5. Variation of the apparent activation energy during annealing of the AA1050 deformed lamellar microstructure at 5 different temperatures [6].

6. Conclusion

The unified coarsening model for deformation microstructures considers an important feature, namely that the apparent activation energy changes during coarsening. The model applicability has been demonstrated in three examples covering a large variation in structural scale. The modelling results show that the apparent activation energies increase during coarsening, leading to a significant decrease in the coarsening rate with increasing time. This change in the apparent activation energy makes coarsening of deformation microstructure significantly different from grain growth in a fully recrystallized coarse-grained material.

Acknowledgement

This project has received funding from the European Research Council (ERC) under the European Union's Horizon 2020 research and innovation programme (grant agreement No 788567). The author is grateful to N. Hansen and D. Juul Jensen for stimulating discussions.

References

[1] Koch, C C 2007 *J. Mater. Sci.* **42** 1403
[2] Estrin Y and Vinogradov 2013 *Acta Mater.* **61** 782
[3] Hansen N and Barlow C 2015 *Physical Metallurgy*, ed E Laughlin and K Hono (Amsterdam: Elsevier) chapter 17 p 1681
[4] Lu K 2016 *Nat. Rev. Mater.* 16019
[5] Humphreys F J and Hatherly M 2004 *Recrystallization and Related Annealing Phenomena* (Oxford: Pergamon)
[6] Yu T and Hansen N 2016 *Acta Mater.* **120** 40
[7] Huang Y and Humphreys 2000 *Acta Mater.* **48** 2017
[8] Yu T, Hansen N and Huang X 2013 *Acta Mater.* **61** 6577
[9] Hegedus Z, Meka S R and Mittemeijer E J 2016 *Acta Mater.* **105** 232
[10] Yu T and Hughes D A 2019 *Philos. Mag.* **99** 869
[11] Mondolfo L F 1975 *Aluminum Alloys: Structure and Properties* (London: Butterworths)
[12] Hood G M 1970 *Philos. Mag.* **21** 305
[13] Beke D L, Godney I, Szabo I A Erdelyi G and Kedves F J 1987 *Philos. Mag. A* **55** 425
[14] Kaur I and Gust W 1989 *Handbook of Grain and Interphase Boundary Diffusion Data*, (Stuttgart: Ziegler Press)

IOP Conf. Series: Materials Science and Engineering **580** (2019) 012055 doi:10.1088/1757-899X/580/1/012055

Influence of rapid annealing on the evolution of fractal abnormal grains in nanocrystalline Pd–10 at% Au

R A Zeller[1,3], H J Fey[1], C Braun[2], R Birringer[2] and C E Krill III[1]

[1] Institute of Functional Nanosystems, Ulm University, Albert-Einstein-Allee 47, 89081 Ulm, Germany
[2] Department of Experimental Physics, Saarland University, Campus D2.2, 66123 Saarbrücken, Germany

[3] E-mail: raphael.zeller@uni-ulm.de

Abstract. During abnormal grain growth, certain grains grow to much larger sizes than their neighbors. Usually, such abnormal grains expand in all directions at roughly the same speed, but in nanocrystalline Pd–10 at% Au, they appear to send forth dendritic offshoots into the surrounding matrix, resulting in highly irregular, fractal-like grain shapes. Exploiting the fast heating capabilities of a salt-bath oven, we have investigated the influence of heating rate on the evolution of fractal grain morphologies in nanocrystalline Pd–10 at% Au. In this material, the nanocrystalline matrix is stable at 170°C, but fractal grains appear within minutes at 200°C and 270°C, followed by fast growth until impingement. Surprisingly, the higher dwell temperature leads to a smaller average grain size in the fully impinged state. For the latter, no additional grain growth is observed at longer dwell times, but the fractal dimension of grain boundaries decreases with time at 270°C. These findings are consistent with abnormal grain formation being governed by a temperature-dependent nucleation process, upon which is superimposed the usual capillary driving force for grain growth; however, the mechanism by which boundaries migrate through the nanocrystalline matrix while retaining fractality remains unclear.

1. Introduction

Like all polycrystals, nanocrystalline (NC) materials manifest grain growth upon thermal treatment [1]. This can happen in a *normal* manner—such that the grain size distribution remains unimodal—or *abnormally,* if a small subpopulation of grains grows much faster than the rest, leading to a bimodal size distribution [2–5]. Recently, when studying nanocrystalline specimens of Pd–10 at% Au prepared by inert gas condensation, we discovered an unexpected growth mode that is, in a certain sense, doubly abnormal [6]: in this material, not only do a few grains suddenly start to grow at a high rate, but they do so in an unusual manner, sprouting "tentacles" that first surround nearby matrix grains and then consume them. The resulting interfaces between large abnormal grains and nanometer-sized matrix grains are highly irregular in shape and surprisingly reminiscent of fractal objects (figure 1); indeed, a box-counting analysis applied to the perimeters of abnormal grains in Pd–Au found a fractal dimension of 1.20 ± 0.05 rather than the smooth-boundary limit of unity [6].

Such growth behavior is difficult to understand from the perspective of the well-established curvature-driven mechanism for grain boundary migration [2], which would be expected to flatten out boundary fluctuations rather than accentuate them. Along the boundary of an abnormal grain, the value of reduced mobility (the product of boundary energy and mobility) likely varies, but as the abnormal

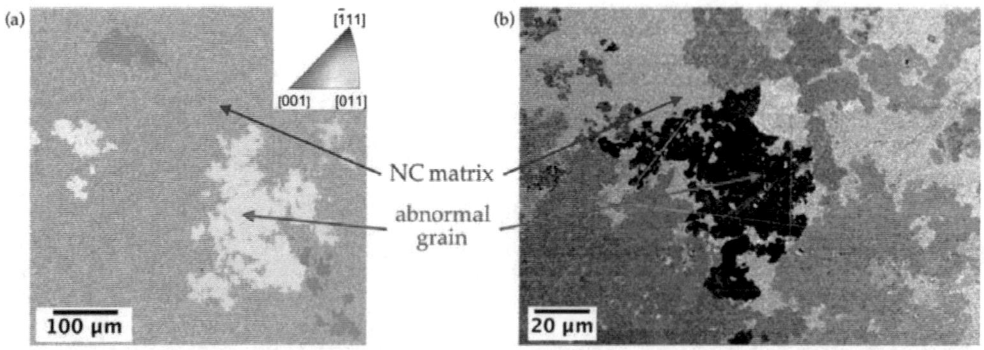

Figure 1. (a) Electron back-scattered diffraction (EBSD) map of AGG occurring in NC Pd–10 at% Au produced by inert gas condensation, with lattice orientation indexed to the sample normal. At an average size of ~10 nm, the matrix grains lie below the resolution limit, resulting in mottled contrast in regions between abnormal grains. (b) Channeling-contrast image of abnormal grains in Pd–Au, obtained by scanning electron microscopy (SEM). The surfaces of abnormal grains are traversed by polishing scratches, whereas vestiges of the NC matrix generate areas with gray/white fluctuations.

grain consumes its NC neighbour grains, the time-averaged boundary migration rate ought to approach the same value everywhere. In NC Pd–10 at% Au, however, this evidently does not occur. We therefore suspect that nonstandard mechanisms might be at least partially responsible for the abnormal grain growth (AGG) observed so frequently in NC systems [7–18]. This suspicion is strengthened by the fact that the technique used to prepare the NC sample shown in figure 1—inert gas condensation—is known to produce specimens of high purity and homogeneity that are free from texture [8]; as a result, we can reasonably exclude particle pinning and texture effects as possible causes of AGG [2].

In order to narrow down the range of possible mechanisms for the fractal AGG seen in NC Pd–10 at% Au, we have varied the rate at which samples prepared in the NC state were heated to a fixed dwell temperature. If fractal AGG is triggered by a "nucleation event" (as in models for AGG in which grain growth kinetics are described by the Johnson-Mehl-Avrami-Kolmogorov equation [2]), then an increase in heating rate might be expected lead to a more refined final microstructure: faster heating would create a greater degree of overheating, thereby raising the rate of nucleation of abnormal grains.

Alternatively, it is conceivable that AGG and/or the emergence of fractality could depend on how the mobility of grain boundaries compares to some threshold mobility value, such as one dictated by a pinning mechanism like solute drag [2]. By heating samples at the same rate to different final dwell temperatures, we can vary the average mobility of the grain boundaries. At high enough temperatures, all boundaries might be extremely mobile (*i.e.,* lie above the threshold value), in which case the interfaces between abnormal and matrix grains would lose their growth advantage over matrix/matrix interfaces, and microstructural evolution would revert to that of normal grain growth, with its characteristic unimodal, lognormal-like grain size distribution [2].

By analyzing the dependence of the final microstructure on both ramp rate and dwell temperature, we aimed in this study to determine the extent to which the AGG observed in NC Pd–10 at% Au can be modeled as a nucleation-and-growth phenomenon, perhaps mediated by a pinning process.

2. Methods

Inert gas condensation [20] was employed to prepare bulk, single-phase specimens of NC Pd–10 at% Au having an average grain size of ~10 nm; additional details are provided in Ref. [6]. This synthesis technique yields disk-shaped samples having a diameter of 8 mm and a thickness of about 500 μm, which were cut into smaller pieces for subsequent processing and characterization.

2.1. Rapid annealing (salt-bath oven)

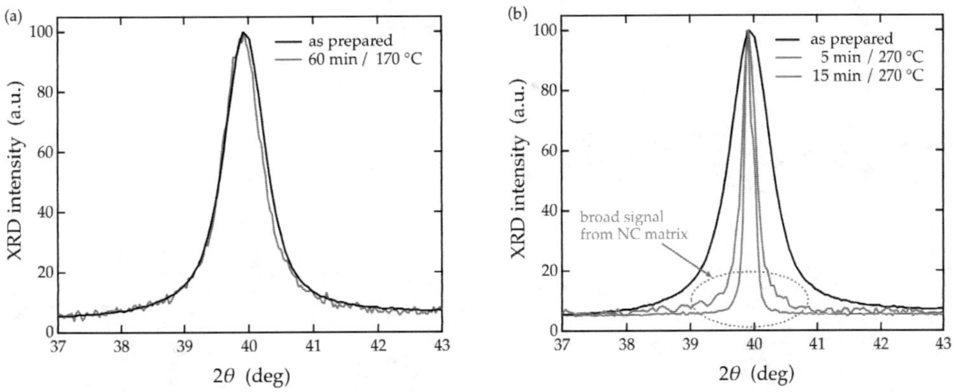

Figure 2. XRD scans of (111) peak intensity (normalized) of NC Pd–10 at% Au in the as-prepared state and after annealing at (a) 170°C and (b) 270°C for various times. Minimal peak narrowing at 170°C is indicative of a stable microstructure; however, at 270°C grain growth is detectable within 5 min, and by 15 min the broad signal arising from the NC matrix has disappeared completely.

At the dwell temperatures of this study (<300°C), the maximum heating rate that could be reached in an evacuated tube furnace was about 11 K/min = 0.18 K/s, even when the furnace had been preheated to the desired temperature prior to introduction of the sample. In contrast, when a specimen was submerged in a molten salt mixture, a heating rate of nearly 100 K/s was achieved, which is some two-and-a-half orders of magnitude higher than in the tube furnace. The salt bath consisted of a commercially available mixture of $NaNO_3$, $NaNO_2$ and KNO_3 salts (AS 140, Durferrit GmbH), which was heated above its melting point of 140–150°C in a custom-built apparatus equipped with a temperature-controlled hot plate and a fixture for rapid removal of the specimen.

2.2. *Microstructural characterization*

The microstructure of annealed specimens was analyzed by x-ray diffraction (XRD) and scanning electron microscopy (SEM). The XRD measurements were carried out in a PANalytical X'Pert Pro diffractometer equipped with a copper tube (λ = 0.15406 nm), scanning the first eight Bragg peaks of the fcc crystal structure. An estimate for the average size of nanocrystalline grains in a given sample could be extracted from measured peak profile widths using the Williamson-Hall method [8]. Because the abnormal grains grew quickly beyond the ~100 nm upper limit of the Williamson-Hall analysis, the size and shape of such grains was assessed using backscattered-electron (BSE) imaging in a Zeiss-LEO 1550 scanning electron microscope. This imaging mode is conducive to channeling contrast, for which the signal intensity depends strongly on the orientation of a grain's crystal lattice with respect to the incident electron beam. Consequently, surface locations belonging to the same micrometer-sized grain share nearly the same gray-scale value, but, for the same reason, an ensemble of NC matrix grains gives rise to rapid gray-scale fluctuations, which take on a characteristic "shiny" appearance (figure 1(b)).

3. Results

Samples were annealed in the salt-bath oven at 170°C, 200°C and 270°C for times ranging from 5 to 60 min, after which they were cooled to room temperature in air and characterized using XRD and SEM. A Williamson-Hall analysis of the Bragg peak profiles of a specimen in the as-prepared state yielded an initial average grain size of 11 ± 1 nm. Heat treatment at 170°C led to minimal narrowing of Bragg peaks (figure 2(a)) up to an annealing time of 60 min, indicating that the NC microstructure was stable at this dwell temperature. In contrast, significant sharpening of peak profiles was observed within 15 min at 200°C and 5 min at 270°C, whereby a broad signal at the base of each Bragg peak suggested that a certain volume fraction of NC matrix grains remained in the specimen. This interpretation was supported

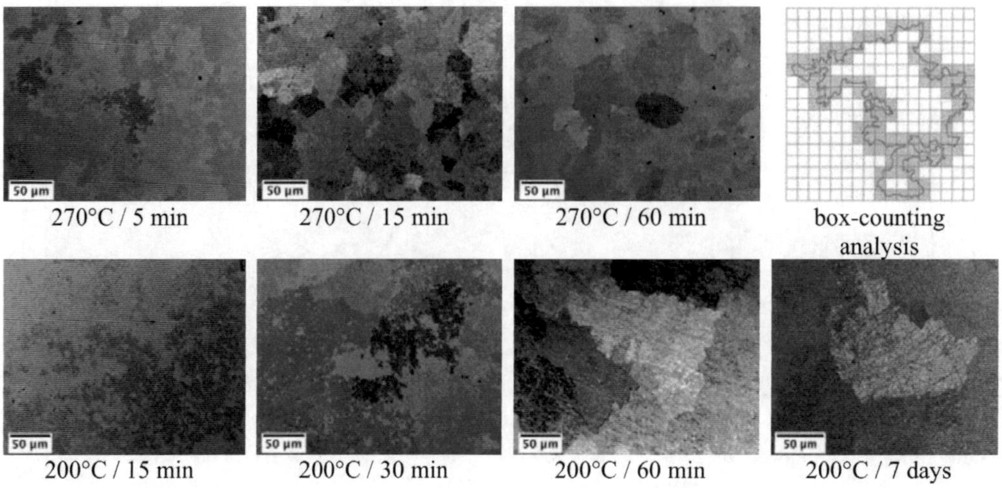

Figure 3. SEM micrographs of NC Pd–10 at% Au following heat treatment in the salt bath oven for various times at (top row) 270°C and (bottom row) 200°C. Image contrast arises from electron channeling, according to which a change in gray scale indicates a rotation of the crystal lattice relative to the direction of the electron beam. (top right) Schematic illustration of the box-counting method applied to the central grain of the micrograph for 270°C / 5 min. (bottom right) The 7-day anneal at 200°C was performed in a tube furnace under vacuum.

by the monotonic decrease in the broad signal's intensity with annealing time (figure 2(b)): within 15 min at 270°C, the Bragg peaks had narrowed in width nearly to the level of the diffractometer's instrumental broadening, whereas the same development took longer to complete at 200°C (30–60 min). From the broad signal's nearly constant width in 2θ, we conclude that the average size of matrix grains did not change over time—in other words, only the abnormal grains underwent coarsening.

The presence of a bimodal grain size distribution during the early stages of annealing at 200°C and 270°C was confirmed by SEM imaging of polished sections (figure 3). Channeling contrast revealed that some grains had grown larger than 50 μm in only 15 min at the lower and 5 min at the higher temperature. The morphology of the abnormal/matrix boundaries was highly irregular, manifesting the same dendritic appearance reported in Ref. [6] for samples prepared under slow-ramping conditions (dT/dt = 1°C/min). A box-counting analysis [6] of the perimeter of the dark grain at the center of the 270°C / 5 min micrograph (top left image in figure 3) yielded a fractal dimension of ~1.3 (figure 4(a)).

With increasing annealing time, fractal abnormal grains impinge upon one another. At 270°C, impingement is accompanied by a marked decrease in perimeter roughness (figure 4(b)), which is reflected in a corresponding drop in fractal dimension from ~1.3 to ~1.1 (figure 4(a)). At 200°C, however, the impinged grain boundaries are still convoluted, with box-counting dimensions in the ~1.2 range. Finally, we subjected a series of NC Pd–10 at% Au samples in the as-prepared state to the slower heating ramp of a tube furnace that had been preheated to 200°C. No significant differences in grain evolution or morphology were noted in comparison to specimens annealed in the salt bath at the same temperature (bottom row of figure 3).

4. Discussion

Comparison of the top and bottom rows of figure 3 reveals that the average size of fractal abnormal grains after impingement is significantly larger at the dwell temperature of 200°C than at 270°C. This finding is consistent with the appearance of fractal grains being governed by a nucleation event, provided that the rate of nucleation increases with temperature. Even if the subsequent growth of fractal grains (*i.e.,* the migration of abnormal/matrix growth fronts) is faster at the higher temperature—as would be

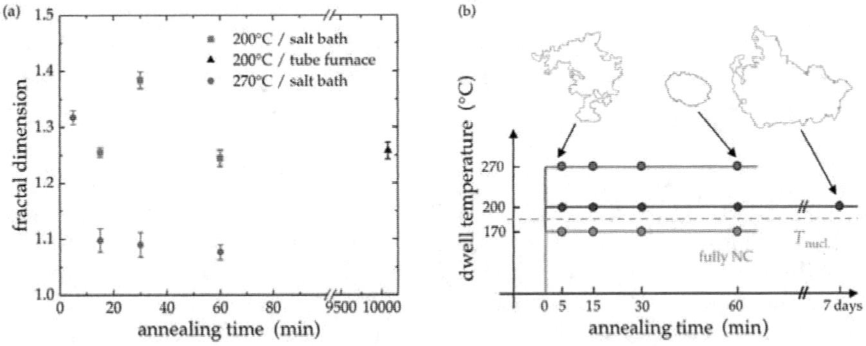

Figure 4. (a) Fractal dimension of the perimeter of selected grains extracted from the SEM images of figure 3, evaluated by the box-counting method of Ref. [6]. (b) Schematic plot of grain morphology as a function of dwell temperature and annealing time. The surmised nucleation temperature $T_{nucl.}$ lies between 170°C and 200°C; smoothing of fractal grain boundaries with time is evident at 270°C but not at 200°C.

expected from the Arrhenius-like temperature dependence of grain boundary mobilities [2]—a faster rate of nucleation at 270°C would lead to the formation of a greater number of abnormal grains per unit volume and, consequently, to a smaller average size at impingement.

Since no fractal grains appear during heat treatment at 170°C, we conclude that the onset temperature for nucleation, $T_{nucl.}$, must lie between 170°C and 200°C. This explains why the fast-ramp and slow-ramp heat treatments performed at 200°C led to qualitatively identical grain morphologies: even if $T_{nucl.}$ takes on its minimum possible value of 170°C, the ramp with dT/dt = 11°C/min would spend only ~3 min between $T_{nucl.}$ and 200°C; once the dwell temperature is reached, the two heat treatments are identical. If (i) the total number of nucleation events during the slow ramp is small in comparison to the number at 200°C or (ii) the nucleation rate increases by only a modest amount between $T_{nucl.}$ and 200°C, then the total number of nucleation events per unit volume should be roughly the same for the two ramp rates, leading to the same average grain size in the impinged state.

Once the NC matrix has been consumed by the abnormally growing grains, there is no evidence for further migration of grain boundaries. At 270°C, the boundaries become noticeably smoother with annealing time (figure 4)—this can be understood as a consequence of the usual capillary driving force for boundary migration, which induces curved boundaries to move toward their respective centers of boundary curvature. As a result, the amplitude of boundary fluctuations should decrease over time. At 200°C, however, the fractal dimension of grain perimeters was found to remain roughly constant, even after a specimen was annealed for 7 days (figure 3 (bottom right micrograph) and figure 4). Apparently, the kinetics of boundary smoothing are so slow at this temperature that no boundary flattening occurs in 7 days, or an unknown pinning force acts on the boundaries to hold them in place.

Finally, we note that the nature of the nucleation event for fractal grains is still unclear, but the behavior seen in NC Pd–10 at% Au is reminiscent of the "particle-assisted" abnormal growth reported by Holm *et al.* in a Monte Carlo Potts model simulation of Zener pinning [21]. There, the polycrystalline microstructure was stable for long stretches of time until one grain suddenly found a way to break free from its pinning particles, after which this grain consumed every other grain in the simulation cell. In NC Pd–10 at% Au, the nucleation of abnormal grains is a similarly rare event: even with a fast ramp to 270°C, for which the nucleation rate appeared to be highest, a single abnormal grain occupied a volume that originally contained roughly $(50\ \mu m\ /\ 11\ nm)^3 \approx 10^{11}$ nanograins! If pinning forces (of unclear origin) play a similar role in fractal AGG as in the simulations of Ref. [21], then candidates for abnormal grain nuclei might include locations in the polycrystalline microstructure at which a specific grain is significantly larger than a majority of its neighbors; such interfaces could conceivably be better able to overcome pinning forces than are boundaries between smaller matrix grains. As a result, the growth of an abnormal grain through the NC matrix would be much faster than migration of a boundary between two nanograins at the same temperature. Once two large, abnormal grains meet at a common boundary, however, the growth rate drops by orders of magnitude, as the curvature-based driving force is now much smaller.

5. Conclusions

The findings reported above are consistent with a nucleation-limited formation of fractal abnormal grains in NC Pd–10 at% Au. Subsequent growth of abnormal grains into the NC matrix proceeds quickly at both 200°C and 270°C, such that the grain size at impingement is a reflection of the relative rate of nucleation rather than growth. The precise nature of the nucleation event has not been determined, but the fact that grain boundaries in the NC matrix were observed to be immobile even at 270°C suggests the presence of some kind of pinning force holding the boundaries in place. The boundary smoothing that occurs in the impinged state at 270°C can be viewed as a consequence of the remaining driving force that is "trapped" in spatial fluctuations of the fractal boundaries. The same driving force for boundary smoothing is present at 200°C: the lack of change in boundary morphology at this temperature can likely be attributed to insufficient kinetics.

References

[1] Krill III C E, Helfen L, Michels D, Natter H, Fitch A, Masson O and Birringer R 2001 *Physical Review Letters* **86** 842–5
[2] Humphreys J, Rohrer G S and Rollett A 2017 *Recrystallization and Related Annealing Phenomena*, 3rd ed (Amsterdam: Elsevier)
[3] Rollett A D, Brahme A P and Roberts C G 2007 *Mater. Sci. Forum* **558-559** 33–42
[4] Kang S-J L 2005 *Sintering: Densification, Grain Growth & Microstructure* (Amsterdam: Elsevier)
[5] Lawrence A, Rickman J M, Hermer M P and Rollett A D 2016 *Acta Mater* **103** 681–7
[6] Braun C, Dake J M, Krill III C E and Birringer R 2018 *Sci. Rep.* **8** 1592–7
[7] Gertsmann V Y and Birringer R 1994 *Scripta Metall. Mater.* **30** 577–81
[8] Ames M, Markmann J, Karos R, Michels A, Tschöpe A and Birringer R 2008 *Acta Mater.* **56** 4255–66
[9] Wang N, Wang Z, Aus K T and Erb U 1997 *Acta Mater.* **45** 1655–69
[10] Dannenberg R, Stach E A, Groza J R and Dresser B J 2000 *Thin Solid Films* **370** 54–62
[11] Hau-Riege S P and Thompson C V 2000 *Appl. Phys. Lett.* **76** 309–11
[12] Greiser J, Müllner P and Arzt E 2001 *Acta Mater.* **49** 1041–50
[13] Hibbard G D, Aust K T and Erb U 2006 *Mater. Sci. Eng.* A **433** 195–202
[14] Hattar K, Follstaedt D M, Knapp J A and Robertson I M 2008 *Acta Mater.* **56** 794–801
[15] Darling K A, VanLeeuwen B K, Koch C C and Scattergood R O 2010 *Mater. Sci. Eng.* A **527** 3572–80
[16] Kacher J, Robertson I M, Nowell M, Knapp J and Hattar K 2011 *Mater. Sci. Eng.* A **528** 1628–35
[17] Dake J M and Krill III C E 2012 *Scripta Mater.* **66** 390–3
[18] Paul H and Krill III C E 2011 *Scripta Mater.* **65** 5–8
[19] Kim S G and Park Y B 2008 *Acta Mater.* **56** 3739–53
[20] Birringer R 1989 *Mater. Sci. Eng.* A **117** 33–43
[21] Holm E A, Hoffmann T D, Rollett A D and Roberts C G 2015 *IOP Conf. Ser.: Mater. Scie. Eng.* **89** 012005

IOP Conf. Series: Materials Science and Engineering **580** (2019) 012056 doi:10.1088/1757-899X/580/1/012056

Investigation of plastic yielding in near-micrometer grain size aluminum using synchrotron microdiffraction

C.L. Zhang[1], A Godfrey[1], W.Q. Gao[1], Y.B. Zhang[2], G.L. Wu[3], W. Liu[4], R.Q. Xu[4], D. Juul Jensen[2]

[1] Key Laboratory of Advanced Materials (MOE), School of Materials Science and Engineering, Tsinghua University, Beijing 100084, China.
[2] Department of Mechanical Engineering, Technical University of Denmark, DK-2800 Lyngby, Denmark.
[3] Joint Laboratory for Light Alloys (Ministry of Education), College of Materials Science and Engineering, Chongqing University, Chongqing 400044, China.
[4] Advanced Photon Source, Argonne National Laboratory, Argonne, IL 60439, USA.

E-mail: awgodfrey@mail.tsinghua.edu.cn

Abstract. The onset of plastic yielding has been studied in samples of aluminum with near-micrometer average grain size using in situ tensile loading combined with synchrotron radiation. A white-beam Laue 3D micro-diffraction technique was used allowing the collection of diffraction signal from a volume with a resolution of 1 μm in all three dimensions. The orientation data obtained from the micro-diffraction are used to estimate the local variation in geometrically necessary dislocation density, and correction factors are suggested for voxels where neighbors are missing in either one or two dimensions. The results demonstrate a considerable heterogeneity in deformation in the early stages of plastic deformation.

1. Introduction

The process by which a polycrystalline metal undergoes plastic yielding, i.e. the transition from elastic to plastic behavior, is of fundamental importance for the understanding of mechanical properties, and yet there are many aspects of this transition that are still only poorly understood [1,2]. The interest in understanding this elastic-plastic transition has intensified in recent years for a number of reasons. These include the observation that even in FCC metals a distinct yield drop is developed for grain sizes in the near-micrometer regime [3,4] (indicating a transition in the way in which plastic strain percolates through a loaded sample), and enhanced combinations of ductility and mechanical strength in samples with a heterogeneous grain size distribution [5].

Digital image correlation (DIC) methods have been used to determine local plastic strain via the deformation gradient tensor, and it has been shown for example that grains with "soft" orientations begin plastic yielding even during the nominally elastic part of the loading curve [6]. Such experiments are however limited to 2D surface observations and it remains very challenging to make such measurements in the elastic-plastic transition region.

An alternative parameter that can be used is the geometrically necessary dislocation (GND) density (ρ_{GND}) [7,8]. Although this is not a direct measure of local plastic strain (for a given stress state different crystal orientations deformed to the same plastic strain will have different ρ_{GND} values), increases in the GND density at a very low strain are a clear indicator of the transition from elastic to plastic deformation.

IOP Conf. Series: Materials Science and Engineering **580** (2019) 012056 doi:10.1088/1757-899X/580/1/012056

The GND density is calculated from local lattice curvature [7,8] and therefore requires spatially resolved measurements of lattice orientation. For study of plastic yielding a technique for measuring crystal orientations with high angular resolution is therefore required, where additionally these measurements should be taken in the bulk with sufficient spatial resolution to allow determination of the local lattice curvature in 3D. These requirements are all satisfied by the X-ray synchrotron facility at beamline 34-ID-E of the Advanced Photon Source, Argonne National Laboratory, where spatially resolved 3D orientation measurements with sub-micrometer spatial resolution can be achieved by use of the differential aperture X-ray microscopy (DAXM) technique [9].

For such an experiment the material for examination should have ideally a fine grain size (to allow observation of a large number of grains while using a high orientation mapping resolution), and be in a fully recrystallized condition (to allow plastic yielding to be studied in an initially deformation-free material). It has been demonstrated elsewhere that these requirements can be achieved in samples prepared by spark plasma sintering (SPS) [4,10,11], where it is possible to achieve an average grain size in the near-micrometer grain size regime. For the present experiment samples with a 5.2 μm average grain size are used, as this grain size is sufficiently large to still exhibit classical yielding behavior, but is fine enough to allow a large number of grains to be mapped using DAXM.

Figure 1. Reconstructed 3D grain maps of the 5 μm sample at strains of (a) 0% and (b) 0.80%. The coloring scheme is based on the Euler angles of each indexed voxel. The mapped volume extends from the sample surface (S) into the interior of the sample

2. Experimental

The SPS technique was used to prepare samples of Al with an average grain size of 5.2 μm. Disks of diameter 20 mm and height 6 mm were sintered from spherical Al powder (average particle size of 5.7 μm) following the procedure described by Le et al. [4]. Dog-bone shape tensile samples were cut from the as-sintered disks by electron-discharge machining with gauge width, length and thickness of 1.8, 12 and 0.6 mm, respectively. Prior to tensile deformation, the specimens were manually ground to 4000-grit silicon carbide paper, and then electrochemically polished in a 10% perchloric acid and 90% alcohol solution to achieve a flat polished surface free of mechanical damage and residual deformation.

For the synchrotron measurements, a polychromatic X-ray beam with energies in the range of 7 - 30 keV was focused using two non-dispersive Kirkpatrick-Baez mirrors, producing a beam of near-Lorentzian profile with a full width at half maximum of ~ 0.3 μm. At the beginning of the experiment DAXM was used to map a volume of dimensions 51 × 5 × 155 μm^3 using a step size of 1 μm along all axes. The volume was located near the center of the gauge length, extending from the sample surface to the sample interior.

For the tensile deformation a purpose-built push-to-pull stage was used, with a nominal strain resolution of ~ 0.05% [12]. The tensile stage was equipped also with a load cell to record the instantaneous force during tensile loading. Full details of the experimental set-up are available in [13]. The specimen was sequentially loaded to three strains of 0.15%, 0.30% and 0.80%. After each deformation step the position of volume of interest was first re-established using a cross-correlation technique based on the white beam Laue patterns [13], after which DAXM was used to map the crystal orientations within the volume, in each case using a step-size of 1 μm in all three dimensions.

3. Results and discussion

Figure 1 shows Euler maps of the investigated volume before and after deformation to a strain of 0.80%. The agreement between the two reconstructed volumes demonstrates successful tracking of a fixed bulk volume, while small differences of colors for the grains in figure 1b qualitatively show development of small rotations inside each grain. The lattice orientation measurement at each voxel can further be used to determine the GND density using either a simple approach based on local kernel average misorientation [14], or a more detailed approach where all dislocations present can be described by the local dislocation density tensor

$$\alpha_{ij} = -\epsilon_{jkl} \frac{\partial \beta_{il}^{el}}{\partial x_k} = -\epsilon_{jkl} \frac{\partial (\varepsilon_{il}^{el} + \omega_{il})}{\partial x_k} \tag{1}$$

where ϵ is the Levi-Civita symbol, β_{il}^{el} is the elastic distortion tensor, which is the sum of the elastic strain tensor ε_{il}^{el} and the lattice rotation tensor

$$\omega_{il} = -\epsilon_{ilm}\theta_m. \tag{2}$$

The latter term in equation 2 is given by the lattice rotation vector $\theta_m = -\epsilon_{mil}\omega_{il}/2$. With the introduction of the lattice curvature tensor

$$\kappa_{ij} = \frac{\partial \theta_i}{\partial x_j} \approx \frac{\Delta \theta_i}{\Delta x_j} \tag{3}$$

and assuming elastic strain is negligible, the relationship between dislocation density tensor and lattice curvature can be established by combining equations (1–3) as

$$\alpha_{ij} = \kappa_{ji} - \delta_{ij}\kappa_{kk} \tag{4}$$

where δ_{ij} is the Kronecker delta, and the scalar measure of GND density is taken as the entrywise one-norm of the dislocation density tensor

$$\rho_{\mathrm{GND}} \approx \frac{1}{b}\|\alpha\|_1 = \frac{1}{b}\sum_i \sum_j |\alpha_{ij}|. \tag{5}$$

For conventional EBSD, since the out-of-plane curvature cannot be measured, the calculated GND density is always an underestimate. One advantage of 3DXRM is it allows the orientation measurement along the missing dimension, and the full dislocation density tensor as well as the GND density can therefore be directly determined. Another advantage of our present data compared to EBSD is a very high angular resolution in orientation determination is accessible, estimated in the present experiment as 0.02°, which also enhances the accuracy of GND density measurement.

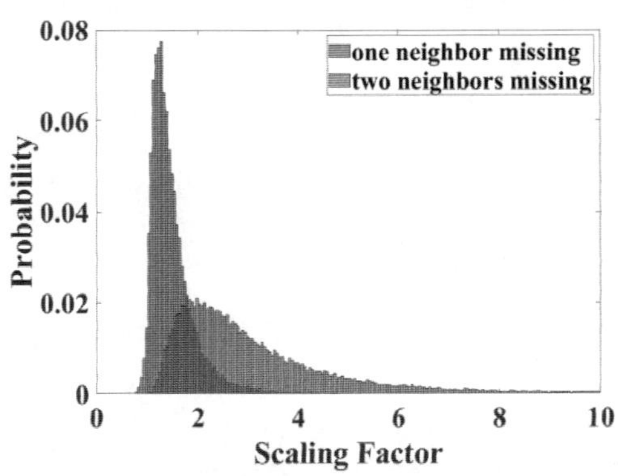

Figure 2. Histograms showing the scaling factor required to compensate for the absence of one or two neighbors in GND density calculations for the 0.80 % strain step. The distributions for the two cases are approximately of Rayleigh distribution shape, with mean values within the 95% confidence interval of 1.45 and 3.14, respectively. A small fraction of voxels with scaling factor larger than 10 is regarded as noise and neglected in calculation of average values.

Nevertheless, even in the 3D data set not every voxel has a full set of 3D neighbors, as some voxels remain unindexed due to low diffraction pattern quality, and some voxels lie either at the end of the examined volume or adjacent to a grain (or subgrain) boundary. As such there are still some places in the data set where neighbors along three positive axes are missing in one dimension, and places where neighbors are missing in two dimensions. To provide an estimation of the effect of such cases on the calculated dislocation density a simulation has been carried out where first the GND density was calculated for all voxels with a full set of neighbors in all three dimensions. Next, for each of these voxels the lattice curvatures along one or two dimensions are set to zero, randomly chosen among the three possible directions for each voxel. Finally, comparing the values allows the difference between the 3D value and the reduced dimension value to be determined.

The results are summarized in figure 2. The mean value within a 95% confidence interval for the one neighbor missing case is 1.45; for the two neighbors missing case the value is 3.14. The mean values are quite close to 1.5 and 3, the reciprocals of which are the inverse ratios of the number of non-zero lattice curvature elements in the GND tensor for each case (6 and 3 out of 9), implying that on average the missing gradient is of similar magnitude as the measured gradient and thus can be extrapolated. For 2D GND density calculations based on EBSD data a correction factor of 3.6 has previously been suggested [15], which as for this study was concluded via simulation. It should be noted that for the present data set, similar distributions and mean correction values for one or two missing neighbors were obtained on data for this sample at other strains, where different sets of randomly chosen missing curvature components were tested.

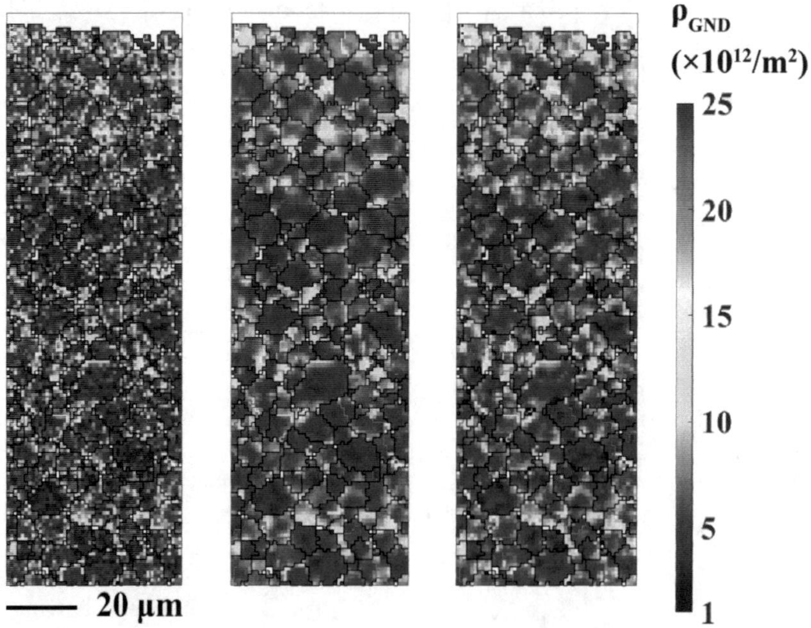

Figure 3. Maps showing the GND density at each voxel in the center layer of the investigated volume before deformation: (a) full 3D data; (b) 3D data after the application of a smoothing filter; (c) GND density calculated considering the center layer as a 2D data set (also with smoothing filter applied). Black and red lines indicate misorientations of 10° and 0.2°, respectively; white lines show features identified as subgrains. The sample surface is at the top of each image.

Based on the above discussion, the scaling factors of 3 and 1.5 were chosen for the GND density correction for all voxels with missing neighbors in the data set. This calculation was carried out within each grain. More precisely, first a grain detection algorithm was applied to undeformed data, using a grain boundary definition of 0.2°, and then a refinement algorithm used to find any subgrains within these grains down to a misorientation angle of less than 0.02°. The determined grain structure (including subgrains) was then mapped on to the data for the deformed samples, so that the voxels inside each grain at each deformation step were identified. For each grain, at each strain, the GND density was calculated for all voxels inside the grain. Note that this is different to the method usually used for GND calculation for EBSD data for example, where an upper cut-off angle is used to ensure that lattice curvature is not calculated across a grain boundary. As the present data set is of high angular resolution, by nature of synchrotron X-ray microscopy, and as the goal is tracking the evolution of GND density, this approach has the advantage of allowing tracking of GND evolution even within volumes identified as subgrains in the initial material.

An example GND density map, showing the center layer of the undeformed material, is given in figure 3a. The low values confirm the recrystallized nature of most grains, though evidence of residual strain, as well as some subgrains, albeit corresponding to very low misorientation angles, is seen, similar to observations of recrystallized Al examined using dark-field X-ray microscopy [16]. Somewhat higher values are seen near the sample surface. As it is expected that the DAXM data quality decreases with increasing depth it is concluded that these higher values are related to the difficulty in removal of all surface damage from sample preparation when a high orientation resolution is used. The variation in GND density is seen more clearly in figure 3b, where 3 × 3 × 3 neighbor smoothing filter has been used.

As it is not always possible to make crystal orientation measurements in 3D, it is also of interest to compare the present 3D results with those that would be obtained if only 2D were available at the same spatial and angular resolution. For this the crystal orientation measurements for the center layer of the mapped volume in the undeformed material were extracted and then the GND density calculated treating the layer as a 2D plane, rather than a 3D volume, applying the same correction factors as described above. The result is shown in figure 3c. A similar overall pattern in the spatial distribution of the GND content is seen, although locally high values in the 2D calculation are typically larger than in the 3D case. This can be understood by the fact that for this sample high in-plane (2D) lattice curvature is in most cases associated with lower out of plane lattice curvature, and the value calculated at each pixel (or voxel in 3D) is the average of the lattice curvature at each position.

Figure 4. Evolution of 3D GND density within each identified grain/subgrain during tensile deformation showing here the center layer of the investigated volume: (a) 0%, (b) 0.15%, (c) 0.30%, and (d) 0.80% deformation. The sample surface is at the top of each image.

Maps showing the evolution of GND density for the volume corresponding to the center layer of the initially mapped volume (calculated using the 3D data set and with smoothing) are shown in figure 4. Only moderate changes are seen at strains of 0.15% and 0.30%, where the average GND density (ρ_{av}) increases to from 5.9 x 10^{12} m^{-2} in the undeformed sample to ρ_{av} = 6.2 x 10^{12} m^{-2} and 6.5 x 10^{12} m^{-2} in samples deformed to strains of 0.15% and 0.30%, respectively. Closer inspection of the data at 0.30% strain shows obvious increases in GND density inside some grains, particularly near grain boundaries, while in a few grains no obvious change in GND density can be seen. At a strain of 0.80% a significant increase in GND density is seen (ρ_{av} = 10.2 x 10^{12} m^{-2}), with a clear heterogeneity in the GND density both between grains and within grains. Preliminary analysis suggests the presence of clusters of volumes with higher GND density providing some evidence of percolation of plastic strain, which will be further analyzed in future work. It is interesting also to note that even at this strain, some larger grains still show a relatively low GND density.

4. Conclusions

A study has been carried out to track the onset of plastic strain in 3D during in situ tensile deformation of aluminum with an average grain size of 5 μm using X-ray synchrotron measurements. The following conclusions from this work can be made:

- The initial sample, produced by spark plasma sintering has a low dislocation density and is in a nearly fully-recrystallized condition. High energy X-ray synchrotron radiation allows the detection of subgrains in the sample down to a misorientation angle of 0.02°.
- The 3D nature of the data provided by DAXM allows calculation of all nine elements of the GND density tensor. For voxels with missing neighbors, average scaling factors are calculated by simulation as 3 and 1.5 for 1D and 2D restricted data, respectively.
- The onset of plastic strain, as indicated by increase in GND density can be tracked using the DAXM technique. Heterogeneity in the onset of plastic deformation is obvious at strains nominally both in the elastic and plastic regimes (based on a conventional 0.2% yield point definition).

Acknowledgements

Financial support from the National Natural Science Foundation of China (project numbers 51471095 and 51671113) is gratefully acknowledged. Use of the Advanced Photon Source was supported by the U.S. Department of Energy, Office of Science, Office of Basic Energy Sciences, under Contract No. DE-AC02-06CH11357. DJJ acknowledges support from the European Research Council (ERC grant agreement No.788567) for part of the work.

References

[1] Hansen N and Barlow C Y 2014 Plastic Deformation of Metals *Physical Metallurgy 5th edition* ed. D E Laughlin and K Hono, Elsevier 1681–764
[2] Thomson R and Levine L E 1998 *Phys. Rev. Lett.* **81** 3884–7
[3] Tsuji N, Ito Y, Saito Y and Minamino Y 2002 *Scr. Mater* **47** 893–9
[4] Le G M, Godfrey A and Hansen N 2013 *Mater. Design* **49** 360–7
[5] Wu X L, Yang M X, Yuan F P, Wu G L, Wei Y J, Huang X and Zhu Y T 2015 *Proc.Natl. Acad. Sci.* **112** 14501–5
[6] Bodelot L, Charkaluk E, Sabatier L and Dufrenoy P 2011 *Mech. Mater.* **43** 654–70
[7] Nye J F 1953 *Acta Metall.* **1** 153–62
[8] Pantleon W 2008 *Scr. Mater.* **58** 994–7
[9] Levine L E, Larson B C, Bennett C, Yang W, Kassner M E, Tischler J Z, Delos-Reyes M A, Fields R J and Liu W 2006 *Nat. Mater.* **5** 619–22
[10] Le G M, Godfrey A, Hansen N, Liu W, Winther G and Huang X 2013 *Acta Mater.* **61** 7072–86
[11] Zhu K N, Godfrey A, Hansen N and Zhang X D 2017 *Mater. Design* 117 (2017) 95–103.
[12] Zhang C L, Zhang Y B, Wu G L, Liu W, Juul Jensen D and Godfrey 2017 *IOP Conf. Ser. Mater. Sci. Eng.* **219** 012050
[13] Zhang C L, Zhang Y B, Wu G L, Liu W, Xu R Q, Juul Jensen D and Godfrey A 2019 accepted for publication *J. Appl. Crystall.*
[14] Zhang Y B, Andriollo T, Fæster S, Barabash R, Xu R, Tiedje N, Thorborg J, Hattel J, Juul Jensen D and Hansen N 2019 *Acta Mater.* **167** 221–30
[15] Ruggles T J and Fullwood D T 2013 *Ultramicrosc.* **133** 8–15
[16] Ahl S R, Simons H, Zhang Y B, Detlefs C, Stöhr F, Jakobsen A C, Juul Jensen D and Poulsen H F 2017 *Scr. Mater.* **139** 87–91

IOP Conf. Series: Materials Science and Engineering **580** (2019) 012057 doi:10.1088/1757-899X/580/1/012057

In-situ study of the effect of strain path change on dislocation boundary evolution in commercial purity aluminum

S.Q. Zhang[1], C.F. Zhang[1], A Godfrey[1]

1 Key Laboratory of Advanced Materials (MOE), School of Materials Science and Engineering, Tsinghua University, Beijing 100084, China

E-mail: awgodfrey@mail.tsinghua.edu.cn

Abstract. An in-situ study of the effect of strain-path change on deformation microstructure and slip activity has been investigated using samples of commercially pure aluminum. Changes in dislocation boundary arrangements and crystal lattice rotations were tracked using CMOS-based electron backscattered diffraction (EBSD) detector combined with in-situ tensile deformation of $\varepsilon = 10\%$. Samples were first cold-rolled to 5% reduction and then deformed by uniaxial tensile along the transverse direction. Despite the use of a CMOS-based detector, analysis of deformation microstructure at such low plastic strains using EBSD is still challenging. Nevertheless in some grains details of the dominant deformation microstructure can be discerned. Slip line observations confirm that overall slip system activity is not influenced by the strain path change, but the EBSD observations show that in some cases dislocation boundary alignment is influenced by the presence of pre-existing boundaries, and in some grains the crystal rotations differ to those expected for tensile deformation of fully recrystallized grains of similar orientation.

1. Introduction

Over the past few decades extensive systematic studies have been carried out to characterize the microstructure evolution in medium-to-high stacking fault energy single-phase face-centered cubic (fcc) metals, such as Al, Cu and Ni [1-10]. Based on these studies it has been established that after deformation the stored dislocations are organized into dislocation rotation boundaries [1-3]. Two categories of dislocation boundaries have been identified, namely geometrically necessary boundaries (GNBs) and incidental dislocation boundaries (IDBs) [2, 3]. Investigations of dislocation structures formed during monotonic deformation (e.g. tension, compression or rolling) have demonstrated that for both the GNBs and IDBs the average boundary spacing decreases and the average misorientation angle across the boundaries increases with increasing strain [4-6]. Additionally it has been found that the dislocation boundary morphology strongly depends on the crystallographic grain orientation, with this relationship extending down in some cases to grain sizes of less than 1 μm [7, 9].

These studies show a clear correlation between dislocation boundary morphology and final grain orientation. However, during manufacturing processes of metal components, the plastic deformation process is often complex including changes of strain path. In contrast to the clearly findings regarding dislocation structure evolution under monotonic deformation, many aspects of the evolution of dislocation structures under strain path change deformation are still unclear [11-14]. For example, Sellars et al. [12] determined GNBs inclination angles in tension followed by compression for a Al-3%

Mg alloy and concluded that the compression step did not change the inclination angle of the GNBs formed in tension, but the misorientation angle across these GNBs decreased, and a new set of GNBs with different angle was formed. Pedrosa et al. [13] examined forward and reverse torsion (using an equal strain value of $\varepsilon = 0.23$ for each direction) on AA1200 Al, and concluded that the reverse strain led to a reduction in the misorientation angle of GNBs formed during the forward deformation, followed by the generation of new GNBs with a lower average misorientation compared with those formed in monotonic torsion to the same strain. Lewandowska et al. [14] studied the influence of both forward/reverse shear and combined rolling-tension-shear strain path changes on the dislocation structures formed in Al 5128 and Al 6016 alloys, observing that reversal of the shear direction partly destroyed the pre-existing dislocation boundary arrangements, while orthogonal deformation promoted an increased dislocation density between the pre-existing GNBs.

A major reason for limited understanding of the microstructural changes that take place during strain path changes is the difficulty of making in-situ observations to follow directly dislocation boundary rearrangements. Examination of thin foils in a transmission electron microscope (TEM) technique is the most reliable and standard method for characterization of dislocation boundary arrangements in deformed metals in metals [15]. This technique only allows, however, in-situ deformation of very small-size samples, with a limited range of both strain and deformation mode. Recently new techniques have been developed for measurement and analysis of crystal orientations using electron backscatter diffraction (EBSD) with improved angular resolution, including a high-sensitivity CMOS-based detector, offering the potential for study of strain path changes on bulk samples deformed in-situ in a scanning electron microscope. The objective therefore of the present work is to explore the possibility of EBSD for the study of the effect of strain path change on dislocation boundary evolution, using as a test material commercial purity Al.

2. Experimental

The material used in this study is a commercial purity Al (AA1200) sheet of thickness 1.2 mm, annealed to a fully recrystallized condition with an average grain size of 35 μm. The Al sheet was first cold rolled to 5% reduction and then dog-bone shaped tensile specimens (gauge width, length and thickness of 10mm, 2mm and 1mm, respectively) were cut from the rolled sheet, with the tension direction taken parallel to the transverse direction of the rolled sheet.

Prior to in-situ tensile deformation, the top surface of the tensile specimen was ground to 4000-grit SiC and then electropolished using 10% $HClO_4$ in ethanol. The in-situ tensile deformation was carried out using a push-to-pull tensile stage at a strain rate of approximately 5×10^{-3} s^{-1}. Before and after the tensile deformation (to a strain of 10%), the microstructure in a selected region on the surface was characterized using EBSD. For the EBSD measurements an Oxford Instruments high sensitivity CMOS-detector was used, with data collected in high-resolution mode. For both the as-rolled and rolled plus tension samples an EBSD step size of 1 μm was used.

3. Results and discussion

Figure1 shows the EBSD results of the same region in the sample rolled to 5% (figure 1(a-c)) and after additional tension deformation of 10% (figure 1(d-f)). The inverse pole figure (IPF) coloring in figure 1(a,d) shows the crystal direction parallel to the extension direction during rolling (i.e., the rolling direction, RD), whereas figure 1(b,e) shows the crystal direction parallel to the tension axis (TA) during in-situ deformation. By comparing the IPF maps it can be seen that, aside from a few grains with orientation close to the cube-orientation (some examples are highlighted with a white ellipse), for most grains the crystal direction parallel to the extension axis is different for the tension step compared to the initial rolling. For example some grains with <001> nearly parallel to RD but with the TA close to either <110> or <111> are marked with a black ellipse. Accordingly the orientation relationship of most grains with respect to the elongation direction changes during the experiment. Figure 1(c,f) also show kernel average misorientation (KAM) maps (constructed using a 3×3 kernel) in which some details of the dislocation boundary microstructure can be seen, even after the low strains used in the present experiment. Comparison of the IPF maps before and after 10% tensile deformation reveals an increase in the orientation variation inside most grains. A similar comparison of the KAM maps reveals some clear changes in the deformation microstructure, including an increase in the misorientation angle of some boundaries, and the appearance of many new dislocation boundaries.

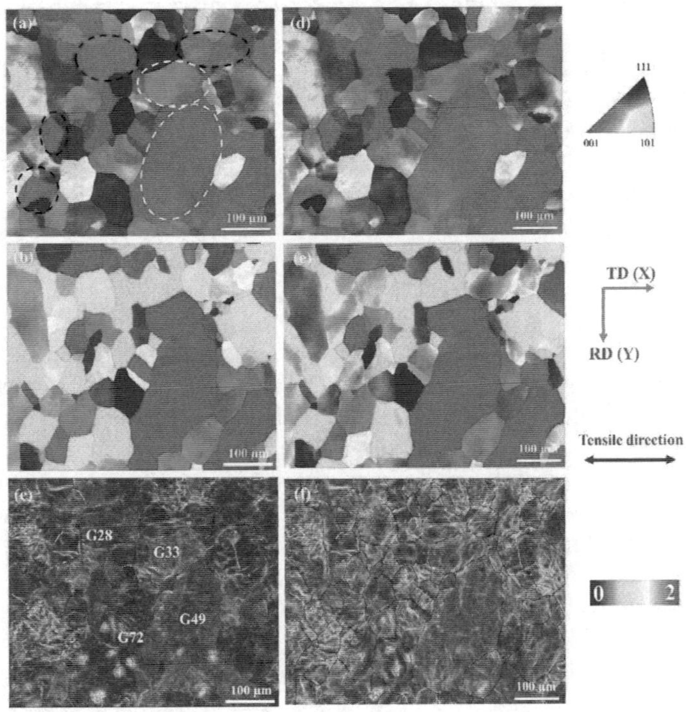

Figure 1. EBSD observations of a selected region in (a-c) the pre-rolled specimen and (d-f) in the same region after 10% tensile deformation (along the transverse direction, TD) : (a, d) and (b, e) are inverse pole figure maps corresponding to the rolling and tensile directions, respectively; (c, f) are kernel average misorientation maps constructed using a 3 × 3 kernel, with an upper cut-off of 5°.

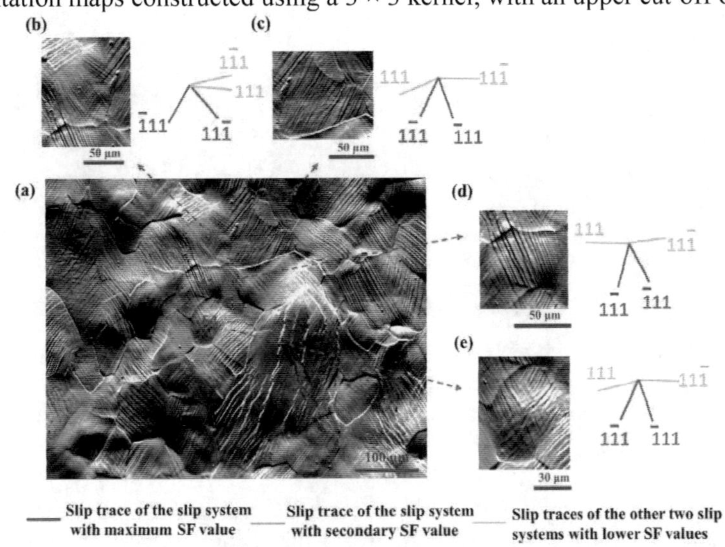

Figure 2. (a) Low magnification forescatter detector (FSD) images of slip bands in the examined region after 10% tension deformation; (b-e) high magnification FSD images together with slip system traces and for selected grains colored based on the Schmid factor (SF) ranking.

It has been reported that slip system activation is a dominant factors in determining the dislocation boundary morphology [16]. In the present experiment the slip traces resulting from the in-situ tensile deformation have been examined from images collected using a forescatter electron detector (FSD). Some examples are shown in figure 2, where it can be seen that slip traces are clearly visible, and well-defined in most of the grains in the examined area. The low magnification FSD image (figure 2a) reveals that one or two set of slip bands are observed in each grain, with most slip bands straight and only a small fraction showing a curved or wavy appearance. It is interesting to note also that many slip bands are observed found to cross the grain boundaries, in agreement with previous studies [17, 18]. Based on the measured orientation of each grain a slip trace analysis has also been carried out where it is found that in each grain the observed slip traces correspond to slip systems with highest (or second highest in the case of grains where two sets of slip traces are visible) Schmid factor [19]. Some examples are shown in figure 2(b-e). Note that the labelling of the {111} planes is arbitrary (depending on the specific crystal symmetry variant reported in the EBSD data) used, but the coloring indicates whether the trace corresponds the highest (red), second highest (blue), or other (yellow) slip system.

It has recently been shown that maps plotting the rotation axis in the specimen coordinate frame of each pixel to the grain average orientation can be useful for highlighting both grain subdivision, and in some cases dislocation boundary arrangements, even in samples deformed only to small plastic strains [20]. Example grain average deviation (GAD) maps are shown for two selected grains in figure 3(a,b), where for comparison KAM maps of the same grains are also shown. The GAD (a1) and KAM (a3) maps of grain labelled G28 reveal the presence of a few dislocation boundaries with relatively high misorientation (marked with red boxes in the figure) after 5% rolling. The GAD (a2) and KAM (a4) maps for the sample after additional tension deformation show these boundaries remain with slightly increased KAM values, indicating additional dislocation storage at these boundaries after the strain path change. In addition the presence of two sets of GNBs formed during tension is seen in both the GAD and KAM maps. This grain has an average tensile axis 6° from [221], for which it is expected that tensile deformation should result in GNBs forming close to the ($44\bar{1}$) plane [16]. The closest set to this trace (marked with a red dashed line in figure 3(a2)) for this grain is approx. 16° from this trace.

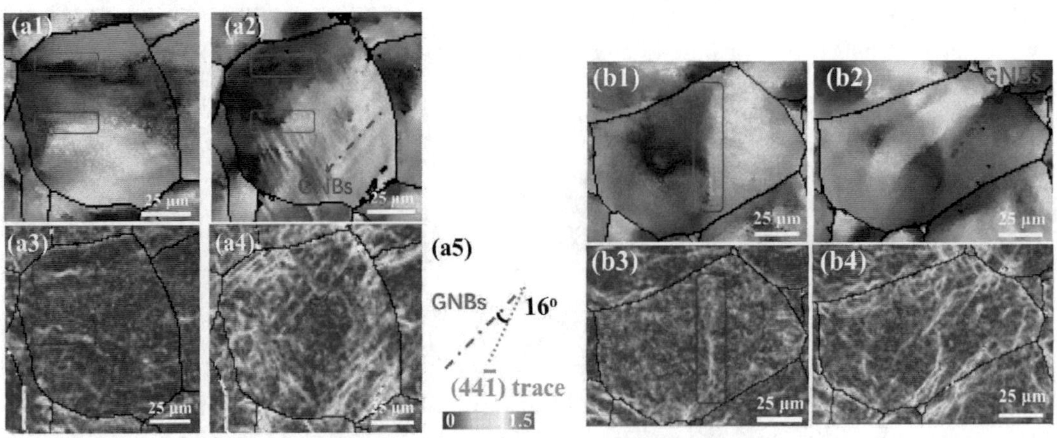

Figure 3. Grain average deviation (GAD) and KAM maps of two typical grains before and after tensile deformation: GAD map of G28 (a1) before tension and (a2) after tension, KAM map of G28 (a3) before tension and (a4) after tension, GAD map of G33 (b1) before tension and (b2) after tension, KAM map of G33 (b3) before tension and (b4) after tension

Fig. 3(b) shows the microstructure before and after tensile deformation for a grain (G33) deviated by 14° from {001}<100>. The GAD (b1) and KAM (b3) maps for this grain before tension (after 5% rolling) show a weak cell-structure with one clearly discernable dislocation boundary traversing the grain (marked with a red box). After additional tension deformation the major boundary is partly preserved with higher local misorientation (KAM value). Additionally, a new set of GNBs (highlighted with a red dashed line in b2) is observed, lying along almost parallel to the ($\bar{1}$ 11) slip trace. For this grain, with the tensile axis just 4° from <001>, it is expected, however, [7] that tension deformation should lead to the formation of a dislocation cell structure, rather than the formation of GNBs.

On account of the in-situ deformation, another aspect of the deformation behavior that can be probed is the lattice rotation of each grain during tension deformation. These rotations can be compared to the rotations both predicted by theory and measured in bulk grains using X-ray synchrotron diffraction [19] for tension deformation of initially recrystallized grains (i.e. without any strain path change). Some examples are shown in figure 4, which also shows the expected rotation paths for tension as a function of the tensile loading axis (figure 4a). It can be seen that for the grains plotted in figure 4(b,c) the rotation paths (as determined from the grain average orientation before and after tension) differ significantly from the expected paths, though some of this difference may be attributed to the difference between the behavior of surface and bulk grains. For grain G28 (figure 4d) a more complex behavior is seen, where part grain rotates along the predicted path (black arrow) while another part of grain rotates along an approximately opposite path (red arrow).

The results presented above show that optimized EBSD observations using a high sensitivity detector are able to at least reveal some features of the changes in deformation microstructure during strain path change, even at the low plastic strains used in this study. One limitation, however, of the present work is that with the current in-situ tension rig, observations have to be made on the rolling-plane of the as-rolled sample, whereas it is known that the deformation microstructure after rolling can be resolved much more easily in the longitudinal plane (formed by the rolling and normal direction). To illustrate the potential for using EBSD to characterize dislocation boundaries in the RD-ND plane at low strain, a map from the same sample (rolled to 5% reduction) in this plane is shown in figure 5. It is seen that at least in some grains a number of dislocation walls are clearly revealed and the microstructure is easier to characterize than in the RD-TD plane. It remains a challenge, however, to design an experiment where in-situ deformation can be carried out allowing viewing of this sample plane.

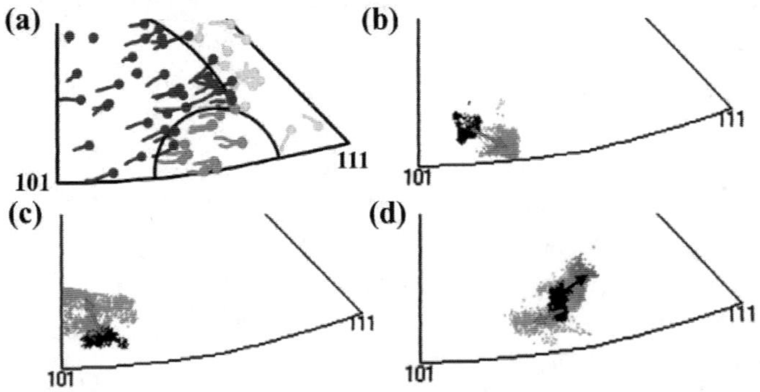

Figure 4. Evolution of crystal direction parallel to the tensile axis during tensile deformation: (a) previously obtained results for bulk recrystallized grains using X-ray synchrotron radiation (image adapted from [19]); (b-d) experimental results for three selected grains. Black pixels represent the initial tensile axis and grey pixels represent the tensile axis after 10% tensile deformation. Red arrows indicate rotations that differ significantly from previous experimental data on fully recrystallized samples.

4. Conclusions

A study has been carried out to assess the potential of using EBSD for in-situ studies of the effects of strain path change on the evolution of the deformation microstructure. The following conclusions from this work can be made:

- Visualization using EBSD of dislocation boundaries in metals deformed to low plastic strain remains a challenge, and a full characterization of the deformation microstructure is still only possible using transmission electron microscopy. Nevertheless in some grains details of the dominant deformation microstructure can be discerned, and EBSD investigation both allows in-situ experiments to be performed and provides reliable data on grain rotations.

- During an orthogonal strain path change from rolling to tension, where the elongation direction is rotated by 90°, some dislocation boundaries formed during the initial 5% rolling remain after 10% tension, with increased local misorientation. Slip line observations after tension deformation suggest that the overall slip system activity is not influenced by the strain path change, but EBSD observations show that in some cases dislocation boundary alignment is influenced by the presence of pre-existing boundaries.
- In-situ deformation allows crystal rotations of surface grains to be tracked and reveals some significant differences to predictions based on tensile deformation of fully recrystallized grains.

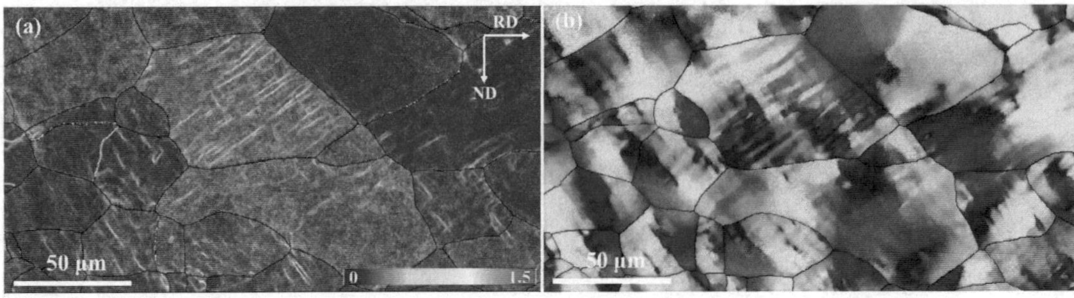

Figure 5. EBSD observations of the longitudinal (RD-ND) plane of a sample rolled to a reduction of 5%: (a) kernel average misorientation (KAM) using a 3 x 3 kernel ; (b) grain average deviation map plotting for each grain the axis component in the sample coordinate frame of the misorientation between each pixel and the grain average orientation.

Acknowledgements
This work was supported by the National Natural Science Foundation of China under project number 51471095.

References
[1] Hansen N and Juul Jensen D 1999 *Phil. Trans. R. Soc. Lond.* A **357** 1447–69
[2] Liu Q and Hansen N 1995 *Phys. Status Solidi A*. **149** 187–200
[3] Kuhlmann-Wilsdorf D and Hansen N 1991 *Scr. Metall. Mater.* **25** 1557–62
[4] Godfrey A and Hughes D A 2004 *Scr. Mater.* **51** 831–36
[5] Godfrey A and Hughes D A 2000 *Acta Mater.* **48** 1897–905
[6] Liu Q and Hansen N 1995 *Scr. Metall. Mater.* **32** 1289–95
[7] Huang X, Winther G 2007 *Phil. Mag.* **83** 5189–214
[8] Le G M, Godfrey A, Hong C S, Huang X and Winther G 2012 *Scr. Mater.* **66** 359–62
[9] Le G M, Godfrey A, Hansen N, Liu W, Winter G and Huang X 2013 *Acta Mater.* **61** 7072–86
[10] Le G M, Godfrey A, Winter G, Hansen N and Huang X 2012 *Proc. Risø International Symposium on Materials Science* eds. S. Fæster et al. **33** 63–78
[11] Rauch E F, Gracio J J, Barlat F, Lopes A B and Duarte J 2002 *Scr. Mater.* **46** 881–6
[12] Zhu Q and Sellars C M 2001 *Scr. Mater.* **45** 41–8
[13] Lopez-Pedrosa M, Wynne B P and Rainforth W M 2006 *J. Microsc.* **222** 97–104
[14] Lewandowska M 2003 *Mater. Chem. Phys.* **81** 555–7
[15] Williams D B and Carter C B 2009 *Transmission Electron Microscopy. A Textbook for Materials Science* (New York: Springer)
[16] Winther G and Huang X 2007 *Phil. Mag.* **83** 5215–35
[17] Davis K G, Teghtsoonian E and Lu A 1966 *Scr. Mater.* **14** 1677–84
[18] Liu W, Wang Z and Xia Y 1989 *Phys. Status Solidi* A **114** 177–84
[19] Winther G, Margulies L, Schmidt S and Poulsen H F 2004 *Acta Mater.* **52** 2863–72
[20] Hong X, Godfrey A, Zhang C L, Liu W and Chapuis A 2017 *Mater. Sci. Eng.* A **693** 14–21

IOP Conf. Series: Materials Science and Engineering **580** (2019) 012058 doi:10.1088/1757-899X/580/1/012058

Exploring high strength metallic materials: a lesson from pearlitic steel wire

X Zhang

Section of Manufacturing Engineering, Department of Mechanical Engineering, Technical University of Denmark, 2800 Kgs, Lyngby, Denmark

Email: xzha@mek.dtu.dk

Abstract. In human history, the development of stronger materials through the ages is reflected with names of eras illustrating our progress. Besides phase transformations, plastic deformation is one of the major methods to produce products with reliable and predictable mechanical properties such as strength. Pearlitic steel wire, the strongest mass-produced steel, shows an excellent combination of formability and strength. The present overview summarizes the investigation of cold-drawn pearlitic steel wires over the last 150 years, covering the pearlite phase transformation, chemical composition design for wires with high strength, microstructure evolution during wire drawing, strengthening mechanisms and structure-strength relationships. By focusing on the structure, challenges and future strategy are outlined to further improve the strength and performance of pearlitic steel wire, where these routes may also applicable to other metals.

1. Introduction

In the history of mankind and civilization, the drive to develop stronger materials through the ages is evidence by with names of eras illustrating our progress [1]. Metals are the major structural material in our society due to their unique properties such as high reliability with predicable strength, excellent fracture toughness compared with other materials (engineering polymetric composites and ceramics) and recyclability for the large scale application. These properties will also make them irreplaceable in the future [2]. However, there exist limitations from case to case in the properties of metals as they are currently used, with a major one being the low specific strength.

Very high strengths, up to 13 GPa have been observed in the tensile and compression testing of microscale and nanocale single crystal metals. These single crystal metals are made by crystal growth or other chemical processes and they are normally quite small (less than 10 micrometer). Moreover, many of these single crystal metals show catastrophic failure without any observable amount of plastic deformation [3]. As a result, such small single crystal metals are rarely used for structural applications.

Metals can be strengthened through controlled introduction of internal defects and boundaries that obstruct dislocation motion. Most of the structural metals used in the industry and our daily lives are polycrystalline and an effective way for improvement of strength is deformation at low temperature ($T \leq 0.35\ T_M$), resulting in an increase of strength by up to two orders of magnitude (where T_M is the melting temperature) due to the increase of internal defects and boundaries [4]. Accompanying the strength increase, the ductility is reduced and the grain size is refined. Very strong steels can also be produced by plastic deformation to very high strain but drawbacks are the large working force required and the poor formability of the final deformed product [5]. These steels owe their strength to a very fine scale microstructure containing a large concentration of defects in the form of dislocations and boundaries. Strong steels, such as bainitic and martensitic steels, can also be produced by phase transformation [6, 7]. One objective of both research and development has therefore been to optimize the different processes and the desired structural refinement.

Nowadays, the maximum flow stress in metals applied in industry (E/100 - E/50) is far from the ideal strength, except for the case of a few metals. Cold-drawn pearlitic wire, with a microstructure resulting from a combination of phase transformation and plastic deformation, widely applied for uses including cables for suspension bridges, steel cords for automobile tires and springs, is one of these. Pearlitic wire is in fact the strongest mass-produced metal, with a maximum strength of 4.8 ~ 5.7 GPa (E/44 ~ E/37), for the grain size around 11 nm [8, 9], and the laboratory record strength around 6.8 GPa (E/31) for a grain size of 10 nm [10]. This high-strength pearlitic wire also excellent properties, with the strength-to-weight comparable to ceramics and carbon fibers but much better fracture toughness (a measure of the energy required for propagating cracks) [11]. These outstanding properties of the nanostructured material guarantee its application in large-scale structures such as the Great Belt Bridge.

The present overview summarizes investigations of cold-drawn pearlitic steel wires in the last several decades, covering topics of of pearlite phase transformation and wire chemical composition design for high strength, microstructure evolution during wire drawing, strengthening mechanisms and structure-strength relationships. Challenges and future strategy are outlined and an outlook to further improve the strength and performance of pearlitic steel wire and other metals is given.

2. Pearlite Phase Transformation and Wire Chemical Composition Design for High Strength

Pearlite is the product of austenite decomposition by a eutectoid reaction. The pearlite phase transformation provides an excellent example of the historical development of phase transformations, where the 'hardening problem' in steel was pursued in the 1880–1925 period using metallography, thermal analysis, and the Gibbs phase rule. Breakthroughs were afforded by X-ray diffraction and other techniques such as transmission electron microscopy regarding metal structure, phase structure and phase transformations in the period of 1925–1970 [12], illustrating the importance of interaction between experimental observations and the development of quantitative models. The pioneering work of Davenport and Bain [13] provided a basis for the development of time-temperature-transformation (TTT) diagrams and the ability to quantify and rationalize the role of alloying elements in terms of hardenability and their influence on structure-property relationships in steels [14]. This was followed by a quantitative description of nucleation and growth rates and interlamellar spacing via a detailed microstructural approach [15], a detailed discussion of analytical and experimental aspects of nucleation and growth kinetics [16], a description of 3D structure from 2D observations [17] and the theory of diffusional growth [18]. The importance of both volume diffusion [19, 20] and diffusion at the advancing interface was first recognized in the 1970s and different models have since been proposed [21, 22] as the rate of motion of the interface in solid state reactions can be controlled by diffusion and thus a large undercooling can occur. Meanwhile, the crystallographic aspects of the pearlite phase transformation including the ferrite-austenite, ferrite-cementite and cementite-austenite orientation relationships have were actively investigated from the 1950s to 1990s [17, 23-27].

Since carbon is the most effective element to harden the steel, hypereutectoid steel with carbon content higher than 0.8 wt% has been investigated for development of high strength steel wires in the industry from the very beginning, and later the combined effect of patenting conditions, cooling rate, chemical composition on pearlite lamellar spacing has been widely investigated [9]. To obtain a high drawability and a high work hardening rate, an appropriate cooling rate corresponding to carbon content should be secured to prevent precipitation of thick proeutectoid cementite. The thickness of cementite lamellae as well as the interlamellar spacing decreases with the increase of carbon content in the practical patenting temperature range. Finally, small-diameter (down to several tens of micrometers) high-strength steel wires with the present strength record have been manufactured on a production basis with a chemical composition typically around 0.96C-0.2Si-0.3Mn-0.2Cr (wt-%) [10].

3. Microstructure Evolution during Wire Drawing

The pearlitic lamellar structure, as shown in figure 1, consists of alternating ferrite and cementite layers as a result of the decomposition of eutectoid austenite, where pearlite colonies with different oriented lamellae nucleate from prior austenite boundaries and form in one austenite grain [14, 28-30]. The first images of pearlite were shown by Sorby at the British Association meeting in September 1864 [31], and drawn pearlitic wires date back to 1886 when wires with a carbon content of 0.828 wt% and a strength not less than 1.65 GPa were produced for military applications [32]. The wire drawing up to medium strains reorients the initial random lamellar directions of different pearlite colonies to the drawing direction [33-36] in the longitudinal direction, while curling of the lamellar structure takes place in the cross-section, as is characteristic [35, 37] for drawing of a bcc metal [4, 38, 39]. Further drawing to large strains continuously thins down the ferrite lamellae with increasing dislocation densities [8, 40] and leads eventually to decomposition of cementite lamellae [8, 10, 41-46]. Detailed microstructural characterization in the last 50 years by transmission electron microscopy (TEM) [40, 47-49], high resolution electron microscopy (HREM) [8] and three dimensional atom probe (3DAP) [45, 46, 50] has quantified the systematic evolution of microstructural parameters with the drawing strain, such as the interlamellar spacing (ILS), the thickness of ferrite (F) and cementite lamellae (C), the dislocation density in the ferrite lamellae and the carbon content in the ferrite lamellae. The major dislocation structure in the ferrite lamellae transforms from threading dislocations and tangles at low and medium strains, to dislocation cells at high strains, see figure 2. The final lamellar size depends on both the drawing strain and the chemical composition, which influences the dynamic recovery during drawing, but can be down to 10 nm in highly drawn wires (see figure 2b) [8, 9].

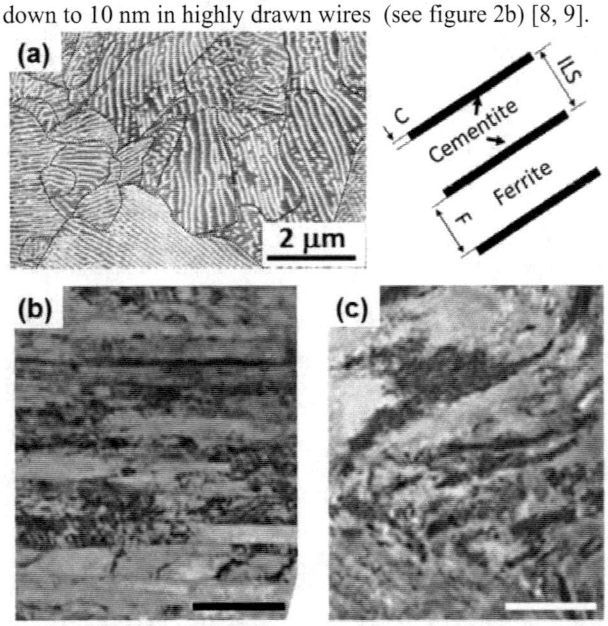

Figure 1. Scanning (a) and transmission (b and c) electron micrographs showing the alternating ferrite and cementite lamellae in the original as-patented state (a), the refined parallel lamellar structure in the longitudinal section (b) and the curling structure in the cross section (c) at a drawing strain of 3.7. The dashed lines in (a) show the pearlite colony boundaries. [30]

IOP Conf. Series: Materials Science and Engineering **580** (2019) 012058 doi:10.1088/1757-899X/580/1/012058

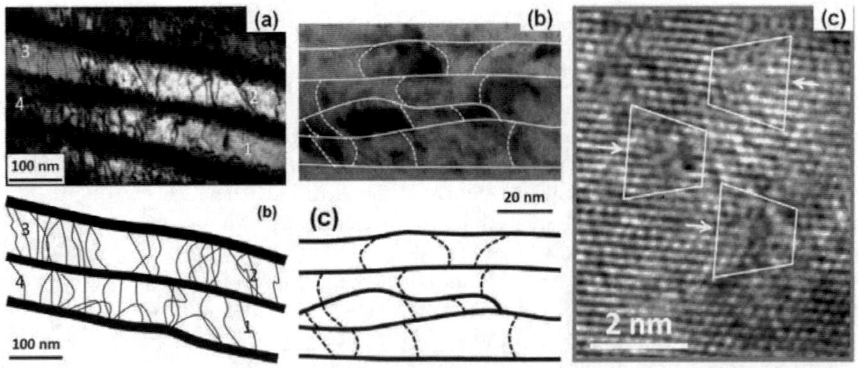

Figure 2. Transmission electron micrographs and sketches showing two major dislocation structures in the deformed ferrite lamellae: threading dislocations (thin lines) at a drawing strain of 0.7 (a), dislocation boundaries (dashed lines) (b) and atomic structure of a low angle dislocation boundary (c) at a drawing strain of 5.4. The thick black lines in the sketches represent cementite lamellae (a) or interfaces between neighboring ferrite lamellae (b). [8, 40]

Table 1. Microstructural parameters of a pearlitic wire containing 0.8 wt-% C in the drawing strain range from 0 to 5.4 [8, 40]

Strain	0	0.7	1.5	2.7	3.7	4.2	5.4
F (nm)	70	56	45	23	18	14	11
C (nm)	19	14	10	5	2	1.4	0.7
ρ (m^{-2})	8×10^{13}	7×10^{14}	2×10^{15}	9×10^{15}	2×10^{16}	3×10^{16}	5×10^{16}

4. Strengthening Mechanisms and Structure-Strength Relationship

Based on the recently elaborate quantification of the microstructural parameters [8, 40, 50], the previous one-parameter (ILS) structure-strength relationship proposed around 1970s [47, 49] has been extended to take into account the thickness of the ferrite (F) and cementite lamellae (C), the dislocation density in the ferrite lamellae, and the carbon content in the ferrite lamellae, as a results of a detailed analysis of the strengthening mechanisms (Table 1). The boundary strengthening is related to the ferrite lamellar thickness as the role of the cementite lamellae is estimated based on a Hall-Petch relation. The dislocation strengthening is related to the dislocation density in the ferrite lamellae through forest hardening proportional to its square root, and the solid solution hardening is related to carbon enrichment in the ferrite lamellae through the cementite decomposition, based on the interaction between dislocation configurations and carbon atoms in solution [8, 50].

Addition of contributions from the above three strengthening mechanisms plus the friction stress, gives a flow stress in good agreement with the experimental values ($\sigma_{0.2\%}$) up to the drawing strain around 2.7. At larger strains beyond 2.7, there exists a difference of around 10% between the experimental and calculated values (Table 2). The overestimation at large strains may have different reasons including: (i) regarding boundary strengthening, the deformation and decomposition of cementite lamellae may reduce the boundary resistance, leading to a reduced Hall-Petch coefficient, however, this reduction may be counteracted by carbon atom segregation to the interfaces between the ferrite lamellae and high angle boundaries/high density of dislocations at the interfaces; (ii) for the calculation of dislocation strengthening contribution using $\alpha MGb\sqrt{\rho}$, a choice of $\alpha = 0.24$, derived for tensile-tested samples with a smaller dislocation density by one to two orders [51], has been used. The value of α is actually proportional to $\ln(R/b)$ where R is the upper cut-off radius and can be taken as the average dislocation spacing ($\rho^{-0.5}$). This means that α may be smaller than 0.24, and a reduced contribution from dislocation strengthening may thus be expected; (iii) for the evaluation of solid solution hardening contributions, models for martensite have been applied [52, 53]. However, the deformation introduced distribution of dislocations and carbon atoms in the heavily drawn wires may be different, which may introduce an overestimation of the flow stress. Additionally, dynamic recovery, which includes dislocation interactions and annihilation, has not been taken into account in the flow stress calculation.

Table 2. Experimental and calculated flow stress in the drawing strain range from 0 to 5.4 [8]

Strain	0	0.7	1.5	2.7	3.7	4.2	5.4
$\sigma_{exp, 0.2\%}$ (MPa)	880	1286	1614	2365	3395	3887	4510
σ_{cal} (MPa)	966	1230	1662	2370	3743	4251	5042
σ_{diff} (MPa)	86	−56	48	5	338 (10%)	364 (9%)	532 (12%)

5. Conclusions and Outlook

The present overview summarizes the investigations of cold-drawn pearlitic steel wires in the last 150 years with the aim to further improve the strength and performance of pearlitic steel wire and other metals.

- Based on the systematic and detailed characterization of drawn wires in the strain range from 0 to 2.7 with a lamellar scale above 20 nm, a good agreement has been found between the experimental flow stresses and calculated flow stresses, taking into account the traditional strengthening mechanisms.
- Challenges and a future research strategy can be: for the structural scale < 20 nm, new advanced microstructural and mechanical characterization techniques [54, 55] including in-situ TEM deformation technique [56] should be further developed. Moreover, the interaction between dislocations and solutes needs further detailed investigation to understand the structure-strength relationship, including the effects of dynamic recovery.

Based on the example of pearlitic steel wire, future development of high strength metallic materials and products, should combine the chemical composition design, phase transformation, dynamic processes and production processes, with the goal of achieving high strength while maintaining formability.

Acknowledgements

X Zhang acknowledge support from the European Research Council (ERC) under the the European Union Horizon 2020 research and innovation programme (grant agreement No 788567-M4D).

References

[1] National Research Council 1975 *Materials and man's needs: materials science and engineering -- volume I* Washington (DC): The National Academies Press
[2] Lu K 2010 *Science* **328** 319
[3] Brenner S S 1956 *J Appl Phys.* **27** 1484
[4] Gil Sevillano J 1980 *Progress in materials science* **25** 69
[5] Tsuji N and Maki T 2009 *Scripta Mater* **60** 1044
[6] Christian J W 2002 *The Theory of Transformations in Metals and Alloys* Pergamon
[7] Bhadeshia H K D H, 2010 *Proc R Soc A* **466** 3
[8] Zhang X, Hansen N, Godfrey A and Huang X 2016 *Acta Mater.* **114** 176
[9] Ochiai M, Nishida S, Ohba H and Kawata H 1993 Tetsu-to-Hagane **79** 1101
[10] Li Y, Raabe D, Herbig M, Choi P P, Goto S, Kostka A, Yarita H, Borchers C and Kirchheim R 2014 *Phy Rev Lett.* **113** 106104
[11] Hohenwarter A, Völker B, Kapp M W, Li Y, Goto S, Raabe D and Pippan R 2016 *Sci Rep.* **6** 33228
[12] Hackenberg R E 2012 *Phase transformations in steels Volume 1*, ed E Pereloma and D V Edmonds (Cambridge: Woodhead Publishing Limited) p 3
[13] Davenport E S and Bain E C 1930 *Trans AIME* **90** 117
[14] Embury D 2012 *Phase transformations in steels Volume 1*, ed E Pereloma and D V Edmonds (Cambridge: Woodhead Publishing Limited) p 276
[15] Mehl R F and Hagel W C 1956 *Progress in Metal Physics* **6** 74
[16] Cahn J W and Hagel W C 1962 *The Decomposition of Austenite by Diffusional Processes*, ed V

F Zackay and H I Aaronson (New York: Wiley) p 131
[17] Hillert M 1962 *The Decomposition of Austenite by Diffusional Processes*, ed V F Zackay and H I Aaronson (New York: Wiley) p 197
[18] Kirkaldy J S 1962 *The Decomposition of Austenite by Diffusional Processes*, ed V F Zackay and H I Aaronson (New York: Wiley) p 39
[19] Brown D and Ridley N 1969 *J Iron Steel Inst* **207** 1232
[20] Puls M and Kirkaldy J S 1972 *Metall Trans* **3** 2777
[21] Hashiguchi K and Kirkaldy J S 1984 *Scan J Metall* **13** 240
[22] Pandit A S and Bhadeshia H K B D 2011 *Proc R Soc A* **467** 508
[23] Bagaryatski Y A 1950 *Dokl Akad Nauk SSSR* **73** 1161
[24] Pitsch W 1962 *Acta Metall* **10** 79
[25] Lupton D F and Warrington D H 1972 *Acta Metall* **20** 1325
[26] Dippenaar R J and Honeycombe R W K 1973 *Proc R Soc A* **333** 455
[27] Zhou D S and Shiflet G J 1992 Metall Trans A 23 1259
[28] Zhang X, Hansen N, Godfrey A and Huang X 2012 *33rd Risø international symposium on materials science: Nanometals – status and perspectives* (Risø, Denmark) p 407
[29] Zhang X, Hansen N, Godfrey A and Huang X 2014 *35th Risø International Symposium on Materials Science: New Frontiers of Nanometals* (Risø, Denmark) p 153
[30] Zhang X, Hansen N, Godfrey A and Huang X 2018 *Mater Sci Technol* **34** 794
[31] Sorby H C 1864 *Report of the 34th meeting of British Assn Adv Science* Part II
[32] Percy J 1886 *J Iron Steel Inst* **29** 62
[33] Zhang X, Godfrey A, Hansen N, Huang X, Liu W and Liu Q 2010 *Mater. Charac* **61** 65
[34] Zhang X, Godfrey A, Liu W and Liu Q 2010 *Acta Metall Sin* **46** 141
[35] Zhang X D, Godfrey A, Liu W and Liu Q 2011 *Mater. Sci. Techno* **27** 562
[36] Zhang X, Godfrey A, Hansen N and Huang X 2013 *Acta Mater* **61**, 4898
[37] Langford G 1977 *Metall Trans A* **8** 861
[38] Hosford W F 1964 *Trans Met Soc AIME* **230** 12
[39] Aernoudt E Van Houtte P and Leffers T 1993 *Plastic deformatio and fracture of materials* (Weinheim: VCH, Newyork; Basel; Cambridge) p 129
[40] Zhang X, Godfrey A, Huang X, Hansen N and Liu Q 2011 *Acta Mater* **59** 3422
[41] Hong M H, Reynolds W T, Tarui T and Hono K 1999 *Metall Mat Trans A* **30** 717
[42] Gavriljuk V G 2003 *Mater Sci Eng A* **345** 81
[43] Languillaume J, Kapelski G and Baudelet B 1997 *Acta Mater* **45** 1201
[44] Hono K, Ohnuma M, Murayama M, Nishida S, Yoshie A and Takahashi T 2001 *Scr Mater* **44** 977
[45] Takahashi J, Tarui T and Kawakami K. 2009 *Ultramicroscopy* **109** 193
[46] Sauvage X, Lefebvre W, Genevois C, Ohsaki S and Hono K 2009 *Scr Mater* **60** 1056
[47] Embury J D and Fisher R M 1966 *Acta Metall* **14** 147
[48] Zhang X, Godfrey A, Huang X, Hansen N, Liu W and Liu Q 2009 *30th Risø international symposium on materials science: nanostructured metals – fundamentals to applications* (Risø, Denmark) p 409
[49] Langford G 1970 *Metall Trans A* **1** 465
[50] Li Y J, Choi P, Borchers C, Westerkamp S, Goto S, Raabe D and Kirchheim R 2011 *Acta Mater* **59** 3965
[51] Hansen N and Huang X 1998 *Acta Mater* **46** 1827
[52] Krauss G 1999 *Mater Sci Eng A* **273-275** 40
[53] Winghell P G and Cohen M 1963 *Electron microscopy and strength of crystals* (Interscience Publishers: New York) p 995
[54] Zhang X, Bech J I and Hansen N 2015 *IOP Conf Ser Mater Sci Eng* **89** 012158
[55] Zhang X, Nielsen C V, Hansen N, Silva C M A and Martins P A F 2019 *Int. J. Plast.* **115** 93
[56] Zhang X, Godfrey A, Hansen N, Huang X, Liu W and Liu Q 2010 *Mater Charac* **61** 65

IOP Conf. Series: Materials Science and Engineering **580** (2019) 012059 doi:10.1088/1757-899X/580/1/012059

Thermal stability of laser shock peening processed Ni-based superalloy DZ17G

Guangni Zhou[1,3], Hao Shen[1], Wenxin Zhu[1], Nobumichi Tamura[2], Kai Chen[1*], Yubin Zhang[3*]

[1] Center for Advancing Materials Performance from the Nanoscale (CAMP-Nano), State Key Laboratory for Mechanical Behaviour of Materials, Xi'an Jiaotong University, Xi'an, Shaanxi 710049, China
[2] Advanced Light Source, Lawrence Berkeley National Laboratory, Berkeley, CA 94720, USA
[3] Department of Mechanical Engineering, Technical University of Denmark, Kongens Lyngby 2800, Denmark

E-mail: kchenlbl@gmail.com; yubz@mek.dtu.dk

Abstract. Ni-based superalloys are widely used in challenging environments due to their exceptional mechanical properties. To enhance their resistance to damage, surface modification by laser shock pinning (LSP) is an approach to improve performance of superalloys. Considering that superalloys are typically used at high temperatures, the thermal stability of the LSP-processed material should be studied. In this article, the hardness and microstructure of a LSP-processed Ni superalloy, DZ17G, are characterized after annealing at various temperatures (up to 1000 °C) for different time periods. The results show that the majority of the strength increase achieved at the sample surface from LSP processing is lost after annealing at 800 °C for 10 h. Detailed microstructural characterization by synchrotron μXRD and metallography shows that the hardness decrease is due primarily to the annihilation of dislocations, and no recrystallization occurs in the temperature range studied.

1. Introduction

Due to their extraordinary mechanical performance, Ni-based superalloys are widely used to fabricate the hot section parts of gas turbine engines [1]. The alloys are mainly strengthened by the ordered γ'-Ni_3(Al, Ti) precipitates, in high coherency with the surrounding FCC solid solution matrix (γ). When the alloys are exposed to a high temperature and high stress environment, various damage in the form of foreign object damage, fretting, corrosion pitting, etc. can cause premature fatigue failure [2]. To extend the service lifetime of such components, mechanical surface treatments are typically used. In recent years, laser shock peening (LSP) has been demonstrated as a promising surface strengthening technique, which can effectively extend the fatigue life of Ni-based superalloys [3]. During the LSP process, a very short and intense laser pulse is used to irradiate the workpiece, see figure 1. Such pulses can instantaneously ablate the absorbing coating surface and induce the formation of a high pressure plasma. A transparent overlay is used to confine the plasma as it expands away from the surface, resulting in a high peak pressure shock wave being launched into the metal. If the pressure exceeds the yield strength of the metal, plastic deformation is generated in the near surface regions [4].

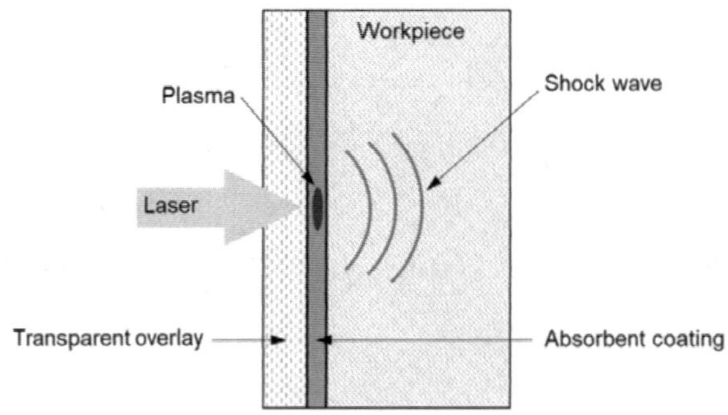

Figure 1. Schematic of LSP processing.

Recently we have conducted systematic work on a LSP-processed Ni-based superalloy DZ17G, and found that the main strengthening mechanism is the introduction of abundant dislocations [5]. Considering the superalloys DZ17G are always used in service at high temperature (below 980 °C), the thermal stability of the LSP-introduced dislocations in material should be clarified. In this study, the annealing behaviour of LSP-processed superalloys is studied. Microhardness is taken as the criterion to evaluate the mechanical properties of the annealed samples, while synchrotron Laue micro X-ray diffraction (μXRD) is applied to investigate microstructural changes during annealing, supplemented with metallographic observations.

2. Experiment

2.1. Material, LSP and annealing

A directionally solidified Ni-based superalloy, DZ17G, composed of γ, together with γ' precipitates, eutectic $\gamma + \gamma'$ and carbides, was used in this study. For LSP processing, a Q-switched Nd laser system was used. An Al foil was used as a protective and absorbing layer, and water was used as a confining overlay to generate shock waves. The LSP treatment was conducted with an overlap rate of 50% and the process was repeated three times on the same sample surface. The LSP direction was perpendicular to the growth direction of directionally solidified DZ17G, which is parallel roughly to the crystal <001> direction. Further details on the sample preparation process can be found in [5]. After LSP processing, the samples were annealed at 600 °C, 800 °C and 1000 °C for 1 h in an Ar gas protected atmosphere. Another sample was annealed at 800 °C for 10 h to study the effect of annealing time.

2.2. Hardness and Laue μXRD

To reveal the effect of annealing, micro Vickers indentation testing was conducted using a HXD-1000TMC/LCD instrument on the sample cross section at a distance of 120 μm away from the LSP-processed surface, to avoid the influence of sample edge on the hardness values. A load of 200 gf and a dwell time of 30 s were used. For the sample annealed at 800 °C for 1 h, the hardness profile in the depth along the LSP direction was also measured to reveal the strength gradient. Synchrotron X-ray microdiffraction was then performed at nominally the same positions at Beamline 12.3.2 of the Advanced Light Source (ALS) at Lawrence Berkeley National Laboratory [6]. For the synchrotron experiment, a polychromatic X-ray beam (5-24 keV) was focused to ~ 1 × 1 μm^2. The Laue diffraction patterns were recorded with a two-dimensional Pilatus-1M detector, mounted in a 90° reflection geometry 150 mm above the sample. The Laue diffraction patterns were processed using the software package, XMAS [7], by which the Laue patterns of the superalloy were indexed and the shape of the 002 spot was quantified (see figure 2). The intensity distribution of the 002 Laue peak in real space was transformed to reciprocal space (see figure 2(b) to 2(c)). By integrating the intensity map along the transverse direction, χ, an intensity profile as a function of 2θ was obtained (figure

2(d)). The full width at half maximum (FWHM) of the profile, δ_{FWHM}, was then determined based on the intensity profile.

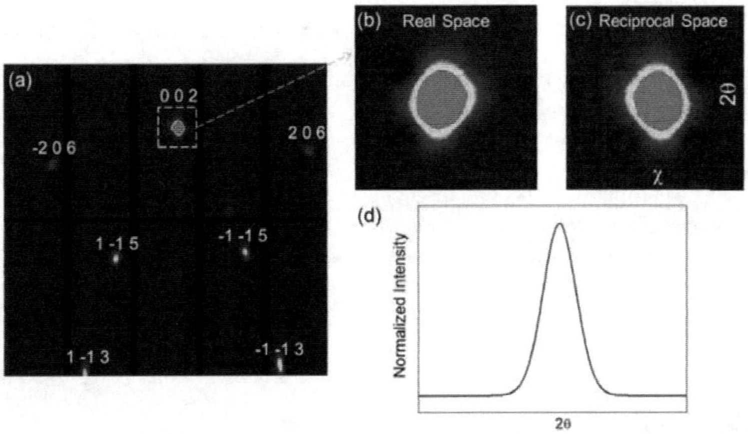

Figure 2. Analysis of the Laue μXRD pattern. (a) is an indexed Laue diffraction pattern of the LSP-processed Ni superalloy DZ17G, (b) is a magnified view of the 002 peak, which is transformed to reciprocal space and shown in (c). (d) is the intensity profile after integrating the intensity along χ.

3. Results

3.1. Hardness

Hardness values of the LSP-processed and annealed samples are shown in figure 3, where the hardness of the as-received Ni-based superalloy without LSP treatment, 384.3 HV is shown by the dashed line as a reference. After LSP, the hardness close to the processed surface (120 μm below) increases about 50%, to 576.5 HV. Upon annealing, the hardness drops monotonically with increasing temperature. After annealing at 1000 °C for 1 h, the hardness close to the processed surface drops to 420.6 HV, which is about 9.4% higher than in the non-treated sample. Similarly longer annealing times lead to further hardness drops, though at a much lower rate (see figure 3). After annealing at 800 °C for 10 h, the hardness drops to 411.4 HV. The remaining hardness increase is still about 14% of the total hardness increase produced by LSP.

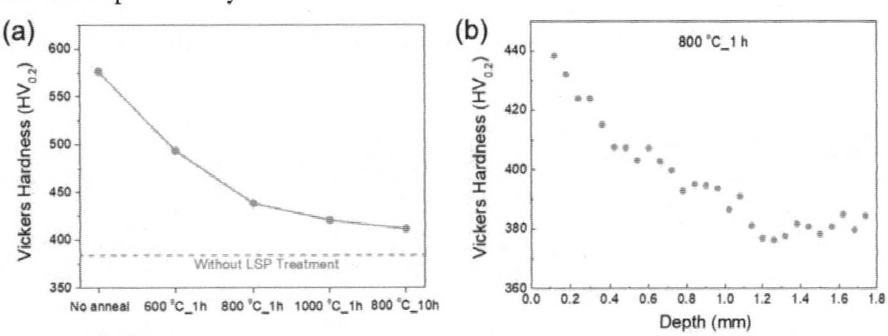

Figure 3. (a) Micro Vickers hardness values measured on the cross section at 120 μm from the LSP-processed surface for deformed and annealed samples. (b) Hardness profile along the depth for the sample annealed at 800 °C for 1 h.

For the sample annealed at 800 °C for 1h, the hardness distribution on the whole cross section is shown in figure 3(b), where a hardness gradient in the depth range 0 - 1.2 mm is seen. Compared to the hardness profile of the LSP-processed sample [5], it can be seen that the hardness drops linearly along the depth.

3.2. Laue μXRD

The 002 peaks for the LSP-processed sample and samples annealed at 800 °C for 1 h and 10 h have been analysed. The 002 spot collected from the sample annealed at 800 °C for 10 h at a region not affected by LSP, which is considered to be a dislocation-free region, is also analysed as a reference. Similar to the diffraction spots seen in figure 2b, the shape of the diffraction spots remains round in reciprocal space, while the size of the spots reduces as a function of annealing time, see figure 4a. Based on the method introduced above, the intensity profiles of the peak 002 along 2θ are determined and shown in figure 4b. The δ_{FWHM} value decreases from 1.68° for the deformed sample to 1.14° and 1.00° after annealing at 800 °C for 1 h and 10 h, respectively, which is about two times larger than that for the reference, 0.53°. It is seen that the majority of the decrease in δ_{FWHM} occurs during the first 1h of annealing, which matches the hardness results.

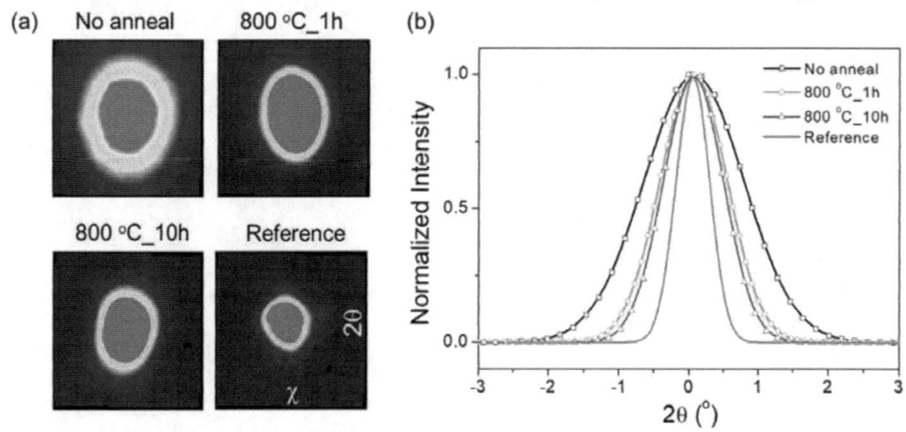

Figure 4. μXRD results. (a) 002 peaks in reciprocal space and (b) intensity profiles along 2θ.

3.3. Metallography

Optical metallography of the shock-peened superalloy without annealing and with annealing at 800 °C for 1 h and 10 h are shown in figure 5, where original grain boundaries developed during solidification are marked by dashed lines. No new grains are observed in any of the three samples, i.e. no recrystallization occurs during annealing (see figure 5).

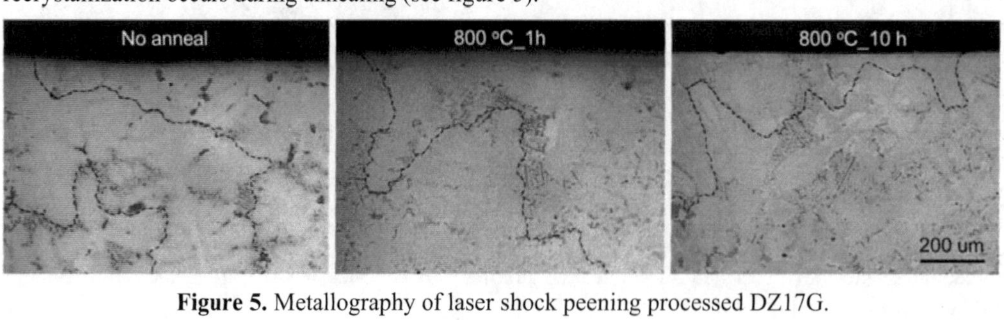

Figure 5. Metallography of laser shock peening processed DZ17G.

4. Analysis

According to Ice and Barabash [8], the broadening of a diffraction peak is typically a result of dislocations from plastic deformation. Additionally, the round shape of a diffraction spot implies that the dislocations are statistically stored dislocations (SSDs), i.e. the dislocations in the illuminated local regions are paired and do not contribute to local lattice misorientation. As discussed in [5], the generation of a high SSD density is primarily related to the high strain rate deformation during LSP.

The dislocation density, ρ, is related to δ_{FWHM} of the intensity distribution by [9]:

$$\rho \propto \delta_{FWHM}^2. \tag{1}$$

As the annealing temperatures user in this study are lower than the solid solution treatment temperature, γ' precipitates are therefore expected neither to coarsen nor refine. The peak width of the diffraction spots is therefore not affected by the precipitates. The observed decrease in peak width is therefore purely a result of reduction in dislocation density. No recrystallization is observed within the temperature range, implying no local boundary migration occurs. The decrease in dislocation density should be a result of the annihilation of SSDs, as paired dislocations can move locally during annealing and annihilate mutually without long range dislocation motion or boundary migration. Annealing at higher temperatures or longer times leads to more dislocation annihilation, while the remaining dislocations are still SSDs, so that the diffraction spots remain round after annealing (see figure 4a).

According to the dislocation strengthening mechanism, the strength (hardness) of a material is related to the dislocation density as [10-12]:

$$HV \propto \sqrt{\rho}. \tag{2}$$

The hardness profile in figure 3b further shows that the rate of dislocation annihilation depends on the dislocation density. Combining equations (1) and (2), the relationship between the measured hardness and the FWHM of the intensity distribution can be obtained as:

$$HV \propto \delta_{FWHM}, \tag{3}$$

or

$$\Delta HV \propto \Delta\delta_{FWHM}. \tag{4}$$

The decrease of δ_{FWHM}, and $\Delta\delta_{FWHM}$, should therefore be attributed directly to the hardness drop, ΔHV, during annealing.

The experimental results have been quantified according to equation (4). The hardness changes between the reference, LSP-processed and annealed at 800 °C for 1 h and 10 h samples, are 195.2 HV, 57.0 HV, and 30.1 HV, while the corresponding $\Delta\delta_{FWHM}$ values are 1.15 °, 0.61° and 0.47°, respectively. The two changes are, however, not found to be proportionally related in this study, with the hardness decreasing more quickly than the decrease of δ_{FWHM}.

5. Discussion

One explanation for the non-proportional relationship could be due to the fact that δ_{FWHM} is determined based on polychromatic X-rays. With polychromatic X-rays, the 004 and 008 peaks are coincident with the 002 peak. Similarly, the 00N (N=1,3,5,7,9) peaks from γ' precipitates are also coincident with the 002 peak. Moreover, the lattice misfit between γ' and γ can also introduce spot broadening in addition to the broadening from size effects and from dislocations. The lattice misfit can change during annealing, and affect the δ_{FWHM} value. Additionally, the hardness values are also affected by the compressive residual stress generated by the LSP process [13]. This stress may be released quickly during the early stages of annealing, leading to a rapid drop of the hardness. This stress release will, however, result only in a shift of the diffraction peak position, which cannot be detected by the polychromatic X-rays, and not in any change to the peak width. To study the relative peak broadening and the peak position from each phase of the superalloy, monochromatic X-ray energy scans around a specific X-ray energy are required [9,14], and such efforts are planned for future work.

6. Conclusions

The thermal stability of LSP treated Ni superalloy DZ17G has been studied. We find that the strength increase from the LSP process is reduced monotonically with increasing annealing temperature and time. After annealing at 800 °C for 10 h, the residual hardness increase is about 14% of the total hardness increase produced by LSP. Characterization by synchrotron micro-diffraction and optical microscopy shows that the strength reduction is due primarily to the annihilation of statistically stored dislocations. No recrystallization is seen in the annealing temperature range.

Acknowledgements

This work is supported by the National Natural Science Foundation of China (Grant No. 9186010 and 51671154), the National Key Research and Development Program of China (Grant No. 2016YFB0700404), the National Basic Research Program of China ("973" Program) (Grant No. 2015CB057400). This research used resources of the Advanced Light Source, which is a DOE Office of Science User Facility under contract no. DE-AC02-05CH11231. G. Zhou thankfully acknowledges financial support from the program of the China Scholarships Council (No. 201806280355).

References

[1] Pollock T M and Tin S 2006 Nickel-based superalloys for advanced turbine engines: chemistry, microstructure and properties *J. Propuls. Power.* **22** 361-74.

[2] Chen X 2005 Foreign object damage on the leading edge of a thin blade *Mech. Mater.* **37** 447-57.

[3] Clauer A H 1996 Laser shock peening for fatigue resistance *Surf. Perf. Ti* 217-30.

[4] Montross C S, Wei T, Ye L, Clark G and Mai Y W 2002 Laser shock processing and its effects on microstructure and properties of metal alloys: a review *Int. J. Fatigue* **24** 1021-36.

[5] Zhou G, Zhang Y, Chen K, Strengthening mechanism in the laser shock peening processed Ni-based superalloy DZ17G, in preparation.

[6] Kunz M, *et al*, 2009 A dedicated superbend x-ray microdiffraction beamline for materials, geo-, and environmental sciences at the advanced light source *Rev. Sci. Instrum.* **80** 035108.

[7] Tamura N 2014 XMAS: A versatile tool for analyzing synchrotron x-ray microdiffraction data, *Strain and dislocation gradients from diffraction: Spatially-resolved local structure and defects,* ed G E Ice and R I Barabash (London: Imperial College Press) pp 125 - 55.

[8] Ice G E and Barabash R I 2007 White beam microdiffraction and dislocations gradients, *Dislocations in Solids* **13** 499-601.

[9] Wilkens M 1970 The determination of density and distribution of dislocations in deformed single crystals from broadened X-ray diffraction profiles *Phys. Status Solidi A* **2** 359-70.

[10] Hansen N and Huang X 1998 Microstructure and flow stress of polycrystals and single crystals *Acta Mater.* **46** 1827-36.

[11] Zhang P, Li S X and Zhang Z F 2011 General relationship between strength and hardness *Mater. Sci. Eng. A* **529** 62-73.

[12] Swadener J G, George E P and Pharr G M 2002 The correlation of the indentation size effect measured with indenters of various shapes *J. Mech. Phys. Solids* **50** 681-94.

[13] Luo K Y, Jing X, Sheng J, Sun G F, Yan Z and Lu J Z 2016 Characterization and analyses on micro-hardness, residual stress and microstructure in laser cladding coating of 316L stainless steel subjected to massive LSP treatment, *J. Alloys Compd.* **673** 158-69.

[14] Levine L E, Larson B C, Yang W, Kassner E M, Tischler J Z, Delos-Reyes M A, Fields R J and Liu W 2006 X-ray microbeam measurements of individual dislocation cell elastic strains in deformed single-crystal copper, *Nat. Mater.* **5** 619.

IOP Conf. Series: Materials Science and Engineering **580** (2019) 012060 doi:10.1088/1757-899X/580/1/012060

Coarsening in nanocrystalline thin films: A 4D problem

A. Öncü[1], D. Zöllner[1,2]

[1]Faculty of Natural Sciences, Otto von Guericke University Magdeburg, Universitätsplatz 2, 39106 Magdeburg, Germany

[2]B CUBE – Center for Molecular Bioengineering, TU Dresden, Tatzberg 41, 01307 Dresden, Germany

dana.zoellner1@tu-dresden.de

Abstract. Materials properties of polycrystalline solids depend strongly on the microstructure. Hence, investigating grain microstructures as they change during grain growth, e.g., by thermal annealing, is of utmost importance. Historically, this has been done by optical microscopy yielding only a two-dimensional characterization. Consequential problems have long been recognized. In particular, a two-dimensional section through a three-dimensional polycrystal yields a rather poor idea about the true size and form of the individual grains, and therefore of the microstructure, and in addition, such a section represents only a snapshot in time and does not give any clues about structural changes during grain growth. In the present study, we investigate coarsening in nanocrystalline thin films using a modified three-dimensional Potts model. By taking the increased importance of higher order grain boundary junctions, i.e. triple lines, at smaller grain sizes into account, it is shown that the microstructure, as well as the associated growth kinetics, undergo distinct changes during coarsening.

1. Introduction

Computational investigations are an integral part of modern materials science and design [1]. They enable not only a very selective manipulation of a broad range of materials parameters that are experimentally inaccessible, e.g., due to technical limitations, but enable also predictions of material behaviour. This holds for processes such as deformation and recrystallization, as well as for the problem of microstructural coarsening during grain growth. In general, grain growth can be understood as a rather simple process: at elevated temperatures the grain boundaries—being two-dimensional defects in the three-dimensional samples—migrate such that the total grain boundary area, and thereby the total boundary energy, is reduced. However, in reality this process is very complex. The grain structure must satisfy a global constraint for volume conservation at all times, while the grain boundaries and their higher order junctions—namely triple lines and quadruple points—strive to preserve a local balance in the network. Understanding this process is of utmost importance as the grain microstructure of a material, characterized for examples by its average grain size and corresponding size distribution, influences strongly material properties. A temporal change in microstructure leads to changes in the properties and may result in materials failure in applications [2].

In contrast to grain growth in bulk materials, where surface effects are usually neglected, they are of great importance for thin films due to the increased material surface to volume ratio [3]. Nevertheless, thin films are still three-dimensional in nature, even though they are commonly treated as two-dimensional objects [4-6]. In such a simplified analysis it is assumed that the microstructure is bamboo-like, i.e. it is fully columnar throughout the film thickness. Although this is indeed the case for some materials, it is not true in general. Specifically, after deposition of a film or after deformation, e.g., by rolling and succeeding recrystallization, the average grain size may be noticeably smaller than the film thickness, and as such the microstructure contains grains of globular nature as in bulk materials. As a result, the associated growth kinetics describing microstructural changes during annealing are distinctively different from samples where the average grain size reaches the size of the film thickness [7].

In the present work, we study grain growth in nanocrystalline thin films, which are of nanocrystalline film thickness, using a modified three-dimensional Potts model. To that aim, the increased importance of the triple lines at smaller grain sizes is taken into account. It is shown that the microstructure as well as the associated growth kinetics undergo distinct changes.

2. Modified Potts model

The Potts model has been used for modelling grain growth for almost 40 years [8,9]. It is a lattice based model, where a given grain microstructure is mapped onto a either a square or hexagonal simulation lattice, such that grain boundaries are defined implicitly as locations between two neighbouring lattice points with different orientations. Triple lines and quadruple points are similarly resulting microstructural features. In typical simulations of bulk materials, periodic boundary conditions are implemented that treat a 3D sample as an infinite continuation in all three dimensions. Also, in the standard version of the Potts model only two materials parameters are taken into account: (i) the specific grain boundary mobility, m, which depends on the crystallographic misorientation angle between two neighbouring grains and can be calculated by the Huang-Humphreys-relation, and (ii) the grain boundary energy per unit length, γ, which can be calculated using the Read-Shockley-equation using also the crystallographic misorientation angle [10]. In addition, a purely technical parameter—namely the simulation temperature—has to be selected carefully [11]. This parameter does not reflect the actual temperature during thermal treatment, but is used to ensure the correct curvature of the grain boundaries as well as the associated correct dihedral angles at triple junctions.

The present implementation combines two previous approaches. On one hand, the thin films are treated as three-dimensional objects by introducing a finite film thickness, h. This is different compared to many previous investigations [4-6], where grain growth in thin films was performed on two-dimensional lattices, but it has already been implemented successfully in Ref. [7]. On the other hand, it has been shown by Gottstein and Shvindlerman [12,13] that a limited triple junction mobility may exert a drag effect on grain boundaries in nanocrystalline materials that changes the grain growth kinetics as observed experimentally [14]. Hence, also in the present work, a limited triple junction mobility is taken into account as in, e.g., [15,16]. The modelling parameters have been selected as described in the following:

- The simulation lattice grid has a size of $500\times500\times h$ lattice points. The film thickness, h, is selected large enough to allow initial three-dimensional growth, but it is also small enough compared to the other two dimensions ($h \ll 500$) [7].
- The sizes of the grains follow a normal distribution, and the grains are equi-axed and randomly distributed in space. Here it should be mentioned that due to the limited film thickness h, 3% of grains span the whole film thickness already at $t = 0$, i.e., are columnar.
- The simulation temperature has been set to $k_B T = 0.8\,\mathrm{J}$ according to Ref. [11] preventing unphysical lattice effects such as, e.g., clinging of boundaries to the underlying grid.
- The mobilities and energies have been selected in such a way that a pure analysis of the effect of triple lines is possible without an overlaying effect of texture. Hence, all grain boundaries are assumed to have identical values for energy, γ_{gb}, and mobility, m_{gb}, whereas the triple lines have significantly lower values for the mobility, m_{tj}, as introduced in [15,16].

3. Results

Already two decades ago, Gottstein and Shvindlerman [12] found that limited triple junction mobility may exert a drag effect on the adjoining grain boundaries that leads to deviations of the grain vertex angles at the junctions from their equilibrium. As a result, grain boundary controlled growth, i.e. normal grain growth follows

$$\langle R\rangle^2 \propto t, \tag{1a}$$

whereas triple junction controlled grain growth results in

$$\langle R\rangle \propto t. \tag{1b}$$

Here t is the annealing time and $\langle R\rangle$ is the average grain radius, where the radius of each grain is calculated as the radius of a grain area equivalent circle. As a consequence of equation (1a) the average grain area, $\langle A\rangle$, e.g., through a 3D section increases linearly with time. The two growth laws can alternatively also be derived by treating grain growth as a dissipative process [17,18].

In the following, at first the thickness of the sample, h, is varied systematically such that $h = \{10,15,20,25\}$ lattice points assuming normal grain growth. As it can be seen in figure 1, the microstructure is characterized early on by a very large fraction of non-columnar grains, which decreases strongly, however, during annealing.

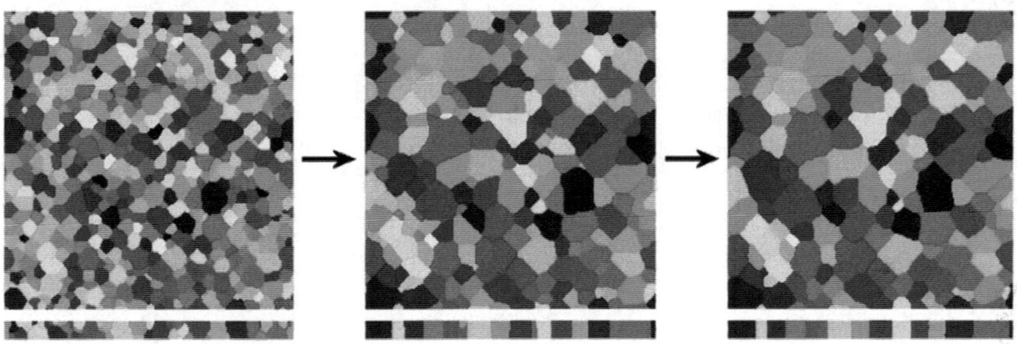

Figure 1. Temporal development of the grain microstructure for normal grain growth with $h = 15$.

This decrease in the number of globular grains is naturally coupled to an increase in the relative number of columnar grains, $\eta = N_{col}\colon N$. In figure 2a, it is interesting to note how the slope of $\eta(t)$ depends on the film thickness, but reaches in each case for long annealing times a steady state, where the relative number is close to 100%, hence forming two main growth regimes. The latter can also be observed with respect to the average growth law, as shown in figure 2b. In contrast to the theoretical description in the form of equation (1), here for reasons of measurements and comparability the average grain area, $\langle A\rangle$, is analysed as the average area of all surface grains. All resulting curves show similar shapes. After an initial period of time (i), which depends on the initial microstructure [19] and is basically non-visible in the presented case, the first growth period (regime I) is coupled to a strong increase in η. A short transition regime (t) follows during which the last 15 to 20% non-columnar grains are transformed. Finally, a period of slow growth follows for long annealing times (regime II).

As expected for normal grain growth, the average grain area increases as a linear function of time. The strong increase in average grain size in regime I is indicated by only one least-squares fit as the four curves yield very similar results. In contrast, the more or less constant number of columnar grains in regime II is associated with a much smaller slopes with similar values [7] of 0.0226 for $h = 10$, 0.0209 for $h = 15$, 0.0271 for $h = 20$, and 0.0267 for $h = 25$. These values are about 15 times smaller than the slope in regime I. It can be concluded that regime I has a larger slope—representing rather three-dimensional coarsening with higher grain curvatures—whereas regime II has a smaller slope—representing rather two-dimensional coarsening with reduced grain curvatures.

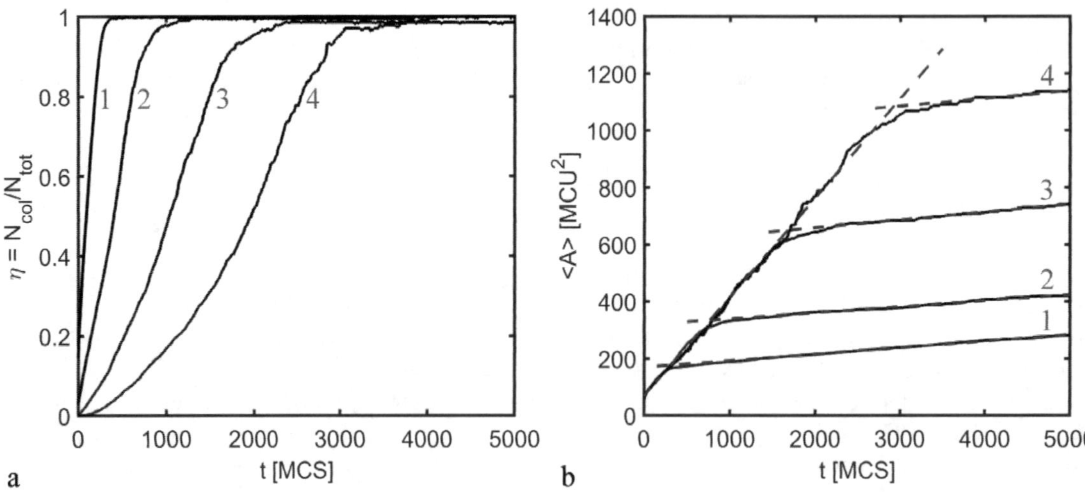

Figure 2. Temporal development of: a – relative number of columnar grains; b – average grain radius; both for different film thicknesses with 1: $h = 10$, 2: $h = 15$, 3: $h = 20$, and 4: $h = 25$. For more details see text.

In contrast, starting the simulations with constant film thickness, $h = 15$, and varying the triple line mobility, m_{tj} with respect to the boundary mobility, m_{gb}, systematically with $m_{tj} < m_{gb}$ yields strong changes in the growth kinetics in agreement with previous 2D and 3D studies [15,16].

This finding is reflected in the average growth law (figure 3b) and with respect to the relative number of columnar grains (figure 3a), when analysing the effect of $m_{tj} = m_{gb}$ in comparison to $m_{tj} = 1/25 m_{gb}$. The linear correlation $\langle A\rangle(t)$ holds only for normal, i.e. grain boundary controlled grain growth, where $m_{tj} = m_{gb}$, in both growth regimes I and II. In contrast, for $m_{tj} \ll m_{gb}$, equation (1b) predicts a linear relation between average radius and annealing time, which means that the average area increases as a quadratic function of time. This is observed to be indeed the case in figure 3b for growth regime I. Due to the much slower coarsening in this case, a longer time period has been considered up to $t = 9000$. This enables not only an observation of the initial coarsening regime (regime i), but also of the longer transition to the second growth regime (regime t).

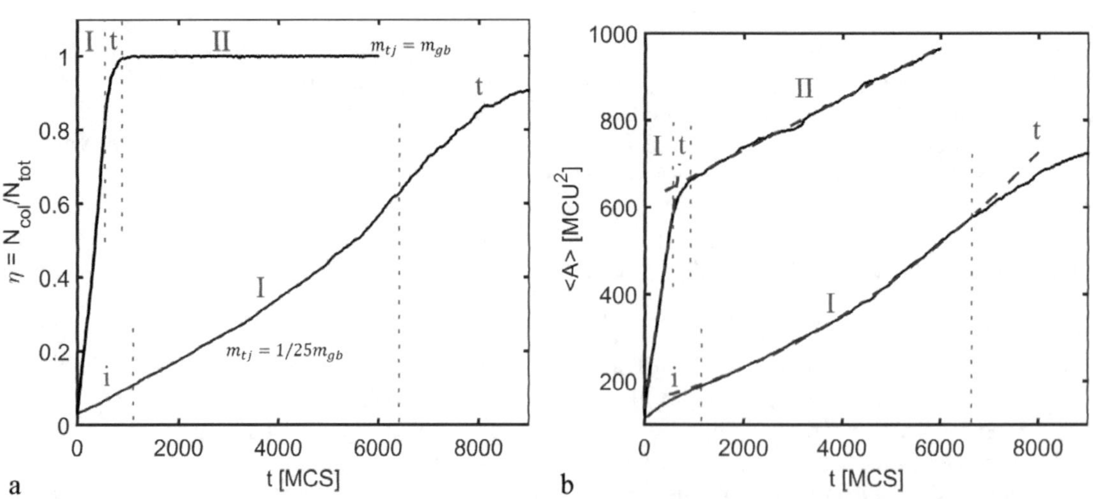

Figure 3. Temporal development of: a – relative number of columnar grains; b – average grain area; for normal grain growth with $m_{tj} = m_{gb}$ and triple junction control with $m_{tj} = 1/25 m_{gb}$. The least-squares fits in figure 3b yield for normal grain growth $\langle A\rangle = 0.8082\, t + 134.4$ for growth regime I and $\langle A\rangle = 0.0592\, t + 613.8$ in regime II in contrast to triple junction control with $\langle A\rangle = 5.5\; 10^{-6}\, t^2 + 0.0277\, t + 153.5$ in regime I.

Overall, a reduced triple junction mobility follows the theoretical prediction in terms of equation (1b), whereas grain boundary controlled growth follows equation (1a). In addition, the results agree well with the experimental measurements of Barmak et al. [6], who investigated grain growth in 100 nm-thin Al films and observed very fast initial coarsening followed by very slow coarsening, cf. figure 9a in [6].

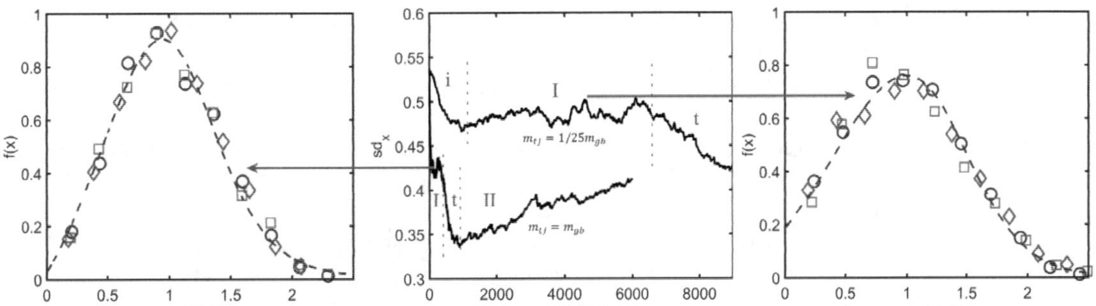

a b c

Figure 4. a – Scaled size distribution for normal grain growth showing time-independence for three random annealing times, i.e., self-similarity in growth regime I together with analytic theory (dashed line) from [17,18]; b – Temporal evolution of standard deviations; c – Scaled size distribution for triple junction controlled grain growth showing self-similarity in growth regime I in comparison with analytic theory (dashed line) from [17,18].

On the other hand, the average growth laws according to equation (1) have furthermore been shown analytically and numerically to be associated with distinct self-similar growth regimes [15-18], where statistical self-similarity is a feature of grain growth in polycrystalline materials [20] that can be confirmed, e.g., with respect to the time-independent scaled size distribution $f(x)$, where the scaled grain size is defined as $x = R/\langle R \rangle$ and the condition $\int f(x)\mathrm{d}x = 1$ holds.

A direct result of the time-independence of $f(x)$ is the fact that the associated standard deviation of the scaled grain size sd_x is constant. In order to check this fact, in figure 4b the standard deviation is plotted as a function of annealing time for both coarsening kinetics. The differences can be spotted easily. In general, triple junction control is associated with larger values describing a broader distribution, where the three growth regimes (initial i, regime I, and transient t) can be detected. In contrast, for normal grain growth the strong changes in the microstructure in the transient and initial regime can be observed, where the latter changes its value from approx. 0.535 to 0.43 within just about 50 time steps. This explains why the initial regime could not be detected in the previous figures. In addition, normal grain growth shows a slow but steady increase of the value within growth regime II, which means that the microstructure is not fully self-similar.

However, when comparing the first growth regimes (I), where the major structural changes happen it can be seen in figure 4a and figure 4c that both show time-independence in $f(x)$, but exhibit clearly different distributions, taking shapes (dashed lines) that have been predicted analytically in [17,18].

4. Summary and conclusions

In the present work, grain growth in nanocrystalline thin films has been studied. To achieve this aim, two approaches have been combined, taking the three-dimensional nature of such films into account, as well as considering the increased importance of triple lines at small grain sizes.

Firstly, the introduction of a finite film thickness yields a temporal evolution of the grain microstructures that can be divided clearly into two main growth regimes. The first growth regime (I) is characterized by a strong increase in average grain size and in the relative number of columnar grains, and represents three-dimensional coarsening with higher grain curvatures. In contrast, the late coarsening regime (II) has a strongly reduced slope regarding the average growth law, and represents quasi-two-dimensional coarsening with reduced grain curvatures. This is true for a range of film thicknesses and in stark contrast to two-dimensional grain growth simulations, which never yield such behaviour.

Secondly, the additional introduction of reduced triple junction mobilities slows the overall coarsening process down visibly and leads to a number of changes:

1. In contrast to normal grain growth, triple junction controlled growth is characterized by a quadratic relation between average grain area and annealing time clearly visible in regime I. This is in agreement with predictions according to Eq. (1b), even though the basic assumption behind this equation, namely purely two-dimensional grain growth, is not fulfilled. Due to the reduction of coarsening speed, growth regime I is ten times longer than for normal grain growth and the initial coarsening regime (i) becomes visible.
2. The two different average growth laws are associated with different grain microstructures:
    a. In regime I, the scaled grain size distribution is narrower for normal grain growth and shifted to a larger number of relatively small grains for triple junction control. This is in agreement with findings on two-dimensional junction controlled grain growth.
    b. The scaled size distributions shown in the present work from growth regime I, where the major changes in the microstructures take place, are self-similar.
    c. Nevertheless, there is an overall temporal change from the initial grain size distribution to long annealing times. The standard deviations—characterizing the width of the distributions—show strong reductions during initial and transient growth regimes. In addition, for long annealing times normal grain growth shows a slow but steady increase in the standard deviation indicating a slight non-self-similarity.

Overall, grain growth in thin films is a four-dimensional problem, for which the introduction of a finite film thickness yields substantial differences in the development of the microstructure compared to purely two-dimensional considerations, even though some analytic theories of 2D grain growth can be applied. In particular, for triple junction controlled grain growth, as may happen in nanocrystalline materials, substantial changes in microstructure and microstructural evolution can be observed.

References

[1] Sinnott S (Editor-in-Chief) 2019 *Computational Materials Science*, Elsevier
[2] Callister WD and Rethwisch DG 2014 *Materials Science and Engineering: SI Version*, Wiley
[3] Mullins WW 1958 *Acta Metallurgica* **6** 414
[4] Hallberg H and Olsson PAT 2016 *Journal of the Mechanics and Physics of Solids* **90** 160
[5] Liu Z-J and Shen YG 2004 *Acta Materialia* **52** 729
[6] Barmak K, Eggeling E, Kinderlehrer D, Sharp R, Ta'asan S, Rollett AD and Coffey KR 2013 *Progress in Materials Science* **58** 987
[7] Zöllner D 2016 *Computational Materials Science* **125** 51
[8] Anderson MP, Srolovitz DJ, Grest GS and Sahni PS 1984 *Acta Metallurgica* **32** 783
[9] Srolovitz DJ, Anderson MP and Sahni PS 1984 *Acta Metallurgica* **32** 793
[10] Zöllner D 2016 *Grain Growth*, In: Saleem Hashmi (editor-in-chief), Reference Module in Materials Science and Materials Engineering, Elsevier
[11] Zöllner D 2014 *Computational Materials Science* **86** 99
[12] Gottstein G and Shvindlerman LS 1998 *Scripta Materialia* **38** 1541
[13] Gottstein G and Shvindlerman LS 2006 *Scripta Materialia* **54** 1065
[14] Krill CE, Helfen L, Michels D, Natter H, Fitch A, Masson O and Birringer R 2001 *Physical Review Letters* **86** 842
[15] Zöllner D 2011 *Computational Materials Science* **50** 2712
[16] Zöllner D 2012 *Scripta Materialia* **67** 41
[17] Streitenberger P and Zöllner D 2011 *Acta Materialia* **59** 4235
[18] Streitenberger P and Zöllner D 2014 *Acta Materialia* **78** 114
[19] Zöllner D, Streitenberger P and Rios PR 2016 *Computational Materials Science* **113** 11
[20] Mullins WW 1986 *Journal of Applied Physics* **59** 1341

AUTHOR INDEX